NOUS·SOMMES·PRETS

SIMON FRASER UNIVERSITY
W.A.C. BENNETT LIBRARY

Biological Approaches
to Sustainable Soil Systems

Organic Chemicals in the Soil Environment, Volumes 1 and 2,
  edited by C. A. I. Goring and J. W. Hamaker

Humic Substances in the Environment, M. Schnitzer and S. U. Khan

Microbial Life in the Soil: An Introduction, T. Hattori

Principles of Soil Chemistry, Kim H. Tan

Soil Analysis: Instrumental Techniques and Related Procedures,
  edited by Keith A. Smith

*Soil Reclamation Processes: Microbiological Analyses and
  Applications,* edited by Robert L. Tate III and Donald A. Klein

Symbiotic Nitrogen Fixation Technology, edited by Gerald H. Elkan

Soil--Water Interactions: Mechanisms and Applications, Shingo Iwata
  and Toshio Tabuchi with Benno P. Warkentin

Soil Analysis: Modern Instrumental Techniques, Second Edition,
  edited by Keith A. Smith

Soil Analysis: Physical Methods, edited by Keith A. Smith
  and Chris E. Mullins

Growth and Mineral Nutrition of Field Crops, N. K. Fageria,
  V. C. Baligar, and Charles Allan Jones

Semiarid Lands and Deserts: Soil Resource and Reclamation,
  edited by J. Skujins

Plant Roots: The Hidden Half, edited by Yoav Waisel, Amram Eshel,
  and Uzi Kafkafi

Plant Biochemical Regulators, edited by Harold W. Gausman

Maximizing Crop Yields, N. K. Fageria

Transgenic Plants: Fundamentals and Applications, edited by
  Andrew Hiatt

*Soil Microbial Ecology: Applications in Agricultural and Environmental
  Management,* edited by F. Blaine Metting, Jr.

Principles of Soil Chemistry: Second Edition, Kim H. Tan

Water Flow in Soils, edited by Tsuyoshi Miyazaki

Handbook of Plant and Crop Stress, edited by Mohammad Pessarakli

Genetic Improvement of Field Crops, edited by Gustavo A. Slafer

Agricultural Field Experiments: Design and Analysis,
  Roger G. Petersen

Environmental Soil Science, Kim H. Tan

*Mechanisms of Plant Growth and Improved Productivity: Modern
  Approaches,* edited by Amarjit S. Basra

Selenium in the Environment, edited by W. T. Frankenberger, Jr.
  and Sally Benson

Plant–Environment Interactions, edited by Robert E. Wilkinson

Handbook of Plant and Crop Physiology, edited by
  Mohammad Pessarakli

Biological Approaches
to Sustainable Soil Systems

EDITED BY

Norman Uphoff

Andrew S. Ball	Cheryl Palm
Erick Fernandes	Jules Pretty
Hans Herren	Pedro Sanchez
Olivier Husson	Nteranya Sanginga
Mark Laing	Janice Thies

Taylor & Francis
Taylor & Francis Group
Boca Raton   London   New York

A CRC title, part of the Taylor & Francis imprint, a member of the
Taylor & Francis Group, the academic division of T&F Informa plc.

Cover picture: Roots of plantain (*Plantago lanceolata*) and fungal hyphae in a grassland soil. © Karl Ritz, National Soil Resources Institute, Cranfield University [k.ritz@cranfield.ac.uk].

Published in 2006 by
CRC Press
Taylor & Francis Group
6000 Broken Sound Parkway NW, Suite 300
Boca Raton, FL 33487-2742

Library of Congress Cataloging-in-Publication Data

Biological approaches to sustainable soil systems / authors/editors, Norman Uphoff ... [et al.].
    p. cm. -- (Books in soils, plants, and the environment ; v. 113)
  Includes bibliographical references and index.
  ISBN-10: 1-57444-583-9 (alk. paper)
  ISBN-13: 978-1-57444-583-1 (alk. paper)
  1. Soil management. 2. Soil biology. 3. Soil fertility. 4. Soil ecology. I. Uphoff, Norman Thomas. II. Series.

S591.B512 2006
621.4--dc22
    2005034628

Taylor & Francis Group
is the Academic Division of Informa plc.

Visit the Taylor & Francis Web site at
http://www.taylorandfrancis.com

and the CRC Press Web site at
http://www.crcpress.com

Foreword

Global agriculture is now at the crossroads. The Green Revolution of the last century that gave many developing countries such as India a breathing spell, enabling them to adjust the growth of their human populations better to the supporting capacity of their ecosystems, is now in a state of fatigue. Average growth rates in food production as well as factor productivity in terms of yield per unit of mineral fertilizer (NPK) are both declining. Yet, India and other developing nations are still confronted with the need to produce more food and other farm commodities under conditions of diminishing arable land and irrigation water resources per capita and with expanding biotic and abiotic stresses, some linked to global climate changes.

In January, 1968, several months before Dr. William Gaud coined the term "Green Revolution," I made the following statement in a presidential address to the Agricultural Sciences Section of the Indian Science Congress held in Varanasi:

> Exploitative agriculture offers great dangers if carried out with only an immediate profit or production motive. The emerging exploitative farming community in India should become aware of this. Intensive cultivation of land without conservation of soil fertility and soil structure would lead, ultimately, to the springing up of deserts. Irrigation without arrangements for drainage would result in soils getting alkaline or saline. Indiscriminate use of pesticides, fungicides, and herbicides could cause adverse changes in biological balance as well as lead to an increase in the incidence of cancer and other diseases, through the toxic residues present in the grains or other edible parts. Unscientific tapping of underground water will lead to the rapid exhaustion of this wonderful capital resource left to us through ages of natural farming. The rapid replacement of numerous locally adapted varieties with one or two high-yielding strains in large contiguous areas would result in the spread of serious diseases capable of wiping out entire crops, as happened prior to the Irish potato famine of 1854 and the Bengal rice famine in 1942. Therefore, the initiation of exploitative agriculture without a proper understanding of the various consequences of every one of the changes introduced into traditional agriculture, and without first building up a proper scientific and training base to sustain it, may only lead us, in the long run, into an era of agricultural disaster rather than one of agricultural prosperity.

Since enhancement in productivity per unit of land is the only pathway available to most population-rich, but land-hungry, countries like India, for the purpose of meeting the food needs of a fast-growing population, I started working on what I termed "an ever-green revolution movement." Ever-green revolution involves the enhancement of farm productivity in perpetuity without associated ecological or social harm. In his 2002 book, *The Future of Life*, E.O. Wilson endorsed this concept:

> The problem before us is how to feed billions of new mouths over the next several decades and save the rest of life at the same time, without being trapped in a

Faustian bargain that threatens freedom and security. No one knows the exact solution to this dilemma. The benefit must come from an ever-green revolution. The aim of this new thrust is to lift food production well above the level obtained by the Green Revolution of the 1960s, using technology and regulatory policies more advanced and even safer than those now in existence.

For an ever-green revolution, we need, first and foremost, to have sustainable soil systems. In spite of all the advances in aquaculture, the soil will remain the main source of food. It is in this context that *Biological Approaches to Sustainable Soil Systems* is an extremely timely and valuable contribution to a wider understanding of what needs to be done for the sake of a truly modern agriculture and for the sake of future generations. We owe a deep debt of gratitude to Dr. Norman Uphoff and all the editors and contributors for their labor of love to the cause of enhancing the productivity and profitability for a wide range of farming systems on an environmentally sustainable basis by sharing both scientific knowledge and practical experience.

There is a growing conflict between the proponents of organic farming and chemical farming. Chapters in this book bring out clearly that there is no need to abandon all external inputs and that there can be synergy between organic and inorganic inputs. The book provides the scientific basis for low external-input agriculture. Soil health is treated in a holistic manner involving attention to the physics, chemistry, and microbiology of soil systems. The various chapters could help scientists chart out and embark on a program of "soil breeding" for high productivity. Such initiatives for sustainable agriculture should receive as much attention as crop breeding has been given if we are to promote advances in productivity in perpetuity without adverse ecological consequences.

Water is becoming a very serious constraint in many countries and of growing concern. This is where Norman Uphoff has rendered a very valuable service by promoting the System of Rice Intensification (SRI) method adapted from agronomic practices developed in Madagascar (Chapter 28). Experience in India has shown that this methodology leads to nearly 40% saving in water without affecting the yield of the crop, indeed, as a rule, increasing crop yields with reduced external inputs. As decisions on land use are invariably water-use decisions, land and water use will in the future have to be dealt with in a more integrated manner. I hope that this ambitious and timely book will be read and used widely in order to ensure the long-term viability and sufficiency of food production systems around the world.

M.S. Swaminathan

Contents

PART III: Strategies and Methods

Introduction

This book has been constructed by the editors and contributors as a report on the state of knowledge and practice for a more biologically oriented and informed agriculture appropriate to a wide range of contemporary agroecosystems. It advances numerous explanations and conclusions for sustainable soil-system management even though we know this is a rapidly expanding area for research and field-based innovation, with many answers still to be found.

The editors and contributors, representing many institutions and diverse disciplines, have worked in dozens of countries. What has brought them together is their involvement with research and/or experience which shows that agricultural production can be increased at the same time that it is made more sustainable by being less dependent on the exogenous resources that have driven the expansion of agriculture in the past century. The book focuses on and illuminates ways in which endogenous processes within soil systems offer opportunities to expand agricultural output with less reliance on external inputs, even while these continue to play an important role in contemporary agriculture.

The book is neither antichemical in orientation nor opposed in principle to the use of external inputs. These will remain important in the decades ahead, and, in fact, the book presents much evidence supporting the optimizing use of mineral fertilizers with biological interventions. However, their pre-eminent role will surely change. For many economic, environmental, and equity reasons, it would be unwise not to consider alternative or complementary approaches to conventional "modern" agriculture. No one can say with certainty what will be the future prices and availability of fossil fuel-based energy and agrochemical inputs. However, it is unlikely that their costs will not increase in real terms in the years ahead.

Fortunately, many opportunities have been emerging in recent years to move agriculture in more biologically driven directions that are less dependent on such inputs. This book shows that there are many ways in which a variety of crops can be produced more abundantly and more cheaply by managing and intensifying endogenous processes in soil systems. Improvements in the growing environments for crops can capitalize on existing genetic potential to achieve substantial increases in output, often 50–100% and sometimes more, with less reliance on external inputs, as summarized in Chapter 49.

Many innovations in agricultural systems have been empirical, not driven by scientific knowledge, but rather by efforts to solve specific problems of soil-system constraints or decline. However, it is desirable that technology does not run too much ahead of science. Improvements in soil-system performance should be explicable in terms of scientific knowledge that can make these innovations more generalizable, more transparent, more replicable, and more optimal and sustainable. Knowing better their limitations and finding additional opportunities can lead to their wider and safer use. This book thus seeks to foster a closer connection between science and practice where now there is often some estrangement.

The 104 contributors come from 28 countries, presently based in Africa (24), the United States (24), Latin America (18), Asia (18), Western Europe (12), and 4 each in the

Middle East and Oceania. Denominating contributors' disciplines is difficult since both original training and current work are important, but these often diverge. Because the contributors represent so many different disciplines and have divergent roles along the research-development continuum, the voices and concerns of the chapters that follow present considerable variety. The different tones and perspectives that readers will find in these chapters mirror the diversity of the underground world about which the authors are seeking to communicate. The variation we trust will make for more interesting reading.

We have included both practitioners and researchers in this review since advancing knowledge in this domain is not a simple linear process of going from science to practice. Practice has often given impetus to scientific inquiry and insights. We look forward to having richer foundations for these various strands of theory and practice in the decades ahead as productive knowledge and experience accumulate. However, we can already identify, assemble, and share enough knowledge so that agriculture in this new century appears positioned for a more successful and sustainable future. There is no need for dire or gloomy predictions.

Both authors and readers of a book that is intended for a multidisciplinary audience face a similar challenge — to optimize between breadth and depth. The editors invited persons with recognized expertise in many different subject areas to contribute from their knowledge. Our request was to focus on the most important things for nonspecialists to know on the respective subjects without compromising substance and scientific rigor. The editorial team, which collectively has broad and deep expertise on soil systems around the world, took responsibility for writing one-third of the book, the remainder being by experts from around the world. The managing editor, himself a social scientist by original training, undertook to ensure coherence in the full set of presentations and to have them presented in language that is broadly accessible.

As seen from the table of contents, this book represents an effort to chart some new directions for agricultural science and practice. It is not, however, "breaking new ground" in that what is presented in Parts I and II has roots in the scientific literature that go back often half a century (e.g., phytohormones) or more (e.g., the effects of mycorrhizal associations).

Much new knowledge is offered through the syntheses and applications presented in Part III, with amplifications in Part IV. Contemporary work shows significant benefits from mobilizing well-documented biological processes within soil systems. Agricultural plants have coevolved with other flora and fauna for several hundred million years. This has led to many productive symbiotic relationships that are severed when crops are treated as separate from and independent of the ecosystems in which they function.

The original idea for this book was suggested by Russell Dekker, senior editor of Marcel Dekker, in a discussion in June 2003 with Norman Uphoff, who became managing editor for this project. Both saw value in bringing together what is known from many countries and disciplines about this "biocentric" way of thinking about agricultural science and practice that is gaining ground. Special thanks go to Virginia Montopoli at the Cornell International Institute for Food, Agriculture and Development (CIIFAD) for her support in producing this book. An undertaking as ambitious and complex as this required continuous and skilled administrative assistance to bring it to fruition.

The concept of an alternative paradigm for crop and soil sciences that is more biologically oriented came into the literature over a decade ago when one member of the editorial team, Pedro Sanchez, presented it in a paper at the 15th World Congress of Soil Science in Mexico (Sanchez, 1994). Since this direction was first articulated, accumulating knowledge and practice have given shape and momentum to this paradigm.

There is still much more to be known and done. However, we expect that readers will agree, after assessing the scientific foundations of this emerging paradigm and its empirical accomplishments to date, that agriculture in the 21st century can be made both more productive and sustainable by accepting this more holistic orientation and associated knowledge than by simply projecting the current paradigm into the future without considering biological factors and actors in more active and central roles.

Reference

Sanchez, P.A., Tropical soil fertility research: towards the second paradigm. In: *Transactions of the 15th World Congress of Soil Science*, Acapulco, Mexico. Mexican Soil Science Society, Chapingo, Mexico, 65–88 (1994).

PART I: OVERVIEW

1

Understanding the Functioning and Management of Soil Systems

Norman Uphoff, Andrew S. Ball, Erick C.M. Fernandes, Hans Herren, Olivier Husson, Cheryl Palm, Jules Pretty, Nteranya Sanginga and Janice E. Thies

CONTENTS

Soil, the foundation for most terrestrial life, has unrivalled complexity. What is seen as mostly mineral and relatively homogeneous material contains uncountable numbers of organisms, as well varying amounts of air and water. A single gram of soil usually contains tens to thousands of millions of fungi and bacteria, plus thousands of diverse plant and animal species. Indeed, the extraordinary diversity of microbes in the soil remains unappreciated because most, perhaps >95% of them, cannot be cultured and thus have never been examined, classified, or recognized for their ecological significance. *Terra incognita* remains a very relevant concept.

Soil scientists have long grappled with the difficulty of discerning the many variations and gradations in soil chemistry and physical structure that have major implications for soil fertility. Yet this fertility is a function also of gases, liquids, organic matter, and myriad organisms that are found along with inert solid mineral material. This book is about soil systems, rather than about soil, to emphasize the contributions that living components make to something that is variously cultivated, grazed, built on, walked on, and utilized in other ways. It explores the implications of a biologically framed understanding of soil systems for making their management more productive and sustainable.

1.1 Components of Soil Systems

Soil systems have four major categories of material content. The first is by far the largest in terms of volume, but the last and smallest category has effects on soil system performance that are far out of proportion to its measurable share.

- *Mineral elements* are usually about half of the soil's volume, even though they can appear to be its totality. The mineral portion of soil, which differs from system to system in its chemical composition and its physical characteristics, has long been the focus of most soil science research. These mineral elements exist in different-sized soil particles, classified (from large to small) as sand, silt, or clay. The mineral composition of soil establishes its physical properties, and it influences and is influenced by the life forms that are present.

- *Water* is usually about a quarter of soil volume, although the actual amount can vary greatly over time and between soil systems. With too little water, soil systems become desiccated, and with too much, they are saturated.

- *Air* in well-aggregated soil can be another quarter of the volume, containing oxygen, hydrogen, nitrogen, and carbon in gaseous forms. The more pore space within the soil, the greater will be its capacity for holding both water and air which benefit plants as well as other flora and fauna in the soil. For any given soil porosity, the amounts of water and air are usually inversely related.

- *Organic material* usually comprises only a small portion of soil by volume, usually between 1 and 6%, although it can be higher than this. This fourth category encompasses (1) nonliving organic matter which is derived from the growth, reproduction, death, and decomposition of plants, animals, and microbes and exists in the soil as humus or as other inanimate material, and (2) an immense variety of living flora and fauna, referred to collectively as the soil biota. This category includes also (3) plant roots, which actively make soil systems more hospitable for the growth of their shoots and other species. The processes involved in these interactions are addressed in Chapter 5 and Chapter 6.

The organic portion of soils includes both soil organisms and the various biological substances and processes that animate soil systems. The connection between the mineral and organic components of soil systems is intimate, converging at the smallest scale of soil structure and function in what are called clay–humus complexes. At the next higher level of structure, in microaggregates, inert and living materials are practically fused. Although this is well known, the biological dimensions of soil systems are too often regarded more as secondary or intervening variables, rather than as central and determining factors.

Here and in the chapters that follow, we present a soil-system perspective that is based on both well-established principles and recent research findings. It gives attention to biological factors without separating them from or opposing them to the soil's physical and chemical aspects, because all three sets of factors are and need to remain integrated. We hope a more biocentric perspective will advance the understanding and management of soil systems so that these can be made more sustainably productive for food production and for their environmental services.

Both above- and belowground biota are the sources of the organic matter that provides necessary energy stores to soil systems as discussed in Chapter 5 and Chapter 6. Microorganisms decompose and mineralize organic materials and accelerate the weathering of inorganic materials that can then be used by plants and other organisms. In addition to solubilizing certain nutrients which then are available for flora and fauna, they also stabilize chemical and physical aspects of the soil, creating a more hospitable habitat for plants as well as for themselves. The activities of all kinds of organisms combine to aggregate the soil, making it possible for water and air to diffuse below the surface. This makes the soil better able to absorb and retain these elements for subsequent use by plants and other organisms.

By breaking down chemical compounds, the soil biota continuously replenish the pool of nutrients that are available for plants and other biota, at the same time that they utilize many of these nutrients for themselves. This latter process known as "immobilization" makes the nutrients absorbed by microorganisms unavailable, at least at the time, for plants and other organisms. However, the process could just as well be regarded as a kind of "banking" of soil nutrients, buffering them in the short run and keeping them in the soil for eventual use by plants, rather than being lost through leaching or volatilization. Collectively, the biological actors in the soil have many and very great consequences for the performance of soil systems.

Plants should themselves be understood as active participants in the continuing ebb and flow of soil processes, rather than just as recipients of water and nutrients, of predation by soil herbivores or infection by pathogens. Plant roots continually adjust their interactions with the soil environment, particularly with the thin layer of biologically rich soil, known as the rhizosphere that envelops the root system and is modified by root exudation. An example of how sophisticated this interaction can be is seen in a recent report on maize plant responses to attacks by the larvae of the western maize rootworm (*Diabrotica virgifera virgifera*), a coleopteran pest that feeds on maize roots. When the roots of certain maize varieties are attacked, they emit a chemical signal, the compound (E)-β-caryophyllene, which attracts a nematode *Heterorhabditis megiditis* which is a natural enemy of the rootworm pest to attack the larvae (Rasmann et al., 2005). Similar strategies are operative above ground as maize leaves, in response to caterpillar herbivory, emit certain volatile compounds which attract parasitic wasps to infest the caterpillars (Turlings et al., 1990; Khan et al., 1997; Degenhardt et al., 2000). Such findings indicate that plants are best understood as active rather than as passive participants in soil systems.

1.2 Understanding Soil System Dynamics

Soil systems contain complex food webs which exhibit contrasting processes of competition, mutualism, predation, and symbiosis. The many and intricate connections of the food web and how organisms support each other in diverse ways, while in turn being consumed by other organisms, are reviewed in Chapter 5. Some of these interactions are unusual, and cannot be simply extrapolated from knowledge about aboveground systems. Thus specific scientific knowledge about belowground actors and processes, reviewed in Part II, is important for understanding means to better manage and improve soil systems, presented in Part III.

Here are some examples of belowground processes which are very different from those observed above ground.

- The fecal pellets (castings) produced by earthworms are populated by microbes that decompose mineral nutrients in these pellets before they are reingested, thus functioning in effect as a kind of "external rumen." This symbiotic relationship supports worms' capacity to enrich and aerate the soil through which they move.
- Symbiotic protozoa that live in the guts of termites break down cellulose that termites have ingested, thereby enabling their hosts to derive energy from wood which no other organisms can digest.
- Some species of ants and termites maintain, in effect, "fungus gardens" in their hills, continuously feeding the resident fungi in order to obtain from them the energy that enables these "ecosystem engineers" to make the soil more permeable to air and water.

The subterranean world thus contains many unexpected actors and relationships, described in informative detail in recent "underground travelogs" by Wardle (2002) and Wolfe (2001). Understanding how this world works requires an integration of chemical, physical, and biological perspectives informed by empirical investigation with a minimum of preconceptions.

1.2.1 Difficulties in Analyzing Biological Components

Studying biological processes and agents is complex, often ambiguous, and invariably more difficult than assessing chemical and physical factors. However, there has been a burgeoning of research on soil biology and ecology in the last 20 years. This book could not have been written two decades ago, and even a decade ago it would have compared poorly with what evidence now available permits us to report.

For the sake of simplifying research design and producing replicable results, much soil research to date has been carried out under what are called axenic conditions, studying soil samples that have been sterilized or fumigated before analysis, thereby eliminating all living organisms. This ensures that no biotic presence or activity will affect the soil being evaluated for its physical or chemical properties and relationships. (Note that etymologically the term "a-xenic" means without anything foreign, implying that the organisms being removed from the soil are alien to it and do not belong there.) Gnotobiotic soil studies are intentionally more biologically oriented, eliminating all but a certain species of organism from the soil to study its effects in isolation. However, when this is done, none of the effects of that species' interactions with other organisms can be considered, even though soil biology research has been demonstrating the ubiquity and significance of such interactions. Having only one species in soil analysis is actually not much better than having none, since there can be opposite effects when other organisms are present in the soil, not just larger or smaller effects.

For example, research on the soil fungus *Trichoderma harzianum* has shown that under axenic conditions it attacks the mycelium of the beneficial mycorrhizal fungus *Glomus intraradices* (Rousseau et al., 1996). However, in a soil-based system where other organisms are present, researchers have found *G. intraradices* unaffected by the presence of *T. harzianum*, and in fact, the Trichoderma fungus appears to be suppressed through nutrient competition (Green et al., 1999, as reported by Whipps (2001)). Thus, not examining relationships with the whole suite of organisms present and active in the soil can give misleading results because *ex situ* studies often give different results from *in situ* investigations.

Soil research that controls for biological factors by removing them or by making them extraneous to the analysis can produce more apparently precise and replicable experimental results. However, this approach excludes important processes that are pervasively influential under field conditions. Conclusions drawn from controlled evaluations are thus only applicable to situations where the same, usually artificial conditions can be found. Conclusions based on axenic or gnotobiotic soil samples, when extrapolated to the real world, should have appropriate qualifications about their limited applicability stated clearly and explicitly.

1.2.2 Methodological Issues and Opportunities

Soil researchers are confronted with their own version of the uncertainty principle that Werner Heisenberg proposed for quantum physics, where the act of measurement itself affects the phenomena under study. Researchers studying living soil relationships have to accept the fact that they cannot produce as exact or as incontestable results as their peers who study soil chemistry and soil physics because reductionist methodologies are less able to illuminate the holistic realm of biology, where emergent properties are particularly important. Because of the dynamism of soil systems, the differing results that are often obtained when analyzing the same soil at two points in time may be due not to measurement error but to variation in the phenomena of interest. This discrepancy does not represent a failure of measurement, but is more likely a realistic representation of diverse and changing soil situations.

Researchers studying soil systems need tolerance for ambiguity, complexity, and uncertainty as they seek to assemble credible and robust explanations of these phenomena. Reductionist methods can and should be used, however, with a sophistication that appreciates (and talks about) what has been left out of the analysis. Sacrificing realism for the sake of precision and replicability is a trade-off that can lead to more misunderstanding than insight. What goes on in soil systems is still incompletely understood because most present scientific knowledge has been based on studies that have underrepresented the biological component.

Fortunately, in recent years, numerous advances have been made in research methods for studying soil biology. Good recent examples of such work are found in van Noordwijk et al. (2004), which goes into considerable scientific detail on many of the subjects addressed in this volume. Indeed, a number of the chapters in this volume report the results of using some of the most technologically advanced investigation techniques, for example, Chapter 8 on natural Rhizobium-cereal associations, Chapter 15 on soil and plant management practices that influence genetic expression for plant senescence and resistance to pests and diseases, Chapter 37 on phosphorus availability, and Chapter 40 on polycropping to control weed and insect pests.

In Chapter 46, we consider some of these advances in measurement capabilities and how they can help us to learn about the functioning of soil systems. Chapter 47 and Chapter 48 review how these advances can be built upon to improve knowledge and practice through monitoring and modeling exercises. Many of the means for measurements now being made of biological actors and activities in soils were not available 10 years ago.

1.3 Soil Systems Analysis and Management

A growing number of researchers in countries around the world have taken on the challenge of developing a new and integrated understanding of soil systems, combining detailed measurement and reductionist analyses with inclusive understandings of plant, soil, microorganism, climate, and other interactions, and then try to apply this knowledge to the improvement of agricultural systems. The principles governing soil-system functioning have been compared with musical themes by Lucien Séguy, one of the contributors to this volume. Themes in the analysis of soil systems are basically the same across different climatic, edaphic, and other conditions, but they will sound somewhat different when played by the different soil-system "orchestras" that differ according to various combinations of biophysical and other factors.

Well-functioning soil systems have the following requirements according to Ana Primavesi, another contributor to this volume:

- A well-aggregated soil structure that has sufficient organic matter in the upper soil horizons so that it can support (1) water and air penetration; (2) aerobic soil life for aggregating the soil and for mobilizing nutrients; and (3) good root development.

- Soil protection against the adverse effects of sun and rain (high temperature and erosion). This can be achieved by (1) mulch from crop residues, green manures, or cover crops that keep the surface covered; (2) intercropping or polycropping with a variety of plants to keep the surface covered; (3) reduced spacing between plants or planting an undercrop such as groundnut; and/or (4) shading where intense sunshine adversely affects both plants and water.
- Diversified and abundant populations of soil organisms to mobilize nutrients.
- Extended and well-functioning root systems with widespread access to soil nutrients.
- Crops adapted to the soil and climatic conditions, or some adaptation of the growing environment to support them, such as provision of nutrients lacking in the soil.
- Windbreaks around cropped areas or reforestation to reduce evapotranspiration, and
- Careful use of machinery to minimize soil compaction, a major hazard for most soil systems.

Many current standard agricultural practices have negative impacts on soil systems. Plowing not only can contribute to soil erosion; it also disturbs and reduces earthworm and other epigeic fauna populations important for soil fertility. By inverting soil layers, plowing kills aerobic organisms on or near the surface by burying them, while eliminating anaerobic organisms living at lower depths by exposing them to the air. The reduced soil porosity that results from compaction creates less favorable conditions for development of roots and many kinds of microbial life. While the application of synthetic fertilizers can have positive effects, this alters soil chemistry, physics, and biology to different extents. Increasing some but not all of the soil nutrients that are in deficit, and possibly creating surpluses of others, can unbalance soil chemistry and biology to the detriment of plant health and growth. The application of some agrochemicals for crop protection also has effects on populations of soil microflora and fauna, diminishing their biodiversity.

Contemporary agriculture has undoubtedly produced many benefits. However, these can come with substantial economic and ecosystem costs (Waibel et al., 1999; Pretty, 2002; 2005; Tegtmeier and Duffy, 2004). Some costs are evident in soil degradation and loss, but often there are greater unseen losses in soil biota. Soils with less diverse microbial communities have less capacity to suppress pathogenic activity, with resulting plant diseases. The use of agrochemical controls can make disease problems worse in the long term.

Agricultural practices, such as deep plowing and fertilizer application that correct some of the symptoms of deteriorating soil systems, do not address the causes of the deterioration and thus do not improve soil conditions in the long run. Biologically-based approaches aim to promote virtuous rather than vicious cycles, improving conditions that will support sustained and beneficial processes. The contributions of biological understanding to soil fertility, discussed in more detail throughout this book, can be summarized as:

- *Soil creation.* Although most soil texts explain this process in terms of physical and chemical processes, biological activity has important roles in soil genesis.
- *Solubilization and recycling of nutrients.* Expanding the available pool of nutrients in the soil is a continuous process, with organisms that operate from micro- to macroscales contributing more to plant nutrition, even more nitrogen, than

is provided through inorganic amendments (Ladha et al., 1998). The role of plant roots in recycling nutrients is generally undervalued.

- *Improvement and stability of soil structure.* By creating soil aggregates and pores, as well as mixing minerals and organic matter through the activities of a wide range of organisms.

- *Detoxification.* By decomposing or neutralizing a wide variety of organic and inorganic substances that would otherwise impede the growth of organisms in and above the ground.

1.4 Soil System Interactions

Soil systems vary widely in their productivity, their stability, and their resilience, with factors and processes varying not just between wet and dry or hot and cold seasons, but changing continuously over time. Despite this variation, however, the factors and processes involved are reasonably similar across all soil systems, and they commonly vary around certain fairly predictable states or cycles.

Certain processes are faster under tropical conditions than in temperate areas, as discussed in Chapter 2 and Chapter 3, but the processes themselves are universal because of the needs of flora and fauna at different scales. The organisms present in soil are never exactly the same in two different locations, yet the functional categories remain similar. Certain functions are often performed by different sets of flora or fauna, often on an as-needed basis. Managing soil systems productively and sustainably requires adapting cropping principles and practices to particular local conditions, capitalizing on the complex systems of interaction that are now coming to be understood better.

A schematic representation of soil system dynamics developed from the work of Séguy and CIRAD researchers in countries including Brazil, Madagascar, France, and Vietnam is presented in Figure 1.1. This analysis underscores the benefits of maintaining permanent cover on fields as discussed in Chapter 22 to Chapter 24. Various means — biological, chemical, and physical — can be used to control weed competition and alter soil structure, the main reasons for plowing over the millennia. Replacing tillage with practices that keep the soil covered and nurture more biotic activity in the upper horizons is one of many biologically-based approaches for better management of soil systems. This kind of analysis and associated changes in standard agronomic practice are opening up new possibilities, with many other examples considered in Part III. These achieve economic production objectives in ways that are more compatible with and supported by natural ecosystem processes and are also more cost-effective.

Soils have often been regarded as a medium for anchoring plants and for receiving fertilizer and other inputs, assuming a fixed endowment of inherent nutrients. Any nutrients that are lost through crop production, according to the standard view, must be replaced whenever exported in order to sustain a given level of fertility. Such a simplification of soil processes has supported some demonstrable successes in agronomic theory and practice. However, a more multifaceted appreciation of how soils function as dynamic systems can help achieve even greater increases in yield and profitability while also contributing to improved soil and water quality.

Biologically-based practices with an agroecological perspective do not focus on single species, respectively. Instead they address the complex interactions of one or more crops with soil biota of many kinds, sometimes even incorporating the management of weeds into cropping systems to take advantage of what such plants can contribute, such as

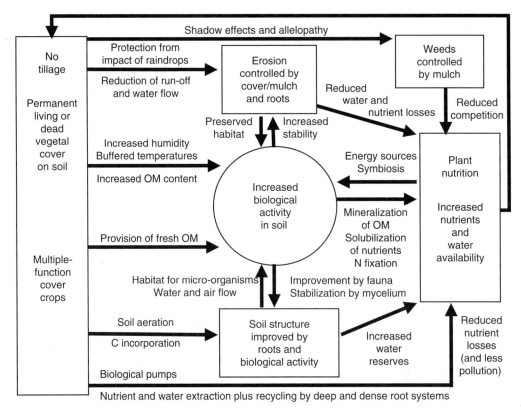

FIGURE 1.1
Schematic diagram of elements and interactions in the management of field cropping in Brazil with direct seeding into permanent vegetative cover to capitalize on plant and soil biota complementarities and synergies. Courtesy of Lucien Séguy, CIRAD.

harboring beneficial insects or putting out more root exudates for the benefit of soil microbes. The purpose is to create more favorable conditions for desired plant growth, through greater soil porosity, canopy humidity, optimum microclimatic temperatures, nutrient mobilization, and other services. Plants have co-evolved over eons with soil organisms, starting with the services of mycorrhizal fungi some 400 million years ago as ascertained from the fossil record (Margulis and Sagan, 1997). Trying to improve the performance of plants with little or no reference to the environment in which they have evolved, and with which they have developed systemic interdependence, will surely limit future success.

1.5 Book Design

The perspective taken here is not a radical departure from present agriculture, though some proponents of more biologically-oriented approaches may prefer to emphasize the differences rather than the similarities. We do not see new approaches as displacing present ones in a zero-sum way. Rather, there is likely to be an accelerated evolution and acceptance of new practices, informed by emerging scientific knowledge and by the success of alternative methods. The impetus for this book arose from the substantial improvements in crop performance that were being achieved through changes in the management of plants, animals, soil, water, and/or nutrients. The innovations and evaluations reported in Chapters 18, 19, 21–23, 28, 29, 31, 34, 41, and 42 are ones with which members of our editorial group were personally acquainted or involved. We were encouraged also by other work that we were learning about, undertaken by colleagues from around the world and reported in other chapters in Part III and Part IV.

This volume is intended to communicate the scope and substance of biological approaches to managing soil systems (a) to scientists who have an interest in innovation and practice, and (b) to practitioners who are concerned with improving agriculture on a broader scale, both spatially and over time, based on better scientific understandings. Part II presents the major important components and aspects of soil systems from multiple disciplinary perspectives. It aims to provide a reasonably inclusive appreciation of what is known about biological factors in soil systems' functioning. Part III seeks to acquaint readers with what is being done around the world with the knowledge that is accumulating for supporting new management approaches. It covers accomplishments and limitations of current efforts to capitalize on biological potential for improving agricultural systems. Part IV puts these approaches into a more comprehensive setting, addressing important complementary issues and opportunities as well as problems and methods of measurement, monitoring, and modeling (Chapter 46 to Chapter 48). Chapter 49 assesses the current and prospective context of agricultural development strategies and then considers what advantages can be achieved with more biologically-driven approaches to cropping system improvement. Chapter 50, in conclusion, considers some issues and opportunities related to these new directions for research and development.

Before reviewing the constituent elements and processes of soil systems in Part II, the remaining chapters in Part I provide three overviews of soil systems and their management, first in tropical settings (Chapter 2), then in temperate ones (Chapter 3), and further, under arid and semi-arid circumstances (Chapter 4). These discussions provide some integrated perspectives on the challenges that farmers and policymakers face for making better use of available soil resources under the different circumstances for their management.

1.6 Challenges for Agricultural Science and Practice

The world in the twenty-first century is going to be quite different from that of the recent past, as seen from the global data and time-series reviewed in Chapter 49. Key trends include the following:

- Arable land per capita by 2050 will be only about one-third of what it was a century earlier, projected to be about 0.08 ha per capita, having been 0.24 ha in 1950 (Worldwatch Institute data). The world's farmers will have to increase their production considerably from the available arable land area. They cannot afford to use nonsustainable practices to reduce arable area or environmental goods and services. Most agricultural land will need to be utilized more intensively as extensive cultivation practices, inherently less productive, become less feasible.

- Production increases will have to be accomplished with diminishing availability of water per capita. More efforts will be needed, for example, to conserve and utilize "green water," i.e. water that is absorbed and stored in the soil and used *in situ* (Savenije, 1998). This is different from "blue water," which is captured in artificial or natural storage facilities or pumped from underground stores and then conveyed to some point of use. Finding ways to induce plants to develop better plant root systems and building up soil organic matter will be important for making better use of potential soil water stores.

- Global climate change will adversely affect the productivity of a large part of the world's farmland, especially in the tropics and subtropics, where many areas will shift from productive to semi-arid status, while other areas will go from being semi-arid to arid. Some agricultural areas at higher latitudes may benefit from gradual warming, but these regions are not where the world's projected needs for food production and consumption are greatest. Moreover, helping plants develop better root systems in association with having more effective communities of supportive soil biota will be important in the future for enabling crops to cope better with climatic fluctuations and extreme events.

- Ways must be found to reduce the loss of existing soil resources through erosion, salinization, and other forms of land degradation, so that present land shortages will not become even more serious. Already about one-quarter of the world's arable land can be classified as degraded, and almost one-third of this vulnerable arable area is in Africa and Central America (Brady and Weil, 2002). As far as is possible, we will need to find ways to restore and improve land previously lost through soil-degrading practices.

For all these tasks, more knowledge of and reliance on biological processes will be part of the process of making soil systems more productive and more sustainable, preserving them and, where they are degraded, rehabilitating them. Fortunately, the contributions to this book provide evidence and reasons to think that these outcomes are attainable. We are pleased that the largest number of the innovative management systems reported in Part III are from Africa, where more productive and more sustainable agriculture is most needed.

Each innovation reported is somewhat different, and where successful it has been adapted to the agroecosystem into which it was introduced, with a view to capitalizing upon biological potentials not just of the crop (or animal) of interest, but of the whole assembly of flora, fauna, and microorganisms in that system. The production benefits from such approaches are summarized in Chapter 49 (Table 49.4), to give readers an overview of the magnitude and range of agronomic advances already being obtained with such methods around the world, under a great variety of climatic, soil, socioeconomic, and other conditions. The wider benefits achievable in terms of natural resource conservation, environmental quality, accessibility for capital-limited farmers, and ultimately human health and well-being are much harder to quantify and compare.

No final or global assessment is presently possible since sustainability can only be validated over many years, and many aspects will probably remain incommensurable. What can be said with some confidence now is that these approaches present great potential that is well worth investigating and pursuing. How far they spread will depend on their utility for farmers and their households, and on the net benefits they produce for larger social and environmental systems.

References

Brady, N. and Weil, R.R., *The Nature and Properties of Soils*, 13th ed., Prentice-Hall, Upper Indian River, NJ (2002).

Degenhardt, J. et al., Attracting friends to feast on foes: Engineering terpene emission to make crop plants more attractive to herbivore enemies, *Curr. Opin. Biotech.*, **14**, 169–176 (2000).

Khan, Z.R., et al., Intercropping increases parasitism of pests, *Nature*, **388**, 631–632 (1997).

Green, H. et al., Suppression of the biocontrol agent *Trichoderma harzianum* by mycelium of the arbuscular mycorrhizal fungus *Glomus intraradices* in root-free soil, *Appl. Environ. Microbiol.*, **65**, 1428–1434 (1999).

Ladha, J.K. et al., Opportunities for increased nitrogen-use efficiency from improved lowland rice germplasm, *Field Crops Res.*, **56**, 41–71 (1998).

Margulis, L. and Sagan, D., *Microcosmos: Four Billion Years of Evolution from Our Microbial Ancestors*, University of California Press, Berkeley (1997).

Pretty, J., *Agri-Culture: Reconnecting People, Land and Nature*, Earthscan, London (2002).

Pretty, J., Ed., *The Pesticide Detox*, Earthscan, London (2005).

Rasmann, S. et al., Recruitment of entomopathogenic nematodes by insect-damaged maize roots, *Nature*, **434**, 732–737 (2005).

Rousseau, A. et al., Mycoparasitism of the extramatrical phase of *Glomus intraradices* by *Trichoderma harzianum*, *Phytopathology*, **86**, 434–443 (1996).

Savenije, H.H.G., The role of green water in food production in sub-Saharan Africa, Paper prepared for FAO program on Water Conservation and Use in Agriculture (WCA) (1998), (http://www.wca-infonet.org).

Tegtmeier, E.M. and Duffy, M.D., External costs of agricultural production in the United States, *Int. J. Agric. Sustainability*, **2**, 155–175 (2004).

Turlings, T.C., Tumlinson, J.F., and Lewis, W.J., Exploitation of herbivore-induced plant odors by host-seeking parasitic wasps, *Science*, **250**, 1251–1253 (1990).

van Noordwijk, M., Cadisch, G., and Ong, C.K., Eds., *Below-Ground Interactions in Tropical Agro-ecosystems: Concepts and Models with Multiple Plant Components*, CAB International, Wallingford, UK (2004).

Waibel, H., Fleischer, G., and Becker, H., The economic benefits of pesticides: A case study from Germany, *Agrarwirtschaft*, **48**, 219–230 (1999).

Wardle, D.A., *Communities and Ecosystems: Linking the Aboveground and Belowground Components*, Princeton University Press, Princeton, NJ (2002).

Whipps, J.M., Microbial interactions and biocontrol in the rhizosphere, *J. Exp. Bot.*, **52**, 487–511 (2001).

Wolfe, D., *Tales from the Underground: A Natural History of the Subterranean World*, Perseus, Cambridge, MA (2001).

2

Soil System Management in the Humid and Subhumid Tropics

Ana Primavesi

University of Santa Maria, Rio del Sul, Brazil

CONTENTS

Although tropical soil systems represent an important part of the world's diversity of soil ecosystems, they have in the past received less scientific attention than temperate zone soils. Soil systems in both categories operate according to the same principles, but climatic differences have put them on divergent paths. Neither should be taken as a norm from which the other deviates. In recent years, research on tropical soil systems has been greatly expanding, helping us to understand better the general principles according to which soil systems function.

While it is true that agricultural productivity has commonly been greater in temperate regions, differences often arise because tropical soils have frequently been managed with assumptions, practices, and technologies transferred from temperate climatic zones. In fact, tropical soil systems can be very productive when their comparative advantages are utilized. Under natural conditions their gross production of biomass above- and

TABLE 2.1

Broad Contrasts of Temperate and Tropical Soil System Characteristics

	Tropical Oxisol/Utisol	**Temperate Mollisol**
Temperature regime	Warm to hot	Cold to warm
Predominant clay form	Variable-charge 1:1 clays, kaolinites (Si–Al)	Permanent-charge 2:1 clays, montmorillonites (Si–Al–Si)
Cation exchange capacities (CEC)	Poor (7–150 mmol dm^{-3})	Rich (500–2200 mmol dm^{-3})
Soil pH	Acidic (5.6–5.8 or lower)	Neutral (6.8–7.0 or higher)
Rooting depth	Shallow to very deep	Moderate to deep
Key compounds	Fulvic acids, leading to leaching of cations	Humic acids, increasing CEC
Biological properties affecting agriculture	Abundant soil biota	Less numerous and diverse soil biota

belowground can be multiples of what is produced in temperate climates, in spite of the poverty of tropical soils when these are assessed only in terms of available mineral nutrients. A tropical Amazonian forest ecosystem, even on poor sandy soil, can produce as much biomass in 18 years as will be created under a boreal northern forest on richer soils in 100 years (Primavesi, 1980).

Even when they are poor in chemical terms, tropical soils can be biologically rich in comparison with the kind of temperate soil systems that are considered in Chapter 3. Whenever soil conditions are warmer and more humid, there will be greater activity of chemical, physical and particularly biological processes. Table 2.1 compares some typical traits of tropical Oxisols and Ultisols with those of a Mollisol more characteristic of the fertile plains of Eastern Europe and North America. The comparisons indicate some contrasting differences between typical humid-tropical and temperate soils.

Tropical soil systems are located in a broad band around the equator between the Tropic of Cancer and Tropic of Capricorn ($23\frac{1}{2}°$ lat. N and S). They differ from temperate systems most obviously in terms of their relatively constant temperatures under widely differing moisture regimes. Some temperature variations naturally occur due to elevation and rainfall regimes, and neither tropical nor temperate regions are homogeneous. Addressing all this variation in a single chapter is impossible, so the focus here will be mostly on the humid tropics, although the discussion is largely applicable to the subhumid tropics as well. Comprehensive reviews of tropical soils have been provided in Sanchez (1976) and Primavesi (1980). Significant differences in soil systems and their management arise once the climatic regimes in tropical areas become arid or semi-arid as reviewed in Chapter 4.

The tropics with their considerable variation in rainfall (with climates ranging from very humid to arid) and vegetation (from dense forest to cleared agricultural lands) encompass a large part of the earth's surface. All together they are home to about one-third of the world's population (Bonell and Hufschmidt, 1993). This region is most noted for its tropical rain forests, but these are diminishing in area under pressures of agriculture, logging, and development.

At the equator, day length is always about 12 h, considerably shorter than the long summer days in temperate regions. The latter can have 50–100% more daily solar radiation during their summer growing season, but year-round exposure to intensive sunlight gives tropical soil systems more total energy to produce biomass above- and belowground. The other main asset in much of the tropics is moisture, although excess rainfall can become a liability. In the humid equatorial zone where rainfall is most abundant and almost continuous, the effects of this resource become dominant. Under native vegetation in the Amazon basin, precipitation can exceed 4000 mm year^{-1}. This includes localized recycling of water transpired by the trees which can be responsible for 50–70% of the local precipitation, accounting for the term rain forest.

There is ongoing debate over whether the clearance of large areas of tropical forest leads to a reduction in rainfall. In the large *cerrado* area of north-central Brazil, for example, where savannahs rather than forests now prevail, decades of monocropping appear to have contributed to smaller rivers drying up and diminished water flow in large ones (Angelo, 2005). It is not agreed yet how much of this trend is attributable to land management practices and how much to climatic changes, but both are related to the functioning of soil systems. The cerrado agroecosystem is disscussed more in Chapters 21 and 22.

The higher temperatures and moisture in tropical areas that are in general favorable for plant growth can create an aggressive climate with rapid mineralization of organic matter and extensive leaching of nutrients, natural or applied, from upper soil horizons. Rainbursts on clean-weeded soil that has been deep-plowed with high application rates of nitrogen fertilizer and of lime to raise pH contribute to rapid decomposition of organic matter, with consequent release of CO_2, loss of soil aggregation, and soil compaction (Primavesi, 1980).

Soil management practices in tropical regions thus need to be carefully chosen and carefully executed so as not to contribute to soil compaction and the worsening of any deficiencies or imbalances in soil nutrients and biota. The challenge in managing tropical soil systems is to achieve sustainable, optimal productivity over time, rather than to maximize highest but unsustainable economic returns that lead to soil and environmental degradation. Tropical areas that have arid or semi-arid hydrological regimes require their own tailored management practices (Chapter 4).

2.1 Soil System Dynamics and Biodiversity in the Humid Tropics

Soil systems with all their biodiversity and continuous interaction of flora and fauna function as a kind of collective entity, not as a collection of isolated factors or subsystems. This multiplicity is the basis of life on our planet (Primavesi, 2003b). Changes in soil moisture drive a kind of subterranean atmospheric circulation that is not unlike the currents in the ocean (Jackson, 2004). Soil moisture changes are part of the overall hydrological cycle that includes not just rain, wind, and extreme storm events, but also evaporation, erosion, and interactions with the aboveground climate. Our scientific knowledge has advanced to the point where soil moisture levels and distribution are now being monitored by two satellite systems, Aqua from NASA and ADEOS II from Japan, as part of weather forecasting, recognizing that the underground supplies and the movement of water substantially affect the wider climate.

In the humid tropics, plant growth can be luxuriant even on predominantly sandy soils with no more than 1% clay content. With a naturally low supply of nutrients per unit volume of soil, plants' growth depends (1) on intense soil biotic activity; (2) on the development of root systems that can explore larger volumes of soil, especially helped by mycorrhizal fungi that substantially increase the root–soil interface; and (3) on the rapid turnover of organic matter. This turnover is usually within 6–8 weeks compared to 3–6 years in some temperate systems. Nutrient recycling is driven by extensive and complex interactions among soil organisms, for example, by the mobilization of nutrients from soil stocks, normally silicates, and nitrogen fixation by aerobic and anaerobic bacteria including rhizobacteria. There is even nitrogen fixation by bacteria that live on the leaves of tropical forest trees, on what is called the phylloplane (Brighigna et al., 2000; Lindow and Brandl, 2003).

Nutrient pollution does not occur in tropical soils under natural vegetation because surpluses get mostly taken up by the abundant vegetation and soil biota. Such systems have little or no leaching of nutrients and there is no pollution of water resources by excess nitrate, phosphate, and other compounds as easily happens when cultivated temperate soils are fertilized. Massive root systems enable many tropical plants to prosper in these environments, as discussed below. It should be noted, however, that root biomass can be even higher in semi-arid savannahs, where plants must adapt to water scarcity, showing how flexible and opportunistic plants' structural and physiological strategies for survival can be.

2.1.1   Three Puzzles of Tropical Soils

2.1.1.1   Nutrient Availability

Luxuriant Amazonian rain forests grow on soils that are considered very poor when assessed by conventional measures. However, when technology is transferred from temperate climatic zones to the tropics, the diversity of flora usually disappears. This does not mean that tropical soils are inherently unproductive, but rather that the practices introduced were inappropriate to the local conditions. While temperate-zone technologies can produce good yields for a while, this usually happens by exploiting soils in ways that contribute to their degradation over time (Senna de Oliveira et al., 2000). This has been seen in some subhumid or semi-arid tropical regions such as the *cerrado* and some other parts of northeast Brazil, in Colombia (the Cauca valley), Cuba, and other tropical areas (Primavesi, 1997). Intensive mechanized, monocrop agriculture runs into difficulties when practiced close to the equator and elsewhere as well. The sustainable productivity of tropical soils depends on the careful exploitation of their intrinsic qualities and processes.

2.1.1.2   Nutrient Replenishment

A tropical soil system that is exhausted and lacking in available nutrients due to continuous cultivation can recover all or most of its exported nutrients within 8–10 years under regrowth of native vegetation in the humid tropics if the tree seed pool still remains (Khatounian, 2001). Where do these nutrients come from? This capacity for tropical soil regeneration has been exploited for centuries by small-scale farmers practicing shifting cultivation (or slash-and-burn) not only in the tropics but also in Europe and North America a few centuries ago. During the fallow period, some nutrients are accumulated in the biomass from rainfall, biological nitrogen fixation, and the dissolution of weatherable minerals. Considerable amounts of nutrients exist in the soil in unavailable forms that can be made available through a combination of chemical, physical, and biological processes, driven by complex interactions among a multiplicity of flora and fauna in the soil. The limitations as well as the prolific nature of such processes need to be appreciated.

2.1.1.3   Organic Matter

At the same time, with monocropped tropical soils, even having a thick ground cover of organic matter, e.g., in cacao plantations, there is no guarantee of crop health and productivity. If organic matter is so important for plants' performance and protection, how can this happen? Quite evidently, organic matter by itself is not enough. To explain the successes or failures of cropping efforts especially in the humid tropics, we need to consider the many interactions among plants, and especially their roots, soil particles, water, microbes, and nutrients. Finding satisfactory answers to the questions posed above requires consideration of the contributions that are made by biodiversity, below as well as above ground.

2.1.2 Biodiversity as a Pervasive Fact of Life in the Tropics

The biodiversity of plant species in the tropics can be 100 times greater than in temperate climates. In Amazonia, indeed, there exist 400,000 different plant species. In general terms, we believe that biodiversity above and below ground are related, even if there have not been sufficient studies for satisfactory quantification. Biodiversity is the base of sustainable agriculture, restoring the productivity of sites when they have been utilized (Primavesi, 2003b). The soil food web, discussed in Chapter 5, more complicated than earlier notions of a soil food chain, is continuously renewed by these processes. Diversified life in the soil depends on similarly diversified plant cover, just as this vegetation benefits from the multiplicity of minerals made available through biotically mediated processes and recycling (Hayman, 1992; Hungria and Urquiaga, 1992; Tsai and Rosetto, 1992).

Plants contribute to the maintenance of biodiversity in part by controlling their own reproduction strategies which, if overly successful, would lead to dominance and homogeneity. A number of tropical plants are known to release substances that are toxic to their own seeds, hindering germination up to a distance of 50 m (Dobremez, 1995). In tropical forests, it is not uncommon to find at most three trees of the same species within an area of 1 ha, for example, rubber, Brazil nut, and mahogany trees. This effect should not be characterized as allelopathy because the chemical inhibition is extended to the plant's own species. This evolutionary strategy benefits plant species both individually and collectively because when roots systems are more diverse, they can exploit a larger volume more thoroughly than if all had the same morphology. This creates more nutrition opportunities and also soil aggregation. Wherever vegetation is homogeneous as with monocropping, plants' root space is restricted and the soil compacts, with plants, especially trees, becoming starved, making it necessary to provide inorganic fertilization. Only with biodiversity can so massive a biomass as the Amazonian forest exist. The biodiversity seen above ground is mirrored below.

2.2 Nutrient Dynamics and Soil Management in Tropical Ecosystems

One can identify at least 18 mineral elements that are necessary for plant growth (Brady and Weil, 2002), but the number may well be higher. Especially in tropical soils, it is easy for fertilization to lead to deficiencies in some nutrients as a consequence of the abundance of others, and also to other problems. In soils that are compacted and chemically depleted, root systems cannot develop properly to access all the nutrients that they need for the plant. Plants lacking complete or balanced supplies of nutrients will become still more unbalanced nutritionally when mineral fertilizers are then applied. Also, such plants need more watering when the soil solution's concentration becomes too high and osmotic pressure is affected. Applying only macronutrients such as N, P, or K alters ratios between macro- and micronutrients (Bussler, 1968; Homés, 1972; Primavesi, 1980), which makes plants more vulnerable to disease (Chaboussou, 2004). We learned long ago that supplying an abundance of mineral nitrogen to rice plants, for example, makes them vulnerable to attacks of *Pyricularia oryzae* (rice blast) due to nitrogen-induced copper deficiency (Primavesi et al., 1971).

Too much of any element and correspondingly too little of others opens the door to a variety of diseases in crops. Further, supplying certain nutrients but not others enables particular species of flora (weeds) and fauna (pests) to prosper at the expense of other species. This creates unfavorable conditions for crop growth. So while it is important to maintain proper balance among nutrients in all soil systems, this is particularly true in the tropics where soil interactions are more dynamic and where imbalances can occur more frequently.

Organic soil amendments have the advantage of providing more or less a full range of nutrients in contrast to mineral fertilizers that supply only a few minerals, and the ratios of nutrients made available are approximately those that plants need. Plant species generally have adapted their growth requirements and patterns to the soils in which they have evolved or been bred.

Chemical analyses to assess the status of individual nutrients seldom give an adequate understanding of the nutrient situation at that location. Analyses of soil nitrogen, for example, reveal only the quantity of nitrogen present at that point in time, whereas nitrogen levels can vary considerably over time. Moreover, such tests do not assess the interrelations of that nitrogen with other elements such as copper or molybdenum. Nor do they include the dynamics of nitrogen fixation by bacteria or losses to the air as N_2 or N_2O when soils are compacted. Phosphorus also presents measurement difficulties because the several methods for extraction used to determine the level of "available P" in soil all give different values. These can differ by as much as seven times depending on the method of extraction used. So phosphorus test results can report spurious precision. (Variability in the measurement of phosphorus is discussed in Chapter 13.)

Not just the absolute amounts of nutrients are important, but also for many nutrients the forms in which they are available. In any soil, the amount of a nutrient that will actually be available to plants is affected also by factors such as soil structure, moisture, and temperature. For their nutrient acquisition, plants depend not simply on the supply of the nutrient in that soil assessed in physicochemical terms, but also on microbial activity, which is affected by root exudation, soil organic matter (SOM), and oxygen supply, factors little considered in soil testing.

Fertilizing the soil with certain elements is not the only way to help plants acquire the nutrients they need. Root systems can be promoted through management practices, such as by alternating the planting of different varieties or by combining certain crops. It is difficult to obtain just the right balance by applying the same amounts of fertilizer to different crops because each crop species and variety has its own optimum requirement.

Where there are naturally-occurring nutrient deficiencies in the soil, soil amendments can be beneficial. In the tropics, boron, for example, is sometimes lacking, and this reduces root growth. Root/shoot ratios improve when sufficient boron is present, so wherever there are boron deficiencies, supplying this nutrient is advisable. Soil amendments with powdered rock phosphate can be used to correct deficiencies of phosphorus. However, because these materials may contain other elements, e.g., fluoride, there can be limits to their utility. No fine-tuning of nutrient supplies is required if plants are enabled to better acquire for themselves the nutrients that they need. Soil and water management practices can allow plants to optimize investments of their own internal nutrients so as to access larger volumes of soil through better root systems.

2.2.1   Interactions between Physical and Biological Factors

Nutrient availability in the soil is thus a necessary but not a sufficient condition for plant growth and health. The uptake of nutrients by plant roots depends on both physical and biological conditions. Favorable soil structure with sufficient porosity is needed for air and water to permeate. SOM is not just a source of nutrients for plants; it provides carbon to feed the soil biota that build up and sustain soil aggregation. In soil with high bulk density, lacking sufficient air, many plant nutrients become reduced, losing their oxygen by exchanging it for hydrogen. This can be detrimental to plant growth and performance.

Aggregation is an all-important process in soil systems, and it can be problematic in tropical soils if biological processes are compromised. Microbes combine the micro-aggregates into bigger crumbs (0.01–0.1 mm) that are built up in soil by trivalent Fe and Al oxides and hydroxides. By producing polyuronic acids, they stick these tiny particles together to form bigger aggregates (0.5–2.0 mm). These aggregates are in turn processed by earthworms and other soil fauna into still larger aggregates, up to 4.0 mm.

The water stability of aggregates, i.e., their resistance to being broken down by water drops, generally lasts only 8–10 weeks. Within this time, the hyphae of fungi exhaust their supply of nutrients and so can no longer hold the aggregates together. Aggregates that have lost their stability must be protected against rain splash to conserve the soil's porous structure until other crumbs can be glued together by microbial activity. Well-aggregated, protected soils resist the force of rain drops, which is especially important in the tropics, allowing infiltration rates that can reach 400 mm h^{-1} (Werner, 1982).

Pores, according to their size, have different capacities: mesopores conserve water within the soil making it available for plant growth; micropores retain water, while macropores drain it into the subterranean watershed, feeding springs and rivers. Pores thus mitigate or avoid soil erosion, flooding and drought, all of which are critical abiotic stresses throughout the tropics, made worse by soil compaction and the consequent run-off of rainwater. Although these soil characteristics are classified as physical, they reflect the biological conditions of the soil (Barnes et al., 1971).

2.2.2 Physical and Chemical Interactions

In the tropics as well as in temperate regions, certain soils compact easily, especially when they are sandy with sufficient silt and clay to promote crusting. Agricultural practices such as plowing and herbicidal weed control, as well as widely spaced crop rows deprive the soil of surface protection and expose it to overheating and rain splash. In Brazil, spacing is now being reduced for almost all crops to give more soil protection. In fruit plantations, interrows are increasingly kept covered with native vegetation. When the organic matter content of the soil decreases and soils get compacted and impermeable to water and air, water absorption diminishes, with increased runoff, erosion, and flooding as a consequence. This can change microclimates and reduce biomass production when soils become salinized (if there is a source of sodium in the subsoils or irrigation water), or when they are exposed to higher temperatures and evapotranspiration with an increase in albedo, i.e., reflection rather than absorption of solar radiation (Senna de Oliveira et al., 2000).

Liming may correct low pH and neutralize Al, Fe, and Mn toxicity. However, this also makes SOM disappear more rapidly due to increased numbers of zymogenic bacteria. This contributes to loss of the aggregated soil structure and to pervasive compaction that is difficult to remedy when some soils are limed to reach a pH of 7. This does not happen when soils are limed at a rate to reach a lower pH (5.6) where exchangeable Al is precipitated. Overliming can make problems worse. Some tropical soils may need the addition of some calcium as a nutrient (Primavesi and Primavesi, 1964), but this is not as beneficial as a corrective to raise soil pH as commonly thought. In the tropics, the most favorable pH which will enable most microbial enzymes to have their maximum activity is 5.6–5.8, lower than the optimum pH in temperate climates, 6.8–7.0. Addition of organic matter can assist in the short-run correction of pH, as well as in amelioration of toxic Al, Mn, or Fe. This also promotes the aggregation, aeration, and oxidation processes (Döbereiner and Alvahydo, 1966).

Soil physics and chemical processes are affected adversely by land clearing. When trees are cut down and winds are free to sweep across the fields, evapotranspiration is increased. Such stress can lower crop yields by as much as 50–70% (Grace, 1977). With no protective measures such as windbreaks, a constant breeze can diminish soil humidity up to an equivalent of 750 mm rainfall year^{-1} (Rodrigues, 2000), making water stress a problem in an area that has otherwise sufficient rainfall.

Irrigation can compensate in the short run for water losses from the plants and soil, but yields will be lower than when wind flow is slowed or impeded by higher plants, shrubs or trees (Grace, 1977). Even though this is known in the literature, such knowledge has seldom been incorporated into management systems. For the sake of expanding field size to accommodate mechanization, the Cuban Government eliminated windbreaks over large areas in the 1960s. This had drastic effects on the soil systems needed to support agricultural and animal husbandry, for example, milk production dropped due to a lack of shade for cows (Funes et al., 2001).

2.2.3   Root Development and Nutrient Access

For plants to absorb water and nutrients through their roots, they must fix CO_2 in their leaves to support the metabolism needed to maintain this absorption. Roots can function as pumps that bring up nutrients from lower horizons, which is especially important in zero-tillage (Chapter 22). Extensive root system development is essential for best nutrient access and absorption, augmented by mycorrhizal fungi that infect plant roots and acquire nutrients for plants from portions of the soil that are not accessed by the roots themselves (Chapter 9). Among other exudates, roots produce substances that chelate (or can produce substances that favor the development of bacteria that do the chelating of) nutrients such as Fe, Cu, and Zn, maintaining these in a soluble state. The fact that root exudates change when plants receive foliar fertilization indicates that exudation processes are regulated by plants' needs for certain nutrients. Heterotrophic soil microorganisms, especially those in the rhizosphere, are continuously active, though at different rates, in mobilizing nutrients.

With their greater soil moisture and temperature, humid tropical soil systems have more physical potential for greater biological activity. The biodiversity in natural ecosystems as well as in the polyculture practiced widely by small farmers contributes to the root development of plants compared with monocultural agriculture (Primavesi, 2003a). Having a diversity of plant species or even of varieties creates a greater multiplicity of root exudates that can support greater soil microbial biodiversity. With this comes more diversity of other soil organisms within the food web. Planting systems with two crops in alternated rows can increase yields of each because the roots of each variety are able to enter the root space of the other, thereby exploring a larger soil volume. This and other mechanisms that can account for the greater productivity of polyculture are reported in Chapters 39 and 40. It is a very fortunate adaptation for tropical agriculture that plants' root development generally increases in nutrient-poor soils (Primavesi, 1980, 2001).

2.3   Managing Soil Systems for Tropical Agriculture

The principles that govern soil system dynamics are the same for all parts of the world. However, how specific soil systems are best managed for productivity and sustainability will vary depending on the conditions created by interaction among climate, edaphic, and biotic factors, as well as human interventions (Primavesi, 2003b). What are regarded as problem soils can have some intrinsic chemical or physical limitations. But most often their problems are a consequence of how they have been used or misused, with adverse effects on their biological components.

Aerobic microbes, particularly cellulolitic ones, are especially important as part of these large-scale soil processes because they are the very efficient decomposers of organic material, are responsible for soil aggregation and structure, and are efficient mobilizers for plant nutrition, while being at the same time food sources for larger organisms in the soil food web (Chapter 5). This makes successful soil-system management in large part management for, if not of, microorganisms. The following practices, some considered in more detail in Part III, can help to achieve more sustainable forms of farming particularly in, but not only for, tropical regions.

2.3.1 Agroforestry

This strategy of integrated land-use management combines perennial trees, shrubs, and grasses, and often livestock, with annual crops, in a purposefully diversified set of plant and animal species that capitalizes on synergies among them. Such systems are evaluated in Chapters 19–21.

2.3.2 Weed Management

Weeds are generally regarded as a major problem in the tropics, where the greater plant growth under favorable temperature and moisture conditions includes that of unwanted plants which can interfere with farming operations. However, under certain conditions, weeds can have positive effects, e.g., for soil protection and soil microbial diversity. Thus, management need not always aim to eliminate all weeds (Tripathi, 1977; Altieri, 1995). In the tropics, where there is an abundance of light and water, there will be times and places where weeds are not serious competitors with crops for these resources, in which case a more tolerant approach can be taken. Where excessive weed growth is a problem, often this can be coped with by management practices such as crop rotation, spacing, and mechanical weed control rather than by chemical applications.

2.3.3 Crop Combinations and Rotations, and Use of Green Manures

In-field diversity of plants commonly gives agronomic and economic advantages, although certain combinations of plants can cause crop losses where species have antagonistic or allelopathic effects. Legumes are usually very productive in combination with cereals but are antagonistic to onion and garlic, for example. Farmers have found that sunflowers should not be used in a rotation with or near potatoes, tomatoes, or tobacco as then none develop normally. However, in general, crop rotations and the use of noncrop plants as green manures can enrich the soil and benefit soil fertility.

Growing different plant species together is often beneficial because different root systems tap different horizons and niches within soil systems, as discussed in Chapters 39 and 40. Because different plants having particular biochemical profiles of exudates are associated with different sets of soil organisms, this strategy can build up more robust biological foundations for ongoing soil fertility. Agricultural researchers would do well to experiment more with various forms of polycropping, moving away from the monocropping preference that has come to dominate cultivation strategies over the past century for the sake of large-scale mechanization.

2.3.4 Maintaining Soil Cover with Diversified Crops

Keeping the soil covered during and also between seasons is important for protecting the life of the soil. This makes the soil less subject to erosion and heat damage and creates a better environment for biodiversity of all kinds. Methods for this that have been developed in Brazil are presented in Chapter 22. Farmers are more often interested in maximizing the economic productivity of their crops rather than in emphasizing sustainability. However, measures favoring the latter contribute to cost-effective economic productivity over time if one looks beyond single seasons, and especially if one wants to make best use of our limited fresh water supplies. Conserving water *in situ*, in the soil where plants and various organisms need them, is the most rational approach. The production and use of more organic matter within a cropping system has many advantages, especially in the humid tropics, for its beneficial effects on ground cover and to have more SOM for promoting soil aggregation.

2.3.5 No-Tillage Farming with Mulch Applications

Alteration of conventional cultivation practices to match natural vegetative systems more closely is gaining popularity in Brazil and other countries especially for avoiding soil erosion and floods. This strategy for soil-system management, discussed in more detail in Chapters 22 and 24, need not be elaborated here, although it should be emphasized that mulch is as important a part of this system as are changes in tillage. One key element in such practice, especially in the tropics, is to keep adding diversified organic matter to the soil, for example, by rotating as many as five different crops. This enhances diverse microbial and mesofaunal life in the soil as well as maintaining a mulch layer that provides soil organisms with a favorable C:N ratio. If soil is compacted to begin, zero-tillage without mulch can be the worst farming system, when neither sufficient air nor water reach plant roots, whereas with mulch it can be the best (Govaerts et al., 2005).

2.4 Discussion

No agricultural technology is sustainable if not consistent with the natural processes and dynamics of a given agroclimatic region. While some highly extractive technologies may yield well in the tropics for a few years, their benefits turn to net costs over time and become eventually unsustainable. The transfer of technologies developed for temperate agricultural production to tropical areas, instead of benefiting the latter, has often contributed to their poverty as tropical soil systems get degraded through uses that are contrary to their natural capabilities and that disregard their constraints.

Soil, water, microbes, nutrients, plants, organic matter, environment, and climate are all intrinsically interrelated and must be seen and managed as a system, i.e., as an agroecosystem, rather than as a collection of units which can be replaced or compensated for by advanced technology. The exploitation of compacted soils in which diverse life forms can no longer live is always unsustainable, and it disrupts hydrologic cycles, creating more arid microclimates that cannot sustain the plant and other growth necessary for healthy ecosystems and healthy people. Tropical soil systems in particular have to be managed according to their needs, keeping all of the relationships discussed here in balance. With appropriate management, these systems can be very productive and can benefit people for many generations to come.

References

Altieri, M., *Agroecology: The Science of Sustainable Agriculture*, 2nd ed., Westview, Boulder, CO (1995).

Angelo, C., Punctuated disequilibrium, *Sci. Am.*, **292**, 22–23 (2005).

Barnes, K.K. et al., *Compaction of Agricultural Soils*, American Society of Agricultural Engineers, St. Joseph, MI (1971).

Bonell, M. and Hufschmidt, M.M., *Hydrology and Water Management in the Humid Tropics*, Cambridge University Press, Cambridge (1993).

Brady, N.C. and Weil, R.R., *The Nature and Properties of Soils*, 13th ed., Prentice-Hall, Upper Saddle River, NJ (2002).

Brighigna, L. et al., The influence of air pollution on the phyllosphere microflora composition of *Tillandsia* leaves (Bromeliaceae), *Rev. Biol. Trop.*, **48** (2000).

Bussler, W., Symptome und Symptomsequenzen bei Ernährungsstörungen von höheren Pflanzen, *Kalibriefe*, **2/3** (1968).

Chauboussou, F., *Healthy Crops: A New Agricultural Revolution*, Jon Carpenter, Charnley, UK (2004).

Döbereiner, J. and Alvahydo, R., Eliminação da toxdes de Mn pela matéria orgânica em solo 'gray hidromorfico', *Pesq. Agropec. Bras.*, **1**, 234–248 (1966).

Dobremez, J.F., Guerre chimique chez les vegeteaux, *La Recherche*, **179**, 912–916 (1995).

Funes, F. et al., Eds., *Transformando el Campo Cubano*, Centro de Estudios de Agricultura Sostenible, Havana (2001).

Govaerts, B., Sayre, K.D., and Deckers, J., Stable high yields with zero tillage and permanent bed planting? *Field Crops Res.*, **94**, 33–42 (2005).

Grace, J., *Plant Response to Wind*, Academic Press, London (1977).

Hayman, M., Phosphorus cycling by soil microorganisms and plant roots, In: *Soil Microbiology: A Critical Review*, Walker, N., Ed., Butterworth, London, 67–91 (1992).

Homés, M.V.L., The effect of completely equilibrated fertilizer on the production of plants cultivated on large scale, *Scient. Varia*, **38**, 469–502 (1972).

Hungria, M. and Urquiaga, S., Transformações microbianas de outros elementos, In: *Microbiologia do Solo*, Cardoso, E.J.B.N., Tsai, S.M., and Neves, M.C.P., Eds., Sociedad Brasileira do Solo, Campinas (1992).

Jackson, T.J., *Hydrologic Remote Sensing, USDA/ARS 3: 20–22*. US Department of Agriculture, Agricultural Research Service, Beltsville, MD (2004).

Khatounian, C.A., *A Reconstrução Ecológica da Agricultura*, Ed., Agroecologia, Botucatú (2001).

Lindow, S.E. and Brandl, M.T., Microbiology of the phyllosphere, *Appl. Environ. Microbiol.*, **69**, 1875–1883 (2003).

Primavesi, A., *Manejo Ecológia do Solo*, Nobel, São Paulo (1980).

Primavesi, A., *Agro-ecologia: Ecosfera, Tecnosfera, Agricultura*, Nobel, Saõ Paulo (1997).

Primavesi, O., *Integração dos sistemas de manejo do solo à ecologia regional e qualidade de vida*, Congr Bras Ci Solo, Ribeirão Prêto, São Paulo (2001).

Primavesi, O., *Fundamentos ecológicos para o manejo efetivo de ambiente rural nos trópicos, educação ambiental e produtividade com qualidade ambiental*. Doc. 33, EMBRAPA/Ministerio de Agricultura, Brasilia (2003a).

Primavesi, O., Biodiversity and sustainability, In: *Encyclopedia of Soil Science*. Marcel Dekker, New York (2003b).

Primavesi, A. and Primavesi, A.M., Relation von Pflanzenernährung und Pflanzenkrankheiten, *Z. Pflanz., Düngung und Bodenkunde*, **105**, 22–27 (1964).

Primavesi, A.M., Primavesi, A., and Veiga, C., Influencia dos equilíbrios nutricionais no arroz irrigaddo, sobre a resistência à brusone (*Piricularia oryzae* Cav.), *Rev. Centro Ci Rur.*, **1**, 101–124 (1971).

Rodrigues, V., Desertificação: Problemas e soluções, In: *Agricultura, Sustentabilidade e o Semi-Árido*, Senna de Oliveira, T. et al., Eds., Federal University of Fortaleza, Ceará (2000).

Sanchez, P.A., *Properties and Management of Soils in the Tropics*, Wiley, New York (1976).

Senna de Oliveira, T. et al., Eds., *Agricultura, Sustentabilidade e o Semi-Árido*, Federal University of Fortaleza, Ceará (2000).

Tripathi, R.S., Weed problems: An ecological perspective, *Trop. Ecol.*, **18**, 138–148 (1977).

Tsai, S.M. and Rosetto, R., Transformações microbianas do fósforo, In: *Microbiologia do Solo*, Cardoso, E.J.B.N., Tsai, S.M., and Neves, M.C.P., Eds., Sociedad Brasileira do Solo, Campinas (1992).

Werner, H., Infiltração de chuva simulada em solos agregados e protegidos, EMBRAPA, Passo Fundo, pers. comm. (1982).

3

Soil System Management in Temperate Regions

G. Philip Robertson and A. Stuart Grandy
*Kellogg Biological Station and Department of Crop and Soil Sciences,
Michigan State University, East Lansing, Michigan, USA*

CONTENTS

The view of soils as principally support media for plants, rather than as complex systems driven by life processes, dominates most thinking about temperate zone agriculture. To a large extent, the success of the last century's Green Revolution was based on new technologies that provided, via inputs external to the system, certain ecological services traditionally supplied by soil — nutrient supply and pest suppression in particular. The result has been an agricultural enterprise that too often values soil largely as a porous medium which supports plants and drains excess rainfall. Not well appreciated are the crucial roles of soil for creating fertility and for buffering the environmental impacts of agricultural production. Nor is enough credit given to the roles that soil systems play as fundamental, interactive components within larger agricultural ecosystems. As a result, the actual and potential contributions of soils to the productivity of intensively managed systems, particularly in temperate regions which rely heavily on exogenous inputs, are undervalued.

Figure 3.1 illustrates how soils are dynamic, living systems that are integral parts of larger ecosystems. Soil subsystems participate fully in the processes that are common to ecosystems as a whole — energy flow; the movement and transformations of water, carbon, and nutrients; and the trophic dynamics that regulate biodiversity and other community characteristics. From a functional standpoint, soil can be regarded essentially in terms of habitat, providing a home to a wide variety of organisms that together provide services critically important to crop productivity and environmental quality: pest protection, pathogen control, nutrient and water availability, water filtration, carbon storage, erosion control, and plant support, among others.

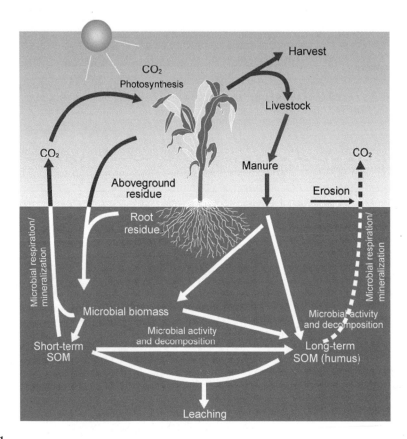

FIGURE 3.1

A maize ecosystem diagram illustrating aspects of the carbon cycle, one of many system-level processes that occur in ecosystems. From Cavigelli, M.A., Deming, S.R., Probyn, L.K., and Harwood, R.R., Eds., In *Michigan Field Crop Ecology: Managing Biological Processes for Productivity and Environmental Quality,* Michigan State University Extension Bulletin E-2646, East Lansing, MI, 1998. With permission.

Are these services in greater demand in temperate than in other soil systems? No. Nor are they more expendable. Rather, 20th-century chemical and mechanical technologies have allowed many of these services to be diminished if not supplanted entirely in management systems dominated by external inputs of energy, nutrients, and pesticides. Although this dominance of input-intensive methods is more common today in temperate regions, it can be found in tropical areas as well (Robertson and Harwood, 2001). In fact, as mechanized, intensive agriculture has moved into the tropics, for example, large-scale soybean production in Brazil, there are the same kinds of problems, though rates of change may differ. The challenge in both temperate and tropical regions is similar: how to maintain, enhance, and restore the contributions of soil biology to the fertility and sustainability of agricultural ecosystems.

3.1   Temperate Region Soil Differences

All generalizations about soil systems have significant exceptions, but two generalizations that differentiate temperate from tropical soils are quite tenable. First, because temperate soils are seasonally cold, during a significant portion of the year, plant growth and soil biological activity are low or nil due to suboptimal or freezing soil temperatures. Seasonality with its temperature fluctuations results in important changes in the chemical and physical soil environment. Freeze–thaw cycles accelerate rock weathering and the breakdown of soil aggregates, for example, and chemical as well as biological reactions that affect mineral weathering, chemical solubility, and other soil chemical properties occur more slowly in winter months. As discussed below, this seasonality provides both challenges and opportunities for effectively managing soil fertility.

A second generalization concerns soil mineralogy and its impact on soil chemistry. Agricultural soils in temperate regions are more likely to be geologically young in comparison to large regions of the tropics. Some tropical soils are also young, especially those developed from geologically recent volcanic and alluvial deposits, but most are not, and this has important implications for soil fertility. In young soils, such as those recently glaciated or formed from windblown loess, primary minerals have weathered little, and the electrical charge system (which confers ion exchange capacity) is largely permanent, with base cations such as Ca^{+2}, K^+, and Mg^{+2} common. In older soils, weathering will have removed most of the 2:1 layer-silicate clays, and electrical charges result mainly from the protonation and deprotonation of surface hydroxyl groups (Uehara and Gillman, 1981).

The ion exchange capacity of older tropical soils therefore depends very much on soil pH. When pH is low, cation exchange will be negligible, and many ions important for plant growth will be in low supply and easily washed from the soil by percolating rainfall. In contrast, the charge system of younger soils is more durable because it mostly results from the crystal lattice structure of 2:1 layer-silicate clays and is thus more impervious to changes in soil pH and soil solution composition. Although all soils contain both permanent and variable charge surfaces, most are dominated by one or the other charge system, and this has a significant impact on nutrient mobility and availability (Sollins et al., 1998). In general, the permanent charge system that dominates most temperate region soils provides these soils with greater chemical and structural resistance to the deleterious effects of chronic disturbance that is typical of mechanized agriculture.

One sometimes reads of other temperate vs. tropical soil differences such as regional differences in soil organic matter (SOM) or soil biodiversity. These are not differences that are regionally inherent. Many tropical soils can have native SOM stores equal to those in temperate regions, so such generalizations do not hold up (Sanchez et al., 1982; Greenland et al., 1992). Likewise, it is difficult to generalize about soil biodiversity because we know so little about it. We know, for example, that 1 g of soil can contain $>10^9$ microbes representing >4000 different, mostly unidentified species (Torsvik et al., 1990), while a liter of soil can contain hundreds of different species of soil fauna (Coleman and Crossley, 2003). On average, about 20% of the organic matter in arable soils is living biomass (Paul and Clark, 1996), yet very little of this can at present be identified by species. In neither temperate nor tropical soils do we know $>1\%$ of the soil biota (Tiedje et al., 1999), so generalizations are hard to substantiate. Moreover, the relationship of this biodiversity to ecosystem functioning is in any case not yet documented. So, scientists and practitioners are both operating with little certain knowledge about the specific organisms present in soil.

3.2 Challenges to Soil Fertility and Management

The inherent fertility of many temperate-region soils is high. In comparison to highly weathered tropical soils, many soils in temperate regions can withstand years of crop production following their conversion from natural vegetation. Eventually, however, soil nutrient stocks decline and soil structure degrades, and most temperate-region cropping systems now owe much of their present productivity to external subsidies, which enhance or compensate for lost ecological services. To bring soil to its full fertility and to sustain this depends on the satisfactory resolution of two major challenges: the restoration and maintenance of SOM, including its all-important living fraction; and the development of nutrient-efficient, and especially nitrogen-efficient, cropping systems. Other challenges are also important — erosion control; water conservation; nutrient losses to groundwater, surface waters and the atmosphere; and pathogen suppression, among others — but in most landscapes they remain secondary. This chapter focuses on these two major challenges, considering secondary challenges within the context of these two chief concerns.

3.2.1   Soil Organic Matter Restoration and Conservation

The loss of SOM, sometimes referred to simply as soil carbon loss, is common to almost all field-crop production systems. The principal cause of SOM loss is accelerated microbial activity as agronomic activities in general, and cultivation in particular, stimulates microbial respiration of soil organic carbon, the foundation of organic matter. This rapid turnover of SOM is the foundation of soil fertility in low-input cropping systems, since as microbes consume carbon they release nitrogen and other nutrients to the soil solution where nutrients become available to plants. Typically 40–60% of a soil's organic carbon stores are lost in the 40–60 years following the initial cultivation of a temperate region soil (Figure 3.2). This occurs even more quickly with cultivation in the tropics. Restoring lost SOM and tempering its turnover is thus a major goal of biologically-based agriculture (Robertson and Harwood, 2001).

The reasons for accelerated microbial activity are complex and related to a number of factors. Chief among them is the breakdown of soil aggregates, small particles of soil (0.05–8 mm) that protect carbon molecules from rapid microbial consumption. Carbon particles inside aggregates exist in an environment very different from the bulk soil environment: certain soil organisms may not be present, and the activity of those that are present is likely be restricted by low oxygen availability. Oxygen diffuses very slowly into

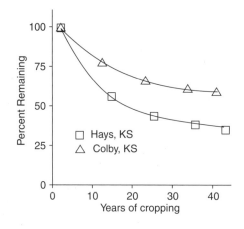

FIGURE 3.2

Soil carbon loss following cultivation at two temperate region sites, Hays and Colby Counties, Kansas, U.S.A. From Haas, H.J., Evans, C.E., and Miles, E.F., In *Nitrogen and Carbon Changes in Great Plains Soils as Influenced by Cropping and Soil Treatments*, USDA Technical Bulletin 1164, Washington, DC, 1957. With permission.

aggregates, with the result that the oxygen consumed by microorganisms is not quickly replaced, and the interiors of aggregates thus tend to be anaerobic to a much greater degree than bulk soil.

Cultivation breaks apart aggregates, especially the larger ones, exposing trapped organic carbon to aerobic microbes that easily respire it to CO_2. Much of the increase in atmospheric CO_2 starting early in the 19th century was the result of pioneer cultivation (Wilson, 1978). This stimulated microbial activity and the turnover of active organic matter pools formerly in aggregates. The basis for soil carbon sequestration as a CO_2 mitigation strategy is recovery of this lost soil carbon (Lal et al., 2004). While this recovery will contribute to greenhouse gas abatement, it will also improve soil productivity and increase microbial biomass and nutrient availability.

Aggregation remains low following cultivation because the microbial processes that stabilize soils are disrupted at the same time that aggregates are more exposed to physically destabilizing processes. Microbial production of polysaccharides, humic substances, and aliphatic compounds that promote particle binding and aggregate stabilization invariably decrease following the start of cultivation. Additionally, extensive networks of fungal hyphae that enmesh soil particles and provide a framework for aggregate stabilization are shattered by tillage. These hyphae are also sensitive to changes in residue placement following agricultural conversion (Jansa et al., 2003).

Increases in the physical forces that destabilize aggregates following cultivation are mostly related to changes in soil water dynamics. Cultivated soils are bare much of the year. During these periods, raindrop impacts result in greater disruption of aggregates at the soil surface, and without transpiration these soils will be wetter for more of the year. Generally, with increasing water content, aggregate structure decreases and dispersed clay increases (Perfect et al., 1990). Bare soils are also more exposed to freezing, which has particularly damaging effects on soil structure since as soil water freezes and expands, it moves into pores and fracture planes between particles, driving them apart. Structural breakdown of cultivated soils leads quickly to wind and water erosion, to substantial and permanent losses of soil carbon, and ultimately to reduced productivity.

Usually not all carbon is lost from soil even after decades or centuries of plowing. However, what remains is carbon that is relatively unavailable to microbes because it is chemically resistant to microbial decomposition or tightly bound to clay particles (Kiem and Kögel-Knaber, 2003) — what soil biologists call slow or passive carbon — plus whatever carbon has been recently added as crop residue (Figure 3.3). These fractions provide a very different soil habitat than before, bereft of many of the benefits of abundant SOM and biota. There is less water-holding capacity, less porosity and aeration, lower infiltration, and a diminished buffer of biologically-available nutrients. Moreover, organic matter itself — even in permanent-charge soils — provides significant cation exchange capacity, which helps to hold biologically important cations against leaching loss.

Soils impoverished in carbon will thus be impoverished in biological activity and in the fertility that this activity confers. Soil nitrogen turnover — the nitrogen-supplying power of the soil — is lower whenever microbial populations are diminished, and there are consequently fewer invertebrates such as earthworms, ground-dwelling beetles, and nonparasitic nematodes. Many of these organisms, discussed in Part II, are needed to promote crop growth by providing services such as decomposing litter, creating soil pores and aggregates, and consuming root-feeding insects, parasites, and plant pathogens (Coleman and Crossley, 2003).

In summary, soil systems that are low in SOM and soil biota, whether for either management or for natural reasons, will be lower in fertility, and for this reason they require substantial external inputs to maintain crop productivity. Restoration and

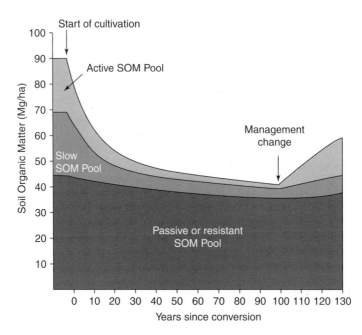

FIGURE 3.3
Changes in soil organic matter fractions following cultivation of a soil profile under native vegetation. After Brady, N.C. and Weil, R.R., In *The Nature and Properties of Soils*, 12th ed., Prentice Hall, Upper Saddle River, NJ, 2002, 523. With permission.

maintenance of SOM in both residual and living forms is thus a crucially important management challenge.

3.2.2   High Nutrient-Use Efficiency

Most cropping systems use and export nutrients at prodigious rates. Some nutrient loss, such as that exported in yield, is unavoidable. Other losses, however, such as nutrients lost via hydrologic and gaseous pathways, are inadvertent. All exported nutrients that can limit crop performance must be replaced for a cropping system to remain productive, from external sources or from within the soil system. Maintaining this nutrient availability in both time and place to match plant needs is one of the toughest of agronomic challenges.

For certain plant nutrients such as calcium and magnesium, most temperate-region soils can maintain a steady supply with little depletion even in the face of significant export. This is because the mineral stores of these nutrients are high in most young soils. For other nutrients, however, particularly nitrogen, phosphorus, and potassium (N, P, and K), the ability of a soil to fully resupply losses is eventually lost. When this occurs, modern cropping systems rely on fertilizers to make up the difference. Nitrogen deficits are especially severe because nitrogen losses can occur via so many different pathways, and it is nitrogen that typically limits the productivity of even natural ecosystems that are not harvested.

The two main strategies for improving nutrient availability in cropped ecosystems are to increase inputs and to reduce losses. Inputs are commonly increased via organic or synthetic fertilizer additions, or specifically for nitrogen, by N_2 fixation (see Chapter 12). Losses can be reduced, on the other hand, by increasing system-wide nutrient-use efficiency. Nitrogen is a case in point. A highly productive maize crop with a yield of

10 tons of grain removes about 260 kg N ha^{-1} (Olson and Kurtz, 1982) or around 5.2 tons of nitrogen over 20 years of cropping. In uncultivated arable soils, organic nitrogen stores can be as high as 10 tons of nitrogen ha^{-1} on average. Continuous cropping of maize thus has the potential to remove, within 20 years, an amount of nitrogen equivalent to 50% of the nitrogen stock in the native SOM, demonstrating the potential for rapid soil nitrogen depletion. Because nitrogen is the most common limiting nutrient in temperate region ecosystems, restoring lost nitrogen is a crucial agronomic goal. Preventing as much nitrogen as possible from inadvertently leaving the system is equally important, from both an agronomic and environmental standpoint.

Improving a cropping system's nutrient-use efficiency requires matching soil nutrient release — whether from organic or inorganic sources — with the demand for nutrients by plants. This matching has to occur both temporally and spatially. In diverse native plant communities and many cropped perennial systems, soil microbial activity will almost always coincide with periods when there is at least some plant need. In native communities, the presence of diverse species having different life cycles means that at least some plants will be actively photosynthesizing whenever temperature and moisture permit. In the annual monocultures typical of temperate-region agriculture, on the other hand, such synchrony is rare.

Most grain crops, for example, are in the ecosystem for only 90–100 days, and only during 30–40 days at midsummer will they be accumulating biomass at a significant rate. In maize, for example, nitrogen uptake rates can reach the astonishing rate of 4 kg N ha^{-1} day^{-1} (contrast this with inputs of nitrogen to the soil from precipitation of 10 kg ha^{-1} year^{-1}). This high rate is sustained for only 3–4 weeks, however, and it falls to nil within the following 2–3 weeks (Olson and Kurtz, 1982). The much longer periods during which atmospheric nitrogen deposition occurs and soil temperature and moisture are sufficient to support microbial nitrogen mineralization do not match crops' peak nutrient demand. This asynchrony creates a huge potential for nutrient loss and for low system-wide nutrient-use efficiency (Figure 3.4).

Spatial symmetry can be as important as temporal synchrony for ensuring that nutrient availability and uptake are well matched (Robertson, 1997). Row crop management, unfortunately, does not often result in well-matched spatial arrangements of plants and

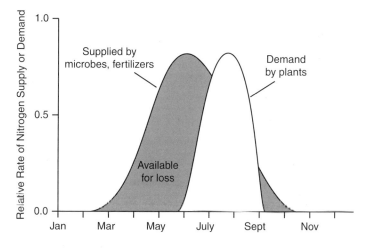

FIGURE 3.4

Asynchrony between nitrogen supply and nitrogen plant demand in a temperate cropping system can lead to periods of nitrogen vulnerability to loss. From Robertson, G.P., In *Ecology in Agriculture*, Academic Press, New York, 1997. With permission.

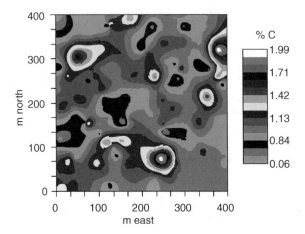

FIGURE 3.5

The variability of soil organic carbon across a 400×400 m^{2}soybean field in southwest Michigan. From CAST, *Climate Change and Greenhouse Gas Mitigation: Challenges and Opportunities for Agriculture*, Council for Agricultural Science and Technology, Ames, IA, 2004, With permission.

resources within a field, and this mismatch also reduces system-wide nutrient-use efficiency. Row vs. between-row differences in soil-nutrient availability have been recognized for decades (e.g., Linn and Doran, 1984), and a number of management strategies, discussed later, can be derived from knowledge about how to increase the water and nutrient-use efficiencies of row crops.

Spatial heterogeneity at larger scales is also emerging as a management issue. Available evidence suggests that soil nitrogen availability is highly variable in natural communities, with variable patches of soil fertility at scales that can affect individual plants (e.g., Robertson et al., 1988). This variability persists after conversion to agriculture (Robertson et al., 1993), so that field-scale soil variability as shown in Figure 3.5 becomes a major factor in most cropping systems. High nutrient-use efficiency from both the spatial and temporal perspectives is thus an important goal of agronomists, and one whose achievement depends on adept combinations of soil and plant management decisions.

3.3   Solutions to the Major Soil System Challenges in Temperate Regions

3.3.1   Restoration of Soil Organic Matter

Decades of research have demonstrated that SOM can be restored and maintained at relatively high levels in most arable soils. Most importantly, those biologically active SOM fractions most rapidly lost following cultivation — such as light-fraction (LF) or particulate organic matter (POM) — can be regenerated. LF has a rapid turnover time of 2–3 years because it is relatively free of mineral material and humification and has high concentrations of carbon and nitrogen (Wander et al., 1994). LF is thus an ideal source of energy and nutrients for microorganisms, and its decomposition releases plant nutrients to the soil solution. Restoring LF and other active SOM pools through strategic crop and soil management thus has the potential to stabilize cropping systems and reduce dependencies on external inputs.

At the simplest level, SOM change is simply the difference between organic carbon added to soil and organic carbon lost via the biological oxidation of SOM carbon to CO_2 carbon. There are thus two ways to build SOM in cropping systems: (1) increase soil carbon inputs via crop residues, cover crops, and soil amendments such as compost and manure, and (2) decrease soil carbon loss by slowing decomposition and (where important) soil erosion.

Carbon inputs to soil are influenced by nearly every facet of agricultural practice (Paustian et al., 1995). These include crop type and productivity, the frequency and duration of fallow periods, and fertilizer and residue management. Organic amendments such as manure, compost, and sewage wastes provide additional management interventions.

High crop productivity based on associated residue inputs does not in itself guarantee higher SOM pools (Paul et al., 1997). Relationships between residue inputs and SOM are complicated by changes in enzyme dynamics and decomposition processes following N-fertilization and other agricultural practices (Fontaine et al., 2003; Waldrop et al., 2004). In the U.S.A.'s corn belt, for example, even though aboveground residues in a maize ecosystem may exceed by a factor of two the amount of litterfall in the forest or native prairie that the agricultural system replaced, SOM levels in the maize system persist at about 50% of the levels in native forest even when maize residues are not exported from the soil system.

This said, substantial residue inputs are still a prerequisite for building organic matter stores in soil. Removing all aboveground residues — as is the case for maize silage, wheat straw production, or biobased fuel production, for example — removes a major source for SOM accumulation. With other factors held equal, in fact, field experiments have generally found a close linear relationship between the rates of residue carbon return and the SOM levels found in temperate agricultural soils (e.g., Rasmussen et al., 1980). Organic amendments also provide a direct and effective means for building SOM. For example, in a long-term continuous wheat experiment at Rothamsted, UK, plots receiving farmyard manure (35 tons ha^{-1} annually) over a 100-year period effectively doubled their SOM levels (Jenkinson, 1982).

Decomposition rates of crop residues and SOM are principally influenced by climate, by the chemical composition or quality of the residue, and by soil disturbance. In general, decomposition occurs faster in warmer, moister (but not saturated) soils, and with management that exposes the soil's surface to greater solar radiation or that uses spring tillage to accelerate soil drying and warming following a winter snow cover so as to promote decomposition. Draining wetland soils for agriculture achieves essentially the same result.

Decomposition is also affected by litter quality. Plant tissues lower in nitrogen and higher in structural compounds such as cellulose, suberin, and lignin decompose more slowly than tissues that are higher in sugars, protein, and nitrogen: for example, soybean leaves decompose much faster than do wheat straw or maize stalks. Few microbes are able to degrade the complex chemical structure of lignin, whereas simple organic compounds can be respired by most soil organisms.

It follows that SOM is likely to accumulate faster with the addition of more structurally complex materials (for example, Figure 3.6), although these relationships may be complicated by interactions between decomposition products and soil physical processes. For example, the rapid production of polysaccharides associated with the decomposition of legumes can facilitate aggregate formation and increased physical protection of SOM. More research is needed, however, to determine how plant and microbial communities interact to control decomposition and, in particular, the formation of particular biochemicals which stabilize SOM in agricultural soils. Manure tends to be more complex

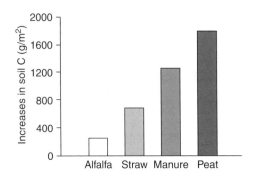

FIGURE 3.6
Soil carbon increases over 20 years following the addition of carbon sources differing in structural complexity or quality to a sandy soil in Canada. Residues were added at the rate of 500 g m^{-2} year^{-1}. From Paustian, K., Elliot, E.T., Collins, H.P., Cole, C.V., and Paul, E.A., *Aust. J. Exp. Agric.*, 35, 929–939, 1995; after Sowden, F.J. and Atkinson, H.J., *Can. J. Soil Sci.*, 48, 323–330, 1968. With permission.

structurally than are uncomposted crop residues because it has already been exposed to microbial attack in the animal gut.

No-till soil management, discussed in Chapter 22 and Chapter 24, and other forms of tillage management that are less destructive than moldboard plowing help to conserve SOM in cultivated soils by helping to maintain soil aggregate stability. Conservation tillage can also conserve SOM by reducing erosion in landscapes subject to wind and water erosion (Lal et al., 2004). Cover crops that maintain plant cover during periods when the primary crop is not present — late fall, winter, and early spring, for example — can also reduce the potential for soil erosion and add additional vegetative residue to the SOM pool.

Restoring SOM in cropping systems can thus be achieved best through some combination of increased organic matter inputs, no-till or other conservation tillage practice, and cover cropping. While any organic matter inputs will help to build SOM, the most effective will be those that are slow to decompose, such as low-nitrogen, high-lignin crop residues or compost and manure. Rotational complexity may also help to restore SOM when one or more crops in the rotation have higher lignin contents, more residue, or a longer growing season than others.

3.3.2 Improving Nutrient Efficiency

Making nutrients available mostly when and where they are needed by the crop improves nutrient-use efficiency. Temporal synchrony is achieved by applying inputs as close as possible to the time required for crop growth. Applying mobile fertilizers, such as nitrogen, in split applications, e.g., 20% at planting and the rest just before the period of greatest crop growth, is common in many temperate systems, although — egregiously — fall application of anhydrous ammonia is still common for maize production in some regions of the Midwestern USA.

Likewise, encouraging decomposition of the previous crop's residue early in a crop's growing season is also beneficial. Spring rather than fall tillage will keep more nutrients in active SOM pools where they are better protected from overwinter leaching and gaseous losses and will serve to stimulate decomposition (and nutrient release) prior to crop growth. Rotary hoeing or some other type of shallow cultivation well into the growing season can stimulate microbial activity just prior to major crop growth. Winter cover crops — particularly fast-decomposing high-nitrogen crops such as legumes — also help to provide active-fraction SOM when the crop most needs it (see Figure 3.4). Cover crops can additionally help to capture nutrients released to the soil solution when the main crop is not present; plants active in the fall and spring when microbes are actively oxidizing SOM can temporarily immobilize nutrients that would otherwise be vulnerable to overwinter or springtime losses from the ecosystem.

Other aspects of crop management that may influence microbial communities and decomposition are nitrogen fertilization, inputs of labile carbon compounds, and irrigation. Many studies have demonstrated that nitrogen or organic matter additions may result in a change in the mineralization of native SOM (Fontaine et al., 2003). This is referred to as the priming effect: a strong change in the turnover of SOM in response to a soil amendment (Kuzyakov et al., 2000). Priming effects may play a critical role in controlling carbon balance and nitrogen turnover in ecosystems. However, our ability to exploit the underlying microbial processes to manage soil fertility is currently limited (De Neve et al., 2004). This is primarily because environmental controls over priming responses are very complex and include interactions between nutrient availability, litter quality, soil texture, and other factors. Despite these challenges, this should remain an area of active research because the potential benefits are great from being able to manipulate SOM turnover and nutrient mineralization when and where it is most needed with relatively modest additions of nitrogen or carbon to the soil.

Spatial synchrony or coincidence can be achieved at two levels. At the row vs. between-row level, inputs such as fertilizers can be applied in bands next to or over the tops of rows using drip irrigation, fertilizer banding, or foliar feeding; or organic amendments or crop residues can be mounded into rows using techniques such as ridge tillage. Ridge tillage, a popular soil management technique for many low-input farmers in the Midwestern USA. (NRC, 1989), minimizes spatial asymmetry by periodically mounding the between-row A_p horizon into semi-permanent ridges on to which the crop is planted. This concentrates the labile organic matter and soil biotic activity within rows, achieving the same effect as fertilizer banding.

At the larger field scale, variability can be addressed by using site-specific application technologies. Many harvest-combines today are sold with global positioning system (GPS) equipment to permit highly-resolved yield mapping. With proper application equipment, these maps can then be used to tailor fertilizer applications to the productive capacity of any given area of the mapped field. Rather than fertilizing an entire field with a single, high rate of application, the highest rates can be applied only where productivity, and therefore plant nutrient uptake, will be high, reducing nutrient losses from low-productivity areas. In effect, this method uses plants in the field as bioassays for the nitrogen made available by soil microbes; it provides additional nitrogen fertilizer in proportion to the plants' abilities to take it up.

In most temperate regions, the current cost of fertilizer is low relative to the marginal increase in productivity that can be gained from applying it at high rates. So from the producer's standpoint, it rarely pays to reduce the inputs of limiting nutrients that are inexpensive, e.g., nitrogen. Thus, socioeconomic influences condition decisions about achieving crop nutrient-use efficiency (Robertson and Swinton, 2005). In many if not most cases we cannot expect improved ecosystem nutrient-use efficiency until policy and other issues affecting farmer decision-making are appropriately resolved (Robertson et al., 2004).

3.4 Discussion

Soil fertility in both temperate and tropical regions is the net result of a complex interplay between the biotic and abiotic components of agricultural ecosystems. The abiotic environment includes both physical and chemical attributes, which can differ between temperate vs. tropical regions, though often less than terminology implies. Important differences are that temperate-region soil systems are usually exposed to a seasonal cold or frozen period; further, temperate region soils are in general developed on geologically younger soils and therefore dominated by a permanent-charge mineralogy. The biological environment includes enormously complex food webs and a truly amazing diversity of microbes and invertebrates. As yet, we still know little about the patterns and importance of this diversity in either temperate or tropical regions, and even less about how to manage it, but this should change rapidly in the coming decades.

The major agronomic challenges related to temperate-region soil system fertility management are SOM restoration and the improvement of ecosystem-level nutrient-use efficiency. Both issues are addressable with current knowledge and technology, and both require active management of the soil biota, either directly or indirectly. To date, methods to manage the soil biota directly through priming or other strategies are often more theoretical than practical in production-oriented systems. A better understanding of the various scientific issues that can illuminate these processes and dynamics is much needed. However, effective utilization of such knowledge will require more attention to social science variables and more integration across disciplines than is currently found in scientific studies or farmer practice.

References

Brady, N.C. and Weil, R.R., *The Nature and Properties of Soils*, 12th ed., Prentice Hall, Upper Saddle River, NJ (2002).

CAST, Climate change and greenhouse gas mitigation: Challenges and opportunities for agriculture, *Council for Agricultural Science and Technology*, Ames, IA (2004).

Coleman, D.C. and Crossley, D.A. Jr., *Fundamentals of Soil Ecology*, 2nd ed., Academic Press, Burlington, MA (2003).

De Neve, S. et al., Manipulating N mineralization from high N crop residues using on- and off-farm organic materials, *Soil Biol. Biochem.*, **36**, 127–134 (2004).

Fontaine, S., Mariotti, A., and Abbadie, L., The priming effect of organic matter: A question of microbial competition, *Soil Biol. Biochem.*, **35**, 837–843 (2003).

Greenland, D.J., Wild, A., and Adams, D., Organic matter dynamics in soils of the tropics: From myth to complex reality, In: *Myths and Science of Soils of the Tropics*, Lal, R. and Sanchez, P.A., Eds., Special Publication 29, Soil Science Society of America, Madison, WI, 17–34 (1992).

Haas, H.J., Evans, C.E., and Miles, E.F., *Nitrogen and Carbon Changes in Great Plains Soils as Influenced by Cropping and Soil Treatments*, USDA, Washington, DC (1957), Technical Bulletin 1164.

Jansa, J. et al., Soil tillage affects the community structure of mycorrhizal fungi in maize roots, *Ecol. Appl.*, **13**, 1164–1176 (2003).

Jenkinson, D.S., The nitrogen cycle in long-term field experiments, In: *The Nitrogen Cycle*, Stewart, W.D.P. and Roswall, T., Eds., The Royal Society, London, 261–270 (1982).

Kiem, R. and Kögel-Knaber, I., Contribution of lignin and polysaccharides to the refractory carbon pool in C-depleted arable soils, *Soil Biol. Biochem.*, **35**, 101–118 (2003).

Kuzyakov, Y., Friedel, J.K., and Stahr, K., Review of mechanisms and quantification of priming effects, *Soil Biol. Biochem.*, **32**, 1485–1498 (2000).

Lal, R. et al., Managing soil carbon, *Science*, **304**, 393 (2004).

Linn, D.M. and Doran, J.W., Effect of water-filled pore space on CO_2 and N_2O production in tilled and non-tilled soils, *Soil Sci. Soc. Am. J.*, **48**, 1267–1272 (1984).

MSU, *Michigan Field Crop Ecology: Managing Biological Processes for Productivity and Environmental Quality*, Cavigelli, M.A. et al., Eds., Extension Bulletin E-2646, Michigan State University, East Lansing, MI (1998).

NRC, *Alternative Agriculture*, National Academy Press, Washington, DC for the National Research Council (1989).

Olson, R.A. and Kurtz, L.T., Crop nitrogen requirements, utilization and fertilization, In: *Nitrogen in Agricultural Soils*, Stephenson, F.J., Ed., American Society of Agronomy, Madison, WI, 567–604 (1982).

Paul, E.A. and Clark, F.E., *Soil Microbiology and Biochemistry*, 2nd ed., Academic Press, San Diego, CA (1996).

Paul, E.A. et al., Eds., *Soil Organic Matter in Temperate Ecosystems: Long-Term Experiments in North America*, Lewis CRC Publishers, Boca Raton, FL (1997).

Paustian, K. et al., Use of a network of long-term experiments in North America for analysis of soil carbon dynamics and global change, *Aust. J. Exp. Agric.*, **35**, 929–939 (1995).

Perfect, E. et al., Rates of change in soil structural stability under forages and corn, *Soil Sci. Soc. Am. J.*, **54**, 179–186 (1990).

Rasmussen, P.E. et al., Crop residue influences on soil carbon and nitrogen in a wheat-fallow system, *Soil Sci. Soc. Am. J.*, **44**, 596–600 (1980).

Robertson, G.P., Nitrogen use efficiency in row-crop agriculture: Crop nitrogen use and soil nitrogen loss, In: *Ecology in Agriculture*, Jackson, L., Ed., Academic Press, New York, 347–365 (1997).

Robertson, G.P. et al., Rethinking the vision for environmental research in US agriculture, *BioScience*, **54**, 61–65 (2004).

Robertson, G.P., Crum, J.R., and Ellis, B.G., The spatial variability of soil resources following long-term disturbance, *Oecologia*, **96**, 451–456 (1993).

Robertson, G.P. and Harwood, R.R., Sustainable agriculture, In: *Encyclopedia of Biodiversity*, Levin, S.A., Ed., Academic Press, New York, 99–108 (2001).

Robertson, G.P. et al., Spatial variability in a successional plant community: Patterns of nitrogen availability, *Ecology*, **69**, 1517–1524 (1988).

Robertson, G.P. and Swinton, S.M., Reconciling agricultural productivity and environmental integrity: A grand challenge for agriculture, *Front. Ecol. Environ.*, **3**, 38–46 (2005).

Sanchez, P.A., Gichuru, M.P., and Katz, L.B., Organic matter in major soils of the tropical and temperate regions, *Trans. 12th Int. Congr. Soil Sci.*, **1**, 99–114 (1982).

Sollins, P., Robertson, G.P., and Uehara, G., Nutrient mobility in variable- and permanent-charge soils, *Biogeochemistry*, **6**, 181–199 (1988).

Sowden, F.J. and Atkinson, H.J., Effect of long-term annual additions of various organic amendments on the organic matter of a clay and a sand, *Can. J. Soil Sci.*, **48**, 323–330 (1968).

Tiedje, J.M. et al., Opening the black box of soil microbial diversity, *Appl. Soil Ecol.*, **13**, 109–122 (1999).

Torsvik, V. et al., Comparison of phenotypic diversity and DNA heterogeneity in a population of soil bacteria, *Appl. Environ. Microbiol.*, **56**, 776–781 (1990).

Uehara, G. and Gillman, G.P., *The Mineralogy, Chemistry and Physics of Tropical Soils with Variable Charge Clays*, Westview Press, Boulder, CO (1981).

Waldrop, M.P., Zak, D.R., and Sinsabaugh, R.L., Microbial community response to nitrogen deposition in northern forest ecosystems, *Soil Biol. Biochem.*, **36**, 1443–1451 (2004).

Wander, M.M. et al., Organic and conventional management effects on biologically active soil organic matter pools, *Soil Sci. Soc. Am. J.*, **58**, 1130–1139 (1994).

Wilson, A.T., The explosion of pioneer agriculture: Contribution to the global CO_2 increase, *Nature*, **273**, 40–41 (1978).

4

Soil System Management under Arid and Semi-Arid Conditions

Richard J. Thomas, Hanadi El-Dessougi and Ashraf Tubeileh

International Center for Agricultural Research in the Dry Areas (ICARDA), Aleppo, Syria

CONTENTS

Soil fertility in systems under arid and semi-arid conditions, hereafter referred to as dry areas or drylands, is constrained by environmental extremes of hot and cold temperatures, as well as by low water availability. With some exceptions, these soils have inherently low fertility, low availability of nitrogen and phosphorus, low water-holding capacity, high pH, low soil organic matter (ranging from 0.1 to 3%), shallowness, stoniness, and other specific problems (Matar et al., 1992). These areas are quite widespread, occupying around 30–40% of the world's terrestrial surface. Given the vulnerability of these lands to degradation, it is estimated that some 44 million km^2 — 34% of the total world's area, supporting 2.6 billion people — is at risk from desertification (Eswaran et al., 2001). Hence, these lands are of great global significance even if their agricultural production potential is relatively low.

There is considerable information on the use of fertilizers and legumes to enhance soil fertility in dry areas, including timing and amount of fertilizers, application methods, crop responses, and effects of crop rotations (e.g., Matar et al., 1992; Ryan, 2004), and also on dryland fallowing and water use (Farahani et al., 1998). However, despite evidence that some nutrient inputs are required to sustain higher agricultural production on these lands, resource-poor land users are reluctant to invest much in inputs such as fertilizers because of the large risks involved and the relatively low economic returns they get from their prevailing cereal–fallow-based systems. Hence, they tend to adopt risk-aversion strategies rather than attempt any maximization of production. Risk-aversion strategies include diversified cropping systems with annuals and perennials, fruits, fuelwood, oilseed, and pharmaceuticals; diversified animal production of cattle, buffalo and other draft animals, goats and sheep, chickens and other fowl; integrated crop–livestock systems; spatial mobility; and flexible livelihood strategies that include off-farm labor and other employment opportunities.

The degradation of drylands continues almost unabated, despite the realization that dry areas need careful management to prevent overexploitation, i.e., through better soil and water management and technologies that are appropriate to their more extensive systems, with due attention to socioeconomic factors. The UN Convention to Combat Desertification (UNCCD) has estimated that perhaps as much as 70% of the world's drylands, some 3.6 million ha, have already been degraded (UNCCD, 2004). However, some so-called "bright spots" of sustainability have been identified, and important lessons are being learned from past experience (see Mortimore, 2004, and Chapter 25).

One lesson is that even though external inputs can increase productivity and contribute to sustainability, the use of high levels of inputs such as fertilizers, even so-called "economic" levels, is usually prohibitively costly in dry areas (e.g., Pender and Gebremedhin, 2004). This consideration, given also poor access to markets and inadequate infrastructure, suggests that low external-input technologies are required for sustainable livelihood improvement for most dryland populations, especially those predominantly dependent on agriculture (Hazell, 2001). It is unlikely that drylands can ever be competitive for grain production with the subsidized producers in North America, Europe, and elsewhere, since dry areas generally have low overall potential given their water and soil constraints.

Soil fertility efforts should aim to enhance both the stabilization and resilience of the production systems, and to support increased income-generation through more diverse, higher-value agricultural products, e.g., organic foods, specialized cereal products, such as durum wheat and couscous, fruits, medicines, cosmetics, and herbs. This is best achieved via low external-input systems with greater attention to what Sanchez (1994) called the "second paradigm of soil fertility" (Chapter 49, section 2). This alternative proposes that crop production enhancement relies more on biological processes, adapting germplasm to adverse soil conditions, enhancing soil biological activity, and optimizing nutrient cycling. The latter will minimize, though not exclude, the use of external inputs and will seek to maximize the efficiency of their use (Sanchez, 1994; Swift, 1999).

This chapter does not attempt a comprehensive review of soil fertility in dry areas, but rather focuses on some pertinent issues and promising options for the future of resource-poor farmers who are heavily dependent on the natural resource base, taking into account recent advances in our knowledge about fertility management in such high-risk environments. As dryland environments are highly variable, we do not attempt to cover all of them but rather consider key issues from the fairly modal arid and semi-arid areas of Central and West Asia and North Africa. These are physical-climatic regions with temperate or steppe areas having typical Mediterranean climates of warm and cold seasons with winter rainfall. These differ from more tropical semi-arid areas or highland regions, which can have quite different rainfall and temperature patterns and different dryland farming systems (Bowden, 1979), e.g., the agro-pastoral millet–sorghum system that is found predominantly in West and Sub-Saharan Africa (Chapter 26) or the extensive commercial cereal–livestock systems in Australia. A more complete discussion on types of drylands is provided in De Pauw (2004).

4.1 Climatic Constraints of Drylands and Coping Strategies

Harsh conditions such as low and erratic rainfall, high temperatures, and strong dry winds are the prevailing climatic features in drylands represented by the large region of Central and Western Asia and North Africa, which is our focus here. Water scarcity rather than soil systems' fertility is often the factor most limiting for crop production and nutrient availability. Thus, conservation and effective utilization of water is a prerequisite for greater availability and more efficient uptake of nutrients. Rainwater harvesting by vegetation barriers, using stones and green mulches to reduce evaporation, digging ditches and ponds are useful low-cost technologies for improving soil moisture content and nutrient availability (Oweis et al., 2004). One such technology is the zaï holes widely used in West Africa, discussed in Chapter 26.

Alternatives include the use of marginal water sources, such as nutrient-rich sewage and wastewater, which in many developing countries is an important source of irrigation water. However, because the use of such water may entail health hazard and environmental risks, it is best suited to nonconsumable products. Management, use and reuse of saline and/or sodic drainage water can also contribute to efficient use of scarce resources. Qadir and Oster (2004) propose developing cropping systems that can endure the stresses caused by the expected levels of salinity and sodicity, and using agriculturally significant plant species that can tolerate salinity and/or sodicity. Such species can also be valuable for bioreclamation of salt-affected soils.

Long-term strategies to cope with climate fluctuations and extremes include the development of more stress- and disease-tolerant germplasm via conventional breeding and biotechnology. Given that current levels of production in drylands are below the existing genetic potential of most crops, a balanced approach will place emphasis on improving agronomic and soil and water management practices to complement the expected advances in breeding and biotechnology, and the former can possibly substitute for them when the latter are not achieved.

4.2 Some Unusual Aspects of Nutrient Cycles in Dryland Areas

The alternative paradigm of soil fertility places increased emphasis on nutrient cycling in soil–plant–farm systems via biological processes coupled with increased nutrient-use efficiency (Sanchez, 1994). We suggest some salient and unusual aspects of nutrient cycles in drylands.

Nutrient cycling in drylands is affected by low and erratic rainfall, wide temperature extremes, alkalinity and/or salinity, and occasionally by relatively high rates of dry deposition of nutrient-enriched soil particles from wind erosion. In extremely dry areas where vascular plants are absent, nutrient cycling through microbial organisms predominates. For example, microbiotic crusts, composed of nitrogen-fixing cyanobacteria, are often found on desert surfaces. These organisms can survive long periods of desiccation with very rapid responses to rehydration. These organisms conserve and cycle water as well as nutrients, increasing water infiltration, slowing evapotranspiration, and reducing wind erosion. Some of the nitrogen they contribute to the soil can be lost, however, through denitrification from underlying anaerobic microsites, which limits nitrogen build-up (Sprent, 1987).

Under certain conditions, nitrate can accumulate in dry soils from repeated rainfall events followed by mineralization of litter and nitrification. This usually occurs in the absence of vascular plants. Spots of higher fertility can occur in the soil that can be exploited temporarily by fodder plants, for example. Where plants grow in arid and semi-arid areas, most of the nitrogen will be in the plants' biomass with efficient internal plant recycling.

Some potentially useful desert plants, such as *Atriplex* and *Artemisia* spp., have negative effects on the nitrogen cycle. Their antimicrobial capacities inhibit nitrogen-fixing cyanobacteria, and their nitrification-inhibiting capacities alter a major part of the nitrogen cycle, especially at high soil pH and temperatures (Sprent, 1987). When these plants impede the oxidation of nitrite to nitrate, this leads to nitrite accumulation and then to denitrification, which results in gaseous nitrogen losses from the soil system (see Sprent, 1987, and references therein). These plants are often particularly efficient in nutrient scavenging and internal nutrient recycling, so they have a competitive advantage. However, their dominance has detrimental effects on nutrient cycling in soil systems that are already constrained by nitrogen limitation. These effects need to be considered when such plants are used in agroecosystems.

The possibilities for enhancing nitrogen in dryland soils by use of legumes are restricted wherever water rather than nitrogen is the most limiting factor. In such situations, plants invest more of their energy and photosynthate into developing more extensive, deeper rooting systems to acquire water rather than into forming root nodules (Sprent, 1985). This may be why there is little evidence that native dryland legumes actually fix significant amounts of nitrogen.

Sprent (1987) discusses the interesting case of certain legume phreatophytes, such as *Prosopis* spp., that have very deep roots to exploit deep water sources in dry areas. They often have nitrogen-fixing activity even when the upper soil layers contain high levels of soil nitrate that would normally inhibit plants' nitrogen fixation. Such capabilities suggest conserving shrubs and trees that retain nitrogen and water in their biomass, a difficult task in marginal areas that are already overexploited and where natural resource managers lack incentives to conserve biodiversity.

More could be said about the diverse and often unusual characteristics of plants that have evolved in arid and semi-arid environments. These examples should suffice to show why one needs to assess nutrient cycling under dryland conditions carefully, in order to avoid unforeseen problems where natural systems have been perturbed by plant introductions or by changes in cropping systems. The evolved relationships among plants and soil organisms in interaction with other components of soil systems, all operating within particular, difficult climatic and edaphic conditions, create both constraints and opportunities for management.

4.3 Agronomic and Soil–Water Management Practices for Improved Soil Fertility

A number of soil fertility management practices under arid and semi-arid conditions are thought to hold great promise. These include soil conservation measures such as cover crops, wind breaks, contour planting, terracing, and reduced tillage to minimize land degradation and erosion, cereal–legume rotations, nitrogen-fixing trees, and better crop–livestock integration.

4.3.1 Conservation Agriculture and Tillage

Conservation agriculture (CA) aims to conserve and improve the natural resource base while using the resources available for agricultural production more efficiently. Several versions of CA are presented in Chapters 22 and 24. CA avoids or minimizes soil tillage, maintains a permanent soil cover of crops and/or residues, and utilizes efficient crop rotations (FAO, 2002). It has been successful in many parts of the world, but has been least successful to date in the dry areas where production of organic matter is too low for permanent soil cover and for significant crop residues because of water shortages. Stewart and Koohafkan (2004) suggest, however, that even small amounts of crop residues can reduce wind erosion considerably and increase soil water storage. Since significant quantities of soil and nutrients are lost by wind and water erosion under arid conditions where the soil remains bare for most part of the year (Zöbisch, 1998), even small savings are worth pursuing.

Interactions between soil nutrients and water are different under conservation tillage compared with conventional tillage systems. For example, nitrogen is more efficiently used under no-till compared to conventional tillage thanks to the higher soil moisture content conserved, especially with legume crops (Pierce and Rice, 1988). This is mainly because no-till reduces water loss in dry years, due to the residue cover, and improves water infiltration during wet years. Somewhat differently, López-Bellido et al. (1996) have reported a positive effect of no-till on wheat yields in dry years, but a negative effect in wet seasons. There is need to further exploit such interactions wherever they can be beneficial in dryland areas.

Stewart and Koohafkan (2004) point out that no dramatic increases in production and soil fertility can be expected in dry areas in the short term, and farmers are unlikely to commit themselves to long-term solutions without adequate incentives. This argues for establishment of policies that will promote better soil and water management under conservation agriculture farming systems.

4.3.2 Legume Rotations and Crop Mixtures

In recent years there has been an encouraging trend away from mainly cereal-based systems in drylands toward cereal–legume and cereal–legume–livestock systems that not only bring economic benefits but also improve soil fertility through the processes discussed above. Examples are found in the drylands of West Asia (Jones and Singh, 1995), Australia (Angus and Good, 2004) and the dry savannas of West Africa (Sanginga et al., 2003a).

Biological nitrogen fixation is the cheapest and most effective management tool for maintaining sustainable yields in low-input agriculture (Dakora and Keya, 1997; see also Chapter 12). Furthermore, this is often the only available source of nitrogen supply for plants in smallholder systems in less-developed countries (Hungria and Vargas, 2000; see also Chapter 27). An alternative to the widely practiced cereal–fallow or cereal monoculture systems is the introduction of nitrogen-fixing legume crops into a rotation with cereals.

The beneficial effects of legumes in crop rotation on soil fertility and subsequently on cereal productivity are well documented (Pierce and Rice, 1988; Robson et al., 2002). There is evidence, for example, that wheat grown in rotations with other plants such as legumes in dry areas gives more efficient water and nitrogen use than does the cropping system of grain followed by bare fallow (Halvorson and Reule, 1994). In dry areas, long-term trials undertaken by ICARDA researchers in northern Syria under rainfed conditions (annual rainfall 250–320 mm) have shown an increase in residual soil nitrogen content, higher wheat and barley yields, and higher water-use efficiency after the legume phase (Keatinge et al., 1988; Jones and Singh, 1995). Similar results have been reported from dry areas of Spain (López-Bellido et al., 1996) and Australia (Pearson et al., 1995).

Legume crops differ in their nitrogen-fixing capacity and yields. ICARDA's long-term trials have showed that fava bean (*Vicia fava*) and lathyrus (*Vicia lathyrus*) can fix more nitrogen than chickpea and lentil (Malhotra et al., 2004). Peoples et al. (1998) report that perennial pastures containing alfalfa (*Medicago sativa*) provide consistently greater annual vegetative production and can fix up to 50% more nitrogen than subterranean clover pastures, especially under drought conditions in Australia.

Considerable efforts have been made during the last two decades to increase the efficiency of legume–rhizobium associations in dry areas under prevailing harsh environmental conditions. For this purpose, Materon and Cocks (1988) identified certain native rhizobium strains that are able to form effective nitrogen-fixing nodules in different annual medic species (of the *Medicago* genus of legumes). Subsequently, Athar and Johnson (1996) have identified two mutant strains of *R. meliloti* that are more effective under drought stress than naturalized alfalfa rhizobium. Guckert et al. (2003) report promising results with nitrogen fixation and dry matter production under cold winter temperatures down to 7°C. In West Africa, Sanginga et al. (2003a) have reported the success of maize–soybean systems as a result of the introduction of promiscuously nodulating soybean varieties in dry savanna areas.

An important additional benefit of the introduction of crop and forage legumes into dryland systems is their apparent ability to utilize relatively inaccessible pools of soil phosphorus. After nitrogen, phosphorus is usually the second most-limiting major nutrient in dryland soils. Lupins were the first crop identified as having a mechanism for enhancing phosphorus availability via the exudation of citric acid (Gardener et al., 1982). Later chickpeas were found to be able to utilize phosphorus from apatites (calcium phosphates) via the exudation of citric acid (Ae et al., 1991). Details on these processes can be found in Randall et al. (2001).

Other beneficial soil fauna, such as fungi and plant growth-promoting rhizobacteria, should be considered in addition to rhizobium in efforts to achieve economically sound sustainable cropping systems for the dry areas. A recent study conducted in southern Australia identified new isolates of *Penicillium* fungi associated with wheat roots having a high potential to mobilize phosphorus originating from phosphate rocks (Wakelin et al., 2004). These fungi are being investigated for their ability to increase crop production on strong phosphorus-retaining soils in dry areas. We are learning that more and more plants have identifiable mechanisms for increasing nutrient availability in soil systems, usually in consort with particular soil organisms.

4.4 Improved Soil Fertility in Dryland Horticultural and Agroforestry Systems

According to Olivares et al. (1988), the plants best able to take advantage of dryland soil and climate conditions are trees, by making better use of limited water and nutrient resources through their deep root systems. Furthermore, many trees, shrubs, and drought-tolerant plants, such as cactus, play an important role in sand stabilization and land protection, in addition to their economic uses for food, fuel, wood, fodder, and other purposes (Nefzaoui, 2002).

4.4.1 Horticultural Systems

Although most cropping systems in arid Mediterranean areas are still based on cereals, fruit trees represent a major component of increasing importance. In Central and West Asia and North Africa, permanent crops, mainly fruit trees, represent 1.5% of the total agricultural area (FAOSTAT, 2004), but a much larger share of agricultural value. Market demand for traditional tree crops is declining, as well as their profitability, so farmers in dry Mediterranean areas are becoming more open to new management practices or even to new crops. In areas receiving more than 400 mm of annual rainfall and where space is available between young or widely spaced trees, farmers are now growing crops that demand little water (lentil, chickpea, or even barley), using moisture not reached by tree roots and at the same time improving soil fertility (Tubeileh, 2004).

With production costs increasing and a trend toward organic farming systems, finding innovative sources of water and nutrients is important to maintain soil fertility and crop yields. Recent research has focused on the use of abundantly available olive-mill wastewater to improve soil fertility in olive-growing areas. The incorporation of such water in sandy soils of Tunisia has increased aggregate stability and reduced evaporation losses from the soil (Mellouli et al., 1998). This is probably because olive-mill wastewater improves soil microbial activity (Kostou et al., 2004), soil organic matter, and levels of inorganic elements in the soil (Paredes et al., 1999). This nutrient-rich water has been shown to result in an improved shoot growth of olive seedlings, although application of large amounts can be phytotoxic (Briccoli-Bati and Lombardo, 1990).

Mycorrhizal fungi provide a very important pathway for improving the water-use efficiency of tree crops. It has been documented that the inoculation of olive trees with *Glomus mosseae* greatly enhances their lateral root development, which in turn stimulates acquisition of water and nutrients (Vitagliano and Citernesi, 1999).

4.4.2 Nitrogen-Fixing Trees

In spite of their multiple uses, relatively little attention has been given to woody leguminous species in the dry areas (Thomson et al., 1994). Leguminous trees can survive with arid soils' low levels of nitrogen due to their nitrogen-fixing capacity. According to Dakora and Keya (1997), legume trees can fix 43–581 kg N ha^{-1} year^{-1}, compared to about 15–210 kg N ha^{-1} year^{-1} from grain legumes. This is why they can be characterized as "fertilizer trees" (Chapter 19). Nitrogen-fixing trees such as Acacia and Prosopis are some of the best sources of this fertilization in arid regions, which is inexpensive and already *in situ* (Zahran, 1999). In addition, most of these species are sources of highly nutritious fodder, fuel, food, charcoal, gums, fiber, and timber (Kang et al., 1990; Fagg and Stewart, 1994).

Acacias are also well adapted to low rainfall and extreme temperatures due to their extremely deep root systems. They include about 1250 species of deciduous or evergreen trees and shrubs widely distributed in the tropics and warmer temperate areas, especially in Australia (Fagg and Stewart, 1994). These species are planted to provide windbreaks, afforest mining and salt-affected areas, stabilize sand dunes, and reduce erosion (Thomson et al., 1994). In addition to their nitrogen-fixing capacity, such species in agroforestry systems are beneficial for maintaining soil fertility due to their efficient nutrient cycling of tree biomass and their uptake of nutrients from deep soil layers (Kang et al., 1990). Peoples and Herridge (1990) report that *Acacia holosericia* trees grown in Senegal are capable of fixing 36–108 kg N ha^{-1}. Seeds of some Acacia species are a traditional food of Australian aboriginal people (Harwood, 1994), who still use Acacia to revegetate or rehabilitate degraded land (Lister et al., 1996).

Leaf pruning of these trees is an important component of sustainability in agroforestry and soil fertility (Zahran, 1999), adding to the biomass available from leaf fall. *Prosopis* spp. are deciduous, thorny shrubs or small trees native to tropical and subtropical regions of the western hemisphere, Africa, the Middle East, and India (Fagg and Stewart, 1994). Their deep rooting systems given them significant tolerance to water stress. This depth allows the roots to achieve good nodulation even under drought conditions, enabling them to survive and contribute to biomass maintenance in the soil (Nilsen et al., 1986; Peláez et al., 1994).

Nutrient levels (N, P, and K), moisture content, and organic carbon of soil are all higher under the canopy of *Prosopis cineraria* compared to the open area (Puri et al., 1994a). The beneficial effects of tree legumes can also be estimated indirectly through their effect on a neighboring or following crop. In the Sahelian region with annual rainfall of 400–600 mm, the yields of millet, groundnuts, and sorghum increased from 500 to 900 kg ha^{-1} when grown under the canopy of *A. albida* (Felker, 1978).

Leguminous trees, which can be grown in hedgerows or intercropped with annual crops, improve soil water conditions as well as enhance soil nitrogen supplies. These trees obtain most of their water from deep soil layers whereas annual field crops take water from the upper layers (Peláez et al., 1994). However, competition between leguminous trees and annual crops has been reported from some dry areas of India (Puri et al., 1994b), as well as East Africa. Optimum spacing and related considerations thus need to be factored into management decisions.

4.4.3 Combining Inorganic and Organic Nutrient Sources

Although it is recognized that the use of fertilizers will continue to be important for agricultural production, there is a need to increase our knowledge of chemical and biological processes that render nutrients more available to plants from soil organic matter and other nutrient sources, such as manures, native vegetation, and crop residues. This will help optimize nutrient cycling and recovery, minimize fertilizer requirements, and maximize the efficiency of nutrient use. Such research is consistent with what farmers actually do with their available nutrient resources. For example, recent surveys in Syria indicate that farmers use crop rotations, manures, and fertilizers, in that order of preference, to maintain their soil fertility (El-Dessougi, unpublished).

While recent research advances have improved guidelines for the effective use of crop and plant residues in tropical regions (Swift, 1999; Palm et al., 2001), similar data for drylands are scarce. Nutrient cycling through the use of low-cost organic inputs such as animal manure and crop residues is known to improve the structural and chemical properties of soil and to replenish macro- and micronutrients. These changes increase soil water-holding capacity and allow for better root penetration and aeration, protecting nutrients against loss by leaching or erosion and enriching soil flora and fauna (Woomer and Swift, 1994). Still, management practices to achieve these effects are not well-worked out for dry areas, where water and temperature factors work differently from humid tropical or temperate conditions.

Increasing nutrient use and uptake efficiencies is important to maximize the use of applied organic and inorganic inputs and to reduce production costs. Careful manage-ment of the methods and timing of fertilizer application will synchronize nutrient release with peak nutrient uptake by crops, offsetting deficiencies in nutrient availability and depletion of soil nutrient stocks (Woomer and Swift, 1994). Proper management of other low-cost sources of organic inputs, such as the collection and composting of household organic waste, as well as the use of agro-industrial waste such as oil mill wastes or sawdust, can further contribute to efficient nutrient cycling, depending on how much immobilization of nutrients accompanies the process.

Swift (1999) has argued that perhaps the most practical means to control and manage biologically the dynamics of soil fertility is via indirect technologies such as manipulation of the quality, timing, and amounts of additions of crop residues, plant litter, and manures. For resource-poor farmers on drylands, this option is likely to be more feasible than direct interventions such as introducing and/or manipulating particular beneficial soil fauna. Our knowledge of the role of soil fauna in soil system fertility in the drylands is meagre compared with the wetter tropical areas. Also there is a lack of cost-effective delivery mechanisms for inocula and soil organisms. This option is discussed in Chapters 32–34, with some promising results; however, the successes reported there are in climates more humid than the dryland areas considered here.

Although there has been optimism that positive interactive effects are attainable from combining inorganic and organic nutrient resources, either by increasing efficiency or by adding value, the evidence on this is sparse. Most recent evidence points to beneficial effects due to mulching, conserving moisture and/or increasing infiltration, rather than from synergistic interactions (Sanginga et al., 2003b). Both are, of course, helpful in dry areas. Other soil management strategies for sustained fertility that can be used in dry areas include planting crops with differing root architecture to achieve more efficient nutrient capture, use of crop residues, improved fallows, and agroforestry (e.g., Jones and Snapp, 1997).

4.4.4 Pest and Disease Management

An integral part of soil fertility management is the control of weeds and diseases. Failure in this can result in crop failure which endangers food security. Manipulating soil microorganisms, soil fauna and plant pathogens as bioherbicides has been proposed for effective control of parasitic weeds and other crop pests and diseases (Liu et al., 1995; Elzein and Kroschel, 2003). This area of research is currently receiving more attention.

Crop rotations can play an important role in pest control as illustrated by maize–soybean rotations in Nigeria where the devastating infestation of maize fields with the parasitic weed *Striga hermonthica* has been reduced (Sanginga et al., 2003a). Intercropping is also emerging as an effective management strategy for striga control (Chapter 40). This is another area of research that requires crossdisciplinary collaboration as discussed in Chapter 41.

4.4.5 Management of Rangelands and Crop–Livestock Interactions

Rangelands covering vast areas of the arid and semi-arid regions are characterized by low levels of productivity, and most are undergoing some degree of degradation as a result of uncontrolled grazing of communal lands and the removal of shrubs and trees for fuelwood. There has been too little effort to establish counteracting strategies that ensure regeneration and conservation of rangeland plants and that avoid the invasion by species of low nutritive value and palatability. Insecure tenure arrangements contribute to declining soil fertility when farmers are not motivated to conserve soils and invest in sustaining fertility where they have no right to the future benefits from their efforts.

Extensive grazing is a system that facilitates the exploitation of nutrients dispersed over wide areas that could not otherwise be profitably used. While there may be some merit in promoting nitrogen-fixing species in rangelands, it is unlikely that they will receive enough other inputs to significantly increase nutrient cycling and accumulation in useful plant or animal products. Water harvesting in rangelands that can lead to the increased production of forage shrubs and herbs, followed by more general improvement in soil fertility, may be the most feasible option for these large areas. The management of rangelands will remain problematic unless supported by adequate technologies, appropriate policy options, information and education, accessibility to markets, and more awareness of environmentally friendly practices.

Crop–livestock interactions constitute an essential component of integrated soil fertility management under marginal dryland conditions, where both extensive and intensive livestock rearing is being practiced to various extents. This is addressed at greater length in Chapter 17. Livestock act as transporters of nutrients from grazed rangelands, crop residues, and native vegetation, and they can supply manure and urine to cropped land, thereby contributing to the cycling of nutrients in the soil and enriching soil organic matter. Substantial amounts of vegetative material containing up to 35–50% organic matter and 80–90% nitrogen and phosphorus are recycled by ruminants (Powell et al., 1999).

However, the interactions between livestock and crops can be competitive. For example, the grazing of crop residues that help replenish soil nutrients or protect bare soils against erosion can result in range deterioration and land degradation if the manure is not returned to the field or there is uncontrolled grazing. Integrated management of range-land, feeding systems, manure management, composting, and application is needed to minimize losses and maximize recycling of nutrients, optimally sustaining soil fertility over time (Tarawali et al., 2001).

4.5 Discussion

Experience shows that soil fertility management technologies developed by researchers and scientists in isolation from resource-poor farmers and that are capital-based have low adoptability, in large part because farmers have to cope with specific microecological or socioeconomic conditions that are not addressed by these technologies. Given the specificity of spatial and temporal microclimatic and socioeconomic conditions, scientific recommendations that are sound and relevant for a particular area can be of no value in other sites. Thus, to ensure adoption and practical applicability, soil fertility management technologies should be developed in participation with farmers, e.g., participatory technology development.

Farmers' thinking about soil fertility must be holistic, considering factors such as the impact of microclimatic variations on crop productivity (Cools et al., 2003) and economic and ethnic influences (Desbiez et al., 2004). Farmers have important context-specific knowledge that can be combined with scientists' detailed knowledge of soil chemical, physical, and biological processes to develop appropriate innovations in fertility management. Innovations developed jointly with farmers stand a better chance of adoption through farmer-led adaptive processes attuned to given agroecologies and to their own specific conditions. Participatory methods can be of value to researchers, not just farmers, enabling them to incorporate into their analysis the indigenous knowledge that farmers have of local soils that has been accumulated and transferred through generations to cope with the adverse conditions of drylands (Ishida et al., 2001).

Recent and on-going studies at ICARDA (e.g., Cools et al., 2003) plus other evidence from arid and semi-arid areas (e.g., Ishida et al., 2001; Reij and Steeds, 2003) have shown that farmers have considerable knowledge about their land and soil fertility patterns. Farmers use criteria such as soil color, texture, and depth to classify their soils as an indication of fertility status and workability (see also Joshi et al., 2004). Farmers generally follow very sound practices to overcome the scarcity of their resources. For example, they know very well the value of animal manure for their soils, but are faced with problems of low quantity and quality of manure, labor availability, other competing claims on productive resources, and infestation by weeds and diseases.

To deal with such constraints, farmers adopt various practices such as confining animals to collect manure; alternating field applications; composting manure and applying it only to high-value crops such as vegetables, or localized application around the base of olive trees; allowing pastoralists and their animals to stay on their land over the dry season to graze crop residues and manure the land; and allocating manure collection and management exclusively to women for more intensive and consistent use.

Under extreme agroclimatic and socioeconomic conditions, farmers are likely to be effective managers of their soils. Reij and Steeds (2003) report that except when facing the most severe natural disasters, people in dry areas are quite capable of sustaining their livelihoods and coping with almost all adverse conditions. These authors documented outstanding cases of farmer innovators who have developed various strategies and techniques to reduce risk, enhance productivity, and improve income sources.

Examples of success stories and innovations that have resulted in significant, positive returns both in terms of soil fertility and farmers' livelihoods include: night paddock manuring systems; construction of stone ponds to collect water and soil; harvesting rain water; inventing simple, inexpensive tools for planting and harvesting; increased diversification of plant biomass; and regeneration of natural vegetation by medicinal and woody tree species in pits filled with manure (Reij and Waters-Bayer, 2001). Locating innovator-farmers and publicizing their successes in coping with adverse situations can be an effective means to help farmers in other similar areas through farmer-to-farmer exchanges.

As external inputs are scarce and costly in dryland areas, management systems that require few external inputs, relying on nutrient cycling and the more efficient use of water and nutrients, are more likely to gain acceptance. These should be more resilient and generate higher-value products than the mainly cereal-based, crop–livestock systems dominant in dryland areas. Promising management options for improving soil fertility in the drylands include conservation agriculture and tillage; crop rotations; low-cost soil and water conservation technologies developed with farmers; higher-value food, medicinal, cosmetic and herb crops; and new tree and livestock options. These options need to be researched and developed by building on the knowledge and experiences of those farmers who have already successfully developed appropriate innovations to cope with the harsh dryland conditions.

Many of these options need an enabling policy environment that removes perverse subsidies and encourages farmers to adopt technologies that conserve their natural resources while providing them with increased income from niche markets for their products. Getting such incentives and policies in place and engaging in participatory technology development are two parallel paths to reverse the current decline in soil fertility and degradation of dryland soil systems. Concurrently, we need to achieve a better understanding of the interactions among plants, animals, and soils that drive nutrient fluxes and cycling under water-scarcity conditions and also soil fertility interactions with pest and diseases. Such knowledge will identify or create opportunities for better, low-cost biological management of soil fertility in these so-called "marginal" areas that are actually extensive and must support large and growing populations.

References

Ae, N., Arihara, J., and Okada, K., Phosphorus response of chickpea and evaluation of phosphorus availability in Indian alfisols and vertisols, In: _Phosphorus Nutrition of Grain Legumes in the Semi-Arid Tropics_, Johansen, C. et al., Eds., ICRISAT, Patancheru, India, 33–41 (1991).

Athar, M. and Johnson, D.A., Influence of drought on competition between selected _Rhizobium meliloti_ strains and naturalized soil rhizobia in alfalfa, _Plant Soil_, **184**, 231–241 (1996).

Angus, J.F. and Good, A.J., Dryland cropping in Australia, In: _Challenges and Strategies for Dryland Agriculture_, CSSA Special Publication 32, Rao, S.C. and Ryan, J., Eds., CSSA/ASA, Madison, WI, 151–166 (2004).

Bowden, L., Development of present dryland farming systems, In: _Agriculture in Semi-Arid Environments_, Hall, A.E. et al., Eds., Springer, Berlin, 45–72 (1979).

Briccoli-Bati, C. and Lombardo, N., Effect of olive oil waste water irrigation on young olive plants, _Acta Hortic._, **286**, 489–491 (1990).

Cools, N., DePauw, E., and Deckers, J., Towards an integration of conventional and evaluation methods and farmers' soil suitability assessment: A case study in northwestern Syria, _Agric. Ecosyst. Environ._, **95**, 327–342 (2003).

De Pauw, E., Management of dryland and desert areas, In: _Encyclopedia for Life Support Systems_. UNESCO/EOLSS Publishers, Paris (2004).

Desbiez, A. et al., Perceptions and assessment of soil fertility by farmers in the mid-hills of Nepal, _Agric. Ecosyst. Environ._, **103**, 191–206 (2004).

Dakora, F.D. and Keya, S.O., Contribution of legume nitrogen fixation to sustainable agriculture in Sub-Saharan Africa, _Soil Biol. Biochem._, **29**, 809–817 (1997).

Elzein, A. and Kroschel, J., Progress on management of parasitic weeds, In: _Weed Management for Developing Countries, addendum I_, Labrada, R., Ed., FAO, Rome, 109–144 (2003).

Eswaran, H., Reich, P., and Beinroth, F., Global desertification tension zones, In: _Sustaining the Global Farm: Proceedings of the 10th International Soil Conservation Organization Meeting_, Stott, D.E. et al., Eds., 24–29 May 1999, West Lafayette, Indiana, 24–28 (2001).

Fagg, C.W. and Stewart, J.L., The value of Acacia and Prosopis in arid and semi-arid environments, _J. Arid Environ._, **27**, 3–25 (1994).

FAO, _Intensifying Crop Production with Conservation Agriculture_, Food and Agriculture Organization, Rome (2002), http://www.fao.rg/ag/ags/AGSE/main.htm

FAOSTAT, _FAO Primary Crops Statistical Database_, FAO, Rome (2004).

Farahani, H.J., Peterson, G.A., and Westfall, D.G., Dryland cropping intensification: A fundamental solution to efficient use of precipitation, _Adv. Agron._, **64**, 197–223 (1998).

Felker, P., _State of the Art: Acacia albida as a Complementary Permanent Intercrop with Annual Crops_, University of California-Riverside, Riverside, CA (1978).

Gardener, W.K., Barber, D.A., and Parberry, D.G., The acquisition of phosphorus by _Lupinus albus_ L. 1: Some characteristics of the soil/root interface, _Plant Soil_, **68**, 19–32 (1982).

Guckert, A. et al., Production et fixation d'azote chez des luzernes annuelles en région méditerranéenne lors de la période froide, _Fourrages_, **173**, 37–47 (2003).

Halvorson, A.D. and Reule, C.A., Nitrogen fertilizer requirements in an annual dryland cropping system, _Agron. J._, **86**, 315–318 (1994).

Harwood, C.E., Human food potential of the seeds of some Australian dry-zone Acacia species, _J. Arid Environ._, **27**, 27–35 (1994).

Hazell, P., _Strategies for the Sustainable Development of Dryland Areas_, International Food Policy Research Institute, Washington, DC (2001).

Hungria, M. and Vargas, A.A.T., Environmental factors affecting N_2 fixation in grain legumes in the tropics, with an emphasis on Brazil, _Field Crops Res._, **65**, 151–164 (2000).

Ishida, F., Tian, G., and Wakatsuki, T., Indigenous knowledge and soil management, In: _Sustaining Soil Fertility in West Africa_, SSSA Special Publication 58, Tian, G. et al., Eds., Soil Science Society of America, Madison, WI, 91–109 (2001).

Jones, M.J. and Singh, M., Yields of crop dry matter and nitrogen in long-term barley rotation trials at two sites in Northern Syria, _J. Agric. Sci._, **124**, 389–402 (1995).

Jones, R.B. and Snapp, S.S., Management of leguminous leaf residues to improve nutrient use efficiency in the sub-humid tropics, In: *Driven by Nature: Plant Litter Quality and Decomposition*, Cadisch, G. and Giller, K.E., Eds., CAB International, Wallingford, UK, 239–250 (1997).

Joshi, L. et al. Locally derived knowledge of soil fertility and its emerging role in integrated natural resource management, In: *Below-Ground Interactions in Tropical Agroecosystems: Concepts and Models with Multiple Plant Components*, van Noordwijk, M., Cadisch, G., and Ong, C.K., Eds., CAB International, Wallingford, UK, 17–39 (2004).

Kang, B.T., Reynolds, L., and Atta-Krah, A.N., Alley farming, *Adv. Agron.*, **43**, 315–359 (1990).

Keatinge, J.D.H., Chapanian, N., and Saxena, M.C., Effect of improved management of legumes in a legume–cereal rotation on field estimates of crop nitrogen uptake and symbiotic nitrogen fixation in northern Syria, *J. Agric. Sci.*, **110**, 651–659 (1988).

Kostou, M. et al., The effect of olive oil mill wastewater (OMW) on soil microbial communities and suppressiveness against *Rhizoctonia solani*, *Appl. Soil Ecol.*, **26**, 113–121 (2004).

Lister, P.R. et al. Acacia in Australia: Ethnobotany and potential food crop, In: *Progress in New Crops*, Janick, J., Ed., ASHS Press, Alexandria, CA, 228–236 (1996).

Liu, L., Kloepper, J.W., and Tuzun, S., Induction of systemic resistance in cucumber against bacterial angular leaf spot by plant growth-promoting rhizobacteria, *Phytopathology*, **85**, 695–698 (1995).

López-Bellido, L. et al., Long-term tillage, crop rotation, and nitrogen fertilizer effects on wheat yield under Mediterranean conditions, *Agron. J.*, **88**, 783–791 (1996).

Malhotra, R.S. et al., Improved livelihoods from legumes: A review of BNF research at the International Center for Agricultural Research in the dry areas, In: *Symbiotic Nitrogen Fixation: Prospects for Enhanced Application in Tropical Agriculture*, Serraj, R., Ed., Oxford and IBH, New Delhi, India, 99–112 (2004).

Matar, A., Torrent, J., and Ryan, J., Soil and fertilizer phosphorus and crop responses in the dryland Mediterranean zone, *Adv. Soil Sci.*, **18**, 81–146 (1992).

Materon, L.A. and Cocks, P.S., Constraints to biological nitrogen fixation in ley-farming systems designed for West Asia, In: *Microbiology in Action*, Murell, I.R. and Kennedy, W.G., Eds., Wiley, New York, 93–106 (1988).

Mellouli, H.J. et al., The use of olive mill effluents ("margines") as soil conditioner mulch to reduce evaporation losses, *Soil Till. Res.*, **49**, 85–91 (1998).

Mortimore, M., Why invest in drylands? Paper Commissioned by the Global Mechanism of the UNCCD, International Fund for Agricultural Development, Rome (2004).

Nefzaoui, A., Cactus to prevent and combat desertification, *Proceedings of the 6th International Conference on the Development of Dry Lands*, Cairo, Egypt, Ryan, J., Ed., ICARDA, Aleppo, Syria, 261–269 (2002).

Nilsen, E.T., Virginia, R.A., and Jarrall, W.H., Water relations and growth characteristics of *Prosopis glandulosa* var. Torreyana in a simulated phreatophytic environment, *Am. J. Bot.*, **73**, 430–433 (1986).

Olivares, J., Herrera, M.A., and Bedmar, E.J., Woody legumes in arid and semi-arid zones: The *Rhizobium–Prosopis chilensis* symbiosis, In: *Nitrogen Fixation by Legumes in Mediterranean Agriculture*, Beck, D.P. and Materon, L.A., Eds., ICARDA and Nijhoff, Dordrecht, The Netherlands, 65–72 (1988).

Oweis, T., Hachum, A., and Bruggeman, A., Eds., *Indigenous Water-Harvesting Systems in West Asia and North Africa*, ICARDA, Aleppo, Syria (2004).

Palm, C.A. et al., Management of organic matter in the tropics: Translating theory into practice, *Nutr. Cycl. Agroecosyst.*, **61**, 63–75 (2001).

Paredes, C. et al., Characterization of olive mill wastewater (alpechin) and its sludge for agricultural purposes, *Bioresour. Technol.*, **67**, 111–115 (1999).

Pearson, C.J., Norman, D.W., and Dixon, J., *Sustainable Dryland Cropping in Relation to Soil Productivity*, FAO Soils Bulletin 72. Food and Agriculture Organization, Rome (1995).

Peláez, D.V. et al., Water relations between shrubs and grasses in semi-arid Argentina, *J. Arid Environ.*, **27**, 71–78 (1994).

Pender, J. and Gebremedhin, B., Impacts of policies and technologies in dryland agriculture: Evidence from Northern Ethiopia, In: *Challenges and Strategies for Dryland Agriculture*, CSSA Special Publication 32, Rao, S.C. and Ryan, J., Eds., CSSA/ASA, Madison, WI, 389–416 (2004).

Peoples, M.B. and Herridge, D.F., Nitrogen fixation by legumes in tropical and subtropical agriculture, *Adv. Agron.*, **44**, 155–223 (1990).

Peoples, M.B. et al., Effect of pasture management on the contributions of fixed N to the N economy of ley-farming systems, *Aust. J. Agric. Res.*, **49**, 459–474 (1998).

Pierce, F.J. and Rice, C.W., Crop rotation and its impact on efficiency of water and nitrogen use, In: *Cropping Strategies for Efficient Use of Water and Nitrogen*, ASA Special Publication 15, Hargrove, W.L., Ed., ASA/CSSA/SSSA, Madison, WI, 21–42 (1988).

Powell, J.M., Ikpe, F.N., and Somda, Z.C., Crop yield and the fate of nitrogen and phosphorus following application of plant material and faeces to soil, *Nutr. Cycl. Agroecosyst.*, **54**, 215–226 (1999).

Puri, S., Kumar, A., and Singh, S., Productivity of *Cicer arietinum* (chickpea) under a *Prosopis cineraria* agroforestry system in the arid regions of India, *J. Arid Environ.*, **27**, 85–98 (1994a).

Puri, S., Singh, S., and Kumar, A., Growth and productivity of crops in association with an *Acacia nilotica* tree belt, *J. Arid Environ.*, **27**, 37–48 (1994b).

Qadir, M. and Oster, J.D., Crop and irrigation management strategies for saline-sodic soils and waters aimed at environmentally sustainable agriculture, *Sci. Total Environ.*, **323**, 1–19 (2004).

Randall, P.J. et al. Root exudates in phosphorus acquisition by plants, In: *Plant Nutrient Acquisition: New Perspectives*, Ae, N., Arihara, J., Okada, K., and Srinivasan, A., Eds., Springer, Tokyo, 71–100 (2001).

Reij, C. and Steeds, D., Success stories in Africa's drylands: Supporting advocates and answering skeptics. Paper commissioned by the Global Mechanism of the Convention to Combat Desertification, Center for International Cooperation, Vrije Universiteit, Amsterdam, The Netherlands (2003).

Reij, C. and Waters-Bayer, A., Eds., *Farmer Innovation in Africa: A Source of Inspiration for Agricultural Development*, Earthscan Publications, London (2001).

Robson, M.C. et al., The agronomic and economic potential of break crops for ley/arable rotation in temperate organic agriculture, *Adv. Agron.*, **77**, 369–427 (2002).

Ryan, J., Ed., Desert and Dryland Development: Challenges and Potential in the New Millennium: *Proceedings of the Sixth International Conference on the Development of Dry Lands, 22–27 August 1999, Cairo, Egypt*. ICARDA, Aleppo, Syria (2004).

Sanchez, P.A., Tropical soil fertility research, towards the second paradigm, In: *Transactions of the 15th World Congress of Soil Science, 10–16 July 1994, Acapulco, Mexico*. International Soil Science Society, Wageningen, The Netherlands, 65–88 (1994).

Sanginga, N. et al., Sustainable resource management coupled to resilient germplasm to provide new intensive cereal–grain–legume–livestock systems in the dry savanna, *Agric. Ecosyst. Environ.*, **100**, 305–314 (2003a).

Sanginga, N. et al., Balanced nutrient management systems for cropping systems in the tropics: From concepts to practice, *Agric. Ecosyst. Environ.*, **100**, 99–102 (2003b).

Sprent, J.I., Nitrogen fixation in arid environments, In: *Plants for Arid Lands*, Wickens, G.E., Goodin, J.R., and Field, D.V., Eds., Allen and Unwin, London, 215–229 (1985).

Sprent, J.I., *The Ecology of the Nitrogen Cycle*, Cambridge University Press, Cambridge, UK (1987).

Stewart, B.A. and Koohafkan, P., Dryland agriculture: Long neglected but of worldwide importance, In: *Challenges and Strategies for Dryland Agriculture*, CSSA Special Publication 32, Rao, S.C. and Ryan, J., Eds., Crop Science Society of America, Madison, WI, 11–23 (2004).

Swift, M.J., Towards the second paradigm: Integrated biological management of soil, In: *Soil Fertility, Soil Biology and Plant Nutrition Interrelationships*, Siqueira, J.O., Moreira, F.M.S., Lopes, A.S., Guilherme, L.R.G., Faquin, V., Neto, A.E.F., and Carvalho, J.G., Eds., Lavras-MG, Brazil, 11–24 (1999).

Tarawali, S.A. et al., The contribution of livestock to soil fertility, In: *Sustaining Soil Fertility in West Africa*, SSSA Special Publication 58, Tian, G. et al., Eds., Soil Science Society of America, Madison, WI, 281–304 (2001).

Thomson, L.A.J., Turnbull, J.W., and Maslin, B.R., The utilization of Australian species of Acacia, with particular reference to those of the subtropical dry zone, *J. Arid Environ.*, **27**, 279–295 (1994).

Tubeileh, A., Bruggeman, A., and Turkelboom, F., *Growing Olives and Other Tree Species in Marginal Dry Environments*, ICARDA, Aleppo, Syria (2004).

UNCCD, *The Causes of Desertification*. Fact Sheet 2. UN Convention to Combat Desertification, Bonn, Germany (2004).

Vitagliano, C. and Citernesi, A.S., Plant growth of *Olea europaea* L. as influenced by arbuscular mycorrhizal fungi, *Acta Hortic.*, **474**, 357–361 (1999).

Wakelin, S.A. et al., Phosphate solubilization by *Penicillium* spp. closely associated with wheat roots, *Biol. Fertil. Soils*, **40**, 36–43 (2004).

Woomer, P.L. and Swift, M.J., Eds., *The Biological Management of Tropical Soil Fertility*, Wiley, Chichester, UK (1994).

Zahran, H.H., Rhizobium—legume symbiosis and nitrogen fixation under severe conditions and in an arid climate, *Microbiol. Mol. Biol. Rev.*, **63**, 968–989 (1999).

Zöbisch, M.A., The West Asia and North Africa region: Some underlying factors of wind erosion and perspectives for research and technology development, In: *Wind Erosion in Africa and West Asia: Problems and Control Strategies*, Sivakumar, M.V.K., Zöbisch, M.A., Koala, S., and Maukonen, T., Eds., ICARDA, Aleppo, Syria, 37–47 (1998).

PART II: SOIL AGENTS AND PROCESSES

5

The Soil Habitat and Soil Ecology

Janice E. Thies and Julie M. Grossman

Department of Crop and Soil Sciences, Cornell University, Ithaca, New York, USA

CONTENTS

This chapter reviews the key functions of soil biota and their roles in maintaining soil fertility. We consider the soil as a habitat for organisms, identifying important sources of energy and nutrients for the soil biota and describing the flow of energy and cycling of materials from above to below ground. A more detailed discussion of energy flows follows in the next chapter. The trophic structure of the soil community, i.e., the organized flow of nutrients within it, and the various interactions among organisms comprising the soil food web are considered here. Linkages between above- and below ground processes are highlighted to illustrate their interconnectedness and to show that soil is not an inert medium, but rather hosts a wide variety of organisms that collectively perform essential ecosystem services.

The functioning of soil systems involves many interactions among plant roots and plant residues, various animals and their residues, a vast diversity of microorganisms, and the physical structure and chemical composition of the soil. To manage soil systems productively, we need to know what practices will help to improve the survival and functioning of beneficial soil organisms while deterring the activity of pathogenic organisms. This volume offers varied examples of how the biological functioning of soil systems can be enhanced to improve their fertility and sustainability.

Here, we present an integrated view of the soil as a fundamental component of terrestrial ecosystems, having a distinct though varying structure and an intricate set of biological relationships. This illustrates how soil organisms contribute to maintaining soil fertility and also how the fertility of soil systems can be improved by managing and enhancing biological interactions. The basic factors and dynamics of soil systems discussed here provide a foundation for understanding the chapters that follow. It is written so that readers not trained in soil science can gain ready access to the subject matter. Persons already familiar with soil science should appreciate the change in perspective that it offers on soil systems, putting living organisms and the organic matter they produce center-stage.

5.1   The Soil as Habitat for Microorganisms

Soil is one of the more complex and highly variable habitats on earth. Any organisms that make their home in soil have had to devise multiple mechanisms to cope with variability in moisture, temperature, and chemical changes so as to survive, function, and replicate. Within a distance of <1 mm, conditions can vary from acid to base, from wet to dry, from aerobic to anaerobic, from reduced to oxidized, and from nutrient-rich to nutrient-poor. Along with spatial variability there is variability over time, so organisms living in soil must be able to adapt rapidly to different and changing conditions. Variations in the physical and chemical properties of the soil are thus important determinants of the presence and persistence of soil biota.

5.1.1   Differences Among Soil Horizons

A typical soil profile has both horizontal and vertical structure. At the base of any soil profile is underlying *bedrock*, or **parent material**, which is the type of geological formation upon which and with which the soil above has been formed. Overlying the bedrock is a *C horizon* that has developed directly from modifications of the underlying parent material. This C horizon remains the least weathered (changed) of the identifiable horizons, accumulating calcium (Ca) and magnesium (Mg) carbonates released from horizons above. Microbial activity in this C horizon is typically very low, in part because of limitations in oxygen (O_2) and organic matter.

Overlying the C horizon is the **subsoil**, or *B horizon*. This is composed of minerals derived from the parent material and of materials that have leached down from the horizons above, including humic materials formed above from the decomposition of organic (plant and animal) matter. Yet, because the B horizon is typically still rather low in organic matter, it supports relatively small microbial populations and has little biological activity. The B horizon is the zone of maximum illuviation, i.e., deposition or accumulation of silicate clays and of iron (Fe) and aluminum (Al) oxides.

The *A horizon*, denoting the upper layers of soil, is usually fairly high in organic matter and often darker in color. This, along with the *O (organic) horizon*, is the horizon in which plant roots and soil organisms are most active. Within the A horizon there are differing extents of leaching and movement of materials from the horizon above to the horizons below. The interface between the A and B horizons is the zone of maximum eluviation, i.e., removal through downward leaching of silicate clays and Fe and Al oxides. The interface between the A horizon and the O horizon above it is where incoming organic residues become incorporated with the mineral soil. Together with incorporated soil organic matter (SOM), the A horizon is often referred to as the **topsoil**.

The O horizon on the surface is the topmost layer, often referred to as the **litter layer**. The largest component of this layer is undecomposed organic matter (OM), and the origins of these organic materials are easy to distinguish — plant litter, manure, or other organic inputs.

5.1.2   Factors in Soil Genesis

In 1941, Hans Jenny (1941) proposed the following soil-forming factors that are still used today:

1. The parent material or underlying geological formation of the region;
2. The climate, referring largely to the temperature and precipitation in the region and to their interaction, which affects soil formation through freezing and thawing cycles;
3. The topography, denoting where soil is located within the landscape, at the top, middle, or bottom of a slope, which has dramatic effects on the outcome of soil formation;
4. Organisms, such as the dominant plant community and associated soil organisms that influence soil formation strongly by depositing OM and aggregating soil minerals; and
5. Time that has passed since the bedrock was laid down in relation to all of the other factors.

These factors combined explain the complex mix of characteristics that differentiate soil types. That soil types can vary considerably over short ranges illustrates the important role of the biota in soil formation because the other factors vary at larger scales both spatially and temporally.

5.1.3   Physical Components of Soil Systems

A typical soil is composed of both a mineral fraction and an organic fraction. These two fractions make up the soil solids, with the remaining soil volume composed of pore space, which at any given time is filled with some combination of air and/or water. When soil is saturated with water, all of the air in its pore spaces will have been displaced; conversely, desiccated soil has only air in the spaces between its soil solids.

The SOM content, the nature of the mineral fraction, and the relative proportions of air and water are critical factors affecting microbial activity and function. Soils with their pore space dominated by water are anaerobic. This condition will limit microbial activity to that of anaerobes and facultative anaerobes, i.e., organisms capable of metabolism in the absence of oxygen (O_2). The anaerobic process of fermentation is energetically less efficient than aerobic metabolism (Fuhrmann, 2005), and its end-products are generally organic acids and alcohols, which can be toxic to plants and many microbes. Hence, a soil with much of its pore space occupied by water much of the time will be a less productive soil, even though water is one of plants' critical needs.

A balance, where about half of the soil's pore space is occupied by air and half by water, is more supportive of both plant growth and microbial metabolism. Roots require O_2 in order to respire, and aerobes (microorganisms capable of aerobic respiration) can derive vastly more energy from this process than can be derived through fermentation or anaerobic respiration.

The nature of the mineral fraction determines the soil texture, content, and concentration of mineral elements as well as the presence of heavy metals, which can have some undesirable effects on plant and/or animal life. Phosphorus (P), potassium (K), and magnesium (Mg) are essential plant macronutrients derived from the soil mineral fraction. Hence, the productive capacity of any soil is very dependent on the composition of its mineral fraction (Brady and Weil, 2002).

5.1.4   Physical Properties and Their Implications for Soil Biology

Other important soil physical properties include texture, bulk density, temperature, aggregation, and structure. Each has important effects on the composition and activity of soil biota.

Texture, which refers to the proportions of sand, silt, and clay in any given soil, will strongly affect the soil's water-holding capacity and its cation- and anion-exchange capacities. The ability of soil to retain water is important because microbes depend on soil water as a solvent for cell constituents and as a medium through which dissolved nutrients can move to their cell surface. Also, water is needed to facilitate the movement of flagellated bacteria, ciliated and flagellated protozoa, and nematodes. Texture thus directly influences biological activity in soil.

Bulk density refers to the weight of soil solids per unit volume of soil. Soils with a bulk density <1 g cm^{-3} are lighter or loose soils, likely to have good aeration and easy for roots to penetrate and for microbes to navigate. Soils with a bulk density >1 g cm^{-3} are considered as increasingly heavier or compacted soils. As bulk density increases, soil porosity decreases, and air and water flows become restricted. This impedes soil drainage and root penetration. Such soils are often prone to waterlogging, creating anaerobic conditions.

Temperature will have varying effects on microbial activity depending on the respective organisms' range of tolerance. Psychrophilic organisms thrive in cold soil, at temperatures $<10°$C; mesophiles have their greatest rates of activity at temperatures between $10–30°$C; while thermophiles are more active at temperatures in excess of $40°$C. Soils in temperate regions experience prolonged periods annually at each of these temperature optima. This leads to marked seasonal shifts in microbial community composition throughout the year and to concomitant changes in the rates of SOM turnover and in the amounts of microbial biomass. Microbial communities in tropical soils also vary seasonally, but this is less determined by temperature.

Soil aggregation is the result of many interacting factors. In their model of soil aggregation, Tisdall and Oades (1982) described the process of aggregation as beginning with the interaction of clay platelets with one another at a scale of 0.2 μm. Microbial colonization of soil particles comes into play at a scale of 2 μm, an order of magnitude greater where bacterial and fungal metabolites serve to glue clay particles together. At a scale of 20 μm, fungal hyphal filaments and various polysaccharides produced by bacteria become the dominant aggregating factors. Then at a 200-μm scale, roots, and fungal hyphae bind these particles together. The resulting soil is a matrix of mineral particles bound together by biological materials at various nested scales to form macroaggregates at the 2-mm scale.

Soil structure describes the extent of micro- and macroaggregation of a soil. A well-aggregated soil is more resistant to erosion from rain and wind. Also, it is generally well drained and more conducive for the growth of aerobic populations. It thus tends to be a more productive soil for plants and the soil biota. The process of aggregation as seen in the preceding discussion is the result of activities of plant roots and soil biota, creating intrinsic bonds between physical and biological characteristics of soil systems.

5.1.5   Influence of Soil Chemical Properties

Soil chemical properties strongly influence the activity of soil organisms, being at the same time themselves affected by such activity. The more important soil chemical properties affecting on biological activity are:

- pH, i.e., the acidity or alkalinity of a soil
- Cation- and anion-exchange capacity
- Mineral content and solubility
- Buffering capacity
- The concentration of nutrient elements in the soil
- The concentration of O_2, carbon dioxide (CO_2), nitrogen (N_2), and other gases in the soil atmosphere
- Soil water content, and
- Salinity or sodicity.

Both plants and soil organisms have varying tolerances to extremes in soil pH. Most organisms prefer near-neutral pH values between 6 and 7.5. Many soil nutrients are most available for uptake by plant roots within this pH range. When soil is more acidic, the metal elements Fe, manganese (Mn), zinc (Zn), and copper (Cu) increase in solubility, while the solubility of most major nutrient elements — nitrogen (N), P, K, Ca, Mg, and sulfur (S) — decreases. The availability of N, K, S, and molybdenum (Mo) is unaffected at high pH; however, that of P, Ca, Mg, and boron (B) decreases above pH 8.0. In general, fungi and actinomycetes (bacteria that resemble fungi in their morphology and growth habits) appear to be relatively tolerant of both high and low pH, whereas many autotrophic and other heterotrophic bacteria are inhibited at low pH. Hence, in acidic soils, fungi and actinomycetes will tend to predominate. Organisms with greater limits of tolerance to changing abiotic conditions will have a competitive edge, which can affect the activity of others through substrate competition and thus inhibit their growth further.

Living organisms require a range of nutrient elements for their survival. Plants obtain their C (from CO_2), hydrogen (H_2) and oxygen (O_2) from the atmosphere, while the remaining elements must be derived from the soil solution. For most soil microbes, the situation is somewhat different as they derive their energy and cell biomass C mainly from decomposing plant and animal residues and from SOM. Notable exceptions include the cyanobacteria and other photosynthetic bacteria that fix CO_2 directly into cell biomass C using light energy, and the chemolithotrophic bacteria that use the bond energy in reduced compounds, such as NH_4, to generate reducing potential to fix CO_2 into cell biomass C chemosynthetically.

There are many pathways by which soil organisms obtain their energy, cell biomass C, and nutrients. Soil microbes obtain many of their other needed elements from the soil solution or soil minerals, which they solubilize to acquire the necessary nutrients, or from the soil atmosphere. Nitrogen is a special case. Almost 80% of the atmosphere is made up of nitrogen (N_2) gas. However, atmospheric N_2 is not available to plants until it has been reduced, either industrially, atmospherically, or through the process of biological nitrogen fixation (BNF). Many bacteria and cyanobacteria have the ability to fix N_2, but the most well-known are the rhizobia that fix atmospheric nitrogen in symbiosis with host legumes (Fred et al., 1932; Giller, 2001). Nitrogen-fixing bacteria, such as Azospirillum and Azotobacter, also form endophytic or associative relationships within or in close association with plant roots (Boddy et al., 2003), and there are many free-living N_2 fixing bacterial species as well (Dobbelaere et al., 2003). BNF is discussed in more detail in Chapter 12. Most soil fauna meet their energy, cell biomass C, and mineral nutrient requirements from consuming other organisms as either grazers or as predators.

The availability of mineral elements is not is the only important aspect; so are the relative proportions or ratios of mineral elements in relation to an organism's needs. A soil may be high in P, Mg, Ca, and S, for example, but if nitrogen availability is low, then the growth of soil organisms will be limited by the lack of this element. This concept is known as Liebig's "Law of the Minimum," where the growth of any organism is restricted by whatever nutrient element is in the shortest supply in its environment relative to its needs (von Liebig, 1843; van der Ploeg, et al., 1999). This concept is important to bear in mind. No matter how much of a given mineral nutrient is added to a soil, this will not improve crop yield or microbial growth if this is not a factor that is limiting production (Thies et al., 1991).

5.1.6 Adaptations to Stress

Given the high spatial variability in soil properties, the microorganisms that live in soil must be capable of rapidly adapting to continually changing surroundings. Soil organisms respond to stress by varying their use of O_2, by forming resting structures, by increasing intracellular solute concentrations, by producing polyols and heat-shock proteins, and/or by altering membrane structure, to name a few of the possible mechanisms.

Microorganisms vary in their need for or tolerance of O_2. We referred above to the two major groups in terms of their functional relationship to O_2: aerobes and anaerobes. Aerobes are species capable of growing at the O_2 concentration found in the atmosphere (21%), and they typically use O_2 as a terminal electron acceptor in the respiratory electron transport chain. There are three main types of aerobes: obligate, facultative, and microaerophilic. Obligate aerobes require the presence of O_2 for their survival; their type of metabolism is aerobic respiration. While facultative aerobes do not require O_2, they grow much better if O_2 is present. These versatile bacteria have the capacity to respire either aerobically or anaerobically. Microaerophiles require O_2, but they can function at much lower levels than atmospheric concentrations. Their form of metabolism is aerobic respiration (Atlas and Bartha, 1998).

Anaerobes, on the other hand, do not or cannot use O_2 as a terminal electron acceptor. There are two basic types of anaerobes: aerotolerant anaerobes and obligate anaerobes. The first do not use O_2 for their metabolism, but they are not harmed by its presence. These organisms depend on a fermentative type of metabolism for their energy. Obligate or strict anaerobes, in contrast, are harmed by the presence of O_2. These organisms metabolize various substrates to derive energy either by fermentation or anaerobic respiration.

Facultative aerobes, microaerophiles, and aerotolerant anaerobes are better able to persist in the soil environment since they have the ability to adapt readily to the often rapid changes in O_2 availability that invariably occur in the soil. The capacity of facultative aerobes for use compounds other than O_2 as terminal electron acceptors in anaerobic respiration, for example, allows them to continue to respire C substrates and to generate the energy-storing molecule ATP via the electron transport chain when O_2 supply is reduced or cut. Nitrate (NO_3^-) and sulfate (SO_4^{2-}) are commonly used as alternative electron acceptors in anaerobic respiration.

The capacity to form spores or cysts is another type of adaptation that can enhance an organism's persistence in soil during periods of low water availability. Bacterial endospores are very durable, thick-walled dehydrated bodies that are formed inside the bacterial cell. When released into the environment, they can survive extreme heat, desiccation, and exposure to toxic chemicals. Bacteria, such as Bacillus and Clostridium that form endospores, and actinomycetes and true fungi, that commonly reproduce by conidia and spores, are well represented in the soil community. Their capacity to form spores gives these species an obvious survival advantage in the soil environment. The much larger protozoa and nematodes (Chapter 10) which feed on bacteria and fungi can both form cysts or thick-walled resting structures that enable them to survive when conditions are not favorable for growth. Once conditions become favorable, such as after a rain or when prey populations increase, the cysts germinate and these protozoa and nematodes then resume feeding, growing, and reproducing.

Other adaptations also enhance the capacity for organisms to survive in the ever-changing soil environment. Examples include producing polyols (alcohols with three or more hydroxyl groups) and heat-shock proteins; increasing intracellular solute concentrations; altering the membrane composition as seen in many Archaea (a prokaryotic lineage distinct from the Bacteria); and producing heat-stable proteins as seen in the thermophiles. In the last two decades, there has been a great increase in our knowledge of the survival strategies and mechanisms of soil biota which make possible the existence of the plethora of species that we are now coming to know, through molecular methods, are present in the soil.

5.1.7 Build It and They Will Come

When the physical and chemical characteristics of a soil are within optimal ranges, biological activity generally follows suit. For example, if soil texture and structure allow for a good balance between adequate drainage vs. moisture retention with sufficient gas exchange, conditions will generally be conducive for microbial growth and activity. If the soil is compacted or water-saturated, it rapidly becomes anaerobic. Under such conditions, fermentative metabolism may predominate, and organic acids and alcohols are produced. Practices that improve SOM content, water-stable aggregation, and drainage, such as growing cover crops and retaining residues (Chapter 30), applying compost (Chapter 31), and reducing tillage (Chapters 22 and 24) all help promote abundant, active soil biological communities.

5.2 Classifying Organisms Within the Soil Food Web

5.2.1 The Soil Food Web as a System

When one thinks of any ecosystem, generally the first things that come to mind are the organisms — plants, animals, and microbes — that live within it and provide a variety of ecosystem services. In ecological terms, these are classified either as producers (plants, algae, and autotrophic bacteria) or consumers (herbivores, predators, and decomposers). The primary producers, most often plants in terrestrial ecosystems, form the base of the food chain, or more accurately, the food web — a vast network of feeding interactions between and among organisms within the system. Primary producers capture energy from sunlight through the process of photosynthesis. This captured energy, stored in chemical bonds, provides the energy for most other organisms within the food web.

Trophic (feeding) interactions can be quite complex, especially below ground. Primary producers, generally plants, are consumed by herbivores, which are the primary consumers. Herbivores are in turn consumed by predators, which are considered secondary consumers within the system. Predators are then consumed by higher-order predators, the tertiary consumers within the system and on upwards. A simplified diagram of the soil food web is given in Figure 5.1.

Consumption is an energetically inefficient process. A rule of thumb is that only 10% of the energy contained at the first trophic level persists as usable energy at the next trophic level. Thus, up to 90% of the energy contained in primary producers, when consumed, becomes unavailable for metabolic work, being mostly lost from the system in the form of heat. This inefficiency of energy flow from one trophic level to the next has important consequences for the structure of ecosystems. The biomass that can be supported at any particular trophic level depends on the amount and availability of biomass in organisms at the trophic level immediately below it, upon which it feeds.

In aboveground systems, the largest biomass will be that of the primary producers. As one moves to higher trophic levels in the food web, both the biomass and often the number of organisms that can be supported decrease. This leads to the concept of a pyramid of biomass, or a pyramid of energy. This shape suggests how the size of successive

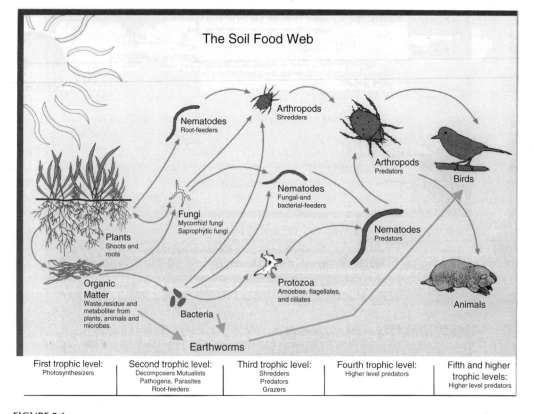

FIGURE 5.1

A simplified soil food web emphasizing trophic (feeding) relationships and functional roles of the soil biota. Adapted from SWCS (2000).

populations in any food web, i.e., their number and biomass, will decrease. Food webs will have, necessarily, a finite number of trophic levels as the total energy available for metabolic work at higher levels is consecutively dissipated as heat.

Organisms in all ecosystems are dependent on a source of energy that can be captured to do metabolic work, discussed in more detail in Chapter 6. Whether they capture it themselves through photo- or chemosynthesis or rely on preformed organic compounds, such as plant or animal tissue from other organisms, is a distinction that becomes very important when we consider the biota within an ecosystem's soil subsystem. The biological system beneath the soil surface operates on the same principles as those above ground, but with some distinct and important differences. The key difference is that primary production is extremely limited below ground since it is not continuously driven by abundant solar energy. This makes the whole subterranean subsystem energy-limited. Root-derived soluble C compounds, sloughing of root cells, and root death below ground, plus litter and animal waste deposited above ground, are the primary sources of energy for the belowground community (Wardle, 2002).

5.2.2 Energy and Carbon as Key Limiting Factors

The necessary goal for any organism is to obtain enough energy, cell biomass C, and mineral nutrients to produce the cellular constituents that are necessary for survival, growth, and reproduction. Metabolism refers to the biochemical processes occurring within living cells that make it possible for organisms to carry out what is necessary to maintain life. Microorganisms can be differentiated, and are categorized, based on three important metabolic requirements: (1) their source of energy; (2) their source of cell biomass C; and (3) their source of electrons or reducing equivalents.

- Phototrophs obtain energy from light, whereas chemotrophs obtain their energy from the chemical bonds in reduced organic or inorganic compounds.
- Autotrophs obtain their cell carbon from either CO_2 or HCO_3, whereas heterotrophs obtain their cell C from organic compounds.
- Lithotrophs derive electrons from reduced inorganic compounds such as NH_4^+, whereas organotrophs derive them from reduced organic compounds.

Four main groups are typically identified based on their sources of energy and cell C: photoautotrophs, photoheterotrophs, chemoautotrophs, and chemoheterotrophs (Atlas and Bartha, 1998). Photoautotrophs, as noted above, include plants, cyanobacteria, and other photosynthetic bacteria that use the process of photosynthesis to convert light energy from the sun into chemical energy. The chemical energy captured is subsequently used for carbon fixation.

Organisms such as the nitrifying bacteria that use ammonium (NH_4) as a source of energy and reducing potential to fix CO_2 into cell biomass are known as chemoautotrophs. Those bacteria and fungi, protozoa and soil fauna that rely on plant and animal residues and SOM as sources of both energy and cell biomass C are classified as chemohetero-trophs, or simply as heterotrophs. Photoheterotrophs are a small and unusual group of photosynthetic bacteria, the green nonsulfur and purple nonsulfur bacteria that use light as a source of energy and organic compounds as their source of cell C.

The activity of heterotrophic soil organisms depends on the availability of degradable organic C compounds. Since primary production below ground is limited by a lack of light, soil heterotrophs must depend on the activity and success of aboveground photoautotrophs, mainly plants, for their survival. In a healthy soil, heterotrophs meet their needs for energy and cell biomass C from the continuous addition of plant and animal residues, from the secretion of organic compounds by plant roots, and from the slow turnover of SOM, which includes the microbial biomass that continually dies off as new microorganisms come to life.

5.3 Primary Producers

5.3.1 Energy Capture in Plants Drives the Soil Community

Plants as primary producers capture energy by in their aerial leaf systems, and much of that energy is transferred below ground to plant roots through the phloem, part of the plant's vascular system specialized for this purpose. Plant roots provide a special, highly energized habitat for microorganisms living next to them in the surrounding soil, referred to as the rhizosphere, discussed below. Some microorganisms are endophytic, inhabiting the interior tissues of roots as mutualists rather than as parasites. Hence, it is sometimes difficult to delineate where the realm of the plant root ends and that of soil organisms begins.

Carbon compounds released by roots serve as the primary source of energy for most heterotrophic soil organisms. Belowground herbivores, plant-parasitic nematodes and pathogenic fungi feed directly on living root tissues, thus reducing plant productivity. However, the vast majority of organisms in the rhizosphere that feed on root-derived compounds are decomposers. In most cases, their presence around the roots is highly beneficial to plant growth, particularly when their activities release mineral nutrients that plants can subsequently acquire, thus creating a positive feedback loop between plants and the rhizosphere microbial community.

Another major source of energy for soil heterotrophs is dead plant material (litter) and animal residues. In woodlands, this would be primarily in the form of leaf fall and tissues of dead plants, plus animal excrement and carcasses. In agricultural systems, much of the plant material is removed during harvests and not returned to the soil. This is an undesirable management practice, however, because it runs down the energy status of the soil, depleting the energy needed by microorganisms to perform their many beneficial functions.

In addition to vascular plants, other primary producers that may be present in surface soil are photosynthetic bacteria, cyanobacteria, and algae. However, their energy contribution to soil is comparatively small. Cyanobacteria, a large and diverse group of photosynthetic bacteria coming in an assortment of shapes and sizes, were previously, mistakenly, called blue–green algae. Ranging from 1 to 10 μm in diameter, they are found as filaments, colonies of numerous shapes, and as single cells. Many of the filamentous cyanobacteria are able to fix atmospheric N_2 within specialized thick-walled cells, called heterocysts. Cyanobacteria, other photosynthetic bacteria, and algae use light energy and generally require high moisture levels; hence, they are not active below the first few millimeters in soil. Some cyanobacteria and algae do, however, form important partnerships with fungi called lichens. Lichens are resistant to desiccation and colonize rock surfaces, tree bark, and other organic and inorganic surfaces. In some ecosystems, such as in the Arctic and very arid environments, lichens and cyanobacterial soil crusts may be the dominant primary producers (Belnap, 2003). Their contribution to soil function in arable lands is not substantial in comparison to vascular plants, however, and we will not consider them further here.

The soil biota are limited mainly by the amount of energy that can be produced and stored by aboveground organisms that is ultimately transferred below ground. Gross and net rates of primary production vary greatly from one plant species to the next due mainly to the photosynthetic pathway used (C3, C4, and CAM) and to abiotic factors such as variations in light, soil moisture, temperature, and nutrient availability. The highest capacities for photosynthesis are seen in plants possessing the C4 photosynthetic pathway such as maize, sorghum, and sugarcane; the lowest capacity is found in plants relying on crassulacean acid metabolism (CAM), such as desert succulents. Variations in photosynthetic capacity have a direct impact on the amount of fixed C that reaches the soil and becomes available for use by heterotrophic soil organisms. Of the total C fixed by photo- or chemosynthetic organisms (gross primary production [GPP]), some portion is used to fuel their own cellular respiration. GPP minus respiration is called net primary production (NPP), or the accumulation of standing plant biomass (and that of other autotrophs). NPP is what fuels the soil subsystem, largely in the form of detritus and root exudates.

5.3.2 Roots

Processes that occur at or near the soil–root interface control the productivity of both plants and soil organisms. This interface is discussed in more detail in Chapter 7. Here, we consider briefly the roles and contributions of root systems as part of the soil food web. We note that roots also offer habitat for bacteria and fungi, referred to as endophytes, living within roots, performing mutualistic services such as documented in Chapter 8, while themselves being benefited by plant roots.

Root systems are composed of long thick roots that provide structural support and shorter, fine roots that are important in the uptake of nutrients and water. Soil biota are not evenly distributed along a single root system. Even though various root types within a single root system support very distinct distributions of both bacterial and fungal species (McCully, 1999), fine roots and root hairs (specialized epidermal cells) have often been neglected in soil ecology studies. Microbial population differences associated with roots of differing size and age need to be taken into account for understanding root–soil dynamics.

Through the roots, plants acquire the water and nutrients that they need for survival. Plant roots are not passive absorbers of nutrients and water, but actually active regulators maintaining complex signaling relationships between roots and shoots (Chapter 15). Features of an actively growing root are shown in Figure 5.2. Root hairs and the root cap are very influential in controlling rhizosphere microbial populations. Root hairs greatly increase the amount of soil that plants can explore and from which they can extract nutrients and water. Root hairs extend into the soil environment usually less than 10 mm and range from 20–70 μm in diameter. They form on both the structural roots, as well as on the finer lateral roots. Root hairs initially grow straight, but when they encounter soil particles they curl, bend, and often develop branches, creating microhabitats in which microbes can reside. Root hairs are often the cells in which mutualistic relationships with mycorrhizal fungi and nitrogen-fixing rhizobia bacteria are initiated, discussed in Chapters 9 and 12.

The growing plant root has three distinct zones: the meristem, or zone of cell division, where new root cells are formed; the zone of elongation where these cells expand and lengthen; and the zone of maturation, or root hair zone, where these cells mature and from whence root hairs originate (Figure 5.2). As roots grow, root cap cells are continuously sloughed off into the soil, being replaced by the dividing meristem cells of the elongating root. Root cap cells secrete a dense mucilage of polysaccharides that serves several significant purposes, including providing a lubricant for the root to grow through the soil and for retaining moisture, thereby guarding root tissues against desiccation (Bengough and Kirby, 1999). Mucilage that undergoes continuous wetting and drying contributes to the formation of soil aggregates, which give the soil better structure and tilth.

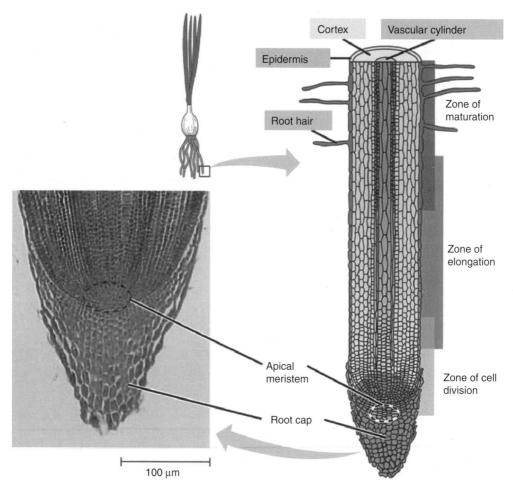

FIGURE 5.2
Cross-section of a typical root showing different zones and organs. From Campbell, N.A. and Reece, J.B., *Biology,* 7th ed., Benjamin Cummings, San Francisco (2005). With permission.

The sloughed-off root cap cells and mucilage remain in the soil, covering the maturing root surface as it continues to grow into the soil environment. Some recent evidence suggests that these sloughed root cap cells may sometimes act as decoys, with potential pathogens colonizing these sloughed cells rather than the intact root cap cells, as the root tip grows away from the area. This process of sloughing off root cells, among other things, thus helps to protect the meristem from pathogen invasion.

5.3.3   The Rhizosphere

The root surface is referred to as the rhizoplane, whereas the rhizosphere is the biologically active area of soil that surrounds the root and is chemically, energetically, and biologically different from the surrounding bulk soil. It is the zone where plants have the most direct influence on their soil environment through root metabolic activities, such as respiring and excreting C-rich compounds, or through nonmetabolically mediated processes that cause cell contents to be released into the surrounding soil, such as cell abrasion or sloughing. The rhizosphere can extend outward up to 1 cm or more from the root surface depending on the plant type and soil moisture and texture (for a comprehensive review, see Pinton et al., 2001). Here we discuss the rhizosphere in terms of the soil food web. A closer look at the components and functions of the rhizosphere is provided in Chapter 7, after considering energy flows in Chapter 6. Practical applications are seen in Chapter 39

Together, the rhizosphere and the rhizoplane provide diverse habitats for a wide assortment of microorganisms. Habitats on root surfaces are affected by differences in moisture, temperature, light exposure, plant age, root architecture, and root longevity. However, the primary way in which plants influence the communities of microorganisms that inhabit the rhizosphere is through their deposition of root-derived compounds. These are classified as root exudates (passive process), secretions (active process), mucigel (root/microbial byproduct mixtures), and lysates (contents of ruptured cells) (Rovira, 1969).

The accumulation of all these various substances put into the soil is called rhizodeposition, and represents the key process by which C is transferred from living plants into the soil subsystem of the larger ecosystem (Jones et al., 2004). Rhizodeposition increases the energy status of the surrounding soil and, consequently, the mass and activity of soil microbes and fauna that are found in the rhizosphere. This is reflected in the R/S ratio, i.e., the biomass of microbes in the rhizosphere (R) in relation to that in the bulk soil (S). This ratio is generally greater than one.

Microorganisms engage in a variety of activities in the rhizosphere. Beneficial interactions include fixing N_2 (Chapters 12 and 27), solubilizing or enhancing uptake of less mobile nutrients (Chapters 13 and 37), promoting plant growth (Chapters 14, 32, 33, and 34), mutualistic symbioses (Chapters 9, 12, and 34), biocontrol (Chapter 41), antibiosis, aggregating and stabilizing soil, and improving water retention. Neutral or variable interactions include free enzyme release, bacterial attachment, competition for nutrients, and nutrient flux. Harmful activities include allelopathy (Chapter 16), phytotoxicity, and infection or pathogenesis. Complementing these positive, neutral, or negative functions are ones that occur within roots, associated with endophytic organisms such as discussed in Chapters 8 and 12.

Many activities of microbes in the rhizosphere are of benefit to plants. Indeed, some research findings have indicated that plants may select for, i.e., support, certain taxonomic or functional groups of organisms present in their rhizospheres; however, laboratory and field experiments have given inconsistent results (e.g., Grayston et al., 1998; Smalla and Wieland, 2001; Singh et al., 2004).

A central interest in soil ecology studies is enhancing or manipulating microbial populations found in the rhizosphere, including abundance and differential distribution of species. Many inoculation programs are aimed at changing species distributions in the rhizosphere either to enhance a particular process or to suppress plant pathogens. Inoculating legumes with specific strains of rhizobia aims to increase BNF (Chapter 12) or provide other benefits (Chapter 8), while inoculating with mycorrhizae is intended to increase plant uptake of poorly mobile nutrients (Chapters 9 and 33). Inoculating with Trichoderma (Chapter 34), plant growth-promoting rhizobacteria (Chapter 32), or applying compost (Chapter 31) may aid in suppressing plant pathogens in many systems.

Part III of this volume provides numerous examples of how managing to enhance beneficial populations in the rhizosphere and improving soil biological activity in general can yield significant benefits to plant productivity and soil quality (see especially Chapter 39). Favorable results, however, are contingent on many factors being aligned in certain ways, so this area of research continues to present many unresolved questions.

5.4 Consumers

The soil biota have a number of important functional roles as consumers which include: C mineralization and OM turnover, nutrient cycling, vital mutualisms with plants and each other, causing and suppressing plant and animal diseases, improving soil structure, bioremediating contaminated soil (Chapter 42), and generating and consuming greenhouse gases (Susilo et al., 2004; and Chapter 43). For many years, the focus has been on measuring pools of nutrients or organic substrates without regard to the organisms responsible for the shifts between one pool and another. This has changed substantially in recent years as the focus has moved toward assessing the abundance, activity, and diversity of communities, populations, individuals, and gene sequences of interest.

5.4.1 Decomposers, Herbivores, Parasites, and Pathogens

The first consumer group of the soil food web, the primary consumers, contains decomposers, i.e., organisms that feed on root exudates and plant and animal residues, and numerous herbivores, parasites, and pathogens that feed on living root tissues. This trophic level encompasses many heterotrophic soil bacteria and fungi. These include the important mycorrhizal fungi and symbiotic rhizobia bacteria discussed in Chapters 8, 9, and 12, as well as several types of pathogenic fungi, oomycetes, and root-feeding nematodes (Figure 5.1). It includes also the larvae and adult stages of insects that feed on the roots and shoots of plants and whose life-cycles are largely carried out in the soil.

Heterotrophic soil bacteria have several functional roles in soil, most importantly as decomposers of dead organic matter. They can also be symbionts that live with plants and other organisms in the soil to mutual benefit, or pathogens that live at the expense of other organisms. Saprophytic bacteria, which feed on dead organic matter, are the most numerous of the decomposers. These bacteria produce, as a group, many different enzymes that give them broad capacities to degrade organic matter, enabling them to metabolize a vast array of C compounds to obtain energy and cell biomass C. Many heterotrophic bacteria facilitate key transformations of various nutrient elements that complete elemental cycling. A prime example is the fixation of N_2 from the atmosphere by nitrogen-fixing bacteria and the return of N_2 to the atmosphere during anaerobic respiration by facultative anaerobes through the process of denitrification, with the sequential reduction of nitrate (NO_3^-) in the soil solution to N_2 gas in the soil atmosphere.

Soil bacteria and fungi are important in developing and maintaining soil structure and aggregation. Bacteria improve soil structure by producing exopolysaccharides and other metabolites that help glue soil particles together. Fungi, by producing a network of hyphal filaments, also help to stabilize aggregates.

Some soil bacteria are important plant pathogens that colonize living plant tissue and cause disease. Common examples are crown gall caused by *Agrobacterium tumefaciens* and the black rot of crucifers caused by *Xanthomonas campestris*. Certain plant-pathogenic bacteria that colonize the rhizosphere produce metabolites that retard plant growth. It is possible for a bacterium that is considered to be plant growth-promoting under some soil conditions to become deleterious to the plant as environmental conditions change. A shift from aerobic respiration to fermentation under O_2-limited conditions, for example, can cause a shift in the endproducts of metabolism from CO_2 to acids and alcohols which may be damaging to roots.

An important mutualism between soil fungi and plants is that of the mycorrhizal fungi. Ectomycorrhizae associate largely with tree species, inhabiting root surfaces and extending their hyphae from there, while endomycorrhizae form associations with most crop plants, actually penetrating and inhabiting their cortical root cells as discussed in Chapter 9. In the relationship between these fungi and a host plant, the plant benefits by enhanced nutrient status, largely from increased uptake of phosphorous and micronutrients, protection from desiccation (through increased water uptake), and protection from pathogens and toxic metals by occupying the same niche or forming a protective layer on the root surface. Mycorrhizal fungi benefit in return by obtaining energy and fixed carbon directly from host plants.

The last of the primary consumers considered here are the plant-feeding nematodes. Infestation by parasitic nematodes causes millions of dollars in crop losses each year (Bird and Koltai, 2000). Most species of plant-feeding nematodes harbor a needle-shaped stylet or mouth part that enables them to pierce the plant cell wall and cell membrane and to feed on the cell contents. Maintaining large populations of beneficial soil organisms — the saprophytic and symbiotic bacteria and fungi, as well as free-living nematode species — is a promising means for reducing and preventing the spread of parasitic nematodes as they all compete for substrates and space within the rhizosphere. Nematodes as primary and secondary consumers within the food web are considered in more detail in Chapter 10.

5.4.2   Organic Matter Decomposition

One of the more important functions of the primary decomposer group of microbes, saprophytic bacteria and fungi, is to break down complex organic materials into their component building blocks by the action of exoenzymes (Reynolds et al., 2003). Enzymes are proteins produced by living cells that facilitate (catalyze) chemical reactions by lowering the energy needed for activating these processes. Most enzymes are characterized by high specificity, which is largely a function of differences in enzyme-active sites.

Different soil bacteria and fungi produce an enormous variety of enzymes that are secreted into the surrounding environment, such as dehydrogenases, proteases, and cellulases. These exoenzymes reduce organic molecules and degrade proteins and cellulose, respectively, into their component parts outside the cells. The products are then taken up through the cell wall and cell membrane for use in metabolic reactions. Producing exoenzymes involves a high carbon cost to bacteria and fungi; hence, they become highly invested in the surfaces that they have colonized. Bacteria often form biofilms on surfaces that enable them to degrade organic compounds more efficiently (Davey and O'Toole, 2000).

Released nutrients are taken up by decomposers, which can result in the immobilization of nutrients within microbial biomass. Inorganic nutrient elements, such as N, P, S, K, and Mg, in excess of their needs, are released back into the soil environment and become available once again for uptake by plants. Since most plants cannot take up nutrients in organic forms, the decomposition of OM is an important source of inorganic nutrients for them. Through their respiration, soil decomposers also release CO_2 back into the atmosphere, making it available once again for plants to capture in the process of photosynthesis, thus completing the C cycle.

The rate and extent of decomposition is directly related to the nature of the OM that is being decomposed. Materials of different composition and energy status will decompose at different rates, and thus there is variation in the length of time that organic materials remain (reside) in the soil before being completely broken down. Many plant and animal residues, such as root exudates, leaf litter, frass (insect excrement), and manure, have very short residence times in soil, being completely decomposed in weeks, months, or at most a few years. Carbon in this form is referred to as part of the labile C fraction. Microbial metabolites, humic acids, and highly lignified materials have lower mineral nutrient contents in relation to the carbon content or require highly specialized enzymes for their decomposition. Carbon in this form has a long residence time in soil on the order of years, decades, or more and is referred to as part of the recalcitrant carbon fraction (Paul and Clark, 1996).

The quality of OM inputs represents a primary limiting factor affecting the growth and reproduction of saprophytic organisms. If the available forms of carbon are high in energy and easily broken down (high quality), as is the case with many plant residues, then decomposers are likely to be both active and abundant. However, where SOM content is low, or when OM inputs consist of more recalcitrant materials, such as lignin and polyphenols (lower quality), microbial activity will be restricted, and the functioning of the whole ecosystem will be affected.

SOM has many key functional roles. Serving as the primary source of carbon and energy for the soil biota, it becomes the primary factor controlling microbial activity. It also influences soil water-holding capacity, air permeability, nutrient availability, and water infiltration rates. SOM content is very sensitive to soil management practices. For example, tillage exposes SOM previously occluded inside aggregates. Once exposed, SOM is rapidly mineralized by colonizing microbes, thus reducing the overall OM content of the soil. Many of the chapters in Part III focus on management practices that can help to conserve and increase SOM quantity and quality as a basic strategy for enhancing soil system functioning and sustainability. The quality, turnover, and functional significance of soil OM inputs are discussed in more detail in Chapters 6 and 18.

5.4.3   Grazers, Shredders, and Predators

The organisms at the next trophic level are the secondary consumers, which include the protozoa, bacterial- and fungal-feeding nematodes, and microarthropods such as mites and collembola. These organisms feed predominantly on soil bacteria and fungi, but also consume SOM. Feeding on live bacteria and fungi is commonly referred to as grazing. Grazers are critically important in the cycling of mineral nutrients since when they feed on nitrogen-rich bacteria, they excrete large amounts of inorganic nitrogen into soil (Bonkowski, 2004).

Grazers have adapted various methods of consuming their prey. Bacteria-feeding protozoa engulf their prey, whereas bacteria- and fungus-feeding nematodes have specialized mouth parts for piercing or penetrating. Those of bacteria-feeding nematodes sweep or suck bacteria off the surfaces of roots and soil particles, while fungus-feeding nematodes often have fine stylets that allow them to pierce the fungal cell walls and consume the cell contents, seen in Figures. 10.1 and 10.2 in Chapter 10. Grazing, no matter the mechanism, results in more rapid nutrient turnover and release because the amount consumed is often in excess of the grazing organism's needs.

Unlike the plant-parasitic nematodes, the bacteria- and fungus-feeding nematodes are very beneficial within soil systems. Their grazing activity helps to regulate the size and structure of bacterial and fungal populations and accelerates nutrient cycling, making them the "good guys" within the soil nematode world. When soil nematicides or fumigants are used, all nematodes can be killed off, the beneficial as well as the deleterious species. This disrupts the functioning of the free-living nematodes and compromises their role in facilitating nutrient turnover (Ibekwe, 2004). More selective ways of dealing with plant-parasitic nematodes are needed, such as developing suppressive soils that enhance the beneficial nematode populations while controlling the plant-feeding species. Enhancing the populations of beneficial nematodes can help keep the deleterious ones in check through several mechanisms, including stimulating induced systemic resistance (ISR) in plants by enhancing nutrient availability and competing for space and other resources. This is an active area of current research in soil ecology. The ecological roles of protozoa and nematodes are discussed further in Chapter 10.

The shredders and predators occupy several trophic levels, depending on the substrates or prey on which they feed. Mites and collembola fragment (shred) and ingest OM and thus are primary consumers, but some also graze on fungi, which makes them secondary consumers. Earthworms and enchytraeids fragment and ingest OM and so are primary consumers, but the OM is often covered with bacteria and fungi, thus they are simultaneously secondary consumers. As we move up the food chain, we find that the feeding relationships are not straightforward or distinct. Many organisms feed at multiple trophic levels, and this contributes to the complexity of trophic relationships and leads to efficient OM turnover and net nutrient release.

Collectively, the shredders are important for controlling microbial populations, shredding organic matter, and cycling nutrients. Shredding, also known as comminution, speeds up residue decomposition as it mixes bacteria and fungi with the residues and increases the surface area available for these decomposers to colonize. Mesofauna (mites, collembola, termites, and enchytraeid worms) and macrofauna (wood lice, millipedes, beetles, ants, earthworms, snails, and slugs) all contribute to the shredding and turnover of organic residues. Shredders also deposit partially digested residues, called frass or insect excrement, in the soil. Frass being very energy rich is an excellent substrate for decomposers. In addition to depositing nutrient-rich casts, earthworm activity also mixes the upper layers of the mineral soil with surface residues (bioturbation) and creates biopores or channels for water and roots to pass through.

The higher trophic levels in the soil food web (tertiary consumers and beyond) contain predatory nematodes and predatory arthropods, such as pseudoscorpions, centipedes, and species of spiders, beetles, and ants. From a soil ecology perspective, the life history and functions of the predators are important because they can help regulate important plant pest populations. Larger soil animals such as moles, while important members of the soil subsystem, are not considered here, but their ecological roles are discussed in Wolfe (2002). Ecological roles and functions of soil fauna are discussed further in Chapter 11.

5.4.4 They All Interact Together

Soil biota, through a continuous and highly interrelated set of feeding relationships, are key to liberating plant-available nutrients in the rhizosphere (Adl, 2003). Without the activities of soil biota, nutrients bound up in organic matter would remain immobilized, and the cycling of nutrients would be greatly limited. Instead, soil organisms mineralize OM, thus facilitating the release of inorganic nutrient elements and their continual cycling.

The effects of this process are not simple because the nutrients liberated are also available for uptake by bacteria, fungi, protozoa, nematodes, and microarthropods living on or in the vicinity of roots. All of these organisms compete with roots for uptake of these mineral nutrients. Soil saprophytes, while important in mineralizing organic matter, are equally important in immobilizing nutrients. Only when these elements are available in excess of what microbial communities need do they become freely available to plants. Immobilization of nutrients in the soil biota is not as negative a process as it sounds. It can actually be quite beneficial by retaining nutrients within the topsoil and rhizosphere, thereby preventing them from leaching into lower soil horizons, beyond the reach of plants unless there is very deep root growth.

Bacteria require more mineral nutrients in relation to their carbon requirements than do fungi or protozoa. Therefore, bacteria are more likely to immobilize mineral nutrients, whereas the activities of fungi, protozoa, nematodes, and microarthropods are likely to result in greater release of available mineral nutrients into the soil solution.

The soil food web is thus an intricate set of interrelationships among a wide diversity of organisms. This web of interactions significantly influences all aspects of the soil environment. Without living organisms and other soil organic materials, the soil would be simply a compilation of minerals, gases, and water. Nutrient elements would not be recycled, and the system would rapidly wind down to lower fertility levels, unless all the elements are constantly replaced from external sources. The strategy adopted by the Green Revolution was essentially indifferent to the roles and contributions of soil biota, and this has contributed to impoverishment of the soil biota and SOM in many areas. To maintain a healthy soil food web is to conserve a self-renewing ecosystem capable of sustaining plant growth for long-term productivity. Ignoring and undermining the rich diversity of life in soil comes at a cost. Better to understand this highly complex community so that soil resources can be managed in more sustainable ways.

5.5 Biological Diversity and Soil Fertility

There is still a continuing debate over whether increasing bacterial and fungal species diversity in the soil environment will lead to longer-term ecosystem sustainability. In particular, questions arise as to how changes in management practices that affect plant community diversity and productivity may have indirect impacts on below ground soil biotic communities and their functioning (Giller et al., 1997; Clapperton et al., 2003). It is still unclear how much soil biotic diversity is required for sustainable soil systems, or if simply having a representative set of organisms that give functional diversity is sufficient (Brussaard et al., 2004). It is well known that plant litter is critical in determining soil physical properties and also the quality and availability of substrates for microorganisms (Wardle et al., 2004). Although strong correlations do exist, many studies have shown that as long as litter quality is maintained, increasing the species richness of plant litter has no predictable effect on decomposition rates or biological activity (Wardle et al., 1997; Bardgett and Shine, 1999).

It will be of great value to determine more conclusively the significance of the operative relationships between soil biodiversity and fertility. Understanding these relationships could allow ecosystem managers to encourage the presence of organisms that are beneficial to soil systems intended for crop and animal production, as well as to overall ecosystem health as discussed in Parts II and III of this volume.

5.6 Discussion

A good summary statement about soil systems by Martius et al. (2001) is cited in Chapter 13: "Soil comes to life through organic matter, which supports highly diverse communities of microorganisms and soil fauna that provide critical ecosystem services, most notably the recycling of nutrients." Soil management clearly needs to focus on managing, directly or indirectly, the soil biological communities for improved soil function and long-term sustainability. Soil is arguably our most precious global resource and one that has been sorely mistreated. This mismanagement has sacrificed millions of hectares of fertile soil through erosion and degradation, which occurred not just because of unwise physical manipulation but due to a loss of the soil life that is needed to maintain its integrity and thus help it resist loss.

Analyzing the chemical and physical aspects of soil systems is much easier than delving into the complex realm of soil biology, and thus the analysis and evaluation of chemical and physical properties has dominated soil science for generations. Today, more soil research is examining soil biology, assisted by new methods for analysis as discussed in Chapter 46. These are overcoming previous limitations to our ability to classify, measure, and assess causal relationships. The chapters that follow in Part II give insights into the various agents and processes composing soil systems, with chapters in Part III then showing how such knowledge is being applied to make soil system management more effective and sustainable.

References

Adl, S.M., Reconstructing the soil food web, In: *The Ecology of Soil Decomposition*, Adl, S.M., Ed., CABI Publishing, Cambridge, MA (2003).

Atlas, R.M. and Bartha, R., *Microbial Ecology: Fundamentals and Applications*, Benjamin Cummings, Menlo Park, CA (1998).

Bardgett, R.D. and Shine, A., Linkages between plant litter diversity, soil microbial biomass and ecosystem function in temperate grasslands, *Soil Biol. Biochem.*, **31**, 317–321 (1999).

Belnap, J., The world at your feet: Desert biological soil crusts, *Front. Ecol. Environ.*, **1**, 181–189 (2003).

Bengough, A.G. and Kirby, J.M., Tribiology of the root cap in maize (*Zea mays*) and peas (*Pisum sativum*), *New Phytol.*, **142**, 421–426 (1999).

Bird, D.M. and Koltai, H., Plant parasitic nematodes: Habitats, hormones, and horizontally-acquired genes, *J. Plant Growth Regul.*, **19**, 183–194 (2000).

Boddy, R.M. et al., Endophytic nitrogen fixation in sugarcane: Present knowledge and future applications, *Plant Soil*, **252**, 139–149 (2003).

Bonkowski, M., Protozoa and plant growth: The microbial loop in soil revisited, *New Phytol.*, **162**, 616–631 (2004).

Brady, N.C. and Weil, R.R., *The Nature and Properties of Soil*, 13th ed., Prentice-Hall, Upper Saddle River, NJ (2002).

Brussaard, L. et al., Biological soil quality from biomass to biodiversity: Importance and resilience to management and disturbance, In: *Managing Soil Quality: Challenges in Modern Agriculture*, Schjønning, P., Elmnolt, S., and Christensen, B.T., Eds., CABI Publishing, Cambridge, MA (2004).

Campbell, N.A. and Reece, J.B., *Biology*, 7th ed., Benjamin Cummings, San Francisco (2005).

Clapperton, M.J., Chan, K.Y., and Larney, F.J., Managing the soil habitat for enhanced biological fertility, In: *Soil Biological Fertility: A Key to Sustainable Land Use in Agriculture*, Abbott, L.K. and Murphy, D.V., Eds., Kluwer, Boston, MA (2003).

Davey, M.E. and O'Toole, G.A., Microbial biofilms: From ecology to molecular genetics, *Microbiol. Mol. Biol. Rev.*, **64**, 847–867 (2000).

Dobbelaere, S., Vanderleyden, J., and Okon, Y., Plant growth-promoting effects of diazotrophs in the rhizosphere, *Crit. Rev. Plant Sci.*, **22**, 107–149 (2003).

Fuhrmann, J.J., Microbial metabolism, In: *Principles and Applications of Soil Microbiology*, Sylvia, D.M. et al., Eds., 2nd edition Prentice-Hall, Upper Saddle River, NJ (2005).

Fred, E.B., Baldwin, I.L., and McCoy, E., *Root Nodule Bacteria and Leguminous Plants*, Studies in Science no 5. University of Wisconsin, Madison, WI (1932).

Giller, K.E., *Nitrogen Fixation in Tropical Cropping Systems*, CABI Publishing, Wallingford, UK (2001).

Giller, K.E. et al., Agricultural intensification, soil biodiversity, and agroecosystem function, *Appl. Soil Ecol.*, **6**, 3–16 (1997).

Grayston, S.J. et al., Selective influence of plant species on microbial diversity in the rhizosphere, *Soil Biol. Biochem.*, **30**, 369–378 (1998).

Ibekwe, A.M., Effects of fumigants on non-target organisms in soils, *Adv. Agronomy*, **83**, 1–35 (2004).

Jenny, H., *Factors of Soil Formation: A System of Quantitative Pedology*, McGraw Hill, New York (1941).

Jones, D.L., Hodge, A., and Kuzyakov, Y., Plant and mycorrhizal regulation of rhizodeposition, *New Phytol.*, **163**, 459–480 (2004).

Martius, C., Tiessen, H., and Vlek, P.L.G., The management of organic matter in tropical soils: What are the priorities?, *Nutrient Cycling Agroecosyst.*, **61**, 1–6 (2001).

McCully, M.E., Roots in soil: Unearthing the complexities of roots and their rhizospheres, *Annu. Rev. Plant Physiol. Plant Mol. Biol.*, **50**, 695–718 (1999).

Paul, E.A. and Clark, F.E., *Soil Microbiology and Biochemistry*, Academic Press, San Diego (1996).

Pinton, R., Varanini, Z., and Nannipieri, P., Eds., *The Rhizosphere: Biochemistry and Chemical Substances at the Soil–Plant Interface*, Marcel Dekker, New York (2001).

Reynolds, H.L. et al., Grassroots ecology: Plant–microbe–soil interactions as drivers of plant community structure and dynamics, *Ecology*, **84**, 2281–2291 (2003).

Rovira, A.D., Plant root exudates, *Bot. Rev.*, **35**, 35–58 (1969).

Singh, B.K. et al., Unraveling rhizosphere–microbial interactions: Opportunities and limitations, *Trends Microbiol.*, **12**, 386–393 (2004).

Smalla, K. and Wieland, G., Bulk and rhizosphere soil bacterial communities studied by denaturing gradient gel electrophoresis: Plant-dependent enrichment and seasonal shift revealed, *Appl. Environ. Microbiol.*, **67**, 4742–4751 (2001).

Susilo, F.X. et al., Soil biodiversity and food webs, In: *Below-Ground Interactions in Tropical Agro-ecosystems: Concepts and Models with Multiple Plant Communities*, van Noordwijk, M., Cadisch, G., and Ong, C.K., Eds., CABI Publishing, Cambridge, MA (2004).

SWCS, *Soil Biology Primer*, rev. ed, Soil and Water Conservation Society, Ankeny, IA (2000).

Thies, J.E., Singleton, P.W., and Bohlool, B.B., Influence of the size of indigenous rhizobial populations on establishment and symbiotic performance of introduced rhizobia on field-grown legumes, *Appl. Environ. Microbiol.*, **57**, 19–28 (1991).

Tisdall, J.M. and Oades, J.M., Organic-matter and water-stable aggregates in soils, *J. Soil Sci.*, **33**, 141–163 (1982).

van der Ploeg, R.R., Böhm, W., and Kirkham, M.B., On the origin of the theory of mineral nutrition of plants and the law of the minimum, *Soil Sci. Soc. Am. J.*, **63**, 1055–1062 (1999).

von Liebig, J., *Die organische Chemie in ihrer Anwendung auf Agriculture und Physiologie (Organic Chemistry and its Application in Agriculture and Physiology)*, Auflage Vieweg, Braunschweig (1843).

Wardle, D.A., *Communities and Ecosystems: Linking the Aboveground and Belowground Components*, Princeton University Press, Princeton, NJ (2002).

Wardle, D.A., Bonner, K.I., and Nicholson, K.S., Biodiversity and plant litter: Experimental evidence which does not support the view that enhanced species richness improves ecosystem function, *Oikos*, **79**, 247–258 (1997).

Wardle, D.A. et al., Ecological linkages between above and belowground biota, *Science*, **304**, 1629–1633 (2004).

Wolfe, D., *Tales From the Underground: A Natural History of the Subterranean World*, Perseus, Cambridge, MA (2001).

6

Energy Inputs in Soil Systems

Andrew S. Ball

School of Biological Sciences, Flinders University of South Australia, Adelaide, Australia

CONTENTS

Many of the most important relationships between living organisms and the environment are ultimately controlled by the amount of available incoming energy received at the Earth's surface from the sun. It is this energy which helps to drive soil systems. The sun's energy enables plants to convert inorganic chemicals into organic compounds. Living organisms use energy in either of two forms: radiant or fixed. Radiant energy exists in the form of electromagnetic energy, such as light, while fixed energy is the potential chemical energy bound in organic substances. This latter energy can be and is released through the biological process known as respiration.

Organisms that take energy from inorganic sources and fix it into energy-rich organic molecules are called autotrophs. They are considered to be "producers." If this energy comes from light, these organisms are photosynthetic autotrophs (or photoautotrophs), and in soil ecosystems plants are the dominant photosynthetic autotrophs. By contrast, organisms that depend for their survival on fixed energy stored in organic molecules are called heterotrophs. They obtain their energy from living organisms and are characterized as "consumers." Those that ingest plants are known as herbivores, while carnivores are those that eat herbivores or other carnivores for their energy supply. Decomposers constitute a third major category of heterotrophs. Often referred to as detritivores, feeding on detritus, these obtain their energy not from consuming living organisms, but from the consumption of dead ones or from ingesting organic compounds dispersed in the environment (Mackenzie et al., 2001). In Chapter 5, the major groups of soil micro-organisms were introduced together with a description of their functional roles within a trophic, food web-based structure. This chapter discusses the flows of energy and nutrients within the soil system and between trophic levels.

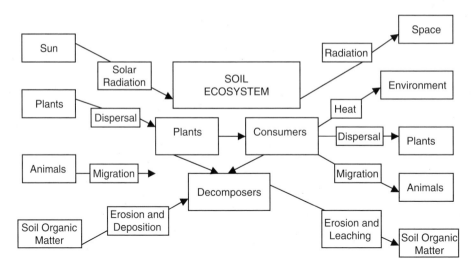

FIGURE 6.1
A conceptual framework illustrating the various inputs and outputs of energy and matter in a typical soil system.

Organic energy, once it has been fixed by plants into organic compounds, moves within soil systems through the consumption of either living or dead organic matter. Through decomposition, the chemicals that were once organized into organic compounds are returned to inorganic forms that can be taken up by plants once again. Organic energy can also move from one soil system to another by a variety of processes that include animal migration, animal harvesting, plant harvesting, plant dispersal of seeds, leaching, and erosion (Figure 6.1). This underscores that soil systems are open systems, both by getting most of their energy from the sun and by having some inflow and outflow of energy in diverse forms (Aber and Melillo, 2001).

Terrestrial ecosystems remove atmospheric carbon dioxide (CO_2) by plant photosynthesis during the day. This process results in the growth of plant roots and shoots and in increased microbial biomass in the soil. Plants release some of their stored carbon (C) back into the atmosphere through respiration. When plants shed leaves and their roots die, this organic material decays, and some of it can become protected physically and chemically as inert organic matter (OM) in the form of humus, which can be stable in soils for even thousands of years.

The decomposition of soil carbon by soil microbes releases CO_2 into the atmosphere. Decomposition also mineralizes OM, thereby making nutrients available for plant growth. The total amount of carbon stored in an ecosystem reflects the long-term balance between plant production and respiration and soil decomposition. Carbon is the essential element for energy storage and transformation in all soil systems, and thus the carbon cycle is one of the most important cycles in soil (Godden et al., 1992).

6.1 Sources of Nutrients in Soil

In natural soil systems, nutrients are derived from one of the three sources, whose relative importance will differ depending on the particular ecosystem. These sources are:

- Inputs from the atmosphere, both with precipitation and in dry form

- Weathering of the parent material underlying the soils
- Decomposition of dead OM.

The earth as a whole is a natural recycling system. No new matter is being added to the earth, so all new biomass must be made from existing matter, including carbon, oxygen (O_2), hydrogen (H), nitrogen (N), phosphorus (P), calcium (Ca), potassium (K), and other chemical elements. All the living things that have ever existed are still here, having been disassembled during decomposition and reassembled during growth. This is in contrast to the fate of energy, given that the earth is an open energy system (Begon et al., 1995).

One result of this process is that in natural ecosystems, nutrient cycling — particularly of nitrogen, which is most commonly a limiting nutrient — is very tightly controlled. Very few nutrients are lost from natural soils, as these processes tend to release nutrients slowly, take them up rapidly, and conserve them. There is thus little loss of nutrients via erosion in natural ecosystems. While some nutrients are lost from soil through such processes as erosion, volatilization to the atmosphere, and leaching with water, these losses are usually minor and often get utilized elsewhere (Aber and Melillo, 2001).

Natural ecosystems generally have a high storage capacity for nutrients. This storage capacity exists largely in and on organic materials that decay slowly. Decomposition of OM may take several months to several years to complete. In tropical regions, the whole process is quite quick because moist conditions and high temperatures enhance biological activity (Aber and Melillo, 2001), and most of the nutrient cycling occurs in the topmost horizons. Under natural conditions, inputs from plants are the most important, including not only nutrients released by organic decomposition but also substances washed from plant leaves (foliar leaching). Losses (system outputs) are by leaching, erosion, gaseous losses like denitrification, and plant uptake. Within the soil, nutrients are stored on the soil particles, in dead OM or in chemical compounds (Foth and Turk, 1972).

6.2 Decomposition

Decomposition has a significant effect on soil structure and fertility by interacting with a number of processes (Ball and Trigo, 1997). Organisms generally die on or in the soil. The breakdown of OM is not a single chemical transformation but a complex process, with many sequential and concurrent steps. These include chemical alteration of OM, physical fragmentation, and finally release of mineral nutrients. Many animals living in the soil contribute to the mechanical decomposition of OM as well as to its chemical decomposition through digestion. Among these species are slugs and snails, earthworms, isopods, millipedes, centipedes, spiders, mites, and ants. Different organisms are involved with the different stages of these processes. Figure 6.2 shows the cycling of nutrients common to all terrestrial systems.

Breakdown starts almost immediately after an organism, or part of it, dies. The OM is quickly colonized by microorganisms that use enzymes to oxidize the OM to obtain energy and carbon. The surfaces of leaves and roots (and often their interiors) are colonized by microorganisms even before they die. Soil animals such as earthworms assist in the decomposition of OM by incorporating it into the soil where conditions are more favorable for decomposition than on the surface. Earthworms and other larger soil animals such as mites, collembola, and ants by fragmenting organic material increase its surface area, enabling still more microorganisms to colonize the OM and decompose it more rapidly (Begon et al., 1995).

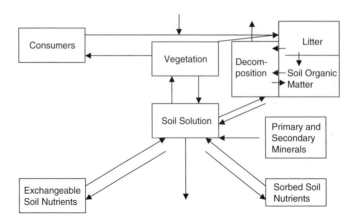

FIGURE 6.2
Common forms and directions of nutrient cycling in terrestrial systems.

6.3   Factors Affecting Rates of Decomposition

6.3.1   Litter Quality and Components

Litters differ in the rate at which they can be decomposed. Litter that decomposes more rapidly is said to be higher quality, while poor-quality litter conversely takes a longer time to decompose (Ball, 1992). Leaves and fine roots in the soil generally decay more rapidly than do suberized (i.e., woody) roots and stems. The quality of different litters reflects how much energy, carbon, and nutrient elements that litter can provide to the microbes involved in its decomposition. Each fraction of the litter is composed of different kinds of molecules that each require different enzymes for their degradation (Ball and McCarthy, 1989). This is one reason why diversity of soil biota contributes to soil fertility. Differences in litter quality have very practical implications in production systems as discussed in Chapter 18 and Chapter 19.

Leaves generally have more cellulose than lignin, and stems generally have more lignin than leaves (Ball et al., 1989). Lignin molecules have a complex, folded structure that makes it difficult for enzymes to release the component parts quickly. When lignin is linked within plant cell walls with cellulose, this makes it harder to degrade the cellulose. Within 10 weeks, most parts of leaves will have been degraded, but it may take up to 30 weeks to degrade the same amount of stem material. Table 6.1 shows the typical concentrations of major carbon compounds present in plant materials.

Litter and soil OM are the resources that drive most microbial growth. Freshly fallen leaf litter usually has a high proportion of readily utilizable energy-rich compounds that enables the soil microbial population to grow quickly. However, should the litter contain little nitrogen, then the microbial growth will be limited. Three general characteristics thus determine the quality of litter in terms of microbial decomposition. First, the type of chemical bonds present and the amount of energy released by their degradation influences litter quality. Second, the size and three-dimensional complexity of the molecules also influences litter quality along with a third consideration, their nutrient content (Betts et al., 1991). The types of C=C bonds present, together with the energy they yield, constitute the carbon quality of the material (Ball, 1993). Nutrient quality describes the nutrient content of the litter together with the ease with which these nutrients can be made available.

TABLE 6.1

Range of Concentrations of the Major Carbon Compounds Found in Woody and Herbaceous Plants (in %)

	Sugars and Starch	Other Solubles	Cellulose	Lignin
Grasses				
Leaves	6.8–12.4	19.6–40.6	44.1–49.5	6.4–10.6
Stems	5.7–7.3	28.9–34.4	52.5–57.7	13.1–14.4
Roots	3.9–5.2	20.0–36.2	41.6–58.0	12.2–22.0
Trees				
Leaves	5.7–7.3	25.1–37.6	43.1–47.4	12.1–22.5
Wood	1.1–4.1	5.9–16.7	40.3–80.5	12.5–38.9
Roots	3.9–5.2	14.6–20.0	47.7–49.5	25.3–33.8

Sources: Ball, A.S. and Trigo, C., The role of actinomycetes in plant litter decomposition, *Recent Developments in Soil Biochemistry*, 1, 9–13 (1997); Ball, A.S. and Bullimore, J., in *Humic Substances, Peats and Sludges*, The Royal Society of Chemistry, Cambridge, U.K., 311–318 (1997); Ball, A.S. and Pocock, S., in *The Environmental, Agricultural and Medical Aspects of Nitrogen Chemistry*, The Royal Society of Chemistry, Cambridge, U.K., 110–118 (1999). With permission.

During decomposition, there are two distinct phases. In the initial stages of decomposition, rates of degradation are determined by the availability of nutrients such as N, P, and sulfur (S), together with the presence of readily decomposable carbon compounds, such as soluble carbohydrates and nonlignified cellulose. Later, rates of decomposition are determined by the ability of the microbial population to degrade lignin.

During decomposition the organic molecules in organic matter get broken down into simpler organic molecules that require further decomposition or into mineralized nutrients (Ball and Allen, 1991). The compounds in organic matter vary in the ease with which microorganisms can break them down (McCarthy and Ball, 1991). The first organic compounds to be broken down are the easiest, simple sugars and carbohydrates. These compounds are also the first products of photosynthesis and are high-quality substrates for decomposition; these molecules are small and their chemical bonds are energy-rich (Ball, 1993). This results in the release of much more energy than would be required to create the enzymes necessary to break down the sugar. Further, small soluble molecules can readily be taken inside the microorganism and metabolized internally (Berrocal et al., 1997).

In plants, simple sugars such as glucose that are not required immediately for respiration and growth are stored as starch, a carbohydrate formed from linked glucose units. The decomposition of starch occurs at a somewhat slower rate than that of the simple sugars because its larger molecule has to be broken down into smaller molecules before the individual glucose units can be metabolized (Aber and Melillo, 2001). Nevertheless, starch represents a relatively degradable, high energy-yielding substrate. Unfortunately for microbes, neither starch nor simple sugars tend to be found in high concentrations in plant litter since plants usually use up these compounds prior to senescence (Table 6.1).

Cellulose also represents a polymer of glucose units linked by various bonds, but this time, glucose units are differently linked than in simple sugars. This different bonding imparts very different properties to the macromolecule, affecting its function in plants. Cellulose is the main component of plants' primary cell walls. Carbohydrates utilized in making cellulose cannot be converted and used in respiration by the plant. Cellulose, the most widely found molecule in terrestrial ecosystems, is more resistant to degradation than starch, and a range of extracellular enzymes, e.g., endoglucanases, are required to cleave this large polymer into smaller chains (Ball and Trigo, 1997). Other cellulolytic enzymes (exoglucanases) cleave smaller oligosaccharide units from the end of the chain.

Other glucosidases separate individual glucose units from the ends of polymers to complete the disassembly (Ball and McCarthy, 1989).

A number of phenolic compounds containing double (unsaturated) C=C bonds are found in plant litter. These compounds, which play very important roles in plant structure and function, also affect decomposition. Two types of these compounds are recognized: polyphenols (or tannins) and small phenol polymers made from a number of phenolic acids. In plants, polyphenols are thought to serve primarily as defense mechanisms against grazing animals and pathogenic microorganisms. The leaching of polyphenols from plant litter into the soil can significantly reduce the rate of decomposition in a soil system because of their inhibiting effects. Although the phenolic ring when decomposed yields less energy than saturated C bonds, these polyphenols can be metabolized by microorganisms. However, their rate of decomposition is difficult to assess due to polyphenols' mobility and their ability to condense with other polyphenols (and proteins) to form lignins (Aber and Melillo, 2001).

Lignins are a second type of phenolic compounds containing C=C bonds, and they are the second most abundant compound present in plant litter (Betts et al., 1991). These large, amorphous and very complex compounds are among the more complex and variable of natural compounds, with quite variable structure and no precise chemical description. Lignin is one of the slowest of the plant components to decay, and its decomposition results in almost no energy gain to the microorganism since large amounts of energy must be expended to complete its decomposition. There is evidence that some energy derived from the decay of higher-quality substrates is necessary to decompose lignin. Certain complex enzyme systems that can decompose the cellulose, hemicellulose, and lignin fractions of plant litter simultaneously have been identified (Trigo and Ball, 1996). The chemical qualities that provide strength and rigidity to the plant are the cause of the very slow rate of decomposition. The implications of these differences in decomposability are discussed further in Chapter 18 and Chapter 19.

6.3.2   The Physical and Chemical Environment

The major requirement for the decay of plant (and animal) residues is an active microbial population in contact with the residue. Soil microbes (bacteria, fungi, and actinomycetes) thrive and are most active under moist, warm conditions. Residue decomposition thus proceeds rapidly in temperate climates during wet spring, summer, and/or autumn days when it is warm and moist (Begon et al., 1995). Decomposition is conversely slow during the winter when it is cold, and water is in frozen form. A wet rainy summer will stimulate greater decomposition than a cool dry summer. In the tropics, conditions for decomposition are generally ideal year-round. In arid and semi-arid areas, higher temperatures would stimulate rapid decay, but a lack of moisture inhibits the growth and activity of microbial populations.

Maximum decomposition occurs in soils that are wet to near field capacity (wet but not muddy, with about 55% water-filled pore space) and at soil temperatures near 30°C. Decomposition proceeds slowly when soil temperatures are below 10°C, and it essentially stops at temperatures near freezing. Decomposition is slow when soil water contents are <40% water-filled pore space (barely moist to the touch), and it stops in soils that are air dry (dusty, hard, and crumbly to the touch). Decomposition is drastically reduced in soils that have become saturated with water. The saturation impedes the diffusion (movement) of oxygen into and within the soil. A sufficient supply of O_2 is required for maximum microbial activity (Aber and Melillo, 2001).

The physical condition of a soil affects the rate of decomposition of plant litter. Severe soil compaction impedes both water and air movement into a soil. If the soil is left in that condition for an extended period, decomposition will diminish. The amount of inorganic nitrogen available in a soil also determines how fast residues will decay. In general, decomposition is greater in soils with high residual inorganic nitrogen and/or high potential for mineralization of inorganic nitrogen from native soil organic matter (humus).

6.4 Roles of Organic Matter: The Humus Connection

Decomposition of organic matter rarely goes to completion, but rather tends to result in the accumulation of very stable, complex substances, collectively known as humus (Waksman, 1932). This is a structureless, dark, chemically complex organic material that has decomposed to a kind of stable equilibrium where further decomposition requires the input of more energy than it yields. It remains a reserve of organic molecules and has physical properties that are very beneficial in soil systems, such as increasing cation exchange capacity (CEC). This is the ability of soil particles to reversibly bind positively charged nutrients ions (cations) that can subsequently become available for plant use.

The distinction between litter and humus is often an operational one based on factors such as root penetration. Humus actually represents a series of high-molecular-weight polymers with a high proportion of phenolic rings having variable side chains. Compared with plant litter, humus is very high in nitrogen and has high-molecular-weight polyphenolic molecules, with less cellulose and hemicellulose than found in litter.

The source of nitrogen in the humus can be difficult to determine since 40% of its nitrogen is not found in amino acids. One possible explanation is that the nitrogen is bound in chitin, a polysaccharide that occurs also in fungal hyphae and in the exoskeletons of insects. Its toughness as a structural compound confers benefits similar to those that plants get from cellulose, to which it is related (Schnitzer, 1978). Humus exists in several forms. Its three fractions — humin, humic acid, and fulvic acid — are differentiated according to their solubility in acid after precipitation in alkaline solution. Both humic and fulvic acids exhibit a variable structure with high content of phenolics.

There is still considerable uncertainty regarding the compounds from which humus is formed. A long-standing hypothesis is that humus is formed from the modification of existing plant residues, with lignin seen as the most important precursor. Humus could be formed from the slow but continual microbial modification of initial lignin molecules. These modifications would include condensation into larger and larger molecules, with the addition of nitrogen through condensation reactions between lignin and proteins.

An alternative hypothesis is that microbes break down all of the large molecules into smaller ones, which are then repolymerized chemically to form the high-molecular-weight humic and fulvic acids. Both these processes may be operating during the formation of humus. That this fundamental process for the functioning of soil systems is still not fully understood indicates how much is still to be learned about their formation, structure, and function (Aber and Melillo, 2001).

While humus is decomposing very gradually, releasing nutrients, more is continuously being created within functioning soil systems. When present in large quantities, humus dominates the chemical and biological dynamics of soils in most terrestrial ecosystems. Humus decomposes very slowly, first, because the carbon compounds present are of low biochemical quality, and second, because OM can form colloids with mineral particles in soils. This association between organic and inorganic materials disrupts the alignment of degrading enzymes with the humus molecules and reduces the ability of these enzymes to degrade the humus. Furthermore, the enzymes that microbes have produced to degrade the humus can become fixed and deactivated by the humus/mineral colloids. The amount of OM that can be stabilized in this way increases in soils with finer textures, as smaller soil particles such as clays bind more effectively with OM. This stabilization of OM is particularly important in tropical soils in which the decomposition of freshly deposited OM is generally rapid (Stevenson, 1979).

Humus is vitally important for soil systems as it serves as an energy source for soil organisms while also providing nutrients during decomposition. It contains large numbers of charged sites — in particular, negatively charged sites to which positively charged nutrient ions such as ammonium (NH_4^+) adhere — so it enhances soil nutrient stores. Humus's fibrous and porous construction increases the water-holding capacity of soil and at the same time improving its aeration. All of this improves the soil environment for microbes and plant roots. Humus also, because of its porosity, increases the infiltration of water into the soil, thereby reducing runoff (Stevenson, 1982).

6.5 Nutrient Cycling: Focus on Carbon and Nitrogen

The importance of making regular OM inputs into soil to maintain or enhance its fertility can be seen from agricultural soils that have lacked regular amendments with OM. In general, we find that where inorganic fertilizers have been used, the OM content of the soils is reduced. When inorganic nutrients are supplied, this enables soil microbes that are nutrient-limited to decompose more rapidly whatever organic matter is available. This faster decay persists at least until structural problems in the fertilized soil outweigh the nutrient advantages provided by the inorganic nutrient supplementation. This is why, at first, adding inorganic nutrient inputs will increase decomposition rates; but then over time this usually reverses.

A dramatic example of this loss of organic material in agricultural soils is found in the Midwestern USA, whose prairie soils have lost approximately 33–50% of their organic material since they began being cultivated 100–150 years ago. Lowered OM in the soil, similar to humus, also results in lowered water-holding capacity and more runoff of water (Pretty et al., 2003).

Mineralization of OM must be considered in terms of the cycling of C, N, P, and S. This is the biological process whereby the organic compounds in OM are chemically converted by soil microorganisms into simpler organic compounds, into different organic compounds, or into mineralized nutrients. Microorganisms release enzymes that oxidize the organic compounds in OM, with this oxidation reaction releasing both energy and carbon, which microorganisms need to live.

The end-products of mineralization processes are nutrients in a mineral form. Unless they are in such a form, very few plants can take them up from the soil. Therefore, all the nutrients in OM must undergo mineralization processes before they can be used again by living organisms. Consider a protein molecule containing C, N, P, and S. When microorganisms mineralize this protein molecule, it undergoes a series of changes into simpler organic molecules until eventually the carbon in the protein is converted into CO_2, the nitrogen into NH_4^+, the phosphorus into phosphate (PO_4^{3-}), and the S into sulfate (SO_4^{2-}).

What is called immobilization of OM is the opposite process from mineralization. In immobilization, mineralized (inorganic) nutrients are incorporated into organic molecules within living cells. This process is very important because it relocates mineral nutrients into pools within the soil that have a relatively rapid turnover time, making them available to plants and preventing their loss by leaching. Plants are generally not efficient at competing with microorganisms for mineral nutrients available in the soil, but with a large microbial population turning over rapidly, dying off but being succeeded by a new generation, there is always a supply of available nutrients with the soil system, although it may be temporarily "immobilized" in living organisms (Mackenzie et al., 2001).

The C-to-N (C:N) ratio of organic matter refers to the amount of carbon present in the soil relative to the amount of nitrogen there. There is always more carbon than nitrogen in organic matter. The ratio C:N is written as a single number, which expresses how much more carbon than nitrogen there is available. If the number is low, the amounts of carbon and nitrogen are reasonably similar. When the ratio is a large number, there is considerably more carbon than nitrogen.

A C:N ratio does not tell us anything about the forms that carbon and nitrogen are in, just how much of each is there. All organisms have an optimal ratio of carbon to nitrogen in their own biomass, which differs between the bacteria, the fungi, and plant and animal cells. Bacteria generally have a low C:N, approximately 5. Fungi have a higher C:N of roughly 15, while plant and animal cells have, on average, a C:N around 10. If the C:N of an added organic material is high, e.g. over 20, then nitrogen will be in short supply for all organisms. In such a case, microorganisms will respire carbon and sequester the nitrogen in their biomass, so the plants growing in such soil will probably be nitrogen-deficient. Some nitrogen addition would be needed to meet the nitrogen requirements of these plants.

This is why incorporating OM into soils can change the amount of nitrogen (and other nutrients) available to plants (Aber and Melillo, 2001). The addition of organic matter with low nitrogen (high C:N ratio) to soil may, therefore, result in low rates of mineralization with increasing rates of immobilization, having an overall effect of reducing the amount of nitrogen available to the plants. In this case, additions of inorganic nitrogen fertilizer may be required to improve plant productivity. Conversely, additions of organic matter rich in nitrogen (low C:N ratio) will increase mineralization rates, making more nitrogen available to the plants. The addition of nitrogen-rich organic material also leads to increases in soil humus content. This is important for sustainable land management practices (Pretty et al., 2003).

6.6   Discussion

In this chapter, the flow of nutrients and energy through the soil ecosystem has been tracked. This has built upon the description of the biotic components found in soil, together with an examination of their roles, as presented in Chapter 5. In examining the flow of energy and nutrients, we have seen that although there are various sources of nutrients, decomposing organic matter forms the most important input. The fate of the decomposing organic matter depends very much on the quality of the material, in particular its lignin content, with the physical and chemical environments also playing significant roles in determining the rate of decomposition and therefore the release of available inorganic nutrients into the soil.

Decomposition rarely results in the complete mineralization of organic compounds. A dark, chemically complex, recalcitrant organic component of soil systems known as humus performs a number of vital functions in these systems, such as increasing water-holding capacity and improving soil aeration. Humus decomposes very slowly because the carbon compounds present are of low biochemical quality, and also because organic matter can form colloids with mineral particles in soils, which retard the decomposing activities of soil microbes.

Although all nutrients are cycled in soil ecosystem, the cycling of carbon and nitrogen are arguably the most important. The release of these elements from organic matter during decomposition (mineralization) is crucial for plant productivity. If the quality of the organic material is low (as reflected in a low C:N ratio), the decomposers may lock away (immobilize) vital nitrogen so that it is unavailable to plants. With an understanding of energy and nutrient cycling, it is clear that the zone of interaction between the micro-organisms involved in the mineralization process and plants' root systems, searching for released inorganic compounds in the rhizosphere, is a vital area that largely determines the fertility of the soil. These important interactions are discussed in more detail in Chapter 7.

References

Aber, J.D. and Melillo, J.M., *Terrestrial Ecosystems*, 2nd ed., Harcourt Academic Press, San Diego, CA (2001).

Ball, A.S., Degradation of plant biomass grown under elevated CO_2 conditions by *Streptomyces viridosporus* T7A, In: *Xylans and Xylanases*, Visser, J., Beldman, G., Kusters van Soneren, M.A., and Vorager, A.G.J., Eds., Elsevier, Amsterdam (1992).

Ball, A.S., Carbohydrates, In: *Biochemistry Labfax*, Chambers, A. and Rickwood, D., Eds., Bios Scientific Publishers, Oxford, 305–315 (1993).

Ball, A.S. and Allen, M., Solubilisation of wheat straw by actinomycetes, In: *Advances in Soil Organic Matter Research: The Impact of Agriculture on the Environment*, Wilson, W.S., Ed., The Royal Society of Chemistry, Cambridge, UK, 275–286 (1991).

Ball, A.S., Betts, W.B., and McCarthy, A.J., Degradation of lignin-related compounds by actinomycetes, *Appl. Environ. Microbiol.*, **55**, 1642–1644 (1989).

Ball, A.S. and Bullimore, J., Decomposition in soil of C-3 and C-4 plant material grown at ambient and elevated atmospheric CO_2 concentrations, In: *Humic Substances, Peats and Sludges*, Wilson, W.B. and Hayes, M., Eds., The Royal Society of Chemistry, Cambridge, UK, 311–318 (1997).

Ball, A.S. and McCarthy, A.J., Comparative analysis of enzyme activities involved in straw saccharficiation by actinomycetes, In: *Energy from Biomass*, 4th ed., Grassi, G., Pirrowitz, D., and Zibetta, H., Eds., 271–274 (1989).

Ball, A.S. and Pocock, S., The effects of elevated atmospheric CO_2 on nitrogen cycling in a grassland ecosystem, In: *The Environmental, Agricultural and Medical Aspects of Nitrogen Chemistry*, Wilson, W.S. and Ball, A.S., Eds., The Royal Society of Chemistry, Cambridge, UK, 110–118 (1999).

Ball, A.S. and Trigo, C., The role of actinomycetes in plant litter decomposition, *Recent Developments in Soil Biochemistry*, **1**, 9–13 (1997).

Begon, M., Harper, J.L., and Townsend, C.R., *Ecology Individuals, Populations and Communities*. 2nd ed., Blackwell Science, Oxford (1995).

Berrocal, M.M. et al., Solubilisation and mineralisation of [^{14}C] lignocellulose from wheat straw by *Streptomyces cyaneus* var. *viridochromogenes* during growth in solid-state fermentation, *Appl. Microbiol. Biotechnol.*, **48**, 379–384 (1997).

Betts, W.B. et al., Biosynthesis and structure of lignocellulose, In: *Biodegradation: Natural and Synthetic Materials*, Betts, W.B., Ed., Springer Verlag, London, 139–156 (1991).

Foth, H.D. and Turk, L.M., *Fundamentals of Soil Science*, Wiley, New York (1972).

Godden, B. et al., Towards elucidation of the lignin degradation pathway in actinomycetes, *J. Gen. Microbiol.*, **138**, 2441–2448 (1992).

Mackenzie, A., Ball, A.S., and Virdee, S.R., *Instant Notes in Ecology*, 1st and 2nd ed., Bios Scientific, Oxford, UK (2001).

McCarthy, A.J. and Ball, A.S., Actinomycete enzymes and activities involved in straw saccharification, In: *Biodegradation: Natural and Synthetic Materials*, Betts, W.B., Ed., Springer, London, 185–200 (1991).

Pretty, J.N. et al., The role of sustainable agriculture and renewable resource management in reducing greenhouse gas emissions and increasing sinks in China and India, In: *Capturing*

Carbon and Conserving Biodiversity: The Market Approach, Swingland, I.R., Ed., Royal Society Press, London, 195–217 (2003).

Schnitzer, M., Humic substances: Chemistry and reactions, In: *Soil Organic Matter*, Schnitzer, M. and Khan, S.U., Eds., Elsevier, Amsterdam, 1–64 (1978).

Stevenson, F.J., Humus, In: *The Encyclopedia of Soil Science*, Pt 1, Dowden, Hutchinson and Ross, Stroudsberg, PA (1979).

Stevenson, F.J., *Humus Chemistry: Genesis, Composition, Reactions*, Wiley, New York (1982).

Trigo, C. and Ball, A.S., Production and characterisation of humic-type solubilised lignocarbohydrate polymer from the degradation of wheat straw by actinomycetes, In: *Humic Substances and Organic Matter in Soil and Water Environments*, Clapp, C.E., et. al., Eds., IHSS Press, Birmingham, UK, 101–106 (1996).

Waksman, S.A., *Humus*, Williams and Wilkins, Baltimore, MD (1932).

7

The Rhizosphere: Contributions of the Soil–Root Interface to Sustainable Soil Systems

Volker Römheld and Günter Neumann

Institute for Plant Nutrition, University of Hohenheim, Hohenheim, Germany

CONTENTS

Changing conditions of local and global markets are pressing farmers to achieve ever-increasing crop productivity together with improved quality of their agricultural products. To reach these objectives, adequate plant acquisition of nutrients is necessary. However, even when nutrients are provided externally, their utilization by plants is highly dependent upon the physical, chemical, and especially biological conditions in the soil that is located in the immediate vicinity of plants' roots, known as the rhizosphere. This layer of soil, just a few millimeters thick, is intimately and continuously affected by roots' metabolic processes, creating a zone of intense activity quite different from the surrounding bulk soil. The rhizosphere's contribution to soil systems' fertility and sustainability and, thus to optimal plant growth, is all out of proportion to its physical volume.

The rhizosphere was described in Chapter 5 as a biologically active zone of soil with a particularly high carbon content, thanks to continuous plant root exudation and rhizo-deposition. Chapter 6 discussed the contribution made to soil fertility by ongoing interactions between these carbon resources and various rhizosphere microorganisms. Here, we consider in more detail the processes that go on in the rhizosphere. This domain serves as a broker between plants and the soil that they inhabit, evoking biological potentials that exist in both.

During preceding decades, our knowledge of the principles that regulate the respective rhizosphere processes has grown enormously (Marschner, 1995), but they still find little application in agricultural practice. There are now some examples of how certain measures can induce changes in rhizosphere processes of practical value (Römheld, 1990; Römheld and Neumann, 2005), so it is clear that this domain is amenable to improved management. This chapter seeks to give readers an appreciation of how the rhizosphere can be integrated within overall strategies of soil-system management, improving nutrient acquisition for better plant growth, while also conferring resistance to biotic and abiotic stress conditions. Chapter 40 provides some evidence from China that rhizosphere knowledge and management are beginning to be combined to good effect.

To understand the causes and extent of changes in certain rhizosphere conditions, it is necessary to consider the genotypes of both plants and associated soil biota, various soil characteristics, and the effects that farmer interventions have on cropping systems under field conditions. We anticipate that in the years ahead, increasingly evident resource limitations such as on water and the higher cost of chemical fertilizer, plus growing sensitivity to the environmental consequences of current agricultural practices, will create incentives for more innovative strategies for more purposeful rhizosphere management.

7.1   Specification of the Rhizosphere

A century ago, the German phytopathologist Lorenz Hiltner mentioned for the first time the rhizosphere as the soil compartment influenced by root activity (Hiltner, 1904). He considered the rhizosphere soil as particularly important for microbial suppression of certain soil-borne diseases (Neumann and Römheld, 2005). As discussed in Section 7.2, the rhizosphere is subject to important gradients in nutrient concentration, pH, redox potential, exudation, and microbial activity. The microbes involved include both noninfecting rhizosphere microorganisms, living freely associated in this zone, and infecting rhizosphere microorganisms which invade the roots. A major category of the latter, arbuscular mycorrhizal fungi, is considered in Chapter 9. Their ectomycorrhizal relatives, which live more superficially on the roots rather than within them, are more common on the roots of trees than in crops. There are also endophytic bacteria that inhabit the roots, creating something referred to as the endorhizosphere. Chapter 8 presents a case study of such organisms and their contributions to plant growth and health.

Depending on the processes and gradients being considered, the spatial extent of the rhizophere can range between less than of one millimeter up to several millimeters, this extent being greatly affected by the length of root hairs, microscopic extrusions from the roots (see Figure 34.1 in Chapter 34). The distance that protons or secreted organic compounds can be diffused from the roots into the soil depends most directly on the amount that is released and on soil factors such as water content and porosity. The spatial extent of the rhizosphere also varies depending on the genotype of the plant and its nutritional status since both affect the amount and kind of root exudation (Neumann and Römheld, 2000).

Often, the reported extent of the rhizosphere is overestimated in model experiments due to their being done under optimal soil conditions, e.g., low buffered, well-watered soils, and with plant species such as white lupin that have roots which are very "efficient" in producing exudates (Dinkelaker et al., 1989). With less ideal soil conditions or with less-efficient plants, the layer of bioactive soil around the roots will be thinner. A typical example of such overestimation are reports of the gradient of pH changes observed in the rhizosphere when it is measured using agar as a medium (Römheld and Neumann, 2005). Because there is less inhibition within an agar sheet of the diffusion of root-released protons, e.g., those associated with ammonium–nitrogen fertilization, a wider gradient is reported than actually operates in a soil system.

7.2   Functions of the Rhizosphere

To understand the contributions of the rhizosphere to plant growth, the extent of the various changes that depend on root metabolic activity as well as the spatial extent and distribution of these biological, chemical, and physical changes need to be appreciated. These changes in the rhizosphere are often restricted to distinct root zones, differing between the area around the root tip, where the root is growing through the soil, and that surrounding the more mature root zones. These zones are indicated in Figure 5.2 in Chapter 5. Usually, the most intense root exudation occurs in the apical root zones that are associated with still low microbial colonization (Neumann and Römheld, 2002). Accordingly, one needs to consider both longitudinal gradients along roots and radial gradients extending from the roots. With an appreciation of these variations in three dimensions, one can better comprehend the various rhizosphere processes that have relevance for plant growth under field conditions.

The uptake of mineral nutrients and of xenobiotics, such as heavy metals, takes place from the water-soluble fraction of minerals in the soil. This is mainly governed by conditions in the rhizosphere via processes such as solubilization or desorption, with less influence exerted by bulk soil conditions (Figure 7.1). While these conditions affect plant growth directly, they also govern the activity of plant growth-promoting rhizobacteria and

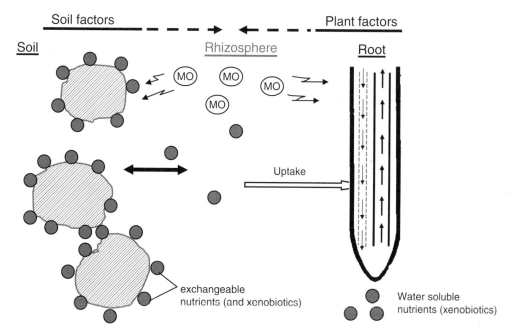

FIGURE 7.1
Interaction of soil and plant root factors in the rhizosphere for the uptake of mineral nutrients and xenobiotics. In a more complete representation, the various interactions of rhizosphere microorganisms (MO) would have to be indicated.

TABLE 7.1

Changes in Selected Rhizosphere Processes Reported in the Literature

Rhizosphere Process	Extent of Change in the Rhizosphere Compared with Bulk Soil	Plant Species (conditions)	Reference
Acidification/alkalinization	Decrease or increase of 2 pH units	Wheat, maize (different form of N)	Römheld (1986)
	Decrease of 3 pH units	White lupin (low P soil)	Dinkelaker et al. (1989)
Redox processes	Reductase activity enhanced by factor > 100	E.g., groundnut; (Fe deficiency)	Römheld (1990)
Root exudates/LMW compounds in the rhizosphere			
	2–3 times increase in sugars	Average values for various plant species	Jones et al. (2003), Jones et al. (2004)
	10 times increase in amino acids	Average values for various plant species	Jones et al. (2004)
	60 times increase in citrate	White lupin (low P soil)	Dinkelaker et al. (1989)
Phosphatase activity	Factor 2–8	Red clover, wheat, rape (low P soil)	Tarafdar and Jungk (1987)
Depletion of extractable P and K	Factor 5–7	Rape, wheat, maize (young seedlings)	Jungk (2002), Vetterlein and Jahn (2004)
Accumulation of Ca	Factor up to 10 with precipitation	Trees, azalea	Jungk (2002)
Accumulation of NaCl	Factor 6	Barley; maize (saline soils)	Schleiff (1986), Vetterlein and Jahn (2004)

of pathogens in the rhizosphere, which affect growth indirectly either by stimulating this or by causing disease.

Many years of research have documented changes in the rhizosphere, such as, those reported in Table 7.1. Such findings make obvious the relevance of the rhizosphere as the continuous close link between bulk soil conditions and the performance of plant roots and populations of associated microorganisms. Conventional soil analyses of bulk soil alone can result in very misleading conclusions regarding nutrient bioavailability for plants and the growth conditions of rhizobacteria because such information ignores the numerous and significant root-induced changes in the rhizosphere (Gobran et al., 1998).

7.2.1 Variability in and within the Rhizosphere

7.2.1.1 Changes in pH

Changes of 2 to 3 pH units are not uncommon within the rhizosphere of distinct root zones (Römheld, 1986). This variation directly affects how much and which mineral nutrients will be available for root uptake. The measured pH of the bulk soil is of minor importance compared to that which prevails in the rhizosphere (see Table 7.1; also Marschner, 1988; Thomson et al., 1993; Neumann and Römheld, 2002). The same kinds of differences are seen in the capacity of root zones to detoxify elements in the soil, such as heavy metals and aluminum (Wu et al., 1989; Ryan et al., 2001; Matsumoto, 2002).

The magnitude of pH changes in the rhizosphere will depend very much on plants' genotype (Römheld, 1986; Neumann and Römheld, 2002), as well as on their nutritional status (Neumann and Römheld, 2000) and on soil conditions, such as soil-buffering

TABLE 7.2

Effect of Nitrogen Form on Rhizosphere pH and Nutrient Uptake in Bean (*Phaseolus vulgaris* L.) Grown on a Sandy Loam pH 6.8 (after Thomson et al., 1993)

N-Form	pH		Uptake (μg/m root length)					
	Bulk	Rhizosphere	P	K	Fe	Mn	Zn	Cu
$Ca(NO_3)_2$	6.6	6.6	123	903	55	8	7	1.4
$(NH_4)_2SO_4$ − N-Serve[a]	5.7	5.6	342	1127	71	20	13	2.0
$(NH_4)_2SO_4$ + N-Serve[a]	6.6	4.5	586	1080	166	35	19	4.6

[a] Commercial nitrification inhibitor for agricultural use in the US.

capacity and free lime (Römheld, 1986; Jungk, 2002). Farmers can manipulate rhizosphere pH to improve crop nutrient acquisition by their selection of crop genotypes, but also by the kind of mineral fertilizer that they apply, in particular by the form of the N fertilizer they use (Römheld, 1986; Thomson et al., 1993). If nitrogen is applied as nitrate, this will promote a pH increase in the rhizosphere, while if nitrogen is applied in the form of ammonium, this promotes a decline. If there is some enhancement of N_2-fixing bacteria in symbiosis with the root system, these can themselves lower rhizosphere pH (Mengel and Steffens, 1982). The magnitudes of effect that different forms of nitrogen can have on rhizosphere pH and on the ensuing uptake of different nutrients are shown in Table 7.2.

7.2.1.2 Changes in Redox Potential

Similar to the changes observable in rhizosphere pH, the soil's redox potential can also vary over a wide range, due to the release of reducing root exudates or by the induction of extracellular reductase; both are affected by plant genotype and nutritional status (Dinkelaker et al., 1995; Neumann and Römheld, 2002). Changes in redox potential in the rhizosphere, which affect reductase activity, have far-reaching consequences for toxicity or for deficiency of Mn for crops as seen in Table 7.3 (also Marschner, 1988).

The changes in Mn redox status that result from increased or decreased microbial activity, which is, respectively, responsible for ensuing Mn reduction or for Mn oxidation, are of particular interest because of the role that Mn plays in plants' own mechanisms for disease suppression *via* the shikimate pathway (through formation of lignin and phytoalexins). There are increasing indications (King et al., 2001; Kremer et al., 2001; Guldner et al., 2005; Römheld et al., 2005) that certain agrochemicals such as glyphosate, a commonly used herbicide, can promote Mn-oxidizing bacteria in the rhizosphere, which in turn decrease Mn acquisition by crop plants. This observation is consistent with reported increases in the severity of certain diseases in areas after repeated use of this herbicide (Sanogo et al., 2000).

TABLE 7.3

Shoot Dry Matter Production and Mn Shoot Concentration of Soybean Cultivars Differing in Fe Deficiency-Induced Reductase Activity on the Rhizoplane as Affected by the Fe EDDHA Supply. Growth in Pots with a Calcareous (Calciaquoll) Soil (after Moraghan and Mascagni, 1991)

Cultivar	Fe EDDHA Supply (2 mg Fe/kg soil)	Relative Reductase Activity	Shoot Dry Matter (g/pot)	Mn Shoot Concentration (mg/kg dry wt)
Bragg (Fe efficient)	0	Very high	1.5	337 (toxicity)
	2	Low	3.5	24
T 203 (Fe inefficient)	0	Medium	1.3	95
	2	Very low	2.2	8 (deficiency)

7.2.1.3 Changes in Microbial Activity

Besides variation in rhizosphere pH and redox potential, there can be comparably high gradients in population density and activity of microorganisms as a consequence of the release of root exudates. There can be 10- to 100-fold differences in the microbial populations found in the rhizosphere and bulk soil (Marschner, 1995; Semenov et al., 1999). Microbial activity within the rhizosphere can be seen as a kind of "fingerprint" of a root system as shown in Figure 7.2, with bacterial colonies growing on agar, covering the roots of a maize plant after 2 weeks' culture in a rhizobox. This figure shows the enhanced microbial activity on and around roots that will affect root growth and also plant health. The effects of various kinds of organisms — associative N_2-fixing bacteria (Chapter 12), plant growth-promoting rhizobacteria (Chapter 14), and rhizosphere pH-dependent root pathogens — all need to be considered.

Differences are observed not only in overall population densities, but also specific changes in the structure of microbial communities due to differences in the composition of root exudates (Marschner et al., 2002; Wasaki et al., 2005). Up to now, this aspect of soil ecology has not been well understood, nor have its possible implications for improving agronomic practices been exploited. One example of the impact that such differences can have on crop performance is presented by results of research indicating significant variations in microbial community structure among different oat cultivars, with associated

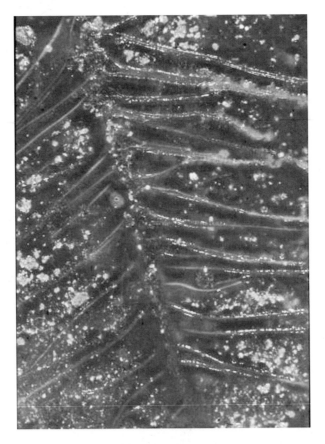

FIGURE 7.2
Fingerprint of the high microbial population density in the rhizosphere of a maize plant after 2 weeks growing in a rhizobox. A prefixed solid agar sheet with a general microbial growth medium was put on the opened rhizobox for 15′ contact and afterward incubated at 25°C in darkness for 72 h (from Römheld, 1990).

impacts on Mn acquisition (Timonin, 1946; Rengel et al., 1996). Such differences could account at least in part for the precrop effects that crop rotation has on the suppression of disease, or on various aspects of soil health in general (Huber and Wilhelm, 1988; Huber and McCay-Buis, 1993).

7.2.1.4 *Changes in Nutrient Availability*

Another change of relevance in the rhizosphere, one that can be visualized with radioisotope tracer techniques, is the accumulation or depletion of specific mineral nutrients and other elements in the rhizosphere. These changes can be of mutual importance for root growth and nutrient uptake. Depending on solubility of a given nutrient in soil solution, its transport to the root surface (*via* the mass flow driven by transpiration) can be higher or lower than the plant's uptake. Whenever it is higher, accumulation can occur up to factor 10 for Ca, Na, or Cl, for example (Marschner, 1995); where it is lower, a depletion of plant-available nutrients such as K and P, or certain micronutrients, occurs, by a factor up to 5 and more in the rhizosphere — and even more on the rhizoplane, i.e., root surface (Fusseder and Kraus, 1986; Jungk, 2002; Vetterlein and Jahn, 2004). Figure 7.3 suggests how differences in genotype can affect nutrient concentrations in the soil solution at different distances from the root's surface; it also lists ways in which nutrient availability and uptake in the rhizosphere can be affected independently of genotype.

Accumulation or depletion of mineral elements will have significant implications for plants' access to water, particularly in semiarid regions or in irrigated fields with salt problems (Schleiff, 1986; Vetterlein and Jahn, 2004). This also affects the acquisition of macronutrients such as K and P, especially under environmental stress conditions, such as drought. This is discussed in Section 7.2.2, in terms of variability in spatial distribution of the rhizosphere.

The importance of changes in these various processes in the rhizosphere is seen to be greater when the interaction among the respective processes is taken into account. For example, low rhizosphere pH promotes greater uptake of Mn by common bean (*Phaseolus vulgaris*) relative to their uptake of Zn, because Mn bioavailability is enhanced directly by

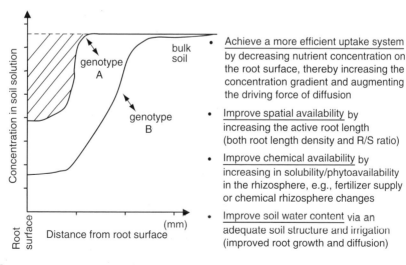

FIGURE 7.3
Possibilities for improving the limiting step of spatial nutrient availability by diffusion and the subsequent uptake of mineral nutrients.

the low pH and additionally by greater Mn^{IV} reduction at low pH levels (Sarkar and Wyn-Jones, 1982).

Another example of complex interactions is the inhibition of the growth of fungal hyphae and the inhibited infection of plant roots by certain root pathogens seen at low rhizosphere pH, e.g., the inhibitory effect that low rhizosphere pH has on *Gaeumannomyces graminis*, the so-called "take-all" fungus that causes root rot in some cereals. It is now evident that this pathogen can be better controlled by soil management practices affecting the growth environment for the pathogens than by employing chemical means (Huber and McCay-Buis, 1993).

Having low rhizosphere pH has other benefits, such as creating favorable conditions for the promotion of plant growth-promoting rhizobacteria (Römheld, 1990). However, rhizosphere management needs to optimize soil conditions, such as pH, because there are often countervailing effects. While low pH diminishes some diseases such as "take-all" in cereals, Verticillium wilt in cotton and potato, and Streptomyces scab in potato, other diseases, such as club rot in cabbage and Fusarium wilt in cotton, are promoted by this soil condition (Huber and Wilhelm, 1988). Efforts to improve crop production by modifying the rhizosphere thus need to take such considerations into account. We know, for example, that high rhizosphere pH can enhance the populations of Mn-oxidizing bacteria, which increase crops' susceptibility to certain plant diseases.

Because various rhizosphere parameters can have contradictory effects, farmers seeking effective rhizosphere management strategies need to consider concurrently and in detail the interplay of rhizosphere changes such as in pH, redox potential, and the structure of microbial populations. Also, we note that these effects should be assessed under realistic conditions, not relying on simplified model experiments or analyses that artificially eliminate this complexity.

7.2.2   Spatial Extent of Rhizosphere Activities

Keeping in mind the large changes in different rhizosphere processes that are common, it is important to consider the rhizosphere's relatively small share of the top soil, or of the top 1 m of the soil profile. As emphasized already, the extent of the rhizosphere depends fundamentally upon the process that is being considered since this extent varies for different processes, as seen below. There is not any single, simple rhizosphere for particular plants, because its coverage varies in both time and space.

The genotype and nutritional status of the plants involved and various soil conditions are the major determining factors for the domain of any given rhizosphere process. In general, various stress factors decrease the effective share of the rhizosphere, mainly due to inhibited root growth and a lower diffusion gradient, e.g., under drought stress (Mackay and Barber, 1985). However, under certain stress conditions, plants contribute more of their carbon and other compounds to the rhizosphere (Sauerbeck and Helal, 1986; Neumann and Römheld, 2000). This elicits increased microbial growth, including mycorrhizas, and also greater microbial activity that can at least partially compensate for stresses such as drought, compaction, low pH, or specific nutrient shortages (Sauerbeck and Helal, 1986; Marschner, 1995; Neumann and Römheld, 2000).

This effect underscores the symbiotic nature of root–rhizosphere interactions. In a world of Darwinian competition where each organism conserves its own resources for its own benefit, plants under stress might be expected to reduce their exudation in a time of stress. Instead, the physiological response of plants to greater stress that has evolved over millions of years is to increase carbon partitioning to the roots and to the rhizosphere. This can result in an increased root/shoot ratio (Anghinoni and Barber, 1980; Stasovski and

Peterson, 1991) and in enhanced support of microorganisms in the rhizosphere when plant growing conditions become suboptimal.

With respect to the capacity for phosphorus acquisition by an annual crop plant or fruit tree, rhizosphere soil can constitute between 0.5 and 20% of a top soil. The roots of plants that hyperaccumulate heavy metals can access these mineral elements from between 0.04 and 2% of the top soil. These estimates, shown in Table 7.4, are mainly affected by root system parameters, such as root length density and root hair length. The data in this table were calculated mainly from Fusseder and Kraus (1986) and from unpublished studies by Margerita Lopez-Lopez and Aiyen on the depletion of heavy metals in the rhizosphere of the hyperaccumulating plant *Thlaspi caerulescens*.

To estimate the rhizosphere share of a 1-m soil profile, the values derived for the total top soil can be divided by a factor of 4 because of the much lower root length density and reduced root activity at deeper soil depths. Differences in the spatial extent of the rhizosphere with respect to acquisition of phosphorus and potassium can be explained by the generally better diffusion of potassium through soil than of phosphorus and by the correspondingly wider depletion gradient (Figure 7.3). The differences in the spatial extent of the rhizosphere with respect to Cd and Zn are mainly due to the higher buffering of Zn in most heavy metal-contaminated soils compared with that of Cd.

Data on the spatial extent of the rhizosphere, deduced from pot experiments, can be expected to be higher than those from field experiments due to the generally better root growth conditions in pot experiments. More sophisticated simulation models that consider root growth parameters such as root hair length and density under various stress conditions such as drought, soil compaction, high salt, or low soil pH will give more diverse figures for the rhizosphere's share of soil. Such models can calculate values for more realistic scenarios by drawing on already available single-factor calculations (Table 7.4).

Knowing differences in the share of rhizosphere soil is of practical relevance for determining best fertilization practices. Even when there is high chemical availability of a particular mineral nutrient in the bulk soil, soil deficiency symptoms can occur in plants where the spatial extent of the rhizosphere is very limited because of restricted

TABLE 7.4

Calculated Share of Rhizosphere Soil within the Top Soil and within Top 1-m Soil Profile for Nutrient Acquisition (K and P) and Heavy Metal Accessibility (Cd and Zn) Showing Effects of Root Growth Characteristics (root length density) and Stress Conditions That Affect Growth and Activity of Roots (e.g., drought, salt, or high Al)

| | | | Top Soil (Ap horizon) | | | | Top 1-m Soil Profile | | | |
| | | | Mineral Nutrients | | Heavy Metals | | Mineral Nutrients | | Heavy Metals | |
	Conditions		K	P	Cd	Zn	K	P	Cd	Zn
Annual plants	High root	No stress	50	20	2	0.4	12.5	5.0	0.5	0.1
	length density	With stress	10	5	0.4	0.08	2.5	1.25	0.12	0.02
Annual plants	Medium root	No stress	25	10	1	0.2	6.25	2.5		
	length density	With stress	5	2.5	0.2	0.04	1.25	0.63		
Fruit trees without fruits		No stress	10	5			2.5	1.25		
		With stress	2	1			0.5	0.25		
Fruit trees		No stress	5	2.5			1.25	0.63		
		With stress	1	0.5			0.25	0.13		

diffusion (e.g., Figure 7.3), or where there is limited root growth, e.g., under various stress conditions, such as drought (Mackay and Barber, 1985). Under such conditions, increasing the total chemical availability of a nutrient by broadcast application of mineral fertilizer, as often recommended by the fertilizer industry, is of limited value. A better and more sustainable technique would be the placement of the fertilizer near the restricted root system in combination with microorganisms that are nutrient-mobilizing (for phosphorus and micronutrients) or N_2-fixing. Such placement would minimize the negative effects of limited spatial access, for example, to phosphorus fertilizer that is applied. Also, to assess the efficacy of phytoextraction as a feasible remediation technology for heavy metal contaminations, one needs to know a realistic figure for what is the share of the heavy metal-depleted rhizosphere.

7.3   Rhizosphere Processes of Importance for Plant Growth

The discussion in the preceding section on different rhizosphere processes that vary in their magnitude and spatial share of top soil illuminates the unique role of the rhizosphere, where plants, microbes, and soil converge to support soil and plant health in general. The rhizosphere constitutes a conjunction of plant, soil, and microbes in an interactive triad, sketched in Figure 7.4. This figure indicates how, given their respective and varied metabolic activities, plant roots and their associated microorganisms are active partners in the rhizosphere. This in turn is a crucial component of the larger soil system within which it exists. The broader soil conditions surrounding the rhizosphere, such as water content, buffering capacity, and nutrient storage capacity modulate these root/microbe-induced changes.

The following rhizosphere processes should receive explicit attention to attain the better understanding of the rhizosphere that is needed for its best-adapted management. These processes are: (1) the various processes factors involved in the mobilization of mineral nutrients to improve nutrient uptake efficiency; (2) root growth promotion by a variety of rhizobacteria; and (3) disease suppression by antagonistic microorganisms as related to micronutrient status. Because there are complex, interactive relationships between root growth characteristics and rhizosphere conditions, in particular affecting the spatial extent of the rhizosphere, the improvement of root growth will be considered first.

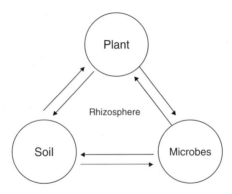

FIGURE 7.4
Interactions of the soil–microbe–plant triad in the rhizosphere.

7.3.1 Improved Root Growth

As noted above, root length density together with the length of root hairs are decisive parameters for the spatial extent of the rhizosphere within the top soil. This directly affects the spatial availability of nutrients and the spatial accessibility of xenobiotics such as heavy metals (Table 7.4). In turn, the availability of certain mineral nutrients, particularly phosphorus and nitrogen, affects root growth characteristics, such as root/shoot ratio and root hair length (Anghinoni and Barber, 1980; Stasovski and Peterson, 1991). This adaptive response enables plants to improve the spatial extent of their rhizosphere in a soil that has, for example, limited phosphorus availability.

In soils with low phosphorus nutritional status, the infection rate by arbuscular mycorrhizas and the growth of their extracellular hyphae often also increase as additional adaptive mechanisms for plants that have symbiotic relationships with microorganisms (Marschner and Dell, 1994; Marschner, 1996; Neumann and Römheld, 2002). This enables them to cope with a situation of low chemical nutrient availability through increased spatial accessibility. Complementary root responses, such as root hair growth, also play a role in this response. Thus, inhibited growth of root hairs often affects the infection rate of legumes by nitrogen-fixing bacteria and the populations of microbes in the root zones of C_4 grasses that have associative N_2 fixation capabilities, such as Azotobacter and *Azospirillum*.

It should be noted that root growth can be severely inhibited on sites that have low pH due to Al toxicity problems. This has severe consequences for the acquisition of nutrients because the spatial extent of the rhizosphere is thereby limited, and so is the infection rate of legumes by nitrogen-fixing bacteria which depend very much on proper root hair formation (Marschner, 1995). Recent work has shown Al-induced secretion of carboxylates, such as malate and citrate, or phenolics that achieve Al detoxification by a chelation process in certain plant species that have evolved as Al-excluders (Matsumoto, 2002). This could explain some well-established genotypical differences in Al-resistance that has been observed in barley, wheat, and soybean (Foy et al., 1969; Marschner, 1995).

Secretion of carboxylates into the rhizosphere is restricted to the most Al-sensitive apical root zones (Kollmeier et al., 2000). This effect has been overlooked for decades because there were no appropriate methods for localized collection of root exudates in the soil–root interface and for sensitive analysis (Engels et al., 2000). Now it will be interesting to see whether management strategies can be developed to support plants' own adaptation mechanisms for Al detoxification to counteract the negative effect of Al on root growth and thus to extend the spatial extent of the rhizosphere. Improving root growth by the detoxification of Al in the rhizosphere could also improve the infection of legumes by N_2-fixing bacteria and the populations of N_2-fixing bacteria associated with C_4 graminaceous species (Marschner, 1995; Kapulnik and Okon, 2002).

There is a largely unexplored potential, in particular for less fertile soils or under adverse soil conditions, to improve root growth and root hair formation by inoculation with certain microorganisms that produce phytohormones, such as Azospirillum species, with the aim of enhancing nutrient acquisition (Chapter 14). Practical applications along these lines are discussed in Chapter 32. In this context, an external supply of bioeffectors or biofertilizers that contain or induce the biosynthesis of phytohormones has been shown to assist in nutrient acquisition, probably *via* improved growth of roots and in particular of root hairs, resulting in increased spatial extent of the rhizosphere (Kapulnik and Okon, 2002; see also Chapters 15 and 37). The fungus Trichoderma is also known to have this effect (see Figure 34.1 in Chapter 34).

7.3.2   Improved Nutrient Acquisition

Besides the spatial availability of mineral nutrients, mainly governed by root growth parameters, the presence of nutrients in a form available for plant uptake plays a decisive role in nutrient acquisition which is affected by plant roots and associated microorganisms. The specific processes involved depend on the mineral nutrient being considered, so there are differences between nitrogen, phosphorus, potassium, and micronutrients in terms of how they are processed in the rhizosphere.

Despite high inputs of industrially synthesized nitrogen mineral fertilizers in many countries, there are still some agroecosystems, e.g., low-input farming systems and organic farms, that operate with limited external inputs of nitrogen. In such systems, the mineralized nitrogen required for a healthy start, for example, after a cold winter period, must come from somewhere. It is apparently delivered by enhanced microbial nitrogen mineralization in the rhizosphere (Parkin et al., 2002).

There are a limited number of studies on nitrogen turnover in the rhizosphere, but we have found greatly enhanced concentrations of amino acids attributable to nitrogen mineralization in the rhizosphere compared with the bulk soil (Römheld, unpublished data). The release of easily degradable root exudates into the rhizosphere could produce a priming effect for mineralization greater than the addition of green manures to a bulk soil. A substantial amount of nitrogen can be sequestered in the rhizosphere microbial biomass. Recent studies suggest an important role of protozoa and free-living nematodes for the liberation and subsequent mineralization of microbial nitrogen by their grazing on populations of bacteria in the rhizosphere (Bonkowski, 2004).

The contribution of associative N_2-fixing bacteria in the rhizosphere of C_4 plants with a high rate of root exudation of carbohydrates as an energy source for rhizosphere microorganisms like Azospirillum can be substantial for the nitrogen budget of sugarcane, millet, sorghum, and maize in Brazil and other subtropical countries (Chapters 12 and 27). Much less is known about the biological N_2 fixation contributions of endophytes (Chapter 12), but these too can be a source of nitrogen for plants.

Another aspect of nitrogen delivery to plants in the rhizosphere could be the exchange in clay minerals of fixed ammonium by the process of rhizosphere acidification (Schneider and Scherer, 1998). This release of fixed ammonium in the rhizosphere can contribute decisive amounts of nitrogen during the growth periods of some plant species (Scherer and Ahrens, 1996).

As discussed in Chapter 13, very little of the soil's total phosphorus is in forms that are readily available for plant uptake. Calculations show that during the main growth stage of an annual crop, the water-soluble phosphorus in the rhizosphere soil has to get replenished 20–50 times a day to guarantee adequate plant growth (Neumann and Römheld, 2002). To meet the requirements for the dynamic delivery of water-soluble phosphorus for plant uptake, various rhizosphere processes, which are often stimulated by plants' low phosphorus status, are operative.

In neutral and alkaline soils, rhizosphere acidification such as is induced by NH_4-supply, can increase the phosphorus availability of acid-soluble Ca phosphates by dissolution as well as the availability of micronutrients (Table 7.2). In acid soils (pH $<$ 5), however, rhizosphere acidification will not increase phosphorus availability. This can occur only *via* secretion of carboxylates. These mediate phosphorus solubilization by mechanisms of ligand exchange, dissolution, and the occupation for phosphorus sorption sites. In contrast, rhizosphere alkalinization as occurs with field-grown pearl millet and some legumes (e.g., cowpeas) can enhance chemical phosphorus availability either by ligand exchange or indirectly by improved microbial phosphorus mineralization and due to an improved spatial availability as a consequence of a better root growth (Neumann and Römheld, 2002).

Secretion of acid phosphatases and phytases by plant roots and also by rhizosphere microorganisms can contribute to phosphorus acquisition by hydrolysis of organic phosphorus esters in the rhizosphere. These compounds can comprise up to 30–80% of the total soil phosphorus. Although such secretion of acid phosphatases into the rhizosphere is often enhanced under low phosphorus nutritional status (Römheld and Neumann, 2005), the availability of organic phosphorus is usually rather limited mainly due to the low solubility of organic phosphorus in most soils. However, a simultaneous release of carboxylates under phosphorus deficiency can increase the efficiency of both carboxylates and phosphohydrolases (Neumann and Römheld, 2000). Soil system processes that govern phosphorus availability are discussed in more detail in Chapter 13.

Rhizosphere acidification can also efficiently enhance potassium release either by desorption or by solubilization of potassium sources with low solubility (Jungk, 2002). Hinsinger et al. (1993) have shown a deep depletion of interlayer potassium in clay minerals in the rhizosphere after acidification by rape plants, with the consequence of enhanced weathering of minerals. As shown for phosphorus, and for micronutrients like Fe, the amount of soluble inorganic Fe in the soil solution comprises less than 0.05% of the daily requirement for adequate plant growth (Kirkby and Römheld, 2004).

All this underscores the importance of plant root-induced mobilization processes that occur in the rhizosphere to enhance the concentration of individual micronutrients in the rhizosphere soil solution. These processes include changes in pH redox potential, chelation by root exudates, and the activity of rhizosphere microorganisms. As summarized for the various micronutrients in Kirkby and Römheld (2004), mobilization processes can be critical for the acquisition and uptake of micronutrients and are often upregulated under low nutritional status (see also Neumann and Römheld, 2002).

7.3.3   Crop Protection against Pests and Pathogens

Adequate and balanced supply of all mineral nutrients is required to achieve a good yield and a high quality product. However, under certain stress conditions, there are clear indications that plants have a higher requirement for distinct mineral nutrients such as potassium and micronutrients, particularly Mn, Zn, and Cu, which are involved as co-factors in various plant defense reactions. This is true for stress factors that are abiotic (drought, salt, low temperature) as well as biotic (pests and diseases) (Vunkova-Radeva et al., 1988; also unpublished data from Ismael Cakmak). Management strategies need to account for enhanced acquisition of all these mineral nutrients. The "take-all" disease of wheat and other cereals caused by the fungus *Gaeumannomyces graminis* is an example where various strategies for rhizosphere management have been developed and applied in practice. The driving force for the rapid development of alternatives has been, until recently, the lack of an effective fungicide. Rhizosphere acidification by ammonium–N fertilizer — [stabilized with a nitrification inhibitor, such as "N-serve" which is nitrapyrin: 2-chloro-6-(trichloromethyl)pyridine or chloride Christensen et al., 1981; Huber and Wilhelm, 1988] — is now a widespread method for suppressing the infection of wheat roots by this fungus. The particular importance of improving Mn availability in the rhizosphere, whether by rhizosphere acidification, by promotion of Mn-reducing rhizosphere microorganisms, or by biofertilizer applications such as Mn-mobilizing Trichoderma strains, is summarized by Huber and McCay-Buis (1993).

However, for many other diseases, alternative measures for suppression are still missing. For example, effective chemical control is still missing for the fast-spreading dieback syndrome of citrus in Brazil, probably caused by a bacterium *Xylella fastidiosa* and a virus. This dieback is observed particularly in conventionally managed citrus orchards where there is regular use of herbicides under the trees to control weeds. In biologically managed orchards that manage weeds by intensive mulching of Brachiaria grass under the trees, with no application of herbicides, the disease is not or at least much less expressed (Guldner et al., 2005).

Plant analysis data have indicated that low Mn and Zn status might be causally involved in the outbreak of this disease by lowering plant resistance to citrus dieback. This is consistent with the role of Mn in "take-all" and other Mn-dependent diseases (Huber and Wilhelm, 1988). It has been shown also that the mulch of Brachiaria grass can release a natural nitrification inhibitor (Subbarao et al., 2004). So the ammonium–N released by mineralization is stabilized with subsequent rhizosphere acidification resulting in increased Mn acquisition by the citrus roots. This assumption is consistent with observations in Indonesia where dieback problems in citrus have been successfully cured by Mn infusion into the tree trunks (Aiyen, personal communication).

Given the predominant role of Mn in plant defense against pathogen attacks via the shikimate pathway (Marschner, 1995), it is evident that the availability of Mn in the rhizosphere, as well as of Cu and B as other protective elements, plays a decisive role in plant health. It appears that the herbicide glyphosate results in increased susceptibility to various diseases and nematodes inasmuch as there is inhibited Mn acquisition after the release of glyphosate from the roots of treated plants into the rhizosphere (Guldner et al., 2005; Römheld et al., 2005). Such observations call attention to the importance of rhizosphere processes which can be affected by agrochemicals with undesirable effects on plant health.

7.4   Discussion

The research reported here shows that the very small proportion of rhizosphere soil is of critical importance for understanding processes such as nutrient and xenobiotic bioavailability, root growth, disease suppression, and stress tolerance. Knowledge of these rhizosphere processes should be used to design the best management practices for the rhizosphere, to achieve desired improvements in plant growth and plant health with less input of agrochemicals (Figure 7.5).

Such knowledge has been put to use in Chinese agriculture by studying and capitalizing on the benefits attainable from intercropping, reported in Chapter 40. In southern Brazil, some cooperatives are now cultivating up to 25% of their land in rotation with black oats (*Avena nuda* L.) as a kind of green manure, plowing it into the soil without harvesting the grain. Farmers consider that such crop management improves soil health (Calagari, 2002). Given what we now know about rhizosphere processes with their complex interplay among soil, microbes, and plants, it appears that black oats may be enhancing soil health by promoting Mn-reducing organisms in the rhizosphere, as described by Timonin (1946). As a consequence of enhanced microbial Mn reduction, the following crop such as winter wheat benefits from higher Mn acquisition and disease resistance.

It is gratifying that, after a long period of neglect, Lorenz Hiltner's very practically oriented research on the rhizosphere about 100 years ago, like that of Professor A.B. Frank on mycorrhizas before him, as discussed in Wolfe (2001), is undergoing a renaissance. The Chinese and Brazilian experiences are demonstrating the possibilities of more sustainable agricultural systems with less unilateral use of mineral fertilizers and other agrochemicals, based on a deeper scientific understanding of the rhizosphere as the bioactive soil/root interface. It will be a challenge to support farmers worldwide, particularly those on less

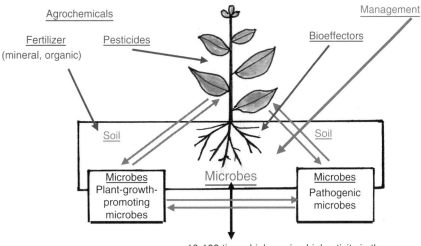

FIGURE 7.5
Management of the rhizosphere for optimizing soil–microbe–plant interplay to achieve improved plant health with less reliance on agrochemicals.

favorable land with adverse soil factors, to become fully aware of this interface and of its implications for their management practices. The best starting point is for soil and plant scientists to collaborate more with soil microbiologists in advancing our knowledge of this sphere and in communicating it broadly.

References

Anghinoni, I. and Barber, S.A., Phosphorus influx and growth characteristics of corn roots as influenced by phosphorus supply, *Agron. J.*, **72**, 685–688 (1980).

Bonkowski, B., Protozoa and plant growth: The microbial loop revisted, *New Phytol.*, **162**, 617–631 (2004).

Calagari, A., The spread and benefits of no-till agriculture in Paraná, Brazil, In: *Agroecological Innovations: Increasing Food Production with Participatory Development*, Uphoff, N., Ed., Earthscan, London, 187–202 (2002).

Christensen, N.W. et al., Chloride effect on water potentials and yield of winter wheat infected with take-all root rot, *Agron. J.*, **73**, 1053–1058 (1981).

Dinkelaker, B., Hengler, C., and Marschner, H., Distribution and function of proteoid roots and other root clusters, *Bot. Acta*, **108**, 183–200 (1995).

Dinkelaker, B., Römheld, V., and Marschner, H., Citric, acid excretion and precipitation of calcium citrate in the rhizosphere of white lupin (*Lupinus albus* L.), *Plant Cell Environ.*, **12**, 285–292 (1989).

Engels, C. et al. Assessment of the ability of roots for nutrient acquisition, In: *Root Methods: A Handbook*, Smit, A.L. et al., Eds., Springer, Berlin, 403–459 (2000).

Foy, C.D., Fleming, A.L., and Arminger, W.H., Aluminum tolerance of soybean varieties in relation to calcium nutrition, *Agron. J.*, **61**, 505–511 (1969).

Fusseder, A. and Kraus, M., Individuelle Wurzelkonkurrenz und Ausnutzung der immmobilen Makronährstoffe im Wurzelraum von Mais, *Flora*, **178**, 11–18 (1986).

Gobran, G.R., Clegg, S., and Courchesne, F., Rhizospheric processes influencing the forest ecosystems, *Biogeochemistry*, **42**, 107–120 (1998).

Guldner, M. et al., Release of foliar-applied glyphosate into the rhizosphere and its possible effects on non-target organisms. In: *Rhizosphere 2004: A Tribute to Lorenz Hiltner*, Hartmann, A. et al., eds, GSF-Report, Munich, Neuherberg, Germany (2005).

Hiltner, L., Über neuere Erfahrungen und Probleme der Bodenbakteriologie unter besonderer Berücksichtigung der Gründüngung und Brache, *Arb. Dt. Landw. Ges.*, **98**, 59–78 (1904).

Hinsinger, P. et al., Root-induced irreversible transformation of a trioctahedral mica in the rhizosphere of rape, *J. Soil Sci.*, **44**, 535–545 (1993).

Huber, H. and McCay-Buis, T.S., A multiple component analysis of the take-all disease of cereals, *Plant Dis.*, **77**, 437–447 (1993).

Huber, D.M. and Wilhelm, N.S., The role of manganese in resistance to plant disease, In: *Manganese in Soils and Plants*, Graham, R.D. et al., Eds., Kluwer, Dordrecht, Netherlands, 155–173 (1988).

Jones, D.L. et al., Organic acid behaviour in soils: Misconceptions and knowledge gaps, *Plant Soil*, **248**, 31–41 (2003).

Jones, D.L., Hodge, A., and Kuzyakov, Y., Plant and mycorrhizal regulation of rhizodeposition, *New Phytol.*, **163**, 459–480 (2004).

Jungk, A.O., Dynamics of nutrient movement at the soil–root interface, In: *Plant Roots: The Hidden Half*, 3rd ed., Waisel, Y., Eshel, A., and Kafkafi, U., Eds., Marcel Dekker, New York, 587–616 (2002).

Kapulnik, Y. and Okon, Y., Plant growth promotion by rhizosphere bacteria, In: *Plant Roots: The Hidden Half*, 3rd ed., Waisel, Y., Eshel, A., and Kafkafi, U., Eds., Marcel Dekker, New York, 869–885 (2002).

King, C.A., Purcell, C.C., and Vories, E.D., Plant growth and nitrogenase activity of glyphosate-tolerant soybean in response to foliar glyphosate applications, *Agron. J.*, **93**, 179–180 (2001).

Kirkby, E.A. and Römheld, V., *Micronutrients in Plant Physiology: Functions, Uptake and Mobility*, Proceeding No. 543. International Fertilizer Society, York, 1–51, UK (2004).

Kollmeier, M., Felle, H.H., and Horst, W.J., Genotypical differences in aluminium-tolerance of maize are expressed in the distal part of the transition zone: Is reduced basipetal auxin transport involved in inhibition of root elongation by aluminum?, *Plant Physiol.*, **122**, 945–956 (2000).

Kremer, R.J. et al., Herbicide impact on *Fusarium* ssp. and soybean cyst nematode in glyphosate-tolerant soybean, *Am. Soc. Agron.*, **503**, 104D (2001), Title summary.

Mackay, A.D. and Barber, S.A., Soil moisture effects on root growth and phosphorus uptake by corn, *Agron. J.*, **77**, 519–531 (1985).

Marschner, H., Mechanisms of manganese acquisition by roots from soils, In: *Manganese in Soils and Plants*, Graham, R.D. et al., Eds., Kluwer, Dordrecht, Netherlands, 191–204 (1988).

Marschner, H., *Mineral Nutrition of Higher Plants*, Academic Press, London (1995).

Marschner, H., Mineral nutrient acquisition in non-mycorrhizal and mycorrhizal plants, *Phyton*, **36**, 61–68 (1996).

Marschner, H. and Dell, B., Nutrient uptake in mycorrhizal symbiosis, *Plant Soil*, **159**, 98–102 (1994).

Marschner, P. et al., Spatial and temporal dynamics of the microbial community structure in the rhizosphere of cluster roots of white lupin (*Lupinus albus* L.), *Plant Soil*, **246**, 167–174 (2002).

Matsumoto, H., Plant roots under aluminum stress: Toxicity and tolerance, In: *Plant Roots The Hidden Half*, 3rd ed., Waisel, Y., Eshel, A., and Kafkafi, U., Eds., Marcel Dekker, New York, 821–838 (2002).

Mengel, K. and Steffens, D., Relationship between the cation/anion uptake and the release of protons by roots of red clover, *Z. Pflanzenernaehr. Bodenk.*, **145**, 229–236 (1982).

Moraghan, J.T. and Mascagni, H.J., *Environmental and Soil Factors Affecting Micronutrient Deficiencies and Toxicities*, Micronutrients in Agriculture, SSSA Book Series No. 4 (1991), 371–425.

Neumann, G. and Römheld, V., The release of root exudates as affected by the plant physiological status, In: *The Rhizosphere: Biochemistry and Organic Substances at the Soil–Plant Interface*, Pinton, R., Varanini, Z., and Nannipieri, Z., Eds., Marcel Dekker, New York, 41–89 (2000).

Neumann, G. and Römheld, V., Root-induced changes in the availability of nutrients in the rhizosphere, In: *Plant Roots The Hidden Half*, 3rd ed., Waisel, Y., Eshel, A., and Kafkafi, U., Eds., Marcel Dekker, New York, 617–649 (2002).

Neumann, G. and Römheld, V., Rhizosphere research: A historical perspective from a plant scientist's viewpoint, In: *Rhizosphere 2004: A tribute to Lorenz Hiltner*, Hartmann, A. et al., Eds., GSF-Report, Munich, Neuherberg, Germany, 35–37 (2005).

Parkin, T.B., Kaspar, T.C., and Cambardella, C., Oat plant effects on net nitrogen mineralization, *Plant Soil*, **243**, 187–195 (2002).

Rengel, Z. et al., Plant genotype, micronutrient fertilization and take-all colonization influence bacterial populations in the rhizosphere of wheat, *Plant Soil*, **183**, 269–277 (1996).

Römheld, V., pH changes in the rhizosphere of various crop plants, in relation to the supply of plant nutrients. *Potash Review*, Berne, Switz, No 12, 6/55, 1–8 (1986).

Römheld, V., The soil–root interface in relation to mineral nutrition, *Symbiosis*, **9**, 19–27 (1990).

Römheld, V. and Neumann, G., The rhizosphere: Definition and perspectives, In: *Rhizosphere 2004: A Tribute to Lorenz Hiltner*, Hartmann, A. et al., Eds., GSF-Report, Munich, Neuherberg, Germany, 47–49 (2005).

Römheld, V. et al., *Plant Nutrition for Food Security, Human Health and Environmental Protection*, Li, C.J., et al. Eds., Tsinghua University Press, Beijing, 476–477 (2005).

Ryan, P.R., Delhaize, E., and Jones, D.L., Function and mechanism of organic anion exudation from plant roots, *Ann. Rev. Plant Physiol. Plant Mol. Biol.*, **52**, 527–560 (2001).

Sanogo, S., Yang, X.B., and Scherm, H., Effects of herbicides *Fusarium solani* sp. *Glycines* and development of sudden death syndrome on glyphsoate tolerant soybean, *Phytopathology*, **90**, 57–68 (2000).

Sarkar, A.K. and Wyn-Jones, R.G., Effect of rhizosphere pH on avialability and uptake of Fe, Mn and Zn, *Plant Soil*, **66**, 361–372 (1982).

Sauerbeck, D. and Helal, H.M., Plant root development and photosynthate consumption depending on soil compaction, *Trans. XIII. Congr. Int. Soil Sci. Soc.*, **3**, Hamburg, Germany, 948–949 (1986).

Scherer, H.W. and Ahrens, G., Depletion of non-exchangeable NH_4–N in the soil–root interface in relation to clay mineral composition and plant species, *Eur. J. Agron.*, **5**, 1–7 (1996).

Schleiff, U., Water uptake by barley roots as affected by the osmotic and matric potential in the rhizosphere, *Plant Soil*, **94**, 354–360 (1986).

Schneider, M. and Scherer, H.W., Fixation and release of ammonium in flooded rice soils as affected by redox potential, *Eur. J. Agron.*, **8**, 181–189 (1998).

Semenov, A.M., van Bruggen, A.H.C., and Zelenev, V.V., Moving waves of bacterial populations and total organic carbon along roots of wheat, *Microb. Ecol.*, **37**, 116–128 (1999).

Stasovski, E. and Peterson, C.A., The effects of drought and subsequent rehydration on the structure and vitality of *Zea mays* seedling roots, *Can. J. Bot.*, **69**, 1170–1178 (1991).

Subbarao, G.V. et al., Can nitrification be inhibited/regulated biologically? *Abstract 3rd International Nitrogen Conference*, Beijing, China, 12–16 October (2004).

Tarafdar, J.C. and Jungk, A., Phosphatase activity in the rhizosphere and its relation to the depletion of soil organic phosphorus, *Biol. Fertil. Soils*, **3**, 199–204 (1987).

Thomson, C.J., Marschner, H., and Römheld, V., Effect of nitrogen fertilizer form on pH of the bulk soil and rhizosphere, and on the growth, phosphorus, and micronutrient uptake of bean, *J. Plant Nutr.*, **16**, 493–506 (1993).

Timonin, M.I., Microflora of the rhizosphere in relation to the manganese-deficiency disease of oats, *Soil Sci. Soc. Am. Proc.*, **11**, 284–292 (1946).

Vetterlein, D. and Jahn, R., Gradients in soil solution composition between bulk soil and rhizosphere — *In situ* measurement with changing soil water content, *Plant Soil*, **258**, 307–317 (2004).

Vunkova-Radeva, R. et al., Stress and activity of molybdenum-containing complex (Molybdenum cofactor) in winter wheat seeds, *Plant Physiol.*, **87**, 533–535 (1988).

Wasaki, J. et al., Root exudation, P acquisition and microbial diversity in the rhizosphere of Lupinus albus as affected by P supply and atmospheric CO_2 concentration, *J. Environ. Quality*, **34**, 2157–2166 (2005).

Wolfe, D.W., *Tales from the Underground: A Natural History of Subterranean Life*, Perseus, Boston (2001).

Wu, Q.T., Morel, J.L., and Guckert, A., Effect of nitrogen source on cadmium uptake by plants, *C. R. Acad. Sci.*, **309**, 215–220 (1989).

8

The Natural Rhizobium–Cereal Crop Association as an Example of Plant–Bacteria Interaction

Frank B. Dazzo[1] and Youssef G. Yanni[2]

[1]*Department of Microbiology and Molecular Genetics, Michigan State University, East Lansing, Michigan, USA*
[2]*Sakha Agricultural Research Station, Kafr El-Sheikh, Egypt*

CONTENTS

High grain production of cereals has only been possible with high inputs of inorganic or synthetic nitrogen fertilizers. This unfortunately creates many environmental, economical, and health risk problems that must be minimized for environmental safety and sustainable agriculture. Although great efforts have been made over the last three decades to increase the productivity of cereals by introducing new varieties and better farm management methods, sufficient production of these crops using external inputs is still a distant target. An alternative, cereal biofertilization, is still constrained by the small number and diversity of candidate microorganisms and by limited extension efforts.

In particular, meeting crops' demand for nitrogen (N) to obtain their maximum yield potential is a challenge. The discovery and development of effective agricultural biofertilizers that can reduce dependence on inorganic N-fertilizer inputs to maximize crop yields is a high priority that can help to achieve the goal of sustainable agriculture worldwide. Our research has shown that the exploitation of naturally selected, beneficial plant–bacteria associations is environmentally sound and has high probability for success.

Growing knowledge about symbiotic relations between cereal crops and certain microbes that exist in the crops' rhizosphere and roots, derived by using some of the most advanced analytical methods, illuminates the complex ways in which plants and soil biota can benefit each other. It expands our understanding of soil systems and also of opportunities for improving agricultural outcomes.

In the mid-1990s, we began a collaborative research project for both basic scientific discovery and agronomic application of novel, beneficial plant–microbe associations. We wanted to assess their potential for promoting crop plant growth and for the eventual development of new, improved biofertilizers. Our work focused originally on Rhizobia and rice with a more recent expansion to include the Rhizobia–wheat association.

Our guiding hypothesis was that natural endophytic associations between Rhizobia and cereal roots would most likely occur where these cereals are successfully rotated with a legume crop that could enhance the soil population of the corresponding Rhizobial symbionts. Such natural Rhizobium–cereal associations would be perpetuated if they were mutually beneficial. If this hypothesis is correct, the cereal roots growing at these sites should harbor, along with other microbes, a high population density of endophytic Rhizobia that are already adapted to be highly competitive for colonization of the interior habitats of crop roots, being protected from stiff competition with other soil-rhizosphere microorganisms under field conditions. This is where endophytic Rhizobia are strategically located because a more rapid and intimate metabolic exchange is possible within host plant tissues rather than just on their epidermal surface.

An ideal place to test this hypothesis was in the Egyptian Nile delta because for more than seven centuries of recorded history, rice there has been rotated with the forage legume, Egyptian berseem clover (*Trifolium alexandrinum* L.). In this area, japonica and (more recently) indica and hybrid rice cultivars are cultivated by transplantation in irrigated lowlands. Currently, about 60–70% of the 500,000 ha of land area used for rice production in Egypt is in rice–clover rotation. Berseem clover's high yield, protein content, and symbiotic N_2-fixing capacity enhance its use as a forage and green manure plant in this region.

An interesting enigma for this successful farming system is that the clover rotation with rice can replace 25–33% of the recommended amount of fertilizer-N needed for optimal rice production. However, N-balance data indicate that this benefit of rotation with clover cannot be explained solely by the increase in available soil nitrogen created by mineralization of the biologically fixed, nitrogen-rich clover crop residues. So we asked: is there a natural endophytic Rhizobium–rice association that has evolved which contributes to this added benefit of clover/rice rotation?

Our studies have indicated that the well-known clover root-nodule occupant *Rhizobium leguminosarum* bv. *trifolii* does indeed participate in such an association with rice, independent of root nodule formation and biological N_2-fixation. Furthermore, we have found that the use of certain strains of endophytic Rhizobia as inoculants for a rice crop can significantly improve its vegetative growth, grain productivity, and agronomic fertilizer-use efficiency (measured as kg grain yield per kg of fertilizer-N applied). Rhizobia can help produce high rice grain yield with less dependence on inorganic fertilizer inputs, fully consistent with sustainable agriculture.

8.1 Confirmation of a Natural Rhizobium–Cereal Association

We undertook an ecological approach which involved multiple cycles of field and laboratory studies to detect, enumerate, and isolate Rhizobial endophytes from surface-sterilized roots of field-grown rice and wheat (Yanni et al., 1997; Yanni et al., 2001). Rice plants were sampled at five different field sites during two rotations with berseem clover in the Nile delta. The first field-sampling site was from vegetative regrowth of "ratooned" rice that remained from the previous growing season, intermingled among clover plants in their current rotation. The second sampling was from four different sites in flooded fields of transplanted rice during the next rice-growing season. The roots were washed and surface-sterilized until bacteria on the root surface could no longer be cultured. Diluted macerates of the surface-sterilized roots were then inoculated directly on axenic seedling roots of berseem clover in enclosed tube cultures containing nitrogen-free medium, and the nodulated plant replicates were scored for most-probable number (MPN) calculations after 1 month of incubation.

This experimental design took advantage of the strong positive selection provided by the clover "trap" host so as to select for the numerically dominant "rice-adapted" clover-nodulating Rhizobia present among the other natural rice endophytes that survived surface sterilization of the roots. It also provided us with an easy means to isolate the dominant strains of endophytic Rhizobia within the clover root nodules that ultimately developed on plants inoculated with the highest dilutions in the MPN series.

The results from all five sample sites provided solid confirmation of our guiding hypothesis, that clover-nodulating Rhizobia intimately colonize the rice root interior in these fields of the Egyptian Nile delta (Yanni et al., 1997; Yanni et al., 2001). The population size of clover-nodulating Rhizobia was 2 to 3 logs higher inside the roots of the ratooned rice that was growing among the clover plants than in the transplanted rice in flooded fields (Figure 8.1). These results suggested that rice root interiors provide more favorable growth conditions for Rhizobia when cultivated in close proximity to clover in the drained, more aerobic soils rather than in the monoculture within flooded soils. This highlighted a long-term benefit of the rice–legume rotation in promoting this intimate plant–microbe association.

Next, we used standard microbiological techniques to isolate into pure culture these Rhizobial nodule occupants representing the numerically dominant endophytes of rice roots. We verified that they were authentic Rhizobia by testing their ability to nodulate berseem clover in gnotobiotic culture and evaluated their N_2-fixing activities on their clover host. These symbiotic performance tests (plus 16S rDNA sequencing) confirmed that the rice-adapted isolates were authentic strains of R. leguminosarum bv. trifolii capable of nodulating berseem clover under gnotobiotic conditions, and that both effective and ineffective Rhizobial isolates were included in the culture collection (Yanni et al., 1997; Yanni et al., 2001). This was followed by tests of Koch's postulates which proved that pure cultures of selected isolates of rice-adapted Rhizobia can invade rice roots under gnotobiotic culture conditions and be isolated back into pure culture as the same authentic inoculant strains (Yanni et al., 1997).

The genomic diversity of the isolates was evaluated to gain a better understanding of the breadth of this ecological niche for Rhizobia and to guide us in selecting isolates that can represent the genomic diversity in various studies on this association. The BOX-PCR and plasmid profiling methods indicated that our culture collection of rice-adapted Rhizobia contained 10 different strain genotypes, representing sufficient variation to define their range in ability to evoke growth responses in cereals. It also indicated that diverse

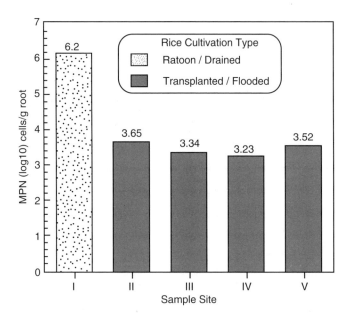

FIGURE 8.1
Most probable numbers of endophyte populations of *R. leguminosarum* bv. *trifolii* in rice roots cultivated in the Egyptian Nile delta. *Source:* From Yanni, Y.G. et al., *Aust. J. Plant Physiol.*, **28**, 845–870 (2001). With permission. Copyright and published by CSIRO Publishing, Melbourne, Australia, 2001.

populations of *R. leguminosarum* bv. *trifolii* colonize rice root interiors in drained and flooded soils used for rice–berseem clover rotation in the Nile delta (Yanni et al., 2001).

In more recent work, we have used a diversity of legumes (berseem clover, alfalfa, soybean, lentil, fava bean, bean) normally cultivated in rotation with wheat as trap hosts in an attempt to better reveal the species/biovar diversity of the numerically dominant Rhizobial endophytes of field-grown wheat in the Nile delta. The result of this study was quite interesting in that the clover symbiont, *R. leguminosarum* bv. *trifolii*, is the dominant Rhizobium forming endophytic associations with wheat roots in fields of the Nile delta, while none of the other Rhizobia within the other legume cross-inoculation groups appear capable of occupying this ecological niche in this same habitat (Yanni and Dazzo, unpublished data).

Since this discovery of a third ecological niche for Rhizobium (Figure 8.2), we have created an international network of collaborators to expand the intrinsic scientific merit of this project in both basic and applied directions of beneficial plant–microbe interactions. As an outcome, many tests of the generality of plant growth-promoting Rhizobia have indicated that this type of association is widespread worldwide rather than being restricted to the Nile delta. Other natural associations of endophytic plant growth-promoting Rhizobia within field-grown roots of wild rice, wheat, barley, sorghum, millet, maize, and rice rotated with legumes have now been described in Senegal, Canada, Mexico, Morocco, South Africa, Venezuela, India, China, and elsewhere (Antoun and Prévost, 2000; Chaintreuil et al., 2000; Matiru et al., 2000; Gutierrez-Zamora and Martinez-Romero, 2001; Hilali et al., 2001; Lupway et al., 2004; Mishra et al., 2004; Y. Jing, personal communication). Thus, despite some initial reservations about this novel finding within the scientific community, there is now no longer any empirical basis upon which to doubt the existence and potential benefits of this plant–microbe association.

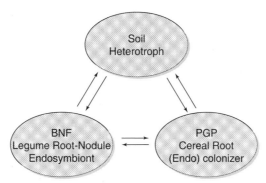

Rhizobium Life Cycle in Legume-Cereal Rotations

FIGURE 8.2

Widespread natural occurrence of three ecological niches for Rhizobium in legume-cereal rotations. *Source:* From Yanni, Y.G. et al., *Aust. J. Plant Physiol.*, **28**, 845–870 (2001). With permission. Copyright and published by CSIRO Publishing, Melbourne, Australia, 2001.

8.2 The Biology of the Rhizobium–Cereal Association

8.2.1 Colonization of Rice Plants by Endophytic Rhizobia

Microscopy has helped us define the infection process in the Rhizobium–cereal association. Early studies on the colonization of rice roots by azorhizobia pointed to the "crack entry" mode of primary host infection in natural wounds of split epidermal cells and fissure sites where lateral roots have emerged, followed by colonization within intercellular spaces and host cells of the outer root cortex (Reddy et al., 1997). Transmission electron microscopy with high magnification/resolution confirmed their endophytic existence within dead host cells adjacent to well-preserved intact cells of the root cortex (Reddy et al., 1997). Other studies using selected strains of our rice-adapted Rhizobia tagged with the green fluorescent protein (gfp) indicated that these bacteria were entering rice roots at lateral root emergence sites, multiplied intercellularly within lateral rootlets, and sometimes migrated from these sites of primary host infection to form long fluorescent rows of bacteria inside roots (Prayitno et al., 1999).

Similar results were reported for the invasion of wheat roots by *Azorhizobium caulinodans* when the cultures were supplemented with the flavonoid naringenin (Webster et al., 1997; Webster et al., 1998). Other recent microscopical studies of rice plants inoculated with gfp-tagged Rhizobia indicate that their infection process is dynamic, beginning with colonization at lateral root emergence, crack entry into the root interior through separated epidermal cells, followed by endophytic ascending migration up into the stem base, leaf sheath, and leaves where they grow transiently to high local population densities (Chi et al., 2005). Thus, this endophytic plant–microbe association is far more invasive than previously thought. Comparisons of viable plate counts vs. direct microscopy indicate that these gfp-tagged bacteria remain active for long periods within the rice plant even while their culturable population densities decline (Chi et al., 2005).

We are using quantitative microscopy to evaluate spatial patterns of Rhizobial distribution on rice roots to better understand their colonization behavior. This work involves scanning electron microscopy of rice roots inoculated with a candidate biofertilizer strain of *R. leguminosarum* bv. *trifolii* analyzed at single cell resolution using our CMEIAS© (Center for Microbial Ecology Image Analysis System) image analysis software developed for these studies in computer-assisted microscopy. We developed new measurement features in CMEIAS, e.g., the Cluster Index (Dazzo et al., 2003), to extract information from digital images of microbial cells on the root surface needed to compute plotless, plot-based, and geostatistical analyses that describe and model spatial patterns of their root surface colonization. This includes statistically defendable interpolation of their distribution and dispersion even within areas of the root that are not sampled (Dazzo and Wopereis, 2000; Liu et al., 2001; McDermott and Dazzo, 2002; Dazzo, 2004).

Typical scanning electron micrographs depicting various morphological features of the colonization of Sakha 102 rice roots by Rhizobial strain E11 used in these studies are illustrated in Figure 8.3. These images reveal that these bacteria: (1) attach in both supine

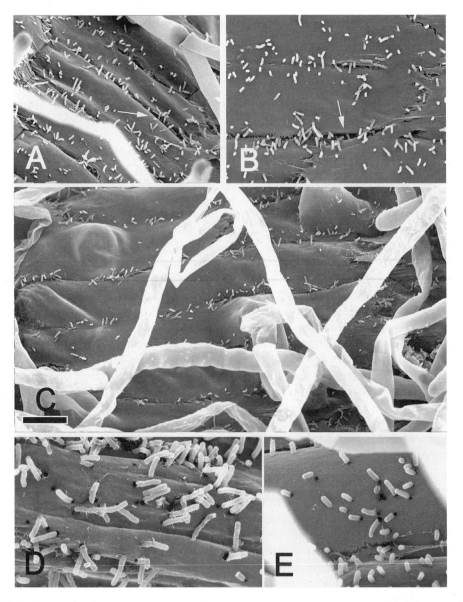

FIGURE 8.3

(a–e) Scanning electron micrographs of the rice epidermal root surface colonized by an endophyte strain of rice-adapted Rhizobia. *Source:* From Yanni, Y.G. et al., *Aust. J. Plant Physiol.*, **28**, 845–870 (2001). With permission. Copyright and published by CSIRO Publishing, Melbourne, Australia, 2001.

and polar orientations preferentially to the nonroot hair epidermis, in contrast to their preferential attachment in polar orientation to root hairs of their host legume; (2) commonly colonize small crevices at junctions between epidermal cells (white arrows in Figure 8.3 A–C), suggesting this route as an alternate portal of entry into the root; and (3) they produce eroded pits on the rice root epidermis (Figure 8.3 D,E). Similarly eroded plant structures are produced in the Rhizobium–white clover symbiosis by plant wall-degrading enzymes bound to the bacterial cell surface, and these pits represent incomplete attempts of bacterial penetration that had only progressed through isotropic, noncrystalline layers of the plant cell wall (Mateos et al., 2001). Consistent with these results, activity gel electrophoresis indicated that the rice-adapted Rhizobia produced a cell-bound CM-cellulase (Yanni et al., 2001). This enzyme likely participates in the invasion and dissemination of the Rhizobial endophyte within host roots.

This type of information derived from microscopy enhances our understanding of root colonization by inoculant strains, the dynamic aspects of Rhizobial dispersion on the root, and how different inoculant delivery systems will ultimately impact on successful application of biofertilizer inoculants.

8.2.2 Plant Growth-Promotion Effects from Rhizobial Endophyte Strains

Our early studies on endophytic colonization of rice by Rhizobia indicated that some strains promote the shoot and root growth of certain rice varieties in gnotobiotic culture (Yanni et al., 1997). Later, more extensive tests established the range of growth responses of japonica, indica, and hybrid rice varieties from Egypt, Philippines, USA, India, and Australia when these cultivars were inoculated with various rice-adapted Rhizobia. The results indicated that the diverse Rhizobial endophytes evoked a full spectrum of growth responses in rice (positive, neutral, and sometimes even negative), often exhibiting a high level of strain-variety specificity (Yanni et al., 1997; Prayitno et al., 1999; Biswas et al., 2000a, 2000b; Yanni et al., 2001). On the positive side, a chronology of PGP$^+$ responses of rice to Rhizobia manifested as increased seedling vigor (faster seed germination followed by increased root elongation, shoot height, leaf area, chlorophyll content, photosynthetic capacity, root length, branching, and biomass). This carries over into increased yield and nitrogen content of the straw and grain at maturity. Similar results were obtained when wheat was inoculated with diverse genotypes of endophytic wheat-adapted Rhizobia in gnotobiotic plant bioassays; there is high strain-variety specificity in Rhizobial promotion of wheat growth.

Numerous field inoculation trials have been conducted to assess the agronomic potential of these Rhizobium–cereal associations under field conditions, with the long-range goal of identifying, developing, and implementing superior biofertilizer inoculants that can promote rice and wheat productivity in real-world cropping systems while reducing their dependence on fertilizer-N inputs. A direct agronomic approach was adopted to address the importance of continued evaluation of the various strain genotypes in our diverse collections of cereal-adapted Rhizobia under experimental field conditions (Yanni et al., 1997; Yanni et al., 2001). This meant first acquiring useful information from laboratory plant growth-promotion (PGP) bioassays. Then this information was applied in designing and implementing small field trials at the Sakha Agricultural Research Station in the Kafr El-Sheikh area. Information from these results is then utilized to conduct scaled-up experiments on large farmers' fields in neighboring areas of the Nile delta where cereal–legume rotations are used, so our results can be compared to real on-farm baselines in grain production.

These experiments included N-fertilizer applications at three rates: 1/3, 2/3, and the full recommended rate, previously assessed without biofertilizer inoculation. The inocula for

TABLE 8.1

Grain Yields of Rice var. Giza 178 in the Best Experimental Treatments vs. Adjacent Farmers' Fields at Different Locations in Kafr El-Sheikh, Nile Delta, Egypt, 2002

Farm Location	Best Experimental Treatment: Inoculated Strains + kg of N Fertilizer[a]	Grain Yield of Best Experimental Treatment (kg ha^{-1})	Yield in Adjacent Field (No Researcher Supervision) (kg ha^{-1})[b]	Increase Over Farmer's Yield (%)
Baltem	E11 + E12 + 96 N	8623	8330	3.5
Beila	E11 + E12 + 96 N	11,309	9520	18.8
Metobas	E11 + E12 + 96 N	12,400	9520	30.3
Sidi Salem	144 N	11,118	9068	22.6

[a] The method of inoculation was direct broadcast of the inoculum (10^9 colony-forming units [CFU] g^{-1}) at the rate of 720 g peat-based inoculum ha^{-1} 3 days after transplanting of the rice seedlings and during a period of calm wind at sunset. N, kg N ha^{-1} was applied as urea (46% N) in two equal doses, 15 days post transplanting and at the mid-tillering stage.

[b] Recommended rate of fertilizer-N for the tested rice varieties when used without inoculation with biofertilizers is 144 kg N ha^{-1}. This rate was used by the farmer in the adjacent field who was not supervised by the research personnel.

Source: Authors' data.

rice and wheat were used as indicated in Table 8.1 and Table 8.2. The field trials were done in subplots each 20 m^2 with four replications. The various trials were supplemented with calcium superphosphate before tillage, with potassium sulfate added 1 month after wheat sowing or rice transplantation. Appropriate broad-spectrum herbicides were applied to control the major narrow and broad leaf weeds.

In total, we have conducted 24 different field inoculation trials of selected endophytic strains of Rhizobia with rice and wheat in the Nile delta. So far, positive increases in grain yield resulting from inoculation with selected strains of our cereal-adapted Rhizobia have occurred in 11 of 13 field tests for rice (85%), and 10 of 11 field tests for

TABLE 8.2

Grain Yields of Wheat in the Best Experimental Treatments vs. Grain Yields in Adjacent Farmers' Fields at Different Locations in Kafr El-Sheikh, Nile Delta, Egypt, 2002–2003

Farm Location	Wheat Variety	Best Experimental Treatment: Inoculated Strain + kg N Fertilizer	Yield of the Best Experimental Treatment (kg ha^{-1})	Yield in the Rest of the Same Farmer's Field (No Researcher Supervision) (kg ha^{-1})[a]	Increase Over Farmer's Yield (%)
1	Sakha-93	EW 54 + 180 N	7112	6120	16.2
2	Sakha-61	EW 72 + 120 N	7382	5712	29.2
3	Sakha-61	EW 72 + 180 N	8247	6936	18.9
4	Sakha-61	EW 72 + 60 N	5802	4896	18.5

EW, rhizobial wheat-root endocolonizer used at the rate of 720 g peat-based inoculum (10^9 colony forming units (CFU)/g) for inoculation of 144 kg wheat seeds for cultivation of one hectare. The method of inoculation involved mixing the seeds with the inoculum in the presence of a suitable quantity of a solution with an adhesive material like pure Arabic-gum, gelatin, or sucrose. The mixed seeds were left for some time in the shade and then planted as fast as possible during sunset followed by irrigation of the field area. N, kg N ha^{-1} was applied as urea (46% N) in two equal doses just before sowing and 75 days later.

[a] Recommended rate of fertilizer-N for the tested wheat varieties when used without inoculation with biofertilizers was 180 kg N ha^{-1}. This rate was used by the farmer in the part of the field that was not supervised by the research personnel.

Source: Authors' data.

wheat (91%) (Yanni et al., 1997; Yanni et al., 2001; Yanni and Dazzo, unpublished data). Table 8.1 and Table 8.2 summarize the data on grain yield from our recent scaled-up experiments on farmers' fields. They illustrate the best performance obtained with field inoculation treatments using our cereal-adapted strains on the scaled-up, researcher-supervised experimental plots vs. yields of the same variety obtained simultaneously by traditional agricultural practices on adjacent fields without inoculation or supervision by research personnel.

The results indicate the beneficial effects of inoculation with our cereal-adapted Rhizobia on grain yields of rice and wheat in three of four cases. Increases ranged between 3.5 and 30.3% and from 16.2 to 29.2% in grain yields of rice and wheat, respectively, using the researchers' package of agronomic treatments rather than farmers' practices. The best inoculation responses for rice occurred with an inoculum combining two strains of rice-adapted Rhizobia (rather than one). It was quite interesting to find that the wheat and rice varieties tested in most of these experiments displayed an increase in agronomic fertilizer-N use efficiency. This indicates that our Rhizobial strains can help these crops to utilize the nitrogen taken up more efficiently to produce grain with less dependence on nitrogen fertilizer inputs.

The results also suggest that even after the recommended amount of nitrogen has been provided, our microbial inoculants facilitate the acquisition of whatever nutrients then become the next limiting factor for rice and wheat productivity in these fields. These results reflect the potential value of Rhizobial biofertilizer inoculants for rice and wheat production in the Nile delta. This knowledge should improve agriculture by advancing basic scientific knowledge on beneficial plant–microbe associations, and also by assisting low-income farmers to increase cereal production on marginally fertile soils using bio-fertilizers consistent with environmental soundness and sustainable agriculture.

8.3 Extensions of Rhizobial Endophyte Effects

8.3.1 Use of Rhizobial Endophytes from Rice with Certain Maize Genotypes

It is of obvious interest to know whether superior Rhizobial endophyte strains that are PGP^+ with rice can also promote the growth of other cereal crops. Field tests conducted in Wisconsin, USA, have provided preliminary evidence for this, at least for certain genotypes of maize (Yanni et al., 2001). In that study, inoculation with one of our rice-endophyte strains resulted in statistically significant increases in dry weight for three of six tested maize genotypes in the greenhouse, and one of seven maize genotypes in experimental field plots receiving no nitrogen fertilizer.

Interestingly, a cross between the high-responding genotype and a different non-responding genotype resulted in a hybrid maize genotype with an intermediate level of growth responsiveness to inoculation with the same Rhizobial strain (Figure 8.4). This suggests the possibility of genetic transmissibility and inheritance in maize of the ability to respond to selected endophytic Rhizobia. This result reinforces the earlier finding that induction of positive growth responses in cereals by Rhizobia is genotype-specific (also in maize). Such experimental results provide leads that may help to identify the genes in cereals necessary for expression of these growth responses to Rhizobia. Also, since recent work indicates that one of our rice endophyte strains of Rhizobia can promote the growth of shoots and roots, root architecture, and uptake of N and Ca^{++} for certain cotton varieties under growth-room conditions (Hafeez et al., 2004), a thorough screen of their potential benefit to a wide variety of nonlegume crop plants should be made.

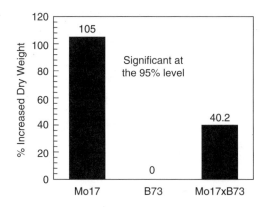

FIGURE 8.4

Genotype-specific inheritance and growth response of maize (*Zea mays*) to inoculation with a rice-endophyte strain of Rhizobia under field conditions in Wisconsin, USA, without fertilizer N application. *Source:* From Yanni, Y.G. et al., *Aust. J. Plant Physiol.*, **28**, 845–870 (2001). With permission. Copyright and published by CSIRO Publishing, Melbourne, Australia, 2001.

8.3.2   Extension of Rhizobium–Rice Associations to Other Rice Varieties

For practical reasons, it is important to examine whether the various desirable interactions of these Rhizobial endophytes can be extended to cereal varieties that are preferred by farmers in cropping systems throughout the world. The number of varieties we have tested so far is too small to make reliable and accurate armchair predictions about genotypes not yet tested. Because many characteristics of this association exhibit high strain/variety specificity, tests of their compatibility at the laboratory bench are necessary before they are tested in the field. Our studies so far have included rice genotypes commonly used in Egypt (Sakha-101, 102, 104, Giza-175, 177, 178, and Jasmine rice), USA (M202 and L204), Australia (Calrose and Pelde), and India (Pankaj).

However, what about rice varieties preferred by low-income farmers who cultivate rice on marginally fertile soils and who cannot afford to purchase fertilizers? To address this question, a study was undertaken to measure how well the Rhizobial-endophyte strain E11 can colonize the root environment of four rice genotypes preferred by Filipino peasant farmers because of their good yielding ability and grain characteristics (Sinandomeng, PSBRC 74, PSBRC 58, and PSBRC 18). Cells were inoculated on axenic seedling roots, then grown gnotobiotically in hydroponic tube culture, and the resultant populations were enumerated by viable plate counts. For comparison, seedlings of equal size received an equivalent inoculum of a local, unidentified isolate BSS 202 from *Saccharum spontaneum* used as a PGP$^+$ biofertilizer inoculant for rice in the Philippines.

The results indicated significant colonization potential of strain E11 on roots of all four rice genotypes (Yanni et al., 2001). In some cases, the populations achieved were even higher than the BSS 202 isolate (Table 8.3). The implications of this experiment are significant: strain E11 exhibits no obvious difficulty in its potential to intimately colonize roots of not only the superior rice varieties that have undergone significant breeding development but require high nitrogen inputs for maximum yield, but also with other rice varieties that perform acceptably on marginally fertile soil without significant N-fertilizer inputs. The latter type of rice cropping could derive significant benefit from the biofertilizer inoculants that our research is intended to develop.

TABLE 8.3

Colonization Potential of *R. leguminosarum* bv. *trifolii* Strain E11 and Isolate BSS 202 from *Saccharum spontaneum* on Four Rice Varieties Preferred by Low-Income Filipino Farmers

Inoculum Strain	Sample Location	Colonization Potential (Viable Plate Count/3 Rice Roots)			
		Sinandomeng	PSBRC 74	PSBRC 58	PSBRC 18
Rlt E11	External rooting medium	1.03×10^7	1.50×10^7	7.16×10^7	8.17×10^7
BSS 202	External rooting medium	9.17×10^7	7.00×10^6	8.00×10^7	1.27×10^7
Rlt E11	Root surface	6.34×10^7	1.05×10^8	6.50×10^7	6.50×10^7
BSS 202	Root surface	1.08×10^7	6.00×10^7	5.60×10^7	5.66×10^7
Rlt E11	Root interior	1.50×10^8	2.19×10^9	1.32×10^8	1.03×10^9
BSS 202	Root interior	7.50×10^7	6.30×10^6	7.34×10^7	1.23×10^8

Source: From Yanni, Y.G. et al., *Aust. J. Plant Physiol.*, **28**, 845–870 (2001). With permission.

8.4 Underlying Mechanisms of Plant Growth Promotion

The ability of some endophytic Rhizobial strains to promote the growth of rice and wheat prompted follow-up studies to identify possible mechanisms operative in this beneficial plant–microbe interaction. These studies have focused primarily on rice, addressing the following possibilities: (1) Rhizobial induction of an expansive root architecture having enhanced efficiency in plant mineral nutrient uptake; (2) Rhizobial production of extracellular growth-regulating phytohormones; (3) Rhizobial solubilization of precipitated inorganic and organic phosphate complexes, thereby increasing the bioavailability of this important plant nutrient; (4) associative nitrogen fixation by Rhizobia; and (5) Rhizobial production of Fe-chelating siderophores.

Although biological control of phytopathogens (via direct antagonism or induction of systemic disease resistance) is traditionally recognized as another important mechanism of plant-growth promotion by rhizobacteria (Kloepper, Schroth, Miller, 1980; Haque and Ghaffar, 1993), we have not tested whether this mechanism is operative in our system because our Rhizobial inoculants can promote rice growth in gnotobiotic culture (independent of phytopathology) and because the (already blast-resistant) uninoculated control plants lack any disease symptoms in our field inoculation trials (Yanni and Dazzo, unpublished observations).

8.4.1 Stimulation of Root Growth and Nutrient-Uptake Efficiency

Responsive rice varieties commonly develop expanded root architectures when inoculated with candidate biofertilizer strains of Rhizobia. This suggests that these Rhizobial endophytes alter root development in ways that could make them better "miners," more capable of exploiting a larger reservoir of plant nutrients from existing resources in the soil. This possibility was suggested in early studies showing significantly increased production of root biomass in plants that had been inoculated (Yanni et al., 1997; Prayitno et al., 1999; Biswas et al., 2000a; Yanni et al., 2001) and by studies using greenhouse potted soil indicating significant increases in N, P, K, and Fe uptake by rice plants inoculated with selected Rhizobia, including our test strains (Biswas et al., 2000b). More recent studies have confirmed this hypothesis using plants grown gnotobiotically with Rhizobia in nutrient-limited medium (50% Hoaglands), followed by measurements of root architecture and mineral nutrient composition using CMEIAS image analysis and atomic absorption spectrophotometry, respectively (Yanni et al., 2001). In these latter studies, inoculated

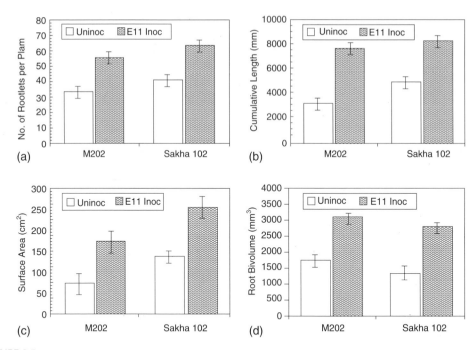

FIGURE 8.5
(a–d) Effect of inoculation with *R. leguminosarum* bv. *trifolii* E11 on root architecture of rice varieties. *Source:* From Yanni, Y.G. et al., *Aust. J. Plant Physiol.*, **28**, 845–870 (2001). With permission. Copyright and published by CSIRO Publishing, Melbourne, Australia, 2001.

plants developed a more expanded root architecture, as seen in Figure 8.5a–d, and accumulated higher concentrations of N, P, K, Ca, Mg, Na, Zn, and Mo than did their uninoculated counterparts (Figure 8.6).

We note also, however, that the levels of Fe, Cu, B, and Mn were not statistically different in the inoculated and uninoculated plants under these microbiologically controlled experimental conditions (Yanni et al., 2001). This selectivity in terms of which plant nutrients exhibit increased accumulation as a result of inoculation indicates that there is no across-the-board, general enhancement of mineral accumulation due just to an expanded

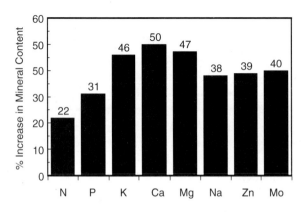

FIGURE 8.6
Effect of inoculation with *R. leguminosarum* bv. *trifolii* E11 on elemental composition of rice plants. *Source:* From Yanni, Y.G. et al., *Aust. J. Plant Physiol.*, **28**, 845–870 (2001). With permission. Copyright and published by CSIRO Publishing, Melbourne, Australia, 2001.

root architecture with increased absorptive biosurface area. The results indicate, quite interestingly, that bacteria can modulate the rice plant's plasticity that enables it to control the adaptability of its root architecture and also physiological processes for more efficient acquisition of selected nutrient resources when they become limiting. This same mechanism is considered one of the major reasons for the beneficial growth-promotion effect on grasses by *Azospirillum brasilense* (Tien et al., 1979; Umali-Garcia et al., 1980; Okon and Kapulnik, 1986; Bashan et al., 1990; Okon and Labandera-Gonzalez, 1994).

These Rhizobium-induced increases in the mineral composition of rice plants raise new possibilities regarding their potential impact on the human nutritional value of this crop. A potential value-added benefit resulting from inoculation could be to increase the nutritional value of resulting grain, not only for increased nitrogen (mostly in the form of protein), but other macro- and micronutrients as well. For instance, rice is an important, indeed even the major bioavailable source of some minerals, e.g., zinc, in the human diet, particularly in developing countries (IRRI, 1999). Zinc is considered an essential micronutrient that is important for maturation of the reproductive organs in women and the developing fetus. Our inoculants have shown a capacity to increase the zinc content of rice grain (Figure 8.6).

SDS-PAGE and RP-HPLC analyses of the protein composition in field-grown Giza 177 rice grains indicated no discernable differences in the ratios of the major nutritionally important storage proteins, particularly glutelin, albumin, and globulin, as a result of inoculation with Rhizobial endophyte strain E11. Field inoculation with a Rhizobial endocolonizer thus did not qualitatively alter rice-grain protein, as all nutritionally important proteins were present in the treated and control samples in similar ratios. However, since inoculation with Rhizobia causes a significant increase in total grain nitrogen per hectare of crop (in protein form), the benefits of inoculation to small farmers will include an increase in the quantity of rice grain protein produced per unit of land used for cultivation. This increases the nutritional value of the harvested grain as a whole in comparison with uninoculated rice produced.

Enhanced uptake of nutrient resources by inoculated rice could be a two-edged sword if accompanied by enhanced bioaccumulation of toxic metals. We therefore analyzed other rice grain samples from this same field inoculation experiment for their heavy metal content (Hg, Se, Pb, Al, and Ag). The results indicated no significant differences in the low levels of these toxic heavy metals in the rice grain of uninoculated vs. Rhizobial endophyte-inoculated treatments. Considered collectively, these studies indicate that rice plants inoculated with our Rhizobial biofertilizer produce rice grain whose human nutritional value is equal to or improved (depending on the nutrient considered) as compared to uninoculated plants.

8.4.2 Secretion of Plant Growth Regulators

Early studies suggested that Rhizobial endophyte strain E11 produced the auxin indoleacetic acid (IAA) in pure culture and in gnotobiotic culture with rice (Biswas et al., 2000a; Biswas et al., 2000b). Further studies indicated that production of IAA-equivalents by strain E11 was tryptophan-dependent. We developed a simple defined medium to optimize production of IAA by fast-growing Rhizobia, and axenic bioassays of filter-sterilized culture supernatant from strain E11 grown in this defined medium showed an ability to stimulate rice root growth (Yanni et al., 2001). These results suggest that endophytic strains of Rhizobia can boost rice growth by producing extracellular bioactive metabolites that promote the development of a more expansive root architecture. The contributions of phytohormones are considered in more detail in Chapter 14.

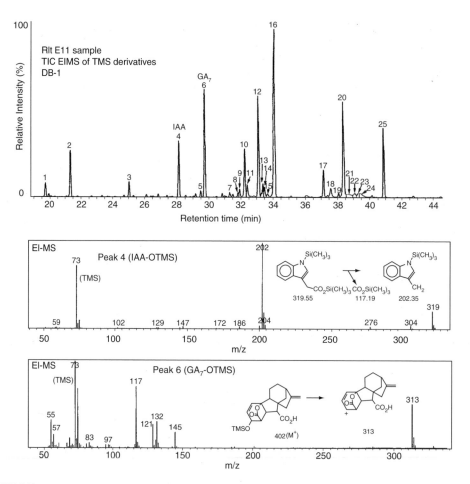

FIGURE 8.7
GC/MS fractionation and identification of phytohormones in the culture supernatant of rice-adapted Rhizobial endophyte strain E11. *Source:* From Yanni, Y.G. et al., *Aust. J. Plant Physiol.*, **28**, 845–870 (2001). With permission. Copyright and published by CSIRO Publishing, Melbourne Australia, 2001.

These results logically led us to identify the growth-regulating phytohormones produced and secreted by strain E11 in pure culture. Analysis of its culture supernatant using electrospray ionization gas chromatography/mass spectrometry (GC/MS) indicated the presence of IAA and a gibberellin (consistent with GA_7) (Figure 8.7) (Yanni et al., 2001). These represent two different major classes of plant growth regulators that play key roles in plant development. This fundamentally new information has increased our understanding of the mechanisms underlying plant growth-promotion in this beneficial Rhizobium–rice association.

8.4.3 Solubilization of Precipitated Phosphate Complexes by Rhizobial Endophytes

Most Nile delta soils used for rice cultivation contain about 1000 Ppm phosphorus, primarily in the unavailable form of precipitated tri-calcium phosphate, $Ca_3(PO_4)_2$. Although waterlogged conditions normally prevail in rice fields, less than 8 Ppm P (Olsen P) is available to rice. Any significant solubilization of precipitated phosphates by rhizobacteria *in situ* would enhance phosphate availability to rice in these soils, representing another possible mechanism of PGP for rice under these field conditions (see Chapter 13). We tested the diversity of rice-adapted Rhizobia for phosphate-solubilizing activity on culture media impregnated with organic and inorganic phosphate complexes. Our improved, double-layer plate assay indicated that some strains are active in solubilizing both inorganic (calcium phosphate) and organic (inositol hexaphosphate = phytate) insoluble P complexes (Yanni et al., 2001). These positive results indicate extracellular acidification and phosphatase enzymes (phytase), respectively. This extracellular PGP[+] activity would potentially increase the availability of phosphorus for rice in rhizosphere soil, and thereby promote rice growth when soil phosphorus is limiting.

8.4.4 Associative Nitrogen Fixation

Rice plants accumulate more shoot-N and grain-N when inoculated with selected strains of Rhizobial endophytes (Yanni et al., 1997; Biswas et al., 2000b; Yanni et al., 2001). However, this additional combined nitrogen is mainly derived from soil mineral nitrogen, not from biological nitrogen fixation (BNF). This conclusion is based on several lines of evidence:

1. Growth benefits by Rhizobia are enhanced rather than suppressed when fertilizer-N is provided (Yanni et al., 1997; Prayitno et al., 1999; Biswas et al., 2000a; Biswas et al., 2000b). Studies have usually shown an inverse relationship between fertilizer-N supply and BNF (Chapter 12).

2. The degree of growth benefit linked to inoculation with Rhizobial endophytes does not correlate with their degree of nitrogen-fixing activity in symbiosis with their normal nodulated legume host (in our case, berseem clover), since some rice-adapted strains of Rhizobia that are Fix-minus on clover are nevertheless PGP[+] on rice.

3. Acetylene reduction tests on rice plants whose growth is promoted by Rhizobial endophytes indicate no associative nitrogenase activity (Yanni et al., 1997; Biswas et al., 2000a).

4. Greenhouse studies using the ^{15}N-based isotope dilution method indicate that the increased nitrogen uptake in inoculated plants is not derived from BNF (Biswas et al., 2000b).

5. Measurements of the natural abundance of nitrogen isotope ratios ($\delta^{15}N$) on field-grown plants indicate that their greater proportion of nitrogen resulting from inoculation with strain E11 is not derived from BNF (Yanni et al., 2001).

Considered collectively, these results indicate that biological nitrogen fixation is not responsible for the positive growth response of rice to inoculation with these rice-adapted Rhizobia.

8.4.5 Production of Fe-Chelating Siderophores

Siderophore production potentially provides a dual mechanism of PGP: enhancing uptake of Fe for the plant, and suppressing rhizosphere pathogens unable to utilize the Fe–siderophore complex. However, none of the genotypes of Rhizobial endophyte strains produced detectable siderophores (Yanni et al., 2001), so this mechanism seems not to be applicable here.

8.5 Discussion

Studies completed so far indicate that we have superior candidate strains of Rhizobial endophytes suitable for use as biofertilizers for rice under field conditions. Information on the spatial distribution of candidate strains is both useful and necessary in order to fully exploit their benefits for sustainable agriculture. Our rationale is that a thorough understanding of their natural spatial distribution within rice agroecosystems should assist our biofertilization strategy program by helping to predict and interpret results of tests to evaluate their efficacy as inoculants.

The cumulative information derived from the studies described here indicates that Rhizobia have evolved an additional ecological niche that enables them to maintain a three-component life cycle that includes a free-living heterotrophic phase in soil, a nitrogen-fixing endosymbiont phase within the root nodules of legumes, and a beneficial growth-promoting endocolonizer phase within cereal roots in the same crop rotation (Figure 8.2). The results further indicate an opportunity to exploit this newly described plant–Rhizobia association by developing biofertilizer inoculants that have potential to increase cereal production, including rice and wheat, with less fertilizer-N inputs, which would be supportive of both sustainable agriculture and environmental safety.

The situation exemplified by the Egyptian Nile delta indicates that inoculation of the cereal crop with the appropriate cereal-endophyte strain(s) of Rhizobia would follow rather than replace a crop rotation with the legume. This way the cereal crop would gain maximum benefit from its biological association with Rhizobia as a symbiotic N_2-fixer in the legume and then as an efficient plant growth-promoting rhizobacterium (PGPR) colonizing within the cereal root interior. Some Rhizobial endophytes can benefit multiple cereal crop species.

Assuming that the past results summarized here accurately reflect the potential benefits of this new agricultural biotechnology based on exploitation of a natural resource (natural Rhizobial endophytes of cereals), the following outcomes can be expected:

1. Increased cereal crop yield above what is reached using inorganic fertilizers alone, with a reduction of up to one-third of the fertilizer input previously recommended and used when there was no biofertilizer inoculation. Economically, our field management packages including this new biofertilization technology can assist farmers to increase their production by 3.5–30.3% for rice and 16.2–29.2% for wheat, with a saving of one-third of their fertilization costs (Table 8.1 and Table 8.2).

2. Decreased environmental pollution and health risks originating from excessive use of inorganic N-fertilizers. However, this needs further study to verify the reduction in costs involved in dealing with diseases associated with the excessive use of agrochemicals and the economical benefits from increasing individuals' work abilities.

3. Decreased energy needed for production, transportation, and distribution of fertilizers, directing this energy to other socioeconomic and industrial uses.

4. A better understanding of how farmers can practice sustainable agriculture by utilizing biofertilizers as a safe and effective alternative to inorganic fertilization.

5. Promotion of cooperation between governmental research institutions, on one side, and private sectors represented by farmers and agricultural biotechnology industries, on the other.

Farmers who hosted our experiments in their fields and were formerly suspicious about the validity of the research practices are now very concerned not only to inoculate their fields with our preparations, but also ask advice about the use of other microbial preparations. Other farmers in the experimentation areas who learned directly or indirectly about our results have become curious about this innovation. However, a well-designed extension program is still necessary to meet the needs of potential users of this technology.

The work reported here complements other discussions in this book on biological nitrogen fixation, P solubilization, mycorrhizal fungi, phytohormones, and related subjects. These all indicate that certain bacteria significantly affect plant growth and development in ways that can be utilized to enhance crop production in sustainable agriculture. One of the many lessons we have learned from our studies on the Rhizobium–cereal association was stated well by Leonardo da Vinci: "Look first to Nature for the best design before invention." This is particularly relevant for the design of strategies for sustainable agriculture, as discussed in the chapters by Primavesi, Séguy, and others, where insights are sought about the processes and potentials of ecosystem dynamics as a starting point for basic and applied research. They then capitalize upon such understanding in efforts to enhance agricultural production in ways that are consistent with complex, evolved biological relationships ascending from the foundation of microbial populations.

Acknowledgments

Parts of the work described here were supported by Projects BIO2-001-017-98-CA 27 and BIO5-001-015-CA 115 from the US-Egypt Science and Technology Joint Fund and by the Long-Term Ecological Research Program at Michigan State University's Kellogg Biological Station. We thank staff members of the Agricultural Research Center (Giza, Egypt), the Soils, Water and Environment Research Institute (Giza, Egypt), and the Sakha Agricultural Research Station (Kafr El-Sheikh, Egypt) for their assistance, and acknowledge the assistance of the Ribosomal Database Project-II and Center for Advanced Microscopy at Michigan State University (East Lansing, MI, USA).

References

Antoun, H. and Prévost, D., PGPR activity of Rhizobium with nonleguminous plants, In: *Proceedings of the 5th International Plant Growth-Promoting Rhizobacteria Workshop*, Villa Carlos Paz, Cordoba Argentina 29 Oct.–3 Nov. 2000, http://www.ag.auburn.edu/argentina/pdfmanuscripts/tableofcontents.pdf.

Bashan, Y., Harrison, S.K., and Whitmoyer, R.E., Enhanced growth of wheat and soybean plants inoculated with *Azospirillum brasilense* is not necessarily due to general enhancement of mineral uptake, *Appl. Environ. Microbiol.*, **56**, 769–775 (1990).

Biswas, J., Ladha, J.K., and Dazzo, F.B., Rhizobia inoculation improves nutrient uptake and growth in lowland rice, *Soil Sci. Soc. Am. J.*, **64**, 1644–1650 (2000a).

Biswas, J.C. et al., Rhizobial inoculation influences seedling vigor and yield of rice, *Agron. J.*, **92**, 880–886 (2000b).

Chaintreuil, C. et al., Photosynthetic Bradyrhizobia are natural endophytes of the African wild rice *Oryza breviligulata*, *Appl. Environ. Microbiol.*, **66**, 5437–5447 (2000).

Chi, F., Shen, S.H., Cheng, H.P., Jing, Y., Yanni, Y.G. and Dazzo, F.B., Ascending migration of endophytic Rhizobia from roots to leaves inside rice plants and assessment of their benefits to the growth physiology of rice. *Appl. Envir. Microbiol.* **71**, 7271–7278 (2005).

Dazzo, F.B., Applications of quantitative microscopy in studies of plant surface microbiology, In: *Plant Surface Microbiology*, Varma, A. et al., Eds., Springer, Germany, 503–550 (2004).

Dazzo, F. and Wopereis, J., Unraveling the infection process in the Rhizobium–legume symbiosis by microscopy, In: *Prokaryotic Nitrogen Fixation: A Model System for the Analysis of a Biological Process*, Triplett, E., Ed., Horizon Scientific Press, Norwich, UK, 295–347 (2000).

Dazzo, F.B. et al., Quantitative indices for the autecological biogeography of a Rhizobium endophyte of rice at macro and micro spatial scales, *Symbiosis*, **35**, 147–158 (2003).

Gutierrez-Zamora, M. and Martinez-Romero, E., Natural endophytic association between *Rhizobium etli* and maize (*Zea mays*), *J. Biotechnol.*, **91**, 117–126 (2001).

Hafeez, F.Y. et al., Rhizobial inoculation improves seedling emergence, nutrient uptake and growth of cotton, *Aust. J. Exp. Agric.*, **44**, 617–622 (2004).

Haque, S.E. and Ghaffar, A., Use of Rhizobia in the control of root rot diseases of sunflower, okra, soybean and mungbean, *J. Phytopathol.*, **138**, 157–193 (1993).

Hilali, A. et al., Effets de l'inoculation avec des souches de *Rhizobium leguminosarum* biovar *trifolii* sur la croissance der ble dans deux sols der Maroc, *Can. J. Microbiol.*, **41**, 590–593 (2001).

IRRI, *More Nutrition for Women and Children, Rice: Hunger or Hope?*. IRRI, Manila, 22–25 (1999).

Kloepper, J.W., Schroth, M.N., and Miller, T.D., Effects of rhizosphere colonization by plant growth-promoting rhizobacteria on potato development and yield, *Phytopathology*, **70**, 1078–1082 (1980).

Liu, J. et al., CMEIAS©: A computer-aided system for the image analysis of bacterial morphotypes in microbial communities, *Microb. Ecol.*, **41**, 173–194 (2001).

Lupway, N. et al., Endophytic Rhizobia in barley, wheat, and canola roots, *Can. J. Plant Sci.*, **84**, 37–45 (2004).

Mateos, P. et al., Erosion of root epidermal cell walls by Rhizobium polysaccharide-degrading enzymes as related to primary host infection in the Rhizobium–legume symbiosis, *Can. J. Microbiol.*, **47**, 475–487 (2001).

Matiru, V., Jaffer, M.A., and Dakora, F.D., Rhizobial colonization of roots of African landraces of sorghum and millet and the effects on sorghum growth and P nutrition. In: *Proceedings of the 9th Congress of the African Association for Biological Nitrogen Fixation: Challenges and Imperatives for BNF Research and Application in Africa for the 21st Century*, 25–29 September, Nairobi, Kenya, 99–100 (2000).

McDermott, T. and Dazzo, F.B., Use of fluorescent antibodies for studying the ecology of soil- and plant-associated microbes, In: *Manual of Environmental Microbiology*, Hurst, C., Eds., American Society for Microbiology Press, Washington, DC, 615–626 (2002).

Mishra, R.N., Singh, R., and Jaiswal, J., The beneficial plant growth-promoting association of endophytic *Rhizobium leguminosarum* bv. *phaseoli* with rice plants. Sixth European N_2-Fixation Conference, Toulouse, France (2004).

Okon, Y. and Kapulnik, Y., Development and function of Azospirillum-inoculated roots, *Plant Soil*, **90**, 3–16 (1986).

Okon, Y. and Labandera-Gonzalez, C.A., Agronomic applications of Azospirillum: An evaluation of 20 years worldwide field inoculation, *Soil Biol. Biochem.*, **26**, 1591–1601 (1994).

Prayitno, J. et al., Interactions of rice seedlings with nitrogen-fixing bacterial isolated from rice roots, *Aust. J. Plant Physiol.*, **26**, 521–535 (1999).

Reddy, P.M. et al., Rhizobial communication with rice: Induction of phenotypic changes, mode of invasion, and extent of colonization in roots, *Plant Soil*, **194**, 81–99 (1997).

Schloter, M. et al., Root colonization of different plants by plant growth-promoting *Rhizobium leguminosarum* bv. *trifolii* R39 studied with monospecific polyclonal antisera, *Appl. Environ. Microbiol.*, **63**, 2038–2046 (1997).

Tien, T., Gaskins, M.H., and Hubbell, D.H., Plant growth substances produced by *Azospirillum brasilense* and their effect on the growth of pearl millet (*Pennisetum americanum* L.), *Appl. Environ. Microbiol.*, **37**, 1016–1024 (1979).

Umali-Garcia, M. et al., Association of Azospirillum with grass roots, *Appl. Environ. Microbiol.*, **39**, 219–226 (1980).

Webster, G. et al., Interactions of Rhizobia with rice and wheat, *Plant Soil*, **194**, 115–122 (1997).

Webster, G. et al., The flavonoid naringenin stimulates the intercellular colonization of wheat roots by *AzoRhizobium caulinodans*, *Plant Cell Environ.*, **21**, 373–383 (1998).

Yanni, Y.G. et al., The beneficial plant growth promoting association of *Rhizobium leguminosarum* bv. *trifolii* with rice roots, *Aust. J. Plant Physiol.*, **28**, 845–870 (2001).

Yanni, Y.G. et al., Natural endophytic association between *Rhizobium leguminosarum* bv. *trifolii* and rice roots and assessment of its potential to promote rice growth, *Plant Soil*, **194**, 99–114 (1997).

9

The Roles of Arbuscular Mycorrihizas in Plant and Soil Health

Mitiku Habte

Department of Tropical Plant and Soil Science, University of Hawaii, Manoa, Hawaii, USA

CONTENTS

Plant species whether in natural ecosystems or in agroecosystems function under the influence of other organisms, particularly soil microorganisms. The roots of plants are inhabited by myriad soil microorganisms that live in parasitic, pathogenic, saprophytic, and/or mutualistic relationships with them. Among the mutualistic associations that plants form with soil microorganisms, giving benefit to both species, the most widespread geographically and botanically are mycorrhizas.

The term "mycorrhiza" was coined by the distinguished forest pathologist Frank in 1885, combining the Greek word *mykēs*, which means fungus, with the Latin word *rhiza*, which refers to roots. This compound word, "fungus–root," stands for the mutualistic interaction that occurs between most plant species and groups of soil fungi that inhabit plant roots, collectively known as mycorrhizal fungi. Note that this refers to the association rather than to either member of the association, even though the term mycorrhiza is commonly used to refer to the fungi themselves.

The seven types of mycorrhizas currently recognized differ from each other in the morphology of the host-endophyte associations that they form, in the range of habitats that they occupy, and in the types of host species involved (Smith and Read, 1997). Despite certain dissimilarities, they are quite similar in the ways that they influence plant life (Smith and Read, 1997). The focus here is on arbuscular mycorrhizas (AM), the most widespread type and one that influences over 80% of known plant species (Giovannetti and Sbrana, 1998; Tawaraya, 2003). These include the vast majority of agronomic, horticultural, silvopastoral, and tropical forest and agroforestry species. The remaining 20% of plant species either do not form mycorrhizas, or they host other types of mycorrhizas, such as ectomycorrhizal fungi, e.g., which are very important in associations with trees. The latter would warrant more attention if this volume dealt more extensively with forestry. This chapter summarizes what is known about the roles that such associations play in plant and soil health, and the key variables that influence practical

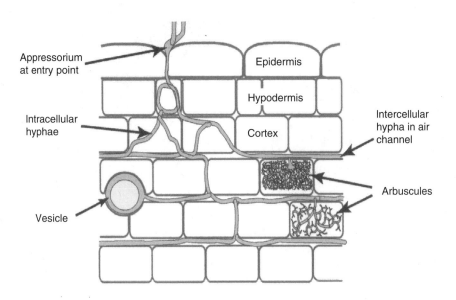

FIGURE 9.1
Diagram of a longitudinal section of a root showing the characteristic structures of arbuscular mycorrhizal fungi. External hyphae exiting the root at top are not shown. Used with permission from Mark Brundrett, INVAM, the International Culture Collection of Arbuscular and Vesicular Mycorrhizal Fungi, West Virginia University, http://mycorrhiza.ag.utk.edu/mimag.htm (accessed January, 2006).

applications of mycorrhizal fungi. Chapter 33 in Part III considers the development and results of such applications and presents their effects on soil system performance. The close association between fungus and root cells is shown in Figure 9.1, indicating the different components.

9.1 The Nature of Arbuscular Mycorrhizas

The fungi that form arbuscular mycorrhiza belong to the order Glomales in the phylum Glomeromycota, which consists of five families and seven genera. As obligate symbionts they are unable to grow and survive in the absence of host plants whose roots they can inhabit. The seven stages in the life cycle of these fungi can be summarized as follows. It parallels the "infection" of leguminous plant roots by rhizobial bacteria described in Chapter 12 (Section 12.1.l).

9.1.1 Spore Germination

The start of the process of forming mycorrhizal associations is not host-dependent (Douds and Schenck, 1991), although in some cases the presence of host plants and/or their exudates stimulates the process (Becard and Piche, 1989). Exudates from nonhost plants or from hosts of other mycorrhizal types do not appear to stimulate spore germination (Giovannetti and Sbrana, 1998). However, there are instances in which the roots of nonhost species or their exudates have inhibited spore germination (Vierheilig and Ocampo, 1990).

9.1.2 Post-Germination Hyphal Growth and Branching

This process occurs in response to volatile and hydrophobic constituents of root exudates (Nagahashi, 2000). This initial growth is a prerequisite for establishing contact with roots. The growth and branching of hyphae, threadlike filaments that extend from the main body of the fungal cell, facilitate the probability of AM fungi making contact with host root surfaces.

9.1.3 Attachment to Root Surfaces

Making a physical connection is facilitated by a specialized structure known as an appresorium (shown in Figure 9.1). The formation of this structure is not host-dependent, although the development of the penetrating hyphae emerging from it may be host-directed (Nagahashi, 2000).

9.1.4 Penetration of Epidermal and Hypodermal Cell Walls

The mechanisms by which fungal hyphae penetrate the surface of roots and their cortical cell walls are not clear, but they appear to involve a combination of hydrolytic enzymes and mechanical forces (Garcia-Garrido et al., 2002).

9.1.5 Spread of Intraradical Hyphae and Penetration of Cortical Cells

After penetrating the root epidermal cells, hyphae develop intraradically, i.e., within the root, and some of these hyphae penetrate the cortical cell wall, though not the plant cell membrane.

9.1.6 Formation of Arbuscules

Next is the essential process whereby structures are formed, known variously as arbuscules, coils or vesicles, within the root. These are essential for AM symbiosis to proceed and constitute its primary diagnostic feature. Arbuscules are formed when the fungal hyphae that have penetrated cortical cells divide dichotomously, producing finely divided, tree-shaped clusters of hyphae with a relatively large surface area. The host plant's plasma membrane becomes modified upon contact with the fungus, and it surrounds the arbuscule with a plant-derived membrane. This portion of the host membrane is analogous to the membrane that envelops rhizobia in the root nodules of leguminous plants.

The arbuscule is the site for nutrient exchange between AM fungi and their associated host plants (Saito, 2000). Vesicles are membrane-bound organelles, round to oval in shape that serve as storage structures, but also serving as reproductive structures when they are older. Because vesicles are absent in two of the seven genera containing these fungi, the term currently preferred to represent this fungus–plant association is arbuscular mycorrhiza, rather than an earlier term still found in the literature, vesicular-arbuscular mycorrhiza (VAM).

9.1.7 The Growth of Extraradical Hyphae and Formation of Spores

The final stage in the AM fungal life cycle is the formation of external hyphae that reach outward from the root, connecting the plant root to the adjacent soil. These hyphae are critical for the transfer of nutrients and water from the soil to plant roots. When spores subsequently develop on these extraradical hyphae, this completes the life cycle of the fungi. Spores are important structures for survival and reproduction, and their physical characteristics are used to establish the taxonomic identity of different AM fungi.

9.2 Mycorrhizal Roles in Plant Nutrition

9.2.1 Nutrient Acquisition

Numerous investigations over the years have shown that AM fungi can absorb nutrients from the soil, including N, P, K, Ca, Mg, S, Fe, Mn, Cu, and Zn, and translocate them to the roots of the associated plants (Auge, 2001). Those nutrients that are more mobile, notably N, Ca, and Mg, are transported to the root surface predominantly by mass flow, making it unlikely that AM fungi play much of a role in their uptake under normal conditions. Mycorrhizal associations become more important for uptake of more mobile nutrients when they are in relatively short supply.

AM fungi play a critical role in the uptake of the relatively immobile forms of nitrogen, notably NH_4^+, both in dry soils and at low root-length densities (George, 2000). Generally speaking, their most prominent and consistent nutritional effect is to improve the uptake of nutrients that are scarce, diffusion-limited, immobile, particularly P, Cu, and Zn. Jones et al. (1998) report that the efficiency with which mycorrhizal plants take up phosphorus, compared with nonmycorrhizal plants, is 3.1 to 4.7 times higher.

One reason for the success of AM fungi in nutrient acquisition is that they have access to phosphate pools that plant roots cannot reach (Yao et al., 2001). In soils not well-supplied with available phosphorus, when the uptake of phosphorus by plants exceeds the rate at which it diffuses into the root zone, zones of P-depletion occur surrounding the roots. AM fungi overcome this problem by extending their external hyphae from root surfaces into areas of soil beyond the zone of phosphorus depletion, accessing a greater volume of the soil than is available to the unassisted root (George, 2000).

The external hyphae of some AM fungi may spread up to 10–12 cm from the root surface (Jacobsen et al., 1992). Sieverding (1991) has calculated that the volume of soil explored by mycorrhizal roots can exceed that explored by the "uninfected" root by a factor of >100. Furthermore, the smaller diameter of AM hyphae compared to plant roots, and hence their greater surface area (Bolan, 1991), as well as their greater affinity for phosphorus (Howeler et al., 1981), make the fungi much more efficient than roots are in the uptake of phosphorus.

An additional consideration is that AM fungi induce changes in root morphology such as root branching and root elongation, something well-documented but as yet not fully explainable. Root size and architecture have a significant influence on nutrient uptake because they affect the volume of soil from which plants can explore for nutrient uptake (Espeleta et al., 1999). So this influence is an obvious mechanism by which AM fungi enhance the uptake of phosphorus and other immobile nutrients (George, 2000).

9.2.2 Nutrient Mobilization

Arbuscular mycorrhizal fungi may also have biochemical and physiological capabilities for increasing the supply of available phosphorus or other immobile nutrients in soil systems. These mechanisms appear to involve acidification of the rhizosphere (Bago and Azcon-Aguilar, 1997), excretion of chelating agents (organic molecules that bind metals, thereby preventing them from rendering phosphorus unavailable), and increases in the activity of phosphatase enzymes in roots (Tarafdar and Marschner, 1994).

Solubilization of phosphorus by microorganisms is discussed more in Chapter 13. The enzymes produced by AM fungi can mobilize phosphate tied up in soil organic matter, making phosphorus available for ready and efficient removal by the external hyphae (Tarafdar and Marschner, 1994). However, the exact role that these enzymes play in phosphorus uptake and the extent of their effect are still not entirely clear (Habte and Fox, 1993; Garcia-Garrido et al., 2002). It has been observed that mycorrhiza-free plants can be as effective and perhaps more effective than AM fungi in mobilizing phosphorus tied up in sparingly soluble or insoluble phosphates of Al, Fe, and Ca, or those tied up in organic compounds (Yao et al., 2001), which suggests that the contribution of AM fungi in this respect may be limited or conditional.

Overall, the results of a large number of field and greenhouse experiments have shown that mycotrophic plant species effectively colonized by AM fungi have significantly higher tissue concentrations of phosphorus than do AM-free plants when both are grown in soil with low to moderate concentrations of soil-solution phosphorus. Such a difference is not seen when soil phosphorus supplies are high, however. This enhanced phosphorus uptake is often accompanied by significant increases in biomass yield (e.g., Miyasaka and Habte, 2001; Douds and Reider, 2003). So some positive relationship between AM fungi and nutrient acquisition is surely operative, although mechanisms and extents remain to be fully established.

9.2.3 Nutrient Amelioration

The role of AM fungi in ameliorating limited nutrient supplies in soil systems emanates largely from their interactions with other soil microorganisms. Among the microorganisms with which AM fungi interact cooperatively are phosphate-solubilizing microorganisms, the root-nodule bacteria Rhizobium, and asymbiotic N_2-fixing bacteria (Catska, 1994). Populations of these microorganisms are known to be stimulated by their proximity to the rhizospheres of plants that have been colonized by AM fungi (George, 2000; Gryndler, 2000). This may help explain why the benefits of inoculum cocktails are greater than applications with individual species as documented in Chapters 32 and 33.

9.2.3.1   Interactions with Phosphorus Solubilizers

Because there is some minimum concentration of inorganic phosphorus (P_i) in the soil solution below which the efficacy of the AM symbiosis will decline (Habte and Fox, 1993), the association of AM fungi with P-solubilizing microorganisms in the rhizosphere can be critical, especially in soils containing suboptimal concentrations of P_i in solution (Yao et al., 2001).

Many soil bacteria and fungi have the ability to solubilize phosphorus compounds, thereby making sorbed phosphorus more accessible and reversing the process of phosphorus fixation. Soil bacteria of the genera Pseudomonas, Enterobacter, and Bacillus, and soil fungi of the genera Penicillium and Aspergillus, have all been observed to be active P-solubilizers (Whitelaw, 2000). Fungi are particularly prominent in this regard (Kim et al., 1998; Osorio and Habte, 2001).

The major mechanisms by which phosphate-solubilizing microorganisms (PSM) increase the supply of plant-available phosphorus is through the production of organic acids (Kim et al., 1998). These acids are products of the fermentative and respiratory metabolism of carbohydrates and other carbonaceous substrates. AM fungi contribute to this pool by stimulating host plants to release root exudates by a mechanism or mechanisms not clearly understood at present. The organic acids thus produced can compete with phosphorus for anion exchange sites on soil colloids and/or form insoluble complexes with cations such as Ca_2^+, Al_3^+, and Fe_3^+, thereby preventing these cations from precipitating out P_i, leaving it in plant-available form (Whitelaw, 2000). AM fungi accelerate the process further by capturing the P_i released in this way before it is reconverted into insoluble phosphate forms (Yao et al., 2001). Efforts to increase the supply of plant-available phosphorus through inoculation of soil with PSM have yielded varied results, however, since they are affected by the soil type, soil treatment, plant species, and type of PSM involved (see Chapters 13, 32, 33, and 37).

9.2.3.2   Interactions with Rhizobium

An increase in nodulation and nitrogen fixation in legumes in the presence of mycorrhizal colonization is a widely reported phenomenon (Azcon-Aguilar and Barea, 1992; Olsen and Habte, 1995). It appears that AM fungi stimulate nodulation and N_2 fixation largely through their effect on improved P, Cu, Zn, Ca, and Fe (Azcon-Aguilar and Barea, 1992; Olsen and Habte, 1995; Ibijbijen et al., 1996). In some low-P soils, leguminous species such as cowpea (*Vigna unguiculata*) and leucaena (*Leucaena leucocephala*) barely form nodules unless the phosphorus status of the soils is first improved, or the soil is inoculated with AM fungi (Mosse, 1986; Habte, unpublished data). The interaction between AM and rhizobium continues to be an important subject for research because of the prevalence and importance of legume-AM fungi-rhizobium symbioses for agriculture, as discussed in Chapters 12 and 27.

9.2.3.3   Maintenance of Nutrient Balance

Enabling associated plants to maintain a more balanced supply of nutrients is another advantage that AM associations can confer when there are nutrient shortages in the soil. This has been seen in plants that are growing in soil with a low P_i concentration but in association with AM fungi, compared to mycorrhiza-free plants that are growing in soil at high P_i concentrations. Having a high P_i concentration in soils can have a negative influence on the availability of other nutrient elements because P_i reacts with them to form insoluble precipitates (Havlin et al., 1999). The nutrient elements that can be most affected in this manner include Fe, Mn, Ca, Cu, and Zn (Jones et al., 1991; Shuman, 1998).

The adverse effect of high P_i on Zn nutrition is well documented, although exactly how P_i fertilization interferes with Zn nutrition is not clearly established (Havlin et al., 1999). High P_i may result in Zn precipitation in soil and/or may lead to antagonistic interactions with Zn in the plant. The occurrence of P-induced micronutrient disorders, particularly Cu and Zn deficiency, has been frequently reported for some tree species (Timmer and Teng, 1990). In addition to causing nutritional disorders because of its direct interactions with other nutrients, high P_i affects the supply of other nutrients by inhibiting mycorrhiza formation (Morghan and Mascagni, 1991).

9.2.3.4　*Transfer of Nutrients between Plants*

The possibility that AM fungal hyphae interconnect root systems of plants was at first hypothesized based on their ability to colonize a wide array of plant species and on the well-known extensiveness of external hyphae compared to root hairs. The actual existence of hyphal interconnections between plants was subsequently demonstrated (Heap and Newman, 1980).

There is clear isotopic evidence for the movement of carbon, nitrogen, and phosphorus between plants interconnected by AM hyphae (Johansen and Jensen, 1996; Waters and Borowicz, 1994). However, the measured quantities of nutrients directly transferred via interconnecting hyphae to the receiving plant have thus far been small (Johansen and Jensen, 1996). There is some evidence, on the other hand, to suggest that nitrogen transfer between intercropped plants can be enhanced through selection of suitable host and AM fungal partners (Martensson et al., 1998).

9.3　Roles in Soil Structure and Plant Protection

Arbuscular mycorrhizal fungi are often implicated in functions that may or may not be related to enhanced nutrient uptake. This section considers the multiple roles that AM fungi play in the structure and functioning of soil systems.

9.3.1　Impacts on Soil Aggregation

The various processes of soil aggregation assemble sand, silt, and clay particles into aggregates of various sizes by organic and inorganic means. The stability of these aggregates when exposed to water, resisting disintegration, and the resulting pore spaces created by the size and strength of these aggregates establish physical soil properties that are critical to soil microbial and macrofaunal activities, to plant growth, and to the movement of materials within soil systems. These physical properties include soil tilth, infiltration, drainage, water-holding capacity, air-filled pore spaces, and resistance to soil loss by erosion (Tisdall, 1994). Soil aggregates less than 250 μm in diameter are considered microaggregates, while those greater than 250 μm are called macroaggregates. Organic residues, bacteria, polysaccharides, and inorganic cementing agents are responsible for creating the stability of microaggregates. Macroaggregates are stabilized in turn by roots, fungal hyphae, and associated mucigels and exopolysaccharides (Tisdall, 1994).

Macroaggregates are the basic units of soil structure, and AM fungi contribute to their stabilization in two ways. They help entangle and enmesh soil particles and micro-aggregates to form macroaggregates. They also produce large quantities of a glycoprotein known as glomalin, which is deposited on hyphal walls and in the adjacent soil. This substance, just discovered within the past 10 years, is a recalcitrant, hydrophobic cement-ing agent believed to reduce the disruption of macroaggregates during wetting and drying (Wright and Upadhyaya, 1998; Miller and Jastrow, 2000). AM fungal hyphae can maintain the stability of macroaggregates up to 22 weeks after the death of associated host plants (Miller and Jastrow, 2000).

9.3.2   Suppression of Phytopathogenic Fungi

Many researchers have studied the interactions between arbuscular mycorrhizal fungi and phytopathogenic fungi in order to understand the resistance of mycorrhizal plants to soil-borne fungal pathogens. These interactions often result in reductions, induced by AM fungi, in disease incidence (Catska, 1994; Matsubara et al., 2001; Thygesen et al., 2004); pathogen development (Cordier et al., 1996); and disease severity (Matsubara et al., 2001). The extent of AM fungal-induced protection of host plants against phytopathogenic fungal attack ranges from complete protection (Torres-Baragan et al., 1996) to the more common partial protection (Matsubara et al., 2001). The degree of partial protection is influenced by the AM fungal species used and the cultivar involved (Yao et al., 2002). At the same time, there are examples where AM fungi either did not protect host species against fungal pathogens (Yao et al., 2002) or even enhanced pathogenicity (Ross, 1972). So the effects of AM fungi in this regard, while generally favorable, are variable and contingent on determinants not yet known.

A number of mechanisms have been proposed to explain AM fungal protection of host plants against soil-borne fungal pathogens (Linderman, 2000; Sylvia and Chellemi, 2001). These are (1) enhanced P nutrition; (2) competition for nutrients and infection sites; (3) changes in the chemical composition of plant tissues; (4) changes in the microbial composition of the mycorhizosphere; (5) biochemical changes in the plant; and (6) allevia-tion of abiotic stress.

AM fungi involvement in regulating microbial populations in the mycorhizosphere may be part of the explanation for phytoprotection against soil-borne fungal pathogens (Catska, 1994; Newsham et al., 1994). According to Linderman (2000), AM fungal regula-tion of plant pathogenic fungi in the mycorhizosphere by antagonistic microorganisms represents the best explanation for the protection of associated hosts against soil-borne fungal pathogens when AM fungi are abundant and active.

9.3.3   Protection against Phytoparasitic Nematodes

Populations of AM fungi and plant parasitic nematodes frequently interact because both are rhizosphere organisms and often occupy the same plant tissue. Nematodes are discussed in the following chapter. The nature of the interactions involving these two populations has received increasing research attention over the past 30 years. Although the results have been varied, in many instances it has been demonstrated that AM fungi can reduce the penetration of nematodes into roots (Smith et al., 1986) and hamper their development subsequent to their penetration into roots (Habte et al., 1999). As a result, AM fungi either reduce or eliminate damage caused by plant parasitic nematodes (Pinochet et al., 1998; Habte et al., 1999). Because of this fact and of the predictability with which AM fungi can be established on host roots, they have the potential for suppressing nematode populations.

Several mechanisms have been proposed to explain the suppression of nematodes by AM fungi. There is some evidence that the two populations may be competing for limited resources such as space and food supply (Tylka et al., 1991). Also, AM fungi may induce changes in roots physiology, altering the characteristics of root exudates, as well as the composition of cell walls, making roots less attractive for colonization, more difficult to penetrate, or less suitable as a niche (Morandi, 1987).

AM fungi could affect nematode activity indirectly by enhancing the efficacy of other soil microorganisms that parasitize nematodes such as *Bacillus penetrans* (Rao and Gowen, 1998). Finally, AM fungi could improve plant vigor through an enhanced uptake of nutrients by nematode-damaged roots, which would otherwise absorb them inefficiently (Pinochet et al., 1995). Improved phosphorus uptake does not consistently explain alteration of nematode damage by AM fungi, however (Strobel et al., 1982; Habte, unpublished data).

9.3.4   Protection against Drought and Salinity

The literature on AM fungi and water relationships is voluminous, but far from conclusive. There is some support for the view that AM fungi have no impact on host water relations (e.g., Syvertsen and Graham, 1990; George et al., 1992). However, other papers show AM fungi improving water relations of associated plants in water-stressed soils (e.g., Estrada-Luna et al., 2000) and in adequately watered soils (Sanchez-Diaz and Honrubia, 1994; Auge, 2001).

According to much of the earlier literature, the influence of AM fungi on water relationships is a function of their impact on improved host P nutrition with a resulting increase in plant size (Sieverding, 1983). However, subsequent studies have shown that AM fungi can also influence the water relations of associated hosts independent of any effect on P nutrition or size of host plants (Goicoechea et al., 1997).

Research has indicated that AM fungi can influence the water relations of associated plants by various means, by: (1) enhancing the efficiency with which plants use water (Al-Karaki, 1998; Kaya et al., 2003); (2) increasing root exploration of the available soil volume and water extraction by external hyphae (Ruiz-Lozano and Azcon, 1995; Espeleta et al., 1999); (3) enhancing the development of water-stable aggregates (Auge, 2001); (4) increasing hydraulic conductivity of host plants by reducing the frequency of nonconductive suberized (matured and hardened) root surface tissues (Hamblin, 1985); and (5) bridging the gap in water flow that can develop when roots extract water from the adjacent soil at a rate much faster than that for replacement by water from the bulk soil farther away from the root (Klepper, 1990).

Currently one cannot say more than that AM fungi affect plants' water relations, but this effect is neither as predictable nor as significant as their well-documented effect on the nutrition and growth of associated plants. The latter effect appears to be the best explanation for how fungi have a favorable impact on the water relations of associated plants (Auge, 2001).

Arbuscular mycorrhizal fungi can completely protect host plants against salinity under moderately saline conditions (electrical conductivity, $4–8\,\mathrm{dSm^{-1}}$) (Pfeiffer and Bloss, 1988). However, in most of the studies reported, their effect has been just to reduce the adverse effects of salinity on host growth. This has been demonstrated for a variety of horticultural and agronomic crop species (e.g., Al-Karaki, 2000; Cantrell and Linderman, 2001; Yano-Melo et al., 2003). In many of the studies in which AM fungi were shown to mitigate the adverse effects of salinity, the effects could be fully explained by improved P nutrition (Ojala et al., 1983; Pfeiffer and Bloss, 1988). There is some indication that AM fungi differ in their ability to protect host plants against salt stress (Yano-Melo et al., 2003).

9.3.5   Protection against Metal Toxicity

9.3.5.1   Aluminum Toxicity

The solubility of aluminum in soil is largely governed by pH, and its toxicity is severe at pH values lower than 5.0, although it may occur at pH values as high as 5.5. Toxic aluminum species in soil, among other things, curtail plant growth by damaging roots and by interfering with the availability of plant nutrients like Ca, P, and Mo (Young and Goulart, 1997). One of the potential means of alleviating this problem is through the use of acid-adapted AM fungi. Cuenca et al. (2001), who grew *Clusia multiflora* in soil irrigated with acid water (pH, 3.0) in the presence and absence of AM fungi isolated from neutral and acid soils, observed that the plants inoculated with the AM fungus isolated from the acid soil grew better than those inoculated with isolates taken from a neutral soil. A number of other plant species have been shown to be protected against aluminum toxicity by AM fungal inoculation (Borie and Rubio, 1999; Cumming and Ning, 2003).

The mechanisms by which AM fungi protect plants against aluminum toxicity are not clearly established. It is possible that aluminum may be complexed by polyphosphates in mycorrhizal roots (Medeiros et al., 1994). The fungi could also absorb and retain aluminum in their hyphae, or produce diffusable low-molecular-weight organic acids that chelate and detoxify it in the root zone (Cuenca et al., 2001). It is also possible that aluminum is being retained by the high cation exchange capacity of roots (Yang and Goulart, 2000).

9.3.5.2   Heavy Metal Toxicity

The heavy metals causing environmental and human health concerns above certain concentrations in soil include Pb and some of the transition elements such Ca, Co, Cu, Cr, Hg, Mn, Ni, and Zn. Some soils are inherently rich in one or more of these metals because of the nature of the parent material from which they are derived. However, the heavy metal content of many soils is due to pollution from mining, smelting, or fossil fuel consumption and/or the addition of sewage sludge and fertilizers to soils (Haselwandter et al., 1994). High levels of heavy metals in soils can inhibit the growth and development of sensitive plants and hamper the formation and function of arbuscular mycorrhizas (Taylor, 2000). There are numerous reports about the occurrence of AM fungi in sites polluted with heavy metals and of the ability of isolates from these sites to enhance the tolerance of plants they colonize (Hildebrandt et al., 1999).

The tolerance of mycorrhizal plants to heavy metals can be accomplished by: (1) increases in the concentration and retention of the metals in plant roots, possibly by mechanisms similar to those described above for aluminum; (2) decreases in their concentration in shoots (Haselwandter et al., 1994; Tonin et al., 2001); or (3) decreases in the concentration of the metals in both roots and shoots (Posta et al., 1994). The mechanisms by which the fungi achieve this accumulation are unclear, however. In cases where the fungi seem to regulate the allocation of heavy metals between roots and shoots, host protection appears to derive from hyphal sequestration of the metals in root tissues, thereby precluding the movement of high concentrations of the metals up into shoots (Bradley et al., 1982).

Manganese and iron have more than one oxidation state, with their reduced forms being associated with phytotoxicity. Another way that AM fungi can protect plants against high concentrations of these metals is by interfering with biochemical processes that favor the reduced forms of the elements. It appears that the fungi can protect associated plants against manganese toxicity by suppressing manganese-reducing bacteria and/or by suppressing the release of manganese-reducing root exudates in the rhizosphere (Posta et al., 1994). These protective functions can be very important for certain soil conditions.

9.4 Determinants of Symbiotic Efficacy

The degree to which mycorrhizal fungi enhance the nutrition and health of associated plants and the quality of soil systems is dependent on a multitude of soil biotic and abiotic factors, as well as on other environmental factors that influence the host, the endophyte, and the endophyte–host association. Factors which help predict the outcomes of the application of AM fungal technology when the number of AMF propagules is adequate are reviewed below.

9.4.1 Soil P Status

There is some optimal concentration of phosphorus in the soil solution at which the endophytic organism and its host maintain a dynamic balance. This concentration varies among plant species. For fast-growing, coarse-rooted plant species like *Leucaena leucocephala* (Lam.) de Wit, this has been measured as 0.02 mg l^{-1} (Habte and Manjunath, 1987). If the phosphorus supply is suboptimal for host plant growth, AM fungal colonization and function will be curtailed because of poor root development and because of competition between the fungus and the plant for limited phosphorus.

At high soil phosphorus concentrations and correspondingly high internal phosphorus concentrations, the host cell membrane is more stable and releases little or no root exudate into the rhizosphere, an apparent conserving mechanism. Such limitation of root exudation will lead then to a reduction in the level of AM fungal colonization of the roots (Graham et al., 1981).

9.4.2 Variations in Plant Dependence on AM Fungi

Mycorrhizal dependence represents the degree to which a plant species relies on the mycorrhizal association for its nutritional status. For most mycorrhizal plants, dependence declines as the concentration of phosphorus in the soil solution is increased, however, some remain more dependent than others at higher levels of phosphorus. It is well established that plant species and cultivars within any given species vary in their dependence on AM fungal colonization for their nutrient uptake (Tawaraya, 2003). In general, species that produce large quantities of fine roots with high frequency of long root hairs have lower external phosphorus requirements, and hence are less dependent on AM fungal colonization, than those with sparse and coarse root systems and few root hairs (Bryla and Koide, 1998).

9.4.3 The Impact of Soil Disturbance and Cropping Sequences

The activities of AM fungi can be severely curtailed by soil disturbance. In native eco-systems, soil disturbance caused by land clearing or mining operations can be so severe that it may not be possible to restore symbiotic function of the fungi merely by reinfesting the affected areas with AM fungi (Habte et al., 1988). In agroecosystems, the activities of AM fungi are generally adversely affected by disturbances such as mechanical plowing or planting operations (McGonigle et al., 1990). The difference that mycorrhizal associations (or their loss) can make in crop productivity is one of the reasons contributing to the acceptance of zero-tillage and conservation agriculture in many countries (Chapters 22 and 24).

The diversity of AM fungal communities tends to decline when native ecosystems are converted into agricultural ecosystems and when agricultural inputs are intensified

(Jansa et al., 2003). Tillage destroys the extraradical hyphal networks of AM fungi that develop in soils in association with the previous mycorrhizal crop (Kabir et al., 1999). In no-till and reduced-tillage systems, preserving the integrity of these hyphal networks contributes to more rapid AM fungal infectivity and to more efficient nutrient uptake than is possible in more severely disturbed soils.

Arbuscular mycorrhizal fungi, being obligate symbionts, are sensitive to cultural practices that hamper or delay their contact with receptive host species. Any cropping system that has a bare fallow period, or one that includes in its rotation a nonmycorrhizal species, or even one that is less hospitable to mycorrhizal fungi, is likely to impede the efficacy of fungi in that ecosystem. The problem is often related to declines in the density of AM fungal propagules that are sustained during a fallow period or during the growth of a nonhost or poorly mycotrophic species (Thompson, 1987). Planting soils with cover-crop species favorable to mycorrhizal associations, such as crotalaria (*Crotalaria juncea*) and rye grass (*Lolium perenne*), will ensure the build-up of AM fungal propagules for the subsequent crop (Boswell et al., 1998).

9.5   Effects of Arbuscular Mycorrhizal Application

The evidence that greater inhabitation of roots by AM fungi gives crops more resistance to biotic and abiotic stresses, and greater nutrient uptake resulting in better health and higher yield, has led to efforts to inoculate roots and rhizospheres. These have had mixed results overall, although the knowledge and means to apply AM fungi effectively are increasing.

AM fungi may not become well established in plant roots if the number of infective propagules contained in an inoculum or in the soil is low. Many instances of poor performance of inocula may be a result of a low level of infective propagules in inoculum formulations. All other things being equal, if a high-quality inoculum is introduced into a low P soil that contains no AM fungi or has a very low density of indigenous AM fungi, the probability of obtaining a positive response to this inoculation is very high (Habte and Fox, 1993; Habte et al., 2001).

If a soil already contains high levels of infective propagules, it is unlikely that plants will respond to inoculation of the soil with AM fungi. However, even if the density of indigenous AM fungal propagules is adequate, there can still be obstacles to effective inoculation. The particular fungi may not be compatible with the target host crop plant (Linderman and Davis, 2004). Or the fungi may be inferior mutualists; or they may lack a desired trait such as disease-suppressing ability (Habte et al., 1999); or still other factors in the soil system may curtail their symbiotic activity. Where such constraints are operative, establishing a more efficacious AM fungus on the host plant prior to planting it in the field, through inoculation, would be advisable.

With a fully functional AM symbiosis, plants can grow effectively with a small fraction of the external phosphorus concentration required for the maximal growth of mycorrhiza-free plants. AM fungal inoculation can make a valuable contribution in (1) native eco-systems such as forest ecosystems; (2) agricultural systems in which production is limited by the high phosphorus-fixing capacities of the soil or supplying phosphorus is not possible or too costly; (3) situations where it is essential to reduce soil fertilizer application rates significantly because of environmental concerns; or (4) where phosphate rock is readily available and can be used instead of more soluble phosphorus sources.

A major constraint is the obligate nature of the AM symbiosis, meaning that these fungi cannot be grown apart from host plants. They cannot readily be multiplied in laboratory media. The current state of technology for producing AM fungal inoculum makes direct application of inoculum to extensive areas of land cumbersome and not cost-effective. However, there have been some advances in this area as reported in Chapter 33. Application of AM fungal inocula during nursery production of seedlings for subsequent outplanting to large areas of land has been the most practical and cost-effective way of utilizing the AM fungal technology so far (Johnson and Pfleger, 1992).

The use of AM fungi for acclimatizing and promoting the subsequent survival and growth of micropropagated species is becoming important. Vegetative micropropagation of agronomic, horticultural, and forest species is used for producing large numbers of genetically homogeneous plants. AM fungal technology is starting to be integrated into protocols for the production of micropropagated plants (Varma and Schuepp, 1994).

9.6 Harmful Effects of AM Fungi

When the host-AM fungus association is truly mutualistic, the benefit that each partner derives from the association outweighs its costs (Fitter, 1991). The primary cost to the host is the photosynthate it provides for the production, maintenance, and respiratory demands of the fungus and for its own growth and development. Under normal conditions, this expenditure is more than compensated by enhanced rate of photosynthesis, due to a fungus-induced increase in leaf area (Harris and Paul, 1987) and perhaps enhanced chlorophyll levels (Tsang and Maun, 1999).

The balance can shift from mutualism to parasitism if the plant allows itself to be colonized by AM fungi under conditions that are optimal for mycorrhiza-free host growth. Because the plant expends as much as 20% of its photoassimilate for production and maintenance of AM fungal biomass and respiration (Johnson et al., 1997), this carbon drain can result in host-growth depression (Harris and Paul, 1987). If the phosphorus supply is suboptimal for mycorrhizal activity, the host and endophyte will compete for the limited amount of phosphorus. Host-endophyte imbalance can also occur when the host plant is unable to produce sufficient photosynthate for itself and the associated AM fungi because of inadequate light supply and/or low temperature (Johnson and Pfleger, 1992).

The host plant is also at a disadvantage when it is colonized by an AM fungus that is an inferior mutualist (Johnson and Pfleger, 1992) or when the fungus stimulates the activities of parasitic angiosperms that co-colonize the host (Sanders et al., 1993). It is also possible that AM fungi can be pathogenic as seen in the relationship between *Glomus macrocarpum* and tobacco, where the fungus induces tobacco stunt disease, although it is not known to induce pathogenicity in any other host plants (Modjo and Hendrix, 1986). AM fungi are known to infect seeds of plants in soil and to kill them, especially seeds of nonmycorrhizal plants (Taber, 1982). They are also known to suppress the growth and development of nonmycorrhizal or weakly mycorrhizal plants when these are grown with strongly mycorrhizal plants (Francis and Read, 1994).

9.7 Discussion

Over the past four decades, our understanding has grown rapidly of how arbuscular mycorrhizal fungi form obligately mutualistic associations with most plant species and influence plant life in numerous important ways, including positive impacts on soil structure. It remains to be clearly established whether the influence of AM fungi on plant health is attributable mostly to their impact on host plant nutrition. Clarifying this will require consistent use of research protocols that measure well-defined phosphorus concentrations present or attained in the soil solution rather than assess the effects of adding certain amounts of phosphorus to the soil. Simply knowing how much phosphorus was added tells us little about the actual conditions in which plant roots, fungi, and other organisms interact. The chemical, physical, and biological influences affecting phosphorus status in soil systems are so many and complex that isolating the influence of mycorrhizal fungi will always be difficult. The obligate dependence of AM fungi on plants for their growth and reproduction means that they cannot be cultivated in conventional media and with standard microbiological methods.

Existing inoculum production techniques are adequate for applying the AM technology cost-effectively to a wide variety of agronomic, horticultural, forest, and agroforestry species that are normally started as transplants, including those multiplied by means of micropropagation. Large-scale field experience of inoculation with mycorrhizal fungi is reported in Chapter 33. Making more predictable applications of AM technology will require a thorough understanding of the main biotic and abiotic variables that govern the infectivity and symbiotic effectiveness of AM fungi which have been reviewed in this chapter.

Acknowledgments

The author acknowledges the support provided by the College of Tropical Agriculture and Human Resources, University of Hawaii, in the preparation of this chapter.

References

Al-Karaki, G.N., Benefit, cost, and water-use efficiency of arbuscular mycorrhizal durum wheat grown under drought stress, *Mycorrhiza*, **8**, 41–45 (1998).

Al-Karaki, G.N., Growth of mycorrhizal tomato and mineral acquisition under salt stress, *Mycorrhiza*, **10**, 51–54 (2000).

Auge, R.M., Stomatal behavior of arbuscular mycorrhizal plants, In: *Arbuscular Mycorrhizas: Physiology and Function*, Kapulnik, Y. and Douds, D.D., Eds., Kluwer Academic Publishers, Boston, 201–237 (2001).

Azcon-Aguilar, C. and Barea, J.M., Interaction between mycorrhizal fungi and other rhizosphere microorganisms, In: *Mycorrhizal Functioning: An Integrative Plant–Fungal Process*, Allen, M.F., Ed., Chapman & Hall, New York, 163–198 (1992).

Bago, B. and Azcon-Aguilar, C., Changes in the rhizosphere pH induced by arbuscular mycorrhiza formation in onion (*Allium cepa*), *Z. Pflanzen. Bodenkunde*, **160**, 333–339 (1997).

Becard, G. and Piche, Y., Fungal growth stimulation by CO_2 and root exudates in vesicular-arbuscular mycorrhizal symbiosis, *Appl. Environ. Microbiol.*, **55**, 2320–2325 (1989).

Bolan, N.S., A critical review on the role of mycorrhizal fungi in the uptake of phosphorus by plants, *Plant Soil*, **134**, 189–293 (1991).

Borie, F. and Rubio, R., Effect of arbuscular mycorrhizae and liming on growth and mineral acquisition of aluminum tolerant and aluminum-sensitive barley cultivars, *J. Plant Nutr.*, **22**, 121–137 (1999).

Boswell, E.P. et al., Winter wheat cover cropping, VA mycorrhizal fungi and growth and yield, *Agric. Ecosyst. Environ.*, **67**, 55–65 (1998).

Bradley, R., Burt, A.J., and Read, D.J., The biology of mycorrhiza in the Ericaceae VIII: The role of mycorrhizal infection in heavy metal resistance, *New Phytol.*, **91**, 197–209 (1982).

Bryla, D.R. and Koide, R.T., Mycorrhizal response of two tomato genotypes relates to their ability to acquire and utilize phosphorus, *Ann. Bot.*, **82**, 849–857 (1998).

Cantrell, I.C. and Linderman, R.G., Pre-inoculation of lettuce and onion with VA mycorrhizal fungi reduces deleterious effects of soil acidity, *Plant Soil*, **233**, 269–281 (2001).

Catska, V., Interrelationships between vesicular-arbuscular mycorrhiza and rhizosphere microflora in apple replant disease, *Biol. Plantarum*, **36**, 99–104 (1994).

Cordier, C., Gianinazzi, S., and Gianinazzi-Pearson, V., Colonization patterns of root tissues by *Pythophthora nicotianae* var. *parasitica* related to reduced disease in mycorrhiza tomato, *Plant Soil*, **185**, 223–232 (1996).

Cuenca, G., Andrade, Z., and Meneses, E., The presence of aluminum in arbuscular mycorrhizas of *Clusia multiflora* exposed to increased acidity, *Plant Soil*, **231**, 233–241 (2001).

Cumming, J.R. and Ning, J., Arbuscular mycorrhiza fungi enhance aluminum resistance of broomsedge (*Andropogon viriginicus* [L]), *J. Exp. Bot.*, **54**, 1447–1459 (2003).

Douds, D.D. and Reider, C., Inoculation with mycorrhizal fungi increases yield of green pepper in a high P soil, *Biol. Agric. Hortic.*, **21**, 91–102 (2003).

Douds, D.D. and Schenck, N.C., Germination and hyphal growth of VAM fungi during and after storage in soil at five matric potentials, *Soil Biol. Biochem.*, **23**, 177–183 (1991).

Espeleta, J.F., Eissenstat, D.M., and Graham, J.H., Citrus root responses to localized drying soil: A new approach to studying mycorrhizal effects on the roots of mature trees, *Plant Soil*, **206**, 1–10 (1999).

Estrada-Luna, A.A., Davies, F.T., and Egilla, J.N., Mycorrhizal fungi enhancement of growth and gas exchange of micropropagated guava plantlets (*Psidium guajava* L.) during ex vitro acclimatization and plant establishment, *Mycorrhiza*, **10**, 1–8 (2000).

Fitter, A.H., Costs and benefits of mycorrhizae: Implications for functioning under natural conditions, *Experimentia*, **47**, 350–355 (1991).

Francis, R. and Read, D.J., The contributions of mycorrhizal fungi to the determination of plant community structure, In: *Management of Mycorrhizas in Agriculture, Horticulture and Forestry*, Robson, A.D., Abbott, L.K., and Malajczuk, N., Eds., Kluwer Academic Publishers, Dordrecht, The Netherlands, 11–25 (1994).

Garcia-Garrido, J.M., Ocampo, J.A., and Garcia-Romera, I., Enzymes in arbuscular mycorrhizal symbiosis, In: *Enzymes in the Environment*, Burns, R.G. and Dick, R.P., Eds., Marcel Dekker, New York, 125–151 (2002).

George, E., Nutrient uptake: Contribution of arbuscular mycorrhizal fungi to plant mineral nutrition, In: *Arbuscular Mycorrhizas: Physiology and Function*, Kapulnik, Y. and Douds, D.D., Eds., Kluwer Academic Publishers, Boston, 307–343 (2000).

George, E. et al., Contribution of mycorrhizal hyphae to nutrient and water uptake of plants, In: *Mycorrhizas in Ecosystems*, Read, D.J. et al., Eds., CAB International, Wallingford, UK, 42–47 (1992).

Giovannetti, M. and Sbrana, C., Meeting the non-host: The behavior of AM fungi, *Mycorrhiza*, **8**, 123–131 (1998).

Goicoechea, N., Antolin, M.C., and Sanchez-Diaz, M., Gas exchange is related to the hormone balance in mycorrhizal or nitrogen-fixing alfalfa subjected to drought, *Plant Physiol.*, **100**, 989–997 (1997).

Graham, J.H., Leonard, R.T., and Menge, J.A., Membrane-related decrease in root exudation responsible for phosphorus inhibition of vesicular-arbuscular mycorrhiza formation, *Plant Physiol.*, **68**, 548–552 (1981).

Gryndler, M., Interactions of arbuscular mycorrhizal fungi with other soil organisms, In: *Arbuscular Mycorrhizas: Physiology and Function*, Kapulnik, Y. and Douds, D.D., Eds., Academic Publishers, Boston, 239–262 (2000).

Habte, M. and Fox, R.L., Effectiveness of VAM fungi in nonsterile soils before and after optimization of P in soil solution, *Plant Soil*, **151**, 219–226 (1993).

Habte, M. and Manjunath, A., Soil solution phosphorus status and mycorrhizal dependency in *Leucaena leucocephala*, *Appl. Environ. Microbiol.*, **53**, 797–801 (1987).

Habte, H., Miyasaka, S.C., and Matsuyama, D.T., Arbuscular mycorrhizal fungi improve forest tree establishment in the field, In: *Food Security and the Sustainability of Agro-Ecosystems*, Horst, W.J., Ed., Kluwer Academic Publishing, Dordrecht, The Netherlands, 644–645 (2001).

Habte, M., Zhang, Y.C., and Schmitt, D.P., Effectiveness of Glomus species in protecting white clover against nematode damage, *Can. J. Bot.*, **77**, 135–139 (1999).

Habte, M. et al., Interaction of vesicular-arbuscular mycorrhizal fungi with erosion in an oxisol, *Appl. Environ. Microbiol.*, **54**, 945–950 (1988).

Hamblin, A.P., The influence of soil structure on water movement, crop root growth, and water uptake, *Adv. Agron.*, **38**, 95–158 (1985).

Harris, D. and Paul, E.A., Carbon requirements of vesicular-arbuscular mycorrhizae, In: *Ecophysiology of VA Mycorrhizal Plants*, Safir, G.R., Ed., CRC Press, Boca Raton, FL, 93–105 (1987).

Haselwandter, K., Levyal, C., and Sanders, F., Impact of arbuscular mycorrhizal fungi on plant uptake of heavy metals and radionuclides from soil, In: *Impact of Arbuscular Mycorrhizas on Sustainable Agriculture and Natural*, Gianinazzi, S. and Schuepp, H., Eds., Berkhauser Verlag, Basel, Switzerland, 179–189 (1994).

Havlin, J. et al., *Soil Fertility and Fertilizers*, 6th ed., Prentice Hall, Upper Saddle River, NJ (1999).

Heap, A.J. and Newman, E.I., The influence of vesicular-arbuscular mycorrhizas on phosphorus transfer between plants, *New Phytol.*, **85**, 173–179 (1980).

Hildebrandt, U., Kaldorf, M., and Bothe, H., The zinc violet and its colonization by arbuscular mycorrhizal fungi, *J. Plant Physiol.*, **154**, 709–717 (1999).

Howeler, R.H., Asher, C.J., and Edwards, D.G., Establishment of an effective mycorrhizal association on cassava in flowing solution culture and its effects on phosphorus nutrition, *New Phytol.*, **90**, 279–283 (1981).

Ibijbijen, J. et al., Effect of arbuscular mycorrhizal fungi on growth, mineral nutrition, and nitrogen fixation of three varieties of common beans (*Phaseolus vulgaris*), *New Phytol.*, **134**, 353–360 (1996).

Jacobsen, I., Abbot, L.K., and Robson, A.D., External hyphae of vesicular-arbuscular mycorrhizal fungi associated with *Trifolium subterraneum* (L): I. Spread of hyphae and phosphorus inflow into roots, *New Phytol.*, **120**, 371–380 (1992).

Jansa, J. et al., Soil tillage affects the community structure of mycorrhizal fungi in maize roots, *Ecol. Appl.*, **13**, 1164–1176 (2003).

Johansen, A. and Jensen, E.S., Transfer of N and P from intact or decomposing roots of pea to barley interconnected by an arbuscular mycorrhizal fungus, *Soil Biol. Biochem.*, **28**, 73–81 (1996).

Jones, M.D., Durall, D.M., and Tinker, P.B., A comparison of arbuscular and ectomycorrhizal Eucalyptus coccifera: Growth response, phosphorus uptake efficiency, and external hyphal production, *New Phytol.*, **140**, 125–134 (1998).

Johnson, N.C., Graham, A.H., and Smith, F.A., Functioning of mycorrhizal association along mutualism-parasitism continuum, *New Phytol.*, **135**, 575–585 (1997).

Johnson, N.C. and Pfleger, F.L., Vesicular arbuscular mycorrhizae and cultural stresses, In: *Mycorrhiza in Sustainable Agriculture*, Bethlenfalvay, G.J. and Linderman, R.R.G., Eds., American Society of Agronomy, Crop Science Society of America, and Soil Science Society of America, Madison, WI, 71–99 (1992).

Jones, J.B., Wolf, B., and Mills, H.A., *Plant Analysis Handbook*, Micro-Macro Publishing, Athens, GA (1991).

Kabir, Z., O'Halloran, I.P., and Hamel, C., Combined effects of soil disturbance and fallowing on plant and fungal components of mycorrhizal corn (*Zea mays* L.), *Soil Biol. Biochem.*, **31**, 307–314 (1999).

Kaya, C. et al., Mycorrhizal colonization improves water-use efficiency in watermelon (*Citrus lantanus* Thumb) grown under well-watered and water-stressed conditions, *Plant Soil*, **253**, 287–292 (2003).

Kim, K.Y., Jordan, D., and McDonald, G.A., Effect of phosphate solubilizing bacteria and vesicular-arbuscular mycorrhizae on tomato growth and soil microbial activity, *Biol. Fert. Soils*, **26**, 79–87 (1998).

Klepper, B., Root growth and water uptake, In: *Irrigation of Agricultural Crops, Agronomy Series 30*, Stewart, B.A. and Nielson, D.R., Eds., Agronomy Society of America, Crop Science Society of America, and Soil Science Society of America, Madison, WI, 281–322 (1990).

Linderman, R.G., Effects of mycorrhizas on plant tolerance to diseases, In: *Arbuscular Mycorrhizas: Physiology and Function*, Kapulnik, Y. and Douds, D.D., Eds., Kluwer Academic Publishers, Boston, 345–365 (2000).

Linderman, R.G. and Davis, E.A., Varied response of marigold (*Tagetes* spp.) genotypes to inoculation with different arbuscular mycorrhizal fungi, *Sci. Hortic.*, **99**, 67–78 (2004).

Martensson, A.M., Rydberg, I., and Vestberg, M., Potential to improve transfer of N in intercropped systems by optimizing host-endophyte combinations, *Plant Soil*, **205**, 57–66 (1998).

Matsubara, Y., Ohba, N., and Fukui, H., Effect of arbuscular mycorrhizal infection on the incidence of Fusarium root rot in asparagus seedlings, *J. Jap. Soc. Hortic. Sci.*, **70**, 202–206 (2001).

McGonigle, T.P. Evans, D.G. and Miller, M.H., Effect of degree of soil disturbance on mycorrhizal colonization and phosphorus uptake by maize in growth chamber and field experiments, *New Phytol.*, **116**, 629–636 (1990).

Medeiros, C.A.B., Clark, R.B., and Ellis, J.R., Effect of excess aluminum on mineral uptake in mycorrhizal sorghum, *J. Plant Nutr.*, **17**, 1399–1416 (1994).

Miller, R.M. and Jastrow, J.D., Mycorrhizal fungi influence soil structure, In: *Arbuscular Mycorrhizas: Physiology and Function*, Kapulnik, Y., Ed., Kluwer Academic Publishers, Boston, 1–18 (2000).

Miyasaka, S.C. and Habte, M., Plant mechanisms and mycorrhizal symbioses to increase phosphorus uptake efficiency, *Commun. Soil Sci. Plant Anal.*, **32**, 1101–1147 (2001).

Modjo, H.S. and Hendrix, J.W., The mycorrhizal fungus *Glomus macrocarpum* as a cause of tobacco stunt disease, *Phytopathology*, **16**, 688–691 (1986).

Morandi, D., VA mycorrhizae, nematodes, phosphorus and phytoalexins on soybean, In: *Mycorrhizae in the Next Decade: Practical Application and Research Priorities, Proceedings of the Seventh North American Conference on Mycorrhizae*, Sylvia, D.M., Hung, L.L., and Graham, J.H., Eds., University of Florida, Gainesville, FL, 212 (1987).

Morghan, J.T. and Mascagni, H.G., Environmental and soil factors affecting micronutrient deficiency and toxicities, In: *Trace Elements in Agriculture*, Mortvedt, J.J., Ed., Soil Science Society of America, Madison, WI, 371–425 (1991).

Mosse, B., Mycorrhiza in sustainable agriculture, *Biol. Agric. Hortic.*, **3**, 191–209 (1986).

Nagahashi, G., In vitro and in situ techniques to examine the role of roots and root exudates during AM fungus–host interaction, In: *Arbuscular Mycorrhizas: Physiology and Function*, Kapulnik, Y. and Douds, D.D., Eds., Kluwer Academic Publishers, Boston, 287–305 (2000).

Newsham, K.K., Fitter, A.H., and Watkinson, A.R., Root pathogenic and arbuscular mycorrhizal fungi determine fecundity of asymptomatic plants in the field, *J. Ecol.*, **82**, 805–814 (1994).

Ojala, J.C. et al., Influence of mycorrhizal fungi on the mineral nutrition and yield of onion in saline soil, *Agron. J.*, **75**, 255–259 (1983).

Olsen, T. and Habte, M., Mycorrhizal inoculation effect on nodulation and N accumulation in *Cajanus cajan* at soil P concentrations sufficient or inadequate for mycorrhiza-free growth, *Mycorrhiza*, **5**, 395–399 (1995).

Osorio, N.W. and Habte, M., Synergistic influence of an arbuscular mycorrhizal fungus and a P solubilizing fungus on growth and P uptake of *Leucaena leucocephala*, *Arid Land Res. Manag.*, **15**, 263–274 (2001).

Pfeiffer, C.M. and Bloss, H.E., Growth and nutrition of Guayule (*Parathenicum argentatum*) in saline soil as influenced by vesicular-arbuscular mycorrhiza and phosphorus fertilization, *New Phytol.*, **108**, 315–321 (1988).

Pinochet, J. et al., Interaction between the root lesion nematode *Pratylenchus vulnus* and the mycorrhizal association of *Glomus intraradices* and Santa Lucia cherry rootstock, *Plant Soil*, **170**, 323–329 (1995).

Pinochet, J. et al., Inducing tolerance to the root lesion nematode *Pratylenchus vulnus* by early mycorrhizal inoculation of micro-propagated Myrobalan 29 C plum rootstock, *J. Am. Soc. Hortic. Sci.*, **123**, 342–347 (1998).

Posta, K., Marschner, H., and Römheld, V., Manganese reduction in the rhizosphere of mycorrhizal and non-mycorrhizal maize, *Mycorrhiza*, **5**, 119–124 (1994).

Rao, M.S. and Gowen, S.R., Bio-management of *Meloidogyne incognita* on tomato by integrating *Glomus deserticola* and *Pasteuria penetrans*, *J. Plant Dis. Prot.*, **105**, 49–52 (1998), Abstract only.

Ross, J.P., Influence of Endogone mycorrhiza on phytophthora rot of soybean, *Phytopathology*, **62**, 896–897 (1972).

Ruiz-Lozano, J.M. and Azcon, R., Hyphal contribution to water uptake in mycorrhizal plants as affected by the fungal species and water status, *Physiol. Plant.*, **95**, 472–478 (1995).

Saito, M., Symbiotic exchange of nutrients in arbuscular mycorrhizas: Transport and transfer of phosphorus, In: *Arbuscular Mycorrhizas: Physiology and Function*, Kapulnik, Y. and Douds, D.D., Eds., Kluwer Academic Publishers, Boston, 85–106 (2000).

Sanchez-Diaz, M. and Honrubia, M., Water relations and alleviation of drought stress in mycorrhizal plants, In: *Impacts of Arbuscular Mycorrhizas on Sustainable Agriculture and Natural Ecosystems*, Gianinazzi, S. and Schuepp, H., Eds., Birkhauser Verlag, Basel, Switzerland (1994).

Sanders, I.R., Koide, R.T., and Shumway, D.L., Mycorrhizal stimulation of plant parasitism, *Can. J. Bot.*, **71**, 1143–1146 (1993).

Shuman, L.M., Micronutrient fertilizers, In: *Nutrient Use and Plant Production*, Rengel, Z., Ed., Haworth Press, New York, 165–195 (1998).

Sieverding, E., Influence of soil water regimes on VA mycorrhizas: II. Effect of soil temperature and water regime on growth, nutrient uptake, and water utilization of *Eupatorium odoratum* L, *J. Agron. Crop Sci.*, **152**, 56–57 (1983).

Sieverding, E., *Vesicular-Arbuscular Mycorrhiza Management in Tropical Agroecosystems*, Deutsche Gesellschaft für Technische Zusammenarbeit, Bremen, Germany (1991).

Smith, G.S., Hussey, R.S., and Roncadori, R.W., Penetration and post-infection development of *Meloidogyne incognita* on cotton as affected by *Glomus intraradices* and phosphorus, *J. Nematol.*, **18**, 429–435 (1986).

Smith, S.E. and Read, D.J., *Mycorrhizal Symbiosis*, 2nd ed., Academic Press, London (1997).

Strobel, N.E., Hussey, R.S., and Roncadori, R.W., Interactions of vesicular-arbuscular mycorrhizal fungi, *Meloidogyne incognita* and soil fertility on peach, *Phytopathology*, **72**, 690–694 (1982).

Sylvia, D.M. and Chellemi, D.O., Interactions among root-inhabiting fungi and their implications for biological control of root pathogens, *Adv. Agron.*, **73**, 1–33 (2001).

Syvertsen, J.P. and Graham, J.H., Phosphorus supply and arbuscular mycorrhizas increase growth and net gas exchange responses of two Citrus spp. grown at elevated CO_2, *Plant Soil*, **208**, 209–219 (1990).

Taber, R.A., Occurrence of Glomus spores in weed seeds in soil, *Mycologia*, **74**, 515–520 (1982).

Tarafdar, J.C. and Marschner, H., Phosphatase activity in the rhizosphere and hydrosphere of VA mycorrhizal wheat supplied with inorganic and organic phosphorus, *Soil Biol. Biochem.*, **26**, 387–395 (1994).

Tawaraya, K., Arbuscular mycorrhizal dependency of different plant species and cultivars, *Soil Sci. Plant Nutr.*, **49**, 655–668 (2003).

Taylor, D.L., A new dawn: The ecological genetics of mycorrhizal fungi, *New Phytol.*, **147**, 236–239 (2000).

Thompson, J.P., Decline of vesicular-arbuscular mycorrhizae in long fallow disorder of field crops and its expression in phosphorus deficiency of sunflower, *Aust. J. Agric. Res.*, **38**, 847–867 (1987).

Thygesen, K., Larsen, J., and Bodker, L., Arbuscular mycorrhizal fungi reduce development of pea root-rot caused by *Aphanomyces euteiches* using oospores as pathogen inoculum, *Eur. J. Plant Pathol.*, **110**, 419–441 (2004).

Timmer, V.R. and Teng, Y., Phosphorus-induced micronutrient disorders in hybrid poplar: II. Response to zinc and copper in greenhouse culture, *Plant Soil*, **126**, 31–39 (1990).

Tisdall, J.M., Possible role of soil microorganisms in aggregation in soils, *Plant Soil*, **159**, 115–121 (1994).

Tonin, C. et al., Assessment of arbuscular mycorrhizal fungi diversity in the rhizosphere of *Viola calaminaria* and effect of these fungi on heavy metal uptake by clover, *Mycorrhiza*, **10**, 161–168 (2001).

Torres-Baragan, R. et al., The use of arbuscular mycorrhizae to control onion white rot (*Sclerotium cepivorum* Berk) under field conditions, *Mycorrhiza*, **6**, 253–257 (1996).

Tsang, A. and Maun, M.A., Mycorrhizal fungi increase salt tolerance of *Strophostyles helvola* in coastal foredunes, *Plant Ecol.*, **144**, 159–166 (1999).

Tylka, G.L., Hussey, R.S., and Roncadori, R.W., Interaction of vesicular-arbuscular mycorrhizal fungi, phosphorus and *Heterodera glycines* on soybean, *J. Nematol.*, **23**, 122–133 (1991).

Varma, A. and Schuepp, H., Infectivity and effectivity of *Glomus intraradices* on micropropagated plants, *Mycologia*, **5**, 29–37 (1994).

Vierheilig, H. and Ocampo, J.A., Effect of isocyanate on germination of spores of *Glomus mosseae*, *Soil Biol. Biochem.*, **22**, 1161–1162 (1990).

Waters, J.R. and Borowicz, V.A., *Effect of clipping benomyl and genet on 14C transfer between mycorrhizal plants*, Oikos, 71, 246–252(1994).

Whitelaw, M.A., Growth promotion of plants inoculated with phosphate solubilizing fungi, *Adv. Agron.*, **69**, 99–151 (2000).

Wright, S.F. and Upadhyaya, A., A survey of soils for aggregate stability and glomalin, a glycoprotein produced by hyphae of arbuscular mycorrhizal fungi, *Plant Soil*, **198**, 97–107 (1998).

Yang, W.Q. and Goulart, B.L., Mycorrhiza infection reduces short-term aluminum uptake and increases root cation exchange capacity of highbrush blueberry plants, *Hortic. Sci.*, **35**, 1083–1086 (2000).

Yano-Melo, A.M., Saggin, O.J., and Maia, L.C., Tolerance of mycorrhizal banana (*Musa* sp. cv *pacovan*) plantlets to saline stress, *Agric. Ecosyst. Environ.*, **95**, 343–348 (2003).

Yao, Q. et al., Mobilization of sparingly soluble inorganic phosphates by the external mycelium of an arbuscular mycorrhizal fungus, *Plant Soil*, **230**, 279–285 (2001).

Yao, M., Twedell, R., and Desilets, H., Effects of two vesicular-arbuscular mycorrhizal fungi on the growth of micropropagated potato plantlets and the extent of disease caused by *Rhizoctonia solani*, *Mycorrhiza*, **12**, 230–242 (2002).

Young, W.Q. and Goulart, B.L., Aluminum and phosphorus interactions in mycorrhizal and nonmycorrhizal highbush blueberry plantlets, *J. Am. Soc. Hortic. Sci.*, **122**, 24–30 (1997).

10

Moving Up within the Food Web: Protozoa and Nematodes

Gregor W. Yeates[1] **and Tony Pattison**[2]

[1]*Landcare Research, Palmerston North, New Zealand*
[2]*Queensland Department of Primary Industries, South Johnstone, Queensland, Australia*

CONTENTS

Within the food web characterized in Chapter 5, there is a large intermediate domain of creatures existing between the microflora and -fauna, on one hand, and the arthropods and other macro-fauna such as earthworms, ants and termites, whose numbers and variety are vast. In the process of feeding on bacteria, fungi, and other organisms and organic material within the web, these mesofauna perform a number of essential functions within soil systems. Their so-called "grazing" on microorganisms, which is a major part of the nutrient cycling in soil systems, tends to maintain these microbial populations in an active state, with higher nutrient content, and increases the availability of nutrients for plant growth. As in other components of the food web, they are interactive and interdependent with a host of other species. This chapter can only sketch, with selected details, this vast domain of soil organisms that is typified by, but not limited to, protozoa and nematodes. These creatures live not separately but fully enmeshed within the biological realm of soil systems, dependent upon the physical and chemical aspects of these systems, but also affecting them in important ways.

10.1   Mesofauna and Their Interactions

Soil protozoa such as ciliates, flagellates, and naked and testate amoebae, occur in thousands per gram of soil. They not only "graze" on microbes, stimulating microbial turnover and making nitrogen available within the root zone, but also their metabolites can stimulate the bacterial populations that provide their sustenance (Clarholm, 1985; Darbyshire, 1994; Bonkowski and Brandt, 2002). Thus, they have the capacity to promote the growth of their own food supply.

Soil nematodes are essentially vermiform animals, commonly 0.3–2.0 mm but up to 12 mm long (Figure 10.1 and Figure 10.2). They are the best-known mesofauna within soil systems because some species, being root-feeders, have detrimental effects on plants. However, most nematodes, being microbial feeders, carnivores or omnivores, are minimally deleterious or even beneficial for plants. There are 20,000 known species of nematodes, and potentially five times as many (Gobat et al., 2004), so they represent a vast underground kingdom, moving through soil voids and feeding on microbes, roots, root hairs, and various soil organisms. Many species are incredibly hardy, able to survive up to 30 years in a cyst phase of existence. Topsoils typically contain 100,000–10,000,000 nematodes m^{-2} representing up to 200 species (Yeates, 2005).

In addition to vast numbers of nematodes, numerous bacterial-feeding rotifers can be extracted from the soil (Donner, 1966), enchytraeids that feed mostly on microbes (Didden, 1993), and tardigrades that feed on the fluids in plant and animal cells or act as predators (Kinchin, 1994). Protozoa are the most difficult to enumerate because of their diversity and huge numbers (50,000 known species). The need to culture protozoa to assess their activity has added to the difficulty of analyzing and evaluating them (Bamforth, 2001).

All of these organisms depend on autotrophs for their survival (living plants, including their above- and belowground exudates, residues and detritus) or on soil microflora, the bacteria and fungi that are primary decomposers of plant material, converting this into energy and nutrient resources. Mesofauna depend on free water in the soil for their activity, lacking significant mechanisms of their own for retaining water. These organisms, which normally move and feed in water films, differ greatly in size, and their ability to enter into soil voids to utilize certain resources varies accordingly. Although plant-feeding nematodes feed on living plants, their net effect is not necessarily negative. Even though they may cause reduced plant yield, their grazing, excretion, and death all transfer energy and nutrients to the rhizosphere and into the decomposer food web (Yeates et al., 1999) (Table 10.1). While herbivorous nematodes are generally considered as plant parasites, nematodes as underground species play many positive roles in soil systems.

The decomposer food web includes: (a) the microflora, (b) the bacterial- and fungal-feeding microfauna that graze on these microflora, and (c) the higher trophic levels that feed in turn on the species that feed on microbes. Various species of protozoa, nematodes, and tardigrades may prey on other microbial feeders as well as on plant-feeding nematodes so that the resulting food webs are very complex, with a wide range of sizes and feeding habits among organisms, both obligate and facultative. This complexity, intensified by temporal and spatial variations in plant input, results in varying soil processes at a range of spatial scales.

The microbe-based food web discussed in Chapter 5 is affected by activities at higher trophic levels involving various microarthropods preying on nematodes, rotifers, and tardigrades. The even larger macrofauna, functioning as "ecosystem engineers", in particular earthworms and termites, affect nutrient distribution patterns on a local scale. Complicating these dynamics and classifications is the fact that the multiplicity of fungi

FIGURE 10.1
Diversity of size and thermal death position of 12 genera of nematodes recovered from soil using a method requiring their active movement. Drawings are at uniform magnification; all but E represent females. A: *Aporcelaimus* (Dorylaimida; omnivorous), B: *Cephalobus* (Rhabditida; bacterial-feeding), C: *Rhabditis* (Rhabditida; bacterial-feeding), D: *Tylenchorhynchus* (Tylenchida; plant-feeding), E: *Heterodera* second stage (Tylenchida; plant-feeding), F: *Paratylenchus* (Tylenchida; plant-feeding), G: *Pratylenchus* (Tylenchida; plant-feeding), H: *Pungentus* (Dorylaimida; plant-associated), I: *Ditylenchus* (Tylenchida; species variously plant-feeding or fungal-feeding), J: *Mononchus* (Mononchida; typically predacious but some have been cultured on bacteria), K: *Anaplectus* (Chromadorida; bacterial-feeding), L: *Helicotylenchus* (Tylenchida; plant-feeding). Scale line 500 μm = 0.5 mm. Modified from Yeates (1978).

may be predators or pathogens of soil animals, as well as providing food. Current knowledge of the links between the above- and belowground components of ecosystems has been well summarized by Wardle (2002).

Studies such as Clarholm (1985), Ingham et al. (1985), and Darbyshire et al. (1994) have demonstrated that grazing by microfauna on microflora increases the availability of nutrients to plants (Table 10.2 and Table 10.3). Further, Elliott et al. (1980) have shown that protozoa can graze on bacteria living in narrower soil pores and are then grazed on in turn by nematodes in wider pores, thereby increasing nematode biomass (Table 10.4).

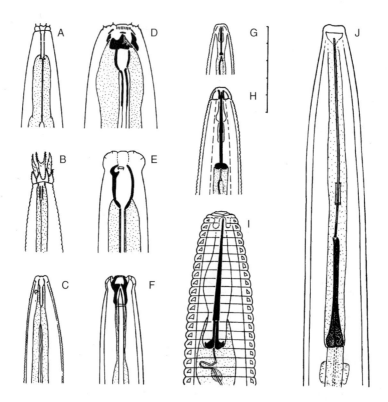

FIGURE 10.2
Diversity of the structure of head and stoma regions of soil-inhabiting nematodes illustrated by females of 10 genera. Drawings are at uniform magnification and a typical body length is given for each genus. Seven of the genera feed on bacteria, fungi, or other microbiota in soil and can generally be regarded as beneficial to plant growth: A: *Rhabditis* (1.0 mm), B: *Acrobeles* (0.7 mm) (Rhabditida), and C: *Plectus* (1.2 mm) (Chromadorida) are solely bacterial-feeding; D: *Diplogaster* (1.0 mm) (Diplogasterida) and E: *Mononchus* (1.5 mm) (Mononchida) are either predators or bacterial-feeders; F: *Actinolaimus* (3.0 mm) (Dorylaimida) is a predator/omnivore; G: *Tylenchus* (1.0 mm) (Tylenchida) is regarded as plant-associated or hyphal-feeding; three of the genera illustrated feed solely on roots of higher plants, and when their populations are combined with other stresses, they may cause pathogenicity and economic crop loss: H: *Rotylenchus* (1.5 mm), I: *Criconema* (0.7 mm) (Tylenchida), and J: *Xiphinema* (3 mm) (Dorylaimida). Scale line 50 μm. Modified from Yeates (1999).

TABLE 10.1

Relative Increase of ^{14}C in Soil Microbial Biomass in the Presence of Nematodes, 15 Days After Pulse-Labeling of Pots of White Clover (*Trifolium repens*)

Nematode	^{14}C in Soil Microbial Biomass
Control	1.17
Heterodera trifolii	1.57
Meloidogyne hapla	1.72
Meloidogyne trifoliophila	1.47
Xiphinema diversicaudatum	2.45
Pratylenchus sp.	2.50
LSD (0.05)	0.49

Note that the increase, which differs among the nematodes tested, is greatest for migratory plant-feeding species.
Source: From Yeates, G.W. et al., *Nematology*, 1, 295–300, 1999. With permission.

TABLE 10.2

Changes in NH_4^+-N Concentration ($mg\ dm^{-3}$) in Pure Cultures of the Bacterium *Arthrobacter* and When Grazed on by the Ciliate *Colpoda steinii* at 10°C for 7 or 14 Days Using Initially Either N-Limited or N-Replete Bacteria

Duration	Initial N Status	7 Days	14 Days
Arthrobacter	Low	+0.15	−0.05
Arthrobacter	High	−0.19	−0.02
Arthrobacter + *Colpoda*[a]	Low	+0.39	+0.50
Arthrobacter + *Colpoda*[a]	High	+0.78	+0.83

[a] Only when *Colpoda* was present were differences in NH_4^+-N between initial N levels after 7 and 14 days significant.

Source: From Darbyshire, J.F. et al., *Soil Biol. Biochem.,* 26, 1193–1199, 1994. With permission.

Jenkinson (1977) has aptly described the soil microbial biomass as "the eye of the needle through which all nutrients must pass." Grazing by protozoa and nematodes not only increases this cycling, but also tends to maintain microflora populations at a higher growth rate with a consequent increase in nutrient levels in the soil. It may also alter the relative abundance of microbial taxa (Sterner and Esler, 2002; Djigal et al., 2004).

With each feeding cycle, not only are nutrients and energy "leaked" back into the soil system through excretion and death to become available to other plant and animal species, but also the C:N and C:P ratios narrow at successive trophic levels. This leads to a surplus of N and P becoming available for plant growth (Sterner and Esler, 2002). Recent work indicates that microfaunal activity continues until the soil approaches its wilting point (Gorres et al., 1999; Yeates et al., 2002). Previously, it was thought that their activities are maximal at field capacity with regard to water saturation, but there can be significant activity over a broad range.

TABLE 10.3

Effect of Increasing Microfloral and Microfaunal Diversity (i.e., food web complexity) on Soil N and Growth of A Grass (*Bouteloua gracilis*) in Microcosms

Organisms Present	NH_4^+-N ($\mu g/g$ soil)[a]	Shoot Biomass (mg)
Plant	33.0	2.1
Plant + bacteria	32.2	2.6
Plant + bacteria + bacterial-feeding nematode	45.5	6.6
Plant + fungus	48.6	7.5
Plant + fungus + fungal-feeding nematode	49.1	8.6
Plant + bacteria + fungus + bacterial-feeding nematode + fungal-feeding nematode	46.4	13.5
HSD ($P < .05$)	10.2	3.0

[a] Note the tendency for NH_4^+-N and shoot biomass to increase with foodweb complexity.

Source: From Ingham, R.E. et al., *Ecol. Monogr.,* 55, 119–140, 1985. With permission.

TABLE 10.4

Effects of Addition of Bacterial-Feeding Amoebae (*Acanthameoba*) on the Abundance and Biomass of Bacterial and Protozoan-Feeding Nematodes (*Mesodiplogsater*) in Microcosms Containing Bacteria (*Pseudomonas*) in Sandy Clay or Clay Soil and Incubated for 33 Days

Nematode Measurement	Soil	Proportional Increase When Amoebae Added
Juvenile number	Sandy clay	1.3
	Clay	2.3
Adult number	Sandy clay	0.6
	Clay	8.7
Total biomass	Sandy clay	1.1
	Clay	2.9

Mesodiplogaster can feed on both *Pseudomonas* and *Acanthamoeba*. The "benefit" to *Mesodiplogaster* of *Acanthamoeba* "mining" *Pseudomonas* from narrower soil pores is greater for adults that are wider and thus more limited in their ability to feed directly on *Pseudomonas*.
Source: From Elliott, E.T. et al., *Okios*, 35, 327–335, 1980. With permission.

While it is known that soil texture, soil structure, mineralogy, and cultivation each has some significant effects on the soil microflora (Jones et al., 1969; Elliott et al., 1980; Hassink et al., 1993; McSorley and Frederick, 2002), these are all soil-specific effects. It is therefore hard to predict what the specific effects of any melioration measure will be when used with a particular soil. Further, similar treatments when applied to different soils can have quite differing outcomes because of the intricate and dynamic relationships among and within these biotic populations. This apparent indeterminacy is not a matter of a lack of causation but due to the complexity of the myriad interactions, such that a specific desired outcome cannot be predicted and guaranteed.

10.2 Applications of Knowledge about Mesofauna

Since soil system dynamics are the focal concern in Part II, rather than their implications for crop performance (Part III), we will not try to assess various practices for managing plant-feeding nematodes (e.g., McSorley, 1998, 2001; Pyrowolakis et al., 1999), which is a frequent concern among agriculturalists. Rather, we are considering how functional groups of nematodes operate within soil systems, examining the possibilities for improving soil fertility and remediating "impaired" soils by affecting or exploiting protozoan and nematode populations. Several case studies will be noted where plant yield has been measured so that the effects of specific interventions can be seen. Not all studies have illuminated the details of nutrient availability, however, which is a key factor affecting nematode populations. Table 10.1, Table 10.2, Table 10.3, and Table 10.4 document links that have been identified among nematode populations, soil conditions, and nutrient availability. They indicate for those unacquainted with the details of this domain how much impact it has on the functioning of soil systems, including the maintenance and enhancement of fertility.

Despite many years of work, there are still no practical ways of manipulating biological control agents in cropping situations so as to have significant impacts on plant-pathogenic nematodes, the unpopular portion of this vast microscopic population (Stirling, 1991).

The above-noted complexity continues to limit what can be done with our present knowledge. However, some steps have been made toward application. It has now been demonstrated, for example, that sheep producers can derive economic benefits from delivering the nematode-trapping fungus *Duddingtonia flagrans* to fecal pats. This fungus captures the free-living, bacterial-feeding stages of trichostrongylid nematodes, thus reducing the reinfection of sheep by gastro-intestinal nematodes, thereby increasing income for sheep-owners (Waller et al., 2004). Other applications of nematode-trapping fungi are still under development.

It is known that nematodes may redistribute growth-promoting bacteria in the soil (Kimpinski and Sturz, 2003), and that protozoa may enhance plant growth by hormonal effects (Bonkowski and Brandt, 2002). Thus, nematodes, protozoa, and similar organisms living in the soil definitely have some impact on agricultural productivity, although measuring this is difficult, and manipulating the organisms is even more of a challenge. Any efforts to do the latter to improve the fertility and sustainability of soil systems must keep in mind the multifunctionality of the system. Some apparent benefits of biological control following the incorporation of organic matter in soil may be due, for example, more to improved nutrient cycling than to any direct biological control of plant pathogens (Yeates and Wardle, 1996).

10.3 Nitrogen Cycling

Nematode and protozoan grazing on microbes increases nutrient mineralization both directly through excretion and indirectly by its effects on microbial activity (Savin et al., 2001). Nitrogen may be mineralized, becoming available for plant growth, if the C:N ratio of any soil additions is less than the microbial C:N ratio, typically 20:1 (Ferris and Matute, 2003). In one recent study, the addition of cephalobid nematodes to microcosms has led to a 12% increase in maize biomass and 16% additional nitrogen uptake (Djigal et al., 2004). This was attributed to nematode activity. Estimated rates of N mineralization due to bacterial-feeding nematodes are $0.0012–0.0058$ μg-N nematode^{-1} d^{-1}, mainly as NH_4^+ (Ferris et al., 1998). Hunt et al. (1987) have estimated that bacterial grazers, i.e., protozoa and bacterial-feeding nematodes, contribute 83% of the N mineralized by fauna, which was 64% as much as that mineralized by microbes. Verhoef and Brussaard (1990) have given a more conservative estimate, suggesting that 30% of N mineralization in the soil could be attributed to the grazing by nematodes on microbes. The contribution of fungal-feeding nematodes may be more important in carbon-rich environments (Couteaux et al., 1991). N mineralization by fungal-feeding nematodes has been estimated at $0.0018–0.0033$ μg-N nematode^{-1} d^{-1} (Chen and Ferris, 2000).

However, there are many restrictions that apply to the rate of nutrient cycling in the soil. Gorres et al. (1999) found significantly higher N mineralization in moist soils at -3 kPa than in drier soils, up to -50 kPa. They attributed this decline in N mineralization in the drier soils to the exclusion of nematode grazers from narrow soil pores that restricted nematode entry under drier soil conditions. Savin et al. (2001) have suggested, on the other hand, that microbial grazing by nematodes stimulates microbial activity more in dry soils than in wet soils. While bacterial activity was stimulated by nematode grazing in dry soils, the grazing of fungal hyphae by nematodes could have decreased fungal biomass and increased microbial respiration in dry soils (Savin et al., 2001).

10.4   Carbon Cycling

The various biochemical processes that cycle carbon in soil systems have been discussed in Chapter 6, addressing their significance for maintaining energy supplies that sustain abundant and diverse biotic populations. Disturbance to soils that are compacted usually results in a short-term increase in microbial biomass, which leads to increased soil respiration and a net loss of carbon from the soil (Ferris and Matute, 2003). The presence of protozoa and nematodes tends to accelerate organic matter decomposition in such cases, and the heterogeneity of the carbon source will influence soil biodiversity (Couteaux et al., 1991; Vetter et al., 2004).

While labile organic sources added to soil promote bacteria-feeding nematodes, the addition of complex carbon compounds tends to increase fungal-feeding nematodes (Ferris and Matute, 2003). Fungal-based decomposition processes are more likely to dominate the decomposition of surface organic matter residues, whereas bacterial-based processes tend to dominate incorporated residues (Neher, 1999). Processing carbon reflects the C:N ratios of substrates, and there is not a linear relation between carbon and nitrogen mineralization. As reviewed by Neher (1999), microfaunal grazing can influence the partitioning of C between roots and shoots, which has a direct effect on crop performance.

10.5   Soil Amendments

Beneficial soil nematodes are more abundant in plots subjected to complex crop rotations, cultivation practices, and organic amendments that enhance soil biological activity (Kimpinski and Sturz, 2003). The addition of plant material significantly raises the rate of soil respiration, with the greatest short-term increases occurring with amendments that have a low C:N ratio (Ferris and Matute, 2003).

There is a succession from bacterial to fungal decomposition of the organic material which occurs more rapidly with amendments having high C:N (Ferris and Matute, 2003). As the organic source becomes more recalcitrant to bacterial decomposition, there is a decline in opportunistic bacterial-feeding nematodes. Bacterial-feeding nematodes make the greatest contribution to the decomposer food web in more intensively managed systems where inputs have high C:N (Kimpinski and Sturz, 2003). Soil amendments such as chitin, paper waste, and pine bark, having a high C:N ratio, may reduce the number of plant-parasitic nematodes, while the number of beneficial nematodes may increase fivefold (Chavarria-Carvajal et al., 2001). With the succession from opportunistic bacterial-feeding nematodes to fungal-feeding species, readily available mineral nitrogen is reduced. This may be advantageous in high rainfall areas to prevent losses of nitrogen from the system, allowing a more sustained release for plant growth.

10.6   Responses to Intensification

The increasing intensification of agriculture has led to increased soil disturbance through cultivation and/or from agrochemical inputs such as fertilizers and biocides (Gupta and Yeates, 1997). The response of soil biotic communities to such disturbance depends on the level of disturbance, the management practices employed, and is often species-specific (Fiscus and Neher, 2002; Wright and Coleman, 2002). Generally, increasing the disturbance of soil decreases the diversity of organisms in the soil and selects for those organisms that can respond most quickly and effectively to disturbance (*r*-strategists). What effect does this disturbance from agricultural intensification have on nutrient cycling and on the maintenance of soil system productivity through disease suppression and nutrient availability?

Populations of herbivorous nematodes respond to changes in the plant community (e.g., Jones et al., 1969; McSorley, 1998; Kimpinski and Sturz, 2003), whereas those of microbivorous nematodes respond to changes in the detritus food web (Mulder et al., 2003). Pankhurst et al. (1995) found that numbers of bacterial-feeding soil protozoa were responsive to changes in soil system management, being more abundant in pasture-wheat rotations with conventional tillage and additional N applications that had increased their food source, predominantly bacteria. Similarly, Mulder et al. (2003) found a shift in the functional groups of nematodes under heavy stocking rates of pastureland, with diversity decreasing as management intensity increased.

It is easily seen that tillage, by affecting soil structure, alters populations of soil biota. Disturbing the soil with tillage increases the decomposition rate of organic matter by 1.4 to 1.9 times, relative to no-tillage situations (Neher, 1999). The plowing of a 5-year-old perennial pasture ley caused an increase in the population levels of the plant-parasitic nematode, Pratylenchus spp., for example, presumably due to the stimulation of "egg hatch" in decomposing roots (Kimpinski and Sturz, 2003). The responses to tillage of a range of plant-feeding nematodes have been shown to differ with soil texture and structure (Jones et al., 1969).

With increasing soil compaction, Foissner (1999) reported a reduction in the number of protozoa and a change in the protozoa community structure. This has been attributed to reduction in pore spaces and moisture content of the soil (Foissner, 1999; Larsen et al., 2004). Neher (1999) suggested that as average pore size decreases, the microfaunal assemblage becomes increasingly dominated by smaller animals. Bouwman and Arts (2000) did not find total nematode abundance affected by increasing soil compaction, but there was a shift among trophic groups, with larger numbers of herbivores and decreased numbers of bacterivores, omnivores, and predators. They attributed this shift in the

TABLE 10.5

Reduction (%) in Simulated Overall Nitrogen Mineralization in 0–10 cm Soil Following the Removal of Various Protozoan and Nematode Groups in Dutch Winter Wheat Fields with Conventional vs. "Integrated" Management

	Conventional	Integrated
Microbial feeders		
Amoebae	39.8	32.4
Flagellates	2.3	1.9
Bacterial-feeding nematodes	10.3	17.3
Fungal-feeding nematodes	1.3	1.2
Enchytraeids	2.1	3.2
Predators		
Predatory nematodes	12.6	19.1
Predatory mites	0.9	0.3
Nematode-feeding mites	0.4	0.2

The effects reported may exceed the direct contribution of certain groups to nitrogen mineralization. Actual mineralization contributions are given in the original paper.
Source: From de Ruiter P.C. et al., *J. Appl. Ecol.*, 30, 95–106, 1993. With permission.

assemblage to changes in the food resources, roots, and microbes, and to the vulnerability of certain species that occupied more stable niches in the soils. Soil compaction may have a double effect on microfauna; by selecting for species that are better able to inhabit the changed soil conditions, this leads to a reduction in the diversity of organisms.

Models help put such data in perspective. Drawing on field data for Dutch winter wheat fields, de Ruiter et al. (1993) showed a marked difference in the impact of protozoan and nematode assemblages on overall nitrogen mineralization between conventional and "integrated" farming systems (Table 10.5). In a California field trial, Ferris et al. (2004) used irrigation to maintain the soil food web in a biologically active state in early autumn, and this enhanced N availability for the vegetative growth of the subsequent summer crop. In plots in which N was marginally limiting, there were positive correlations between autumn food web activity (largely measured by nematode establishment and channel indices using bacterial- and fungal-feeding nematodes, and N availability) and subsequent crop yield.

10.7   Soil Restoration through Fallow

Leaving fallow a soil that has been cultivated for some time provides an opportunity for the reconstitution of soil biological communities, though this can be a transitory effect. When millet was grown on soil in Senegal that had been fallow for 21 years, the composition of the nematode assemblage fauna rapidly approached that of continuously cultivated soil (Villenave et al., 2001). There was a decrease in the ratio of fungal-feeding to bacterial-feeding nematodes that reflected the decreasing importance of fungal-mediated decomposition once cultivation was resumed. Although soil quality fell soon after cultivation resumed, it was better on the soil that had been fallow for 21 years compared to that with only 10 years of fallow. This rapid shift in the makeup of the nematode assemblage when cultivation was resumed is comparable with the disappearance of long-term crop management effects on nematode trophic structure just 1 year after disruptive soil management in Californian farming systems (Berkelmans et al., 2003). In both ecosystems, the pattern of abundance of plant-feeding nematodes was more complex, related in part to having better plant growth resulting from more organic inputs to the soil.

10.8   Incorporation of Crop Residues

The amount and composition of soil organic matter can be altered by incorporating crop residues. Under the Mediterranean climate of West Australia, decomposition of crop residues, i.e., availability of plant nutrients, is strongly moisture-dependent, with measurable populations of protozoa and nematodes which were initially greater on legume than on wheat residues being detected before microarthropods (Van Vliet et al., 2000). In both laboratory and field studies, Fu et al. (2000) found that bacterial-feeding nematodes responded to the addition of maize (*Zea mays*) residue earlier than did fungal-feeding nematodes; predators and omnivores did not show any response until toward the end of the 40-day experiment. The impacts of crop residues, apart from those of soil moisture retention and soil structure, are best measured in terms of the availability of nutrients such as nitrogen and carbon, as described in Sections 10.3 and 10.4.

10.9   Interactions with Macrofauna

Because of the intricacy and interdependence of subsurface species, it is often hard to determine their respective contributions. It is well known that the activities of termites can lead to an accumulation of organic matter in termitaria. This is discussed more in the next chapter. A recent study of Natal sugarcane fields by Cadet et al. (2004) found that cane yield was 5 times greater in areas where termite mounds had been broken up than elsewhere along a transect. However, the nematode assemblage when measured related better to the limits of these termite-affected areas than to either abiotic or bacterial parameters. The greater abundance of plant-feeding nematodes reflected the greater root biomass with the sugarcane crop, yet overall cane yield corresponded well with soil characteristics at 30–60 cm depth. Earthworms, the other main "ecosystem engineers," have also been found to have a significant impact on populations of soil microfauna. Analysis suggests a more rapid loss of labile plant material and a consequent relative increase in the contributions of fungal-feeding nematodes.

10.10   Plant-Pathogenic Nematodes

Agricultural intensification in terms of tillage and external inputs is conducive to the development of economically significant, pathogenic populations of nematodes and other microfauna. The presence of continuous areas of similar plant materials, rather than natural spatial and temporal heterogeneity, may lead to large populations of microfauna adapted to use these uniform resources. Such dominance and lack of heterogeneity may result in loss of functionality. Certain species, and even races, of plant-feeding nematodes may come to dominate; a uniform, recalcitrant residue may cover the surface; and disease and predators that normally contribute to the control of microfaunal populations may cease to be effective.

In general, extreme intensification is likely to result in reduction of the ecosystem services provided by the interactions described in the preceding sections. Strategic management, such as the autumn irrigation of Ferris et al. (2004), may ameliorate this loss. Any agricultural system that increases plant growth provides a greater resource for potential pathogens. Interventions to maintain the desired balance between "good guys" and "bad guys" will be part of a cost-effective soil system management strategy.

10.11   Discussion

Through their grazing activities, protozoa and bacterial- and fungal-feeding nematodes contribute significantly to nitrogen mineralization and other processes that make nutrients available to plants from plant exudates, residues, and mulches. For more readily decomposable organic materials, the bacterial-based chain of organisms is more important. Over time and with reduced litter quality, the process tends to become more fungal-dominated. Recalcitrant organic matter contributes to maintaining soil structure. In addition, improved plant growth associated with the addition of organic matter may provide a greater root biomass that in turn supports more plant-feeding nematodes. This need not cause economic problems unless these populations get out of control, expanding at the expense of other species, because root-feeders contribute to the nutrient endowment of the rhizosphere, providing as well as extracting nutrients.

While there are few cases where the direct links between protozoan and nematode-induced nutrient availability have been demonstrated to contribute to improved plant yield, the indications for such links are strong. While these effects may be enhanced by encouraging bacterial-dominated decomposition processes, i.e., with higher N organic matter and with higher soil moisture, this must be matched to plant uptake to avoid increased leaching of N and other nutrients into ground water. Organic substrates, microfaunal populations, and soil structure exhibit a wide range of interactions in a given soil system. Their spatial heterogeneity is influenced by ecosystem engineers, e.g., earthworms and termites (Chapter 11), and by management practices. This chapter has reviewed some ways in which microfaunal activity can promote production while maintaining sustainable soil systems.

References

Bamforth, S.S., Proportions of active ciliate taxa in soil, *Biol. Fertil. Soils*, **33**, 197–203 (2001).

Berkelmans, R. et al., Effects of long-term crop management on nematode trophic levels other than plant feeders disappear after 1 year of disruptive soil management, *Appl. Soil Ecol.*, **23**, 223–235 (2003).

Bonkowski, M. and Brandt, F., Do soil protozoa enhance plant growth by hormonal effects?, *Soil Biol. Biochem.*, **34**, 1709–1715 (2002).

Bouwman, L.A. and Arts, W.B.M., Effects of soil compaction on the relationships between nematodes, grass production and soil physical properties, *Appl. Soil Ecol.*, **14**, 213–222 (2000).

Cadet, P., Guichaoua, L., and Spaull, V.W., Nematodes, bacterial activity, soil characteristics and plant growth associated with termitaria in a sugarcane field in South Africa, *Appl. Soil Ecol.*, **25**, 193–206 (2004).

Chavarria-Carvajal, J.A. et al., Changes in populations of microorganisms associated with organic amendments and benzaldehyde to control plant-parasitic nematodes, *Nematropica*, **31**, 165–180 (2001).

Chen, J. and Ferris, H., Growth and nitrogen mineralization of selected fungi and fungal-feeding nematodes on sand amended with organic matter, *Plant Soil*, **218**, 91–101 (2000).

Clarholm, M., Interactions of bacteria, protozoa and plants leading to mineralization of soil nitrogen, *Soil Biol. Biochem.*, **17**, 181–187 (1985).

Couteaux, M.-M. et al., Increased atmospheric CO_2 and litter quality: Decomposition of sweet chestnut leaf litter with animal food webs of different complexities, *Oikos*, **61**, 54–64 (1991).

Darbyshire, J.F., Ed., *Soil Protozoa*, CAB International, Wallingford, UK (1994).

Darbyshire, J.F. et al., Excretion of nitrogen and phosphorus by the soil ciliate Colpoda steinii when fed the soil bacterium *Arthrobacter* sp, *Soil Biol. Biochem.*, **26**, 1193–1199 (1994).

de Ruiter, P.C. et al., Simulation of nitrogen mineralization in the belowground food webs of two winter wheat fields, *J. Appl. Ecol.*, **30**, 95–106 (1993).

Didden, W.A.M., Ecology of terrestrial Enchytraeidae, *Pedobiologia*, **37**, 2–29 (1993).

Djigal, D. et al., Influence of bacterial-feeding nematodes (*Cephalobidae*) on soil microbal communities during maize growth, *Soil Biol. Biochem.*, **36**, 323–331 (2004).

Donner, J., *Rotifers*, Warne, London (1966).

Elliott, E.T. et al., Habitable pore space and microbial trophic interactions, *Oikos*, **35**, 327–335 (1980).

Ferris, H. and Matute, M.M., Structural and functional succession in nematode fauna of a soil food web, *Appl. Soil Ecol.*, **23**, 93–110 (2003).

Ferris, H., Venette, R.C., and Scow, K.M., Soil management to enhance bacterivore and fungivore nematode populations and their nitrogen mineralization function, *Appl. Soil Ecol.*, **25**, 19–35 (2004).

Ferris, H. et al., Nitrogen mineralization by bacterial-feeding nematodes: Verification and measurement, *Plant Soil*, **203**, 159–171 (1998).

Fiscus, D.A. and Neher, D.A., Distinguishing nematode genera based on relative sensitivity to physical and chemical disturbances, *Ecol. Appl.*, **12**, 565–575 (2002).

Foissner, W., Soil protozoa as bioindicators: Pros and cons, methods, diversity, representative examples, *Agric. Ecosyst. Environ.*, **74**, 95–112 (1999).

Fu, S. et al., Responses of trophic groups of soil nematodes to residue application under conventional tillage and no-till regimes, *Soil Biol. Biochem.*, **32**, 1731–1741 (2000).

Gobat, J.-M., Aragon, M., and Matthey, W., *The Living Soil: Fundamentals of Soil Science and Soil Biology*, Science Publishers, Enfield, NH (2004).

Gorres, J.H. et al., Grazing in a porous environment: 1. The effect of soil pore structure on C and N mineralization, *Plant Soil*, **212**, 75–83 (1999).

Gupta, V.V.S.R. and Yeates, G.W., Soil microfauna as bioindicators of soil health, In: *Biological Indicators of Soil Health*, Pankhurst, C.E., Doube, B.M., and Gupta, V.V.S.R., Eds., CABI Publishing, Wallingford, UK, 201–233 (1997).

Hassink, J. et al., Relationships between soil texture, physical protection of organic matter, soil biota, and C and N mineralization in grassland soils, *Geoderma*, **57**, 105–128 (1993).

Hunt, H.W. et al., The detrital food web in a shortgrass prairie, *Biol. Fertil. Soils*, **3**, 57–68 (1987).

Ingham, R.E. et al., Interactions of bacteria, fungi, and their nematode grazers: Effects on nutrient cycling and plant growth, *Ecol. Monogr.*, **55**, 119–140 (1985).

Jenkinson, D.S., The soil biomass, *N. Z. Soil News*, **25**, 213–218 (1977).

Jones, F.G.W., Larbey, D.W., and Parrott, D.M., The influence of soil structure and moisture on nematodes, especially *Xiphinema*, *Longidorus*, *Trichodorus* and *Heterodera* spp, *Soil Biol. Biochem.*, **1**, 153–165 (1969).

Kimpinski, J. and Sturz, A.V., Managing crop root zone ecosystems for prevention of harmful and encouragement of beneficial nematodes, *Soil Till. Res.*, **72**, 213–221 (2003).

Kinchin, I.M., *The Biology of Tardigrades*, Portland Press, London (1994).

Larsen, T., Schjonning, P., and Axelsen, J., The impact of soil compaction on euedaphic Collembola, *Appl. Soil Ecol.*, **26**, 273–281 (2004).

McSorley, R., Alternative practices for managing plant-parasitic nematodes, *Am. J. Altern. Agric.*, **13**, 98–104 (1998).

McSorley, R., Multiple cropping systems for nematode management: A review, *Soil Crop Sci. Soc. Fla Proc.*, **60**, 132–142 (2001).

McSorley, R. and Frederick, J.J., Effect of subsurface clay on nematode communities in a sandy soil, *Appl. Soil Ecol.*, **19**, 1–11 (2002).

Mulder, C.H. et al., Observational and simulated evidence of ecological shifts within the soil nematode community of agroecosystems under conventional and organic farming, *Funct. Ecol.*, **17**, 516–525 (2003).

Neher, D.A., Soil community composition and ecosystem processes, *Agroforest. Syst.*, **45**, 159–185 (1999).

Pankhurst, C.E. et al., Evaluation of soil biological properties as potential bioindicators of soil health, *Aust. J. Exp. Agric.*, **35**, 1015–1028 (1995).

Pyrowolakis, A., Schuster, R.P., and Sikora, R.A., Effect of cropping pattern and green manure on the antagonistic potential and the diversity of egg pathogenic fungi in fields with *Heterodera schachtii* infection, *Nematology*, **1**, 165–171 (1999).

Savin, M.C. et al., Uncoupling of carbon and nitrogen mineralization: Role of microbivorous nematodes, *Soil Biol. Biochem.*, **33**, 1463–1472 (2001).

Sterner, R.W. and Esler, J.J., *Ecological Stoichiometry: The Biology of Elements from Molecules to the Biosphere*, Princeton University Press, Princeton, NJ (2002).

Stirling, G., *Biological Control of Plant Parasitic Nematodes: Progress, Problems and Prospects*, CAB International, Wallingford, UK (1991).

Van Vliet, P.C.J., Gupta, V.V.S.R., and Abbott, L.K., Soil biota and crop residue decomposition during summer and autumn in southwestern Australia, *Appl. Soil Ecol.*, **14**, 11–124 (2000).

Verhoef, H.A. and Brussaard, L., Decomposition and nitrogen mineralisation in natural and agroecosystems: The contribution of soil animals, *Biogeochemistry*, **11**, 175–211 (1990).

Vetter, S. et al., Limitations of faunal effects on soil carbon flow: Density dependence, biotic regulation and mutual inhibition, *Soil Biol. Biochem.*, **36**, 387–397 (2004).

Villenave, C. et al., Changes in nematode communities following cultivation of soils after fallow periods of different length, *Appl. Soil Ecol.*, **17**, 43–52 (2001).

Waller, P.J. et al., Evaluation of biological control of sheep parasites using *Duddingtonia flagrans* under commercial farming conditions on the island of Gotland, Sweden, *Vet. Parasitol.*, **126**, 199–315 (2004).

Wardle, D.A., *Communities and Ecosystems: Linking the Aboveground and Belowground Components*, Princeton University Press, Princeton, NJ (2002).

Wright, C.J. and Coleman, D.C., Response of soil microbial biomass, nematode trophic groups, N-mineralization, and litter decomposition to disturbance events in the southern Appalachians, *Soil Biol. Biochem.*, **34**, 13–25 (2002).

Yeates, G.W., Populations of nematode genera in soils under pasture. I: Seasonal dynamics in dryland and irrigated pasture on a southern yellow–grey earth, *N. Z. J. Agric. Res.*, **21**, 321–330 (1978).

Yeates, G.W., Effects of plants on nematode community structure, *Annu. Rev. Phytopathol.*, **37**, 127–149 (1999).

Yeates, G.W., Diversity of nematodes, In: *Biodiversity in Agricultural Production Systems*, Benckiser, G. and Schnell, S., Eds., Marcel Dekker, New York (2005).

Yeates, G.W. and Wardle, D.A., Nematodes as predators and prey: Relationships to biological control and soil processes, *Pedobiologia*, **40**, 43–50 (1996).

Yeates, G.W., Dando, J., and Shepherd, T.G., Pressure plate studies to determine how soil moisture affects access of bacterial-feeding nematodes to food in soil, *Eur. J. Soil Sci.*, **53**, 355–365 (2002).

Yeates, G.W. et al., Increase in ^{14}C-carbon translocation to the soil microbial biomass when five plant-parasitic nematodes infect roots of white clover, *Nematology*, **1**, 295–300 (1999).

11

Soil Fauna Impacts on Soil Physical Properties

Elisée Ouédraogo,[1] Abdoulaye Mando[2] and Lijbert Brussaard[3]

[1]*Agroecology Department, Albert Schweitzer Center for Ecology, Ouagadougou, Burkina Faso*
[2]*International Center for Soil Fertility and Agricultural Development (IFDC), Lomé, Togo*
[3]*Department of Soil Quality, Wageningen University, Wageningen, The Netherlands*

CONTENTS

The physical degradation of soils is a major problem in many countries, and especially in areas where the efficiency of rainwater use determines the success of crop production (Kiepe et al., 2001; Ouédraogo et al., 2006). The deterioration of the physical properties of soil can lead to the collapse of the soil's structure (Lal et al., 1989; Pieri, 1989). After this, soil system recovery is no longer possible without the action of flora or fauna (including humans) that create the conditions necessary for restoring the structural integrity of the soil.

Plant roots in association with soil microbes and other organisms can reconstruct soil structure over time, as seen in slash-and-burn systems with a long fallow period. Human interventions such as ploughing make changes more quickly, although some plants with aggressive rooting capabilities such as Brachiaria and finger millet (*Eleusine coracana*) are able to have quick and beneficial effects on soil structure, even in very degraded situations. Various different organisms, such as basidiomycete fungi, can stabilize and enhance soil structure, while root systems play an important complementary role by "injecting" carbon into the soil.

In parallel, soil structure can be improved by the activities of various soil fauna which create a more hospitable environment for plant growth, with this growth itself contributing to the process of soil improvement. A number of species, particularly earthworms, termites, beetles, and ants, collectively known as macrofauna, are an important part of the soil food web and play crucial roles as "ecosystem engineers" (Jones et al., 1994).

To have productive soil systems, it is necessary to have soil physical properties that are conducive to plant root growth and enhance the efficiency of rainwater use, especially in areas facing climatic constraints (Ouédraogo et al., 2006). Trying to improve soil physical properties with the use of heavy machinery is often inappropriate, whether due to the high cost of this technology, which in developing countries is seldom within the financial reach of farmers, or due to the weak structure of certain soils given their low organic matter and clay content (Mando, 1997).

In low-input agricultural systems, because of the increased demographic pressure and decreasing yields, earlier successful practices for restoring and maintaining soil fertility with shifting cultivation and long-term fallows have been replaced by exploitative continuous cropping. As farmers' yields decrease, area expansion is often the only means available to increase the absolute amount of food produced, and "marginal" lands are increasingly brought under cultivation (Mukwunye et al., 1996). The physical properties of these soils are often a major constraint to their productivity. This chapter analyses the role of soil faunal activity in improving these properties and biological alternatives for achieving such improvement.

The contribution of earthworms to enhancing soil quality is widely recognized, even if few soil management strategies systematically aim to enhance their populations and performance. Termites can be similarly effective; however, their roles in the functioning of soil systems are less well known. Earthworms are also considered in Chapter 31 in connection with vermicomposting and in Chapter 37 on improving phosphorus availability, so we focus here more on termites, appreciating their prominence in tropical soil systems, many of which are constrained physically in ways that termites as well as earthworms can mitigate.

11.1   Parameters of Soil Structure That Affect Crop Production

Soil structure has numerous effects on soil fertility through different processes:

- Physical processes: erosion, runoff, infiltration, aeration, drainage, water retention, soil evaporation, and thermal and mechanical properties of soil.
- Nutrient cycling: mineralization, immobilization, and ion exchange, with physical benefits from capillary action and biological contributions from root activity.
- Carbon cycling: respiration, organic inputs, root and microorganism turnover, decomposition, humification, and physical protection of organic matter.
- Biological activities: movement of soil fauna, and microorganism activity (Blanchart et al., 1999).

In semiarid areas such as the Sahelian zone of Africa, the combined effect of organic matter depletion due to overgrazing, continuous cultivation, and adverse climatic conditions has resulted in the marked deterioration of soil physical properties. This is seen in low-rain infiltration capacity, nutrient imbalance, reduced biodiversity, and low primary production. Mando (1997) found that runoff on these soils may be close to 100%, and the deteriorated soil structure does not permit establishment of any vegetation. Efforts to rehabilitate these soils by constructing bands of stone lines, sowing grass, or planting trees were often hampered by the crusted state of the degraded soils, which limits the infiltration that is necessary to achieve land rehabilitation. The most common response of farmers to this phenomenon has been to abandon cultivated land, although, as shown in Chapter 26, an indigenous technology in the West African Sahel (zaï holes) can reclaim infertile soil through a physical intervention that is augmented by stimulation of biological activity.

11.2   Soil Fauna That Improve Soil Physical Properties

Various soil fauna have important direct effects on soil physical properties by being able to dig into, ingest, and/or transport soil materials. Their indirect contribution derives from the way they affect the dynamics of soil organic matter and from the effects of the structures that they create (mounds, casts, burrows, etc.), or from their impacts on the population dynamics of other decomposer biota (Beare et al., 1997; Hendrix et al., 1998).

Termites and earthworms are the most important soil fauna affecting soil physical properties. Their direct contribution to the improvement of soil physical properties comes from opening of voids, their burrowing activities, and litter decomposition (Marinissen and Hillenaar 1997; Blanchart et al., 1997; Van Vliet et al., 1998; Ouédraogo et al., 2004). Termites are considered first as they are less well known than earthworms.

11.2.1   Termites

Termites, belonging to the order Isoptera, are polymorphic social insects which live in nests (termitaria) of their own construction (Lee and Wood, 1971). They are the major decomposers in most tropical terrestrial ecosystems and are responsible for the mineralization of up to 30% of net primary production and for the breakdown of up to 60% of litter fall (Bouillon and Malhot, 1965; Bachelier, 1978; Mando, 1997; Brussaard et al., 1997). Their intense interaction with the soil makes them the most important soil fauna in the arid and semiarid tropics (Lee and Wood, 1971; Bachelier 1978; Lobry de Bruyn and Conacher, 1990). Both biotic and abiotic conditions in the soil affect termite populations and activity. Soil water content is a key element affecting termite activity, given that termites are very susceptible to desiccation because their soft cuticle has poor water-retaining properties (Moore, 1969).

Three main functional groups of termites have been defined on the basis of their respective food sources.

1. Herbivorous termites feed on plant material, either fresh or dead (Wood, 1996);
2. Fungivorous termites ingest fungi, either in the soil, at the soil/wood interface, in wood, and/or in litter (Jones and Eggleton, 2000);
3. Humivorous termines depend on humus for their energy and nutrition.

Concurrently, two main groups are distinguished according to their nesting behavior:

1. Soil-nesting species, mound-building, and subterranean species all have below-ground nests; while
2. Nonsoil nesting species include arboreal nesting species in Amazonia (Martius, 1994).

Termites live in nests, in gallery systems, or in sheetings made from soil or from a mixture of soil and other material, established either within the soil horizons or on the soil's surface (sheetings are thin layers of fine soil transported to the surface and constructed around herbaceous or woody plants). Construction processes affect the physical and chemical status of both the material used for the construction and the surrounding soil from which the materials were taken.

The soil structure and its structural stability, porosity, and chemical status are all considerably altered by termite activity (Lee and Wood, 1971; Elkins et al., 1986; Lobry de Bruyn and Conacher, 1990; Mando, 1991; Lee and Foster, 1991; Humphreys, 1994). Termites' behavior in selecting, transporting, and manipulating soil particles and cementing them together with saliva brings immediate and important changes in soil structure and properties (Wood, 1996; Mando, 1997). Although most termites do not prefer live materials for food, there are a number of termite species that may become crop pests. It has been noted that this problem is the most severe when crops are already under stress due to drought and/or nutrient deficiencies (Wood, 1996). In fact, the activities of "ecosystem engineers" generally prevent such stress conditions by promoting greater water infiltration and by increasing water- and nutrient-use efficiencies, as discussed in Section 11.4.

11.2.2 Earthworms

Earthworms affect soil physical properties similarly through the galleries they create in the soil and by their upward and downward transportation of soil. Soil aggregation is one of the important contributions that earthworms make to soil structure modification. Blanchart et al. (1999) have reported that, after 14 months of experimentation, the soils in a treatment without earthworms had a much smaller percentage of aggregates >2 mm (5%) compared with soils that were treated with earthworms (45%).

An important part of earthworms' contribution to the formation of soil aggregates is by creating unique structures (casts) as they move through the soil. Fragmentation of organic residues, clay dispersion, addition of polysaccharides, and intense mixing of mineral and organic materials in the earthworms' guts all stimulate the formation of microaggregates in earthworm casts (Pulleman et al., 2005a). The size of microaggregates is influenced by the fact that worms can only ingest aggregates of the size of their mouths, or it is the result of splitting old casts into smaller ones, as done by small filiform earthworms.

Earthworms' impact on soil structure is greatly affected by soil management practices (Brussaard et al., 1990). Using thin sections, Pulleman et al. (2005b) showed that the volume (%) of worm-worked soil of undisturbed soil samples was 49–54% in permanent pasture, 23–33% in an organic arable system, and only 5–10% in a conventional arable system. The reduced impact of earthworms on soil structure in conventionally-cultivated soil compared to permanent pasture arises from the negative effects of tillage and soil compaction and from reduced organic inputs. Manure application and the absence of pesticides in an organic arable system account for the more beneficial effect of earthworms on its soil structure.

The impact of earthworms on soil physical properties relies, in large part, on the huge amount of reconstituted soil deposited during their casting activity. In subhumid West African savannas, deposits on the soil surface of certain earthworms (*Millsonia inermis*) have been measured at 30–74 ton ha^{-1} dry weight by Ouédraogo et al. (2006). In subhumid savannas, the amount of soil that annually passes through the guts of earthworms has been estimated at 250–1250 ton ha^{-1} (Lavelle et al., 1994).

The impact of earthworms on soil physical properties also depends on earthworm species and the type of soils. Lavelle and co-workers have distinguished two types of earthworm casts: globular casts that are created by compacting earthworms (their casting activity increases the bulk density of soil in their casts), and granular ones produced by small filiform earthworms (Chuniodrilus spp.). The latter are sometimes called decompacting earthworms because they break down the large casts produced by compacting earthworms (Lavelle et al., 1999).

Soil texture also has an influence on earthworm cast constitution. Certain earthworm species ingest soil particles selectively, but the texture of other earthworm species' casts is similar to the surrounding soil (Henrot and Brussaard, 1997). The effects of earthworms, as in efforts to revegetate inert soils, are discussed in Chapter 42. The contributions that earthworms make to improving the effectiveness of compost, by production of vermi-compost, are considered in Chapter 31, and to P availability in section 37.1.1.

11.2.3 Other Fauna

Other organisms also play roles as "ecosystem engineers," for example, the *Scarabaeoidea*, better known as scarab beetles. Some of these are root-feeding, as discussed in Chapter 41, but others, which are more numerous in undisturbed (untilled) soil with high biodiversity, construct large numbers of tunnels (some $>70 \, \text{m}^{-2}$). These can extend >1 m into the soil from the surface and permit the beetles to take plant litter or dung down into the soil, where they (or their larvae) ingest it, contributing to decomposition and mineralization (Brown and Oliveira, 2004).

Many other species play similar roles, contributing to the energy and mineral stores in the soil by playing complementary parts in the process of nutrient cycling (Lavelle et al., 1998). Each has its own life cycle and fits somewhat differently into the soil food web, discussed in Chapter 5. In the next section we will elaborate on what is known about termites as ecosystem engineers, appreciating that their contributions are only some of those made in a continuous drama that has a cast not of thousands but of billions and trillions.

11.3 How Termites Rehabilitate Degraded Crusted Soils

Termite-mediated processes are traditionally important in many land management practices in the Sahel. These practices include zaï holes (Kaboré, 1994) and mulching (Mando, 1997). Before termites will rework degraded soil and restore its physical properties so that it becomes suitable for vegetative rehabilitation, they need to be stimulated and supported by some food source, i.e., by organic resources.

Not surprisingly, the quality of organic resources affects the contribution that soil fauna can make to decomposition and to the improvement of soil physical properties. Tian et al. (1993) showed that in Nigeria, termite populations increase with mulching, and termite mulch consumption is higher on plant residues that have low nutritional quality. The plant residue quality index developed by Tian et al. (1995), based on the relationship between the chemical composition of organic material (C:N ratio, lignin content, and polyphenolic content) was negatively correlated with termite densities.

Ouédraogo et al. (2004) have shown that the decomposition of organic resources by abiotic and microorganism influences only, i.e., in the absence of soil macrofauna, was very slow in an annual crop production cycle under semiarid conditions. Soil macrofauna mediated the disappearance of these organic materials depending on their quality; termites actually preferred recalcitrant organic material over high-quality organic material. This suggests that application of low-quality organic material mulch can be a good way to trigger termite activity (Figure 11.1).

Mando (1997) has investigated the role played by soil fauna in the rehabilitation of crusted soils with special attention to termites, the most common genera being Microtermes, Macrotermes, and Odontotermes. The study area was Bourzanga, northern Burkina Faso (13° 26′ 15″ W, 1° 37′30″ N), with a Sahelian–Sudanian climate. Rainfall is unimodal, occurring from June to September, with vegetation of the steppe

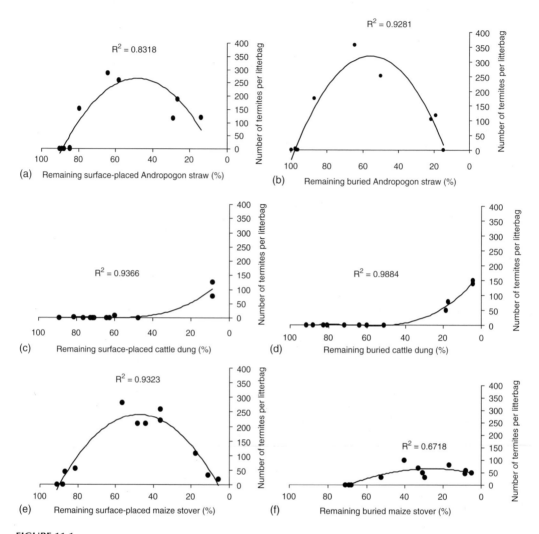

FIGURE 11.1

Correlations between termite density and the remaining surface-placed or buried organic materials, August to October 2000, in semiarid Burkina Faso. The decline in termite number is likely due to resource depletion. *Source:* Ouédraogo, E, Mando, A., and Brussaard, L., Soil macrofaunal-mediated organic resource disappearance in semi-arid West Africa, *Appl. Soil Ecol.*, **27**, 259–267 (2004).

type, according to UNESCO's classification (1973), and also with large bare areas. The dominant grasses are annual therophytes, and perennial grasses are rare (Mando, 1991). Woody vegetation is mostly shrubs of the Mimosaceae and Combretaceae families. Prevailing soil types in the area are Lixisols and Cambisols. Annual runoff is very high, 60–80% of annual rainfall on Cambisols and 50–90% on Lixisols, because of their degraded status caused by continuous cultivation, overgrazing, trampling by cattle, and rainfall intensity. While soil crusting is the major problem, these soils also suffer from nutrient deficits.

An experiment was conducted on wasteland consisting of completely bare crusted soil without any vegetation. Mulch treatments were randomly applied in these plots. The straw used was *Pennisetum pedicellatum* applied at 3 ton ha^{-1}; woody material of *Pterocarpus lucens* was applied at 6 ton ha^{-1}, and a composite mulch (woody material and straw) at 4 ton ha^{-1}. In addition, there was a control plot with no mulch (bare plot). Surface-placed organic material on crusted soil was used to attract termites (Odonto-termes and Microtermes spp.) to test the hypothesis that the presence of dry vegetal material on structurally-crusted soil can trigger termite activity and may thereby improve soil infiltration sufficiently to activate vegetation establishment. Termite activity in the mulched plots was excluded with the application of insecticide (dieldrin) and compared to the bare soil and to mulched soil without insecticide.

The subterranean Macrotermitinae Odontotermes *smeathmani* (Fuller) and Microtermes *lepidus Sjöstadt* were the dominant species found. On all mulched plots, termite activity could be seen from the presence of open macropores, up to 1 cm in diameter. Such voids are made by Odontotermes *smeathmani* (Fuller) according to Kooyman and Onck (1987) since Microtermes voids are less than 2 mm in diameter. Sheetings of organic material, with bridged grain structures, were constructed by termites for protection (Table 11.1) (Mando and Miedema, 1997). Mulched plots (using woody material + straw) had both a large number of macropores (86 ± 20 ha^{-1}) and a large amount of sheetings (10.7 ± 3.9 ton ha^{-1}). Bare plots showed no traces of termite-made features in the topsoil. Within 2 weeks, the termites dug out over 10 ton ha^{-1} from the soil and had reworked it into sheetings.

In the profiles of bare plots, the soil at 0–10 cm depth showed a compact grain microstructure with primary coatings and a locally more dense grain microstructure. No aggregates were found in this layer, and the voids at the top of the thin section were mostly packing voids. At depths of 30 cm and 100 cm, the pedofeatures and structure of bare plots give a clear indication of previous termite activity, whereas any features such as chambers and channels in the topsoil had been completely destroyed by the degrading processes acting on the soil. Thin coatings of silt suggest that sediment yield in water flow may be responsible for pores being filled up. Profiles where termite activity was present differed basically from those without current termite activity in that they have large voids (channels and chambers) in the 0–10 cm layer (Table 11.1).

Water infiltration, measured as the difference between rainfall and runoff, showed that infiltration (expressed in % of annual rainfall) increased in both with-fauna and without-fauna plots that had mulch application. The increase was significantly larger in termite plots, however (Table 11.2). A multiple regression analysis which included soil cover data,

TABLE 11.1

Termite-Made Voids and Infillings in Mulched Termite Plots and Bare Plots (NT)

Parameters		Depth (cm)	% Thin Section Area	ECD[a] (mm)	Number of Voids
Voids	T[b]	0–7	11.8 ± 2.7[c]	1.1–12	11 ± 3
		30–37	3.44 ± 1.6	1.2–4.1	14 ± 3
	NT[d]	0–7	0.0 ± 0.0	N/A	0 ± 0
		30–37	3.6 ± 0.4	1.1–9.2	12 ± 2
Infillings	T	0–7	1.2 ± 1.2	1.3–13.2	2 ± 0
		30–37	17.5 ± 3.5	1.01–18.5	8 ± 5
	NT	0–7	0.0 ± 0.0	N/A	0 ± 0
		30–37	15.5 ± 2.5	0.8–9.2	14 ± 3

[a] ECD = equivalent circle diameter (results reported are the smallest and biggest ECDs).

[b] T = termite plots (mulched).

[c] $\pm$ standard deviation ($n = 3$).

[d] NT = non-termite plots (bare).

Source: From Mando A. and Miedema R., *Appl. Soil Ecol.*, **6**, 261–263 (1997). With permission.

TABLE 11.2

Ponding Time (PT), Runoff Time (RT), and Cumulative Infiltration (Ic) after 30 min of Simulated Rain

Treatment[a]	PT (min)	RT (min)	I_C (mm)
Termite plots: 1st rain	8'15" a	52'25" a	12.9 a
Termite plots: 2nd rain	6'20" a	50'12" ab	11.89 a
Termite plots: 3rd rain	7'35" a	46'32" ab	9.2 ab
Non-termite plots: 1st rain	4'13" a	49'7" ab	5.28 b
Non-termite plots: 2nd rain	5'13" a	45'57" b	5.28 b
Non-termite plots: 3rd rain	5'13" a	45'20" b	4.51 b

[a] Treatments having the same letter(s) in the same row are not significantly different.

Source: From Mando, A., Stroosnijder, L., and Brussaard, L., Effects of termites on infiltration into crusted soil, *Geoderma,* **74**, 107–113 (1996). With permission.

sediment accumulation, and termite-made voids pointed to these voids as the only factor that had significantly influenced runoff on the plots. This was due to the fact that the burrowing activity of soil fauna loosened the soil and provided voids for water flow into the soil and throughout the soil profile. If runoff is controlled, it is evident that erosion will also be reduced and soil water content will be increased. Voids increased the time until ponding and delayed the onset of runoff (Mando et al., 1996).

The results on vegetation rehabilitation highlighted the fact that non-termite plots responded weakly to mulch-only treatments, but even in the first year, vegetation established well on termite + mulch plots. Termite activity resulted in an increase of plant cover, plant species number, phytomass production, and rainfall use efficiency. Bare plots remained bare throughout the experiment. Analysis of the termite and mulch interaction indicated that mulch plots without termites did not perform better than bare plots, especially in the case of woody plant regeneration (Mando et al., 1999).

11.4 Soil Macrofauna Diversity and Improvement of Soil Physical Properties

The relationships between soil biodiversity and soil function are complex and poorly understood. When discussing the implications of changes in soil biodiversity for ecosystem function, it is important to identify the level of taxonomic resolution being used to describe the diversity and the spatial scale at which diversity–function relationships are considered (Vandermeer et al., 1988). The functional attributes of species are a more important consideration than a fine level of taxonomic resolution for assessing impact on ecosystem processes (Beare et al., 1997; Brussaard, 1998).

A study of soil macrofaunal diversity in the central plateau of Burkina Faso by Ouédraogo et al. (2002) has indicated that organic resources and soil management can influence soil macrofaunal diversity, and thereby soil physical properties, with a subsequent impact on crop performance. The experiment evaluated organic resources of different qualities, singly or combined, used with mineral fertilizer (or none) under a till (or no-till) system. Sorghum (*Sorghum bicolor* L. Moench) was sown in all the plots. Organic materials and fertilizer (urea) were applied before sowing and before the plots were tilled with animal power. No pesticide was used during the cropping period or in the surrounding area.

The methods of the Tropical Soil Biology and Fertility (TSBF) Program were used to estimate soil faunal numbers in the experimental plots after harvest. A soil core

$(30 \times 30 \times 30 \text{ cm}^3)$ was taken in each plot after harvest. The core was divided into two layers, 0–10 cm and 10–30 cm. All soil fauna were collected and identified. Diversity in this analysis was regarded as diversity in the different groups of soil fauna, following the classical "diversity index" suggested by Hill (1973).

The results, some of which are summarized in Figure 11.2, showed that soil macrofauna composition is indeed affected by soil management practices and by organic resource allocations. Predators (ants and spiders) were more important in no-till plots than tilled plots probably because access to prey is easier in surface-placed organic resources. Termites were more numerous in the 10–30 cm layer in tilled plots. This may be attributed to the nesting behavior of the termites species found (Odontotermes spp.). These are beneath the tilled layer (0–12 cm) and are not destroyed by tillage. In the 0–10 cm layer, a high diversity of soil macrofauna was observed in the sheep-dung treatments and the maize straw treatments (Ouédraogo et al., 2002).

Easily decomposable organic material (e.g., sheep dung with C:N = 16) promoted more soil fauna diversity as availability of nutrients allowed many soil organisms of the food web to contribute to decomposition. In slowly decomposable organic material such as maize straw, however, a few species of soil organisms dominate the decomposition and predation processes. Termites are known to feed on low-quality organic material, and this explains their dominance in maize straw treatment (Ouédraogo et al., 2004).

As ecosystem engineers, termites engage in organic material breakdown, soil bioturbation, and habitat creation for other soil organisms. Ants were obviously attracted

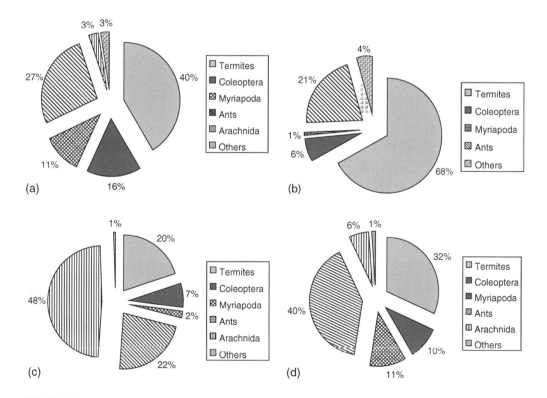

FIGURE 11.2

(a,b) Soil fauna components in tilled plots. (a) = 0–10 cm layer; (b) = 10–30 cm layer. (c,d) Soil fauna components in non tilled plots, (c) = 0–10 cm, (d) = 10–30 cm.

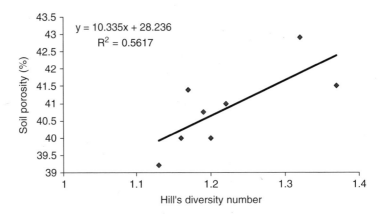

FIGURE 11.3
Correlation between soil macrofaunal diversity and soil porosity for 0–10 cm layer in semiarid Burkina Faso.

by termites' activity as the numbers of termites and ants were correlated. Ants are representatives of predators, granivores, and bioturbators, which also bring about important changes in the physical and chemical properties of soil (Brussaard et al., 1997). Coleoptera and Myriapoda are representatives of detritivore groups which live in rich litter environments and in soil with relatively high organic matter (Bachelier, 1978; Chinery, 1986).

The addition of urea increased nutrient availability for soil organisms in maize straw, leading to higher soil fauna diversity than in the single maize straw application. In the sheep-dung treatment, urea addition led to the reverse. Correlation between Hill's diversity number and crop performance showed a strong association in maize straw + urea treatments (Ouédraogo et al., 2002). Although the relationship between soil macrofaunal diversity and physical soil properties may be site-specific and linked to the composition of the soil fauna, Ouédraogo et al. (2002) found that soil porosity was correlated with Hill's diversity number (Figure 11.3). These results demonstrate that changes in management can create conditions that allow "the soil to work for us" (Elliott and Coleman, 1988).

11.5 Discussion

Maintenance of soil functions is related to the maintenance of beneficial effects from soil fauna communities. Soil fauna are a key element to be considered in soil management systems for land rehabilitation in countries where water infiltration on degraded soil is a problem. Mechanical methods such as soil tillage have been reported to be unsustainable to control crusting because rainfall builds up a new crust when soil structure has been degraded (Stroosnijder and Hoogmoed, 1984). Management practices that eliminate soil macrofauna also eliminate their beneficial contribution to soil quality maintenance. Soil biodiversity may be indirectly promoted by manipulating the quality of organic materials applied for the improvement of both soil physical properties and crop performance. Enhancing populations of soil fauna represents a biological strategy for the alleviation of soil physical constraints.

References

Bachelier, G., *La Faune du Sol: Son Action*, Editions de l'ORSTOM, Paris (1978).

Beare, M.H. et al., Agricultural intensification, soil biodiversity and agroecosystem function in the tropics: The role of decomposer biota, *Appl. Soil Ecol.*, **6**, 87–108 (1997).

Blanchart, E. et al., Regulation of soil structure by geophagous earthworm activities in humid savannas of Côte d'Ivoire, *Soil Biol. Biochem.*, **29**, 431–439 (1997).

Blanchart, E. et al., Effects of earthworms on soil structure and physical properties, In: *Earthworms Management in Tropical Agroecosystems*, Lavelle, P., Brussaard, L., and Hendrix, P., Eds., CAB International, Wallingford, UK, 149–197 (1999).

Bouillon, A. and Malhot, G., *Quel est ce termite Africain?*, Université Leopoldville, Congo (1965).

Brown, G.G., and Oliveira, L.J., White grubs as agricultural pests and as ecosystem engineers, *Abstract for XIV International Colloquium on Soil Zoology and Ecology*, 30 August–3 September, Rouen, France (2004).

Brussaard, L., Soil fauna, guilds, functional groups and ecosystem processes, *Appl. Soil Ecol.*, **9**, 123–135 (1998).

Brussaard, L. et al., Biodiversity and ecosystem functioning in soil, *Ambio*, **26**, 563–570 (1997).

Brussaard, L. et al., Biomass, composition and temporal dynamics of soils organisms of a silt loam soil under conventional and integrated management, *Neth. J. Agric. Sci.*, **38**, 283–302 (1990).

Chinery, M., *Le multilingue nature des insectes d'Europe en couleurs*, Bordas, Paris (1986).

Elkins, N.Z. et al., The influence of subterranean termites on the hydrological characteristics of a Chihuahuan desert ecosystem, *Oecologia*, **68**, 521–528 (1986).

Elliott, E.T. and Coleman, D.C., Let the soil work for us, *Ecol. Bull.*, **39**, 23–32 (1988).

Hendrix, P.F., Long-term effects of earthworms on microbial biomass nitrogen in coarse and fine textured soils, *Appl. Soil Ecol.*, **9**, 375–380 (1998).

Henrot, J. and Brussaard, L., Abundance, casting activity, and cast quality of earthworms in an acid Ultisol under alley-cropping in the humid tropics, *Appl. Soil Ecol.*, **6**, 169–179 (1997).

Hill, M.O., Diversity and evenness: A unifying notation and its consequences, *Ecology*, **72**, 1374–1382 (1973).

Humphreys, G.S., Bioturbation, biofabrics and biomantle: An example from the Sydney Basin, In: *Soil Micromorphology: Studies in Management and Genesis*, *Dev. Soil Sci.*, 22. Ringrose-Voase, A.J. and Humphreys, G.S., Eds., 421–436 (1994).

Jones, D.T. and Eggleton, P., Sampling termite assemblages in tropical forests: Testing a rapid biodiversity assessment protocol, *J. Appl. Ecol.*, **37**, 191–203 (2000).

Jones, C.G., Lawton, J.H., and Shachak, M., Organisms as ecosystem engineers, *Oikos*, **69**, 373–386 (1994).

Kaboré, V., Amélioration de la production végétale des sols dégradés (zippélla) du Burkina Faso par la technique des poquets (Zaï), Thèse de doctorat, École Polytechnique Fédérale de Lausanne, Lausanne, Switzerland (1994).

Kiepe, P. et al., Dynamics of soil resources in Sahelian villages, In: *Agro-silvo-pastoral Land Use in Sahelian Villages*, *Advances in Geoecology*, 33, Stroosnijder, L. and Van Rheenen, T., Eds., Catena Verlag, Reskirchen, 131–141 (2001).

Kooyman, C. and Onck, R.F.M., The interactions between termite activity, agricultural practices and soil characteristics in Kisii district, Kenya, *Wageningen Agricultural University Papers*, **87** (3), 1–120 (1987).

Lal, R., Hall, F.J., and Miller, F.P., Soil degradation: Basic processes, *Land Degrad. Rehab.*, **1**, 51–69 (1989).

Lavelle, P., Brussaard, L., and Hendrix, P., *Earthworm management in Tropical Agroecosystems*, CAB International, Wallingford, UK (1999).

Lavelle, P. et al., The relationship between soil macrofauna and tropical soil fertility, In: *The Biological Management of Tropical Soil Fertility*, Woomer, P.L. and Swift, M.J., Eds., Wiley-Sayce, Sussex and Exeter, UK, 137–139 (1994).

Lavelle, P. et al., Soil function in changing world: The role of invertebrate ecosystem engineers, *Eur. J. Soil Biol.*, **33**, 159–193 (1998).

Lee, K.E. and Foster, R.C., Soil fauna and soil structure, *Aus. J. Soil Res.*, **6**, 745–775 (1991).

Lee, K.E. and Wood, T.G., *Termites and Soils*, Academic Press, London and New York (1971).

Lobry de Bruyn, L.A. and Conacher, A.J., The role of termites and ants in soil modification: A review, *Aus. J. Soil Res.*, **28**, 55–93 (1990).

Mando, A., *Impact de l'activité des termites sur la dégradation de la biomasse végétale et quelques propriétés physiques des sols dégradés: Etude menée a Zanamogo (Burkina Faso), Mémoire de fin d'études*, Université de Ouagadougou, Ougadougou, Burkina Faso (1991).

Mando, A., *The Role of Termites and Mulch in the Rehabilitation of Crusted Sahelian Soils, Tropical Resource Management Papers No 16*. Wageningen University, Wageningen, Netherlands (1997).

Mando, A. and Miedema, R., Termite-induced change in soil structure after mulching degraded (crusted) soil in the Sahel, *Appl. Soil Ecol.*, **6**, 261–263 (1997).

Mando, A., Brussaard, L., and Stroosnijder, L., Termite- and mulch-mediated rehabilitation of vegetation on crusted soil in West Africa, *Restor. Ecol.*, **7**, 33–41 (1999).

Mando, A., Stroosnijder, L., and Brussaard, L., Effects of termites on infiltration into crusted soil, *Geoderma*, **74**, 107–113 (1996).

Marinissen, J.C.Y. and Hillenaar, S.I., Earthworm-induced distribution of organic matter in macro-aggregates from differently managed arable fields, *Soil Biol. Biochem.*, **29**, 391–395 (1997).

Martius, C., Diversity and ecology of termites in Amazonian forests, *Pedobiologia*, **38**, 407–428 (1994).

Moore, B.P., Biochemical studies, *Biology of Termites*, Vol. 1. Kirshma, K. and Weesner, F.M., Eds., Academic Press, New York, 407–432 (1969).

Mukwunye, A.U., Jager, A., and Smaling, E.M.A., Restoring and maintaining the productivity of West Africa soils: Key to sustainable development, *Miscellaneous Fertiliser Studies no 14*. International Fertilizer Development Centre, Mussel Shoals, AL, 61–72 (1996).

Ouédraogo, E. et al., Organic resources and earthworms affect phosphorus availability to crop after phosphate rock addition in semi-arid West Africa, *Biol. Fertil. Soils*, **41**, 458–465 (2005).

Ouédraogo, E., Mando, A., and Brussaard, L., Effect of organic resources management on soil biodiversity and crop performance under semi-arid conditions in West Africa, In: *Proceedings of the Workshop on Sustaining Agricultural Productivity and Enhancing Livelihoods through Optimization of Crop and Crop-Associated Biodiversity with Emphasis on Semi-Arid Tropical Agroecosystems, 23–25 September 2002, Patancheru, Andra Pradesh, India*, Waliyar, F., Colette, L., and Kenmore, P.E., Eds., International Crop Research Institute for the Semi-Arid Tropics/UN Food and Agriculture Organization, Patancheru, India, 89–102 (2002).

Ouédraogo, E., Mando, A., and Brussaard, L., Soil macrofaunal-mediated organic resource disappearance in semi-arid West Africa, *Appl. Soil Ecol.*, **27**, 259–267 (2004).

Ouédraogo, E., Mando, A., and Brussaard, L., Soil macrofauna affect crop nitrogen and water use efficiencies in semi-arid West Africa, *Eur. J. Soil Biol.* (2006), accepted for publication.

Pieri, C., *Fertilité des terres de savane: Bilan de trente années de recherche et de développement agricole au sud du Sahara*, Institut des Recherches Agronomique Tropicales, Paris (1989).

Pulleman, M.M. et al., Soil organic matter distribution and microaggregate characteristics as affected by agricultural management and earthworm activity, *Eur. J. Soil Sci.*, **56**, 453–467 (2005a).

Pulleman, M.M. et al., Earthworms and management affect organic matter incorporation and micro-aggregate formation in agricultural soils, *Appl. Soil Ecol.*, **29**, 1–15 (2005b).

Stroosnijder, L. and Hoogmoed, W.G., Crust formation on sandy soils in the Sahel, II: Tillage and its effects on the water balance, *Soil Till. Res.*, **4**, 321–337 (1984).

Tian, G., Brussaard, L., and Kang, B.T., Biological effects of plant residues with contrasting chemical compositions under tropical conditions: Effects on soil fauna, *Soil Biol. Biochem.*, **25**, 731–737 (1993).

Tian, G., Brussaard, L., and Kang, B.T., An index for assessing the quality of plant residues and evaluating their effects on soil and crop in the (sub-)humid tropics, *Appl. Soil Ecol.*, **2**, 25–32 (1995).

UNESCO, International classification and mapping of vegetation, *Ecol. Cons.*, **6**, UNESCO, Paris (1973).

Van Vliet, P.C.J. et al., Hydraulic conductivity and pore size distribution in small microcosms with and without enchytraeids (Oligochaeta), *Appl. Soil Ecol.*, **9**, 277–282 (1998).

Vandermeer, J. et al., Global change and multi-species agroecosystems: Concepts and issues, *Agric. Ecosyst. Environ.*, **67**, 1–22 (1998).

Wood, T.G., The agricultural importance of termites in the tropics, *Agric. Zool. Rev.*, **7**, 17–150 (1996).

12

Biological Nitrogen Fixation in Agroecosystems and in Plant Roots

Robert M. Boddey, Bruno J.R. Alves, Veronica M. Reis and Segundo Urquiaga
EMBRAPA Agrobiologia Center, Seropédica, Rio De Janeiro, Brazil

CONTENTS

Nitrogen is considered a key factor in agricultural production and in soil systems not only because most crops require and accumulate more of this nutrient than any other, but also because N can be lost from agroecosystems by various paths, due to natural processes but also through agronomic practices. These can contribute not only to erosion and leaching, which affect also other important nutrients, but losses also occur in gaseous forms (NH_3, N_2O, and N_2) via volatilization of ammonia and denitrification. Providing and replacing nitrogen in various inorganic forms is a costly matter for farmers, and often there are supply, logistic, or other constraints also making it difficult for farmers to provide sufficient N for their growing crops.

Many farming systems throughout less-developed countries are mainly dependent on biological nitrogen fixation (BNF) associated with legumes as their source of N for forage and grain crops, whether planted as monocrops, intercrops, green manures, or tree crops. Other cropping systems derive nitrogen from associative N_2-fixation by soil microorganisms living in, on, or around the roots of nonleguminous plant species. This chapter considers BNF within soil systems and farming systems, focusing first on the interactions between leguminous plants and microbes. These are widely known about although their intricacy and robustness are not so widely understood. The chapter then considers the less-known associations between other plant species and endophytic microorganisms in more detail. The latter contribute N to plants' nutrition in return for biochemical products from the plants conferring energy and other benefits. In a subsequent section, tradeoffs between inorganic and organic sources of nitrogen in soil systems are discussed as well as implications for soil system management. A follow-on discussion in Part III (Chapter 27) addresses ways in which the benefits of leguminous BNF can be mobilized to support more sustainable small-farm agricultural production.

12.1   Legume-Based Nitrogen-Fixing Systems

On a global scale, the largest inputs of BNF to terrestrial ecosystems come from nodulated legumes. Approximately 16,500 species of legumes have been catalogued worldwide, but not all of these are able to form N_2-fixing nodules. Less than 24% of known legume species have been studied for their capacity to form nodules on their roots. Studies to date have shown that among leguminous species, most Caesalpinioideae ($>$75%) are unable to nodulate; however, most Mimosoideae ($\sim$89%) and Papilionoideae ($\sim$96%) can (Faria et al., 1999). Most Caesalpinioideae and Mimosoideae are perennial woody legumes, tropical in origin. However, nearly all of the legume species that are used in conventional agriculture, i.e., grain, forage, and green-manure legumes, are Papilionoideae.

For many years the bacteria that infect legume roots to form N_2-fixing nodules were classified together in the genus Rhizobium. They were subsequently divided into species which showed either slow or fast growth on a solid culture medium, and Jordan (1982) suggested removing the slow-growing strains from the genus Rhizobium to create a new genus Bradyrhizobium for these strains. The latter included most of the bacterial strains that infect tropical grain species such as soybean (*Glycine max.*), cowpea (*Vigna unguiculata*), and forage and green-manure legumes, e.g., Stylosanthes, Centrosema, and Crotalaria spp.

Since the end of the 1980s, however, with the advent of molecular biology techniques to investigate the taxonomy of these bacteria (e.g., 16S RNA sequence analysis and DNA:DNA hybridization), there has been a proliferation of genera now identified, six at present, and of species, over 30 (Young and Haukka, 1996). The consensus today is that all bacteria capable of forming N_2-fixing nodules on the roots (and occasionally the stems) of legumes should be called rhizobium (neither italicized nor capitalized) or bacteria of the rhizobium group (Giller, 2001).

Nodules formed on legume roots differ widely in size and in many aspects of their structure, but all of these nodules have central (internal) zones that are infected by the rhizobium to form symbiosomes, or bacteroids. These include the cells containing the enzyme nitrogenase, which converts gaseous N_2 initially into ammonium and then into other compounds such as ureides, glutamine, and/or asparagine. These are the most important compounds that transport N in the xylem tissues to the aerial tissues of the plant (Table 12.1).

Much information on the diversity of nodule structures, processes of bacterial infection, and the probable evolutionary development of legume nodules can be found in Sprent (2001). There seems to be little correlation between the evolutionary development of nodule structure and the quantitative capacity of the legume to obtain N from BNF. Some nodules have relatively primitive structures, for example, such as those encountered on the legume *Mimosa scabrella*. This tree lacks fully developed bacteroids, having just "persistent infection threads" (Faria et al., 1987), yet these are able to provide the host plant with all of the N that it needs for rapid growth. This provision is carried out in the same manner as in legumes, such as soybean, that have determinate nodules and completely developed bacteroids within their nodules that exhibit extremely sophisticated systems for the control of oxygen diffusion. So the process is more important than structure when it comes to BNF though some structure is always necessary.

TABLE 12.1

Effect of Inoculation of *Gluconacetobacter diazotrophicus* and *Paenobacillus azotofixans* on Micro-Propagated Sugarcane var. NA 5679 Grown in Pot Soil and Harvested 7 Months after Transplanting and Acclimatization

Treatments	Fresh Biomass (g plant^{-1})	Dry Biomass (g plant^{-1})	Total N (mg plant^{-1})
Control	434.2b	189.2	304b
G. diazotrophicus PAL5	614.7a	220.4	388a
P. azotofixans 2RC1	490.3ab	244.4	349ab
Mixture PAL5 + 2RC1	506.9ab	229.2	360ab
CV %	13.1	14.4	

Source: Fabio L. Olivares, Universidade Estadual Norte Fluminese, Campos dos Goytacazes, State of Rio de Janeiro, Brazil.

12.1.1 Nodule Formation

The process of nodule formation has become a topic of intense interest to molecular biologists who are now able to track the exchange of molecular signals between the legume and rhizobium that guide and control the biological construction of root architecture for purposes of BNF. The exact mechanisms can differ somewhat between different legume/rhizobium symbioses, but the basic process is as follows:

1. Root exudates such as sugars and amino acids attract rhizobium (through chemotaxis) into the rhizosphere.

2. At this point, specific compounds of the flavonoid group which are produced by the plant roots reach the rhizobium and induce its genes to produce so-called polysaccharide "nod factors," officially known as lipo-chitin oligosaccharides (LCOs).

3. These LCOs produced by the bacteria then induce molecular activity in the *nod* and *nol* genes in the plant which start the process of nodulation. Different rhizobia produce different LCOs, and these differences in molecular structure largely determine which rhizobium bacteria will form nodules on which legume, controlling host-plant specificity.

4. As the rhizobia multiply, they become attached (usually in a polar manner) to the root surface. Examples of polar vs. supine attachment can be seen in Figure 3 of Chapter 8.

5. At this point the bacteria start to infect the plant, usually via root hairs, wounds or cracks, or between the cells of intact epidermises (Sprent, 2001).

6. The most studied system is root hair infection, where the root hair cell wall is breached (the exact enzymatic process is not known), and a tubular structure called the infection thread is formed from plant tissues.

7. Simultaneously, or even before infection of the root tissues, changes occur in the cortical tissues of the root to initiate the infrastructure of the nodule. Vascular tissues, both phloem and xylem, surround the infected cells and eventually provide them with carbon substrates from the host plant (through the phloem) and export the products of nitrogen fixation from the nodule to the rest of the plant (through the xylem).

8. This infection thread penetrates the root hair, and some or all of the infected cells become transformed into bacteroids. These specialized cells, which are formed

jointly by the genomes of the two organisms (legume and rhizobium), contain the nitrogenase and leghaemoglobin essential to the biochemical processes of N-fixation and transport.

Once the nodule is formed, nitrogen is continuously fixed in the bacteroids. These cells must, however, stay at a low oxygen concentration (low partial pressure) because nitrogenase, the enzyme driving the N-fixing process, will only function under anaerobic conditions. At the same time, aerobic respiration is necessary to provide the large quantity of energy (in the form of ATP, etc.) required to reduce N_2 to NH_3. The leghemoglobin allows a high flux of oxygen through the nodule to be sustained while maintaining a low oxygen partial pressure. The ammonia produced by nitrogenase is quickly converted into amino compounds or other compounds such as allantoin that are transferred to host plant tissues via the xylem.

Apart from hosting the nitrogenase and providing C substrates from plant photosynthesis and an export channel for the fixed N, nodules provide other services, especially that of controlling the oxygen diffusion rate so that the supply of O_2 to the nodules does not exceed the demand of the nitrogenase complex for this gas. Some symbioses have nodules equipped with very sophisticated and rapidly responding oxygen diffusion barriers (Minchin, 1997). This may help to explain why nodulated legumes, as well as nodulated actinorhizal symbioses, are the N_2-fixing associations that contribute such large amounts of N to both agricultural and natural ecosystems.

There is a very large amount of literature on the physiology and especially on the molecular biology of different legume/rhizobium symbioses. One of the most interesting points is that the genetic information for nodule formation and infrastructure is found on the host plant's genome. These genes are activated by molecular signals from the rhizobium bacteria, and the bacteria also provide the genes for production of nitrogenase. This means that bacterial infection of nonleguminous plants with a strain of rhizobium, or even infection of legumes that do not possess genes for nodulation (most Caesalpinioideae), will not lead to the formation of nodules. This makes it very difficult to create, using molecular biological techniques, new symbioses of rhizobium with nonlegumes because it will be necessary to transfer a great number of interacting plant genes from an N_2-fixing legume to the genetically modified plant species.

12.1.2   Alternative Modes of Biological Nitrogen Fixation

Some non-nodulating species of plants, particularly gramineae, are able to obtain N from associated N_2 fixation by bacteria in their rhizosphere as well as in and on plant tissues, particularly endophytically in plant roots as discussed in the next section. Because these systems of N-fixation do not involve nodules, it has been difficult to get consistent and repeatable evidence of significant amounts of BNF associated with these plant species, including tropical grasses such as sugar cane and rice.

Knowledge on this parallel form of BNF is just beginning to accumulate, and it may turn out to be much more important than presently appreciated. Because much research has been done on leguminous N-fixation, we can utilize this with considerable confidence in trying to enhance the nutrient status of soil systems. In Part III, we consider the practical contributions that legume-based systems for BNF can make to sustainable tropical agroecosystems (Chapter 27). The remainder of this chapter considers processes of endophytic N-fixation in nonlegumes that are attracting growing scientific attention and that could have a large impact on how we understand and manage soil systems.

12.2 Endophytic Biological Nitrogen Fixation

Over the last 30 years, BNF associated with plants of the Poaceae family, formerly referred to as Gramineae, has been the subject of considerable research, although only a few practical applications of this knowledge have emerged to date. It has been well established in the literature that BNF by various microorganisms living within plant tissues, particularly in roots, is an important element in soil systems and plant nutrition. Considerable efforts have been made to identify and localize the organisms responsible for the observed BNF activity, to quantify their contribution of fixed N_2, and to select bacteria and plant genotypes that can improve plant productivity. Efforts to exploit BNF associated with cereal and grass crops are motivated in part by a desire to reduce the agricultural use of N fertilizer given that its price often puts it beyond the means of poorer farmers and also its production requires significant expenditures of fossil energy. However, promoting endophytic BNF remains a challenge.

The most significant contribution of such BNF to crop production documented thus far is for Brazilian sugar cane (Boddey et al., 2003). This crop has been planted in Brazil for over 400 years, and, nowadays, Brazil has the lowest production cost of sugar in the world, partly due to its low N fertilizer input compared to that of other cane-producing countries. N fertilization by farmers in the state of São Paulo, where 60% of Brazilian sugar cane is grown, averages just 65 kg N ha^{-1}, with productivity of around 80 tons fresh stems per ha, one of the best levels anywhere. Producers in the USA, India, Australia, Colombia, and Mexico usually apply between 150 and 300 kg N ha^{-1} year^{-1}. In Australia, for example, with applications of 200 kg N ha^{-1} and irrigation on much of the area, productivity is 84 t ha^{-1}, only slightly above that of São Paulo. In these countries, owing partly to high N fertilizer input, the energy balance for bio-ethanol production from sugar cane as a fuel for motor vehicles is much less favorable than in Brazil, even negative in some cases, making it not really economically advantageous given the high energy input costs (Pimentel, 1991; Macedo, 1998).

Nutritional limitation, especially nitrogen depletion, has contributed to the degradation process that has lowered the productivity of Brachiaria pastures that cover a large area of Brazil, probably more than 80 million ha (Boddey et al., 2004). Agroforestry interventions for upgrading the productivity of these pasture lands are discussed in Chapter 21. Evidence from ^{15}N isotope studies has shown that under controlled conditions, species of this grass were able to obtain the equivalent of up to 30–40 kg N ha^{-1} year^{-1} from plant-associated BNF (Boddey and Victoria, 1986). With good grazing management and the addition of modest annual additions of P and K fertilizer, annual live weight gains of beef cattle between 200–300 kg could be obtained for many years without any input of fertilizer N (Lascano and Euclides, 1996; Boddey et al., 1996). Another tropical forage grass that has been shown to obtain some benefit from BNF is elephant grass (*Pennisetum purpureum*), which achieves modest but sustainable yields without any N fertilizer application (Reis et al., 2001). Several different species of N_2-fixing bacteria have been isolated from the rhizosphere and root tissues of finger millet (*Eleusine coracana*) which suggests that this crop also may benefit from plant-associated BNF (Vijayakumar and Narasimham, 1991; Loganathan et al., 1999).

There is a general consensus that plant-associated BNF is strongly controlled by plant genotype (Miranda et al., 1990; Urquiaga et al., 1992; Reis et al., 2000b), although it is possible that the selection and introduction of more efficient bacterial strains could have a significant impact, given that this is a thoroughly mutualistic and symbiotic relationship between plants and microorganisms. In Chapter 8, Dazzo and Yanni documented a strongly mutualistic relationship of rhizobia with rice plant roots. This research is all the more significant because rhizobia contribute significant BNF within the nodules on leguminous plant roots, as discussed in this section. In the rhizobium–rice association, however, these bacteria enhance crop productivity and health in other ways.

The most documentation of bacterial BNF with any crop so far has been with sugar cane, as noted above. A recent survey in Brazil using the ^{15}N natural abundance technique indicated that under actual on-farm conditions, contributions from bacterial processes ranged from zero to 60% of the crop's total N (Boddey et al., 2001). Large and diverse populations of N_2-fixing bacteria are associated with the roots and rhizosphere of sugar cane as microscopic evidence has shown that many species can infect the internal tissues of sugar cane. Because we are concerned with soil systems in this book, our focus in this chapter is on root endophytes, but it should be noted that N-fixing bacteria also inhabit plants' above-ground shoot and leaf tissues in the phyllosphere.

To date, no active N_2-fixing association between a monoxenic cane plant and a single bacterial strain or species has been convincingly demonstrated, although Sevilla et al. (1998) recorded a considerable response to inoculation with *Gluconacetobacter diazotrophicus* in greenhouse-grown cane plants that did not occur when a non-N_2-fixing mutant strain of the bacteria was used. Evidence from using ^{15}N-enriched N_2 gas has indicated that some fixed N was incorporated into the plant, although it was not clear whether the fixed N was still within the bacterial cells or had actually been assimilated by plant tissues. The exploitation and improvement of BNF in sugar cane and other grass-family species has so far been hindered by the fact that there is no clear candidate for any single N_2-fixing bacteria that is responsible for the observed N_2 fixation. Probably, the process is contributed to by a number of species, which makes reductionist scientific conclusions extremely difficult. While such research needs to continue, it is more likely that an ecological perspective on plant–soil–microbial relationships will bring advances in knowledge and practice.

12.3   Plant–Diazotrophic Bacteria Relationships

Once N balance and ^{15}N dilution studies had shown that some Brazilian sugar cane varieties were able to obtain agronomically significant contributions from plant-associated BNF (Lima et al., 1987; Urquiaga et al., 1992), efforts were redoubled to investigate the populations of N_2-fixing bacteria, collectively known as diazotrophs, associated with this crop. Attempts were made to isolate N_2-fixing bacteria from sugar cane juice, and almost immediately a novel diazotroph, *G. diazotrophicus* (formerly known as *Acetobacter diazotrophicus*) was discovered.

This bacterium has many interesting properties that differentiate it from other diazotrophs studied earlier. In common with several other N_2-fixing bacteria, the organism appears as small, gram-negative rods and is aerobic, showing pellicle formation in N-free semisolid medium. While it will grow on modest concentrations of sucrose or glucose (e.g., 5 g l^{-1}), its maximum growth rate occurs at approximately 100 g sucrose or glucose l^{-1}, which indicates that it is adapted to sugar-rich cane stems. At this concentration of sucrose, the bacteria grows while producing gluconic acid to an extent that pH falls to less than 3.0 while active N_2 fixation continues (Stephan et al., 1991). It lacks nitrate reductase and can fix N_2 in the presence of >20 mM NO_3^-. Further genotypic and phenotypic details are given in Gillis et al. (1989), Reis et al. (1994), James et al. (1994), Baldani et al. (1997).

This bacterium is very interesting because, while it is ideally suited for survival and growth within cane tissues, it survives very poorly in soil, does not easily infect intact sugar cane plants, and has a narrow range of hosts (Baldani et al., 1997). So far, it has only been isolated from sugar cane, elephant grass (*P. purpureum*) (Döbereiner, 1988), sweet potato (Paula et al., 1989), coffee (Jiménez-Salgado et al., 1997), and pineapple (Tapia-Hernandez et al., 2000), and recently from other genera of the sugar cane family (Gonzalez and Barraquio, 2000). Its strong preference for the internal tissues of plants led Döbereiner (1992) to classify these diazotrophs as endophytic, and more recently several other endophytic species have been discovered.

The existence of bacteria living inside plants without causing disease symptoms was proposed over 50 years ago by Tervet and Hollis (1948), Hollis (1951). The definition for bacterial endophytes proposed by Kado (1991) is now generally accepted: endophytes are bacteria that live in plant tissues without doing substantive harm or gaining benefit other than residency. This definition excludes bacteria that cause disease symptoms or detrimental effects on host plants, and it includes organisms living in plant parts above-ground as well as in root systems.

In recent years, several genera and bacterial species have been added to the list of endophytes that can improve plant growth by internal localization, e.g., two species of Herbaspirillum: *H. seropedicae* (Baldani et al., 1986) and *H. rubrisubalbicans* (Gillis et al., 1991; Baldani et al., 1996). Only a few phenotypic properties differentiate between these two species, such as their ability to grow on certain substrates and their different behavior in infecting plants. They have in common that both species survive poorly in soil, but *H. seropedicae* is found in a wide range of grasses and cereals (Baldani et al., 1997). *H. seropedicae* has never been isolated from naturally occurring cane leaves, although cane roots and stems are often highly infested. When a suspension of *H. rubrisubalbicans* was injected into the leaves of a cane variety susceptible to mottled stripe (B-4362), the bacteria completely blocked some of the xylem vessels and colonized the intercellular space of the mesophyll cells. At this particular site, the presence of nitrogenase was observed in the center of the microcolonies (James et al., 1997). In sorghum, the metaxylem was also colonized by this bacterium, and the nitrogenase antigen was observed to associate with this bacterium (Olivares et al., 1997). These comparisons show that what appear to be small differences in miniscule organisms nevertheless have functional consequences.

The best-characterized endophytic diazotrophic bacteria are those belonging to the genus Azoarcus. Inoculation experiments under gnotobiotic conditions showed that the strain BH72 of Azoarcus sp. can colonize the root cortex both inter- and intra-cellularly, as well as the aerenchyma and the xylem vessels of Kallar grass and rice, in a similar manner with no apparent pathogenic reaction. Similar bacterial colonization has been seen in the field (Hurek et al., 1994). The colonization of rice xylem by the N_2-fixing *Azorhizobium caulinodans* has been demonstrated by Gopalaswamy et al. (2000).

Isolation of bacteria from plant tissue does not necessarily prove that they are endophytes. To demonstrate this, it is necessary to inoculate and localize the bacterium inside the plant tissue. This can be difficult, but modern analytical techniques are advancing researchers' capability to do this. For example, Chelius and Triplett (2000) confirmed the endophytic localization of *Klebsiella pneumoniae* strains 2028 and 342[B-N1] using green fluorescent protein (gfp) as a reporter gene, in intercortical layers of the stem and within the region of maturation in the root; also, the nitrogenase enzyme was localized inside the plant tissue by polyclonal antiserum. *Rhizobium leguminosarum* bv. *trifolii* has been isolated in wetland rice in Egypt (Yanni et al., 1997, 2001), with conclusive evidence that this bacteria established endophytic colonization. As seen in Chapter 8, this rhizobial association with rice increases both rice yield and protein produced ha^{-1}.

12.4 Contributions to Plant Growth

The use of endophytic diazotrophic bacteria in field experiments has been limited because appreciation of their agricultural relevance is fairly recent. Experiments of inoculation using rice planted in pots showed positive results with increments of N derived from BNF around 17–19% for *Herbaspirillum seropedicae* and around 11–20% for *Burkholderia brasilensis* strains (Baldani et al., 2000). In the field, depending on the variety used, the increment in the production could reach 50% using a selected strain (Baldani et al., 2000). Using micropropagated sugar cane inoculated with *G. diazotrophicus*, a 28% increase in shoot yield was recorded under controlled conditions (Baldani et al., 2000).

In a gnotobiotic study on micropropagated sugar cane with a strain of *G. diazotrophicus* combined with increasing applications of N fertilizer, it was shown that N application could be reduced by half in the presence of the bacteria without observing any decrease in the production of fresh stems (Moraes and Tauk-Tornisielo, 1997). More recently, Oliveira et al. (2000) have documented a synergistic effect when micropropagated sugar cane plants were inoculated with a mixture of endophytic bacteria when compared to single inoculation treatments.

In Chapter 28, a strong association is reported between increased populations of endophytic Azospirillum and higher rice yields when using the plant, soil, water, and nutrient management practices of the System of Rice Intensification (SRI). Of course, not all of the yield effect should be attributed to this single species, nor just to the BNF that this rhizobium species can contribute. Such results, however, suggest that endophytic relationships and their effects warrant more systematic evaluation.

12.5 Effects of N Fertilizer on Diazotrophic Associations

In sugarcane, Vose et al. (1981) demonstrated that high levels of mineral N caused a significant decrease in the acetylene reduction activity (ARA). This effect is probably due to the inhibition of the synthesis by bacteria and/or plant root cells of nitrogenase, the enzyme essential for BNF. Fuentes-Ramírez et al. (1993) reported that the association between *G. diazotrophicus* and sugarcane can be limited severely by high N-fertilization, which would account for the decrease in ARA. In their study, they found that they could isolate *G. diazotrophicus* much more frequently from crops fertilized with 120 kg N ha^{-1} than from those fertilized with 300 kg N ha^{-1}.

Muthukumarasamy et al. (1999) obtained similar results in India for *G. diazotrophicus* and *Herbaspirillum* spp. in sugar cane. They suggested that this effect is not a direct negative relationship between the presence of the bacteria and the high levels of nitrogen, since *G. diazotrophicus* continues to fix nitrogen in culture media with high concentrations of NO_3^- (60 mM). It is more probable that the physiological state of the plant is altered by nitrogen, and this subsequently affects the association of the plant with the microorganism.

In a field experiment, Rivera et al. (1991) found that high concentrations of mineral N decreased the population of diazotrophic microorganisms in sugarcane, and these populations were restored when the mineral N content of the soil decreased. Reis et al. (2000a, 2000b) have confirmed the results obtained by Fuentes-Ramírez and Muthuku-marasamy in an experiment that used two sugarcane varieties planted in a sandy soil, comparing the growth of those fertilized with 300 kg of N ha^{-1} with that of unfertilized controls.

In an evaluation of Azospirillum in the roots of rice plants grown in replicated trials with different plant, water, soil, and nutrient management practices in Madagascar, it was observed that while the addition of NPK fertilizer to plots cultivated with SRI practices and no other soil nutrient amendments added 50% to yield, there was a 60% reduction in the populations of Azospirillum in the plant roots. Most of the N for this increased yield would presumably have come from inorganic sources. When, in other trials, compost was used as a source of nutrients instead of fertilizer, there was a further 16% increase in yield as the Azospirillum population mg^{-1} of root went from 45×10^4 to 14×10^5 (Chapter 28, Table 3).

Using DNA-analysis techniques, Tan et al. (2003) have documented a rapid change in the diazotrophic population structure within 15 days after application of N fertilizer, also documenting effects of environmental conditions and plant genotype. They profiled diazotrophic microbial communities using terminal restriction fragment length polymorphism (T-RFLP) analysis of nitrogenase gene (*nifH*) fragments to evaluate the impact of N-fertilization on the diazotrophic populations associated with the roots of rice (Oryza species), both modern (IR-72) and "wild" (*O. longistaminata*), under field conditions in France, the Philippines, and Nepal. The detailed analyses showed that both the abundance and diversity of diazotrophs in terms of the expression of their *nifH* genes, necessary to produce the nitrogenase needed for BNF, were diminished with N-fertilizer applications. Gene activity was suppressed when N was abundant in inorganic form, making the production of fixed N$_2$ less or not necessary.

Earlier research in Brazil had already found that BNF is reduced in soil that has been N-fertilized in recent cropping seasons, and in cultivars that have been fertilized in recent generations (Döbereiner, 1987). This kind of variability in plant–microbial interaction and in response to environmental conditions, while beneficial for the micro- and macroflora involved, makes it very difficult for scientists to measure and demonstrate these complex relationships to their own and others' satisfaction.

12.6 Discussion

Many important questions remain to be answered before these beneficial plant/microbe interactions can be exploited on a large scale to improve crop production in the field. More is known about relationships with leguminous plants, but the record of success with soil, plant, or seed inoculation is still mixed. Even more questions arise with regard to nonleguminous plant species: what proportion of the endophytic bacteria found in these plants are diazotrophic? How many of them are necessary to have a positive effect on plant growth and N nutrition? What about efficiency versus numbers of these bacteria? Is the fixed nitrogen transferred directly to plant cells or does it reach them only after the death and mineralization of the bacteria? Where do the bacteria get their nutrients? Is nitrogenase always active? What regulates the supply response of bacterial and/or plant root cells to N insufficiency by producing the nitrogenase that promotes BNF? How reliably and for what period of time will supply respond to demand? To what extent is the response a collective, ecological one, i.e., more than just an individual organism or species response? In Chapter 32, a number of significant crop yield responses in different countries to applications of Azotobacter are reported. This is an area of practice and of application of microbiological theory where little is yet known definitively. However, much useful knowledge can be expected over the next decade or two, building on several decades of prior research and field evaluation.

Many other questions arise that go beyond those related to nitrogen fixation considering that most of these organisms also interact with the plant in other ways such as production of phytohormones, siderophores, and biocontrol of diseases (Steenholdt and Vanderleyden, 2000). It appears anomalous that the benefits which *R. leguminosarum* bv. *trifolii* can confer on a rice crop, for which it is endophytic when rice is alternated with a clover crop whose leguminous root nodules it inhabits, do not come from BNF but rather from other nutritional and prophylactic services. This finding, discussed in Chapter 8, strengthens the case for expanded research on endophytes and on their roles in plant–microbe relationships.

Intensive studies should be carried out on the host plant side of the interaction, as several reports have shown that plant genotype favoring BNF is a trait that can be selected for improvement (Urquiaga et al., 1992; Wu et al., 1995; Shrestha and Ladha, 1996). This is supported by the research reported in Chapter 32. Perhaps both plant genotypes and bacterial strains need to be selected together to find the "best" combination(s). This can best be done through an integrated approach that involves laboratory, greenhouse, and field experiments. It will be necessary to deal not only with the plants and bacteria directly involved, but also with the characteristics and dynamics of the associated soils and other organisms in the soil.

Acknowledgments

The authors thank Dra. Maria Cristina P. Neves for her critical reading of the text and suggestions for modifications.

References

Baldani, J.I. et al., Characterization of *Herbaspirillum seropedicae* gen. nov. sp. nov: A root-associated nitrogen-fixing bacterium, *Int. J. Syst. Bacteriol.*, **36**, 86–93 (1986).

Baldani, J.I. et al., Emended description of Herbaspirillum: Inclusion of (*Pseudomonas*) *rubrisubalbicans*, a mild plant pathogen, as *Herbaspirillum rubrisubalbicans* comb. nov. and classification of a group of clinical isolates (EF group 1) as Herbaspirillum species 3, *Int. J. Syst. Bacteriol.*, **46**, 802–810 (1996).

Baldani, J.I. et al., Recent advances in BNF with non-legume plants, *Soil Biol. Biochem.*, **29**, 911–922 (1997).

Baldani, J.I. et al., Biological nitrogen fixation (BNF) in non-leguminous plants: The role of endophytic diazotrophs, In: *Nitrogen Fixation: From Molecules to Crop Productivity, Proceedings of the 12th International Congress on Nitrogen Fixation*, Pedrosa, F.O. et al., Eds., Kluwer, Dordrecht, The Netherlands, 397–400 (2000).

Boddey, R.M., Rao, I.M., and Thomas, R.J., Nutrient cycling and environmental impact of Brachiaria pastures, In: *Brachiaria: The Biology, Agronomy and Improvement*, Miles, J.W., Maass, B.L., and Valle, C.Bd., Eds., CIAT, Cali, Colombia, 72–86 (1996).

Boddey, R.M. and Victoria, R.L., Estimation of biological nitrogen fixation associated with Brachiaria and Paspalum grasses using ^{15}N-labelled organic matter and fertilizer, *Plant Soil*, **90**, 265–292 (1986).

Boddey, R.M. et al., Use of the ^{15}N natural abundance technique for the quantification of the contribution of N_2 fixation to sugar cane and other grasses, *Aust. J. Plant Physiol.*, **28**, 889–895 (2001).

Boddey, R.M. et al., Endophytic nitrogen fixation in sugar cane: Present knowledge and future applications, *Plant Soil*, **252**, 139–149 (2003).

Boddey, R.M. et al., Nitrogen cycling in Brachiaria pastures: The key to understanding the process of pasture decline, *Agric. Ecosyst. Environ.*, **103**, 389–403 (2004).

Chelius, M. and Triplett, E.W., Immunolocalization of dinitrogenase reductase produced by *Klebisiella pneumoniae* in association with *Zea mays* L, *Appl. Environ. Microbiol.*, **66**, 783–787 (2000).

Döbereiner, J., *Nitrogen-Fixing Bacteria in Non-Leguminous Crop Plants*, Springer, Berlin (1987).

Döbereiner, J., Isolation and identification of root associated diazotrophs, *Plant Soil*, **110**, 207–212 (1988).

Döbereiner, J., History and new perspectives of diazotrophs in association with non-leguminous plants, *Symbiosis*, **13**, 1–13 (1992).

Faria, S.M.de., McInroy, S.G., and Sprent, J.I., The occurrence of infected cells, with persistent infection threads, in legume root nodules, *Can. J. Bot.*, **65**, 553–558 (1987).

Faria, S.M.de. et al., Nodulação em espécies florestais, especificidade hospedeira e implicações na sistemática de Leguminosae, In: *Soil Fertility, Soil Biology and Plant Nutrition Interrelationships*, *Brazilian Society of Soil Science and Department of Soil Science*, Siqueira, J.O. et al., Eds., Federal University of Lavras, Lavras, Brazil, 667–686 (1999).

Fuentes-Ramirez, L.E. et al., *Acetobacter diazotrophicus*, an indoleacetic acid producing bacterium isolated from sugarcane cultivars of Mexico, *Plant Soil*, **154**, 145–150 (1993).

Giller, K.E., *Nitrogen Fixation in Tropical Cropping Systems*, CABI, Wallingford, UK (2001).

Gillis, M. et al., *Acetobacter diazotrophicus* sp. nov.: A nitrogen-fixing acetic acid bacterium associated with sugar cane, *Int. J. Syst. Bacteriol.*, **39**, 361–364 (1989).

Gillis, M. et al., Taxonomy relationship between '*Pseudomonas*' *rubrisubalbicans*, some clinical isolates (EF group 1), *Herbaspirillum seropedicae* and '*Aquaspirillum*' *autotrophicuim*, In: *Nitrogen Fixation*, Polsinelli, M., Materassi, R., and Vicenzini, M., Eds., Kluwer, Dordrecht, The Netherlands, 292–294 (1991).

Gonzalez, M.S. and Barraquio, W.L., Isolation and characterization of *Acetobacter diazotrophicus* (Gillis) in *Saccharum officinarum* L., *S. spontaneum* L., and *Erianthus sp*, *Philipp. Agric. Sci.*, **83**, 173–181 (2000).

Gopalaswamy, G. et al., The xylem of rice (*Oryza sativa*) is colonized by *Azorhizobium caulinodans*, *Proc. R. Soc. London Ser. B*, **267**, 103–107 (2000).

Hollis, J.P., Bacteria in healthy potato tissue, *Phytopathology*, **41**, 350–366 (1951).

Hurek, T. et al., Root colonization and systemic spreading of Azoarcus sp. strain BH72 in grasses, *J. Bacteriol.*, **176**, 1913–1923 (1994).

James, E.K. et al., Infection of sugar cane by the nitrogen-fixing bacterium *Acetobacter diazotrophicus*, *J. Exp. Bot.*, **45**, 757–766 (1994).

James, E.K. et al., Herbaspirillum, an endophytic diazotroph colonising vascular tissue in leaves of *Sorghum bicolor* L. Moench, *J. Exp. Bot.*, **48**, 785–797 (1997).

Jiménez-Salgado, T. et al., *Coffea arabica* L. a new host plant for *Acetobacter diazotrophicus*, and isolation of other nitrogen-fixing Acetobacteria, *Environ. Microbiol.*, **63**, 3676–3683 (1997).

Jordan, D.C., Transfer of *Rhizobium japonicum* Buchanan 1980 to *Bradyrhizobium* gen nov: A genus of slow-growing, root nodule bacteria from leguminous plants, *Int. J. Syst. Bacteriol.*, **32**, 136–139 (1982).

Kado, C.I., Plant pathogenic bacteria, In: *The Prokaryotes*, Balows, A., et al., Eds., Springer, New York, 659–674 (1991).

Loganathan, P. et al., Isolation and characterization of two genetically distant groups of *Acetobacter diazotrophicus* from a new host plant *Eleusine coracana* L, *J. Appl. Microbiol.*, **87**, 167–172 (1999).

Lascano, C.E. and Euclides, V.P.B., Nutritional quality and animal production of Brachiaria pastures, In: *Brachiaria: The Biology, Agronomy and Improvement*, Miles, J.W., Maass, B.L., and Valle, C.B.de., Eds., CIAT, Cali, Colombia, 106–123 (1996).

Lima, E., Boddey, R.M., and Döbereiner, J., Quantification of biological nitrogen fixation associated with sugar cane using a ^{15}N aided nitrogen balance, *Soil Biol. Biochem.*, **19**, 165–170 (1987).

Macedo, I.C., Greenhouse gas emissions and energy balances in bio-ethanol production and utilization in Brazil, *Biomass Bioenergy*, **14**, 77–81 (1998).

Minchin, F.R., Regulation of oxygen diffusion in legume nodules, *Soil Biol. Biochem.*, **29**, 881–888 (1997).

Miranda, C.H.B., Urquiaga, S., and Boddey, R.M., Selection of ecotypes of *Panicum maximum* for associated biological nitrogen fixation using the ^{15}N isotope dilution technique, *Soil Biol. Biochem.*, **22**, 657–663 (1990).

Moraes, V.A. de. and Tauk-Tornisielo, S.M., Efeito da inoculação de *Acetobacter diazotrophicus* em cana-de-açúcar (Sacharum spp.) variedade SP 70-1143, a partir de cultura de meristemas. [Abstract] XIX Congresso Brasileiro de Microbiologia, 215 (1997).

Muthukumarasamy, R., Revathi, G., and Lakshminarasimhan, C., Influence of N fertilisation on the isolation of *Acetobacter diazotrophicus* and *Herbaspirillum* spp. from Indian sugarcane varieties, *Biol. Fertil. Soils*, **29**, 157–164 (1999).

Olivares, F.L. et al., Infection of mottled stripe disease susceptible and resistant varieties of sugar cane by the endophytic diazotroph Herbaspirillum, *New Phytol.*, **135**, 723–737 (1997).

Oliveira, A.L.M. et al., Biological nitrogen fixation (BNF) in micropropagated sugarcane plants inoculated with different endophytic diazotrophic bacteria, In: *Nitrogen Fixation: From Molecules to Crop Productivity, Proceedings of the 12th International Congress on Nitrogen Fixation*, Pedrosa, F.O., et al., Eds., Kluwer, Dordrecht, The Netherlands, 425 (2000).

Paula, M.A.de., Döbereiner, J., and Siqueira, J.O., Nutrição e produção de batata-doce micro-propagada e inoculada com fungo micorrízico VA e bactérias diazotróficas. [Abstract] III Reunião Brasileira sobre Micorrizas, Piracicaba, São Paulo (1989).

Pimentel, D., Ethanol fuels: Energy security, economics, and the environment, *J. Agric. Environ. Ethics*, **4**, 1–13 (1991).

Reis, F.B. et al., Influence of nitrogen fertilisation on the population of diazotrophic bacteria Herbaspirillum spp. and *Acetobacter diazotrophicus* in sugarcane (Saccharum spp.), *Plant Soil*, **219**, 153–159 (2000a).

Reis, F.B. et al., Ocorrência de bactérias diazotróficas em diferentes genótipos de cana-de-açúcar, *Pesquisa Agropecuária Brasileira*, **35**, 985–994 (2000b).

Reis, V.M., Olivares, F.L., and Döbereiner, J., *Acetobacter diazotrophicus* and confirmation of its endophytic habitat, *World J. Microbiol. Biotechnol.*, **10**, 101–104 (1994).

Reis, V.M. et al., Biological nitrogen fixation associated with tropical pasture grasses, *Aust. J. Plant Physiol.*, **28**, 837–844 (2001).

Rivera, R., Velazco, A., and Treto, E., La fertilizacion (^{15}N), nutricion nitrogenada y actividad de los microorganismos nitrofijadores en la caña de azucar, cepa de caña planta, cultivada sobre suelo ferralitico rojo, *Cultivos Tropicales*, **12**, 21–28 (1991).

Sevilla, M.de. et al., Contributions of the bacterial endophyte *Acetobacter diazotrophicus* to sugarcane nutrition: A preliminary study, *Symbiosis*, **25**, 1283–1288 (1998).

Shrestha, R.K. and Ladha, J.K., Genotypic variation in promotion of rice dinitrogen fixation as determined by nitrogen-15 dilution, *Soil Sci. Soc. Am. J.*, **60**, 1815–1821 (1996).

Sprent, J.I., *Legume Nodulation*, Royal Botanic Gardens, London (2001).

Steenholdt, O. and Vanderleyden, J., Azospirillum, a free-living bacterium closely associated with grasses: Genetic, biochemical and ecological aspects, *FEMS Microbiol. Rev.*, **24**, 487–506 (2000).

Stephan, M.P. et al., Physiology and dinitrogen fixation of *Acetobacter diazotrophicus*, *FEMS Microbiol. Lett.*, **77**, 67–72 (1991).

Tan, Z., Hurek, T., and Reinhold-Hurek, B., Effect of N-fertilization, plant genotype and environmental conditions on *nifH* gene pools in roots of rice, *Environ. Microbiol.*, **5**, 1009–1015 (2003).

Tapia-Hernandez, A. et al., Natural endophytic occurrence of *Acetobacter diazotrophicus* in pineapple plants, *Microb. Ecol.*, **39**, 49–55 (2000).

Tervet, I.W. and Hollis, J.P., Bacteria in the storage organs of healthy plants, *Phytopathology*, **42**, 960–967 (1948).

Urquiaga, S., Cruz, K.H.S., and Boddey, R.M., Contribution of nitrogen fixation to sugarcane: Nitrogen-15 and nitrogen-balance estimates, *Soil Sci. Soc. Am. J.*, **56**, 105–114 (1992).

Vijayakumar, B.S. and Narasimham, A.V.L., Microbial ecology of rhizosphere of *Saccharum officinarum* L. and *Eleusine coracana* L. grown in semi-arid tropical soils of Anantapur District (A.P.). India, *Geobios*, **18**, 145–149 (1991).

Vose, P.B. et al., Potential N_2-fixation by sugarcane, Saccharum sp. in solution culture — I. Effect of NH_4^+ vs. NO_3^-, variety and nitrogen level, In: *Associative N_2-Fixation*, Vose, P.B. and Ruschel, A.P., Eds., CRC Press, Boca Raton, FL, 119–123 (1981).

Wu, P. et al., Molecular-marker-facilitated investigation on the ability to stimulate N_2 fixation in the rhizosphere by irrigated rice plants, *Theor. Appl. Genet.*, **91**, 1177–1183 (1995).

Yanni, Y.G. et al., Natural endophytic associations between *Rhizobium leguminosarum* bv. *trifolii* and rice roots and assessment of its potential to promote rice growth, *Plant Soil*, **194**, 99–114 (1997).

Yanni, Y.G. et al., The beneficial plant growth-promoting association of *Rhizobium leguminosarum* bv. *trifolii* with rice roots, *Aust. J. Plant Physiol.*, **28**, 845–870 (2001).

Young, J.P.W. and Haukka, K.E., Diversity and phylogeny of rhizobia, *New Phytol.*, **133**, 87–94 (1996).

13

Enhancing Phosphorus Availability in Low-Fertility Soils

Benjamin L. Turner,[1] **Emmanuel Frossard**[2] **and Astrid Oberson**[2]

[1]*Smithsonian Tropical Research Institute, Panama City, Republic of Panama*
[2]*Institute of Plant Sciences, Swiss Federal Institute of Technology, Zurich, Switzerland*

CONTENTS

Phosphorus deficiency is a widespread constraint on agronomic productivity. This chapter describes biological processes that enhance phosphorus availability in phosphorus-deficient soils, which can form the basis for sustainable approaches to maintaining phosphorus nutrition. We focus on tropical soils, which are notoriously poor in plant-available phosphorus, with almost half being considered phosphorus-deficient for agricultural production (Fairhurst et al., 1999). More than 70% of these are highly-weathered Oxisols and Ultisols (Soil Survey Staff, 1999), termed Ferralsols in the FAO taxonomy scheme. These acidic soils are low in exchangeable bases, but rich in well-weathered material such as 1:1 clay minerals (mainly kaolinite) and sesquioxides of iron and aluminum (Tiessen and Shang, 1998). This confers a considerable capacity to retain phosphate, but means that the concentrations available to plants are naturally very low. Increasing phosphorus availability in such soils is therefore essential for improving agricultural productivity.

The Green Revolution was based on new crop varieties suitable for greater use of chemical fertilizers and pesticides. However, the strong sorption of phosphate in many tropical soils means that conventional attempts to improve phosphorus fertility by adding readily-soluble mineral fertilizer increase phosphorus availability slowly at best (Smithson and Giller, 2002). Water-soluble mineral fertilizers are often financially or logistically unavailable to farmers in tropical regions, and the quality of rock phosphate available in African markets is often dubious. The price of mineral fertilizer is also several times greater for African farmers than those in Europe, Asia, and North America, and increases markedly from ports to the interior (Sanchez, 2002). Mineral fertilizer application is therefore a relatively inefficient way to improve fertility in tropical soils, although large applications of rock phosphate can raise phosphorus availability in some soils, notably those with pH <6 (Fardeau et al., 1988). Also, the use of small amounts of fertilizer in conjunction with plant germplasm that is adapted to low-fertility soil can result in significant improvement in crop and pasture production, as discussed in Chapter 37 in Part III. The discussions of benefits from bio-char in Chapter 36 and from intercropping in Chapter 39 also show how phosphorus can be mobilized through root-soil-microbial interaction.

Is it possible to improve phosphorus availability in tropical agroecosystems without the intensive use of processed mineral fertilizer? This is an important question given the urgent need to improve productivity in developing regions (Sanchez, 2002). The question is also of interest for organic agriculture, which requires that no processed mineral fertilizers be used, although fairly large amounts of phosphorus can be used if not in a processed form. Research on a variety of agricultural systems suggests that soil phosphorus availability can be improved and maintained by biological processes, but this requires a shift away from conventional approaches to raising soil fertility by adding large amounts of water-soluble mineral fertilizer. This prevailing philosophy, reflecting experience and research in temperate climates, is less applicable for tropical soils that have a high capacity to sorb phosphate.

13.1   Focusing on Organic Phosphorus

Ever since the German chemist Justus von Liebig demonstrated the value of chemical fertilizers in the mid-19th century, agricultural practice has focused on *inorganic* phosphorus (i.e., phosphate) because this is the form of phosphorus taken up directly by the crop. This thinking is ingrained so deeply that even studies of biological mechanisms that improve phosphorus availability to plants, such as root-induced pH changes (e.g., Hinsinger, 2001) and phosphate solubilizing bacteria (e.g., Kucey et al., 1989) investigated only the acquisition of inorganic forms of phosphorus.

However, there is growing recognition of the importance of *organic* forms of phosphorus, especially in tropical soils (Tiessen et al., 1994; Nziguheba and Bünemann, 2005; Organic Phosphorus Workshop, 2005). This chapter describes how the availability and turnover of organic phosphorus can be improved by promoting biologically active soil that is rich in organic matter and contains an abundant microbial community. This is fundamental for maintaining phosphorus availability in strongly phosphorus-fixing soils and, therefore, for the long-term sustainability of tropical agriculture.

The biological approaches to improving soil phosphorus availability considered here do not include strategies such as inoculation of soils with phosphate-solubilizing bacteria. There is little evidence for the effectiveness of such practices (Richardson, 2001), although some applied research has shown encouraging results (Chapter 32). Here, we focus on improvements in biological soil health and on the exploitation of natural processes that improve phosphorus fertility without degrading the soil. Such methods can be adopted immediately by farmers to improve phosphorus fertility. While we recognize that socio-economic factors also contribute to problems with poor fertility in tropical soils (Smithson and Giller, 2002), considering these is beyond the scope of this chapter. Some applications of the concepts presented here are given in Part III.

Improving and maintaining organic matter is fundamental to sustaining agriculture in tropical soils (Coleman et al., 1989; Woomer and Swift, 1994). This cannot be underestimated because organic matter supports diverse biological communities that sustain a multitude of soil functions, including nutrient supply (Lal, 2004; Wardle et al., 2004). This chapter reviews these processes from the perspective of phosphorus nutrition and describes the biological mechanisms involved in maintaining and enhancing phosphorus availability in a healthy soil. We discuss the phosphorus paradox, whereby plentiful concentrations of phosphorus in tropical soils coexist with only small concentrations of plant-available phosphate, and we then describe the importance of organic phosphorus and its turnover via microbial processes for supplying phosphorus to plants. Finally, we discuss options for managing agroecosystems to enhance phosphorus availability to crops.

13.2 Phosphorus in Tropical Soils

13.2.1 The Phosphorus Paradox

The large capacity of most tropical soils to sorb phosphate means that the concentrations of phosphate available to plants are typically low compared to the values considered necessary for successful agricultural production. For example, Smithson and Giller (2002) report that more than 80% of farmers' fields in the densely populated Lake Victoria region of western Kenya contained plant-available phosphate concentrations below 5 mg P kg^{-1}, as determined by bicarbonate extraction. This means that they are classified as severely deficient in phosphorus.

Paradoxically, however, tropical soils often contain relatively large concentrations of total phosphorus, typically orders of magnitude greater than concentrations of plant-available phosphate. In a review of organic phosphorus in tropical soils, Nziguheba and Bünemann (2005) found that *total* phosphorus concentrations ranged between 61 and $1780 \text{ mg P kg}^{-1}$, showing that limited bioavailability can coexist with large potential supply. This phenomenon is not restricted to tropical soils, because almost all soils contain much lower plant-available phosphate than total phosphorus (Russell, 1988). An important and ongoing agronomic question is therefore whether part of the "recalcitrant" phosphorus pool can be made available to crops (Richardson et al., 2005).

13.2.2 The Importance of Organic Phosphorus

Some of the "recalcitrant" soil phosphorus is occluded within minerals and will only become available to plants through slow weathering over long time-scales, although some organisms can access this phosphorus more rapidly (Kucey et al., 1989). However, a large proportion of the soil phosphorus occurs in organic forms (Harrison, 1987). Organic phosphorus is conventionally regarded as relatively unavailable to plants, because it must be hydrolyzed before the phosphate moiety can be taken up by roots. This has impeded efforts to understand its behavior in soil and its role in phosphorus nutrition. However, it is now clear that plants can indeed use organic phosphorus compounds, even as effectively as inorganic forms (Tarafdar and Claassen, 1988). In particular, studies of phosphorus depletion close to roots have revealed that plants do not discriminate among phosphorus forms of different chemical extractability, depleting even the apparently stable pools of inorganic and organic phosphorus (Chen et al., 2002).

As tropical soils tend to have a large capacity to sorb phosphate, organic phosphorus is of considerable importance in supplying phosphorus to plants in tropical agriculture. Indeed, organic phosphorus turnover is probably more important than phosphate desorption in supplying phosphorus to crops (Tiessen and Shang, 1998; Nziguheba and Bünemann, 2005). Widening the scope of what we consider to be potentially available to plants, from a small pool of phosphate to most of the organic fraction, allows the development of new strategies for improving phosphorus availability to crops without the need for high doses of mineral phosphate fertilizer (refer to section 8.4.3 above).

13.2.3   Forms and Amounts of Soil Organic Phosphorus

Organic phosphorus is defined here as phosphorus that is covalently bound to organic molecules, although the very definition of what constitutes an organic molecule remains somewhat controversial. Our definition does not include inorganic polyphosphates as these contain no carbon. This is a strict chemical definition and does not include phosphorus inputs that are made to the soil. For example, plant residues and manures are often described as being organic phosphorus fertilizers, although they contain *both* organic and inorganic phosphorus compounds. We refer to manures and plant residues as *organic amendments* or *organic residues* to avoid implying that all the phosphorus within such material is organic.

Typically, around two-fifths of the total phosphorus in surface layers of tropical soils is in organic forms, although reports range from 2% in a Nigerian sand to more than 90% in some Indian soils (reviewed in Harrison, 1987). Unfortunately, there is considerable analytical uncertainty associated with measurements of organic phosphorus in highly-weathered tropical soils because the most common procedure for its determination, the ignition method, tends to overestimate the likely true values (Condron et al., 1990b).

There was relatively little information on organic phosphorus compounds in tropical soils until the application of solution ^{31}P NMR spectroscopy to soil phosphorus. An example of solution ^{31}P NMR spectroscopy of a tropical Oxisol from Madagascar is shown in Figure 13.1. A wide variety of organic phosphorus compounds occur in tropical soils, which are classified by the nature of their phosphorus bond into phosphate esters, phosphonates, and anhydrides. Phosphate esters are subclassified according to the number of ester groups that are linked to each phosphate. Thus, phosphate monoesters have a single ester bond to a carbon moiety per phosphorus moiety, while phosphate diesters have two bonds. Phosphate monoesters in soil occur mainly as inositol phosphates, a widespread family of phosphoric esters of hexahydroxycyclohexane (inositol) (Turner et al., 2002). Other phosphate monoesters present in small concentrations include sugar phosphates, phosphoproteins, and mononucleotides.

Phosphate diesters include nucleic acids (DNA and RNA), phospholipids, and teichoic acids. They occur in much smaller proportions than phosphate monoesters, typically less than 10% in agricultural soils, but are of agricultural interest because they are readily degraded when soils are brought under cultivation (Condron et al., 1990a). Phosphonates contain a direct carbon–phosphorus bond, which makes them markedly different from other soil organic phosphorus compounds, but they are rarely detected in tropical soils. Organic phosphoanhydrides (organic polyphosphates), such as adenosine triphosphate (ATP), are a source of energy in most metabolic processes, but are rarely detected in soils. Inorganic polyphosphates are often present in tropical soils, usually as pyrophosphate, although their precise origin and function remain unclear.

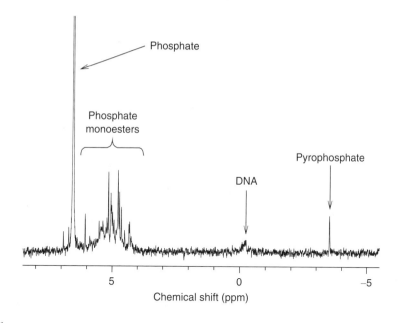

FIGURE 13.1

The phosphorus composition of a tropical soil, determined by extraction in sodium hydroxide and EDTA (ethylenediaminetetraacetate) and solution ^{31}P nuclear magnetic resonance spectroscopy (B.L. Turner, unpublished). The soil was an Oxisol (pH 4.79) from Ambaiboho, Madagascar, that had been managed according to the System of Rice Intensification (Chapter 28) for 4 years. The spectrum is truncated vertically to show the organic phosphate signals in detail. One-third of the extracted phosphorus was in organic form (phosphate monoesters and DNA), of which around 16% was phytic acid. See Section 13.2.3 for details of the identified phosphorus compounds.

We are thus dealing with a complex and previously elusive topic, although the central importance of phosphorus to plant growth and production makes its study highly relevant. A detailed treatment of the organic phosphorus composition of tropical soils is not necessary here. However, a comprehensive collection of data from the older literature can be found in Harrison (1987), while Nziguheba and Bünemann (2005) reviewed more recent information based on analysis by solution ^{31}P NMR spectroscopy.

13.3 Microbes and Phosphorus Turnover in Soil

There is now an increasing appreciation that the management of soil microorganisms will be fundamental in sustainable agricultural practices (Richardson, 2001; Oberson and Joner, 2005). However, soil microorganisms affect phosphorus availability through complex processes that remain poorly understood, especially in tropical soils. To simplify discussion, we separate these processes into three categories:

1. Acquisition of phosphorus from "stable" forms in soil, such as recalcitrant organic phosphorus or primary phosphate minerals. This "microbial mining" operates in the long-term to maintain equilibrium phosphorus concentrations in plant-soil systems.

2. Turnover of organic residues added to soil. These include compost, animal manure, and fresh plant material. The nutrient composition and the carbon quality of such materials (i.e., their suitability as a substrate for decomposition),

determine whether nutrients are released immediately into the soil solution or are retained in microbial cells.

3. Short-term turnover of microbial phosphorus through the soil solution. This occurs through environmental perturbation (e.g., wetting and drying), or through microfaunal grazing by protozoa and nematodes (Chapter 10). This third process renders the phosphorus taken up by microbes through the first two processes available to plants.

These processes are depicted in Figure 13.2, which emphasizes the central role of the microbial biomass in the soil phosphorus cycle. In tropical soils, microbes intercept phosphate released from decomposing organic residues and retain it in their cells. This

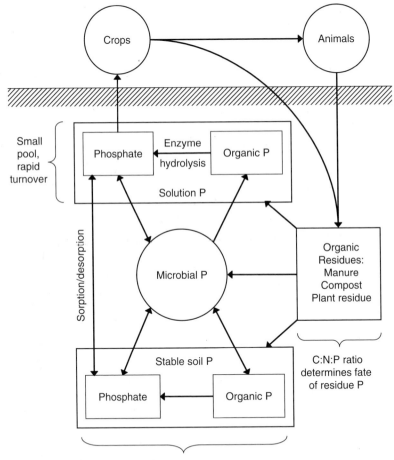

FIGURE 13.2
Conceptual model of phosphorus turnover in tropical soils, emphasizing the central role of the microbial biomass. Microbes can access stable pools of soil phosphorus through solubilization and/or enzymatic hydrolysis, and regulate the fate of phosphorus in organic residues returned to soil from plants and animals. Microbial phosphorus is released into solution predominantly through environmental changes such as wetting and drying (mainly organic phosphorus) and microfaunal grazing (mainly phosphate). The organic phosphorus released to the soil solution becomes available for uptake by plants or microbes following enzymatic cleavage of the phosphate moiety. The pathway of phosphate desorption from the stable phosphate pool to solution, which is important in supplying phosphate to crops in heavily fertilized temperate agriculture, is limited in tropical soils due to their strong capacity to sorb phosphate.

process, often referred to as *immobilization*, is extremely important because it prevents the loss of phosphorus from the biological cycle by sorption in the soil. The immobilized phosphorus becomes available to plants following subsequent microbial turnover.

Organic phosphorus must be converted to phosphate before it can be taken up by plant roots. This is commonly referred to as *mineralization*, an ambiguous term that refers to the release of phosphate from organic amendments or soil organic phosphorus, microbial biomass. In all cases, mineralization is mediated by phosphatase enzymes at some stage, with microbes playing a central role. Microbial processes involved in the mineralization of organic phosphorus are discussed in detail below.

13.3.1 Long-Term Acquisition of Stable Phosphorus

Even when phosphorus concentrations are extremely low, it is rare to find a soil microbial community that appears limited in its activity by the availability of phosphorus (Cleveland et al., 2002). This means that microbes, at least at the community level, must be extremely efficient at acquiring phosphorus from the soil irrespective of the availability of the other major nutrients. Indeed, numerous soil microbes, including bacteria and fungi, can solubilize primary phosphate minerals, or acquire phosphorus from recalcitrant organic phosphorus compounds such as inositol phosphates (Richardson, 2001; Turner et al., 2002). Various mechanisms are employed, including the synthesis of phosphatase enzymes and the secretion of protons, hydroxide ions, or organic anions (Tarafdar and Claassen, 1988; Hinsinger, 2001; Hocking, 2001). An example with particular relevance for tropical soils is the ability of some bacteria to solubilize iron oxides (Stemmler and Berthelin, 2003). As phosphorus is commonly associated with iron oxides in tropical soils, the action of these bacteria may also release phosphorus for biological uptake.

As the activity of soil microbes is almost always limited by the availability of degradable organic carbon, this also regulates the microbial acquisition of phosphorus from recalcitrant compounds (Chauhan et al., 1981; Thien and Myers, 1992). In a healthy soil, carbon is supplied by the addition of organic residues, the secretion of organic anions by plant roots, and by the slow turnover of soil organic matter, including microbial biomass. The secretion of organic anions by roots is especially relevant, because it forms the basis for symbiotic relationships between plants and microbes. Plants supply energy to belowground heterotrophic microbes, which in turn enhance the supply of nutrients from the soil to the root. Such symbiotic relationships are typified by mycorrhizal associations (Smith and Read, 1997), while relationships with bacteria are more indirect.

This dynamic equilibrium indicates why inoculations with specific organisms often fail (Richardson, 2001). Given the complexity and diversity of soil habitats and microorganisms, it is likely that the organisms present are highly adapted to the soil in question, so organisms added by inoculation have difficulty establishing or finding a niche (Powlson et al., 2001). The most promising approach to promoting the functions provided by such organisms as phosphate-solubilizing bacteria may be to ensure that conditions are suitable for their indigenous populations.

Biological acquisition of phosphorus from mineral phosphates is relatively well-studied in comparison to acquisition from organic phosphorus compounds. Inositol phosphates are the most abundant form of organic phosphorus in most soils, but their availability to plants is conventionally thought to be poor, because they bind strongly in soil to form insoluble complexes and few plants produce the phytase enzyme necessary to dephosphorylate them (Turner et al., 2002). In contrast, some microorganisms can readily access inositol phosphates in soil (Richardson et al., 2001), notably ectomycorrhizal fungi and some bacteria (Antibus et al., 1992; Richardson and Hadobas, 1997). There is also evidence that organisms capable of accessing phytic acid are favored in soils with limited phosphorus availability (Turner et al., 2003b). Clearly an active and diverse microbial community is essential for promoting the turnover of recalcitrant soil organic phosphorus.

Microbes in tropical soils might be expected to be strongly adapted to phosphorus limitation, being efficient at acquiring phosphorus despite its limited availability in the soil. There is no direct evidence for this, although fertilization with mineral phosphate altered the microbial community composition in a Kenyan Oxisol (Bünemann et al., 2004a). Also in Kenya, conversion from continuous maize to a nitrogen-fixing fallow system increased soil organic matter and caused marked changes in microbial community composition (Bossio et al., 2004). Additional studies are required to assess whether such changes involve functional groups of organisms involved in acquisition of soil phosphorus from stable compounds.

13.3.2   Microbial Decomposition of Fresh Organic Amendments

A second key process for microbial turnover of phosphorus is the degradation of organic amendments, which can be of major importance in supplying phosphorus to crops (Oberson and Joner, 2005). Such amendments include plant remains, compost, animal manure, and biosolids and are often the sole source of external phosphorus on smallholder farms that typify many tropical regions. The use of organic nutrient sources on small farms has been promoted heavily in recent years, notably through the use of green manures, managed fallows, agroforestry systems, and the recycling of animal manures (Drechsel et al., 1996; Fischler and Wortmann, 1999; Sanchez, 1999), with the addition of mineral phosphorus fertilizer to raise initial soil phosphorus status sometimes emphasized (Smithson and Giller, 2002).

Nutrient supply to crops from decomposing organic residues depends in part on the quality of the residue. Of particular importance are the concentration of readily-degradable carbon and the ratios of carbon-to-phosphorus and carbon-to-nitrogen in the residue (Bünemann et al., 2004a). Residues rich in phosphorus are decomposed quickly and some phosphate may be released into solution, from where it can be taken up by plants. Conversely, residues containing little phosphorus decompose slowly, and the phosphorus will be retained almost completely within microbial cells (Kwabiah et al., 2003; Bünemann et al., 2004b).

The precise pattern, duration, and quantity of phosphorus turned over during residue decomposition will depend also on soil properties and the size, activity, and composition of the microbial biomass (Nziguheba and Bünemann, 2005; Oberson and Joner, 2005). Complete immobilization of phosphorus is only expected during decomposition of residues that have carbon-to-phosphorus ratios greater than approximately 200 (Dalal, 1977). This prevents phosphorus from being immediately available to plants, but is important in tropical soils because it also prevents rapid and potentially long-term sorption to iron and aluminum oxides in the soil (Myers et al., 1997). Phosphorus is thus protected from sorption in the soil by conversion to a form from which it can become available to crops during subsequent turnover. This is discussed in the next section.

A key process during the decomposition of organic residues is the "priming effect," whereby microbes acquire phosphorus from the soil to support their degradation of organic amendments. This phosphorus is often derived from stable pools in the soil, as indicated by a study involving the addition of radio-labeled *Crotalaria grahamiana* residue to a Kenyan Oxisol (Bünemann et al., 2004b). The priming effect is likely to be significant only for organic amendments with relatively high carbon-to-phosphorus ratios. However, it is an important process by which stable soil phosphorus is converted to microbial tissue, which then contributes to plant uptake after subsequent microbial turnover.

13.3.3 Short-Term Turnover of Microbial Phosphorus in Soil

13.3.3.1 Turnover Rates

Despite the ability of microbes to take up soil phosphorus and protect it from sorption, plants can only benefit following microbial turnover and the release of cellular phosphorus into the soil solution. An active microbial community is therefore fundamental to the supply of phosphorus to plants. Most microbial phosphorus is in organic forms, mainly simple phosphate monoesters (e.g., sugar phosphates) and diesters (DNA and phospholipids) (Oberson and Joner, 2005). Unlike free phosphate, these types of organic compounds do not sorb strongly to soil constituents. Once in solution, they become available to plants following cleavage by extracellular phosphatase enzymes released from both microbes and plant roots (Quiquampoix and Mousain, 2005). Through this process, phosphorus is converted from organic structures within microbial cells into free phosphate that can be directly taken up by plant roots, as indicated in Figure 13.2.

The importance of microbes in regulating phosphorus turnover through the soil solution was demonstrated by studies of nutrient leaching that manipulated soil microbial biomass. Elimination of microbial activity by soil sterilization caused organic phosphorus to become undetectable in leachate water in incubated columns, even though it was the dominant fraction in leachate from a corresponding unsterilized soil (Seeling and Zasocki, 1993). Similarly, when laboratory columns of a calcareous soil were amended with organic carbon sources, including crop residues and sucrose, phosphorus in leachate increased 38-fold compared to unamended soil and was almost all in organic forms (Hannapel et al., 1964). In both these studies, elevated biological activity maintained organic phosphorus turnover through the soil solution. In tropical soils this probably prevents phosphorus from becoming tied up by strong sorption in the soil.

Microbes constitute a large fraction of the total phosphorus in most soils (Oberson and Joner, 2005) so their impact on soil phosphorus turnover is unsurprising. Microbial nutrients have a short turnover time relative to those in more stable organic matter, so the continuous release of low concentrations of microbial organic phosphorus to soil solution can be considerable on an annual basis (Jenkinson and Ladd, 1981). The greatest turnover rates are in grassland soils, with annual phosphorus fluxes estimated to be in the order of 40 kg ha^{-1} (Cole et al., 1977; Brookes et al., 1984). These values are probably under-estimations, because microbial turnover rates are now considered to be much more rapid than those used to calculate phosphorus fluxes in the earlier studies (Oehl et al., 2001). Microbial phosphorus concentrations tend to be smaller in tropical soils than in temperate soils (Oberson et al., 2001; Bünemann et al., 2004a), although microbial turnover is much more rapid in tropical soils. Phosphorus flux through the microbial biomass in tropical soils may therefore far exceed the quantity of phosphorus held in this pool at any one time (Oberson and Joner, 2005).

Annual estimates of microbial phosphorus turnover can conceal strong seasonal differences, which in the tropics are usually determined by rainfall (Nziguheba and Bünemann, 2005). Greatest microbial biomass concentrations were measured during the dry season in a tropical dry forest (Campo et al., 1998), whereas studies of agroforestry systems in Nigeria (Wick et al., 2002) and tropical pastures in Colombia (Oberson et al., 1999) reported minima during the dry season.

Seasonal fluctuations in microbial phosphorus almost certainly influence phosphorus availability. Oberson et al. (1999) observed an increase in microbial phosphorus and a decrease in plant-available phosphate when tropical pastures resumed their growth at the onset of the rainy season, suggesting that seasonal fluctuations in microbial phosphorus may influence phosphorus availability to plants. It is now established that rewetting dry soil can release considerable pulses of phosphorus into solution (Campo et al., 1998; Turner and Haygarth, 2001) as discussed below.

The importance of microbial turnover in maintaining phosphorus turnover through the soil solution is further highlighted by measurements of organic phosphorus mineralization. Daily rates, measured using various techniques, range between 0.2 and 2.5 mg P kg^{-1} soil (e.g., Harrison, 1982; Oehl et al., 2004), with the differences appearing to depend in part on assay temperature. Measurements were taken under steady-state conditions without organic amendments, although the highest mineralization rate was detected in a soil that received regular organic amendments and contained an active microbial community (Oehl et al., 2004). It is known that organic phosphorus mineralization is closely related to soil carbon turnover, which is greater under tropical than temperate conditions (Tiessen and Shang, 1998). Therefore, basal organic phosphorus mineralization is probably greater in tropical soils than in comparable temperate soils. Unfortunately, it is currently impossible to confirm this, because radioisotope methods are difficult to use in tropical soils, which strongly sorb phosphate (Bühler et al., 2003).

13.3.3.2   *Mechanisms of Microbial Phosphorus Turnover*

Which mechanisms drive the turnover of microbial phosphorus in soil? Changing environmental conditions are among the most important processes, because considerable amounts of phosphorus can be released by sudden changes in moisture or temperature. It has been known for some time that soil drying can render soil organic carbon soluble in water (Birch, 1958) and influence the solubility of inorganic nutrients (e.g., Sparling et al., 1985). The effect on organic phosphorus can also be considerable. In a range of pasture soils in the U.K., 7 days of air drying at 30°C from approximate field moisture capacity increased the concentrations of water-extractable organic phosphorus by between 185 and 1900%, with organic phosphorus accounting for up to 100% of the increase (Turner and Haygarth, 2001). The released organic phosphorus was derived at least partly from microbial cells (Turner et al., 2003a), which was not unexpected because rapid rehydration can kill between 17 and 58% of soil microbes through osmotic shock and cell rupture (Kieft et al., 1987).

The physical stress induced by soil drying also disrupts organic matter coatings on the surfaces of clays and minerals (Bartlett and James, 1980), which may contribute to the solubilization of organic phosphorus following rewetting (Turner and Haygarth, 2003). Increases in soluble inorganic phosphorus were reported for tropical soils (Srivastava, 1997; Campo et al., 1998), although the effect on organic phosphorus was not determined.

The degree of microbial death upon rewetting is primarily determined by the moisture content of the soil prior to extraction, although significant death only occurs after drying to $< 10\%$ gravimetric moisture content (West et al., 1988). The susceptibility of microbes to drying and wetting is also influenced by the age and type of organisms and the degree of physical protection offered by the soil matrix (Van Gestel et al., 1993). It might be expected that microbial communities would become resistant to desiccation and rewetting in soils where this is common, but the impact of drying on microbes appears similar across a wide range of soils and moisture regimes (Sparling et al., 1989). For example, rewetting killed more than half of the microbial biomass in xeric soils of the western U.S.A. that undergo intense summer drying (Kieft et al., 1987), while irrigation caused substantial shifts in microbial community composition within a single growing season in semi-arid soils where tolerance to extreme fluctuations in moisture stress might be expected (Lundquist et al., 1999). Such changes clearly demonstrate that nutrient release from microbes following the rewetting of dry soils is important even in soils that experience intense seasonal drying.

Management practices such as liming and nitrogen fertilization can influence soluble organic matter in soils (Chantigny, 2003). This may affect the balance between mineralization and immobilization of phosphorus (Nziguheba and Bünemann, 2005), because microbial activity is closely related to the availability of mineralizable organic carbon (Chen et al., 2002). Changes induced by management practices such as adding organic residues are generally of short duration, whereas long-term effects are related more to changes in vegetation type, land use, and the amount of organic material being returned to the soil (Chantigny, 2003; Oberson and Joner, 2005).

13.3.3.3 *Microfaunal Grazers and Phosphorus Turnover*

Soil fauna that graze the microflora on and around plant roots are an additional biological factor that can release microbial phosphorus into solution. The organisms involved include protozoa, microarthropods, and nematodes. They are better known with reference to their influence on nitrogen availability (e.g., Clarholm, 1985), but can also have a considerable impact on the availability of phosphorus (Cole et al., 1978).

Phosphorus is excreted by microfauna mainly as soluble phosphate, although they may strongly influence organic phosphorus concentrations through indirect effects on microbial numbers and activity (Clarholm, 1985; Ingham et al., 1985). Microfaunal grazers also respond rapidly to fluctuations in microbial biomass; for example, protozoan populations can double their size in just a few hours given sufficient numbers of bacteria (Persmark et al., 1996).

Several authors have stressed the importance of microfaunal grazers in tropical soils (e.g., Lavelle et al., 1997; Tiessen and Shang, 1998) and it seems likely that they play a key role in regulating nutrient availability. Additional studies are clearly warranted, notably on the link between microfaunal diversity and nutrient availability (Bradford et al., 2002).

13.4 Implications for Soil Management and Improvement

This volume examines ways in which nurturing more biologically-active soils can open new and better avenues for improving agricultural productivity. This chapter has focused on how appropriate plant, soil, and nutrient management can improve phosphorus nutrition through microbially-mediated processes. Our approach does not preclude the use of mineral phosphate fertilizer, because low doses can raise yields in some circumstances, and can be important for initially improving fertility in soils with very low concentrations of total phosphorus. However, the importance of enhancing soil organic matter in tropical agriculture is clear. Soil comes to life through organic matter, which supports highly diverse communities of microorganisms and soil fauna that provide critical ecosystem services, most notably the recycling of nutrients (Martius et al., 2001).

Careful management of soil organic matter is fundamental for successful agriculture in the tropics, because fertility depends to a large degree on soil organic matter turnover (Tiessen and Shang, 1998). However, soil organic matter can be depleted rapidly in tropical soils, which has adverse effects on soil fertility. An unfortunate example comes from Brazil, where soil organic matter degradation following the cultivation of a forest soil transformed its phosphorus into unextractable forms. This process was irreversible, despite fallow periods and a return of soil organic matter to former concentrations (Tiessen et al., 1992). There is also some evidence that loss of phosphorus fertility precedes the full decline of soil organic matter, which may explain why land becomes unproductive prior to a complete depletion of mineralizable organic carbon (Tiessen and Shang, 1998).

Phosphorus availability to plants in a biologically-healthy soil is promoted by an active microbial community, which performs various functions critical to the maintenance of phosphorus availability in tropical soils. In particular, it converts phosphorus contained within organic amendments into cellular material, which prevents the loss of phosphorus from the biological cycle by strong sorption to soil constituents. A diverse microbial community also contains specialized organisms with the capacity to access phosphorus from stable pools in the soil. The success of such organisms is probably best promoted as part of a healthy microbial community rather than by inoculation of cultured species (Powlson et al., 2001; Richardson, 2001).

Detailed information on many of the processes involved in organic phosphorus turnover is limited, but is important if we are to effectively evaluate and improve agricultural practices that promote biological soil health. Accepting such practices requires a shift from conventional approaches to improving fertility that are based on the application of water-soluble phosphate fertilizer, to approaches based on the maintenance of an active pool of soil organic matter and a diverse biological community. The assessment of phosphorus fertility in such systems is not straightforward, because conventional measurements of plant-available phosphate are unsuitable for tropical soils with a high capacity to sorb phosphate. Novel methods are therefore required for determining the ability of the soil in biologically-managed systems to supply phosphorus to the crop. Such methods should include measurements of organic phosphorus turnover, although these are currently extremely difficult to make.

Once the ability of biologically-managed systems to supply phosphorus to crops can be accurately assessed, we can undertake experiments that illuminate the mechanisms involved in organic phosphorus turnover, and field trials to assess the suitability of biological approaches in a wide range of agroecosystems. Given the urgent need to improve productivity in many regions where phosphorus-deficiency limits crop yield, such studies are of considerable importance for the success of our global future. In Part III, with other colleagues, we report on successful interventions along these lines in Colombia and Kenya (Chapter 37).

References

Antibus, R.K., Sinsabaugh, R.L., and Linkins, A.E., Phosphatase-activities and phosphorus uptake from inositol phosphate by ectomycorrhizal fungi, *Can. J. Bot.*, **70**, 794–801 (1992).

Bartlett, R. and James, B., Studying dried, stored soil samples: Some pitfalls, *Soil Sci. Soc. Am. J.*, **44**, 721–724 (1980).

Birch, H.F., The effect of soil drying on humus decomposition and nitrogen availability, *Plant Soil*, **10**, 9–31 (1958).

Bossio, D.E. et al., Soil microbial community response to land use change on model grassland ecosystems, *Microb. Ecol.* 49, 50–62 (2004).

Bradford, M.A. et al., Impacts of soil faunal community composition on model grassland ecosystems, *Science*, **298**, 615–618 (2002).

Brookes, P.C., Powlson, D.S., and Jenkinson, D.S., Phosphorus in the soil microbial biomass, *Soil Biol. Biochem.*, **16**, 169–175 (1984).

Bühler, S. et al., Isotope methods for assessing plant available phosphorus in acid tropical soils, *Eur. J. Soil Sci.*, **54**, 605–616 (2003).

Bünemann, E. et al., Microbial community composition and substrate use in a highly weathered soil as affected by crop rotation and P fertilization, *Soil Biol. Biochem.*, **36**, 889–901 (2004a).

Bünemann, E. et al., Phosphorus dynamics in a highly weathered soil as revealed by isotopic labeling techniques, *Soil Sci. Soc. Am. J.*, **68**, 1645–1655 (2004b).

Campo, J., Jaramillo, V.J., and Maass, J.M., Pulses of soil phosphorus availability in a Mexican tropical dry forest: Effects of seasonality and level of wetting, *Oecologia*, **115**, 167–172 (1998).

Chantigny, M.H., Dissolved and water-extractable organic matter in soils: A review on the influence of land use and management practices, *Geoderma*, **113**, 357–380 (2003).

Chauhan, B.S., Stewart, W.B., and Paul, E.A., Effect of labile inorganic phosphate status and organic carbon additions on the microbial uptake of phosphorus in soils, *Can. J. Soil Sci.*, **61**, 373–385 (1981).

Chen, C.R. et al., Phosphorus dynamics in the rhizosphere of perennial ryegrass (*Lolium perenne* L.) and radiata pine (*Pinus radiata* D. Don.), *Soil Biol. Biochem.*, **34**, 487–499 (2002).

Clarholm, M., Interactions of bacteria, protozoa and plants leading to mineralization of soil nitrogen, *Soil Biol. Biochem.*, **17**, 181–187 (1985).

Cleveland, C.C., Townsend, A.R., and Schmidt, S.K., Phosphorus limitation of microbial processes in moist tropical forests: Evidence from short-term laboratory incubations and field studies, *Ecosystems*, **5**, 680–691 (2002).

Cole, C.V. et al., Trophic interactions in soils as they affect energy and nutrient dynamics. V. Phosphorus transformations, *Microb. Ecol.*, **4**, 381–387 (1978).

Cole, C.V., Innis, G.S., and Stewart, J.W.B., Simulation of phosphorus cycling in semi-arid grassland, *Ecology*, **58**, 1–15 (1977).

Coleman, D.C., Oades, J.M., and Uehara, G., Eds., *Dynamics of Soil Organic Matter in Tropical Ecosystems*, University of Hawaii Press, Honolulu, HI (1989).

Condron, L.M. et al., Chemical nature of organic phosphorus in cultivated and uncultivated soils under different environmental conditions, *J. Soil Sci.*, **41**, 41–50 (1990a).

Condron, L.M. et al., Critical evaluation of methods for determining total organic phosphorus in tropical soils, *Soil Sci. Soc. Am. J.*, **54**, 1261–1266 (1990b).

Dalal, R.C., Soil organic phosphorus, *Adv. Agron.*, **29**, 85–117 (1977).

Drechsel, P., Steiner, K.G., and Hagedorn, F., A review on the potential of improved fallows and green manure in Rwanda, *Agroforest. Syst.*, **33**, 109–136 (1996).

Fairhurst, T. et al., The importance, distribution and causes of phosphorus deficiency as a constraint to crop production in the tropics, *Agroforest. Forum*, **9**, 2–8 (1999).

Fardeau, J.C., Morel, C., and Jahiel, M., Does long contact with the soil improve the efficiency of rock phosphate? Results of isotopic studies, *Fertil. Res.*, **17**, 3–19 (1988).

Fischler, M. and Wortmann, C.S., Green manures for maize-bean systems in eastern Uganda: Agronomic performance and farmers perceptions, *Agroforest. Syst.*, **47**, 123–138 (1999).

Hannapel, R.J. et al., Phosphorus movement in a calcareous soil: I. Predominance of organic forms of phosphorus in phosphorus movement, *Soil Sci.*, **97**, 350–357 (1964).

Harrison, A.F., ^{32}P-method to compare rates of mineralization of labile organic phosphorus in woodland soils, *Soil Biol. Biochem.*, **14**, 337–341 (1982).

Harrison, A.F., *Soil Organic Phosphorus: A Review of World Literature*, CAB International, Wallingford, UK (1987).

Hinsinger, P., Bioavailability of soil inorganic P in the rhizosphere as affected by root-induced chemical changes: A review, *Plant Soil*, **237**, 173–195 (2001).

Hocking, P.J., Organic acids exuded from roots in phosphorus uptake and aluminum tolerance of plants in acid soils, *Adv. Agron.*, **74**, 63–97 (2001).

Ingham, R.E. et al., Interactions of bacteria, fungi, and their nematode grazers: Effects on nutrient cycling and plant growth, *Ecol. Monogr.*, **55**, 19–140 (1985).

Jenkinson, D.S. and Ladd, J.N., Microbial biomass in soil: Measurement and turnover, *Soil Biochemistry*, Vol. 5, Paul, E.A. and Ladd, J.N., Eds., Marcel Dekker, New York, 415–471 (1981).

Kieft, T.L., Soroker, E., and Firestone, M.K., Microbial biomass response to a rapid increase in water potential when dry soil is wetted, *Soil Biol. Biochem.*, **19**, 119–126 (1987).

Kucey, R.M.N., Janzen, H.H., and Leggett, M.E., Microbially mediated increases in plant-available phosphorus, *Adv. Agron.*, **42**, 199–228 (1989).

Kwabiah, A.B. et al., Response of soil microbial biomass dynamics to quality of plant materials with emphasis on P availability, *Soil Biol. Biochem.*, **35**, 207–216 (2003).

Lal, R., Soil carbon sequestration and impacts on global climate change and food security, *Science*, **304**, 1623–1629 (2004).

Lavelle, P. et al., Soil function in a changing world: The role of invertebrate ecosystem engineers, *Eur. J. Soil Biol.*, **33**, 159–193 (1997).

Lundquist, E.J. et al., Rapid response of soil microbial communities from conventional, low input, and organic farming systems to a wet/dry cycle, *Soil Biol. Biochem.*, **31**, 1661–1675 (1999).

Martius, C., Tiessen, H., and Vlek, P.L.G., The management of organic matter in tropical soils: What are the priorities?, *Nutr. Cycl. Agroecosys.*, **61**, 1–6 (2001).

Myers, R.J.K., van Nordwijk, M., and Vityakon, P., Synchrony of nutrient release and plant demand: Plant litter quality, soil environment and farmer management options, In: *Driven by Nature: Plant Litter and Decomposition*, Giller, G.C. and Giller, K.E., Eds., CAB International, Wallingford, UK, 215–229 (1997).

Nziguheba, G. and Bünemann, E.K., Organic phosphorus dynamics in tropical agroecosystems, In: *Organic Phosphorus in the Environment*, Turner, B.L., Frossard, E., and Baldwin, D.S., Eds., CAB International, Wallingford, UK, 243–268 (2005).

Oberson, A. et al., Phosphorus transformations in an Oxisol under contrasting land-use systems: The role of the soil microbial biomass, *Plant Soil*, **237**, 197–210 (2001).

Oberson, A. et al., Phosphorus status and cycling in native Savanna and improved pastures on acid low-P Colombian Oxisol, *Nutr. Cycl. Agroecosys.*, **55**, 77–88 (1999).

Oberson, A. and Joner, E.J., Microbial turnover of organic phosphorus in soils, In: *Organic Phosphorus in the Environment*, Turner, B.L., Frossard, E., and Baldwin, D.S., Eds., CAB International, Wallingford, UK, 133–164 (2005).

Oehl, F. et al., Basal organic phosphorus mineralization in soils under different farming systems, *Soil Biol. Biochem.*, **36**, 667–675 (2004).

Oehl, F. et al., Kinetics of microbial phosphorus uptake in cultivated soils, *Biol. Fert. Soils*, **34**, 31–41 (2001).

Organic Phosphorus Workshop, Synthesis and recommendations for future research, In: *Organic Phosphorus in the Environment*, Turner, B.L., Frossard, E., and Baldwin, D.S., Eds., CAB International, Wallingford, UK, 377–380 (2005).

Persmark, L., Banck, A., and Jansson, H., Population dynamics of nematophagous fungi and nematodes in an arable soil: Vertical and seasonal fluctuations, *Soil Biol. Biochem.*, **28**, 1005–1014 (1996).

Powlson, D.S., Hirsch, P.R., and Brookes, P.C., The role of soil microorganisms in soil organic matter conservation in the tropics, *Nutr. Cycl. Agroecosys.*, **61**, 41–51 (2001).

Quiquampoix, H. and Mousain, D., Enzymatic hydrolysis of organic phosphorus, In: *Organic Phosphorus in the Environment*, Turner, B.L., Frossard, E., and Baldwin, D.S., Eds., CAB International, Wallingford, UK, 89–112 (2005).

Richardson, A.E., Prospects for using soil microorganisms to improve the acquisition of phosphorus by plants, *Aust. J. Plant Physiol.*, **28**, 897–906 (2001).

Richardson, A.E. et al., Utilization of organic phosphorus by higher plants. In: *Organic Phosphorus in the Environment*, Turner, B.L., Frossard, E., and Baldwin, D.S., Eds, CAB International, Wallingford, UK, 165–184 (2005).

Richardson, A.E. and Hadobas, P.A., Soil isolates of *Pseudomonas* spp. that utilize inositol phosphates, *Can. J. Microbiol.*, **43**, 509–516 (1997).

Richardson, A.E. et al., Utilization of phosphorus by pasture plants supplied with *myo*-inositol hexaphosphate is enhanced by the presence of soil micro-organisms, *Plant Soil*, **229**, 47–56 (2001).

Russell, E.W., *Soil Conditions and Plant Growth*, 11th ed., Longman Scientific and Technical, Harlow, UK (1988).

Sanchez, P.A., Improved fallows come of age in the tropics, *Agroforest. Syst.*, **47**, 3–12 (1999).

Sanchez, P.A., Soil fertility and hunger in Africa, *Science*, **295**, 2019–2020 (2002).

Seeling, B. and Zasocki, R.J., Microbial effects in maintaining organic and inorganic solution phosphorus concentrations in a grassland topsoil, *Plant Soil*, **148**, 277–284 (1993).

Smith, S.E. and Read, D.J., *Mycorrhizal Symbiosis*, 2nd ed., Academic Press, San Diego, CA (1997).

Smithson, P.C. and Giller, K.E., Appropriate farm management practices for alleviating N and P deficiencies in low-nutrient soils of the tropics, *Plant Soil*, **245**, 169–180 (2002).

Soil Survey Staff, *Soil Taxonomy: A Basic System of Soil Classification for Making and Interpreting Soil Surveys*, 2nd ed., U.S. Department of Agriculture, Natural Resources Conservation Service, Washington, DC (1999).

Sparling, G.P., West, A.W., and Reynolds, J., Influence of soil moisture regime on the respiration response of soils subjected to osmotic stress, *Aust. J. Soil Res.*, **27**, 161–168 (1989).

Sparling, G.P., Whale, K.N., and Ramsay, A.J., Quantifying the contribution from the soil microbial biomass to the extractable P levels of fresh and air dried soils, *Aust. J. Soil Res.*, **23**, 613–621 (1985).

Srivastava, S.C., Microbial contribution to extractable N and P after air-drying of dry tropical soils, *Biol. Fert. Soils*, **26**, 31–34 (1997).

Stemmler, S.J. and Berthelin, J., Microbial activity as a major factor in the mobilization of iron in the humid tropics, *Eur. J. Soil Sci.*, **54**, 725–733 (2003).

Tarafdar, J.C. and Claassen, N., Organic phosphorus compounds as a phosphorus source for higher plants through the activity of phosphatases produced by plant roots and microorganisms, *Biol. Fert. Soils*, **5**, 308–312 (1988).

Thien, S.J. and Myers, R., Determination of bioavailable phosphorus in soil, *Soil Sci. Soc. Am. J.*, **56**, 814–818 (1992).

Tiessen, H., Cuevas, E., and Chacon, P., The role of soil organic matter in sustaining soil fertility, *Nature*, **371**, 783–785 (1994).

Tiessen, H., Salcedo, I.H., and Sampaio, E.V.S.B., Nutrient and soil organic matter dynamics under shifting cultivation in semi-arid northeastern Brazil, *Agric. Ecosyst. Environ.*, **38**, 139–151 (1992).

Tiessen, H. and Shang, C., Organic-matter turnover in tropical land-use systems, In: *Carbon and Nutrient Dynamics in Natural and Agricultural Tropical Ecosystems*, Bergström, L. and Kirchmann, H., Eds., CAB International, Wallingford, UK, 1–14 (1998).

Turner, B.L. et al., Potential contribution of lysed bacterial cells to phosphorus solubilisation in two rewetted Australian pasture soils, *Soil Biol. Biochem.*, **35**, 187–189 (2003a).

Turner, B.L. and Haygarth, P.M., Phosphorus solubilization in rewetted soils, *Nature*, **411**, 258 (2001).

Turner, B.L. and Haygarth, P.M., Changes in bicarbonate-extractable inorganic and organic phosphorus following soil drying, *Soil Sci. Soc. Am. J.*, **67**, 344–350 (2003).

Turner, B.L., Mahieu, N., and Condron, L.M., Quantification of *myo*-inositol hexakisphosphate in alkaline soil extracts by solution ^{31}P NMR spectroscopy and spectral deconvolution, *Soil Sci.*, **168**, 469–478 (2003b).

Turner, B.L. et al., Inositol phosphates in the environment, *Philos. Trans. R. Soc. London, Ser. B*, **357**, 449–469 (2002).

Van Gestel, M., Merckx, R., and Vlassak, K., Microbial biomass responses to soil drying and rewetting: The fate of fast- and slow-growing microorganisms in soils from different climates, *Soil Biol. Biochem.*, **25**, 109–123 (1993).

Wardle, D.A. et al., Ecological linkages between above and belowground biota, *Science*, **304**, 1629–1633 (2004).

West, A.W. et al., Comparison of microbial C, N-flush and ATP, and certain enzyme activities of different textured soils subject to gradual drying, *Aust. J. Soil Res.*, **26**, 217–229 (1988).

Wick, B. et al., Temporal variability of selected soil microbiological and biochemical indicators under different soil quality conditions in south-western Nigeria, *Biol. Fert. Soils*, **35**, 155–167 (2002).

Woomer, P.L. and Swift, M.J., Eds., *The Biological Management of Soil Fertility*, Wiley, Chichester, UK (1994).

14

Phytohormones: Microbial Production and Applications

Azeem Khalid, Muhammad Arshad and Zahir Ahmad Zahir

Institute of Soil and Environmental Sciences, University of Agriculture, Faisalabad, Pakistan

CONTENTS

Plants contain a wealth of compounds which function as coordinators of plant growth and development and are referred to as phytohormones, plant hormones, or plant-growth regulators (PGRs). These naturally occurring organic compounds influence various physiological processes in plants such as cell elongation and cell division. They do this at concentrations far below the levels at which nutrients and vitamins normally affect plant processes.

Plant physiologists have recognized five major classes of phytohormones, namely, auxins, gibberellins, cytokinins, ethylene, and abscisic acid. Although it is now well established that a great many soil microorganisms can produce these plant growth-regulating substances (Table 14.1), little has been done to exploit the influence of microbially-produced phytohormones on plant growth and development, partly because our knowledge is still so incomplete, but even more so because the processes involved are so complex. This chapter reviews what is known about microbial biosynthesis of phytohormones and their influence on the growth and development of plants. This will provide a better understanding of how soil systems function, as complexes of plant, microbe, and other biotic and abiotic interactions.

The plant rhizosphere is a remarkable ecological environment as myriad microorganisms colonize in, on, and around the roots of growing plants. Distinct communities of beneficial soil microorganisms, often referred to as plant growth-promoting rhizobacteria (PGPR), that are capable of synthesizing various concentrations of phytohormones are

TABLE 14.1

Production of Phytohormones by Microorganisms and Their Influence on Plant Growth

Phytohormones Detected	Microorganisms	Plants	Responses	References
Auxin-indole-3 -acetic acid (IAA)	Rhizobacteria	Wheat	Rhizobacterial strains active in IAA production had relatively more positive effects on inoculated seedlings	Khalid et al. (2003)
	Rhizobacteria	*Brassica juncea*	Significant correlation observed between auxin production by PGPR *in vitro* and growth-promotion of inoculated seedlings	Asghar et al. (2002)
	Pseudomonas putida GR12-2 and an IAA-deficient mutant	Canola and mungbean	Primary roots of canola seeds treated with wild-type strain 35–50% longer than roots from seeds treated with the IAA-deficient mutant and roots from uninoculated seeds. Exposing mung bean cuttings to high levels of IAA by soaking in a suspension of wild-type strain stimulated formation of many adventitious roots	Patten and Glick(2002)
	Azotobacter	Maize	Inoculation with strains efficient in IAA production had significant growth-promoting effects on maize seedlings	Zahir et al. (2000)
	Rhizobium, Azospirillum	Rice	Inoculation with diazotrophs had significant growth-promoting effects on rice seedlings	Biswas et al. (2000)
	Rhizobium legumino-sarum (strain E11)	Rice	Growth-promoting effects upon inoculation on axenically-grown rice seedlings were observed	Dazzo et al. (2000)
Auxin (IAA) and ethylene	*Arthrobacter mysorens* 7, *Flavobacterium* sp. L30, *Klebsiella mobilis* CIAM880	Barley	All the PGPR actively colonized barley root system and rhizosphere, and significantly stimulated root elongation up to 25%	Pishchik et al. (2002)

(continued)

TABLE 14.1 (Continued)

Phytohormones Detected	Microorganisms	Plants	Responses	References
	Pseudomonas putida GR 12-2 (wild type), GR12-2/acd36 (ACC-deaminase minus mutant), GR 12-2/aux1 (IAA-over producers)	Mung bean	Only the wild-type strain influenced the development of longer roots	Mayak et al. (1999)
Auxin (IAA) and gibberellin-like substances	*Azotobacter chroococcum*	Wheat	A 5-day-old culture increased root and shoot length, most likely through production of PGRs	Pati et al. (1995)
	Azotobacter spp.	Barley	Bacterial metabolites showed a stimulatory effect on plant height and dry weight of barley	Mahmoud et al.(1984)
Auxin (IAA), gibberellin-like substances, and cytokinin-like substances	*Azospirillum brasilense*	Pearl millet	Combination of IAA, GA_3, and kinetin produced changes in root morphology similar to those produced by the inoculum	Tien et al. (1979)
	Azospirillum brasilense	Pearl millet and sorghum	Combination of IAA, GA_3, and kinetin produced changes in root morphology similar to those produced by the inoculum	Hubbell et al. (1979)
Auxin (IAA) and abscisic acid	*Azospirillum brasilense*	*Beta vulgaris* (spp. *cicla*), and wheat	Root elongation of *Beta vulgaris* spp. *cicla* was stimulated, and the number of lateral roots was increased in response to inoculation. Exogenous application of IAA to wheat plants caused a similar response	Kolb and Martin (1985)
Gibberellins	*Bacillus pumilus, Bacillus licheniformis*	Alder	Inoculation showed strong growth-promoting activity on alder	Gutiérrez-Mañero et al. (2001)
ACC-deaminase activity and reduced concentration of C_2H_4	*Enterobacter cloacae, Pseudomonas putida, Pseudomonas fluorescens*	Canola	Inoculation significantly enhanced root elongation of canola under gnotobiotic conditions	Penrose and Glick (2003)

(continued)

TABLE 14.1 (Continued)

Phytohormones Detected	Microorganisms	Plants	Responses	References
	Pseudomonas putida Am2, *Pseudomonas putida* Bm3, *Alcaligenes xylosoxidans* cm4, *Pseudomonas* sp. Dp2	Spring rapeseed	Significant increase in root elongation of phosphorus-sufficient seedlings of rapeseed in a growth-pouch culture experiment was observed in response to inoculation	Belimov et al. (2002)

associated with the root systems of all higher plants. Many of these, having mutualistic relationships with plant roots, are highly dependent for their survival on preformed substrates excreted by plant roots as exudates: amino acids, organic acids, carbohydrates, nucleic acid derivatives, vitamins, and other growth substances. In turn, the soil microflora inhabiting the rhizosphere can cause dramatic changes in plant growth and development by contributing to the host plants' endogenous pools of phytohormones and/or by facilitating the supply and uptake of nutrients and providing other services.

Soil systems present a complex matrix of organic and inorganic constituents being combined in diverse conditions. These create a unique and dynamic environment for the microorganisms that are releasing metabolites which affect plants and other organisms. Even very low concentrations of phytohormones in the vicinity of roots can have pronounced effects on the growth and development of plants. The amounts of a hormone in the soil atmosphere can vary greatly from soil to soil depending upon many biotic and abiotic factors. In the rhizosphere, different crops and varieties or species can produce quite different types and profiles of root exudates, which can support the activity of inocula and/or serve as substrate(s) for the formation of biologically-active substances by the inocula (Frankenberger and Arshad, 1995).

Plant hormones of microbial origin are a class of agrochemicals that could be of great significance in promoting agricultural production. Microbially-released phytohormones not only provide a continuous supply, at no financial cost, but may be more useful than the one-time application of chemically-synthesized PGRs. Moreover, synthetic compounds have narrow thresholds between their inhibitory and stimulatory levels, so a continuous release of PGRs at low concentration (as a result of microbial activity) in the rhizosphere may be more beneficial to modify plant growth in a desired direction. Stimulatory or inhibitory effects of PGRs on plants are concentration-dependent. Contrary to usual expectations, in general, low concentrations are stimulatory, whereas a high concentration may be inhibitory. Plants exposed to higher amounts of PGRs could possibly shift excess amounts of the phytohormone supplied exogenously by the microorganisms into storage forms for later use.

Since synthesized hormones in soil undergo various decomposition processes, the amounts of phytohormones that can be detected in the soil represent net balances of production minus decomposition, so the levels detected are usually less, even much less, than the amount that is actually synthesized by microorganisms and plants. Although the latter can synthesize phytohormones by themselves, these may not be sufficient for plants' optimum growth and development.

Identifying the ability of microorganisms to produce physiologically-active concentrations of PGRs opens up a new field in soil science and allied disciplines, with many possibilities for enhancing soil system performance with reduced dependence on external inputs. This chapter will not go into the most technical aspects of this subject, attempting to give readers an idea of the kinds of research findings that are illuminating this aspect of soil system dynamics.

14.1 Functions and Impacts of Phytohormones

Phytohormones perform many functions in plants, and a function attributed to one hormone group may overlap with the function of other groups. Mostly, various groups of hormones function in plants in coordination with each other. Each group of hormones is known to perform many functions, but in general, auxins are primarily involved in cell enlargement, cytokinins in cell division, and gibberellins in stem elongation by stimulating cell division and cell elongation, while ethylene and abscisic acid are mostly involved in ripening and senescence.

The exogenous provision of phytohormones as a result of microbial activity affects plants' endogenous hormonal levels, either by supplementing the plants' own suboptimal levels or by interacting with the synthesis, translocation, or inactivation of existing hormone levels. The presence and rate of uptake of physiologically-active concentrations of these hormones in the rhizosphere determines actual plant responses. Although plants themselves synthesize a diversity of growth-regulating phytohormones, they also respond to exogenous applications of these PGRs which, after being taken up by the plant, may affect its growth directly as growth stimulants or indirectly by modifying the rhizosphere.

Under suboptimal climatic and environmental conditions, plants are unlikely to synthesize sufficient endogenous concentrations of these phytohormones to sustain optimal growth and development, so exogenous sources often contribute to better plant growth and health. Phytohormone effects to date have been demonstrated largely from exogenous applications. Since specific phytohormones typically do not act alone, the final outcome in terms of plant growth suppression or enhanced development represents the net effect of various hormonal balances. This, of course, makes measurement and assessment difficult.

Recently, scientists have shown that plant growth can be improved when seeds or roots of agricultural crops are inoculated with specific microbial strains that are known to be active in the production of PGRs. The use of microbial inoculants has been successful in a number of countries for a range of field, garden, and tree crops, as seen in Chapter 32 and Chapter 33. In practice, it has often been difficult to achieve consistent and reproducible increases in growth and yield under field conditions, due to the lack of understanding of the mechanisms by which particular microorganisms promote plant growth. Similarly, not enough is known about how the soil type, host plant genotype, and environmental conditions affect the ability of the inocula to colonize and persist on plant roots. We do know, however, that inocula in the presence of a specific physiological precursor of a PGR and/or inocula that produce physiologically-active concentrations of a phytohormone can be highly effective in promoting plant growth and in enhancing consistency and reproducibility (Frankenberger and Arshad, 1995; Arshad and Frankenberger, 1998, 2002; Zahir et al., 2004).

Recent developments in molecular biology and biotechnology reveal that phytohormones play critical roles in gene regulation mechanisms and can modify or enhance the rate of genetic transformation and gene expression in an organism (Chateau et al., 2000; Walz et al., 2002; Jakubowska and Kowalalczyk, 2004). This is explored with research evidence in Chapter 15.

14.2   Sources of Phytohormones

Phytohormones could be either natural, i.e., synthesized by plant/microorganisms, or synthetic, i.e., chemically synthesized, compounds that, after their application, target a plant tissue to alter its life processes or its structure to improve quality, increase yields, or facilitate harvesting.

14.2.1   Natural Sources

14.2.1.1   Plants

Higher plants are capable of synthesizing all the five major classes of phytohormones listed above. Endogenously-produced plant hormones which regulate intracellular processes are distributed within tissues from cell to cell as in the case of auxin, or via vascular bundles as in the case of cytokinin, or through intercellular spaces as occurs with ethylene. Any excessive amounts may be stored in plant tissues as conjugates for later use. It remains open whether they are stored in one or the other plant compartments, and whether they become biologically active by being set free from such compartments. More details about these endogenous plant hormones can be found in several published reviews and books (Moore, 1989; Davies, 2005).

14.2.1.2   Microorganisms

Soil microorganisms, particularly rhizosphere microflora, are a major natural source of PGRs. Very small quantities of PGRs released by microorganisms can have pronounced effects on plant growth and development. The production of PGRs in pure culture and in soil has been often demonstrated (Frankenberger and Arshad, 1995; Arshad and Frankenberger, 1998, 2002; Khalid et al., 2004). However, the soil pool of these PGRs may originate in part from plants that have released these hormones into the rhizosphere as root exudates. How large a proportion this is is not known, and it surely varies over time and in space. Certainly much of the pool is synthesized by soil microbiota *in situ*. Microflora that are known to be able to produce PGRs *in vitro* are present in appreciable numbers in the rhizosphere of practically all plants. The type and amount of PGRs produced by various microorganisms are variable, however. The presence of suitable substrate(s) or precursor(s) also affects PGR production by microorganisms.

Barea et al. (1976) reported that among 50 bacterial isolates collected from the rhizosphere of various plants, 86, 58, and 90% produced auxins, gibberellins, and kinetin-like substances, respectively. Similarly, Mansour et al. (1994) tested 24 isolates of thallobacteria belonging to the genus *Streptomyces* for their potential to produce PGRs. It was seen that all the isolates were capable of producing auxins, gibberellins, and cytokinins in liquid medium. Khalid et al. (2004) reported that different isolates of rhizobacteria varied greatly in their efficiency for producing auxins in a broth medium; among 30 isolates tested, 22 (73%) produced auxins. They also found that in the presence of the auxin precursor L-tryptophan (L-TRP), bacterial efficiency in synthesizing auxins was enhanced several-fold.

Different amounts of cytokinin produced by rhizobacteria have also been confirmed by Garcia de Salamone et al. (2001). They reported that five strains of PGPR produced a cytokinin, dihydrozeatin riboside (DHZR), in pure culture. *Pseudomonas fluorescens* G20-18 produced higher amounts of three cytokinins (isopentyl adenosine [IPA], trans-zeatin ribose [ZR], and DHZR) than did three selected mutant strains. Plant-associated methylotrophs (Long et al., 1996) and phototrophic purple bacterium (Serdyuk et al., 1995) have been reported capable of producing cytokinins *in vitro*.

Among the natural sources, soil microbiota are the most potent producers of the plant-growth regulator ethylene *in vitro* and *in vivo*. Many studies have revealed that diverse groups of soil microbiota are active in producing ethylene (Sato et al., 1997; Weingart et al., 1999). Akhtar et al. (2005) isolated several fungi from rhizosphere soils of various crops through enrichment with L-methionine (L-MET); of the fungi isolated from the rhizosphere soil of wheat, maize, potato, and tomato, 78, 83, 89, and 72%, respectively, were found to produce ethylene from L-MET.

Similarly, production of gibberellins (GAs) and abscisic acid (ABA) by several soil and rhizosphere microflora has been reported by many scientists (Okamoto et al., 1988a, 1988b; Janzen et al., 1992; Gutiérrez-Mañero et al., 2001).

14.2.2 Synthetic Sources

Many synthetic compounds with auxin-like activity have been developed that are used for a variety of purposes, for example, indole-3-butyric acid (IBA) for root initiation, and naphthalene acetic acid (NAA) for apple fruit thinning. Similarly, other synthetic auxins (2,4 dichloro-phenoxy acetic acid, 2,4,5-trichlorphenoxy acetic acid) are well-known for their strong herbicidal effects. Commercially available ethylene gas or ethylene-releasing compounds such as ethephon/ethrel and retprol are used as important synthetic sources to promote the ripening of fruits or vegetables in various parts of the world. Kinetin, a synthetic cytokinin, has also been employed to promote fruit size and other desirable characteristics.

14.3 The Biochemistry of Microbial Biosynthesis of Phytohormones

The biochemical processes whereby different microbes synthesize one or more forms of the five categories of phytohormones are very complex, which may explain in part why much of our knowledge about phytohormones, though many decades old, remains on the margins of plant and soil sciences. It is not necessary to understand all the details of these biochemical processes to appreciate the contributions that microbes make to soil systems through their production of phytohormones, which is presented in detail in Frankenberger and Arshad (1995).

Microbial isolates have been found to synthesize phytohormones from a variety of compounds. Microbial production of ethylene, for example, can be derived from a number of structurally-unrelated compounds. Further, in the microbial biosynthesis of both gibberellins and abscisic acid for which mevalonic acid acts as substrate, the formation of different intermediate compounds during the growth phase of a microorganism determines whether GA or ABA is the final compound synthesized.

To make matters more complicated, phytohormone synthesis can take place either in the presence or in the absence of certain biochemical precursors. Although the presence of tryptophan (TRP) stimulates auxin production, its synthesis in microorganisms can occur in the absence of TRP. This means that TRP is a precursor but not a requirement for biosynthesis. We know that zeatin (Z) acts as the central point for the synthesis of other cytokinins in plants; however, in microorganisms it is not clear whether or not the cytokinins identified are derived from Z or directly from a precursor.

Often PGR production by microorganisms proceeds via more than one biochemical pathway. The detection of various intermediates (indole pyruvic acid, indole acetamide, indole aldehyde, tryptamine) in a TRP-supplemented medium shows that microorganisms often follow different routes for their formation of auxins (Arshad and Frankenberger, 1998). While L-methionine (L-MET) has been identified as a common physiological precursor for ethylene biosynthesis in both higher plants and microorganisms, the biosynthetic pathway found in higher plants is substantially different from the pathways in lower plants (Jia et al., 1999; Arshad and Frankenberger, 2002; Nazli et al., 2003).

The level of expression of any particular phytohormone by certain bacteria depends on the interaction of several different factors, in particular, the biosynthetic pathways being followed, the kind and location of the genes involved and their regulatory sequences, and the presence of enzymes that can convert the active form of a phytohormone into an inactive, conjugated form (Patten and Glick, 1996, 2002).

All of these observations underscore the complexity and contingency of these myriad biochemical processes. Demonstrating these relationships between microorganisms and plant growth, let alone measuring them, has been very difficult; however, these relationships have been well, if not fully, established by microbiologists for several decades now. So it is appropriate to begin factoring their presence and potentials into our understanding and management of soil systems.

14.4   Applications in Agriculture

In recent years, there has been a renewed interest in the screening of microorganisms that are efficient in producing biologically-active substances including PGRs to promote plant growth and yields in various parts of the world. There is no question that microbial production of PGRs is one of the major mechanisms for altering the growth and yield of plants. Many scientists have demonstrated that inoculation with specific microorganisms can affect plant growth via production of PGRs (Table 14.1).

Recently, Khalid et al. (2004) have reported that the potential of rhizobacteria for auxin biosynthesis can be used as a tool for screening effective PGPR strains. A series of laboratory experiments conducted on two cultivars of wheat under gnotobiotic conditions revealed a linear positive correlation ($r = 0.99$) between *in vitro* auxin production and an increase in growth parameters of inoculated seeds. Results of field trials also demonstrated that the PGPR strain which produced the highest amount of auxins in nonsterilized soil also caused maximum increase in growth and yield of two wheat cultivars. Under gnotobiotic conditions, Noel et al. (1996) have shown the direct involvement of IAA and cytokinin production by PGPR in the growth of canola and lettuce.

Glick and coworkers have suggested the involvement of an enzyme, 1-aminocyclopropane-1-carboxylic acid (ACC) deaminase produced by *P. putida* GR12-2, in modifying the root growth of different plants (Glick et al., 1998; Belimov et al., 2002). They found that this bacterium hydrolyzes ACC, the immediate precursor of ethylene in higher plants. ACC deaminase might act to stimulate plant growth by sequestering and then hydrolyzing ACC from germinating seeds, thereby lowering the endogenous levels of ACC, which subsequently results in plant growth promotion. Involvement of phytohormones in inoculation-evoked plant responses has also been well-documented by various authors (Frankenberger and Arshad, 1995; Arshad and Frankenberger, 1998, 2002; Vessey, 2003; Zahir et al., 2004).

Inoculation of plants has proven beneficial for improving plant growth and yield; however, there is a lack of consistency and reproducibility in the effectiveness of inocula.

To achieve consistent yield increases and reproducibility, an approach has been developed that was based on the hypothesis that inoculation in the presence of a specific precursor was more effective than the inoculation alone (Frankenberger and Arshad, 1995; Arshad and Frankenberger, 1998). Using this approach, the production of a particular phytohormone in the rhizosphere can be controlled by providing suitable concentrations of a precursor that can evoke the desired physiological response. Several studies have indicated that the effectiveness of inoculation can be enhanced by the simultaneous applications of one or more physiological precursors of particular PGRs (Table 14.2).

Another novel and innovative approach takes microbial metabolites (not living cells) that contain desired concentrations of a biologically-active substance or PGR. Using this material for seed or root treatment has been shown to be effective for promoting the growth and yield of various plants/crops (rice, maize, wheat, chilies, and tomatoes). After testing in the laboratory/growth room and wire house trials, extensive work on rice was undertaken on farmers' fields to test the validity of this approach. Very encouraging results were obtained: an average increase in yield up to 20%. This approach is different from the conventional biofertilizer approach which uses formulations consisting of living cells of microbial organisms and has often given inconsistent results (Arshad et al., 2003).

14.5 Discussion

Enhancement and use of PGRs is one of the newly-emerging options for meeting challenges to the agricultural sector: raising production to meet the still-growing aggregate demand for food; producing food free from synthetic chemicals; protecting plants against abiotic stresses: frost, excessive heat, etc.; repairing damaged plants; and enhancing the conservation of endangered species and environmental safety. However, before PGRs can contribute to such benefits, scientists must learn more about them and explore ways and means for their better utilization.

Phytohormones coordinate information from many sources through many points of regulation, from biosynthesis to transduction. Phytohormones produced by rhizosphere microorganisms or provided by inocula in the root vicinity certainly affect the growth and development of plants after being taken up by the roots. Future research should focus on managing plant-microbe interactions, particularly with respect to the production and effective utilization of PGRs for the benefit of plants. Scientists need to address agronomic impacts, rates and timing of application, stability and bioavailability in soil systems, and nutritional and root exudation aspects for getting maximum benefit from microbially-released PGRs.

Biotechnological and molecular approaches could possibly develop some genetically-engineered microorganisms that have more successful plant-microbe interaction and/or that are able to synthesize more of particular hormones. Efforts could also focus on whether the use of phytohormones could reduce pesticide use and whether certain plant diseases could be cured or prevented by management practices based on a better understanding of the mechanism(s) of signaling and action. More explicit comprehension of these aspects/disciplines could enable researchers to develop practical models for predicting and assisting growth patterns. Although the use of PGRs is presently small in volume and value compared to chemical pesticides, herbicides, insecticides, and fungicides, it is possible that in the future, the rate of growth of use of naturally- or industrially-synthesized PGRs will increase, making them a rapidly expanding segment of the agricultural service sector.

TABLE 14.2

Precursor-Inoculum Interaction and Physiological Plant Responses

Precursor	Hormone Released as a Metabolite	Inocula	Effects	References
L-Tryptophan (L-TRP)	Auxin (indole-3-acetic acid)	*Pisolithus tinctorius*	*Pisolithus tinctorius* stimulated the growth of potted seedlings of Douglas fir only in the presence of L-TRP. There was basically no difference in growth between the inoculated and noninoculated treatments in the absence of L-TRP	Frankenberger and Poth (1987)
		Azotobacter	Azotobacter inoculation plus L-TRP was more effective than their application alone, increasing the tuber and straw yield of potato	Zahir et al. (1997)
			Azotobacter (capable of producing auxins) plus L-TRP ($10^{-3} M$) increased wheat yield (21.3%) compared to the untreated control	Khalid et al. (1999)
			Maximum length and weight of roots/shoots of maize were observed where Azotobacter inoculum was amended with L-TRP at $10^{-3} M$ concentration	Zahir et al. (2000)
		Rhizobium	Rhizobium inoculation alone increased the grain yield of lentil by 20.9% compared to the non-inoculated control, while application of 1.7 mg L-TRP kg^{-1} soil along with the Rhizobium inoculation increased the yield by 30.6%	Hussain et al. (1995)
		Enterobacter taylorae	*Enterobacter taylorae* with L-TRP inhibited the root growth of field bindweed seedlings, with positive influence on plants of red clover, wheat, pigweed, green foxtail, morning glory, maize, and soybean, most likely due the production of auxins as a result of the precursor-inoculum interactions	Sarwar and Kremer (1995)

(continued)

TABLE 14.2 (Continued)

Precursor	Hormone Released as a Metabolite	Inocula	Effects	References
Adenine (ADE) and isopentyl alcohol (IA)	Cytokinins	*Azotobacter chroococcum*	The combined application of ADE, IA, and the bacterial inoculum enhanced the dry weight of root and shoot tissues, leaf area and chlorophyll a content of radish to a much greater degree than in the presence or absence of the cytokinin precursors (ADE or IA) or the bacterium alone	Nieto and Frankenberger (1990)
			A combined treatment of ADE, IA, and the bacterial inoculum enhanced the vegetative growth of maize to a greater degree than did the application of ADE plus IA; ADE plus *Azotobacter chroococcum*; or ADE, IA, or *A. chroococcum* alone	Nieto and Frankenberger (1991)
L-Methionine (L-MET) and L-ethionine	Ethylene	Indigenous soil microflora	Soil application of L-MET affected the vegetative growth and resistance to stem breaking (lodging) of two cultivars of maize. L-Ethionine application resulted in a significant epinastic response, enhanced fruit yield, and early fruit formation and ripening in tomato	Arshad and Frankenberger (1990)
			Albizia lebbeck L. responded positively to low to moderate concentrations of L-MET applied to soil	Arshad et al. (1993)
			A significant yield response was observed with soybean exposed to L-MET applied to soil	Arshad et al. (1995)

Acknowledgments

This work was funded by various donors: the Gatsby Charitable Foundation (U.K.), the Rockefeller Foundation, the Department for International Development (U.K.), the Global Environmental Facility, Farm Africa, and the Biotechnology and Biological Sciences Research Council (U.K.).

References

Akhtar, M.J. et al., Substrate-dependent biosynthesis of ethylene by rhizosphere soil fungi and its influence on etiolated pea seedlings, *Pedobiologia* **49**, 211–219 (2005).

Arshad, M. and Frankenberger, W.T. Jr., Response of *Zea mays* L. and *Lycopersicon esculentum* to the ethylene precursors, L-methionine and L-ethionine, applied to soil, *Plant Soil*, **122**, 219–227 (1990).

Arshad, M. and Frankenberger, W.T. Jr., Plant growth regulating substances in the rhizosphere: Microbial production and functions, *Adv. Agron.*, **62**, 46–151 (1998).

Arshad, M. and Frankenberger, W.T. Jr., *Ethylene: Agricultural Sources and Applications*, Kluwer Academic/Plenum Publishers, New York (2002).

Arshad, M. et al., Effect of soil applied L-methionine on growth, nodulation and chemical composition of *Albizia lebbeck* L, *Plant Soil*, **148**, 129–135 (1993).

Arshad, M., Hussain, A., and Zahir, Z.A., Effect of soil applied precursors of phytohormones on growth and yield of soybean, *PGRSA Q.*, **23**, 63–69 (1995).

Arshad, M., Khalid, A., and Zahir, Z.A., *Rice-Biofert: Technology Development for Sustainable Rice Production. Annual Report*, University of Agriculture, Faisalabad, Pakistan (2003).

Asghar, H.N. et al., Relationship between *in vitro* production of auxins by rhizobacteria and their growth-promoting activities in *Brassica juncea* L, *Biol. Fert. Soils*, **35**, 231–237 (2002).

Barea, J.M., Navarro, E., and Montoya, E., Production of plant growth regulators by rhizosphere phosphate-solubilizing bacteria, *J. Appl. Bacteriol.*, **40**, 129–134 (1976).

Belimov, A.A., Safronova, V.I., and Mimura, T., Response of spring rape (*Brassica napus* var. Olifera L.) to inoculation with plant growth promoting rhizobacteria containing 1-aminocyclopropane-1-carboxylate deaminase depends on nutrient status of the plant, *Can. J. Microbiol.*, **48**, 189–199 (2002).

Biswas, J.C. et al., Rhizobial inoculation influences seedling vigor and yield of rice, *Agron. J.*, **92**, 880–886 (2000).

Chateau, S., Sangwan, R.S., and Sangwan-Norreel, B.S., Competence of *Arabidopsis thaliana* genotypes and mutants for *Agrobacterium tumefaciens*-mediated gene transfer: Role of phytohormones, *J. Exp. Bot.*, **51**, 1961–1968 (2000).

Davies, P.J., Ed., *Signal Transduction, Action!*, 3rd ed., Kluwer, Dordrecht (2005).

Dazzo, F.B. et al., Progress in multinational collaborative studies on the beneficial association between *Rhizobium leguminosarum* bv. *trifolii* and rice, In: *Nitrogen Fixation in Rice*, Ladha, J.K. and Reddy, P.M., Eds., International Rice Research Institute, Los Baños, Philippines, 167–189 (2000).

Frankenberger, W.T. Jr. and Arshad, M., *Phytohormones in Soil: Microbial Production and Function*, Marcel Dekker, New York (1995).

Frankenberger, W.T. Jr. and Poth, M., Biosynthesis of indole-3-acetic acid by pine ecto-mycorrhizal fungi *Pisolithus tinctorius*, *Appl. Environ. Microbiol.*, **53**, 2908–2913 (1987).

Glick, B.R., Penrose, D.M., and Li, J., A model for lowering plant ethylene concentration by plant growth promoting rhizobacteria, *J. Theor. Biol.*, **190**, 63–68 (1998).

Gracía de Salamone, I.E., Hynes, R.K., and Nelson, L.M., Cytokinin production by plant growth promoting rhizobacteria and selected mutants, *Can. J. Microbiol.*, **47**, 404–411 (2001).

Gutiérrez-Mañero, F.G. et al., The plant-growth-promoting rhizobacteria *Bacillus pumilus* and *Bacillus licheniformis* produce high amounts of physiologically active gibberellins, *Physiol. Plant.*, **111**, 206–211 (2001).

Hubbell, D.H. et al., Physiological interactions in the Azospirillum-grass root association, In: *Associative N$_2$-Fixation*, Vose, P.B. and Ruschel, A.P., Eds., CRC Press, Boca Raton, FL, 1–6 (1979).

Hussain, I. et al., Substrate-dependent microbial production of auxins and their influence on the growth and nodulation of lentil, *Pak. J. Agric. Sci.*, **32**, 149–152 (1995).

Jakubowska, A. and Kowalalczyk, S., The auxin conjugate 1-O-indole-3-acetyl-{beta}-D-glucose is synthesized in immature legume seeds by IAGlc synthase and may be used for modification of some high molecular weight compounds, *J. Exp. Bot.*, **55**, 791–801 (2004).

Janzen, R.A. et al., *Azospirillum brasilense* produces gibberellin in pure culture and on chemically-defined medium in co-culture on straw, *Soil Biol. Biochem.*, **24**, 1061–1064 (1992).

Jia, Y.J. et al., Synthesis and degradation of 1-aminocyclopropane-1-carboxylic acid by *Penicillium citrinum, Biosci. Biotechnol. Biochem.*, **63**, 542–549 (1999).

Khalid, A., Arshad, M., and Zahir, Z.A., Growth and yield response of wheat to inoculation with auxin producing plant growth promoting rhizobacteria, *Pak. J. Bot.*, **35**, 483–498 (2003).

Khalid, A., Arshad, M., and Zahir, Z.A., Screening plant growth-promoting rhizobacteria for improving growth and yield of wheat, *J. Appl. Microbiol.*, **96**, 473–480 (2004).

Khalid, M. et al., Azotobacter and L-tryptophan application for improving wheat yield, *Pak. J. Bio. Sci.*, **2**, 739–742 (1999).

Kolb, W. and Martin, P., Response of plant roots to inoculation with *Azospirillum brasilense* and the application of indole acetic acid, In: *Azospirillum III: Genetics, Physiology, Ecology*, Klingmüller, W., Ed., Springer, Berlin, 215–221 (1985).

Long, R.L.G. et al., Evidence for cytokinin production by plant associated methylotrophs, *Plant Physiol.*, **111**, 89 (1996).

Mahmoud, S.A.Z. et al., Production of plant growth promoting substances by rhizosphere microorganisms, *Zentrabl. Mikrobiol.*, **139**, 227–232 (1984).

Mansour, F.A., Ildesuguy, H.S., and Hamedo, H.A., Studies on plant growth regulators and enzyme production by some bacteria, *Qatar Univ. Sci. J.*, **14**, 281–288 (1994).

Mayak, S., Tirosh, T., and Glick, B.R., Effect of wild type and mutant plant growth promoting rhizobacteria on the rooting of mung bean cuttings, *J. Plant Growth Regul.*, **18**, 49–53 (1999).

Moore, T.M., *Biochemistry and Physiology of Plant Hormones*, 2nd ed., Springer, New York (1989).

Nazli, Z.H., Arshad, M., and Khalid, A., 2-Keto-4-methylthiobutyric acid-dependent biosynthesis of ethylene in soil, *Biol. Fert. Soils*, **37**, 130–135 (2003).

Nieto, K.F. and Frankenberger, W.T. Jr., Influence of adenine, isopentyl alcohol and *Azotobacter chroococcum* on the growth of *Raphanus sativus* (radish), *Plant Soil*, **127**, 147–156 (1990).

Nieto, K.F. and Frankenberger, W.T. Jr., Influence of adenine, isopentyl alcohol and *Azotobacter chroococcum* on the vegetative growth of *Zea mays, Plant Soil*, **135**, 213–221 (1991).

Noel, T.C. et al., *Rhizobium leguminosarum* as a plant-growth rhizobacterium: Direct growth promotion of canola and lettuce, *Can. J. Microbiol.*, **42**, 279–283 (1996).

Okamoto, M., Hirai, N., and Koshimizu, K., Biosynthesis of abscisic acid in *Cercospora pinidensiflorae, Phytochemistry*, **27**, 2099–2103 (1988a).

Okamoto, M., Hirai, N., and Koshimizu, K., Biosynthesis of abscisic acid from α-ionylideneethanol in *Cercospora pinidensiflorae, Phytochemistry*, **27**, 3465–3469 (1988b).

Pati, B.R., Sengupta, S., and Chandra, A.K., Impact of selected phyllospheric diazotrophs on the growth of wheat seedlings and assay of the growth substances produced by the diazotrophs, *Microbiol. Res.*, **150**, 121–127 (1995).

Patten, C.L. and Glick, B.R., Bacterial biosynthesis of indole-3-acetic acid, *Can. J. Microbiol.*, **42**, 207–220 (1996).

Patten, C.L. and Glick, B.R., Role of *Pseudomonas putida* indoleacetic acid in development of the host plant root system, *Appl. Environ. Microbiol.*, **68**, 3795–3801 (2002).

Penrose, D.M. and Glick, B.R., Methods for isolating and characterizing ACC deaminase-containing plant growth-promoting rhizobacteria, *Physiol. Plant*, **118**, 10–15 (2003).

Pishchik, V.N. et al., Experimental and mathematical simulation of plant growth promoting rhizobacteria and plant interaction under cadmium stress, *Plant Soil*, **243**, 173–186 (2002).

Sarwar, M. and Kremer, R.J., Enhanced suppression of plant growth through production of L-tryptophan-derived compounds by deleterious rhizobacteria, *Plant Soil*, **172**, 261–269 (1995).

Sato, M. et al., Detection of new ethylene producing bacteria, *Pseudomonas syringae* pvs. *Cannabina* and *Sesami*, by PCR amplification of genes for the ethylene forming enzyme, *Phytopathology*, **87**, 1192–1196 (1997).

Serdyuk, O.P. et al., 4-Hydroxyphenethyl alcohol-a new cytokine-like substance from the phototrophic purple bacterium *Rhodospirillum rubrum* IR, *FEMS Lett.*, **365**, 10–12 (1995).

Tien, T.M., Gaskins, M.H., and Hubbell, D.H., Plant growth substances produced by *Azospirillum brasilense* and their effect on the growth of pearl millet (*Pennisetum americanum* L.), *Appl. Environ. Microbiol.*, **37**, 1016–1024 (1979).

Vessey, J.K., Plant growth promoting rhizobacteria as biofertilizers, *Plant Soil*, **255**, 571–586 (2003).

Walz, A. et al., A gene encoding a protein modified by the phytohormone indoleacetic acid, *Plant Biol.*, **99**, 1718–1723 (2002).

Weingart, H., Volksch, B., and Ullrich, M.S., Comparison of ethylene production by *Pseudomonas syringae* and *Ralstonia solanacearum*, *Phytopathology*, **63**, 156–161 (1999).

Zahir, Z.A. et al., Substrate-dependent microbially derived plant hormones for improving growth of maize seedlings, *Pak. J. Biol. Sci.*, **3**, 289–291 (2000).

Zahir, Z.A. et al., Effect of an auxin precursor L-tryptophan and *Azotobacter* inoculation on yield and chemical composition of potato under fertilized conditions, *J. Plant Nutr.*, **20**, 745–752 (1997).

Zahir, Z.A., Arshad, M., and Frankenberger, W.T. Jr., Plant growth promoting rhizobacteria-perspectives and application in agriculture, *Adv. Agron.*, **81**, 96–168 (2004).

15

Crop Genetic Responses to Management: Evidence of Root–Shoot Communication

Autar K. Mattoo and Aref Abdul-Baki

Sustainable Agricultural Systems Laboratory, Henry A. Wallace Agricultural Research Center, U.S. Department of Agriculture, Beltsville, Maryland, USA

CONTENTS

A number of successful sustainable agriculture systems for vegetable production have been developed in recent years for the mid-Atlantic states of the USA. Of particular interest is the use of an annual legume, hairy vetch, which has proved to be a beneficial cover crop because it fits well into different cropping rotations, is capable of high nitrogen fixation, and produces substantial biomass (Kelly et al., 1995; Teasdale and Abdul-Baki, 1997; Araki and Ito, 2004). In the UK, an assessment of agricultural practices recently concluded that no-till agriculture is potentially more beneficial than others (Trewavas, 2004). Thanks to methods for molecular biological investigation, we are beginning to understand how and why such practices are contributing to better crop performance and more sustainable agricultural production, by affecting the expression of favorable genetic potentials in crop plants.

This chapter reviews some recent research that indicates some previously unrecognized ways in which the management of plants, soil, and nutrients affects crop growth. These connections have been apparent in gross observational terms and have been reported in measured correlations, but without a good understanding of the mechanisms involved. More research is needed to establish the extent and malleability of the mechanisms now being mapped out. However, work so far illuminates part of the complex web of plant–soil interactions that determine agricultural success. Although microbial partners in these interactions have not been studied yet, they are likely to have some roles in this dynamic process as well. Here we report what is presently known, expecting that further research in this area will expand this knowledge.

15.1   Hairy Vetch-Vegetable Associations

When planted in mid-September and grown through the winter, hairy vetch is ready to be terminated before summer vegetables such as tomatoes and peppers are transplanted in the field in late spring (Teasdale and Abdul-Baki, 1997). We note that for fall vegetables such as broccoli, a mixture of forage soybean (*Glycine max* L.) and foxtail millet (*Setaria italica* L. P. Beaver) that can be grown over the summer has proved to have similar effects (Abdul-Baki et al., 1997).

Yields of tomatoes that are grown in hairy vetch mulch over several years have averaged 20–40% higher than those produced by conventional methods, which usually include the use of black polyethylene mulch (Teasdale and Abdul-Baki, 1997; Whitehead and Singh, 2003). Further, when mulching with hairy vetch, soil fertility has substantially improved over time, even with lower input of chemical fertilizer. This finding has presented a challenge to agricultural scientists to explain why hairy vetch mulch accompanied by less inorganic fertilization gives better results than the "modern" system that employs polyethylene mulch and high levels of synthetic fertilizer application. If this alternative system warrants confidence, there are significant environmental benefits to be gained from greater reliance on organic inputs to expand horticultural production.

Various benefits seen from using cover-crop mulches in cropping systems can directly enhance crop productivity and quality — the retention of longevity and greenness in the leaves (Teasdale and Abdul-Baki, 1997; Kumar et al., 2004), less pest infestation, and enhanced tolerance or resistance to disease (Mills et al., 2002). Economic studies have supported the sustainability and superiority of this alternative system over the conventional system (Kelly et al., 1995; Lu et al., 2000). However, despite these evident advantages of alternative farming practices, their spread has been slow. This may be partly because it is thought that the scientific basis for alternative agriculture is weak (Trewavas, 2001). Until more is known about the mechanism(s) that underlie these beneficial aspects of legume cover crops, many producers may be hesitant to adopt such alternative farming practices. The desirable performance of cover-crop, mulch-grown vegetables has presented an opportunity to explore how the environment can control gene function. Recent research (Kumar et al., 2004, 2005) has shown that in tomatoes grown after a hairy-vetch cover crop, a highly organized and specific network of genes is modulated in the crop, making it possible for the same or reduced inputs to produce greater outputs, something that agriculturalists generally desire.

15.2   Activation of Specific Genes in Crop Associations

Long duration of foliage greenness in plants is an indication of delayed senescence (aging) of their leaves. In all plants' development, a process of leaf senescence will at some point invariably follow that of leaf expansion; however, the timing can vary. Senescence is generally thought to result from the rapid degradation of certain essential proteins in the plant, which accumulate during leaf development and expansion. These include, particularly, rubisco (ribulose-1,5-bisphosphate carboxylase/oxygenase), a protein essential for photosynthesis and N remobilization; glutamine synthetase (GS), which plays a major role in N conversion and C utilization for the synthesis of amino acids, the building blocks for all proteins; the enzyme ATPase, and chlorophyll-binding proteins in the chloroplast, cytb559, and plastocyanin, all important for photosynthesis and converting harvested chemical energy into food (Roberts et al., 1987; Mehta et al., 1992). Further description of their functions is provided below.

The onset of degradation of rubisco and other proteins during leaf senescence is thought to provide nitrogen in the form of amino acids that are released to support the growth of other plant tissues, for example, young or developing leaves and fruits (Thimann, 1980). The stable accumulation and retention of these compounds in a developed leaf is thus associated with leaf longevity. Of special interest is the persistence in leaf tissues of rubisco and GS, which are critical for photosynthesis and the carbon–nitrogen (C/N) metabolism of any crop (Makino et al., 1984).

It is therefore significant that in the leaves of tomato plants grown in hairy-vetch mulch, we find that the transcripts as well as protein levels of rubisco and cytosolic GS remain higher and stable for a longer duration than in plants grown in the conventional black-polyethylene mulch. On the other hand, the levels of gene transcripts for a nitrogen-responsive transporter, ATPase, chlorophyll-binding proteins, and many other chloroplast-localized proteins were not found to be much different (Kumar et al., 2004). Thus, the hairy-vetch-based production system specifically induces the expression of genes that are associated with prolonged plant growth.

Similarly, it has been found that genes contributing to plants' defense against pests and diseases are activated in tomatoes grown with vetch mulch. Basic chitinase (Broglie et al., 1986) and osmotin (Liu et al., 1994) are among the defense proteins that enable plants to resist attacks by pests including pathogenic fungi. The gene transcripts for these two proteins are highly expressed — and their proteins accumulate for a more prolonged period — in tomatoes that were grown in hairy-vetch mulch compared to those grown in black polyethylene. It is interesting that tomatoes grown in either hairy vetch or with black polyethylene were found (Kumar et al., 2004) not to be different in the plant disease-resistance pathways signaled via nitric oxide, a previously documented mechanism for defense against pathogen infection (Durner and Klessig, 1999).

Chaperones are a class of proteins that stabilize native proteins, prevent the aggregation of denatured proteins, and catalyze proper folding of proteins (Pelham, 1986; Hammond and Helenius, 1995). Two such proteins — heat shock protein-70 (hsp70) (Li et al., 1999), and ER binding protein (BIP) (Kalinski et al., 1995) — which are normally activated during periods of plant stress were found to be present at higher steady-state levels (both for transcripts and for proteins) in tomatoes grown in the hairy-vetch mulch compared to those grown in black-polyethylene mulch. The higher stability of hsp-70 and BiP in vetch-mulched tomato plants suggests that these proteins continued to be recruited and maintained in plant tissues longer, which helps prolong the longevity of these plants compared to those grown with black polyethylene (Kumar et al., 2004).

One of the promoters of aging and senescence in plants is the hormone ethylene, which also regulates defensive processes in some plants (Boller, 1991; Lund et al., 1998; Mattoo and Handa, 2004). A key enzyme that enables a plant to synthesize ethylene is known as 1-aminocyclopropane-1-carboxylate (ACC) synthase. There are several forms of this enzyme (Rottmann et al., 1991; Fluhr and Mattoo, 1996). We have found that the gene transcripts for one of them, ACS6 — and for a senescence-associated gene, SAG12 (Noh and Amasino, 1999) — accumulate slowly in hairy-vetch-grown tomatoes compared to the black-polyethylene-grown plants (Kumar et al., 2004). This contrasts with the opposite pattern of accumulation for a receptor protein kinase (CRK) that senses (Papon et al., 2002) the level of cytokinin, an anti-senescence and pro-growth hormone. That CRK is higher in the hairy-vetch-grown tomato suggests that hormonal "cross-talk" and more signaling are key factors associated with the beneficial attributes we find characteristic of plants grown on hairy-vetch mulch.

The activation and retention of gene transcripts for the following proteins define the signature of tomato plants cultivated in the hairy-vetch-based alternative agriculture (Kumar et al., 2004, 2005). We note what each contributes to plant metabolism and growth to indicate the intricacy and complexity of the various processes that plant and soil management practices are affecting:

- N-responsive GS1, important for C/N metabolism and signaling;
- rubisco, one of the most abundant proteins in the biosphere, critical for C fixation;
- nitrite reductase, an enzyme that converts nitrite to ammonia, facilitating N utilization and reducing nitrite toxicity;
- glucose-6-phosphate dehydrogenase, an enzyme that contributes to N-use efficiency;
- two chaperone proteins, heat shock protein-70 and ER lumenal binding protein, that are important for keeping other proteins in an active state;
- anti-fungal proteins, chitinase and osmotin, that provide a defense against pests and are binding partners for cytokinin, a growth-regulating hormone;
- cytokinin-responsive receptor kinase, a protein that determines cytokinin import into the leaf tissue;
- GA_{20} oxidase, an enzyme involved in signaling via another class of plant-growth hormones known as gibberellins (GA), discussed in the preceding chapter.

These studies have shown the existence of a specific interface between hairy-vetch mulch and the tomato plant that results in a fundamentally distinct expression-profile of gene transcripts and proteins in the tomato leaf which affect the growth and longevity of plants being raised on this leguminous mulch.

15.3   More Than Just Nitrogen Is Involved

An easy assumption could be that hairy-vetch residue has its beneficial effect on the growth of tomato plants, and even on the differential gene expression presented in the previous section, through an increased provision of nitrogen, since hairy vetch as a legume has known capacity to fix N (Araki and Ito, 2004). However, this is not an adequate explanation. First, the hairy-vetch-grown tomato plants received less total N input, 100 kg ha^{-1} (in the form of urea) compared to 200 kg ha^{-1} supplied to those plants grown in black polyethylene beds (Kumar et al., 2004; see also Abdul-Baki et al., 1997). Second, there was no significant correlation between total leaf N content and the differences observed in disease onset (or severity) and the senescence index as a function of the mulch used.

Although it is clear that differential gene expression described in the previous section is reminiscent of C/N signaling, because of the types of genes involved and the nature of their expression, hairy-vetch-based effects on tomato leaves represent a different biochemical mode of regulation. For example, short-term exposure of plants to applied N normally results in the activation of senescence-associated protein (SAG12) and the nitrate transporter (Wang et al., 2000); but this did not occur with hairy-vetch-grown tomato (Kumar et al., 2004). Similarly, exposure to a high level of N causes a reduction in osmotin transcripts in Arabidopsis (Wang et al., 2000), whereas in the hairy-vetch production system their transcripts actually accumulate.

The question arises: what are the other factors that work in conjunction with C/N signaling to produce the demonstrably beneficial effects in crops grown on a cover-crop mulch? Another contributing factor could be the type of soil microbes that are enriched in the cover-crop/tomato root association. It is interesting to note that some of the agronomic and physiological effects we have seen with hairy-vetch/tomato production system have been documented in Egypt with rice that is grown in association with berseem clover, also a legume (Chapter 8). The observations of Dazzo and Yanni, based on N-balance data, suggested that the benefit of rotating rice with clover was not due to an increased availability of soil N, from mineralization of the biologically-fixed, N-rich clover crop residues. As in our studies with tomato grown with a lower N fertilizer input, they obtained higher rice grain yields with less dependence on chemical fertilizer inputs, depending instead on certain strains of endophytic rhizobia as inoculants for the rice crop.

Of further significance for our discussion here, Dazzo and Yanni found that rice plants developed expanded root architectures when inoculated with rhizobial endophytes, which could make them better "miners" of other nutrients. They report higher concentrations of N, P, K^+, Na^+, Ca^{+2}, Mg^{+2}, and Zn^{+2} in rice inoculated with the endophytes as compared to plants that remained uninoculated (see Figure 8.6 and Section 15.6 below).

We have found that cover crops differ considerably in their ability to "mine" soil nutrients. For instance, a comparison of hairy-vetch residue vs. rye residue in terms of their respective abilities to accumulate minerals showed the former to be richer in P, B, Mn^{+2}, Mg^{+2}, Ca^{+2}, Fe^{+2}, and Zn^{+2} as well as in N (Abdul-Baki and Mattoo, unpublished data). At the same time, no significant differences were found in K, S, and Cu. What effect, if any, these differences in mineral nutrient availability between the two cover crops have on the ability of tomato plants to take these up and accumulate them in the foliage and fruit has not been assessed.

Dazzo and Yanni suggest an interesting scenario based on these findings where the nutritional value of grain could be enhanced, for example, for an essential micronutrient such as zinc. Whether specific microorganisms associated with the rhizosphere of the cover crop production system are responsible for the robust root architecture of tomato plants and/or assist in the acquisition of selected mineral nutrients as in the case of rice–endophyte association is an important subject that remains to be studied.

15.4 Root-to-Leaf Signals via Hormones

Our findings suggest that root physiology and hormonal signaling contribute to the delayed senescence of plants cultivated under leguminous cover-crop mulch. Hairy-vetch-based cultivation leads to more robust and larger spread of roots (Sainju et al., 2000), which contributes to higher yields of the crop. Of relevance for the mechanisms involved in hairy-vetch-based cropping is the fact that root growth regulates the synthesis of cytokinin, a plant hormone whose decrease is associated with initiation of senescence in plant organs (Nooden et al., 1997; Smart et al., 1991; Sakakibara et al., 1998; Hwang and Sheen, 2001).

Conversely, a continued supply of cytokinin from the roots to the upper parts of a plant should delay senescence. This seems to be the case with the tomato plants cultivated under hairy-vetch mulch, indirectly indicated by higher measured levels of an indicator gene, the cytokinin receptor protein kinase, CRK (Papon et al., 2002; Kumar et al., 2004). Cytokinin is known to inhibit the accumulation of senescence-enhancing genes (Noh and Amasino, 1999).

Another facet of interactive C/N metabolism and cytokinin signaling may be a factor in the coordination of delayed senescence of plants cultivated under hairy-vetch mulch with their greater tolerance to disease (Mills et al., 2002; Kumar et al., 2004). Cytokinin signaling has been implicated in plant–microbe interactions involving rhizobacteria (Ryu et al., 2003), while it is known that the engineered accumulation of cytokinins induces several defense-related genes including basic chitinase (Memelink et al., 1987) and osmotin (Thomas et al., 1995). Higher and steady accumulation of transcripts and protein of these two antifungal defense proteins, chitinase and osmotin, in hairy-vetch-grown tomato (Kumar et al., 2004) indicates that cytokinin signaling not only regulates senescence but also disease tolerance. Chitinase is an enzyme that degrades chitin, a polysaccharide component of the gut linings of insects and in the cell walls of fungi. Chitinolytic activity makes these enzymes potent anti-insect and anti-fungal agents, and their occurrence may have implications in the organism-to-organism interactions in the ecosystem (www.glfc. forestry.ca/frontline/bulletins/bulletin_no.18_e.html).

Our studies thus raise some important larger questions. How do hairy-vetch-grown tomatoes coordinate enhanced cytokinin and C/N signaling to cause delayed leaf senescence and greater tolerance of pests? In these plants, what induces and maintains cytokinin synthesis in the roots, and what regulates the root-to-leaf and leaf-to-root communications? Research is only beginning to get some grasp on these issues.

Another possible factor for some of the observations made with hairy-vetch-grown tomato could be their difference from the black-polyethylene-grown plants in the pace of growth. In the field in spring, soils stay cooler under vetch mulch than under polyethylene, which results in slower growth of tomatoes initially, in regard to root length and aboveground. Eventually they develop a more robust leaf area around the beginning of fruit maturity. Rapid tomato growth in polyethylene mulch could lead to faster accumulation of cytokinins in leaf tissue, and this can signal a feedback mechanism to shut down further cytokinin production, particularly when the levels of the cytokinin-binding partners is low in black-polyethylene-grown tomato. In turn, this would bring the polyethylene-grown tomato into reproductive mode sooner and cause earlier senescence under black-polyethylene mulch.

15.5   Possible Mechanisms Involved

The following proposed outline could account for the observed metabolic and genetic regulation in hairy-vetch-based alternative agriculture. For reasons yet to be fully understood, hairy-vetch cultivation of tomatoes generates a healthier and more robust root system as well as a higher steady-state level of cytokinins. It is understandable that a steady flow of cytokinins from the root to the leaf would signal for enhanced and stable levels of basic chitinase and osmotin. Both these proteins provide a defense against insect pests by binding to actin, thereby causing cytoplasmic aggregation (Takemoto et al., 1997). It is also known that basic chitinase and osmotin are binding partners for cytokinins (Kobayashi et al., 2000).

As a result of the tripartite combination of cytokinins, basic chitinase and osmotin, the latter two defense proteins would remain stable for a longer duration, which would maintain the level of free cytokinins at a minimum. Having a minimal threshold level of cytokinins in their free form could create a situation where biochemical communication between the leaf and the root is delayed. We know that surpassing a high threshold level of cytokinins in the leaf can send a feedback signal to the roots, which inhibits cytokinin transport into the leaf. This means that a low level of free cytokinin in the leaves could induce a continuing flow of cytokinins from the roots to the leaves. This could result in the continuing active expression of key genes, for example, accumulation of rubisco, cytoplasmic glutamine synthetase, nitrite reductase, glucose-6-phosphate dehydrogenase, and chaperones, all of which would enable efficient interactions between C and N signaling pathways and would delay the onset of senescence.

C/N signaling and cytokinins in the leaves will also activate defense and receptor kinase proteins that promote disease-tolerance. Thus, the expression of genes associated with the promotion of senescence would remain underrepresented and at low levels, prolonging plant survival in its productive rather than aging phase. These genes include those for the ethylene biosynthesis enzyme-ACC synthase, cysteine protease, and the senescence-associated SAG12 gene (Kumar et al., 2004). These latter enzymes are underrepresented and remain at low levels in hairy-vetch-grown tomatoes, indicating interactions and connections between hairy-vetch residue and the growth of tomato plants.

15.6 Discussion

Crop productivity and protection are major concerns in a world that has become aware of the harmful effects to the environment and to animal and human health from excessive use of chemicals in agriculture. Reducing the use of chemicals in agriculture without having an adverse impact on the yield or quality of the crop is a reasonable objective for current agricultural research. Agriculturists will be more willing to adopt alternative practices if a strong scientific understanding has been developed of the reasons why crops cultivated under cover-crop mulches offer more net benefits. Venturing into a field situation to identify operative genetic mechanisms is daunting from a reductionist point of view because one has to deal not only with the plant itself but also with, among other things, the cover-crop residue chemistry, root physiology, and interactions of the root systems with a wide variety of microorganisms in the rhizosphere. However, unless we look at whole organisms in a field setting, as summarized here, we cannot recognize what is and what is not happening in the real world.

The employment of genetic screening, gene expression studies, and quantified protein levels using specific and sensitive methods can unambiguously demonstrate, as seen in this chapter, the existence of certain specificity inherent within the dynamic metabolic and signaling pathways offered by cover-crop-based, alternative agriculture. It is becoming apparent that many routes of communication exist between different plant organs, which are tied to receptor proteins and downstream signaling pathways. Included in these are: metabolic signals via specific metabolites when their intra- or extra-cellular levels reach some threshold, hormonal signaling via hormone receptor kinases specific to each hormone involved, and environmental signaling due to changes in the environment (Mattoo and Handa, 2004). In the real world, however, what balances these and which ones override which others still remains a matter for future discovery and definition.

Much of the work to improve N-use and C-use efficiency in crops has been carried out in isolated systems, although engineering the genes of specific proteins/enzymes, e.g., the C-fixing enzyme rubisco, glutamate-metabolizing enzymes, nitrate reductase, and photosynthetic phosphoenolpyruvate carboxylase, has generated considerable information (see Foyer and Noctor, 2002). Whatever the merits of these approaches, it is important to remain aware of the impacts that taking a more holistic view can have, focusing on common features that are defined in test tube-level experiments and at the same time reproducible in the field.

It is encouraging to see certain ideas about inter-organ signaling, e.g., root-to-leaf-to-root (Krapp et al., 2002; Sugiyama and Sakakibara, 2002) and source-to-sink for young leaves, roots, fruits (Nooden et al., 1997) — that were developed from physiological and biochemical experiments including studies on split roots, leaf discs, photosynthetic organelle, and genetic mutants — being reinforced now by our studies of tomato plants cultivated on legume cover-crop-based, alternative agriculture (Kumar et al., 2004).

Use of hairy vetch as a cover crop leads to improved vigor, increased fruit yield, delayed senescence, and suppression of weeds and pathogens when used for growing vegetables. Until recently we have known little about the mechanisms that are involved in these economically important responses. The activation of specific genes as identified above brings to the fore the importance of $G \times E$ (gene $\times$ environment) interactions in determining beneficial attributes in plants. It opens possibilities for evoking from crop plants a more productive response, not by changing genetic characteristics or by adding external inputs, but by altering plants' growing conditions.

Acknowledgments

We wish to thank Norman Uphoff and John Teasdale for their constructive comments on the manuscript.

References

Abdul-Baki, A.A., Teasdale, J.R., and Korcak, R.F., Nitrogen requirements of fresh-market tomatoes on hairy vetch and black Polyethylene mulch, *HortScience*, **32**, 217–221 (1997).

Araki, H. and Ito, M., Decrease of nitrogen fertilizer application in tomato production in no-tilled field with hairy vetch mulch, *Acta Hort.*, **638**, 141–146 (2004).

Boller, T., Ethylene in pathogenesis and disease resistance, In: *The Plant Hormone Ethylene*, Mattoo, A.K. and Suttle, J.C., Eds., CRC Press, Boca Raton, FL, 293–314 (1991).

Broglie, K.E., Gaynor, J.J., and Broglie, R.M., Ethylene-regulated gene expression: Molecular cloning of the genes encoding an endochitinase from *Phaseolus vulgaris*, *Proc. Natl Acad. Sci.*, **83**, 6820–6824 (1986).

Durner, J. and Klessig, D.F., Nitric oxide as a signal in plants, *Curr. Opin. Plant Biol.*, **2**, 369–374 (1999).

Fluhr, R. and Mattoo, A.K., Ethylene: Biosynthesis and perception, *Crit. Rev. Plant Sci.*, **15**, 479–523 (1996).

Foyer, C.H. and Noctor, N., *Photosynthetic Nitrogen Assimilation and Associated Carbon and Respiratory Metabolism*, *Advances Photosynthesis and Respiration 12*. Kluwer, Netherlands, 284 (2002).

Hammond, C. and Helenius, A., Quality control in the secretory pathway, *Curr. Opin. Cell Biol.*, **7**, 523–529 (1995).

Hwang, I. and Sheen, J., Two-component circuitry in *Arabidopsis* cytokinin signal transduction, *Nature*, **413**, 383–389 (2001).

Kalinski, A. et al., Binding-protein expression is subject to temporal, developmental and stress-induced regulation in terminally differentiated soybean organs, *Planta*, **195**, 611–621 (1995).

Kelly, T.C. et al., Economics of a hairy vetch mulch system for producing fresh-market tomatoes in the mid-Atlantic region, *J. Am. Soc. Hort. Sci.*, **120**, 854–860 (1995).

Kobayashi, K. et al., Cytokinin-binding proteins from tobacco callus share homology with osmotin-like protein and an endochitinase, *Plant Cell Physiol.*, **41**, 148–157 (2000).

Krapp, A., Ferrario-Mery, S., and Touraine, B., Nitrogen and signaling, In: *Photosynthetic Nitrogen Assimilation and Associated Carbon and Respiratory Metabolism*, Foyer, G.H. and Noctor, G., Eds., Kluwer, Netherlands, 205–225 (2002).

Kumar, V. et al., An alternative agriculture system is defined by a distinct expression profile of select gene transcripts and proteins, *Proc. Natl Acad. Sci.*, **101**, 10535–10540 (2004).

Kumar, V. et al., Cover crop residues enhance growth, improve yield and delay leaf senescence in greenhouse-grown tomatoes, *HortScience* **40**, 1307–1311 (2005).

Li, Q.B., Haskell, D.W., and Guy, C.L., Coordinate and non-coordinate expression of the stress 70 family and other molecular chaperones at high and low temperature in spinach and tomato, *Plant Mol. Biol.*, **39**, 21–34 (1999).

Liu, D. et al., Osmotin overexpression in potato delays development of disease symptoms, *Proc. Natl Acad. Sci.*, **91**, 1888–1892 (1994).

Lu, Y.C. et al., Cover crops in sustainable food production, *Food Rev. Int.*, **16**, 121–157 (2000).

Lund, S.T., Stall, R.E., and Klee, H.J., Ethylene regulates the susceptible response to pathogen infection in tomato, *Plant Cell*, **10**, 371–382 (1998).

Makino, A., Mae, T., and Ohira, K., Relationship between nitrogen and ribulose-1,5-bisphosphate carboxylase in rice leaves from emergence through senescence, *Plant Cell Physiol.*, **25**, 429–437 (1984).

Mattoo, A.K. and Handa, A.K., Ethylene signaling in plant cell death, In: *Plant Cell Death Processes*, Nooden, L.D., Ed., Elsevier, San Diego, CA, 125–142 (2004).

Mehta, R.A. et al., Oxidative stress causes rapid membrane translocation and *in vivo* degradation of ribulose-1,5-bisphosphate carboxylase/oxygenase, *J. Biol. Chem.*, **267**, 2810–2816 (1992).

Memelink, J., Hoge, J.H.C., and Schilperoort, R.A., Cytokinin stress changes the developmental regulation of several defence-related genes in tobacco, *EMBO J.*, **6**, 3579–3583 (1987).

Mills, D.J. et al., Factors associated with foliar disease of staked fresh market tomatoes grown under differing bed strategies, *Plant Dis.*, **86**, 356–361 (2002).

Noh, Y.S. and Amasino, R.M., Identification of a promoter region responsible for the senescence-specific expression of SAG12, *Plant Mol. Biol.*, **41**, 181–194 (1999).

Nooden, L.D., Guiamet, J.J., and John, I., Senescence mechanisms, *Physiol. Plant.*, **101**, 746–753 (1997).

Papon, N. et al., Expression analysis in plant and cell suspensions of *CrCKR1*, a cDNA encoding a histidine kinase receptor homologue in *Catharanthus roseus* (L.) G. Don., *J. Exp. Bot.*, **53**, 1989–1990 (2002).

Pelham, H.R.B., Speculations on the functions of the major heat shock and glucose regulated proteins, *Cell*, **46**, 959–961 (1986).

Roberts, D.R. et al., Differential changes in the synthesis and steady-state levels of thylakoid proteins during bean leaf senescence, *Plant Mol. Biol.*, **9**, 343–353 (1987).

Rottmann, W.E. et al., 1-Aminocyclopropane-1-carboxylate synthase in tomato is encoded by a multi-gene family whose transcription is induced during fruit and floral senescence, *J. Mol. Biol.*, **222**, 937–961 (1991).

Ryu, C.M. et al., Bacterial volatiles promote growth in *Arabidopsis*, *Proc. Natl Acad. Sci.*, **100**, 4927–4932 (2003).

Sainju, U.M., Singh, B.P., and Whitehead, W.F., Cover crops and nitrogen fertilization effects on soil carbon and nitrogen and tomato yield, *Can. J. Soil Sci.*, **80**, 523–532 (2000).

Sakakibara, H. et al., A response-regulator homolog possibly involved in nitrogen signal transduction mediated by cytokinin in maize, *Plant J.*, **14**, 337–344 (1998).

Smart, C.M. et al., Delayed leaf senescence in tobacco plants transformed with *tmr*, a gene for cytokinin production in *Agrobacterium*, *Plant Cell*, **3**, 647–656 (1991).

Sugiyama, T. and Sakakibara, H., Regulation of carbon and nitrogen assimilation through gene expression, In: *Photosynthetic Nitrogen Assimilation and Associated Carbon and Respiratory Metabolism*, Foyer, G.H. and Noctor, G., Eds., Kluwer, Netherlands, 227–238 (2002).

Takemoto, D. et al., Identification of chitinase and osmotin-like protein as actin-binding proteins in suspension-cultured potato cells, *Plant Cell Physiol.*, **38**, 441–448 (1997).

Teasdale, J.R. and Abdul-Baki, A.A., Growth analysis of tomatoes in black polyethylene and hairy vetch production systems, *HortScience*, **32**, 659–663 (1997).

Thimann, K.V., Ed., *Senescence in Plants*, CRC Press, Boca Raton, FL (1980).

Thomas, J.C., Smigocki, A.C., and Bohnert, H.J., Light-induced expression of *ipt* from *Agrobacterium tumefaciens* results in cytokinin accumulation and osmotic stress symptoms in transgenic tobacco, *Plant Mol. Biol.*, **27**, 225–235 (1995).

Trewavas, A., Urban myths of organic farming, *Nature*, **410**, 409–410 (2001).

Trewavas, A., A critical assessment of organic farming-and-food assertions with particular respect to the UK and the potential environmental benefits of no-till agriculture, *Crop Prot.*, **23**, 757–781 (2004).

Wang, R. et al., Genomic analysis of a nutrient response in *Arabidopsis* reveals diverse expression patterns and novel metabolic and potential regulatory genes induced by nitrate, *Plant Cell*, **12**, 1491–1509 (2000).

Whitehead, W. and Singh, B., Cover crop mixtures and components vs. synthetic nitrogen effect on above ground biomass and yields of tomato, *HortScience*, **38**, 807 (2003).

16

Allelopathy and Its Influence in Soil Systems

Suzette R. Bezuidenhout[1] and Mark Laing[2]

[1]*KwaZulu-Natal Department of Agriculture and Environmental Affairs, Pietermaritzburg, South Africa*

[2]*Department of Plant Pathology, University of KwaZulu-Natal, Pietermaritzburg, South Africa*

CONTENTS

Plants and microbes release various metabolites into the soil environment via leaching, exudation, decomposing plant material, and volatilization. The study of allelopathy is concerned with the very detrimental effects that some of these metabolites can have on other plants growing in the same soil — what might be called "the dark side" of plant–plant and plant–microbial relationships in the soil. Because allelochemicals are subjected to various ecological processes in the soil, these processes affect the production of these compounds and their impact on susceptible plants. The allelochemicals that are produced by plant roots or soil organisms which have been identified so far belong to at least 16 chemical groups, so this subject is chemically as well as biologically complex.

　　The success of any individual plant in its environment is determined by its ability to acquire the resources essential for its growth and to avoid negative growth factors. There are numerous physical or chemical mechanisms by which certain plants inhibit the growth of neighboring plants, either by competition, whereby plants vie with each other for available water, nutrients, light and space, or by allelopathy, whereby certain chemical compounds are produced and released into the soil environment. Some confusion has been caused by the consideration of allelopathy as part of or a kind of competition.

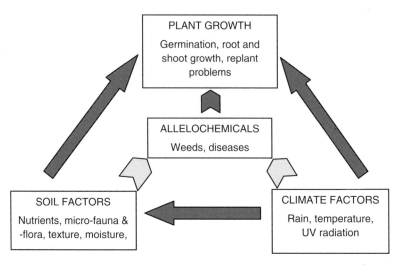

FIGURE 16.1
Heuristic model of allelopathy.

However, competition involves the removal of a shared resource, while allelopathy involves the addition of a chemical compound to the environment through different processes (Rice, 1984; Putnam, 1985). Further, an allelochemical may remain a residual problem after the plant dies, so it is giving no competitive advantage to the parent plant.

Allelopathy is derived from the Greek words *allelon* "of each other" and *pathos* "to suffer" (Rizvi et al., 1992). Researchers agree that allelopathy not only includes the effect of plants on one another, but also the adverse effects that microorganisms can have on plants. The retention, transformation, and transport of allelochemicals are influenced by the nature of the chemical, the organisms present, the properties of the soil, and environmental conditions (Figure 16.1).

Allelopathy is not a new concept (Willis, 1985). Theophrastus (300 BC) first noticed the deleterious effect that cabbage had on grapevines when these plants were grown together and suggested that the impact was due to odors. A common problem in both Greek and Roman times was so-called "soil sickness," associated with declining yields of agricultural fields. In the 17th and 18th centuries, botanists relied strongly on a comparative approach. Stephen Hales believed that root exudates facilitated the excretion of used compounds. This theory of root excretions was a basis for the concept of allelopathy. Swiss botanist Auguste Pyrame de Candolle developed a theory of plant interaction via root excretions. He was influenced by the increasing information about phytochemistry and the effects, positive and negative, of diverse compounds on plant growth.

16.1   Allelopathic Influences on Plant Growth

The negative effects of certain plants on others' growth in agriculture has been noted and studied for some time (Rice, 1984, 1995; Putnam and Tang, 1986). The mechanisms include delayed or inhibited germination, the stimulation or inhibition of root and shoot growth, and induced failures in plant establishment (Rizvi et al., 1992). These effects are often observed in the field or under controlled conditions. Sometimes allelopathic effects are obvious and startling, but mostly the effects are subtle and more difficult to assess.

16.1.1 Impairment of Certain Growth Processes

The biological activity of allelochemicals is concentration-dependent. The growth inhibition threshold of a specific chemical depends on the receiving species, the environmental conditions, and the sensitivity of the receiving plant. Rizvi et al. (1992) point out that allelopathy deals with

- Biochemical interactions and their effects on physiological processes, and
- The mode of action of allelochemicals as specific biochemical sites and the molecular level.

The mode of action of allelochemicals may be direct or indirect. The indirect action represents the effects that occur as a result of soil properties, nutrient status of the soil, and microorganism populations. The direct action involves the biochemical and physiological effects of allelochemicals on various important processes of plant growth and metabolism. Processes influenced by allelochemicals include the following (Rice, 1984; Putnam and Tang, 1986):

- Protein synthesis, e.g., gougerotin inhibits protein synthesis in pro- and eukaryotes;
- Specific enzyme activity, e.g., caffeic acid inhibits phenylalanine ammonia-lyase;
- Membranes and membrane permeability, e.g., salicylic acid inhibits $K+$ transport;
- Phytohormones, e.g., tannins inhibit gibberelic acid in cucumbers;
- Cytology and ultrastructure, e.g., phenolic compounds alter radicle cells of radish;
- Water relationships, e.g., caffeic acid lowers water potential in *Glycine max*;
- Respiration, e.g., juglone inhibits oxidative phosphorylation;
- Photosynthesis, e.g., vanillic acid reduces chlorophyll content in *G. max*; and
- Mineral uptake, e.g., inhibition of $K+$ intake by *Avena sativa*.

16.1.2 Forestry Examples

One of the best-known cases of allelopathy is the way black walnut trees (*Juglans nigra*) inhibit the growth of various vegetables and other plants (Rice, 1984; Willis, 2000). Sensitive species such as tomatoes, peas, potatoes and rhododendrons planted in the vicinity of *J. nigra* display signs of wilting, browning of vascular tissue, necrosis, and eventually death. Toxicity is associated with tree root secretion of juglone, a napthoquinone that becomes harmful when exposed to air as this oxidizes the substance into a toxic form (Rietveld, 1983). The degree of toxicity is affected by root morphology and environmental conditions.

The failure of seedling establishment of patula pine (*Pinus patula*) on old agronomic sites in KwaZulu-Natal, South Africa, prompted investigation. The cause of this problem was attributed to allelopathic influence of a weed, yellow nutsedge (*Cyperus esculentus*) (Bezuidenhout, 2001). Analysis revealed that aqueous extracts of *C. esculentus* tubers and foliage significantly inhibited the growth of the ectomycorrhizal species *Boletus maxima*, which is essential for pine seedling growth (Reinhardt and Bezuidenhout, 2001).

16.1.3 Cover Crop Examples

Cover crops as components of no-tillage systems help reduce erosion, fix atmospheric nitrogen, and improve the physical and chemical characteristics of the soil (Mitchell and

Teel, 1977; Worsham, 1990). Masiunas et al. (1995) found that chemical as well as biological and physical factors contributed to the success of certain cover crops. Stooling rye (*Secale cereale*) planted as a winter cover crop resulted in a 50–75% reduction in weed biomass during the following summer season when its residues were left on the soil surface. Zasada et al. (1997) concluded that a cover crop of stooling rye, without use of additional herbicides, performed adequately to reduce weed populations of common lambsquarter (*Chenopodium album*) when planted at low densities, although not at high densities. The phytotoxicity of *S. cereale* is mainly attributed to the cyclic hydroxamic and phenolic acids that it produces (Shilling et al., 1985; Barnes and Putnam, 1986).

The degree of toxicity from such cover crops is influenced by environmental and production factors such as ultraviolet (UV) radiation and soil fertility (Mwaja et al., 1995). The allelopathogenic effect of a rye grass (*Lolium perenne* or *L. multiflorum*) cycle routinely suppresses maize yields in a maize–rye grass rotation, commonly used in South Africa. Similarly, when perennial rye grass (*L. perenne*) is overseeded into kweek grass (*Cynodon dactylon*) in recreational turf environments, a standard practice in South Africa, the *C. dactylon* is eliminated over the winter season and has to be completely replanted in summer when the rye grass dies off in the heat.

16.1.4 Crop Opportunities

Rice (*Oryza sativa*) is the world's most important cereal grain in terms of production area and consumption. Weed control in rice is a challenge since most rice is planted in developing countries where there is little access to expensive herbicides, and yield is compromised. Researchers have noticed that some rice cultivars suppress the barnyard grass (*Echinochloa crus-galli*) and other annual weed species (Kim and Shin, 1998; Kuk et al., 2001). A number of allelochemicals have been identified in rice root exudates and in decomposing rice straw. These include phenolic and aromatic acids, benzene derivatives, fatty acids, and sterols (Rimando and Duke, 2003). This particular set of allelopathic effects could be utilized positively to protect rice crops from deleterious weed impacts once more is known about how and under what conditions these biochemical compounds are produced.

16.1.5 The Case of Eucalyptus

Eucalyptus tree species have been particularly controversial for their reported allelopathic effects on soil organisms and the growth of other plant species (e.g., May and Ash, 1990; Sanginga and Swift, 1992; Shivanna and Prasana, 1992; Florentine and Fox, 2003). Leaf extracts from Eucalyptus have been found to have a significant inhibiting effect on germination of several plants tested (yarrow, brome and wild rye), more than extracts from oak leaves and the control (Watson, 2000). Leachates of Eucalyptus, as well as Acacia, reduce seedling emergence and other growth parameters for maize (*Zea mays* L.) and kidney bean (*Phaseolus vulgaris* L.) (El-Khawas and Shehata, 2005). So definite allelopathic effects can be demonstrated.

However, evaluating these effects within soil systems is difficult because of the operation of other modes of influence on other flora and fauna, such as the slow decomposition of eucalyptus leaf litter that inhibits the emergence of other tree and herbaceous plant species, something different from allelopathy (Reis and Reis, 1993). The adverse impacts on biodiversity that have been observed with eucalyptus plantations are in part those of monoculture, ones that would occur with any single-species agroecosystem (Couto and Betters, 1995).

It has been shown that the concentration of nitrifying bacteria was very low under eucalyptus plantation litter (Florenzano, 1956). However, other research has shown long-term improvement in soil fertility under eucalyptus plantings (Philliphis, 1956; Kerson, 1961; Ricardo and Madeira, 1985). A 25-year Eucalyptus plantation in Brazil had 27 t ha^{-1} of ground litter compared with an average of 12 t ha^{-1} in nearby native forests and pastures, and more microorganisms and nutrients in the soil (Fonseca, 1984). Chemical allelopathic effects are only one factor among many kinds of interaction going on among flora and fauna within soil systems. This is why it is important to take a systems perspective that enlarges upon, and puts into context, what is observed from investigations of individual or just a few species.

16.2 The Nature and Production of Allelochemicals

Most allelochemicals are secondary metabolites produced as by-products of primary metabolic pathways (Rice, 1984; Putnam and Tang, 1986; Rizvi et al., 1992). According to Aldrich (1984), secondary compounds have no physiological function essential for the maintenance of life. Allelochemicals are present in virtually all plant tissue, i.e., leaves, fruit, stems, and roots (Rice, 1984, 1995; Putnam, 1985). They are released by processes such as volatilization, root exudation, leaching, and decomposition of plant residues. Although allelochemicals appear to be present in all plant tissues, their presence does not guarantee an allelopathic effect (Putnam and Tang, 1986; Heisey, 1990).

16.2.1 Categories of Allelochemical Compounds

Allelochemicals represent a great range of chemical compounds, from simple hydrocarbons and aliphatic acids to complex polycyclic structures. Table 16.1 presents the most prominent secondary products classified in biochemical categories, recognizing that it is impossible to enumerate and include each and every chemical identified as an allelochemical. Rice (1984, 1995) and Putnam and Tang (1986) have grouped allelochemicals into various chemical groups, with phenols, benzoic acids, and cinnamic acids being the most common categories identified thus far. According to Einhellig (1986), terpenoids will be released first, followed by water-soluble phenols and alkaloids; decomposing material will then release phytotoxic phenols. The relative importance of any of these actions depends on the specific allelochemical involved, associated environmental stresses, and which aspect of plant growth is being inhibited.

16.2.2 Allelochemical Production

A variety of environmental conditions influence the quantity of chemicals produced, according to Aldrich (1984) and Rice (1984):

- Light: Some allelochemicals are influenced by the amount, intensity and duration of light, with the greatest quantities produced during exposure to ultraviolet and long-day photoperiods. Thus, understory plants will produce fewer allelochemicals because overstory plants filter out ultraviolet rays.

- Mineral deficiency and drought stress: More allelochemicals are produced under these conditions. It is inferred that allelochemical production gives a source plant some advantage in the competition with other plants for scarce nutrients and/or water.

- Temperature: At lower temperatures, greater quantities are produced. The location and effects of allelochemicals within the plant seem to vary.

TABLE 16.1

Categories and Sources of Allelochemicals

Chemical Group	Allelochemicals	Source Organisms
Simple water-soluble organic acids, straight chain alcohols, aliphatic aldehydes, and ketones	Malic and tartaric acid Fumaric acid Acetic and butyric acid	Fruits *Pinus resinosa*
Unsaturated lactones	Patulin	*Penicillium expansum*
Long-chain fatty acids and polyacetylenes	Oleic and stearic acids Phenylhepatryne	*Polygonum aviculare* *Bidens pilosa*
Naphthoquinone, anthroquinones, and complex quinines	Juglone Helminthosporin	*J. nigra* *Helminthosporium graminium*
Simple phenols	Vanillin and vanillic acid	*Zea mays*
Benzoic acid and derivates	*p*-hydroxybenzoic acid	*A. sativa*
Cinnamic acid and derivates	Chlorogenic acid Ferulic and *p*-coumaric acids	*Helianthus annuus* *Z. mays*
Coumarins	Novobiocin	*Streptomyces niveus*
Flavonoids	Kaempferol	*Selenastrum capriconutum*
Tannins	Tannic acid	*Nicotiana tabacum*
Terpenoids and steroids	Camphor, alternaric acid	*Salvia leucophylla* *Alternaria solani*
Amino acids and polypeptides	Various	Multiple
Alkaloids and cyanohydrins	Ergothioneine	*Claviceps purpurea*
Sulphides and glucosides	Allyl isothiocyanate	*Brassica nigra*
Purines and nucleotides	Caffeine	*Coffea arabica*

Source: Compiled from Rice (1984) and Putnam and Tang (1986).

Aldrich (1984) stated that, as a rule, environmental conditions that restrict plant growth tend to increase the production of allelochemicals. Rice (1984) found that the phenolic compounds released from certain grasses inhibited the nitrification rate in the rhizosphere and consequently the invasion of nitrophilous species. Much of the evidence indicates that several chemicals are often released together and may exert toxicities that are additive or synergistic (Rice, 1995).

16.3   Allelochemicals and the Soil Environment

Understanding the interactions between allelochemicals and soil systems is essential for assessing plant–plant dynamics. Once a chemical enters the environment, a number of interacting processes such as retention, transformation, and transfer take place (Cheng, 1992). Retention involves retardation of the movement of the chemical from one location to another through soil, water, and/or air. Transformation changes the form or structure of the chemical, leading to partial change or total decomposition of the molecule. Transport covers the chemicals' movement in the environment. These processes are influenced by the nature of the chemical, the species present, the properties of the soil, and environmental conditions.

Whether they originate in the rhizosphere or on the soil surface from decomposing plant material, allelochemicals must move through the top of the soil profile to have an allelopathic effect on other plants (Schmidt and Ley, 1999). The complex physical, chemical, and biological characteristics of soil systems will influence the quantitative and qualitative availability of, and response to, allelochemicals (Dao, 1987; Blum and Shafer, 1988; Cheng, 1995). Of the many allelopathic compounds that enter the soil, only a few are absorbed by plants. The size, structure, charge, and polarity of allelochemical molecules are likely to affect their rate of diffusion. Chemicals interact with the surfaces of soil particles and micropores, and this increases the chances of the chemical being sorbed or sequestered before it reaches potential target organisms.

Most allelochemicals have been shown to become bound by humic material in the soil which presumably inactivates them. Equilibria between the absorption and desorption processes determine the concentration of allelochemicals in the soil solution and consequently their bioactivity, movement, and persistence in the soil (Dao, 1987). Phenolic acids are subjected to sorption onto both mineral and organic-matter fractions in the soil. Soils with high levels of organic matter or clay content generally retain phenolic compounds more than sandy soils (Dalton et al., 1989). The type of clay, aluminum, and iron hydroxides in the soil can directly influence the ability of the soil to adsorb phenolic compounds. Many phenolic acids are ionized at the pH value found for most soils, and this makes them more likely to interact with positive sites on clay minerals (Dalton et al., 1989).

Allelopathic effects can be influenced by the nutrient status of the soil. Stowe and Osborn (1980) found that phenolics were more phytotoxic when the levels of nitrogen and phosphorus in the soil were low. Hall et al. (1982) have reported that the phenolic content in sunflower (*Helianthus annuus*) increased as nutrient stress increased. It also had a significant inhibitory effect on the germination of pigweed (*Amaranthus retroflexus*). Phenolic compounds may influence the accumulation and availability of phosphates as they compete for anion absorption sites on clay and humus, resulting in their binding to aluminum, iron, and manganese, which might otherwise bind to phosphate, thereby affecting its availability (Tan and Binger, 1986; Appel, 1993). Raising the nutrient status of the soil, however, does not necessarily alleviate this problem (Bhowmik and Doll, 1984).

The significance of soil texture and moisture in explaining allelopathy has been highlighted by Kuiters and Denneman (1987), who found larger amounts of mild alkaline-extractable phenolics in sandy soil than in loam soil. Fisher (1978) reported that juglone excreted from *J. nigra* resulted in more damage to red and white pine trees when these were grown on wet sites than on dry sites. This could be attributed to the fact that juglone was more available in solution for absorption.

Many allelochemicals are decomposed in soil, either abiotically or by microorganisms. De Scisciolo et al. (1990) found that juglone is more rapidly degraded in soils with higher microbial populations; however, microbial decomposition does not necessarily result in a decrease in allelopathic activity. Willis (1999) reported that mature *Eucalyptus pilularis*, *E. delegetensis*, and *E. regens* all accumulate allelochemicals from the soil. This results in a soil environment that favors pathogens which eventually cause the death of mature trees.

Given the slow movement of chemicals in the soil and the high probability of sorption and/or destruction of most organic chemicals released in the soil, it is likely that true allelopathic interactions take place only when chemicals are released in close proximity to certain species that are susceptible to them. It is unlikely, for example, that allelochemicals leached from plant litter affect the growth of roots very deep in the soil profile. Ponder and Tadros (1985) found that juglone concentration, for example, decreased with soil depth and distance from the source. It is more likely that seedling growth under and through plant litter will be affected by the release of allelochemicals. It has been shown that sometimes direct physical root-to-root contact is necessary for allelopathic effects to become manifest (Dao, 1987; Mahall and Callaway, 1992).

16.4 Discussion

It is only relatively recently that allelopathy has become recognized as an important component of plant science. There are still more questions than answers, and only by improving techniques and by studying the phenomena closely can their complexities be unraveled. Research areas of greatest interest are:

1. Production and release of allelochemicals in response to abiotic and biotic stress;
2. Influence of allelochemicals on soil nutrient dynamics;
3. Roles of soil ecology and their effects on allelochemicals;
4. Influence of soil texture on allelopathic expression; and
5. Transportation of allelochemicals from the donor plant to the recipient.

The soil system's parameters and outcomes derive from multiple interactions among its mineral and organic fractions and its biotic residents. Once in the soil, allelochemical compounds are subjected to many complex reactions that can render them inactive or, conversely, can increase their toxicity. The rhizosphere and plant-root ecology remain areas that are relatively understudied in the literature on allelopathy, which is still itself relatively small. Understanding the interactions of allelochemicals with soil microorganisms and macrobiota are essential in order to understand how plants affect one another, possibly positively through mycorrhizal connections, or negatively through biochemical impediments. Competition and cooperation are woven throughout the fabric of soil systems and their food webs. Allelopathy is a reminder that the mechanisms affecting the balances among plant and animal species are many and often subtle.

References

Aldrich, J.D., *Weed–Crop Ecology: Principles and Practices*, Breton Publishers, N. Scituate, MA (1984).

Appel, H.M., Phenolics in ecological interactions: The importance of oxidation, *J. Chem. Ecol.*, **19**, 1521 (1993).

Barnes, J.P. and Putnam, A.R., Evidence for allelopathy by residues and aqueous extracts of rye (*Secale cereale*), *Weed Sci.*, **34**, 384 (1986).

Bezuidenhout, S.R., Allelopathic effect of the weed *Cyperus esculentus* on the growth of young *Pinus patula* plantations. MSc thesis, University of Pretoria, Pretoria, South Africa (2001).

Bhowmik, P.C. and Doll, J.D., Allelopathic effects of annual weed residues on growth and nutrient uptake of corn and soybeans, *Agron. J.*, **76**, 383 (1984).

Blum, U. and Shafer, S.R., Microbial populations and phenolic acids in soils, *Soil Biol. Biochem.*, **20**, 793 (1988).

Cheng, H.H., A conceptual framework for assessing allelochemicals in the soil environment, In: *Allelopathy: Basic and Applied Aspects*, Rizvi, S.J.H. and Rizvi, H., Eds., Chapman & Hall, London, 21–29 (1992).

Cheng, H.H., Characterization of the mechanism of allelopathy: Modelling and experimental approaches, In: *Allelopathy: Organisms, Processes and Applications*, Dakshini Inderjit, K.M.M. and Einhellig, F.A., Eds., American Chemical Society, Washington, DC, 132–141 (1995).

Couto, L., Betters, D.R., Short-rotation eucalypt plantations in Brazil: Social and environmental issues (http://bioenergy.ornl.gov/reports/euc-braz/eucaly6.html) (1995).

Dalton, B.R., Blum, U., and Weed, S.B., Plant phenolic acids in soils: Sorption of ferulic acid by soil and soil component sterilized by different techniques, *Soil Biol. Biochem.*, **21**, 1011 (1989).

Dao, T.H., Sorption and mineralization of phenolic acids in soils, In: *Allelochemicals: Role in Agriculture and Forestry*, Waller, G.R., Ed., American Chemical Society, Washington, DC, 358–370 (1987).

De Scisciolo, B., Leopald, D.J., and Walton, D.J., Seasonal patterns of juglone in soil beneath *Juglans nigra* (black walnut) and influence of *J. nigra* on understory vegetation, *J. Chem. Ecol.*, **16**, 1111 (1990).

Einhellig, F.A., Mechanisms and mode of action of allelochemicals, In: *The Science of Allelopathy*, Putnam, A.R. and Tang, C.S., Eds., Wiley, New York, 171–188 (1986).

El-Khawas, S.A. and Shehata, M.M., The allelopathic potentialities of *Acacia nilotica* and *Eucalyptus rostrata* on monocot (*Zea mays* L.) and dicot (*Phaseolus vulgaris* L.) plants, *Biotechnology*, **4**, 23–34 (2005).

Fisher, R.F., Juglone inhibits pine under certain moisture regimes, *Soil Soc. Am. J.*, **42**, 801 (1978).

Florentine, S.K. and Fox, J.E.D., Allelopathic effects of *Eucalyptus victrix* L. on Eucalyptus and grasses, *Allelopathy J.*, **11**, 77–84 (2003).

Florenzano, G., *Richerche sui terreni coltivati ad eucalitti. Il Recherche Microbiologiche e Biochimiche*, Centro di Sperimentazione Agricola e Forestale, Laimburg, Italy, 133–152 (1956).

Fonseca, S., Propriedades físicas, quimicas e microbiológicas de um latossolo vermelho-amarelo sob eucalipto, mata natural e pastagem. MS thesis, Universidade Federal de Viçosa, Imprensa Universitária, Viçosa (1984).

Hall, A.B., Blum, U., and Fites, R.C., Stress modifications of allelopathy of *Helianthus annuus* L. debris on seed germination, *Am. J. Bot.*, **69**, 776 (1982).

Heisey, R.M., Evidence for allelopathy by tree-of-heaven (*Ailanthus altissima*), *J. Chem. Ecol.*, **16**, 239 (1990).

Kerson, R., *Soil evolution as affected by Eucalyptus*, Second World Conference on Eucalyptus, Vol. 2. Food and Agriculture Organization, Saõ Paulo, 897–904 (1961).

Kim, K.U. and Shin, D.H., Rice allelopathy research in Korea, In: *Allelopathy in Rice: Proceedings of the Workshop on Allelopathy in Rice*, Olofsdotter, M., Ed., International Rice Research Institute, Manila, 39–43 (1998).

Kuiters, A.T. and Denneman, A.J., Water-soluble phenolic substances in soils under several coniferous and deciduous tree species, *Soil Biol. Biochem.*, **19**, 765 (1987).

Kuk, Y.I., Burgos, N.R., and Talbert, R.E., Evaluation of rice by-products for weed control, *Weed Sci.*, **49**, 141 (2001).

Mahall, B.E. and Callaway, R.M., Root communication mechanism and intercommunity distribution of two Mojave desert shrubs, *Ecology*, **73**, 2145 (1992).

Masiunas, J.B., Weston, L.A., and Weller, S.C., Impact of rye cover crops on weed populations in a tomato cropping system, *Weed Sci.*, **43**, 318 (1995).

May, F.E. and Ash, J.E., *An Assessment of the Allelopathic Potential of Eucalyptus*, Forestry Paper 59. Food and Agriculture Organization, Rome (1990).

Mitchell, W.H. and Teel, M.R., Winter-annual cover crops for no-tillage corn production, *Agron. J.*, **69**, 569 (1977).

Mwaja, V.N., Masiunas, J.B., and Weston, L.A., Effects of fertility on biomass, phytotoxicity, and allelochemical content of cereal rye, *J. Chem. Ecol.*, **21**, 81 (1995).

Philliphis, A., *Protección de los cultivos y defense del suelo*, First World Conference on Eucalyptus. Food and Agriculture Organization, Rome (1956).

Ponder, F. and Tadros, S.H., Juglone concentration in soil beneath black walnut interplanted with nitrogen-fixing species, *J. Chem. Ecol.*, **11**, 937 (1985).

Putnam, A.R., Weed allelopathy, In: *Weed Physiology, Reproduction and Ecophysiology*, Vol. 1. Duke, S.O., Ed., CRC Press, Boca Raton, FL, 131–155 (1985).

Putnam, A.R. and Tang, C.S., Allelopathy: State of the science, In: *The Science of Allelopathy*, Putnam, A.R. and Tang, C.S., Eds., Wiley, New York, 1–19 (1986).

Ricardo, R.P. and Madeira, M.A.V., Relações solo-eucalyto. Universidade Técnica de Lisboa unpublished paper (1985).

Reinhardt, C.F. and Bezuidenhout, S.R., Growth stages of *Cyperus esculentus* influences its allelopathic effect on ectomycorrhizal and higher plant species, *J. Crop Prod.*, **4**, 323 (2001).

Reis, M.C.F. and Reis, G.G., *A contribuiçaõ da pesquisa florestal para a reduçaõ de impactos ambientais do eucalypto*, Anais do I simpósio Brasiliero de Pesquisa Florestal. Society for Forest Research 119–135 (1993).

Rice, E.L., *Allelopathy*, 2nd ed., Academic Press, NY (1984).

Rice, E.L., Allelopathy in forestry, In: *Biological Control of Weeds and Plant Diseases: Advances in Applied Allelopathy*, Rice, E.L., Ed., University of Oklahoma Press, Norman, OK, 317–378 (1995).

Rietveld, W.J., Allelopathic effects of juglone on germination and growth of several herbaceous and woody species, *J. Chem. Ecol.*, **9**, 295–308 (1983).

Rimando, A.M. and Duke, S.O., Studies on rice allelochemicals, In: *Rice: Origin, History, Technology and Production*, Smith, C.W., Ed., Wiley, Hoboken, NJ, 221–244 (2003).

Rizvi, S.H.J. et al. A discipline called allelopathy, In: *Allelopathy: Basic and Applied Aspects*, Rizvi, S.J.H. and Rizvi, V., Eds., Chapman and Hall, London, 1–8 (1992).

Sanginga, N. and Swift, M.J., Nutritional effects of Eucalyptus litter on the growth of maize (*Zea mays*), *Agric. Ecosyst. Environ.*, **41**, 55–65 (1992).

Schmidt, S.K. and Ley, R.E., Microbial competition and soil structure limit the expression of allelochemicals, In: *Principles and Practices in Plant Ecology*, Dakshini Inderjit, K.M.M. and Foy, C.L., Eds., CRC Press, Boca Raton, FL, 339–351 (1999).

Shilling, D.G., Liebl, R.A., and Worsham, A.D., Rye (*Secale cereale*) and wheat (*Triticum aestivum*) mulch: The suppression of certain broad-leaved weeds and the isolation and identification of phytotoxins, In: *The Chemistry of Allelopathy: Biochemical Interactions among Plants*, Symposium Series 268, Thompson, A.C., Ed., American Chemical Society, Washington, DC (1985).

Shivanna, L.F. and Prasana, K.T., Allelopathic effects of Eucalyptus: An assessment on the response of agricultural crops, *Myforest*, **18**, 131–137 (1992).

Stowe, L.G. and Osborn, A., The influence of nitrogen and phosphorus levels on the phytoxicity of phenolic compounds, *Can. J. Bot.*, **58**, 1149 (1980).

Tan, K.G. and Binger, A., Effect of humic acid on aluminum toxicity in corn plants, *Soil Sci.*, **141**, 20 (1986).

Watson, K., *The Effect of Eucalyptus and Oak Leaf Extracts on California Native Plants*, Environmental Sciences Program, College of Natural Resources, University of California, Berkeley (2000).

Willis, R.J., The historical basis of the concept of allelopathy, *J. Hist. Biol.*, **18**, 71–102 (1985).

Willis, R.J., Australian studies on allelopathy in Eucalyptus: A review, In: *Principles and Practices in Plant Ecology*, Dakshini Inderjit, K.M.M. and Foy, C.L., Eds., CRC Press, Boca Raton, FL, 201–219 (1999).

Willis, R.J., Juglans spp., juglone and allelopathy, *Allelopathy J.*, **7**, 1–55 (2000).

Worsham, A.D., Weed management strategies for conservation tillage in the 1990s, In: *Conservation Tillage for Agriculture in the 1990s*, Special Bulletin 90-1, Mueller, J.P. and Wagger, M.G., Eds., North Carolina State University, Raleigh, NC, p. 42 (1990).

Zasada, I.A., Linker, M.H., and Coble, H.D., Initial weed densities affect no-tillage weed management with a rye (*Secale cereale*) cover crop, *Weed Technol.*, **11**, 473 (1997).

17

Animals as Part of Soil Systems

Alice N. Pell

Department of Animal Science, Cornell University, Ithaca, New York, USA

CONTENTS

Manure is often considered to be the primary or only contribution of domestic livestock to maintaining soil fertility. Little attention is paid to the many other ways in which animals affect soil health. Domestic and wild animals can alter soil microbial populations (Bardgett and Wardle, 2003), compact soils (Ritz et al., 2004), influence soil pH (Powell et al., 1998), affect carbon and nitrogen dynamics (Abril and Bucher, 2001), and influence the amount of land cultivated and the intensity with which it is cultivated (Wilson, 2003). Because animals are used to till more than half of the cultivated land in the world, they have a tremendous impact on soil texture, erosion potential, and compaction (Wilson, 2003). As livestock often represent the most valuable, easily convertible assets owned by rural households, they buffer farming communities against inflation, political instability, and crop failures, indirectly affecting how much stress is put on soil systems. Profits realized from the sale of livestock and their products can be used for natural resource investments, education, or health expenses. Investments in all of these areas have impacts on the status of the soil.

The use of manure as a soil amendment, with emphasis on the amount and availability of nutrients and organic matter, has been well researched (Brouwer and Powell, 1998; Mafongoya et al., 2000; Harris, 2002), and its value is well-known. Some advantages of manure as a soil amendment are obvious; it is locally produced and widely available. Less apparent effects of adding manure to soils, beyond provision of nutrients, are often overlooked. These effects include changes in soil texture, soil organic matter, microbial populations, carbon and nitrogen dynamics, and plant communities. Some of the effects of animals on soils are indirect; for example, herbivory alters both the nutrient composition of individual plants and species diversity, which in turn affect the availability of carbon and soil microbial populations. Foliar herbivory stimulates the release of carbon into the

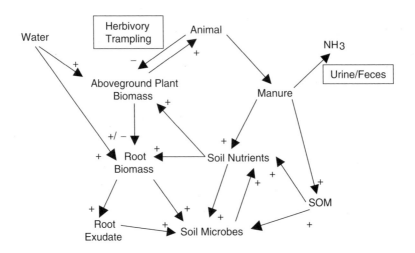

FIGURE 17.1
Interactions among plants, soils and animals.

rhizosphere which stimulates microbial growth and increases nitrogen available to the plant (Wardle et al., 2004).

This chapter will consider some of the many effects, both positive and negative, that livestock can have on soils and plants. In the discussion of nutrient cycling and soil fertility, integrated crop-livestock systems will be the focus. These are defined by Seré and Steinfeld (1996) as systems in which at least 10% of the total agricultural production comes from crops. In dry areas such as Mauritania, Botswana, and Namibia, where livestock represent more than 80% of the agricultural gross domestic product (GDP) (Winrock International, 1992), livestock production eclipses crop production. However, maintaining soil fertility is essential to such dryland systems, heavily dependent on livestock, as well as to those more crop-based systems with a more abundant water supply.

The task of fully understanding crop–soil–animal–human interactions is immense, and there is a risk of meaningless generalities, e.g., both plants and animals need water. Water availability, plot history, soil type and nutrient content, crop choices, indigenous vegetation, and livestock species all influence the functioning of integrated systems. Some of the interactions and feedbacks among system components are indicated in Figure 17.1. In this chapter, animal–soil interactions are considered at three different scales. First, the focus is on animal manure and soil amendments and on the labor-saving contributions of livestock, taking a farmer's view of the system. Then, aboveground changes that affect what happens to soil microbes and other populations belowground in the short and medium term are reviewed. Finally, a more encompassing look is taken at the effects of livestock on soils at a broader scale over longer periods of time.

17.1   Animal Manure and Soil Amendments

Much has been written about the effects of soil degradation on decreasing crop yields, especially in densely populated areas with continuous cultivation. As population has increased, farm size has decreased, and farmers, in order to feed their families, have been forced to till their land continuously, without fallow breaks for soil regeneration. In Sub-Saharan Africa (SSA), the amount of cultivatable land per capita decreased by one-third between 1970 and 2000, from 0.53 to 0.35 ha (Place et al., 2003). Land-poor farmers cannot afford to leave any land "idle" without income, but declining crop yields are intolerable as well. This dynamic, and ways it has been reversed in some parts of Africa, are discussed in Chapter 25.

The best predictor of whether farmers will use improved fallows is whether some of the plant biomass from a fallow plot can be used for forage, assuming that the farmer has livestock to feed. Provision of fodder from fallow plots provides incentive to allow the land to regenerate. Similarly, perennial plant species characterized now as "fertilizer trees" (Chapter 19) also can serve as fodder trees, provided that the foliage of the trees does not contain high levels of secondary compounds. Fertilizer is expensive in Africa, costing as much as six times more in SSA than in Europe or North America (Sanchez, 2003). Because a bag of fertilizer can cost a month's cash earnings, locally produced alternatives to inorganic fertilizer are attractive and even essential, e.g., the use of "fertilizer trees" and animal manures.

Using biomass from fallows or crop residues to enrich the soil rather than as animal feed often involves a direct trade-off. If more organic matter is returned to the soil, with residues incorporated directly, farmers suffer some loss of income and other benefits from livestock ownership, but this zero-sum situation assumes that the animals in question are being fed from cropland only. It is common for animals to be fed from off-farm fodder or forage sources, either through grazing or a cut-and-carry system. This makes farms into open rather than closed systems, and when animals function as nutrient movers, they are indeed transporting valuable nutrients onto the farm.

17.1.1 Manure Soil Amendments

Animal manure is the most commonly used soil amendment in many parts of the world. Over 34.4 million tons of N and 8.8 million tons of P are provided from manure globally each year (Sheldrick et al., 2003). The organic matter content of manure is highly variable, depending on animal diet, storage conditions, and the amount of bedding included with the feces and urine. Cow manure collected from manure piles on Spanish farms was found to contain 25–67% organic matter on a dry matter basis (Moral et al., 2005), while sheep and goat manure contained more organic matter with somewhat less variation (51.3 and 54.6%, respectively). When the contributions of nutrients by animal species are considered, cattle provided 47% of the N from manure, while pigs supplied an additional 24%. Cattle contributed 37% of the P, and pigs accounted for 32%. In Nepal, more than 80% of the N applied to amend the soils was as manure-compost (Thorne and Tanner, 2002).

Manure or manure-compost is locally available, requires limited cash expenditure, and provides needed nutrients and organic matter for the soil. However, the supply of animal manure is not sufficient in most places to meet crop requirements and maintain soil productivity (Fernández-Rivera et al., 1995). As much as 35 ton ha^{-2} of manure on a wet weight basis may be required to maintain initial soil organic carbon levels, significantly more than is practical (Nandwa, 2001). Assuming manure application rates of 3 ton ha^{-1}, an amount barely sufficient to replenish nutrients removed by grain and stover, only 33–42% of the fields in Mali, Chad, and the Gambia and only 8% of the cropland in Niger would receive this much manure (Fernández-Rivera et al., 1995). Countries with many animals and relatively small areas devoted to crops fare better in this comparison than do regions with extensive cropping without complementary animal husbandry. The regions with more crops and fewer animals are those most likely to suffer from serious soil degradation.

Manure quality is both variable and important. Prediction of the quality and quantity of manure as it is excreted from the animal is relatively straightforward provided needed information on the animal and its diet is available. Assessing the availability of manure nutrients to plants is complicated by losses from manure during storage and the variable efficiencies of manure nutrient recovery. Volatilization, leaching, and transport losses mean that the manure applied to the soil differs from what comes out of the animal. Fernández-Rivera et al. (1995) assumed that only half of the manure excreted would be actually available for spreading on crops, a sobering but realistic assumption. Storage losses are difficult to predict because environmental and storage conditions are influential, and the amount of loss varies by nutrient. Nitrogen is especially vulnerable to loss because of NH_3 volatilization, while the P applied to fields through manure is relatively immobile and more likely to be retained in the field.

Diet, animal requirements, and storage practices all contribute to the highly variable nutrient contents and quantities of manure. Nitrogen content of manure can vary from 0.5 to 2.0% of dry matter (Murwira et al., 1995), and a similar fourfold variation in P levels is common. Season affects the amount of manure excreted; during the wet season when feed is abundant, the amount of manure produced may be double the quantity available in the dry season, and the amounts of nitrogen and phosphorus may be three times as high (Powell et al., 2004). Fortunately, the season when manure is abundant coincides with high plant demands for nutrients. Because of the change in the amount and quality of manure available by season, the common reliance on annual averages regarding manure composition and quality is ill-advised.

Partition of N excretion between urine and feces is dictated by the amounts and forms of N and polyphenolics, especially tannins, consumed by the animals. Although urine raises soil pH and improves phosphorus availability in the short-term (Powell et al., 1998), approximately two-thirds of urinary N is in the form of urea, most of which is lost to the atmosphere as NH_3. Fecal nitrogen, which consists largely of nitrogen associated with the plant and microbial cell walls, is much less susceptible to atmospheric loss, but is slowly mineralized. When animals consume diets that are high in condensed tannins, as may be the case when fodder from leguminous trees is used as a protein supplement, indigestible complexes of tannins and dietary proteins are formed. Feces of animals consuming tannins contain more N than from those animals whose diets do not contain tannins. The tannin–protein complexes are more slowly mineralized than are the original forage proteins, which will be advantageous, or not, depending on the timing and amount of plant nutrient needs.

Reliance on animal manure as the sole soil amendment may be problematical because the N:P ratio in manure is typically lower than the ratio required by plants. If enough manure is applied to meet the plants' N requirements, P will be provided in excess (Powell et al., 2004). By combining the use of manure with green manures that decompose at different rates, nutrient availability can be synchronized with plant demand (Palm et al., 2001b). The greatest crop responses to animal manures have been obtained when these are supplemented with inorganic fertilizers or in combination with green manures. The adequacy of nutrient supply and the timing of release must both be considered to ensure that nutrient recovery is optimal.

Thus far, we have focused on nutrient supply but have not considered one of the most important contributions of manure: organic matter. Reliance on inorganic fertilizers alone to provide supplementary soil nutrients is unaffordable for many farmers and can reduce soil pH, which has adverse effects on the availability of some scarce soil nutrients. Soil organic matter is both a sink and a source of plant nutrients, a nutrient "storage tank" that can be called upon when nutrients are needed. The storage tank analogy is imperfect because, after application of organic matter with high levels of polyphenols or lignin, N may be immobilized, making it unavailable to plants and microbes for some time after application.

The organic matter in manure has already been subjected to extensive digestion in the animals' gut, so it is more lignified than the original forage, and it decomposes slowly. In a short-term pot experiment, Powell et al. (1999) compared the effects of applying the green manures of six plant species or the manure of animals fed those same plants on millet growth and soil parameters. Predictably, the manure had 1.4 times more lignin than did the original plant material (20.6 vs. 15.1%), but both amendments exceeded the 15% lignin threshold identified by Palm et al. (2001a) for organic resources whose N mineralizes slowly because of immobilization prior to decomposition.

There was approximately 13% more N in the manure than in the leaves, but more important than differences in the quantity of N were changes in quality. Approximately 73% of the manure N was associated with the fiber fraction (neutral detergent-insoluble nitrogen, NDIN), and thus was slowly released compared to 55% NDIN in the plant material. The efficiency with which N was used was higher in the manure treatments than with the leaf applications, an effect that was attributed to better synchrony with plant nutrient demand.

17.1.2 Management Considerations

Addition of organic matter and other nutrients from different sources must be synchronized to ensure that plants can meet their requirements throughout the growing cycle (Palm et al., 2001b). This requires an understanding of the amount of nutrients present in the soil and from soil amendments, and of the rates at which they will be released. How farmers can gain access to sufficient information about the quality of their manure is problematical given the wide variation in the amount and quality of animal manure produced and in storage losses. Input data bases (Palm et al., 2001a) are a starting point, but tabular values do not reflect local conditions. Mineral values given in tables are often unreliable because soils largely determine the mineral concentration of plants grown on them.

As farmers in areas with higher agricultural potential adopt zero-grazing systems, stall-feeding their animals, they will be able to exert more control over the diets of their animals and can more easily collect manure. They can optimize manure storage conditions and also times and places of application. This practice is gaining acceptance in many local contexts. Tanner et al. (1995) described a system in Java in which excess bedding was provided for stall-fed sheep purposefully to generate superior soil amendments by absorbing feces and urine. Although there were marginal gains in animal productivity under this management system, the primary benefit was in having more and better quality compost. When questioned about the practice, these farmers valued the production of the manure from their sheep as highly as the meat produced. Whether farmers are willing to invest the amount of labor required in a cut-and-carry system depends on the availability of land and labor and on potential markets for their products. The benefits of cut-and-carry systems in providing better compost to maintain soil health may be offset by transferring the laborious task of nutrient transport from foraging animals to people.

17.2 Animal Traction

One of the most significant impacts of animals on soil systems is through their provision of power for traction. Farmers owning draft animals can plow more land than those without. Zambian farmers who cultivated with a hoe are able to till 0.8 ha while those with a single pair of oxen can plow 2.4 ha, and those with more than two oxen have 4.8 ha of cropped land (Wilson, 2003). So, to the extent that increased or more extensive tillage has adverse effects on soil structure and fertility, there can be a negative contribution from animals to soil system sustainability, although their positive contribution through manure, discussed above, can more than compensate for liabilities.

Elsewhere in this volume, the benefits of no-till systems are outlined. These suggest that traction will detract from soil health. However, it can improve labor efficiency and income sufficiently to permit the adoption of natural resource management strategies that enhance soil systems. If farmers can till more land, for example, they could be inclined to use fallow periods to restore soil fertility. Alternatively, draft animals may help to reduce pronounced soil fertility gradients that frequently occur within farms (B. Vanlauwe, personal communication) by reducing the labor required to carry manure and household wastes to distant fields that otherwise receive no soil amendments. In Ethiopia, adoption of no-till cultivation resulted in more productive herds that included more cows and fewer steers because oxen were no longer required for plowing (Benin et al., 2002). So livestock populations and soil biotic communities interact in varied and usually complex ways that depend on biological principles and human behavior.

17.3   Animals, Soil Function, and Belowground Biodiversity

One of the most intriguing new areas of research is on how animals and herbivory affect soils and belowground biodiversity. Elsewhere in this book the case is made for studying plant roots as seriously as aboveground biomass, for assessing soil microbes as indicators of soil health, and for including the impacts of human decision-making and actions within the purview of biological systems analysis. Similarly, the effects of animals, large and small, warrant consideration within a soil systems perspective. Wardle et al. (2004) have proposed that an understanding of soil food webs must encompass both above- and belowground components, including the herbivory and excretion of animals.

Herbivory by animals affects which plants will be present and the chemical composition of these plants. Changes in plant species and quality in turn influence the quality of litter and manure and soil fertility. Populations of soil macro-, meso-, and microfauna are influenced by the availability of nutrients essential for their survival, and these belowground fauna influence what is happening aboveground as well. If conditions shift so that populations of plant pathogens are favored, the costs are obvious. Likewise, the rate at which soil fauna decompose organic matter, varying the nutrient supply available to plants, has important consequences.

Animals figure into these dynamics in numerous ways. Historic observations and experimental data show that deer and goats in New Zealand forests (Wardle et al., 2001), sheep in the north of England (Bardgett et al., 2001), and wildlife in the Kenyan savannas (Augustine and Frank, 2001; Sankaran and Augustine, 2004) affect belowground microbial and mesofaunal populations. Understanding the underlying mechanisms of these interactions requires thinking across temporal scales, from the life cycle of a bacterium (hours to days) to the time required for soil formation (eons). Likewise, very diverse spatial scales are involved, ranging from bacterial chemotaxis (microns) to the amount of range land covered by pastoralists and their animals (tens to hundreds of square kilometers).

Herbivory influences how plants allocate above- and belowground the carbon that they fix through photosynthesis (Table 17.1) (Bardgett and Wardle, 2003). These responses vary based on plant species, physiological state and age of the plant, and environmental conditions (Bardgett et al., 1998). Immediately after being grazed, plants may transfer nutrients belowground to protect themselves from more herbivory. Somewhat longer-

TABLE 17.1

Dynamics of Herbivory, Decomposition, Soil Fauna, and Fertility

Ecosystem Productivity	High	Limited
Plants	↑ Growth rate Nutrient-rich biomass ↑ C allocation to growth	↓ Growth rate Nutrient-poor biomass ↑ C to secondary compounds
Herbivores	↑ Manure deposition Nutrient-rich manure ↑ Productivity + reproduction	↓ Manure deposition Nutrient-poor manure ↓ Productivity + reproduction
Litter	↓ Litter deposition ↑ Litter% N ↓ Litter phenolics ↓ Litter fiber + lignin	↑ Litter deposition ↓ Litter% N ↑ Litter phenolics ↑ Litter fiber + lignin
Soil Fauna	↑ Earthworms ↑ Soil microbes	↓ Earthworms ↓ Soil microbes
Soil Processes	↑ Nutrients for plant growth ↑ Decomposition rate ↑ Mineralization rate ↓ C sequestration in soil ↑ Soil perturbation	↓ Nutrients for plant growth ↓ Decomposition rate ↓ Mineralization rate ↑ C sequestration in soil ↓ Soil perturbation

Source: From Bardgett, R.D. Wardle, D.A., and Yeates, G.W., *Soil Biochem.*, 30, 1867–1878 (2003) Wardle, D.A., Bardgett, R.D., Klironomos, J.N., Setälä, H., van der Putten, W.H., and Wall, D.H., *Science*, 304, 1629–1633, 2004.

term responses include new shoot development to maintain plants' photosynthetic capability, synthesis by plants of defensive compounds that deter herbivores, or additional root growth. Grazing by animals may induce more root exudation, more root formation, or root decay, all of which in turn influence populations of soil fauna by affecting carbon supply. Even longer-term considerations are how much organic matter is returned to the soil, and whether the functional plant and mammalian communities are altered. The time frame imposed or assumed for any study will greatly influence how the effects of herbivory on plants are perceived.

The effects of grazing are not always the same for soil biotic populations. In the northwestern U.K., microbial populations increased in response to low to moderate herbivory due to changes in the plant community, although when there was intensive grazing, fewer microbes were present (Bardgett et al., 2001). On the other hand, in the semi-arid Kenyan savannas, herbivory by cattle demonstrably decreased soil microbial populations (Sankaran and Augustine, 2004). In New Zealand, the effects of grazing by deer on microbial populations were variable (Wardle et al., 2001).

These diverse responses to grazing occur possibly because competing processes are at work (Sankaran and Augustine, 2004):

1. Grazers either use or relocate plant nutrients within the ecosystem, altering the quality and quantity of nutrients supplied to microbes, and

2. Herbivory induces plant responses that change the quantity and quality of the biomass produced.

In nutrient-limited systems, the relationship between grazers and soil microbes appears to be antagonistic, while in systems with more abundant resources the relationship is beneficial (Sankaran and McNaughton, 1999; Wardle et al., 2004).

The effects of nutrient redistribution by animals can persist for decades (Augustine, 2003). Abandoned bomas or kraals used to confine animals at night and glades, previous sites of human habitation, have been found to have significantly higher levels of N and P than in the surrounding land for as much as 40 years after the land was used. Grasses predominated in the bomas and glades, while bush vegetation was found in the adjacent area.

Grazing or browsing affects individual plants. Regrowth vegetation usually has higher nutritive value than the initial plant growth under similar conditions (Van Soest, 1994). While some pot experiments have suggested that herbivory decreases root biomass and function, field experiments in the Serengeti and elsewhere have not supported this conclusion (McNaughton et al., 1998). Initial soil nutrient conditions, especially soil organic matter content, are usually more important predictors of plant responses to grazing than is the number of animals being supported, unless grazing pressure is very heavy (Sankaran and Augustine, 2004).

Animals alter plant populations as well as the composition of individual plants. The classic argument has been that animals preferentially select palatable species, leaving those that are less desirable to go to seed and reproduce. The consequence of this preference is that less desirable and invasive species become dominant, to the detriment of the more desirable, grazing-intolerant species. In evaluations over 14 years, however, moderate grazing had little effect on biodiversity because of plants' adaptation to herbivory (Hiernaux, 1998). Except where there was persistent heavy grazing, climatic variations between years were more important for explaining species richness, particularly of herbaceous plants, than was grazing (Oba et al., 2000).

Many plant species tolerate moderate grazing, and the harvesting of some plants by animals enhances their productivity. Plants that have evolved in the presence of herbivory usually regrow after intense, but short-lived, grazing (Frank et al., 1998). Grazing-tolerant species can decrease their root biomass as they increase shoot growth, while species not adapted to herbivory often reallocate resources belowground (Guitian and Bardgett, 2000). These plant responses affect carbon allocation within the plant, and thus the amounts of energy and nutrients available to the soil fauna.

The grazing habits of different species of animals affect plant diversity. In the Laikipia region of Kenya, wildlife enclosure–exclosure experiments with impala, dik-dik, and elephants, chosen to represent different foraging strategies, and shrubs of different sizes clearly showed that foraging strategy profoundly altered plant composition. When only browsers like dik-dik were present, greater twig removal and less recruitment of saplings reduced bush encroachment (Augustine and McNaughton, 2004); large, bulk-eating animals, on the other hand, reduced shrub cover and also biomass accumulation. When both species were present, bush encroachment was less likely. With domestic species, similar complementarities have been observed when sheep and cattle were grazed together in Australia (Tainton et al., 1996). Grazing with multiple domestic and wild animal species may be a viable tactic for preventing bush encroachment which lowers plant diversity and reduces soil fertility.

In their outline of the effects of herbivores on soils and belowground biodiversity, Wardle et al. (2004) suggested that the above- and belowground linkages seen in well-functioning grazing ecosystems differ from what happens in nutrient-limited environments. They demonstrated how the various parts of their proposed framework could fit together at the landscape level, but they agree that there are still many "black boxes" in their construct. Both the spatial heterogeneity of soil systems and our lack of knowledge of most microbes (some estimate that <1% of all bacteria have been cultured) are serious constraints to resolving the apparently contradictory results from analyses of the effects of grazing and animals on soil microbes and soil organic matter.

Sources of heterogeneity are easy to identify. Variable rainfall patterns, seasonal differences, variations in soil types, plant, animal and microbial species, and human interventions are some of the major variables that influence soil fauna. Explaining how these factors interact to predict what will happen to soils and their productivity is a more difficult task. The laborious methods currently available to microbial ecologists for assessing population changes preclude their examining large numbers of samples, making it difficult to capture and interpret spatial heterogeneity.

Equally important, we do not understand how shifts in microbial species or reduced microbial diversity affects ecosystem function (Bengtsson, 1998). Redundant functions are common in many ecosystems to ensure that if one species is knocked out of the system, its functional niche is occupied. We still do not know which microbes perform which functions or when thresholds are reached that impair ecosystem function. The effects of macro- and mesofauna also must be considered. Our inability to capture the effects of spatial heterogeneity is especially serious for evaluations of grazing lands that are known for their patchiness (Augustine and Frank, 2001).

Soil macrofauna, those ecosystem engineers including termites, earthworms, and ants discussed in Chapter 11, play important roles in transformation of organic matter and soil health. They devote as much as half of the energy they consume to burrowing and soil perturbation. In the process, they create more favorable environments for other soil fauna and plant roots. They are sensitive to disruptions of their environment and favor soils that contain plentiful organic matter. As a result, they favor pastures over cropped land (Lavelle et al., 2001) and thrive in areas where manure has been applied, suggesting that their contributions to soil systems can be enhanced by herbivorous animals. The interactions between soil macro- and mesofauna, microbes, plants, and large herbivores are in any situation extremely complex and highly dependent on initial conditions. Understanding these relationships better should engage the efforts of a wide range of biological scientists in the next decade.

17.4 Overgrazing and Soil Erosion

For many years, a debate has raged about overgrazing, land degradation, and soil erosion (Scoones, 1993; Illius and O'Connor, 1999; Oba et al., 2000; Rowntree et al., 2004).There is no disagreement that animal stocking rates under smallholder management are high; in South Africa, the number of animals per hectare on communal grazing areas is twice that found on commercial farms (Rowntree et al., 2004), and similar densities are maintained in Zimbabwe. There are significant disagreements whether large numbers of animals per hectare cause irreparable damage to soil and vegetation systems, or whether this high animal density is an efficient strategy to maximize resource use. The resilience of semi-arid ecosystems to sustained grazing pressure is an on-going debate among range ecologists (Scoones, 1993; Oba et al., 2000).

Appropriate management strategies that retain spatial flexibility so that animals can track rainfall and follow forage production are essential for animal productivity and to ensure that overgrazing does not lead to soil erosion. In dry areas where the coefficient of variation for rainfall can exceed 50% (Lal, 1987; McPeak, 2003), the environments are often characterized by "patchiness" with productive areas adjacent to ones that yield little. If animals are mobile, large concentrations of animals do not occur except for short periods of time. This usual pattern of dispersion reduces stress on the environment.

Adaptable strategies in which animals and their owners seek sources of water and forage where they are available have merit over ones that are keyed to raising predetermined numbers of animals on specified areas without consideration of variations in rainfall or biomass production. Two sustainable dryland animal systems, based on wildebeests in the Serengeti and pastoralists' cattle, are both characterized by animal migration. A strategy of movement ensures that animals have access to essential nutrients including minerals, protein, and energy while plants are not repeatedly grazed, avoiding depletion of essential reserves.

When animals are constantly on the move, overgrazing, soil erosion, and destruction of savanna vegetation are unlikely. In areas with flexible grazing systems where feed supplies are adequate, the need to "jump start" the grazing season through use of fire to eliminate old vegetation and to stimulate nutritious new growth is reduced. Unfortunately, population pressures and limited land availability impose constraints on grazing management options, which means that fire is used both for land clearing and for pasture "improvement," with serious environmental consequences.

Changes in land tenure arrangements, adverse economic circumstances, growth of animal and human populations, and droughts that cause a loss of migratory flexibility, threaten both natural and managed animal-based systems with environmental and economic collapse. Serious environmental degradation with widespread soil erosion and bush encroachment is evident around human settlements where pastoralists have become partially sedentary while areas with traditional migrations remain in good health.

In environments with sufficient rainfall to sustain crop production, animals often range freely in common areas. Aside from the inevitable, unintended consumption of crops by errant goats, a second drawback to this management is lack of control of grazing, especially when feed is in short supply because of poor rainfall or limited area for pasture (Husson et al., 2004). Competitive grazing with farmers trying to use feed resources before their neighbors' animals consume them results in overgrazing and soil compaction. Even considerable manure deposition cannot offset the effects of heavy, sustained grazing. Increased bulk density of the soil from compaction affects soil structure, water retention, pasture productivity, plant rooting depth, and soil microbes and fauna. The extent to which compaction reduces soil system productivity depends on whether the grazing area was initially grassland or was deforested specifically for grazing, on the size of the grazers (large cattle have more impact than small antelopes), on initial style, type and texture, and on grazing intensity and the plant species present.

17.5 Discussion

In the richer countries of the North, animal wastes are becoming an environmental hazard leading to either accumulation of N and P in the soil or to run-off and leaching of these nutrients into surface and ground water supplies. Nitrate contamination of water and ecosystem-altering algal blooms result. In poorer countries of the South, most farm families would be more than pleased to have access to such waste materials for the N, P, and organic matter that they contain. As long as many millions of farming households in this world who operate mixed farming systems continue to rely on livestock for a substantial part of their income and nutrition, their soils will also depend on animal wastes for sustained productivity.

Little attention has been given to making technological improvements for the collection, conservation, and application of manure-composts for the enrichment and maintenance of soil systems. More scientifically-based methods and practices for handling animal wastes and for grazing animals as part of integrated nutrient management could make an important contribution to an agriculture in the 21st century that is biologically informed and that capitalizes on the multiple facets of agroecological production systems.

In dryland areas, soil erosion due to concentrated overgrazing is the primary threat to soil fertility posed by animals. Avoidance of sustained overgrazing by retaining flexibility and mobility of where animals graze is essential to prevention of soil erosion "hot spots" (McPeak, 2003). For arid systems and for those with adequate rainfall, the next generation of research should focus on both the function of components of the system and how one subsystem affects other parts of the system. We need to understand functional biodiversity better at the microbial, plant and animal scales, and how changes in this functional biodiversity affect ecosystem functioning.

References

Abril, A. and Bucher, E.H., Overgrazing and soil carbon dynamics in Chaco, Argentina, *Appl. Soil Ecol.*, **16**, 243–249 (2001).

Augustine, D.J., Long-term, livestock-mediated redistribution of nitrogen and phosphorus in an East African savanna, *J. Appl. Ecol.*, **40**, 137–149 (2003).

Augustine, D.J. and Frank, D.A., Effects of migratory grazers on spatial heterogeneity of soil nitrogen properties in a grassland ecosystem, *Ecology*, **82**, 3149–3162 (2001).

Augustine, D.J. and McNaughton, S.J., Regulation of shrub dynamics by native browsing ungulates on East African rangeland, *J. Appl. Ecol.*, **41**, 45–48 (2004).

Bardgett, R.D. et al., Soil microbial community patterns related to the history and intensity of grazing in sub-montane ecosystems, *Soil Biol. Biochem.*, **33**, 1653–1664 (2001).

Bardgett, R.D. and Wardle, D.A., Herbivore-mediated linkages between aboveground and belowground communities, *Ecology*, **84**, 2258–2268 (2003).

Bardgett, R.D., Wardle, D.A., and Yeates, G.W., Linking above-ground and below-ground interactions: How plant responses to foliar herbivory influence soil organisms, *Soil Biol. Biochem.*, **30**, 1867–1878 (1998).

Bengtsson, J., Which species? What kind of biodiversity? Which ecosystem function? Some problems in studies of relations between biodiversity and ecosystem function, *Appl. Soil Ecol.*, **10**, 191–199 (1998).

Benin, S., Pender, J., and Ehui, S., Policies for sustainable land management in the East African highlands, *EPTD paper 13*. International Food Policy Research Institute, Washington, DC (2002).

Brouwer, J. and Powell, J.M., Increasing nutrient use efficiency in West-African agriculture: The impact of micro-topography on nutrient leaching from cattle and sheep manure, *Agric. Ecosyst. Environ.*, **71**, 229–239 (1998).

Fernández-Rivera, S. et al., Faecal excretion by ruminants and manure availability for crop production in semi-arid West Africa, In: *Livestock and Sustainable Nutrient Cycling in Mixed Farming Systems in Sub-Saharan Africa*, Powell, J.M., Fernández-Rivera, S., Williams, T.O., and Renard, C., Eds., International Livestock Centre for Africa, Addis Ababa (1995).

Frank, D.A., McNaughton, S.J., and Tracy, B.F., The ecology of the earth's grazing ecosystems, *BioScience*, **48**, 513–521 (1998).

Guitian, R. and Bardgett, R.D., Plant and soil microbial responses to defoliation in temperate semi-natural grassland, *Plant Soil*, **220**, 271–277 (2000).

Harris, F., Management of manure in semi-arid farming systems in West Africa, *Exp. Agric.*, **38**, 131–148 (2002).

Hiernaux, P., Effects of grazing on plant species composition and spatial distribution in rangelands of the Sahel, *Plant Ecol.*, **138**, 191–202 (1998).

Husson, O. et al., Diagnostic agronomique des facteurs limitant le rendement du riz pluvial de montagne dan le nord du Vietnam, *Cahier Agric.*, **13**, 421–428 (2004).

Illius, A.W. and O'Connor, T.G., On the relevance of nonequilibrium concepts to arid and semiarid grazing systems, *Ecol. Appl.*, **9**, 798–813 (1999).

Lal, R., *Tropical Ecology and Physical Edaphology,* Wiley, Chichester, UK (1987).

Lavelle, P. et al., SOM management in the tropics: Why feeding the soil macrofauna?, *Nutr. Cycling Agroecosyst.,* **61,** 53–61 (2001).

Mafongoya, P.L., Barak, P., and Reed, J.D., Carbon, nitrogen and phosphorus mineralization of tree leaves and manure, *Biol. Fertil. Soils,* **30,** 298–305 (2000).

McNaughton, S.J., Banyikwa, F.F., and McNaughton, M.M., Root biomass and productivity in a grazing ecosystem: The Serengeti, *Ecology,* **79,** 587–592 (1998).

McPeak, J.G., Analyzing and addressing localized degradation in the commons, *Land Econ.,* **79,** 515–536 (2003).

Moral, R. et al., Characterisation of the organic matter pool in manures, *Bioresour. Technol.,* **96,** 153–158 (2005).

Murwira, K.H., Swift, M.J., and Frost, P.G.H., Manure as a key resource in sustainable agriculture, In: *Livestock and Sustainable Nutrient Cycling in Mixed Farming Systems in Sub-Saharan Africa,* Powell, J.M., Fernández-Rivera, S., Williams, T.O., and Renard, C., Eds., International Livestock Centre for Africa, Addis Ababa (1995).

Nandwa, S.M., Soil organic carbon (SOC) management for sustainable productivity of cropping and agro-forestry systems in Eastern and Southern Africa, *Nutr. Cycling Agroecosyst.,* **61,** 143–158 (2001).

Oba, G., Stenseth, N.C., and Lusigi, W.J., New perspectives on sustainable grazing management in arid zones of Sub-Saharan Africa, *BioScience,* **50,** 35–51 (2000).

Palm, C.A. et al., Organic inputs for soil fertility management in tropical agroecosystems: Application of an organic resource database, *Agric. Ecosyst. Environ.,* **83,** 27–42 (2001a).

Palm, C.A. et al., Management of organic matter in the tropics: Translating theory into practice, *Nutr. Cycling Agroecosyst.,* **61,** 63–75 (2001b).

Place, F.M. et al., Prospects for integrated soil fertility management using organic and inorganic inputs: Evidence from smallholder African agricultural systems, *Food Policy,* **28,** 365–378 (2003).

Powell, J.M., Ikpe, F.N., and Somda, Z.C., Crop yield and the fate of nitrogen and phosphorus following application of plant material and feces to soil, *Nutr. Cycling Agroecosyst.,* **54,** 215–226 (1999).

Powell, J.M. et al., Urine effects on soil chemical properties and the impact of urine and dung on pearl millet yield, *Exp. Agric.,* **34,** 259–276 (1998).

Powell, J.M., Pearson, R.A., and Hiernaux, P.H., Crop-livestock interactions in the West African drylands, *Agron. J.,* **96,** 469–483 (2004).

Ritz, K. et al., Spatial structure in soil chemical and microbiological properties in an upland grassland, *FEMS Microbiol. Ecol.,* **49,** 191–205 (2004).

Rowntree, K. et al., Debunking the myth of overgrazing and soil erosion, *Land Degradation Dev.,* **15,** 203–214 (2004).

Sanchez, P.A., Soil fertility and hunger in Africa, *Science,* **295,** 2019–2020 (2003).

Sankaran, M. and Augustine, D.J., Large herbivores suppress decomposer abundance in a semiarid grazing ecosystem, *Ecology,* **85,** 1052–1061 (2004).

Sankaran, M. and McNaughton, S.J., Determinants of biodiversity regulate compositional stability of communities, *Nature,* **401,** 691–693 (1999).

Scoones, I., Why are there so many animals? Cattle population dynamics in the communal areas of Zimbabwe, In: *Range Ecology at Disequilibrium,* Behnke, R.H., Scoones, I., and KIerven, C., Eds., Overseas Development Institute, London (1993).

Seré, C. and Steinfeld, H., *World livestock production systems: Current status, issues and trends,* FAO, Rome (1996).

Sheldrick, W., Syers, J.K., and Lingard, J., Contribution of livestock excreta to nutrient balances, *Nutr. Cycling Agroecosyst.,* **66,** 119–131 (2003).

Tainton, N.M., Morris, C.D., and Hardy, M.B., Complexity and stability in grazing systems, In: *The Ecology and Management of Grazing Systems,* Hodgson, J. and Illius, A.W., Eds., CAB International, Wallingford, UK (1996).

Tanner, J. et al., Feeding livestock for compost production: A strategy for sustainable upland agriculture on Java, In: *Livestock and Sustainable Nutrient Cycling in Mixed Farming*

Systems in Sub-Saharan Africa, Powell, J.M., Fernández-Rivera, S., Williams, T.O., and Renard, C., Eds., International Livestock Centre for Africa, Addis Ababa (1995).

Thorne, P.J. and Tanner, J.C., Livestock and nutrient cycling in crop–animal systems in Asia, *Agric. Syst.*, **71**, 111–126 (2002).

Van Soest, P.J., *Nutritional Ecology of the Ruminant*, 2nd ed., Cornell University Press, Ithaca, NY (1994).

Wardle, D.A. et al., Ecological linkages between aboveground and belowground biota, *Science*, **304**, 1629–1633 (2004).

Wardle, D.A. et al., Introduced browsing mammals in New Zealand natural forests: Aboveground and belowground consequences, *Ecol. Monogr.*, **71**, 587–614 (2001).

Wilson, R.T., The environmental ecology of oxen used for draught power, *Agric. Ecosyst. Environ.*, **97**, 21–37 (2003).

Winrock International, *Assessment of Animal Agriculture in Sub-Saharan Africa*, Winrock International Institute for Agricultural Development, Morrilton, AK (1992).

PART III: STRATEGIES AND METHODS

18

Integrated Soil Fertility Management in Africa: From Knowledge to Implementation

Bernard Vanlauwe, Joshua J. Ramisch and Nteranya Sanginga
Tropical Soil Biology and Fertility Institute, CIAT, Nairobi, Kenya

CONTENTS

Sustainable management of soil, water, and other natural resources is the most critical challenge confronting agricultural research and development in sub-Saharan Africa (SSA). Soil fertility decline is a multi-faceted problem and, in ecological parlance, a "slow variable," one that interacts pervasively over time with a wide range of other factors, biological, and socio-economic. Sustainable agroecosystem management is not just a matter of remedying deficiencies in soil nutrients. Impediments include mismatched germplasm and faulty cropping system design, the multiple interactions of crops with pests and diseases, reinforcing feedback effects between poverty and land degradation, institutional failures, and often perverse incentives that stem from national policies and global dynamics. Dealing with soil fertility issues in cost-effective and sustainable ways thus requires a long-term perspective and a holistic approach such as embodied in the concept of integrated soil fertility management (ISFM).

The concepts of ISFM grew out of a series of paradigm shifts generated through experience in the field and from changes in the overall socio-economic and political environments faced by the various stakeholders, in particular, by farmers and researchers. In retrospect, the need for and elements of this integrated strategy should have been obvious much sooner than they were, but this is true for many advances in thinking and practice. We now understand better how the judicious use of mineral fertilizers together with organic sources of nutrients for plants and soil organisms supported by appropriate soil and water conservation and land and crop management measures can counteract the agricultural resource degradation that results from nutrient mining, the exploitation of fragile lands, and associated losses in biodiversity. Appropriate soil fertility management will produce benefits that reach beyond the farm, serving whole societies through the various ecosystem services associated with the soil resource base, e.g., provision of clean water, erosion control, and support for biodiversity.

Part III of this book presents a series of cases and analyses where new as well as often old knowledge is being drawn on to inform and formulate improved practices that can achieve more productive and more sustainable soil systems. In this chapter, after highlighting some of the problems underlying declining soil fertility in SSA, the region where we have been working, we briefly review some shifts in paradigms related to tropical soil fertility management. Several examples are then considered of how science has been translated into practice, with some discussion in conclusion of the challenges that persist and how we envisage addressing them.

18.1 Problems Driving Research and Development for Sustainable Soil Systems in Africa

The fertility status of most soils in SSA is generally poor due to low inherent quality and inappropriate management practices, the latter being the result of various other secondary and tertiary causes. This dynamic is seen from a number of observations that have specified the nature of soil systems' deficiencies and vulnerabilities in the region:

- Sharply negative soil nutrient balances at the regional and national scale for the major plant nutrients, with annual losses of NPK estimated at 8 million tons (Stoorvogel and Smaling, 1990). These negative balances reflect the very low use of mineral inputs across SSA, although they also show the effects of climatic and other conditions discussed in Chapter 2. How nutrient limitations can be mitigated through changes in soil system management is a principal focus of this and following chapters.

- Average crop yields on smallholder farms in many countries are generally around 30% of the yields obtained on research farms (Tian et al., 1995). Closing this yield gap is a major challenge to researchers and farmers.

- Moisture stress affects over two-thirds of all soils. While this often reflects adverse rainfall patterns, much is attributable to the soils' poor water husbandry. Their low levels of organic matter (living and dead) and their unfavorable topsoil structure exacerbate water shortages.

- It is estimated that nearly 500 million ha of land are degraded, approximately 40% of the total arable area, due principally to the forces of water and wind erosion (Oldeman, 1994), which have more adverse effects on soils that have diminished biological integrity.

All these processes have led to declining per capita food production in SSA, which has resulted in over 3 million tons of food aid yearly (Conway and Toenniessen, 2003). Inadequate and inappropriate soil systems management has exacerbated these problems to an alarming extent.

18.2 From an External-Input Paradigm to an Integrated Soil Fertility Management Paradigm

During the past three decades, the ideas that have shaped soil fertility management research and development efforts in SSA have undergone substantial change. During the 1960s and 1970s, an external-input paradigm was framing the research and development agenda. Appropriate use of certain external inputs, whether fertilizers, lime, or irrigation water, was believed to be able to alleviate any constraints to crop production. Organic resources were seen as only playing a minor role (Table 18.1). By working within this paradigm, and benefiting from the development and use of improved cereal germplasm, bolstered by extensive fertilizer demonstrations and subsidization, what became known as the Green Revolution boosted agricultural production in Asia and Latin America in ways not seen before. Seeking similar yield enhancement, subsidies together with government distribution schemes were introduced in many African countries to promote fertilizer use by farmers. However, while some of these met with success, overall they did not come close to overcoming the estimated nutrient depletion rates in SSA or in matching the use rates of farmers in Asia and Latin America. By the early 1980s, these programs became mostly financially unsustainable as costs rose and productivity gains were not achieved (Kherallah et al., 2002).

TABLE 18.1

The Changing Role of Organic Resources in Tropical Soil Fertility Management

Period	Soil Fertility Management Paradigm	Role of Organic Resources
1960s/1970s	External-input paradigm	Organic matter plays a minor role
1980s	Biological management of soil fertility as part of low-external-input sustainable agriculture	Organic matter is mainly a source of nutrients and especially N
1994	Second paradigm — combined application of organic resources and mineral fertilizer	Organic matter fulfils other important roles besides supplying nutrients
Today	Integrated soil fertility management (ISFM) as a part of integrated natural resource management (INRM)	Organic matter management has social, economic, and political dimensions, with multiple stakeholders' interests

18.2.1 The Search for Less Input-Dependent Agricultural Systems

During the 1980s, exclusive reliance on chemical fertilizers for soil fertility enhancement was challenged by proponents of low-external-input sustainable agriculture (LEISA) who correctly argued that organic inputs were viewed as essential to sustainable agriculture (Okigbo, 1990). Further, it was argued that LEISA was preferable because it was more accessible to low-income rural households, who could afford little fertilizer and few agrochemicals. Organic resources were considered to be the major sources of nutrients (Table 18.1) and substitutes for mineral inputs. Additionally, the logistical problems of acquiring and transporting fertilizer, the uncertainty and unevenness of its supply in rural areas, and frequent issues of quality and efficacy reinforced the concern. However, LEISA approaches had little widespread acceptance, in large part because of technical and socio-economic constraints, e.g., insufficient training, lack of sufficient organic resources to apply in the field, and the labor-intensity of these technologies (Vanlauwe et al., 2001a, 2001b).

In this context, Sanchez (1994) proposed an alternative, second paradigm for tropical soil fertility research and remediation: "Rely more on biological processes by adapting germplasm to adverse soil conditions, by enhancing soil biological activity and by optimizing nutrient cycling to minimize external inputs and maximize the efficiency of their use." This paradigm, discussed more in Chapter 49, recognized the need for judiciously combining both mineral and organic inputs to sustain crop production and soil system fertility. The need for both organic and mineral inputs was advocated because (i) both resources fulfill different functions related to crop growth, (ii) under most small-scale farming conditions, neither is available and/or affordable in sufficient quantities to be applied alone, and (iii) for reasons still not fully researched, there were often added benefits when applying both inputs in combination, reflecting a degree of synergy. The alternative paradigm also highlighted the need for improved germplasm well-adapted to local conditions and able to give the most output from the available land, labor, water and nutrient inputs

As in the first paradigm, the LEISA approach put more emphasis on the quantity and quality of nutrient supply than on managing the demand for these nutrients. Obviously, optimal synchrony or use-efficiency requires that both supply and demand be coordinated. While organic resources were initially seen as complementary inputs to mineral fertilizers, over time, as seen in Table 18.1, their role has been seen as more than a short-term source of N, evolving to emphasize a wide array of benefits that can be derived from organic inputs to soil systems, both in the short and long term.

18.2.2 The Search for Optimizing Strategies

From the mid-1980s to the mid-1990s, the shift in thinking toward a more combined use of organic and mineral inputs was accompanied by a movement toward more participatory involvement of various stakeholders in the research and development process. One of the important lessons learned was that farmers' decision-making processes are not driven primarily by variations in soil and climate but by a whole set of factors encompassing the biophysical, socio-economic, and political domains (DFID, 2000).

18.2.2.1 Integrated Soil Fertility Management

The ISFM paradigm shown in Figure 18.1 goes beyond Sanchez's second paradigm to recognize the important roles that social, cultural, and economic processes play in soil fertility management strategies and also the many interactions that soil fertility has with other ecosystem services. ISFM presents a holistic approach to soil fertility research and

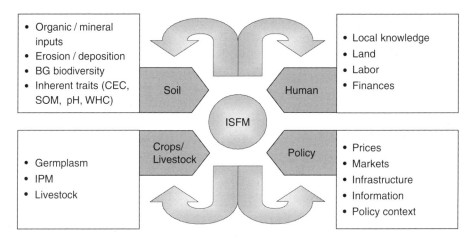

FIGURE 18.1
The processes and components of integrated soil fertility management (ISFM). BG, belowground; CEC, cation exchange capacity; SOM, soil organic matter; WHC, water-holding capacity; IPM, integrated pest management.

practice that embraces the full range of driving factors and consequences related to soil degradation — biological, physical, chemical, social, economic and political. Organic resource use has many social, economic, and policy dimensions besides biological and technical aspects reflected in belowground relationships.

The emergence of the ISFM paradigm parallels the development and spread on a wider scale of concepts of integrated natural resource management (INRM). It is increasingly recognized that natural capital (soil, water, atmosphere and biota) not only creates services that generate goods having market value, e.g., crops and livestock, but also services that are essential for the maintenance of life, e.g., clean air and water. Organic resource management is viewed as the link between soil fertility and broader environmental benefits, particularly ecosystems services such as carbon sequestration and biodiversity protection (Swift, 1997). Due to the wide array of services accruing from natural capital, different stakeholders may have conflicting interests in natural capital, and thus thinking has to extend into social and even political domains. INRM aims to develop policies and interventions that take both individual well-being and broader social needs into account (Izac, 2000). Soil system management is one component, but a basic component, of larger INRM strategies.

18.2.2.2   *Tropical Soil Biology and Fertility Research*

The Tropical Soil Biology and Fertility (TSBF) Institute, initially a program of UNESCO, was founded in 1986 to promote and develop capacities for soil biology as a research discipline benefiting the tropical regions. For over a decade, the program worked closely with the International Center for Agroforestry Research in Nairobi. However, since 2001 it has operated as an institute within the International Center for Tropical Agriculture (CIAT) based in Colombia, while remaining based in Kenya.

The biological management of soil fertility is held to be an essential component of sustainable agricultural development. The program's mission is directed toward four goals:

1. Improve understanding of the role of biological and organic resources in tropical soil fertility and their management by farmers to improve the sustainability of land-use systems.

2. Enhance the research and training capacity of national institutions in the tropics in the fields of soil biology and management of tropical ecosystems.

3. Provide land users in the tropics with methods for soil management that improve agricultural productivity while conserving soil resources.

4. Increase the carbon storage equilibrium and maintain the biodiversity of tropical soils in the face of global changes in land-use and climate.

The implementation strategy for achieving these goals has evolved along with the changes in soil fertility management paradigms described above. In the following section, this will be seen from two case studies examining the contributions that scientific investigations have made to better soil system management.

18.3   Translating Science into Practice

Despite the inherent complexity of the problems underlying the widespread decline in soil fertility in SSA, the good news is that progress is being made. At a 2002 meeting organized by the Rockefeller Foundation to take stock of progress with soil fertility research for development, advances were identified in three areas: (i) number and range of stakeholders influenced, (ii) soil management principles identified or clarified, and (iii) methodological innovations (TSBF, 2002a). National and international research and development organizations, networks, NGOs, and extension agencies working in SSA are increasingly using ISFM approaches (e.g., World Vision, 1999). There has been a rapid increase of membership and activities of the African Network for Tropical Soil Biology and Fertility (AfNet) coordinated by TSBF, with growing agreement on how soil systems can be better managed (Bationo, 2004).

International agricultural research has contributed significantly to the development of sound soil management principles that can help achieve sustainable crop production without compromising the ecosystem service functions of soil systems. Examples of such principles are:

- Application of organic resources in optimizing combinations with mineral inputs so as to maximize input-use efficiencies and farmers' return to their investment.

- Integration of multiple-purpose woody and herbaceous legumes into existing cropping systems to increase the supply of organic resources, crop yields, and farm profits (e.g., Sanginga et al., 2003).

- Enhancement of the soil organic carbon pool as an integrator of various soil-based functions that are related to production and ecosystem services (Swift, 1997).

- Improved sustainability of nutrient cycles through the integration of livestock with arable production activities.

- Soil conservation methods to control soil loss and improve water capture and use-efficiency.

Due to the complex and interactive nature of the major factors that promote poverty and act at different scales, it has been necessary to develop approaches that deal with such a complex environment:

- Pro-poor participatory research approaches that increase the appreciation and use of local knowledge systems in the development of improved soil management interventions and principles have been developed (e.g., Defoer and Budelman, 2000).

- Tools for scaling-up improved soil management practices, including GIS spatial analysis to better characterize problems and target interventions and to obtain a better understanding of information flow pathways, are emerging.

- Rapid assessment techniques using diagnostic indicators of land quality, e.g., spectrometry techniques such as in Shepherd et al. (2005), are now available.

- Molecular tools are being used to study soil biodiversity and pest population dynamics.

The following two sections describe areas where scientific principles have been translated into practice. They also illustrate how the dominant soil fertility management paradigm has shifted.

18.3.1   The Organic Resource Quality Concept and Organic Matter Management

Although use of organic inputs is hardly new to tropical agriculture, the first seminal analysis and synthesis on the decomposition and management of organic matter (OM) was contributed by Swift et al. (1979). Between 1984 and 1986, a set of hypotheses was formulated in terms of two broad themes for soil system management: synchrony, and soil organic matter (SOM) (see Swift, 1984, 1985, and 1986). These two focuses built upon the concepts and principles presented in 1979.

Under the first theme, the organisms-physical environment-quality (OPQ) framework for understanding OM decomposition and nutrient release, formulated by Swift et al. (1979), was elaborated and translated into specific hypotheses. These could explain the efficacy of management options that improved nutrient acquisition and crop growth with an explicit focus on organic resource quality. Under the second theme, the role of OM in the formation of functionally-different SOM fractions was stressed. It should be noted, however, that during this period, organic resources were still mainly regarded as sources of nutrients, and specifically of N (Table 18.1). Their multiple functions within soil systems were not much considered.

During the 1990s, the formulation of research hypotheses related to residue quality and N release led to many research efforts to validate these hypotheses, both within TSBF and other research groups that dealt with tropical soil fertility. Results from these activities were entered in the Organic Resource Database (ORD) (ftp://iserver.ciat.cgiar.org/webciat/ORD/) (Palm et al., 2000). This database contains extensive information on organic-resource quality parameters, including macronutrient, lignin, and polyphenol contents of fresh leaves, litter, stems, and/or roots from almost 300 species utilized in tropical agroecosystems. Data on the soil and climate from where the material was collected are also included, as are decomposition and nutrient-release rates for many of the organic inputs.

Analysis of N-release dynamics revealed four classes of organic resources having different rates and patterns of N release associated with varying organic resource quality assessed in terms of their N, lignin, and polyphenol content (Palm et al., 2000). Based on this analysis and information, a decision support system (DSS) for management of organic N was formulated (Figure 18.2a). This system distinguishes four types of organic resources, suggesting how each can be managed optimally for short-term N release to immediately enhance crop production. Materials with lower N and higher lignin and/or polyphenol contents are expected to release less N and thus they require supplementary N in the form of fertilizer or higher-quality organic resources to maintain nutrient supply at comparable levels.

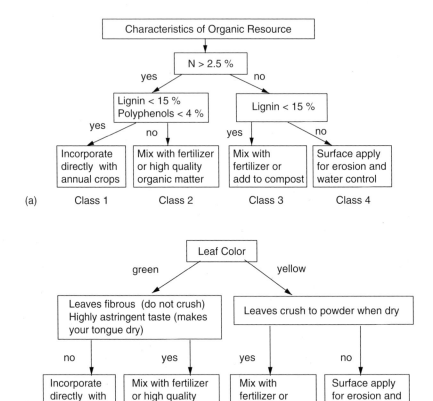

FIGURE 18.2

A decision tree to assist management of organic resources in agriculture. (a) is based on Palm et al. (2000); (b) is a "farmer-friendly" version of the same from Giller (2000).

Being based on laboratory incubations, the DSS needed to be tested under field conditions and was assessed in western, eastern, and southern Africa, using biomass transfer systems with maize as a test crop. The results clearly indicated that (i) the N content of the organic resources is an important factor affecting maize production, (ii) organic resources with a relatively high polyphenol content result in relatively lower maize yields for the same level of N applied, (iii) manure samples do not observe the general relationships followed by the fresh organic resources, and (iv) N fertilizer equivalency values of organic inputs often approach or even exceed 100% of what would be supplied from inorganic sources.

These results gave strong support for the DSS constructed by Palm et al. (2000), except for manure samples. Manure behaves differently from plant materials since it has already gone through a decomposition phase when passing through the digestive system of cattle, rendering the C less available and thus resulting in relatively less N immobilization, as discussed in the preceding chapter. The observation that certain organic resources have fertilizer equivalency exceeding 100% indicates that these organic materials can alleviate other constraints to maize production besides low soil-available N. In the short-term, organic resources not only release nutrients; they can enhance soil moisture conditions or improve the available P in the soil (Nziguheba et al., 2000). In the long term, continuous inputs of OM influence the levels of incorporated SOM and the quality of some or all of its nutrient pools (Vanlauwe et al., 1998; Cadisch and Giller, 2000).

Following field-level testing of the DSS, it has been applied and adapted in a variety of farmer learning activities. These give farmers the knowledge they need to identify and evaluate the potential use of organic resources in their environment. Because there is so much diversity of such resources in any given context, the elements of the DSS provide a generic, easy-to-use tool for farmers to use when confronted with resources that scientists have not themselves evaluated.

Farm-level adaptation of the DSS began with exercises where researchers and farmers in selected communities identified all the organic resources available locally as potential soil inputs. The quality analysis of these materials in one setting (Table 18.2) shows that among

TABLE 18.2

Organic Resources (leaf residues) and Their Chemical Composition, Identified in Farms Around Emuhaya Division, Vihiga District, Western Kenya

Genus and Species Name	Common Name or Local Name	N	P	K	Lignin	PP[a]	Class[b]
		% Dry Matter					
Markhamia lutea		3.20	0.24	1.77	21.21	3.99	1
Psidium guajava		2.32	0.19	1.50	11.20	14.35	3
Persea americana	Avocado	2.07	0.12	0.82	20.25	10.90	4
Not identified	Not known	4.98	0.44	6.66	14.93	3.27	1
Bridelia macrantha		2.37	0.17	1.13	18.53	8.31	4
Vernonia spp		4.88	0.42	4.72	11.31	2.44	1
Croton macrostachyus		4.33	0.38	1.75	10.25	8.42	2
Not identified	*Esikokhakokhe*	3.84	0.39	6.59	9.07	1.32	1
Solanum aculeastrium	Sodim apple	2.87	0.21	1.25	13.70	2.39	1
Erythrina exselsa		4.99	0.33	2.42	6.63	2.26	1
Buddleja davidi		3.30	0.27	1.46	7.94	6.20	2
Senna didymobotra		5.23	0.39	2.13	4.62	4.08	2
Vernonia auriculifera		3.65	0.35	5.25	14.86	4.93	2
Hurungania madagascariensis		3.21	0.18	1.04	13.31	12.70	2
Spathodea campanulata	Nandi flame	3.09	0.21	1.76	17.34	8.58	2
Erythrina abyssinica		2.66	0.20	1.70	11.21	3.36	1
Morus alba	Mulberry	2.86	0.43	2.16	4.28	4.62	2
Acanthus pubescens		3.30	0.30	2.11	5.17	7.56	2
Ricinus commus	Castor plant	4.21	0.30	2.34	3.39	5.27	2
Maesa lanceolata		2.78	0.22	2.06	10.37	12.04	2
Mangifera indica	Mango plant	1.52	0.12	1.00	11.15	12.43	3
Teclea nobilis		3.15	0.22	1.57	9.05	4.83	2
Not identified	*Libinzu*	3.91	0.29	3.28	12.27	5.67	2
Sapium elliptian		3.11	0.18	0.77	6.34	11.73	2
Vangneria apiculata		3.67	0.23	1.76	4.91	4.27	2
Ficus spp		2.55	0.20	2.62	9.55	5.76	2
Ipomoea potatus	Sweet potato	5.07	0.34	2.56	4.34	8.81	2
Not identified	*Omuterema*	3.85	0.34	5.27	2.85	1.20	1
Plectranthus barbutus		3.87	0.28	4.01	16.11	4.98	2
Maesa lanceolata		3.80	0.28	3.92	10.70	6.65	2
Vernonia spp		4.26	0.37	3.67	9.80	5.09	2

[a] PP, polyphenols.

[b] Class refers to classes 1 to 4 indicated in Figure 18.2.

Source: Authors' data.

the plant resources that farmers would consider incorporating into their soils, the large majority were class 2 resources. Of the 38 organic resources assessed, only eight belonged to class 1 and could be classified as equivalent to N fertilizer. *Tithionia diversifolia* had already been identified as a high-quality organic resource during a previous hedgerow survey in the same area, also belonging in class 1 (Gachengo et al., 1999).

When these results were presented and discussed with farmers in a second step, the decision-tree criteria proposed by Palm et al. (2000) were translated into a more farmer-friendly version, using locally-acceptable criteria that do not require scientific equipment (Figure 18.2b). This locally-adapted decision tree was then used by local farmer field schools to design their own experimental trials that tested the validity of the claims that scientists were making regarding the use and management of organic resources (TSBF, 2002b).

These trials, conducted at a variety of sites and through several seasons, provided many opportunities for farmers to compare the effects of these organic inputs under different conditions. During evaluation activities, farmers ranked the classes of organic resources in terms of effect on maize yield as: Tithonia (Class 1) > manure > Calliandra (class 2) > maize stover (Class 3). They confirmed the hypothesis that the differing quality of organic materials would have a demonstrable impact on crop yields.

Scientists also drew many valuable lessons from this exercise. They found, for example, that farmers considered the biomass transfer technology being tested to be less practical and cost-effective than using compost, a common local practice. Their interest in adding their organic resources to compost heaps before application to the soil has stimulated new joint research activities between farmers and scientists on how to improve compost quality (TSBF, 2002b). (Benefits of composting are discussed in Chapter 31.) A second line of experimentation used the resource-quality concept to assess the use of organic materials, especially comparatively-scarce, high-quality Tithonia residues, on high-value crops such as kale rather than on maize (TSBF, 2002b).

18.3.2 Exploring Positive Interactions between Mineral and Organic Inputs

The paucity of class 1 resources at the farm level, and the consequent advice to mix class 2 or 3 resources with minimal amounts of fertilizer N, has led to a diversification of the research agenda toward the combined application of organic and mineral inputs. As mentioned above, such a strategy is consistent with the ISFM paradigm and can potentially lead to added benefits in terms of extra crop yield and/or extra soil fertility enrichment where there are positive interactions between both inputs, as illustrated in Figure 18.3.

Although the concept of interaction between two plant growth factors was already implied in Liebig's Law of the Minimum, it has recently received new attention in work dealing with the combined application of fertilizer and organic inputs. Besides adding nutrients, organic resources also provide C as a substrate for soil organisms and may interfere with pests and diseases when the plants are grown *in situ*.

Two sets of hypotheses can be formulated, based on whether the interactions between fertilizer and organic matter are direct or indirect. Since fertilizer N is susceptible to substantial losses if not used quickly and efficiently by a crop, direct interactions result from microbially-mediated changes in the availability of the fertilizer N when there is an increase in available C. Further, the addition of fertilizer N may also affect the availability of soil-derived N, although this will be less important whenever the bulk soil is C-limited. Indirect interactions are the result of a general improvement in plant growth and demand for nutrients by alleviation, through the addition of organic matter, of another growth-limiting factor.

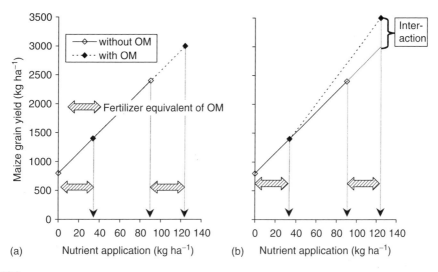

FIGURE 18.3

Theoretical response of maize grain yield to the application of certain levels of nutrients as fertilizer in the presence or absence of organic matter (a) without interaction, and (b) with positive interaction between the fertilizer nutrient and organic matter. *Source:* Vanlauwe et al. (2001a, 2001b).

The direct hypothesis regarding N fertilizer can be stated as: temporary immobilization of applied fertilizer N may improve the synchrony between the supply of and demand for N and also reduce losses to the environment. Observations made under controlled conditions justify this hypothesis, showing interactions in decomposition or N mineralization between different organic materials (Vanlauwe et al., 1994) or between organic matter and fertilizer N (Sakala et al., 2000).

The indirect hypothesis may be formulated for a certain plant nutrient X supplied by fertilizer amendments as: any organic matter-related improvement in soil conditions affecting plant growth (except that attributable to nutrient X) may lead to better plant growth and consequently to enhanced efficiency of the applied nutrient X. The growth-limiting factor can be located in the domain of plant nutrition, soil physics or chemistry, or soil (micro)biology.

Most of the mulch effects or benefits of crop rotation could be classified under the indirect hypothesis. Positive interactions based on the indirect hypothesis may be immediate through direct alleviation of growth-limiting conditions after applying organic matter, e.g., improvement of the soil moisture status after surface application of organic matter as a mulch, or delayed through the improvement of the SOM status after continuous application of organic matter and an associated better crop growth, e.g., improvement of the soil's buffering capacity.

Under on-station conditions, positive interactions can often be observed and measured. However, explaining the mechanisms underlying these interactions is often more problematic:

- In a field study in West Africa, Vanlauwe et al. (2002) observed positive interactions, likely caused by higher soil moisture retention in treatments where organic resources were applied (Figure 18.4).
- Bationo et al. (1995) observed a doubling of the fertilizer N-use efficiency after application of crop residues in Sahelian conditions, attributable to much less wind erosion on treatments when crop residues were applied.

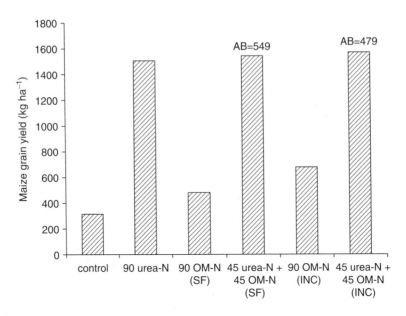

FIGURE 18.4
Maize grain yields in Sekou, southern Benin Republic, as affected by the application of urea, organic materials, or the combination of both. SF, surface-applied; INC, incorporated; OM, organic matter; AB, added benefits. Numerical values for treatments are expressed as kg N ha^{-1}. Adapted from Vanlauwe et al. (2001a, 2001b).

- In Zimbabwe, added benefits ranging between 663 and 1188 kg maize grains ha^{-1} were observed by Nhamo (2001), possibly because the supply of cations contained in the manure alleviated constraints to crop growth caused by the low cation content of the very sandy sites where clay content ranged between 2 and 10% and CEC varied between 1.2 and 2.5 cmol kg^{-1}.

Translating these principles into cropping systems that are adaptable by farming communities has resulted in a series of development innovations, e.g., rotations of maize with promiscuously-nodulating soybean that combine high N-fixation and the ability to kill large numbers of *Striga hermonthica* seeds in the soil; and rotations of millet and dual-purpose cowpea that greatly enhance the productivity and sustainability of integrated livestock systems (Sanginga et al., 2003).

These two systems are effectively used for the replenishment of soil nutrients and organic matter. They contribute positive residual soil N for the following crops while at the same time providing farmers with seeds for food and fodder for feed, as well as income from marketing these farm products. Another option offered to any farmers who have manure available is the opportunity to derive benefits from the combined application of manure and fertilizer to maize. This practice allows farmers to complement the modest fertilizer quantities that they can afford with high-quality organic nutrients, thereby benefiting from the synergism that occurs when combining the two sources of nutrients. Currently, Sasakawa Global 2000 is testing the above options in Northern Nigeria with promising results.

18.4   Challenges and the Way Forward

Although soil fertility replenishment has had a prominent position on the research and development agenda in SSA for decades with tangible progress as seen above, widespread adoption of ISFM strategies is lacking. A full discussion of the reasons for this is beyond the scope of this chapter, but certain issues that have hampered large-scale adoption of ISFM options can be singled out.

18.4.1 Adjusting to Variability at the Farm and Community Levels

Farmers' production objectives are conditioned by a complex set of biophysical as well as social, cultural, and economic factors. One must also take account of the fertility gradients existing within farm boundaries. Most soil fertility research has been targeted at the plot level, but decisions are made at the farm level, considering the production potential of all plots. In Western Kenya, farmers will preferentially grow sweet potato on their most degraded fields, while bananas and cocoyam occupy the most fertile fields (Tittonell et al., 2005). Current recommendations for use of organic resources and mineral inputs do not take into account these gradients in soil fertility status. On the contrary, recommendations are often formulated at the national level and disregard the much greater variations that exist between regions in terms of inherent soil properties and access to input and output markets (Carsky and Iwuafor, 1999).

18.4.2 Use of Adapted Germplasm to Overcome Abiotic and Biotic Constraints and Create More Resilient Cropping Systems

Breeding and biotechnology can help small farmers to sustainably increase their productivity through improved drought-tolerance, soil acidity-tolerance, pest-resistance, and increased efficiency of N-fixation. ISFM acknowledges the importance of the interaction between new crop germplasm and more efficient natural resource management for intensifying food and forage crop systems. Such a combination would utilize the best variety for a given environment when grown in an improved soil using appropriate crop management technologies. Interactions between adapted germplasm and key inputs such as organic residues, mineral fertilizers, and water can lead to improved use-efficiency of nutrients and water at a system level. ISFM bridges a commodity focus and an eco-regional approach, working alongside germplasm development and integrated pest and disease management.

18.4.3 Market-Led Integrated Soil Fertility Management

ISFM practices require some additional inputs of resources, whether minimal amounts of mineral fertilizer, more organic matter, improved germplasm, or greater labor. As most of these inputs require access to financial resources, implementing ISFM strategies will often require farmers to have access to local or national markets so that they can acquire more resources to reinvest in improved soil fertility management. It has been hypothesized that improved profitability and access to markets will motivate farmers to invest in new technology, particularly to integrate use of new varieties with improved soil management options (John Lynam, 2004, personal communication).

Some current evidence does not show conclusive support for this hypothesis, however. For instance, the increased movement of bananas to urban markets in Uganda without replenishment of the soil resource base could lead to a faster degradation of banana-based systems within the production areas. It is also important to consider nutritional consequences. Farmers who sell most of their produce could use the money received for other uses rather than ensuring sufficient and nutritious food for the household. This could lead to poorer health status with unfavorable consequences for household labor availability and quality.

18.4.4 Scaling Up

The knowledge-intensive nature of ISFM means that the kind of simplistic extension methods such as "training and visit" promoted by the World Bank in the 1980s and 1990s are not suitable for disseminating soil management technologies. This lack of suitability accounted in part for the collapse of training-and-visit extension in the mid-to-late 1990s (e.g., Gautam, 2000, on Kenya experience). Since then, the move in many countries toward the decentralization of government services, the improved capacity of NGOs in service delivery, and the beginnings of farmer groups and collective action have created the preconditions for greater innovation and for the redesign of extension and dissemination systems.

  Recognizing the wide diversity in agroecological and socio-economic conditions under which most farmers work has led to a general realization that research and extension agencies do not have the capacity to fine-tune their technological recommendations to the level required by farmers. As extension services have become increasingly marginalized and nonfunctional, the gaps in knowledge-dissemination and technological improvement have been largely filled by a variety of NGOs and, in some cases, community-based organizations. Scaling up information dissemination requires the reinforcement of communication networks and strengthening of information centers (agricultural input suppliers, community centers, field schools), as well as supporting farmers in various ways to transfer knowledge farmer-to-farmer across communities.

18.4.5 Policy Changes

Since the 1980s, most countries in SSA have initiated extensive agricultural market reforms (Kherallah et al., 2002). The expectation of agricultural market reform is that increasing crop prices and improving markets will generate a positive supply response, increasing both agricultural output and income levels. However, the average growth of agricultural production per capita has been negative in SSA since the 1970s. In many countries, reform has meant the elimination of government input and credit subsidies. This has kept yields stagnant or reduced them, or has made input supplies irregular or completely absent, undermining the stability of local prices. What production growth has occurred has often been due either to expansion of crop area rather than increases in productivity per unit area, or to the output of cash-crop farmers still operating within systems who have good access to credit and inputs.

  For ISFM to operate on a broader scale, there is a need for (i) regional policy harmonization and policy reform frameworks for improved management within sub-regional areas, (ii) development of appropriate partnerships to facilitate efficient input-output markets and strengthen their links to ISFM, (iii) identification of marketing opportunities through participatory research within a comprehensive, resource-to-consumption framework, and (iv) development of appropriate seed supply systems and resilient germplasm. Since not all farmers have the capacity to buy themselves out of poverty, there is a major need for a series of "stepping stones" that enable poor farmers to have access to inputs, services, and markets so that they can "climb out of poverty" as their agricultural productivity increases.

Acknowledgments

The Rockefeller Foundation and the Belgian Directorate-General for Development Co-operation are gratefully acknowledged for their continued financial support of this work.

References

Bationo, A., *Managing Nutrient Cycles to Sustain Soil Fertility in Sub-Saharan Africa*, Academy Science Publishers, Nairobi (2004).

Bationo, A. et al., A critical review of crop-residue use as soil amendment in the West African semi-arid tropics, In: *Livestock and Sustainable Nutrient Cycling in Mixed Farming Systems of Sub-Saharan Africa, Technical Papers*, Vol. 2, Powell, J.M. et al., Eds., International Livestock Centre for Africa, Addis Ababa, 305–322 (1995).

Cadisch, G. and Giller, K.E., Soil organic matter management: The role of residue quality in carbon sequestration and nitrogen supply, In: *Sustainable Management of Soil Organic Matter*, Rees, R.M. et al., Eds., CAB International, Wallingford, UK, 97–111 (2000).

Carsky, R.J. and Iwuafor, E.N.O., Contribution of soil fertility research and maintenance to improved maize production and productivity in sub-Saharan Africa, In: *Strategy for Sustainable Maize Production in West and Central Africa*, Badu-Apraku, B. et al., Eds., International Institute for Tropical Agriculture, Ibadan, 3–20 (1999).

Conway, G. and Toenniessen, G., Science for African food security, *Science*, **299**, 1187–1188 (2003).

DFID, *Sustainable Livelihoods Guidance Sheets*, Department for International Development, London (2000).

Defoer, T. and Bundelman, A., *Managing Soil Fertility in the Tropics: A Resource Guide for Participatory Learning and Action Research*, KIT Publishers, Amsterdam, Netherlands (2000).

Gachengo, C.N. et al., Tithonia and senna green manures and inorganic fertilizers as phosphorus sources for maize in western Kenya, *Agroforest. Syst.*, **44**, 21–36 (1999).

Gautam, M., *Agricultural Extension: The Kenya Experience: An Impact Evaluation*, World Bank, Operation Evaluation Department, Washington, DC (2000).

Giller, K.E., Translating science into action for agricultural development in the tropics: An example from decomposition studies, *Appl. Soil Ecol.*, **14**, 1–3 (2000).

Izac, A.-M., What paradigm for linking poverty alleviation to natural resource management? In: *Proceedings of an International Workshop on Integrated Natural Resource Management in the CGIAR: Approaches and Lessons, 21–25 August, 2000, Penang, Malaysia*, CGIAR Secretariat, Washington, DC (2000).

Kherallah, M. et al., *Reforming Agricultural Markets in Africa*, John Hopkins University Press, Baltimore, MD (2002).

Nhamo, N., An evaluation of the efficacy of organic and inorganic fertilizer combinations in supplying nitrogen to crops, M Phil thesis, University of Zimbabwe, Zimbabwe (2001).

Nziguheba, G. et al., Organic residues affect phosphorus availability and maize yields in a Nitisol of western Kenya, *Biol. Fert. Soils*, **32**, 328–339 (2000).

Okigbo, B.N., Sustainable agricultural systems in tropical Africa, In: *Sustainable Agricultural Systems*, Edwards, C.A. et al., Eds., Soil and Water Conservation Society, Ankeny, IA, 323–352 (1990).

Oldeman, L.R., The global extent of soil degradation, In: *Soil Resilience and Sustainable Land Use*, Greenland, D.J. and Szabolcs, I., Eds., CAB International, Wallingford, UK, 99–118 (1994).

Palm, C.A. et al., Organic inputs for soil fertility management in tropical agroecosystems: Application of an organic resource database, *Agric. Ecosyst. Environ.*, **83**, 27–42 (2000).

Sakala, W.D., Cadisch, G., and Giller, K.E., Interactions between residues of maize and pigeonpea and mineral N fertilizers during decomposition and N mineralization, *Soil Biol. Biochem.*, **32**, 679–688 (2001).

Sanchez, P.A., *Tropical Soil Fertility Research: Towards the Second Paradigm, Transactions of the 15th World Congress of Soil Science, Acapulco, Mexico*. Mexican Soil Science Society, Chapingo, Mexico, 65–88 (1994).

Sanginga, N. et al., Sustainable resource management coupled to resilient germplasm to provide new intensive cereal-grain legume-livestock systems in the dry savanna, *Agric. Ecosyst. Environ.*, **100**, 305–314 (2003).

Shepherd, K.D., et al., Decomposition and mineralization rates of organic residues predicted using near infrared spectroscopy. *Soil Biol. Biochem.* in press (2005).

Stoorvogel, J.J. and Smaling, E.M.A., *Assessment of Soil Nutrient Depletion in sub-Saharan Africa 1983–2000*, Winand Staring Center, Wageningen, Netherlands (1990).

Swift, M.J., *Soil Biological Processes and Tropical Soil Fertility: A Proposal for a Collaborative Programme of Research, Biology International special issue 5*. International Union of Biological Sciences, Paris, France (1984).

Swift, M.J., *Tropical Soil Biology and Fertility: Planning for Research, Biology International, Special issue 9*. International Union of Biological Sciences, Paris, France (1985).

Swift, M.J., *Tropical Soil Biology and Fertility: Inter-regional Research Planning Workshop, Biology International Special Issue 13*. International Union of Biological Sciences, Paris, France (1986).

Swift, M.J., Special issue: Soil biodiversity, agricultural intensification and agroecosystem function, *Appl. Soil Ecol.*, **6**, 1 (1997).

Swift, M.J., Heal, O.W., and Anderson, J.M., *Decomposition in terrestrial ecosystems*, Studies in Ecology, Vol. 5, Blackwell Scientific Publications, Oxford, UK (1979).

TSBF, *Soil Fertility Degradation in sub-Saharan Africa: Leveraging Lasting Solutions to a Long-Term Problem: Conclusions from a Workshop held at the Rockefeller Foundation Bellagio Centre, March 4–8, 2002*, Tropical Soil Biology and Fertility Institute, Nairobi (2002a).

TSBF, *Demonstration Plot: Harvesting and Evaluations. A Report on August 16 and 22, 2002 Events at the Farmer–Researcher Demonstration Site in Emuhaya*, Tropical Soil Biology and Fertility Institute, Nairobi (2002b).

Tian, G. et al., Food production in the moist savanna of West and Central Africa, In: *Moist Savannas of Africa Potentials and Constraints for Crop Production*, Kang, B.T. et al., Eds., International Institute of Tropical Agriculture, Ibadan, Nigeria, 107–127 (1995).

Tittonell, P., et al., Exploring diversity in soil fertility management of smallholder farmers in western Kenya. I. Heterogeneity at region and farm scale. *Agric., Ecosyst. Envir.* **110**, 149-165.

Vanlauwe, B. et al., Maize yield as affected by organic inputs and urea in the West-African moist savanna, *Agron. J.*, **93**, 1191–1199 (2001a).

Vanlauwe, B., Dendooven, L., and Merckx, R., Residue fractionation and decomposition: The significance of the active fraction, *Plant Soil*, **158**, 263–274 (1994).

Vanlauwe, B. et al., Organic resource management in sub-Saharan Africa: Validation of a residue quality-driven decision support system, *Agronomie*, **22**, 839–846 (2002).

Vanlauwe, B., Sanginga, N., and Merckx, R., Soil organic matter dynamics after addition of ^{15}N labeled Leucaena and Dactyladenia residues in alley cropping systems, *Soil Sci. Soc. Am. J.*, **62**, 461–466 (1998).

Vanlauwe, B., Wendt, J., and Diels, J., Combined application of organic matter and fertilizer, In: *Sustaining Soil Fertility in West Africa*, Tian, G., Ishida, F., and Keatinge, J.D.H., Eds., SSSA Special Publication 58, Madison, WI (2001b).

World Vision, *Food Security Program Newsletter 4:2*, World Vision International, Accra, Ghana (1999).

19

Managing Soil Fertility and Nutrient Cycles through Fertilizer Trees in Southern Africa

Paramu L. Mafongoya, Elias Kuntashula and Gudeta Sileshi
World Agroforestry Centre (ICRAF), Lusaka, Zambia

CONTENTS

Low soil fertility is increasingly recognized as a fundamental biophysical cause for declining food security among small-farm households in sub-Saharan Africa (SSA) (Sanchez et al., 1997). Because maize is the staple food crop in most of southern Africa, it will be our focus in this chapter. In 1993, SSA produced 26 million metric tons of maize on approximately 20 m ha; approximately 54 million metric tons is expected to be needed by 2020. Meeting this maize production goal will depend on sustaining and improving soil fertility levels that have been declining in recent years.

Soil fertility is not the only significant constraint; lack of appropriate, high-quality germplasm, unsupportive policies, and inadequate rural infrastructure also limit maize production. However, protecting and enhancing soil fertility is the most basic requirement for achieving production goals. As discussed in Chapters 40 and 41, even controlling the parasitic weed Striga hinges on this fundamental factor.

In most cases, nitrogen is the main nutrient that limits maize productivity, with phosphorus and potassium being occasional constraints. Although inorganic fertilizers are used throughout the region, the amounts applied are seldom sufficient to meet crop demands due to their high costs and uncertain availability. Most countries in southern Africa have formulated fertilizer recommendations for all their major crops, sometimes with regionally specific adaptations. However, the amount of fertilizer used in southern Africa is very small in comparison to other parts of the world. For most smallholders, fertilizer use averages as low as $5 \, kg \, ha^{-1} \, year^{-1}$ (Gerner and Harris, 1993).

While the need for increasing the availability of soil nutrients in southern Africa is quite apparent, increasing their supply is very challenging. A high-external-input strategy cannot rely on standard fertilizer-seeds-credit packages without addressing other requirements for successful uptake of Green Revolution technologies, including reliable irrigation, credit systems, infrastructure, fertilizer manufacture and supply, and access to markets. Most African conditions differ starkly from those in the prime agricultural regions of Asia. Approaches that produced successes in Asia are not readily transferable to the African continent. Considering the acute poverty and the limited access to mineral fertilizers in SSA, therefore, an ecologically robust approach of promoting "fertilizer trees" is discussed here. This is a product of many years of agroforestry research and development by the International Center for Research on Agroforestry (ICRAF), now called the World Agroforestry Center, working with various partners in eastern and southern Africa.

19.1   Fertilizer Trees and a Typology of Fallows

Improved fallows involve the deliberate planting of fast-growing species, usually woody tree legumes, referred to here as fertilizer trees, for the rapid replenishment of soil fertility. Improved fallows were not a major area for research during the Green Revolution due to its focus on eliminating soil constraints by use of mineral fertilizers. Biological approaches to soil fertility improvement began to receive attention in connection with the articulation of a second soil-fertility paradigm based on adaptability and sustainability considerations (Sanchez, 1994). Research on fertilizer trees had begun increasing from the mid-1980s, so by the mid-1990s they had growing justification in research (e.g., Kwesiga and Coe, 1994; Drechsel et al., 1996; Rao et al., 1998; Snapp et al., 1998). Large-scale adoption of fertilizer trees by farmers is now taking place across southern and eastern Africa. A more general consideration of fallows is presented in Chapter 29.

19.1.1   Use of Non-Coppicing Fertilizer Trees

Non-coppicing species do not resprout and regrow when cut at the end of the fallow period, typically after 2 years of growth. Non-coppicing species include *Sesbania sesban, Tephrosia vogelii, Tephrosia candida, Cajanus cajan,* and Crotalaria spp. Since the work of Kwesiga and Coe (1994) on Sesbania fallows, much has been learned about the performance of improved fallows using tree species that do not coppice. There has been extensive testing of various species and fallow length on-farm to determine their impact on maize productivity and to assess the processes that influence fallow performance. The performance of Sesbania and Tephrosia under a wide range of biophysical conditions is shown in Table 19.1.

Trials at Msekera Research Station, Zambia, have shown that natural regeneration of Sesbania fallows is possible through self-reseeding, but it is highly erratic. Improved fallows of 2-year duration using either Tephrosia or Sesbania significantly increased maize yields well above those of unfertilized maize, the most common farmer practice in the region. While it was true that fertilized maize usually performed better than improved fallows in most cases, this required a greater cash outlay, so improved fallows could be more profitable. The problem demonstrated in these trials was that the residual effects of these improved fallows on maize yield declined after the second year of cropping (Table 19.1). In a third year of cropping, maize yields following fallow were similar to those of unfertilized maize. The marked decline of maize yields two or three seasons after a non-coppicing fallow is probably related to depletion of soil nutrients and/or to deterioration in soil chemical and physical properties.

19.1.2   Use of Coppicing Fertilizer Trees

Coppicing species include *Gliricidia sepium*, *Leucaena leucocephala*, *Calliandra calothyrsus*, *Senna siamea*, and *Flemingia macrophylla*. Fallowing with a coppicing species, in contrast to a non-coppicing species, shows increases in residual soil fertility beyond 2–3 years because of the additional organic inputs that are derived each year from coppice regrowth that is cut and applied to the soil. An experiment was established in the early 1990s at Msekera Research Station to examine these relationships. These plots have now been cropped for 9 years during which time both maize yields and coppice growth were monitored.

The species evaluated showed significant differences in their coppicing ability and biomass production, with Leucaena, Gliricidia, and *Senna siamea* having the greatest coppicing ability and biomass production, while Calliandra and Flemingia performed poorly. The trends in maize yields have been tracked carefully. In the plots with Sesbania fallow, while maize yields were high for the first three seasons, they then declined to the same level as control plots. Flemingia and Calliandra showed low maize yields over all years. There were no significant differences in maize grain between Gliricidia and Leucaena fallows over the seasons.

The effects of different fallow species on maize yield can be explained partly by the different amounts of biomass added and the quality of the biomass and coppice regrowth. Species such as Leucaena and Gliricidia, which have good coppicing ability, produce large amounts of high-quality biomass with high nitrogen content and low contents of lignin and polyphenols, thereby contributing to higher maize yields (Mafongoya and Nair, 1997; Mafongoya et al., 1998). While Sesbania produces high quality biomass, its lack of coppice regrowth means that it cannot supply nutrients for an extended period of cropping. Species such as Flemingia, Calliandra, and *Senna siamea*, on the other hand, produce low-quality biomass, high in lignin and polyphenols and low in nitrogen. Their use as fallow species leads to N immobilization and reduced maize yields.

TABLE 19.1

Effect of Fallows on Maize Grain Yield Across 18 Locations in Zambia

	Maize Grain Yield (t ha^{-1})		
Land Use	Year 1	Year 2	Year 3
Sesbania sesban fallow	3.9	1.7	1.1
Tephrosia vogelii fallow	2.4	0.8	0.9
Traditional grass fallow	1.1	0.7	0.7
Unfertilized maize	1.0	0.7	0.6
LSD	0.8	0.6	0.6

Source: Authors' data.

Both Gliricidia and Leucaena have shown good potential as coppicing fallows. Over 9 years of cropping, cumulative maize yield of these fallows is greater than unfertilized maize, maize grown after Sesbania, and traditional grass fallow. Continuous nutrient replenishment is achieved by applying the coppice regrowth as mulch to the soil. This trial will be continued for another three seasons to test the sustainability of coppicing fallows in terms of nutrient budgets such as for NPK. On-farm trials have already been established to evaluate responses more widely and to screen more coppicing fallow species.

19.1.3   Mixed-Species Fallows

Improved fallow practices using shrub legume species such as Sesbania have become popular agroforestry systems for soil fertility management in southern Africa and western Kenya. Large increases in maize yields have been reported following short-duration fallows of 9–24 months with single species. Sesbania has been the main focus for these improved fallows given its ability to provide large amounts of high-quality biomass and fuel wood. Dependence upon a few successful fallow species has revealed some drawbacks, however. Sesbania is susceptible to root nematodes and the Mesoplatys beetle. The introduction of any new species can lead to an outbreak of new pests and diseases, as was observed with *Crotalaria grahamiana* in western Kenya (Cadisch et al., 2002). Thus, there is an urgent need to diversify the fallow species and types offered to farmers. Mixing species with compatible and complementary rooting or shoot-growth patterns in fallow systems should lead to more diverse systems and maximize growth and resource utilization above- and belowground. Sowing herbaceous legumes under open-canopy tree species can increase the use of photosynthesis radiation by the whole canopy and thus enhance the system's primary production.

Mixing shallow-rooted species with deep-rooted species can enhance the soil-water and nutrient-uptake zone within the soil profile. More important, it enhances the utilization of subsoil nutrients such as the nitrate that is otherwise lost through leaching. Mixing species in fallows may also reduce the risks with fallow establishment, e.g., if one species is susceptible to water stress, diseases or pests, another can survive and even prosper. Obtaining multiple products from mixed fallows as well as increasing the biodiversity of the system makes the whole system more robust. We have assessed a variety of mixed fallows of tree legumes or tree legumes with herbaceous legumes to test these hypotheses.

Mixing a coppicing fallow species such as *Gliricidia sepium* with a non-coppicing species like Sesbania (Chirwa et al., 2003) significantly increased maize yields compared to single-species fallows (Table 19.2). However, mixtures of non-coppicing species did not increase maize yield compared to sole species (Table 19.3). Mixing coppicing and non-coppicing species reduces the level of subsoil nitrate, and we found that it reduces Mesoplatys beetles (Sileshi and Mafongoya, 2002). We have found also that mixing Gliricidia, Tephrosia, or Sesbania with herbaceous legumes such as Mucuna or Dolichos reduces tree growth, and hence maize yield. Such mixtures also lead to a build-up of the Mesoplatys beetle, which can cause more damage (Sileshi and Mafongoya, 2002).

19.1.4   Biomass Transfer Using Fertilizer-Tree Biomass

Traditionally, resource-poor farmers in parts of Southern Africa have collected leaf litter from secondary forest, called miombo, as a source of nutrients for their crops. In the long term, this practice is not sustainable because it mines nutrients from the forest ecosystems in order to enhance soil fertility in croplands. Also, the miombo litter is of low quality and may immobilize N instead of supplying N immediately to the crop (Mafongoya and Nair, 1997). An alternative means of producing high-quality biomass is through the

TABLE 19.2

Maize Grain Yield (t ha^{-1}) from 3-Year Coppicing Mixed-Fallow Species Treatments at Msekera, Eastern Zambia

Species	2003	2004
Fertilized maize	5.9	3.4
Acacia angustissma (34/88)	3.7	1.3
Acacia angustissma + *Sesbania sesban*	4.6	2.2
Gliricidia sepium (Retalhuleu)	4.1	2.9
Gliricidia sepium + *Sesbania sesban*	4.6	2.7
Gliricidia sepium + *Tephrosia vogelii*	3.3	2.1
Leucaena diversfolia	3.6	1.5
Leucaena diversfolia + *Sesbania sesban*	4.3	2.0
Sesbania sesban	3.9	1.9
Tephrosia vogelii	4.3	2.6
Tephrosia vogelii + *Sesbania sesban*	4.3	2.0
Traditional grass fallow	2.5	1.3
Unfertilized maize	1.7	1.4
SED:	0.5	0.8
F probability	< 0.001	< 0.05

establishment of on-farm "biomass banks" from which the biomass is cut and transferred to crop fields in different parts of the farm. In western Kenya, for example, the use of *Tithonia diversifolia*, *Senna spectabilis*, *S. sesban*, and *Calliandra calothyrsus* planted as farm boundaries, woodlots, and fodder banks has proven to be beneficial as a source of nutrients for improving maize production (Palm, 1995; Palm et al., 2001). A study by Gachengo (1996) found that Tithonia green biomass grown outside a field and transferred into a field was quite effective in supplying N, P, and K to maize, equivalent to the amount of commercial NPK fertilizer recommended. In some cases, maize yields were higher with Tithonia biomass than with commercial mineral fertilizer.

Biomass transfer using fertilizer-tree species is a more sustainable means for maintaining nutrient balances in maize and vegetable-based production systems, as the tree leafy materials are able to supply to the soil N (Kuntashula et al., 2004). Synchrony between nutrient release from tree litter and crop uptake can be achieved with well-timed

TABLE 19.3

Maize Grain Yield (t ha^{-1}) from 2-Year noncoppicing Mixed-Fallow Species Treatments at Msekera, Eastern Zambia

Species	2002	2003
Maize with fertilizer	4.7	4.3
Tephrosia vogelii + *Cajanus cajan*	4.7	2.0
Sesbania sesban + Tephrosia	4.4	1.3
Sesbania sesban + *Cajanus cajan*	4.0	1.8
Tephrosia vogelii alone	3.9	1.6
Sesbania sesban alone	3.4	1.0
Cajanus cajan alone	2.7	0.9
Maize without fertilizer	1.3	0.4
SED	0.9	0.4
F probability	< 0.001	< 0.001

biomass transfer. The management factors that can be manipulated to achieve this are litter quality, rate of litter application, and method and time of litter application (Mafongoya et al., 1998).

Biomass transfer technologies require more labor for managing and incorporating the leafy biomass, however. If used for the production of low-value crops such as maize, the higher maize yield from biomass-transfer technologies may not be enough to compensate for the higher labor cost. Most economic analyses have concluded that it is unprofitable to invest in biomass transfer when labor is scarce and its cost is thus high. However, when prunings are applied to high-value crops like vegetables, the technology becomes profitable (ICRAF, 1997). This practice has been found quite suitable for vegetable production in dambo areas of southern Africa (Kuntashula et al., 2004).

Dambos are shallow, seasonally or permanently waterlogged depressions at or near the head of a natural drainage network, or alternatively they can occur independently of a drainage system. All together, dambos serve approximately 240 million ha in all of sub-Saharan Africa (Andriesse, 1986), of which 16 million ha are in southern Africa. Though dambos are extremely vulnerable to poor agricultural practices, rising population pressure has caused their agricultural use to become increasingly important (Kundhlande et al., 1995). Without applying fertilizers or cattle manure, smallholder farmers cannot produce vegetables successfully in dambos that are degraded due to their continuous cultivation for over 25 years (Raussen et al., 1995). Inorganic fertilizer is not always available to smallholder farmers, and cattle manure is accessible only to those with animals. This calls for alternatives such as biomass transfers for fertilizing vegetables in dambos of southern Africa. Additional results of such evaluations are given in Section 29.3.2.1.

Farmer participatory experiments conducted in 2000–2004 by Kuntashula et al. (2004) have shown that biomass transfer using *Leuceana leucocephala* and *Gliricidia sepium* is tenable for sustaining vegetable production in dambos. In addition to increasing yields of vegetables such as cabbage, rape, onion, tomato, and maize grown after vegetable harvests, biomass transfer has shown potential to increase yields of other high-value crops such as garlic (Table 19.4). Our studies suggest that biomass transfer has greatest potential when (a) the biomass is of high quality and it rapidly releases nutrients, (b) when the opportunity costs of labor are low, (c) when the value of the crop is high, and (d) when the biomass does not have other, valued uses apart from being a reliable source of nutrients.

TABLE 19.4

Selected Vegetable Yields (t ha^{-1}) in Dambos Using Inorganic Fertilizers or Organic Inputs from Manure or Tree Leaf Biomass in Chipata District, Zambia

Treatments	Cabbage Yield (n = 31) (2000)	Green Maize Yield After Onion (t ha^{-1})	Onion Yield (n = 12) (2001)	Green Maize Yield After Cabbage (t ha^{-1})	Garlic Yield (n = 6) (2004)
Manure 10 t + 1/2 rec. fertilizer	66.8	11.6	96.0	11.7	9.1
Recommended fertilizer	57.6	8.4	57.1	10.4	7.2
Gliricidia sepium (12 t)	53.6	12.4	79.8	17.3	- -
Gliricidia sepium (8 t)	43.1	10.9	68.3	14.9	10.3
Leucaena leucocephala − 12 t	32.6	- -	- -	13.0	- -
Nonfertilized	17.0	6.4	28.1	7.8	4.2
SED	5.3	2.06	11.2	3.04	1.2
F probability	< 0.001	< 0.001	< 0.05	< 0.05	< 0.05

- -, treatment not evaluated.

19.2 Mechanisms for Improved Soil Fertility and Health

19.2.1 Biomass Quantity and Quality

The success of maize crop rotations with fertilizer trees depends very much on processes for pruning biomass and on their nutrient yields. Analysis of maize yields across several sites with different fertilizer trees shows that maize yield is most closely correlated with the N content of prunings, with rainfall, and with the quantity of biomass applied. Low and insufficient biomass yields, combined with low quality of prunings in most instances, have contributed to frequent low performance of the technology. The low production of biomass for pruning may result from the use of unsuitable species, poor tree growth due to low soil fertility, soil acidity, moisture stress, or poor management of the species.

Work carried out for many years has shown how organic decomposition and nutrient release are affected by the levels of polyphenol, lignin, and nitrogen content of the organic inputs (Mafongoya et al., 1998). Recently, we have also found that maize yields after fallows with various tree legumes were negatively correlated with the (L + P) to N ratio and positively correlated with recycled biomass. Fallow species with high N, low lignin, and low polyphenols such as Gliricidia and Sesbania gave higher maize yields compared to species such as Flemingia, Calliandra, and Senna. This work has shown that it is not the quantity of polyphenols that is critically important, but rather their quality as measured by their protein-binding capacity (Mafongoya et al., 2000). Legume species for improved fallows can be screened for their suitability based on the above characteristics.

19.2.2 Biological Nitrogen Fixation and N Cycles

The contribution of leguminous trees to crop yield through N_2 fixation is well recognized, although not all legumes fix N_2. Numerous nonleguminous species have N fixed in their roots and root zones through associations with N-fixing bacteria (Chapter 12). Nitrogen fixation in alley cropping systems in the humid and subhumid zones of Africa has been reviewed by Sanginga et al. (1995). There has been little work carried out quantifying N_2 fixation by trees in southern Africa, however. Such analysis has been difficult due to constraints in the methodologies for measuring the N_2 fixed. A series of multi-location trials has been set up to measure the amount of N_2 fixed by different tree genera and provenances using the ^{15}N natural abundance method. The data on percent N derived from atmospheric N_2 fixation (Ndfa) shows high variability among species and provenances of the same species. Greater variation was also recorded for the same species across different locations. So the measurement task is a challenging one.

Sanginga et al. (1990) found that the Ndfa ranged from 37 to 74% for different provenances of *Leucaena leucocephala*. The initial data show a huge potential of trees to fix N_2 and increase N inputs in N-deficient soils. In future analysis we will focus on factors responsible for the variability in N_2-fixation across sites and on how to optimize N_2 fixation under field conditions.

An estimated value of the level of inorganic N in soil before a cropping season begins is an accepted test for assessing prospective soil productivity. Results of studies in Southern Africa show that preseason inorganic N can also be an effective indicator of the N that is plant-available after fallow with different species (Barrios et al., 1997). Studies we conducted at 18 locations in eastern Zambia have indicated that in a tropical soil with a pronounced dry season, total preseason inorganic N (i.e., $NO_3 + NH_4$) is more closely related to maize yield ($R^2 = 0.62$; b = 0.27, se = 0.03) than to preseason NO_3 alone. While large amounts of NH_4 can accumulate during a dry season, it may not be nitrified when the soil is sampled at the beginning of the rainy season. We have concluded that preseason inorganic N is a relatively rapid and simple index that is related fairly well to maize yield on N-deficient soils, and hence it can be used to screen fallow species and management practices.

19.2.3 Deep Capture of Soil Nutrients

The retrieval and cycling of nutrients from soil below the zone exploited by crop roots is referred to as nutrient pumping (Van Noordwijk et al., 1996; also Chapters 20 and 21). Soil nutrients not accessible to annual crops such as maize can be extracted by perennial trees through deep capture. The distributions and density of roots, the demand of plant for nutrients, and the distribution and concentration of plant-extractable nutrients and water will influence deep capture of nutrients by fertilizer trees (Buresh et al., 2004). Deep capture is favored when perennials have a deep rooting system and a high demand for nutrients, when water or nutrient stress occurs in the surface soils, and/or when considerable extractable nutrients or weatherable minerals occur in the subsoil (Buresh and Tian, 1997). These conditions were observed in eastern Zambia where nitrate accumulated in the subsoil during periods of maize growth, and fertilizer trees grown in rotation with maize could then effectively retrieve the nitrate in the subsoil that had been "lost" to maize.

Intercropping rather than rotating fertilizer trees with crops appears to improve the long-term efficiency of nutrient use in deep soils. When perennials such as *G. sepium* are intercropped with maize, they remain always present in the agroecosystem compared with non-coppicing trees such as *S. sesban*. In a mixed fallow, Gliricidia provides a safety-net function to reduce nitrate leaching. In the Sesbania-maize rotation, there is no active perennial legume. Therefore, nitrate leaches into deep soil below the effective rooting depth of maize. Intercropping with fertilizer trees such as Gliricidia may thus be more effective for pumping of soil nutrients than a Sesbania-maize rotation. In base-rich deep soils of Msekera, eastern Zambia, there is potential for subsoil accumulation of highly mobile cations such Ca, Mg, and K, due to the weathering of minerals and leaching of cations that accompany NO_3 leaching in fully fertilized maize crops without any trees present. The introduction of Gliricidia with maize rotation has a great potential for deep capture of Ca and Mg compared to continuously fertilized monoculture maize.

19.2.4 Soil Acidity and Phosphorus

Acidic soils cover approximately 27% of the land in tropical Africa. Acidic soils are characterized by low pH, deficiencies of phosphorus, calcium, and magnesium, and toxic levels of aluminum. This is why finding strategies that offset soil acidity and low P availability is so important. Here, we discuss how agroforestry systems can address these two related constraints. In Chapter 37, there is a more detailed consideration of such a strategy, focused in Western Kenya.

Lime application is the most widely used remedy for high acidity in countries such as in Brazil and U.S.A., but it is financially prohibitive for resource-poor farmers in southern Africa and cannot be considered a viable solution to the problem. Numerous laboratory experiments have recorded increased soil pH, decreased Al saturation, and improved conditions for plant growth as a result of the addition of plant materials to acid soils such as tree prunings, which also supply base cations such as Ca, Mg, and K. The value of tree prunings as a "liming" material for acid soils is related in their cation content (Wong et al., 2000). There is evidence from field experiments (see Wong et al., 1995) that the lateral transfer of alkalinity can be achieved by pruning pure stands of agroforestry trees and applying their pruned biomass to a maize crop.

Several mechanisms contribute to the increase in soil pH through such measures (Wong and Swift, 2003). These processes depend on the ash alkalinity of the organic inputs and on organic anion content. Leguminous materials are particularly useful in this respect because they have high ash alkalinity and offer the benefit of N_2 fixation, while providing cash-limited farmers with inexpensive biological means of liming acid soils without having to buy costly inorganic lime.

In small-scale farming systems in Africa, crop harvesting removes almost all of the P accumulated by cereal crops (Sanchez et al., 1997). In agroforestry systems, root systems may account for as much as 80% of the primary production. Application of plant biomass as green mulch can contribute to P availability, either directly by releasing tissue P during decomposition and mineralization (biological processes) or indirectly by acting on chemical processes that regulate P adsorption-desorption reactions.

Soil organic matter contributes indirectly to raising P in soil solution by complexing certain ions such as Al and Fe that would otherwise constrain P availability. Decomposing organic matter also releases anions that can compete with P for fixation sites, thus reducing P adsorption. In agroforestry development, we have focused on the enhancement of the use-efficiency of soil P, i.e., on increasing the amount of biomass production for a set amount of P, as a more cost-effective means of improving P availability to crops. The more extensive root systems that trees and shrubs have compared to crops increase the exploration of larger soil volumes which results in enhanced P uptake. Recycled tree biomass is an important source of available P (Jama et al., 1997).

Perennial tree species also produce organic anions. However, this production has not been as well studied as that achieved by annual crops (Grierson, 1992). As a result of mycorrhizal "infections," trees can readily produce organic anions that increase P availability through chelation or solubilization mechanisms (Chapter 9). Of these mechanisms, reactions involving metal chelates are most important in tropical acid soils (Gardner et al., 1992). Plant-microbial mechanisms that enhance P bioavailability can be incorporated into tropical agroforestry in the same way that N_2-fixing trees have been integrated in agroforestry systems to increase N availability. However, much more research is needed on factors that control organic anion release from tree roots, their longevity in the soil, and different effects on P mobilization in different soils. Other strategies for managing tropical acid soils and increasing their availability of phosphorus are considered in Chapters 23 and 37.

19.2.5   Soil Physical Properties

The ability of trees and biomass from trees to maintain or improve soil physical properties has been well documented. Alley-cropping, for example, can definitely improve the soil physical conditions on alfisols (Hullugalle and Kang, 1990). Plots alley-cropped with four hedgerow species showed lower soil bulk density, higher porosity, and greater water infiltration rates compared with a no-tree treatment (Mapa and Gunasena, 1995). Tree fallows can also improve soil physical properties due to the addition of large quantities of litter fall, root biomass, root activity, biological activities, and roots leaving macropores in the soil following their decomposition (Rao et al., 1998).

In our studies, we have seen that Sesbania fallow increases the percentage of water-stable aggregates with a diameter >2 mm compared with continuous maize cultivation without fertilizer. After 6 months of cropping, the decrease in water-stable aggregates was highly significant under Sesbania (18%) compared with a traditional grass fallow, which did not lose its aggregate stability. A decrease in aggregate stability was more pronounced under Sesbania followed by maize without fertilizer compared with pigeon pea (*Cajanus cajan*) followed by maize with fertilizer (Chirwa et al., 2004). Under a Sesbania fallow, the improvement in soil structure was evident, as reflected by the results from our time-to-runoff studies. Time-to-runoff after fallow clearing followed this order: traditional grass fallow >Sesbania > fertilized maize (Phiri et al., 2003). After one season of cropping, time-to-runoff decreased in all treatments, except that the traditional grass fallow maintained longer time-to-runoff, reflecting its good maintenance of aggregate stability.

Through rainfall simulation studies, Nyamadzawo et al. (2005) evaluated the effects of improved fallows on runoff, infiltration, and soil and nutrient losses under improved fallows. Tree fallows of Sesbania and Gliricidia mixed with Dolichos increased infiltration rates significantly compared with continuously fertilized maize plots (Nyamadzawo et al., 2005). Tree fallows also significantly reduced soil loss compared to no-tree plots.

That fertilizer trees improve soil physical properties is seen from measured increases in infiltration rates, increased infiltration decay coefficients, and reduced runoff and soil losses. However, these benefits are short-lived and decline rapidly during the first year of cropping where non-coppicing species are used. This is consistent with an increase in soil loss in the second year and a decrease in infiltration rates as well. Mixing a coppicing species like Gliricidia with a herbaceous legume like Dolichos maintains high infiltration rates and reduced soil loss over 2 years of cropping (Mafongoya et al., 2005). In agroforestry as in other agriculture we see repeated advantages of polycropping over use of single species.

19.3 Effects on Soil Biota

Soil biological processes, mediated by roots, flora, and fauna, are an integral part of the functioning of natural and managed fallows (Sanginga et al., 1992; Adejuyigbe et al., 1999). As discussed in the preceeding chapter, this plays a key part in regulating the productivity of ecosystems (see also Sanginga et al., 1992). Among the soil biota essential in soil processes in agroforestry, probably the most important ones are the so-called ecosystem engineers, e.g., termites, earthworms, and some ants, and the litter transformers including millipedes, some beetles, and many other soil-dwelling invertebrates. Sileshi and Mafongoya (2005) compare the population of various soil macro-invertebrates under maize grown in an agroforestry system and monoculture maize. In five separate experiments conducted at Msekera and Kalunga, the number of invertebrate orders per sample and the total macrofauna recorded were higher under maize grown in coppicing fallows than under fully fertilized monoculture maize (Figure 19.1a and b).

Similarly, the population density of total macrofauna (all individuals per square meter) under maize grown in coppicing fallows was higher than those under fully fertilized monoculture maize in all experiments at Msekera. Earthworm, millipede, and centipede populations under maize grown in coppicing fallows were also higher than under monoculture maize. Millipedes were absent from monoculture maize at both Msekera and Kalunga sites during most of the sampling periods. At Msekera, the population density of beetles was also higher under legume fallows compared to monoculture maize. Clearly, what is grown aboveground affects the flora and fauna belowground.

We also noted differences according to fertilizer-tree species used for fallows. Cumulative litter fall, tree leaf biomass, and resprouted biomass under the respective legume species appeared to influence macrofauna populations. Macrofauna diversity (number of orders) was positively associated with total recycled biomass. The litter biomass under the tree species at fallow termination also influenced populations of beetles and earthworms. The tree-leaf biomass incorporated into the soil at fallow

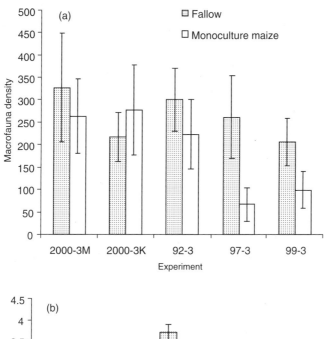

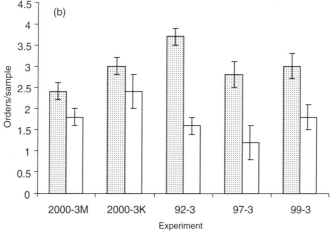

FIGURE 19.1

Macrofauna density (number of all individuals per square meter) and number of orders per sample under improved fallows and fully fertilized monoculture maize in various experiments. 2000-03M = Experiment established in 2000 at Msekera; 2000-3K = Experiment established in 2000 at Kalunga; 92-3 = Experiment established in 1992; 97-3 and 99-3 = experiments established in 1997 and 1999, respectively, at Msekera.

termination was positively correlated with populations of beetles, earthworms, and millipede in the wet season. Among the fallows species, litter transformer populations were higher under *G. sepium*, which produced good quality organic inputs. On the other hand, a higher population of ecosystem engineers was found under trees that produce poor quality organic inputs (Sileshi and Mafongoya, 2005).

The soil under fertilizer trees also harbors plant pathogens and soil insects that can adversely affect the crop and trees in agroforestry. Among the major soil pests of fertilizer trees are plant-parasitic nematodes (Meloidogyne and Pratylenchus spp.) and termites. Root knot nematodes seriously affect the planting of *S. sesban*, pigeon pea (*Cajanus cajan*), and *T. vogelii* in southern Africa (Karachi, 1995; Shirima et al., 2000). In Tanzania, Meloidogyne infections were consistently highest when tobacco was planted after 2 years of *S. sesban* fallow (Shirima et al., 2000). So plants' interactions with the soil systems that support them can have some undesirable effects.

Although termites are generally essential ecosystem engineers in fallows, some are also crop pests. In parts of Kenya, Tanzania, Zambia, Malawi, Zimbabwe, and South Africa, 20–30% of preharvest loss in maize is said to be due to termites (Nkunika, 1994; Munthali et al., 1999; Riekert and Van den Berg, 2003; Van den Berg and Riekert, 2003). Termites are estimated to affect maize production on approximately 80,000 ha in the arid north and northwestern parts of South Africa (Riekert and Van den Berg, 1999).

Few, if any, effective methods exist to control termite species that have subterranean nests such as Microtermes and Odontotermes (Van den Berg and Riekert, 2003). In a study conducted in eastern Zambia, Sileshi and Mafongoya (2003), recorded lower termite damage (% lodged plants) on maize planted after *T. vogelii* + pigeon pea, *S. sesban* + pigeon pea, and pure *S. sesban* compared with maize grown after traditional grass fallow. Monoculture maize grown after traditional grass fallow had approximately 11 and 5 times more termite damage compared to maize grown after *T. vogelii* + pigeon pea and *S. sesban* + pigeon pea, respectively. In another set of experiments, Sileshi et al. (2005) monitored termite damage on maize grown in coppicing fallows. Those studies showed that fully fertilized monoculture maize and maize grown in *S. siamea* and *F. macrophylla* fallows suffered higher termite damage compared to maize grown in *G. sepium* and *L. leucocephala*.

Soil-dwelling insects such as white grubs (*Schizonycha* spp.) and snout beetles (*Diaecoderus* sp.) also affect trees and associated crops. Larvae of *Diaecoderus* spp. develop in the soil, and these later attack maize roots. Emerging adults also attack pigeon pea, *Crotalaria grahamiana*, *Gliricidia sepium*, and *Tephrosia vogelii*. Adult populations building up on these legumes during the fallow phase later infest maize plants (Sileshi and Mafongoya, 2003). In an experiment involving pure fallows and mixtures of these legume species, the density of Diaecoderus beetles was found to be significantly higher in maize planted after *S. sesban* + *C. grahamiana* compared with maize planted after traditional grass fallow. The population of snout beetles was significantly positively correlated with the amount of nitrate and total inorganic nitrogen content of the soil and with cumulative litter fall under fallow species (Sileshi and Mafongoya, 2003).

Other soil biota can play vital roles in controlling plant pathogens and soil-dwelling insect pests. Kenis et al. (2001) and Sileshi et al. (2001) reported some natural enemies of pests that affect fallow species. The entomopathogenic nematode, *Hexamermis* sp., the braconid wasp *Perilitus larvicida*, the carabid beetle *Cyaneodinodes fasciger*, and ants *Tetramorium sericeiventre* and *Pheidole* sp., all live in the soil and are natural enemies of *M. ochroptera* (Kenis et al., 2001; Sileshi et al., 2001). Most of these natural enemies were found to be more abundant in the improved fallows compared to monoculture maize.

Although it is known that soil biota are the major determinants of soil processes and that pest management is an integral part of crop production, studies on soil biota have rarely been undertaken in conjunction with the design of agroforestry practices in southern Africa. The few studies cited above have shown that fertilizer trees can increase the diversity and function of soil biota compared to a continuously cropped, fully fertilized monoculture maize. This subject is discussed further in Chapter 41.

Maintaining active soil invertebrate communities in soils could considerably improve the sustainability of cropping systems through regulation of the soil process at different scales of time and space. To increase the activity of natural enemies and reduce pest problems, fallow management practices that provide habitat, cover, and refuge for natural enemies and reduce build-up of pestiferous species need to be adopted. As experience and knowledge increase, we expect that in the future, routine fallow management practices will be manipulated to meet pest management objectives (Sileshi and Kenis, 2001).

19.4 Sustainability of Fertilizer Tree-Based Land Use Systems

Improved fallows with Sesbania or Tephrosia have been shown to give subsequent maize grain yields of $3-4$ t ha^{-1} without any inorganic fertilizer addition. Palm (1995) showed that organic inputs of various tree legumes applied at 4 t ha^{-1} can supply enough nitrogen for maize grain yields of 4 t ha^{-1}. However, most of these organic inputs could not supply enough phosphorus and potassium to support such maize yields over time.

The question for sustainability is: can improved fallows potentially reduce soil stocks of P and K over time while maintaining a positive N balance? To answer this question we have conducted nutrient balance studies on improved fallow trials at Msekera Research Station. These plots were maintained under fallow-crop rotations for 8 years. The studies on nutrient balances addressed the following questions: (1) Can nutrient balances be used as land quality indicators? (2) Can they be used to assess soil fertility status, productivity, and sustainability? (3) Can they be used as a policy instrument for determining the types of fertilizers to be imported or distributed to farmers?

The nutrient balance studies considered the nutrients added through leaves and litter fall, which were incorporated after fallows as inputs. The nutrients in maize grain harvested, in maize stover removed, and in fuelwood taken away at end of the fallow period were then considered as nutrient exports. For all the land use systems, there was a positive N balance in the 2 years of cropping after the fallow. Fertilized maize had the highest N balance due to the annual application of 112 kg N ha^{-1} for the past 10 years. Unfertilized maize had lower balances due to low maize grain and stover yields over time. The tree-based fallows had a positive N balance due to BNF and deep capture of N from depth. These results are consistent with those of Palm (1995) showing that organic inputs can supply enough N to support maize grain yields of $3-4$ t ha^{-1}.

However, we note that in the second year of cropping, the N balance became very small. This is consistent with our earlier results which showed a decline of maize yields in the second year of cropping after 2-year fallow. The large amount of N supplied by fallow species could be lost through leaching beyond the rooting depth of maize. Our leaching studies have shown substantial inorganic N at some depths under maize after improved fallows. This implies that if cropping goes beyond 3 years after fallowing, there will be a negative N balance. Thus, the recommendation of 2 years of fallow followed by 2 years of cropping is supported by both N balance analyses and maize grain yield trends.

Most of the land use systems showed a positive P balance. This can be attributed to low off-take of P in maize grain yield and stover. However, it should be noted that this site had a high phosphorus status already. The trees could have increased P availability through the secretion of organic acids and increased mycorrhizal populations in the soil. These issues are under investigation at our site. In general, we have observed positive P balances over 8 years. However, this result needs to be tested on-farm where the soils are inherently low in P.

Most land-use systems showed a negative balance for K. For tree-based systems, Sesbania showed a higher negative K balance compared to pigeon pea. This is attributed to the higher fuelwood yield of Sesbania with subsequent higher export of K compared to pigeon pea. The higher negative K balance for fully fertilized maize is due to higher maize and stover yields which export a lot of potassium. This implies that the K stocks in the soil were very high and that K mining has not reached a point where it negatively affects maize productivity. However, in sites with low stocks of K in the soil, maize productivity may become adversely affected.

Overall, the tree-based fallows maintained positive N and P balances. However, on low-P-status soils, a negative P balance would be expected. There was a negative K balance with most land-use systems. It can be hypothesized that as improved fallows are scaled up on depleted soils on farmers' fields, the K and P balances would be, or become, negative. This has implications for fertilizer policy. In Zambia, a mixture called "compound D" containing N, P, and K is the currently imported basal fertilizer for maize. If farmers adopt improved fallows on a wider scale, these will meet their N requirement for maize. Where there is K and P deficit, farmers may not need to buy "compound D" because N is adequately supplied by fallows since they need only K and P as nutrients to supplement their N from the fallow. This may require a shift in government policy on the type of fertilizer imported. There is an urgent need to conduct nutrient budget analyses at a landscape level on farmers' fields to test the validity of our findings.

19.5 Discussion

This chapter has described the progress that has been made during the past decade in research efforts to understand the mechanisms involved in how fertilizer trees work. A large amount of knowledge has been generated. However, some aspects of improved fallows have received little evaluation, and these will be highlighted here.

Work on improved fallows has focused on just a few genera of fertilizer trees such as Sesbania, Tephrosia, Crotalaria, and Gliricidia. Further work is needed to identify more species for improved fallows. Given a large number of potential genera and species of legumes, the selection process could be accelerated by creating a database containing information on fallow performance in relation to environmental factors such as rainfall, soil type and chemistry, and incidence of pests and diseases. Our recent trials across sites have shown a great potential for *Tephrosia candida* as an alternative species to Sesbania and *T. vogelii*, and equally for *Leucaena collinsii* and *Acacia angustissima* as alternative coppicing fallow species to *G. sepium*.

The biophysical limits of improved fallows need to be assessed and extended to facilitate scaling up with minimum research efforts. Simulation modeling, both as a tool for research and for extrapolation, has potential for integrating research results, identifying key components or processes that merit greater research attention, and also ecozones where appropriate fallow species and management techniques have a good chance of success.

Agroforestry land-use systems have been reported to have large potentials to sequester soil carbon. However, there are few studies, if any, in southern Africa that have measured C sequestration in improved fallows. The relationship between increased soil aggregation and carbon storage also needs further research.

As noted earlier, the interaction of pests with soil fertility is gaining attention due to wider interest in scaling-up of improved fallows. So far, most research efforts have concentrated on insect pests and nematodes. Equally important for farmers, however, are plant diseases and weeds. Little effort has been invested in these issues. With scaling-up across many ecozones, the incidence of new pests and diseases is likely to increase. This means there will be a need to monitor pests and diseases with farmers to determine which economic pests need to be dealt with in a concerted research program. Such work is now beginning in southern Africa.

Many of the species currently used in improved fallows are prolific seed producers. If not managed well, these species can become invasive weeds and become a menace to natural ecosystems such as the miombo woodlands. To date there has been no concerted research effort to determine the invasiveness of introduced fertilizer-tree species. There is an urgent need to use current models to predict the potential of new species to become invasive, while at the same time studying the reproductive biology and design management practices that will mitigate potential invasions of natural ecosystems.

Research during the last decade has established the main mechanisms explaining how improved fallows work. Despite significant progress in biophysical research in improved fallows in southern Africa, the application of that scientific knowledge by small-scale farmers is still minimal. The main challenge now is to increase the generation of viable and acceptable fallow options that can make improved fallows more productive so that they markedly increase the income and food security of small-scale farmers. Future research issues on biomass transfer will involve the residual effect of low- and high-quality biomass, combinations of organic and inorganic sources of nutrients, the effects of biomass banks on nutrient mining, agronomic research of biomass transfer possible with different leguminous species, and economic analysis of the systems.

Acknowledgments

The authors are grateful to the Swedish International Development Agency (SIDA) and Canadian International Development Agency (CIDA) for their continued financial support for agroforestry research for over 10 years.

References

Adejuyigbe, C.O., Tian, G., and Adeoye, G.O., Soil microarthropod populations under natural and planted fallows in southern Nigeria, *Agroforest. Syst.*, **47**, 263–272 (1999).

Andriesse, W., *Area and Distribution: The Wetlands and Rice in Sub-Saharan Africa, Proceedings of a Workshop, 4–8 November, 1985.* International Institute of Tropical Agriculture, Ibadan, Nigeria, 15–30 (1986).

Barrios, E. et al., Light fraction soil organic matter and available nitrogen following trees and maize, *Soil Sci. Soc. Am. J.*, **61**, 826–831 (1997).

Buresh, R.J. and Tian, G., Soil improvement by trees in sub-Saharan Africa, *Agroforest. Syst.*, **38**, 51–76 (1997).

Buresh, R.J. et al., Opportunities for capture of deep soil nutrients, In: *Below-ground Interactions in Tropical Agroecosystems*, van Noordwijk, M., Cadisch, G., and Ong, C.K., Eds., CABI Publishing, Wallingford, 109–125 (2004).

Cadisch, G. et al., Resource acquisition of mixed species fallows: Competition or complementarity?, In: *Balanced Nutrient Management Systems for the Moist Savanna and Humid Forest Zones of Africa*, Vanlauwe, B., Sanginga, N., and Merckx, R., Eds., Kluwer Academic Publishers, Dordrecht, The Netherlands, 143–154 (2002).

Chirwa, T.S., Mafongoya, P.L., and Chintu, R., Mixed planted-fallows using coppicing and non-coppicing tree species for degraded Acrisols in eastern Zambia, *Agroforest. Syst.*, **59**, 243–251 (2003).

Chirwa, T.S. et al., Changes in soil properties and their effects on maize productivity following *Sesbania sesban* and *Cajanus cajan* improved fallow systems in eastern Zambia, *Biol. Fert. Soils*, **40**, 28–35 (2004).

Drechsel, P., Steiner, K.G., and Hagedorn, F., A review on the potential of improved fallows and green manure in Rwanda, *Agroforest. Syst.*, **33**, 109–136 (1996).

Gachengo, C.N., *Phosphorus Release and Availability on Addition of Organic Materials to Phosphorus Fixing Soils.* M.Phil. Thesis, Moi University, Kenya (1996).

Gardner, W.K., Parbery, D.G., and Barber, D.A., The acquisition pf phosphorus by Lupinus albus L.I. Some characteristics of the soil-root interface, *Plant Soil*, **68**, 19–32 (1992).

Grierson, P.F., Organic acids in the rhizosphere of *Banksia integrifolia* L.F, *Plant Soil*, **144**, 259–265 (1992).

Gerner, H. and Harris, G., The use and supply of fertilizers in sub-Saharan Africa, In: *The Role of Plant Nutrients for Sustainable Food Crop Production in Sub-Saharan Africa*, Reuler, H.V. and Prins, W., Eds., Vereniging van Kunstmest Producenten, Leidschendam, The Netherlands (1993).

Hullugalle, N.R. and Kang, B.T., Effect of hedgerow species in alley cropping systems on surface soil physical properties of an Oxic paleustalf in southwestern Nigeria, *J. Agric. Sci.*, **114**, 301–307 (1990).

ICRAF., *Annual Report 1996*. International Centre for Research in Agroforestry, Nairobi, Kenya (1997).

Jama, B.A., Swinkels, R.A., and Buresh, R.J., Agronomic and economic evaluation of organic and inorganic sources of phosphorus in western Kenya, *Agron. J.*, **89**, 597–604 (1997).

Karachi, M., Sesbania species as potential hosts to root-knot nematode (*Meloidogyne javanica*) in Tanzania, *Agroforest. Syst.*, **32**, 119–125 (1995).

Kenis, M., Sileshi, G., and Bridge, J., Parasitism of the leaf beetle *Mesoplatys ochroptera* Stal (Coleoptera: Chrysomelidae) in eastern Zambia, *Biocontrol Sci. Technol.*, **11**, 611–622 (2001).

Kundhlande, G., Govereh, J., and Muchena, O., Socioeconomic constraints to increased utilisation of dambos in selected communal areas, In: *Dambo Farming in Zimbabwe: Water Management, Cropping and Soil Potentials for Smallholder Farming in the Wetlands*, Owen, R., Verbeek, K., Jackson, J., and Steenhuis, T., Eds., CIIFAD/University of Zimbabwe, Ithaca, NY/Harare, 87–96 (1995).

Kuntashula, E. et al., Potential of biomass transfer technologies in sustaining vegetable production in the wetlands (dambos) of eastern Zambia, *Exp. Agric.*, **40**, 37–51 (2004).

Kwesiga, F. and Coe, R., Potential of short-rotation Sesbania fallows in eastern Zambia, *Forest Ecol. Manag.*, **64**, 161–170 (1994).

Mafongoya, P.L., Barak, P., and Reed, J.D., Carbon, nitrogen and phosphorus mineralization from multipurpose tree leaves and manure from goats fed these leaves, *Biol. Fert. Soils*, **30**, 298–305 (2000).

Mafongoya, P.L., Giller, K.E., and Palm, C.A., Decomposition and nitrogen release patterns of tree prunings and litter, *Agroforest. Syst.*, **38**, 77–97 (1998).

Mafongoya, P.L. and Nair, P.K.R., Multipurpose tree prunings as a source of nitrogen to maize under semiarid conditions in Zimbabwe: Nitrogen recovery rates in relation to pruning quality and method of application, *Agroforest. Syst.*, **35**, 47 (1997).

Mafongoya, P.L. et al., The effects of mixed planted fallows of tree species and herbaceous legumes on soil properties and maize yields in eastern Zambia, *Exp. Agric.* (2005), in press.

Mapa, R.B. and Gunasena, H.P.M., Effect of alley cropping on soil aggregate stability of a tropical Alfisol, *Agroforest. Syst.*, **32**, 237–245 (1995).

Munthali, D.C. et al., Termite distribution and damage to crops on smallholder farms in southern Malawi, *Insect Sci. Appl.*, **19**, 43–49 (1999).

Nkunika, P.O.Y., Control of termites in Zambia: Practical realities, *Insect Sci. Appl.*, **15**, 241–245 (1994).

Nyamadzawo, G. et al., Soil and nutrient losses from two contrasting soils under improved fallows in eastern Zambia, *Adv. Soil Sci.* (2005).

Palm, C.A., Contribution of agroforestry trees to nutrient requirements of intercropped plants, *Agroforest. Syst.*, **30**, 105–124 (1995).

Palm, C.A. et al., Organic inputs for soil fertility management: Some rules and tools, *Agric. Ecosyst. Environ.*, **83**, 27–42 (2001).

Phiri, E., Verplancke, H., and Mafongoya, P.L., Water balance and maize yield following improved Sesbania fallow in eastern Zambia, *Agroforest. Syst.*, **59**, 197–205 (2003).

Raussen, T., Daka, A.E., and Bangwe, L., *Dambos in Eastern Province: Their Agroecology and Use*, Department of Agriculture, Chipata, Zambia (1995).

Rao, M.R., Nair, P.K.R., and Ong, K., Biophysical interactions in tropical agroforestry systems, *Agroforest. Syst.*, **38**, 3–49 (1998).

Riekert, H.F. and Van den Berg, J., *Final Report: Research on Lodging of Maize in North West Province*. Unpublished research report, ARC-Grain Crops Institute, Potchefstroom, South Africa (1999).

Riekert, H.F. and Van den Berg, J., Evaluation of chemical control measures for termites in maize, *S. Afr. J. Plant Soil*, **20**, 1–5 (2003).

Sanchez, P.A., Tropical soil fertility research, towards the second paradigm. *Transactions, 15th World Congress of Soil Science*, Vol. 1, Acapulco, Mexico, 65–88 (1994).

Sanchez, P.A. et al., Soils fertility replenishment in Africa: An investment in natural resource capital, In: *Replenishing Soil Fertility in Africa*, Buresh, R.J., Sanchez, P.A., and Calhoun, F., Eds., Soil Science Society of America Special Publication 51. SSSA/ASA, Madison, WI, 1–46 (1997).

Sanginga, N., Mulongoy, K., and Swift, M.J., Contribution of soil organisms to the sustainability and productivity of cropping systems in the tropics, *Agric. Ecosyst. Environ.*, **41**, 135–152 (1992).

Sanginga, N., Vanlauwe, B., and Danso, S.K.A., Management of biological N_2-fixation in alley cropping systems: Estimation and contribution to N balance, *Plant Soil*, **174**, 119–141 (1995).

Sanginga, N. et al., Effect of successive cuttings on uptake and partitioning of nitrogen 15 among plant parts of *Leucaena leucocephala*, *Biol. Fert. Soils*, **9**, 37–42 (1990).

Sileshi, G. and Kenis, M., Survival, longevity and fecundity of over-wintered Mesoplatys ochroptera Stål (Co.: Chrysomelidae) defoliating *Sesbania sesban* (Leguminosae) and implications for its management in southern Africa, *Agric. Forest Entomol.*, **3**, 175–181 (2001).

Sileshi, G. and Mafongoya, P.L., Incidence of Mesoplatys ochroptera Stål (Coleoptera: Chrysomelidae) on *Sesbania sesban* in pure and mixed species fallows in eastern Zambia, *Agroforest. Syst.*, **56**, 225–231 (2002).

Sileshi, G. and Mafongoya, P.L., Effect of rotational fallows on abundance of soil insects and weeds in maize crops in eastern Zambia, *Appl. Soil Ecol.*, **23**, 211–222 (2003).

Sileshi, G. and Mafongoya, P.L., Variation in macrofauna communities and functional groups under contrasting land-use systems in eastern Zambia, *Appl. Soil Ecol.*, **31**, in press (2005).

Sileshi, G. et al., Predators of Mesoplatys ochroptera Stal in Sesbania-planted fallows in eastern Zambia, *BioControl*, **46**, 289–310 (2001).

Sileshi, G. et al., Termite damage to maize grown in agroforestry systems, traditional fallows and monoculture on nitrogen-limited soils in eastern Zambia, *Agric. Forest Entomol.*, **7** (2005) 61–69.

Shirima, D.S. et al., Effect of natural and Sesbania fallows and crop rotation on the incidence of root-knot nematodes and tobacco production in Tabora, Tanzania, *Int. J. Nematol.*, **10**, 49–54 (2000).

Snapp, S.S., Mafongoya, P.I. and Waddington, S., Organis matter technologies for integrated nutrient management in smallholder cropping systems in Southern Africa., *Agric., Ecosyst., Envir.,,* **71**, 183–200 (1998).

Van den Berg, J. and Riekert, H.F., Effect of planting and harvesting dates on fungus-growing termite infestation in maize, *Suid-Afrikaanse Tydskrif Plant Grond*, **20**, 76–80 (2003).

Van Noordwijk, M. et al., Root distribution of trees and crops: Competition and/or complementarity, In: *Tree-Crop Interactions: A Physiological Approach*, Ong, C.K. and Huxley, P., Eds., CAB International, Wallingford, UK, 319–364 (1996).

Wong, M.T.F. and Swift, R.S., Role of organic matter in alleviating soil acidity in farming systems, In: *Handbook of Soil Acidity*, Rangel, Z., Ed., Marcel Dekker, New York, 337–358 (2003).

Wong, M.T.F. et al., Initial responses of maize and beans to decreased concentrations of monomeric inorganic aluminium with application of manure or tree prunings to an Oxisol in Burundi, *Plant Soil*, **171**, 275–282 (1995).

Wong, M.T.F. et al., Measurement of the acid neutralizing capacity of agroforestry tree prunings added to tropical soils, *J. Agric. Sci.*, **134**, 269–276 (2000).

20

Biological Soil Fertility Management for Tree-Crop Agroforestry

Götz Schroth[1] and Ulrike Krauss[2]

[1]Conservation International, Washington, DC, USA
[2]CAB International, Turrialba, Costa Rica

CONTENTS

Tree crop-based agroforestry systems, which function as multistrata systems, are an ecologically and economically important group of land-use systems in the humid and subhumid tropics. They are also found in dry climates in places with high water availability; however, this aspect is not focused on here. Multistrata systems are widespread in lowland and mountainous areas, often surrounding homesteads and thus referred to as homegardens or forming a transitional zone between cultivated land into forests (Murniati et al., 2001; Schroth et al., 2004a).

By definition, multistrata agroforestry systems are composed of several strata of trees and tree crops. While the simplest systems consist of only two strata — a lower stratum of tree crops such as coffee (Coffea spp.), cocoa (Theobroma cacao), or tea (Camellia sinensis), and an upper, equally monospecific stratum of shade trees — the most complex systems approach the structural complexity of natural forest and may harbor more than a hundred plant species and varieties, being "agroforests," as defined by Michon and de Foresta (1999).

Groups of plant species typically found in multistrata systems include: shade-tolerant understory tree crops such as coffee, cocoa, and tea; overstory tree crops such as, rubber trees (Hevea brasiliensis), large fruit trees such as Brazil nut (Bertholletia excelsa) and durian (Durio zibethinus), and overstory palms; timber trees, often in a dominant position, and often remnants of previous forest; smaller "service" trees including leguminous shade trees; and midstory species such as citrus (Citrus sp.) and avocado trees (Persea americana), bananas (Musa sp.) and smaller palms. Especially in younger or more open systems, there may also be annual food crops in the understory. Moreover, there may be substantial amounts of spontaneous vegetation in all strata from herbs to emergent trees, depending on the history of the system (e.g., whether it was established on a cleared field or under-planted into thinned forest) and the intensity of management, both present and past.

20.1   Variation in Multistrata Agroforestry Systems

Multistrata systems can be broadly classified according to their diversity of products and the degree of domestication of the system. An example for a fully domesticated system with a single dominant product would be an intensively-managed coffee plantation with a monospecific, planted shade canopy of "service trees," i.e., trees whose principal purpose is to create a suitable environment for the coffee. The service trees in such systems are usually legumes such as Inga spp., *Gliricidia sepium*, and Erythrina spp. These are capable of fixing atmospheric nitrogen and are tolerant of frequent, intensive lopping, although some coffee farms in Costa Rica have recently adopted Eucalyptus spp., chiefly because these trees provide suitable shade without the need for regular pruning, therefore economizing on labor (Schaller et al., 2003b). These systems may receive high levels of mineral fertilizer, herbicides, and pesticides.

Other multistrata systems combine a single dominant product with a much lower degree of domestication of the system. For example, some Amazonian rubber agroforests are essentially "weedy plantations" in the sense that seeds of a single tree-crop species (rubber), plus eventually a few fruit trees, have been sown into slash-and-burn fields, which through extensive management, tolerance of useful forest regeneration, and periodic abandonment have developed into secondary forests enriched with rubber trees (Schroth et al., 2003a). In several regions, coffee, cocoa, and tea systems were traditionally established by underplanting thinned primary or secondary forest; but over the past decades these systems have often been intensified and simplified through substitution of forest remnant trees with planted tree shade (e.g., Johns, 1999).

While these systems are managed, more or less intensively, for a single dominant product (and eventually a number of secondary products such as fuelwood), in other systems management aims at balancing the needs of a number of different tree-crop species that are planted together to make better use of land, labor, and other inputs. Homegardens, widespread in all tropical regions, are a case in point. Other examples include the fruit tree agroforests in Southeast Asia that may contain commercial fruit species as well as coffee, tea, and food plants for home consumption.

As long-term rotational or (semi)permanent, tree-based land-use systems, multistrata systems, especially the more complex ones, help to retain forest functions within agricultural landscapes, including retention of carbon stocks and substantial levels of biodiversity (Schroth et al., 2004a). They offer conservative yet productive land-use options for sensitive landscape positions, such as river banks and steep slopes where annual cropping would lead to soil loss; they reduce the use of fire in the landscape and provide a number of products that might otherwise need to be extracted from natural forests (Murniati et al., 2001).

Economically, multistrata systems that are based on market crops, such as coffee or cocoa, offer income opportunities for farmers, but they expose rural livelihoods to fluctuations of commodity prices and consequent availability of labor. Systems depending on a single commodity are particularly susceptible to such risks. For example, the planted rubber agroforests of the Tapajós region in the Brazilian Amazon have a history of extensive management and periodic abandonment at times of low rubber prices, culminating in a widespread abandonment of the practice in the early to mid-1990s when the internal rubber market disintegrated in the region (Soares, 2003). The extensive management of the groves, a logical response to the unreliability of the national rubber market, sustained their forest character and biodiversity but not the livelihoods of their owners, who abandoned (and often converted) them for slash-and-burn cultivation of cassava (*Manihot esculenta*) and who are only now, owing to more favorable prices, slowly returning to their traditional rubber production (Schroth et al., 2003a). There are similar stories of systems based on cocoa or coffee, especially where market and environmental (especially disease) shocks coincided. Diversified systems that produce a range of products are less sensitive to shocks and a more reliable basis for farmers' livelihoods.

20.2 Soil Fertility Management in Tree-Crop Agroforestry

Despite the economic importance of tree crops in tropical agriculture and commentaries on their ability to maintain and regenerate soil fertility (Sanchez et al., 1985; Schroth et al., 2001a), much less information is available on the biological mechanisms of soil fertility for multistrata systems than for annual crop-based systems, with most research being focused on mineral availability and balances. It is becoming clear, however, that despite many very old and apparently sustainable multistrata systems, such as homegardens (Kumar and Nair, 2004), tree crop-based systems are not safe from soil degradation. The productivity of their soil systems can progressively be undermined, especially over a number of production cycles. This is especially true where, during peaks in commodity prices, tree crops were established on marginally suitable soils, e.g., for cocoa (Ayanlaja, 1987).

Owing to their higher biomass and litter inputs and stronger root systems, tree crop-based systems are better able than some other production systems to regenerate soil conditions, such as soil organic matter levels and soil structure, of land that was previously under annual crops or pasture (see Chapter 21). This is less true for mineral nutrient levels, as evident from the low nutrient levels even in black earth soils under some Amazonian rubber agroforests (Schroth et al., 2004b). However, when tree-crop systems are established on previously forested land, soil characteristics such as organic matter levels and soil structure often decline over time (see review in Schroth et al., 2001a).

Whether a new equilibrium is reached that is still adequate for sustained production depends, among other factors, on soil and climatic conditions, crop species, management practices, and levels of external inputs (the latter three obviously influenced by economic factors). The alternative is reduced soil fertility and crop production, leading to a downward spiral that ends with the abandonment or conversion of the land into less demanding uses (often pasture). The latter scenario is particularly likely on sandy soils where loss of organic matter rapidly leads to degradation of soil structure, which in turn feeds back into diminished plant growth via reduced root development and lessened soil faunal and microbial activity.

Gradual fertility loss under tree crops is not always easy to recognize, especially as it affects soil organic matter. For example, in an experiment conducted over 15 years on a Ferralsol in the central Amazon which was previously under forest cover, it was found that African oilpalm (*Elaeis guineensis*) did not respond to nitrogen fertilizer amendments despite reasonably high production levels (Schroth et al., 2000a). This could have been interpreted as a comparative advantage of the site for oilpalm production; however,

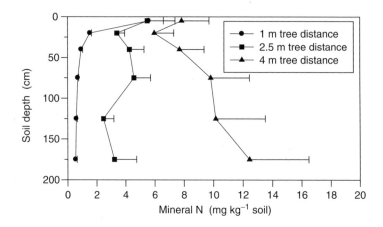

FIGURE 20.1

Nitrate concentrations in the soil of an oil palm (*Elaeis guineensis*) plantation in the central Amazon that has never been fertilized with nitrogen, measured at 1, 2.5 and 4 m distances from the oil palm trees. The nitrate accumulation in the subsoil at larger tree distances shows the leaching of nitrate derived from soil organic matter loss. *Source:* Schroth et al., *Soil Use Manage.*, **16**, 222–229 (2000a). With permission.

another study suggested that primary forest growth in the area responded positively to higher nitrogen levels in this soil (Laurance et al., 1999). The apparent contradiction was resolved by showing that the evident nitrogen sufficiency of the palms was linked to progressive soil organic matter loss (with concomitant nitrogen release) in the topsoil; surplus nitrogen was being leached as nitrate into the subsoil between the palms (Figure 20.1). The comparative advantage for the area was being lost over time and this would presumably make future rotations less profitable.

From an agronomic viewpoint, the objective of biological soil fertility management in tree-crop agroforestry is essentially to create a favorable environment for the acquisition of soil water and nutrients by the crop plants through a combination of favorable soil structure, high nutrient availability, thorough exploration of the soil by root systems and their mycorrhizas, and absence of disease. The principal biological agents that bring about these conditions, and that are managed either directly or indirectly by farmers, are the plants themselves with their root systems and litter, soil organisms (soil fauna and microbes, including antagonists of soil pests and pathogens), and the supply of soil organic matter. In the following discussion, the principal management interventions will be briefly reviewed, with a focus on recent results from Amazonian tree-crop agroforestry systems.

20.3   Root Processes and Their Management

Root systems and their mycorrhizas are the interface through which plants explore the resources (water, nutrients) in the accessible soil. They are also means through which the vegetation influences the soil beneath it by releasing carbon sources, which feed soil biota and increase soil organic matter, and by building and stabilizing soil structure. A study in West Africa showed a linear increase of nitrogen mineralization in the soil with increasing root mass of different legume tree species, although the relationship did not hold for nonleguminous tree species (Schroth et al., 2003b).

While strong *seasonal dynamics* are an important feature of the root systems of annual crops, requiring the close *synchronization* of supply and demand of soil water and nutrients through planting date, timed application of fertilizer, irrigation water, and weed control, a major characteristic of the root systems of tree-crop systems, even mature ones, is their pronounced *spatial variability,* or patchiness. This makes the *synlocation* of nutrient and water supply with plants' demands an equally important concept in managing tree-crop systems than their synchronization. The concept of synlocation not only means it is necessary to apply fertilizer or manure in zones of high root activity, but also to space and arrange trees so as to optimize the exploration of the soil by the root systems of trees (and associated crops or cover crops) as they develop over time. Figure 20.1 illustrates a situation of incomplete soil exploration in a mature monoculture plantation, where nutrients and water were unproductively lost, suggesting that additional crops could have been grown between the palms without affecting their resource use.

The discontinuous root distribution of the oilpalm monoculture in Figure 20.1 contrasts with that in a coffee plantation with *Eucalyptus deglupta* shade trees in Figure 20.2. Although the spacing between the coffee rows was only 2 m, the soil was not fully permeated by the coffee roots. Instead, these were concentrated in the proximity of the coffee rows, pushing the tree roots into the central part of the interrow spaces. Despite very fast growth of the eucalypts, no competitive effects on the coffee were observed at this high-fertility site. The application of fertilizer just along the coffee rows, an example for synlocation, had previously been questioned by scientists, but this study confirmed the

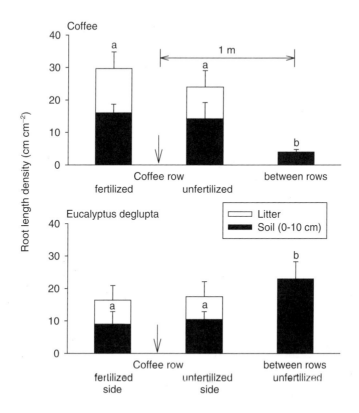

FIGURE 20.2
Root density of coffee (*Coffea arabica*) and shade trees (*Eucalyptus deglupta*) across the coffee rows in a plantation in Juan Viñas, Costa Rica, showing division of the soil space between the two species (similar letters indicate that mean values for these positions are not significantly different). *Source:* Schaller et al., *For. Ecol. Manage.,* **175,** 205–215 (2003b). With permission.

validity of this practice (Schaller et al., 2003b). While the root distribution of the individual species in this system was patchy, the patches were contiguous and gave practically the same total root density in the topsoil in all positions.

This study revealed a process of self-organization where competition between the root systems of two species led to an equilibrium situation with homogeneous root density throughout the topsoil, without any specific management intervention. Root overlap (and thus interspecific competition) was reduced, and the exploration of soil parcels where rooting densities were low was increased. In this system this exploration included the interrow spaces, while in other multispecies systems the same mechanism has been shown to stimulate exploration of the subsoil (Schroth, 1999).

Patchiness, the self-organization of competing tree root systems, and the lower sensitivity of tree crops to root competition compared with annual crops, which results from the longer period of water and nutrient uptake of tree crops over the year and their larger root volume, help in understanding why tropical farmers often plant different tree crops at surprisingly close spacing, giving the appearance of true "agroforests." While the often wider spacing of plants in researcher-managed multistrata plots may give higher yields per plant, the exploration of the soil by the patchy tree root systems risks being incomplete so that resources are wasted (Schroth et al., 1999) or taken up by weeds. However, it should be noted, not all tree root systems respond in the same flexible way to the presence of other root systems (Schroth, 1999). Such differences could be an underlying factor for classifications of tree species according to their suitability for multistrata associations in local knowledge systems (Joshi et al., 2004).

20.4   Managing Soil Microbes and Fauna in Tree-Crop Agroforestry

Root systems represent the "demand side" for soil resources, while soil microbes and fauna contribute to the "supply side" by influencing the decomposition of litter and other organic materials in the soil with associated release of nutrients and both the degradation and stabilization of soil organic matter. Especially fungi and larger soil fauna, the "ecosystem engineers" discussed in Chapter 11, also play important roles in creating and stabilizing soil structure, complementing and interacting with the activities of roots, although these interactions have been rarely studied. The role of these organisms in the control of soil pests and diseases in multistrata agroforestry systems is discussed further below.

Much like root distribution and processes, the distribution and activity of microbes and fauna in the soil and litter of multistrata systems tend to be strongly patchy, and the analysis of these patterns provides clues about design and management factors that influence biological processes on and in the soil. The mineralization of soil nitrogen, one of the soil microbial processes that most directly influences plant growth, is a case in point. At different sites in the central Amazon, microbial nitrogen mineralization in the topsoil of tree-crop systems was found to be substantially higher in the interspaces between trees, which were covered with ground vegetation, than in the regularly weeded soil close to the tree crops. Consequently, the vegetation-covered spaces had higher soil moisture and lower bulk density, which in concert with a larger pool of mobile nitrogen in the soil organic matter led to significantly higher nitrogen mineralization rates than in uncovered soil (Schroth et al., 2000a; Schroth et al., 2001b).

These differences demonstrate the benefits of a management strategy that maintains permanent soil cover and thereby high microbial activity and rapid nutrient turnover in the soil. In contrast, clean-weeding, even if restricted to circles around trees for localized fertilizer application, may result in markedly reduced nutrient mobilization in the soil, which may then have to be compensated for through mineral fertilizer. This means that fertilization may become, to some extent, self-perpetuating (Schroth et al., 2000b).

To create soil cover in plantations, agronomists often recommend leguminous cover crops, which were also used in this experiment. However, this advice has seldom been taken up by smallholder farmers who find it difficult to invest labor in an unproductive crop. Instead, farmers usually establish trees and tree crops in fields of annual or semiperennial food crops. These crops cover the soil while reducing the opportunity cost of having planted tree crops not yet in production. Research has shown that this practice of interplanting food crops may lead to similar or even better tree growth than the use of cover crops (Schroth et al., 2001a). Later in the development of the plantation, when shade prevents successful growth of food crops, aggressive ground-cover species such as grasses can be weeded out to select for a soil cover of unaggressive, broadleaved "noble weeds" (Baker, 2001; Pohlan, 2002).

Even more than microbial activity in the soil, the fauna in soil and litter of multistrata plots responds to small-scale variations in microclimate and litter quantity and quality, which are created by plant cover and management practices, with the potential to enhance litter decomposition, nutrient release, and soil structure (Lavelle et al., 2003). The influence that soil cover (and thus microclimate) can exert on soil and litter fauna was highlighted by a baiting experiment for wood-feeding termites in a multistrata system and monoculture of peach palm (*Bactris gasipaes*) in the central Amazon. After 5 months in the field, the percentage of wood-stick baits (buried in the topsoil), which had been infested by termites, decreased significantly in this order: soil under cover crop (*Pueraria phaseoloides*) in the multistrata plots (45%) > uncovered soil in multistrata plots at about 2 m distance from the previous location (18%) > peach palm monoculture where all soil was uncovered because of the intensive shade (2.5%) (Hanne, 2001).

Another study at this site showed that several litter-dwelling invertebrate groups formed a diversified landscape on a plot with four local tree crops and ground vegetation of legumes and grasses; the area had distinct patches of more or less suitable habitat depending on the group of fauna (Vohland and Schroth, 1999). Several fauna groups including earthworms and millipedes showed a preference for the fleshy litter of peach palm and the soft, nutrient-rich litter of the legume cover crop and annatto (*Bixa orellana*) trees, while avoiding the hard, recalcitrant litter of Brazil nut and cupuaçu (*Theobroma grandiflorum*) trees. Certain other species preferred the litter of Brazil nut trees (caterpillars) or grasses (bugs and thrips).

Overall, faunal density and biomass were strongly determined by the quantity of litter produced by the different plant species. From the data it appears that a minimum quantity of 3 t ha^{-1} of litter, but preferably twice that amount, is necessary to maintain an active and diverse litter fauna in an agroforestry plot (Figure 20.3). Based partly on these data, Lavelle et al. (2003) have suggested that, for biological soil fertility management, agroforestry practices should ensure a supply of ~2 t ha^{-1} yr^{-1} of labile organic matter from litter for digestive assimilation by the soil fauna, while maintaining 3–6 t ha^{-1} of litter on the soil surface. Research is needed to determine the quantity and quality mix of litter that will achieve these values at a range of sites.

The patchiness of faunal distribution in tree-crop agroforestry systems provides further arguments for high planting densities, and for intimate mixing of species with less favorable litter characteristics with species that produce high litter quantity and quality, so that larger patches with hostile conditions for microflora and fauna, and the soil biological processes they drive, do not develop.

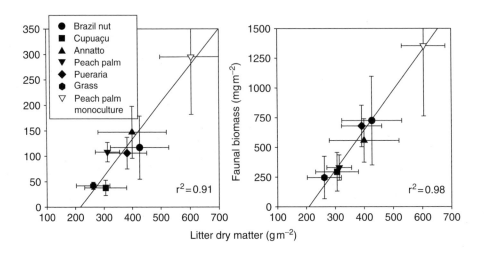

FIGURE 20.3
Faunal density and biomass in the litter layer under different tree and ground cover species in a multistrata system and a monoculture in the central Amazon. The close correlations illustrate a strong positive relationship between litter quantity and litter fauna. *Source:* Vohland, K. and Schroth, G., *Appl. Soil Ecol.*, **13**, 57–68 (1999). With permission.

20.5   Soil Organic Matter Management in Tree-Crop Agroforestry

Soil organic matter has a profound influence on chemical, physical, and biological soil properties, and its loss from cultivated soil is a widely used indicator of soil degradation. Compared with other forms of agriculture, tree-crop agroforestry systems offer far better opportunities to conserve soil organic matter since they maintain permanent soil cover, which also stimulates faunal and microbial activity; produce more biomass and thus above- and belowground litter, especially at high planting densities, which also increase the efficiency of soil exploration by root systems; seldom require the use of fire, which destroys biomass and litter; and do not employ soil tillage, which stimulates soil organic matter breakdown and destroys faunal populations and structures in the soil.

However, tree-crop systems usually lose soil organic matter if established on forest soil, as shown above, but the extent of loss differs between tree species. By associating different kinds of trees and tree crops, land users can capitalize on the ability of some species to retain higher levels of organic matter in the soil than do other, perhaps commercially, more valuable species. This buffers the system against organic matter and fertility loss. For example, in a 40-year-old multistrata system in Nigeria, cola trees (*Cola nitida*) retained higher organic matter levels in the soil than did the cocoa trees with which they were interplanted (Ekanade, 1987). Similarly, in a 7-year-old association of four local tree crops and a legume ground cover in the central Amazon, two tree species with recalcitrant litter, Brazil nut and cupuaçu, had similar organic matter levels in the top 10 cm of soil as adjacent rainforest, while the species with higher litter quality with which they had been interplanted (peach palm, annatto, and a leguminous cover crop) showed a tendency for lower topsoil carbon levels even after this relatively short time period (Schroth et al., 2002). This suggests that tree crops with recalcitrant litter can serve, to a certain extent, as "insurance" against soil organic matter loss in multistrata systems, in addition to their production role (Schroth et al., 2002), while other species with more nutrient-rich, high-quality litter will have a more beneficial effect on litter (and soil) fauna and essential soil microbial processes, as shown above.

20.6   Managing Soil Pests and Diseases in Tree-Crop Agroforestry

Soil-borne pests and diseases often prevent plants from developing healthy, fully functional root systems, thereby reducing plants' ability to absorb soil water and nutrients, sometimes causing plant death. There is some evidence to suggest that through the suitable design and management of tree-crop agroforestry systems, soil-borne plant diseases may be reduced.

Soil-borne pathogens frequently spread from infected to uninfected plants through direct root contact. Consequently, the alternation of plants susceptible and resistant to certain root diseases in a diversified agroforestry system may slow disease spread. For example, Kapp and Beer (1995) have suggested that the higher mortality of *Acacia mangium* trees, due to the soil fungus Rosellinia sp., in pure stands than in a mixed system with the fruit tree *araçá* (*Eugenia stipitata*), maize, and ginger (*Zingiber officinalis*), was due to this phenomenon. On the other hand, verticillium wilt, which affects avocado, is often aggravated if these trees are intercropped with other hosts of this soil fungus, such as olives (*Olea europaea*) and certain stone fruits (Ploetz et al., 1994), because this increases overall connectivity of susceptible root systems and facilitates disease spread.

In diversified systems, the physical distance between host plants is often greater than in monospecific plantations. In addition, nonhost plants will reduce the propagation of root diseases if they reduce lateral root spread of host plants, thereby further reducing intermingling of host root systems. While root systems of closely-spaced trees of different species do, of course, intermingle (Kummerow and Ribeiro, 1982), they also influence each other's distribution, often substantially so, presumably through a mix of resource competition, physical effects (soil drying around roots), and perhaps allelopathic interactions (Schroth, 1999). The confining effect on eucalypt root systems exerted by dense coffee root systems, even under conditions of high resource availability in the soil, has been mentioned before (Figure 20.2). Schaller et al. (2003a) also found that some grasses tend to repel tree roots, thereby acting as partial barriers to their lateral spread. The importance of such "biological barrier" effects in mixed tree-crop systems for root-to-root disease propagation, and whether they can be strategically employed to reduce root contacts between host plants, are good topics for research.

As seen above, tree-crop agroforestry systems allow growers to maintain relatively high levels of organic matter in the soil, although not as high as in tropical forest. Disease incidence and severity, including that caused by soil-borne pathogens, are often reduced with higher soil organic matter levels. One reason is probably the favorable effect that soil organic matter has on plant vigor, reducing susceptibility to disease. In addition, pathogens tend to be less efficient than are soil saprophytes in utilizing carbon sources in the soil, such as decomposing plant materials; they thus tend to decline in soils that are well supplied with organic matter. For example, saprophytic fungi of the genus Trichoderma have a strongly antagonistic effect on a wide range of fungal soil pathogens on which they feed and with which they compete for carbon sources. Certain strains are being used in the biological control of diseases such as witches' broom (*Crinipellis perniciosa*) in cocoa. In Chapter 34 there are further reports on this remarkable fungal genus with the potential for biocontrol and other benefits.

It should be noted that while carbon-rich organic materials tend to favor antagonists over pathogens, nitrogen-rich organic materials may have the opposite effect. In a study in

California, Casale et al. (1995) found a negative correlation between the nitrogen content of various mulches and the growth of *Trichoderma harzianum*, as well as the percentage of healthy roots of avocado and citrus trees to which they were applied. Similarly, Mendoza et al. (2003) observed that the fungal pathogen Rosellinia sp. made more efficient use of *bokashi*, a high-nitrogen organic soil amendment, than did its nonpathogenic antagonists, leading to increased root rot and mortality of cocoa seedlings. The significance of these recent findings for the widespread use of high-nitrogen leguminous materials in more intensively managed plantations of coffee, cocoa, and other tree crops has not yet been well researched.

Overall, it appears that if certain precautions are taken, diversified tree-crop systems, which limit root contacts between plants that are hosts to the same diseases, which enhance plant vigor, and which increase antagonistic microbial populations in the soil through the management of organic matter, may help to control soil-borne plant diseases and prevent the disastrous disease outbreaks known to so many annual crops, especially if grown in monoculture.

20.7   Diversifying Multistrata Systems

As seen in the preceding sections, the association of different plant species with complementary properties is not only an essential element in the biological management of soil fertility in multistrata systems, but also has economic advantages. Broadening the economic basis of multistrata systems that are too dependent on a single commodity will make them less vulnerable to ecological and market shocks, and thus more reliable contributors to farmers' livelihoods. Systems that produce several commodities, and also produce food and help to meet subsistence needs, are less likely to be converted to other land uses at times when market prices for one of their commercial products are low.

While the main focus of this chapter has been the design and management of biological factors that help to maintain high levels of soil fertility in diversified tree-crop systems, an important complementary approach to improving soil systems' fertility is to use tree-crop species and varieties adapted to conditions of low soil fertility and extensive management. Adaptation and combination of species for low external-input conditions has not been a typical focus of selection and breeding programs for tree crops. However, tree crops combining reasonable productivity with adaptation to low-input conditions already exist in various wild and semidomesticated populations, including traditional agroforests.

For example, recent research in traditional rubber agroforests in the Brazilian Amazon found individual, seed-grown (i.e., not grafted) rubber trees that were highly productive despite infertile soil and very limited management activity. They have coevolved with the biotic pressures of a region where the South American leaf blight fungus (*Microcyclus ulei*) is endemic and a constant threat to conventional plantation agriculture (Schroth et al., 2004b). This shows the high value of maintaining traditional agroforests as genetic reservoirs since certain tree individuals could be a basis for a specific "agroforest breed" of undemanding yet relatively productive rubber trees for smallholder systems.

Further work on the domestication and improvement of such species should take place in a participatory way under conditions representative of those of smallholder fields (including low soil fertility and extensive management). Assessments should apply criteria determined together with the farmers who will be the final users of these trees (Simons and Leakey, 2004). For example, fruit-tree species are usually selected for small size to facilitate harvest, but this is not necessarily an advantage in smallholder agroforestry systems. Other considerations, such as access to light, multifunctionality (timber as a secondary product of fruit and nut trees), and the advantage of high biomass production for maintaining soil fertility and a forest microclimate, will be important under less intensive management. For example, small-sized, grafted Brazil nut trees have not been adopted in the Amazon, in part because their timber is also valued locally.

20.8   Discussion

The maintenance of multiple biological soil functions in tree-crop agroforestry systems derives from the choices made about (or tolerance of) functionally complementary plant species. Decisions must weigh close spacing to intensify resource exploration and interactions with the soil through plant root systems and litter, and the most favorable management of ground vegetation (intercrops, cover crops, and weeds) to have permanent soil cover that feeds and protects soil organisms. These measures need to be complemented with the application of fertilizers, manures, and crop residues.

Farmers' decisions about which tree species to integrate into their land use systems are, of course, mainly driven by markets and consumption preferences, within the limits set by species requirements and site conditions. Choices will not always be driven by considerations of soil fertility management as examples of long-term soil degradation under tree crops testify. Many multistrata systems contain sizable amounts of spontaneous forest vegetation, which may help to balance limitations of the dominant crop species, such as weak root systems, and little or recalcitrant litter.

In more intensively managed multistrata systems, farmers may either choose service trees with abundant, easily decomposable litter, such as legume shade trees, to complement their main crop species, or ideally, they may find valuable trees and tree crops that complement each other in their effect on biological soil fertility and mineral cycling. With the traditional focus of soil research in agroforestry on leguminous multipurpose species, relatively little has been written on the interactions with soil fertility and its biological agents of the many nonleguminous fruit and other tree species that typically compose smallholder multistrata systems, such as homegardens and fruit tree agroforests, throughout the tropics. This is an important field for future agroforestry research with immediate relevance to tropical farmers.

References

Ayanlaja, S.A., Rehabilitation of cocoa (*Theobroma cacao* L.) in Nigeria: Physical and moisture retention properties of old cocoa soils, *Trop. Agric., (Trinidad)*, **64**, 237 (1987).

Baker, P.S., *Coffee Futures: A Source Book of Some Critical Issues Confronting the Coffee Industry.* CABI-FEDERACAFE-USDA-ICO, Chinchiná, Colombia (2001).

Casale, W.L. et al., Urban and agricultural wastes for use as mulch on avocado and citrus and for delivery of microbial biocontrol agents, *J. Hortic. Sci.*, **70**, 315–332 (1995).

Ekanade, O., Spatio-temporal variations of soil properties under cocoa interplanted with kola in a part of the Nigerian cocoa belt, *Agrofor. Syst.*, **5**, 419–428 (1987).

Hanne, C., Die Rolle der Termiten im Kohlenstoffkreislauf eines amazonischen Festlandregenwaldes. Ph.D., thesis, University of Frankfurt/Main, Frankfurt/Main (2001).

Johns, N.D., Conservation in Brazil's chocolate forest: The unlikely persistence of the traditional cocoa agroecosystem, *Environ. Manage.*, **23**, 31–47 (1999).

Joshi, L. et al., Soil and water movement: Combining local ecological knowledge with that of modellers when scaling up from plot to landscape level, In: *Below-ground Interactions in Tropical*

Agroecosystems, van Noordwijk, M., Cadisch, G., and Ong, C.K., Eds., CABI Publishing, Wallingford, UK, 349–364 (2004).

Kapp, G.B. and Beer, J., A comparison of agrosilvicultural systems with plantation forestry in the Atlantic lowlands of Costa Rica, *Agrofor. Syst.*, **32**, 207–223 (1995).

Kumar, B.M. and Nair, P.K.R., The enigma of tropical homegardens, *Agrofor. Syst.*, **61**, 135–152 (2004).

Kummerow, J. and Ribeiro, S.L., Fine roots in mixed plantations of Hevea (*Hevea brasiliensis* H.B.K.Müll.Arg.) and cacao (*Theobroma cacao* L.), *Rev. Theobr.*, **12**, 101–105 (1982).

Laurance, W.F. et al., Relationships between soils and Amazon forest biomass: A landscape-scale study, *For. Ecol. Manage.*, **118**, 127–138 (1999).

Lavelle, P., Senapati, B.K., and Barros, E., Soil macrofauna, In: *Crops and Soil Fertility: Concepts and Research Methods*, Schroth, G. and Sinclair, F.L., Eds., CAB International, Wallingford, UK, 303–323 (2003).

Mendoza, R.A. et al., Evaluation of mycoparasites as biocontrol agents of Rosellinia in cocoa, *Biol. Control*, **27**, 210–227 (2003).

Michon, G. and de Foresta, H., Agro-forests: Incorporating a forest vision in agroforestry, In: *Agroforestry in Sustainable Agricultural Systems*, Buck, L.E., Lassoie, J.P., and Fernandes, E.C.M., Eds., Lewis Publishers, Boca Raton, 381–406 (1999).

Murniati, Garrity, D.P., and Gintings, A.N., The contribution of agroforestry systems to reducing farmers dependence on the resources of adjacent national parks: A case study from Sumatra, Indonesia, *Agrofor. Syst.*, **52**, 171–184 (2001).

Ploetz, R.C. et al., *Compendium of Tropical Fruit Diseases*, American Phytological Society Press, St. Paul, MN (1994).

Pohlan, J.H.A., Manejo integrado de malezas, In: *México y la Cafeticultura Chiapaneca*, Pohlan, J., Ed., Shaker, Aachen, Germany, 215–222 (2002).

Sanchez, P.A. et al., Tree crops as soil improvers in the humid tropics?, In: *Attributes of Trees as Crop Plants*, Cannell, M.G.R. and Jackson, J.E., Eds., Institute of Terrestrial Ecology, Huntingdon, UK, 327–358 (1985).

Schaller, M. et al., Root interactions between young *Eucalyptus deglupta* trees and competitive grass species in contour strips, *For. Ecol. Manage.*, **179**, 429–440 (2003a).

Schaller, M. et al., Species and site factors that permit the association of fast-growing trees with crops: The case of *Eucalyptus deglupta* as coffee shade in Costa Rica, *For. Ecol. Manage.*, **175**, 205–215 (2003b).

Schroth, G. A review of belowground interactions in agroforestry, focussing on mechanisms and management options, *Agrofor. Syst.*, **43**, 5–34 (1999).

Schroth, G. et al., Rubber agroforests at the Tapajós river, Brazilian Amazon: Environmentally benign land use systems in an old forest frontier region, *Agric. Ecosys. Environ.*, **97**, 151–165 (2003a).

Schroth, G. et al., conversion of secondary forest into agroforestry and monoculture plantations in Amazonia: Consequences for biomass, litter and soil carbon stocks after seven years, *For. Ecol. Manage.*, **163**, 131–150 (2002).

Schroth, G. et al., Subsoil accumulation of mineral nitrogen under polyculture and monoculture plantations, fallow and primary forest in a ferralitic Amazonian upland soil, *Agric. Ecosys. Environ.*, **75**, 109–120 (1999).

Schroth, G. Harvey, C.A., and Vincent, G., Complex agroforests: Their structure, diversity, and potential role in landscape conservation, In: *Agroforestry and Biodiversity Conservation in Tropical Landscapes*, Schroth, G. et al., Eds., Island Press, Washington, DC, 227–260 (2004a).

Schroth, G., Lehmann, J., and Barrios, E., Soil nutrient availability and acidity, In: *Trees, Crops and Soil Fertility: Concepts and Research Methods*, Schroth, G. and Sinclair, F.L., Eds., CAB International, Wallingford, UK, 93–130 (2003b).

Schroth, G. et al., Plant-soil interactions in multistrata agroforestry in the humid tropics, *Agrofor. Syst.*, **53**, 85–102 (2001a).

Schroth, G., Moraes, V.H.F., and da Mota, M.S.S., Increasing the profitability of traditional, planted rubber agroforests at the Tapajós river, Brazilian Amazon, *Agric. Ecosys. Environ.*, **102**, 319–339 (2004b).

Schroth, G., Rodrigues, M.R.L., and D'Angelo, S.A., Spatial patterns of nitrogen mineralization, fertilizer distribution and roots explain nitrate leaching from mature Amazonian oil palm plantation, *Soil Use Manage.*, **16**, 222–229 (2000a).

Schroth, G., Salazar, E., and da Silva, J.P., Soil nitrogen mineralization under tree crops and a legume cover crop in multi-strata agroforestry in central Amazonia: Spatial and temporal patterns, *Expl. Agric.*, **37**, 253–267 (2001b).

Schroth, G. et al., Nutrient concentrations and acidity in ferralitic soil under perennial cropping, fallow and primary forest in central Amazonia, *Eur. J. Soil Sci.*, **51**, 219–231 (2000b).

Simons, A.J. and Leakey, R.R.B., Tree domestication in tropical agroforestry, *Agrofor. Syst.*, **61**, 167–181 (2004).

Soares, A.T., Situação econômica e perspectiva do extrativismo de borracha na Amazônia, In: *Seringueira na Amazônia — Situação e Perspectivas*, Frazão, D.A.C., Cruz, E.S., and Viégas, I.J.M., Eds., Embrapa Amazônia Oriental, Belém, Brazil, 23–54 (2003).

Vohland, K. and Schroth, G., Distribution patterns of the litter macrofauna in agroforestry and monoculture plantations in central Amazonia as affected by plant species and management, *Appl. Soil Ecol.*, **13**, 57–68 (1999).

21

Restoring Productivity to Degraded Pasture Lands in the Amazon through Agroforestry Practices

Erick C.M. Fernandes,[1] Elisa Wandelli,[2] Rogerio Perin[2] and Silas Garcia[2]
[1]*The World Bank, Washington, DC, USA*
[2]*EMBRAPA Western Amazon Agroforestry Center (CPAA), Manaus, Brazil*

CONTENTS

The conversion of primary forest for subsistence agriculture, industrial logging, and pasture establishment continues to be the predominant cause of tropical deforestation (Laurance, 1999). In the last 30 years, an estimated 58.8 million ha of primary forest have been cleared in the Brazilian Amazon alone (INPE, 2004). Of this, 24 million ha were converted to pastures during the 1970s and 1980s, making this the most common land-use change in the Amazon region (Serrão et al., 1995).

Although pastures for beef cattle continue to dominate deforested landscapes in the Brazilian Amazon, many formerly productive pastures have become degraded and are now abandoned, left to be colonized by secondary vegetation (Fearnside and Guimaraes, 1996; Silver et al., 2000). Based on an analysis of Amazonian land-use data, Fearnside (1996) estimated that 47% of all deforested land in the Amazon is currently in some form of regenerating forest on degraded or abandoned pastures. Owing to the large area of degraded pastures in the Amazon, a number of local research and development agencies are investigating the possibility of rehabilitating the productivity of degraded pasturelands as a means of deflecting the continuing pressure to establish agricultural lands at the forest frontier. Multispecies agroforests have been identified as promising alternatives for rehabilitation of degraded pasturelands (Fernandes and Matos, 1995; Parrotta et al., 1997).

Many of the original pastures in the Amazon were established with revenues from the sale of high-value timber trees. Mahogany as the most valuable species has been severely overexploited (Verissimo et al., 1995). This chapter reviews the successful reintroduction of large-leafed mahogany to degraded pasture lands in the Amazon via an agroforestry approach that harnessed local agroecological knowledge together with scientific information on integrated nutrient management (INM) and integrated pest management (IPM) appropriate for the Amazon.

21.1 Characteristics of Pasture Systems and Abandoned Pastureland in the Amazon

Despite their previous support of lush rainforest, plant growth on deforested soils can be severely constrained by generally low levels of available soil nutrients and high acidity. Deficiencies of available calcium (Smyth and Cravo, 1992), available phosphorus (Gehring et al., 1999), and nitrogen (Davidson et al., 2004) are commonly reported. Cochrane and Sanchez (1982) estimated that only 7% of the land area in the Brazilian Amazon is free from major limitations for plant growth. Soil phosphorus deficiencies (<7 mg kg^{-1}) are said to constrain productivity in 90% (436 million ha) of Brazilian Amazonia, and aluminium toxicity (Al saturation of $>60\%$) occurs over 73% of this area. Another significant problem is the high amount and intensity of rainfall (1800–3000 mm yr^{-1}) which facilitates the loss of nutrients via leaching and surface runoff from bare soil.

In the Brazilian Amazon, pastures are established by felling primary forest, burning the forest biomass to release the nutrients it contains, and planting pasture grasses (Brachiaria spp.). Once established, poor management of both livestock and pastures usually results in pasture degradation within 7 to 10 years. Rueda (2002) found that, paradoxically, low livestock stocking rates, resulting in low grazing pressure, has led to a rapid decline in pasture palatability and productivity. Ranchers rejuvenate the pasture by burning the poorly palatable forage, and this leads to the direct loss of nutrients (especially nitrogen) and facilitates the loss of nutrients in the residual ash via surface runoff and leaching. The disruption of communities of soil macrofauna, microfauna, and microflora, and of their functions, plus the depletion of forest-species seed pools, are key biological reasons for pasture degradation and poor subsequent forest regeneration (Fernandes et al., 1997).

Degraded pastures are characterized by depletion of available soil nutrient stocks, low vegetation biomass, depleted seed banks of forest species, high seed predation, and low stump sprouting (Nepstad et al., 1990), as well as soil surface sealing and compaction (Eden et al., 1991). The speed with which abandoned pastures are colonized by tree and shrub vegetation is directly related to the intensity of their use as pastures. The greater the grazing pressure and the longer the period grazed, the slower is the development of the fallow vegetation and the time for recovery of site productivity (Uhl et al., 1988; Nepstad et al., 1990). The principal pasture grasses (Brachiaria spp.) are aggressive C4 plants that are easily able to dominate establishing seedlings of forest species (mostly C3 species). Feldpausch et al. (2004) found that regenerating secondary forests on abandoned pasturelands could rapidly accumulate biomass, but this resulted in an equally rapid depletion of soil phosphorus and calcium stocks.

Although many degraded pastures are currently at different stages of secondary forest regeneration, there is an absence of long-term empirical data on the quality and dynamics of the regenerating secondary vegetation on abandoned pastures in the Amazon. A study near Manaus showed that regenerating forest vegetation is unable effectively to capture leaching soil nitrogen (Schroth et al., 1999). Similar to forest fallows that follow several cycles of cropping (Chapter 29), the emerging research data from abandoned pasture sites suggest that the regeneration may be limited by critically low levels of available soil nutrients, soil compaction, and low seed pools (Keller et al., 2004). An alternative pathway to natural regeneration on degraded pastures is the establishment and management of biologically diverse, integrated tree-, crop-, and livestock-based systems that are modeled upon native and migrant farmer systems found in the region. Such systems generally involve agroecological approaches to land management that optimize biological processes and, wherever possible, the use of locally available organic inputs (Pretty, 1995; Fernandes et al., 1993b).

21.2   Indigenous Knowledge and Scientific Data for Developing a Sustainable Timber–Pasture System for Degraded Pasturelands

We adopted the following sequential approach to develop appropriate cropping systems to rehabilitate degraded pasture land in the Amazon:

- A review of the Amazonian literature on indigenous technical knowledge and on local agroecosystems;
- A review of the international literature on conceptual approaches to develop improved natural resource management models for the humid tropics and the Amazon;
- A survey of approximately 30 local communities in settlement areas to identify farmers' priorities and approaches to develop productive agroecosystems; and
- Use of the knowledge and information gained from the reviews and surveys to generate specifications of adapted and biologically-robust production systems with potential to rehabilitate the productivity of degraded pasture lands.

21.2.1   Key Characteristics of Robust and Productive Agroecosystems

Our reviews of the literature suggested that both biological diversity and INM are consistent practices in long-lived, integrated farming systems. Biological diversity is required in a structural as well as functional sense. Native stocks of available plant nutrients need to be managed to avoid outputs exceeding inputs and, where necessary, these stocks need to be supplemented from external (organic and/or chemical) sources in order to sustain system function and productivity.

21.2.1.1 *Structural and Biological Diversity of Integrated Farming Systems*

The great majority of productive and long-lived, traditional farming systems have high species diversity and species associations of different age classes spread over several sites (Chang, 1977; Clawson, 1985; Thrupp, 1998). In Latin America, much of the production of staple crops occurs in polycultures. More than 40% of the cassava, 60% of the maize, and 80% of the beans are intercropped with each other or other crops (Francis, 1986).

The strategy of reducing risk by planting several species and varieties of crops stabilizes yields over the long term, provides a range of dietary nutrients, and maximizes returns under low levels of technology and limited resources (Harwood, 1979). These system characteristics maximize labor efficiency per unit area of land, minimize risk of catastrophic crop failure due to drought or severe pest attack, and guarantee the availability of food at medium to high levels of species productivity. In most multiple-cropping systems developed by small farmers, yields per unit area are often 20 to 60% higher than under sole cropping with the same level of management (Beets, 1982). These differences can be explained by a combination of factors that include the reduction of losses due to pests and disease and a more efficient use of the available resources of water, light, and nutrients.

Another benefit of multiple species associations is the creation of additional niches for pollinators, decomposers, and natural enemies of crop pests (Andow, 1991). The plant diversity provides alternative habitat and food sources such as pollen and nectar, and alternative hosts to predators and parasites (Altieri, 1995; Ackerman et al., 1998). Both above- and belowground species and processes are affected and can contribute to agroecosystem productivity and stability (Tillman et al., 1996; Giller et al., 1997). Such integrated farming systems sustain a higher level of agrobiodiversity than intensively managed, monoculture crop systems (Perfecto et al., 1996; Power, 1996).

Turner et al. (1995) have suggested that there exists a three-way interaction among biodiversity, ecosystem processes, and landscape dynamics. Any land management practices that increase biodiversity at a landscape level are also likely to benefit ecosystem services, such as nutrient, water, and soil conservation, biological pest control, and efficient nutrient cycling (Cullota, 1996; Tilman et al., 1996). Although it appears that obtaining appropriate species mixtures rather than maximizing species numbers is more important in the provision of ecosystem services, high species richness may increase agroecosystem resilience following any disturbance by increasing the number of alternative pathways for the flow of resources (Silver et al., 1996). Table 21.1 lists a range of species encountered in traditional systems in the Amazon.

21.2.1.2 *Elements of Integrated Nutrient Management*

A common feature of many traditional and other farming systems that have sustained populations over several decades and, in some cases centuries, is the continuous use of locally-available, biological and organic resources to minimize nutrient losses from the system. Plant nutrients are usually removed from the system via harvests of grain, tubers, fruit, and wood, and by surface erosion and subsurface leaching. The literature on INM documents the following key requirements for effective nutrient management and sustainable cropping.

TABLE 21.1

Tree and Crop Species Encountered in Surveys of Agroforests in the States of Acre, Amazonas, Para, Rondonia, and Roraima

Common Name	Scientific Name	Uses
Avocado	*Persea americana*	Fruit, cash crop
Coconut	*Cocos nucifera*	Food, oil, cash crop
Guaraná	*Paulinia cupana*	Drink, cash crop
Tucumã	*Astrocaryum aculeatum*	Fruit, fibre
Breadfruit	*Artocarpus altilis*	Seeds
Jackfruit	*Artocarpus heterophyllus*	Fruit, seeds
Guava	*Psidium guajava*	Fruit
Lime	*Citrus aurantifolia*	Fruit, cash crop
Mango	*Mangifera indica*	Fruit, cash crop
Peach palm	*Bactris gassipaes*	Fruit, palm heart
Cashew	*Anacardium occidentale*	Fruit, nut cash crop
Pineapple	*Ananas comosus*	Fruit, cash crop
Cupuaçu	*Theobroma grandiflorum*	Fruit, cash crop
Annatto	*Bixa orellana*	Seeds for dye
Acerola	*Malpigia glabra*	Fruit, cash crop
Black pepper	*Piper nigrum*	Spice, cash crop
Cacao	*Theobroma cacao*	Beverage, cash crop
Banana	*Musa* spp.	Fruit, cash crop
Coffee	*Coffea canephora*	Beverage, cash crop
Hog plum	*Spondias mombin*	Fruit, juice
Ingá	*Inga edulis*	Fruit, fuel wood
Biriba	*Rollinia mucosa*	Fruit, cash crop
Soursop	*Anona muricata*	Juice, cash crop
Açai	*Euterpe oleracea*	Fruit, heart of palm, cash crop
Araça boi	*Eugenia stipitata*	Juice
Jambu	*Eugenia jambos*	Fruit
Pitanga	*Eugenia uniflora*	Fruit, juice
Papaya	*Carica papaya*	Fruit, cash crop
Caimito	*Pouteria caimito*	Fruit
Sapotilla	*Manilkara zapota*	Fruit, chewing gum
Bacuri	*Platonia insignis*	Fruit
Genipapo	*Genipa americana*	Fruit, wood
Araticum	*Anona montana*	Fruit
Bacaba	*Oenocarpus bacaba*	Wine, wood, leaf baskets
Cassava	*Manihot esculenta*	Tubers for starch, cash crop
Passion fruit1	*Passiflora nitida*	Fruit
Passion fruit2	*Passiflora macrocarpa*	Fruit
Umari	*Poraqueiba sericea*	Fruit
Rubber	*Hevea brasiliensis*	Latex, cash crop
Mapati	*Pourouma cecropiaefolia*	Fruit
Cubiu	*Solanum sessiliflorum*	Fruit
Pitomba	*Talisia esculenta*	Fruit
Carambola	*Averrhoa carambola*	Fruit
Buriti	*Mauritia flexuosa*	Fruit

Source: Fernandes and Matos, 1995

21.2.1.2.1 Eliminate Soil Erosion and Leaching

The most effective way to reduce soil erosion and leaching is to maximize soil cover via the use of cover crops and mulches and by integrating perennials in vegetative strips along the contours to further stabilize the soil (Fernandes et al., 1993a). Where soil depth is adequate, contour vegetative strips permit interstrip erosion and result in a gradual leveling of the slope and terrace formation without the need for labor-intensive, manual terrace formation.

21.2.1.2.2 Cycle All Flows of Organic Nutrients

One method is to return all crop residues to the field of origin. In many cases, however, crop residues are fed to livestock. Ideally, the livestock should be fed the residues in the field so that the manure goes directly on to the soil. If the residues are removed and fed to livestock elsewhere, then the manure should be returned to the field as soon as possible. The transport and spreading of manure on fields, however, is often a problem owing to labor constraints.

Many farmers combine crop residues with manure to "make the manure go further." In scientific terms, this represents the use of the lignin and polyphenol compounds in crop residues to tie up the nitrogen that would have otherwise been lost through volatilization and leaching. The composting of vegetable residues and animal manure is an efficient way to conserve farm nutrients. Making this knowledge available to all farmers could make a significant improvement in nitrogen/nutrient budgets of small farms.

21.2.1.2.3 Enhance Biological Sources of Nutrients

Nitrogen-fixing trees, shrubs, herbaceous, and crop species can fix nitrogen from the atmosphere and make it available to subsequent crops via biological or associative nitrogen ha^{-1} fixation (Chapter 12). Data from several studies show that it is possible to contribute between 15 and 200 kg of nitrogen to cropping systems via biological nitrogen fixation (Peoples and Herridge, 1990).

21.2.1.2.4 Compensate for Nutrient Exports by Adding Nutrients First as Green or Animal Manure, and if Necessary Supplement with Inorganic Fertilizers

Where soil nutrients have been severely depleted, it is often necessary to restore the minimum levels required for adequate plant growth and yield. Sanchez et al. (1997) have argued for phosphorus replenishment in sub-Saharan Africa as a means of priming the biological nitrogen-fixation process and improving crop productivity. Animal manure and plant litters are generally low in phosphorus, and unlike nitrogen, phosphorus cannot be fixed from the atmosphere. Phosphorus deficiency is a major constraint to effective nitrogen fixation because phosphorus is an important nutrient in the process of nodulation and nitrogen fixation. Guano and rock phosphate can be good sources for phosphorus (and guano also for nitrogen) where such materials are available locally (e.g., Peru, Madagascar, Zaire, West Africa).

21.2.1.2.5 Select and Use Adapted Efficient Species as Components of Improved Systems That Are Designed to Take Advantage of the INM Concept

Some leguminous tree species (e.g., Inga spp.) are able to fix nitrogen at very low levels of available soil phosphorus (Fernandes, 1998). Interestingly, many Inga species are used to provide shade and mulch in traditional integrated farming systems (Pennington and Fernandes, 1998). The grain legume *Cajanus cajan* has been shown to be able to absorb phosphorus from insoluble calcium–phosphorus complexes in high pH soils (Ae et al.,

1990). Leguminous crops that combine some grain yield with high root and leaf biomass, and thus have a low nitrogen harvest, offer a useful compromise of meeting farmers' food security concerns and improving soil fertility (Snapp et al., 1998). Promising genotypes include Arachis, Cajanus, Dolichos, and Mucuna spp. On-farm nitrogen budgets indicate that legumes with high-quality residues and deep root systems are effective ways of improving nutrient cycling.

In addition to soil protection, the fibrous root systems of leguminous cover crops, such as *Centrosema macrocarpum*, *Desmodium ovalifolium*, and *Pueraria phaseoloides*, can also improve soil physical properties (Broughton, 1977). In the Peruvian Amazon, Arevalo et al. (1998) measured improvements in soil physical properties and increased livestock weight gains when cattle were managed in a silvopastoral system with peach palm and *Centrosema macrocarpum* compared with a traditional grass-based pasture. In the western Amazon, Rueda et al. (2003) reported significantly improved pasture and livestock productivity in Brachiaria pastures that contained the herbaceous legume *Pueraria phaseoloides* (see also Chapter 37).

Farmers who have no access to markets and limited capital to invest in synthetic fertilizers or pesticides can rely more on biological methods and synergies to minimize their pest problems, reduce nutrient losses, and enhance nutrient inputs, e.g., via nitrogen fixation. Thurston (1997) has identified many traditional systems in the tropics that rely upon INM approaches.

Our review of findings suggested that the presence of one or more leguminous species in managed systems can significantly enhance not only the protection of existing soil productivity functions but can also provide critical biological leverage points to improve the resilience of the managed systems against ecological and climatic shocks. The several leguminous crop species available to farmers are generally short-duration species (3 to 15 months). As long-term components of managed systems, multipurpose, leguminous tree species that do not need replanting every season can help to sustain the critical biodiversity and INM functions as the system evolves.

21.3 Traditional Tree–Crop Systems as Models for Sustainable Farming Systems on Degraded Pasturelands

Amerindian peoples of the Amazon have long planted and managed trees for a variety of products and services in close association with annual and perennial food crops (Posey, 1985; Denevan and Padoch, 1987). The Kayapó create "resource islands" of trees, shrubs, herbs, and root crops at the forest margin and also in open grasslands. These species are generally collected as seedlings in the forest and transplanted to clearings and campsites. Over a hundred species have been encountered in these "agroforestry" islands (Kerr and Posey, 1984; Posey, 1984).

In many traditional systems, farmers crop the deforested land for a few years and then allow the regeneration of forest species during a fallow period of between 5 and 50 years (Chapter 29). Most of the Amazonian trees (*Bactris gassipaes*, *Euterpe* spp., Theobroma spp., Inga spp.) and other food crops (e.g., cassava, capsicum, native solanaceae) that are in use today were probably domesticated via these traditional fallow-based, mixed-species systems (see Table 21.1). In addition to high species diversity, such indigenous agroforests are characterized by a variety of species associations of different age classes spread over several sites. These system characteristics maximize labor efficiency per unit area of land, minimize the risk of food crop failure due to drought or severe pest attack, and guarantee the availability of food even at relatively modest levels of species productivity.

Tree-based homegardens (agroforests) include both native and exotic species for fruit, timber, shade, medicines, spices, and forage. In the Amazon, agroforests have been reported to involve around 30 perennial and annual plant species in Para, Brazil (Subler and Uhl, 1990) and over 70 species in Peru (Paddoch and de Jong, 1991). Agroforests are not unique to the Amazon and as many as 190 plant species at various stages of domestication have been recorded in tropical agroforests (Fernandes and Nair, 1986). The spatial and temporal associations of components are highly dynamic, and although the plant diversity may be low at any given point in time, agroforests can be very biodiverse over their rotations of 50 to 100 years. In Africa, where deforestation is resulting in significant loss of biodiversity, agroforests have been identified as important *in situ* germplasm banks of food, fruit, and medicinal species, whose wild relatives are fast-disappearing as primary forest is cut down (Okafor and Fernandes, 1987).

Our reviews of reports on indigenous technical knowledge from the Amazon showed that the native leguminous genus Inga was a common component of traditional agroforestry systems. This genus, with 258 described species (Pennington, 1997), is used in managed fallow systems as a trap crop for edible caterpillar species, and as a shade tree for perennial crops such as cacao, coffee, and tea (León, 1966; Carrasco, 1971). Lawrence (1995) reported that the main reasons why farmers like Inga as a shade tree are because "the leaves are a good fertilizer, the shade is perennial, the litter and shade provide good weed control, and the shade keeps the soil humid."

Of the 258 species described, the most commonly used species is *Inga edulis*, which occurs as a component in traditional Amazonian agroforestry systems (León, 1966; Pennington and Fernandes, 1998). The qualities that make *I. edulis* an ideal species to facilitate INM in managed systems include:

- Fast growth and ability to tolerate two to three shoot prunings a year that yield between 8 and 10 ton of biomass ha^{-1} yr^{-1} (Szott et al., 1991b).

- Nitrogen fixation potential of 10 to 50 kg N ha^{-1} yr^{-1} depending on soil conditions and plant management (Fernandes et al., 1997).

- Good adaptation to acid, infertile soils and capacity to nodulate effectively with native Rhizobia in soils with high aluminum saturation and low available phosphorus. Studies also show that Inga is a good host for native mycorrhizal fungi that enable the trees to exploit low levels of available soil phosphorus (Fernandes, 1990).

- Moderate to high nutrient concentrations in leafy biomass (e.g., 2.0–3.5% N, 0.2–0.3% P, 1–3% K, and 0.5–1.5% Ca). Leafy biomass derived from vigorously growing trees of *Inga edulis* grown on an Ultisol contained the following concentration of micronutrients: Mn 112, Cu 13, Zn 35, and Fe 95 mg kg^{-1} (Fernandes et al., 1993a).

- Recalcitrant litter that decomposes slowly over 3 to 5 months and thus forms an effective mulch layer on the soil surface that is beneficial for soil protection, soil moisture conservation, and a microhabitat for soil invertebrate populations.

Our community surveys revealed that although Inga is greatly appreciated for its edible fruit (often called "ice cream bean") and its soil-improving properties, many respondents also indicated that the tree attracts undesirable insects, such as stinging ants, as well as more desirable ones, such as spiders as insect predators. There was good reason to anticipate that Inga would be a good species to facilitate IPM in managed systems in the Amazon.

In the 1990s, ranchers not only deforested large tracts of the Amazon to establish extensive pastures, but also used the revenues from the sale of high-value timber species to finance the deforestation and pasture establishment. One timber species that has been aggressively harvested, to the brink of extinction (CITES Appendix II 2003), is large-leafed mahogany (*Swietenia macrophylla* King.). To reverse the negative ecological impacts of the first-generation Amazonian pastures, and the growing local and international demand for range-fed beef, we decided to develop a mahogany-based improved pasture system to restore the productivity of degraded pastures in the Amazon.

21.4 Accounting for Plant–Insect Relationships in the Design of a Biodiverse Mahogany–Pasture System

Previous attempts to manage mahogany in plantations in the neotropics have not been successful due to constant attacks by the mahogany shoot borer, *Hypsipyla grandella* Zell, a lepidopteran (moth) pest (Patiño Valerra, 1997). Repeated attacks by the shoot borer can kill young trees because they thwart the growing shoots (meristems) of the canopy. If the tree survives, the attacks produce a witch's broom effect and excessive branching that adversely affects the commercial quality of the timber.

Given the scientific knowledge on Inga's biological nitrogen-fixation potential and the numerous reports of Inga as a host plant for potential predators of the mahogany moth, we reviewed the literature to determine the potential usefulness of Inga as a nurse species for mahogany, protecting it against the mahogany moth. This showed that:

1. Many of Inga's more than 280 described species have extrafloral nectarines, i.e., plant glands located outside the flowers (Koptur, 1984). The secretion of nectar occurs as the leaf unfolds, and continues through its mature and expanded state. Since new leaves are produced year-round, this extrafloral nectar is almost always available (Koptur, 1984).

2. Many species of ants, while visiting the foliar nectaries of Inga, provide protection of both young and mature leaves against a variety of insect herbivores by predation or by disturbing the herbivores until they leave (Koptur, 1984). Leston (1973) suggested that shade trees that support an "ant mosaic" in commercial plantations may increase protection of the crop plant by natural enemies.

3. In addition to ants, Inga attracts a variety of other insect species. Our own field observations revealed the presence of several insect predators (spiders, wasps) in Inga canopies.

We expected that by using Inga as a nurse species for mahogany we could reduce shoot borer attacks because of the following three factors:

1. Inga has a dense canopy that would physically shield the mahogany trees and make it difficult for *H. grandella* to locate them.

2. Inga attracts a variety of insect species including ants, which defend their territories against other insects, and could thus prevent *H. grandella* from reaching the mahogany.

3. The insects that frequent the foliar nectaries of Inga attract predators such as spiders and birds, which are also potential predators of *H. grandella*.

21.5 Design and Establishment of the Components of a Biodiverse Mahogany–Pasture System

Degraded pastures near Manaus, Brazil were identified and characterized for soil, vegetation, and above- and belowground biodiversity. The average aboveground biomass was 17 ton ha^{-1}, the majority of which was found in the standing litter and tree components. Species richness on the pastures was low, with the majority of the biomass being represented by many individuals of a few species. The most important species on the site included *Brachiaria humidicola* Rendle, the original introduced pasture grass; *Borreria verticillata* (L.) G.F.W. Mey. and *Rolandra fruticosa* (L.) Kuntze, both invasive weeds; plus the tree species *Laetia procera* (Poeppig) Eichler, *Vismia amazonica* Ewan, *Vismia lateriflora* Ducke, and *Vismia cayennensis* Jacq (McKerrow, 1992).

The degraded pasture biomass contained 150 kg N, 4.8 kg P, 87 kg K, 20 kg Ca, and 83 kg Mg per hectare. Slashing and burning the vegetation, which is normal practice in the region for establishing pastures and other cropping systems, resulted in a loss of 132 kg N, 43 kg K, 29 kg Ca, and 6 kg Mg, all per hectare. Since pastures take several years to establish before they can be productively grazed, and the mahogany timber trees would be grown on a long rotation (>50 years), we designed the system to produce rice (*Oryza sativa*), maize (*Zea mays* L.), cowpea (*Vigna unguiculata*), and cassava (*Manihot exculenta* Crantz.) for the first 4 years. Also, because the slashing and burning of the degraded pasture vegetation resulted in a low net contribution of nutrients to the soil, and the soils showed significant surface compaction, we compared two soil amendment treatments:

1. Slashing and burning of the natural regeneration on the degraded pasture, followed by a single application of 20 kg ha^{-1} P (as triple superphosphate) and then cropping with rice, cowpea, and cassava.

2. Slashing and burning of the natural regeneration on the degraded pasture, an application of lime (1 ton ha^{-1} CaCO$_3$ equivalent), mechanization to break up surface compaction and to incorporate the lime, followed by applications of 50 kg N, 20 kg P, and 70 kg K ha^{-1} and then cropping with maize, cowpea, and cassava.

Our strategy for protecting mahogany from the shoot borer moth was targeted to providing maximum physical shielding and niches for moth predators during the first 5 to 6 years of mahogany growth. The design involved two guard rows of Inga (6 m apart) and a central row of mahogany. To obtain overhead protection we interplanted the mahogany with a fast-growing native tree species, *Schizolobium amazonicum*. The design resulted in a vegetative tunnel with sideways protection of the mahogany by dense canopies of Inga loaded with foliar nectaries and mahogany moth predators, with overhead protection by a sparse canopy of Schizolobium. Since mahogany is a light-demanding species, it is important to provide adequate overhead light while retaining a physical shield against the moth. The dense lateral shade and dappled overhead shade mimics conditions in forest gaps, where mahogany seedlings regenerate in the primary forest, and quick upward growth of the mahogany is forced with a minimum of lateral branching.

21.6 Results of the Inga-Mahogany Experiment

During the first 5 years following establishment, the low-phosphorus-input system produced 0.8 ton of rice and 16 tons of cassava, and approximately 4 tons Inga fuelwood per hectare. The mechanized, lime plus NPK, moderate-input system produced 2 tons maize, 0.5 ton of cow pea, 20 tons cassava, and 7 tons Inga fuelwood per hectare. Thus, priming the systems with adequate nutrients (lime + NPK) significantly improved the crop productivity and growth of the Inga and mahogany trees. Over 10 years of growth, the aboveground biomass in the system accumulated per hectare 260 kg N, 24 kg P, 195 kg K, 173 kg Ca, and 36 kg Mg (McCaffery, 2003).

Pasture productivity, especially that of grasses, was significantly enhanced by the initial application of moderate amounts of lime and NPK relative to only phosphorus applications (Table 21.2). A major benefit of improved herbage productivity was the reduction of weed invasions in the pasture. As most of the invading weeds are not grazed or only poorly palatable, the initial nutrient and lime inputs had significant direct and indirect impacts on pasture productivity and palatability.

Results after 4 years of growth showed that the Inga nurse trees significantly delayed the onset of *H. grandella* attacks on interplanted mahogany. Mahogany trees in adjacent plots growing without the protection of Inga were attacked at heights of around 2 m by *H. grandella.* in the second year after field establishment. In the Inga-mahogany association, however, the attacks took place largely in the fourth year and at heights of 6 to 7 m. Interestingly, 90% of the trees in the low-input, sparse-canopy Inga-mahogany system were attacked vs. 70% in the high-input, denser Inga canopy system.

The attack of *H. grandella* on mahogany increased once the mahogany trees grew taller than the Inga (6 m) as evidenced by the bushy growth and bifurcation of the mahogany stems. The delayed attack by *H. grandella* on mahogany that was interplanted with Inga is significant because the older trees were better able to survive the attack than the younger, open-grown mahogany trees. In addition, the mahogany trees that were not attacked in the first 3 years developed a single stem, which makes the tree more valuable for sawn timber than stems that are branched. In the Ecuadorian Amazon, a similar approach of interplanting mahogany in groves of *I. edulis* or *I. ilta* has also resulted in significantly reduced attacks by *H. grandella* compared with control plots where mahogany was not protected by Inga (Niell and Revello, 1998).

Tapia-Coral et al. (2004) have reported on the development of a substantial and persistent litter layer in Inga-mahogany systems. Although naturally-regenerating secondary vegetation control plots had a significantly higher litter layer than the mahogany-pasture system, the latter maintained a good litter layer in both dry and wet seasons (Figure 21.1). In a separate study, Barros et al. (2003) reported that the build-up of the litter layer in the mahogany–pasture system resulted in significantly improved soil invertebrate populations and soil structure. Nine years after its establishment, the mahogany–pasture system accumulated 16 tons of aboveground C ha^{-1} ($\sim$1.8 ton ha^{-1} yr^{-1}), which is significantly better than the carbon losses (-0.2 to -0.6 ton ha^{-1} yr^{-1}) reported for tropical pastures (Sanchez, 2000) (Table 21.3).

TABLE 21.2

Biomass of Pasture Components (grass, legume, weeds) in 5-Year-Old, Low Phosphorus Input (ASP1) vs. Mechanized and Limed (ASP2) Inga-Mahogany Systems Established on Degraded Pastures Near Manaus, Brazil

	Planted Species			Natural Regeneration			Biomass	
Treatment	Legumes	Pasture Grass	Subtotal	Unpalatable Weeds	Other Weeds	Subtotal	Total	Litter
Dry matter (t ha^{-1})								
ASP1	3.89 a[a]	0.90 b	4.79 b	1.76 a	1.30 a	3.06 a	7.85 a	6.40 b
ASP2	3.36 a	3.38 a	6.74 a	0.60 b	0.58 b	1.17 b	7.91 a	7.29 a
Total dry matter (%)								
ASP1	49.58 a[a]	11.40 b	60.98 b	22.43 a	16.60 a	39.02 a	100	—
ASP2	42.46 a	42.70 a	85.16 a	7.55 b	7.29 b	14.83 b	100	—

[a] Numbers in each column with the same letter are not significantly different at $P > 0.01$ by Tukey test.

Source: Authors' data.

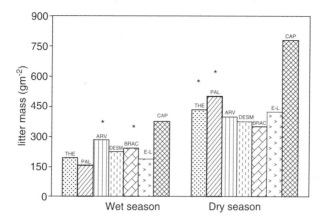

FIGURE 21.1

Litter-layer mass in the wet and dry seasons under agroforestry systems and secondary forest in central Amazonia, Brazil. THE, Theobroma system; PAL, Palm system; ARV, Mahogany trees; DESM, Desmodium; BRAC, Brachiaria; E–L, inter-rows; and CAP, second growth. *Source:* From Tapia-Coral, S.C., Luizao, F., Wandelli, W., Fernandes, E.C.M., *Agroforest. Syst.*, **65**, 33–42 (2004).

21.7 Discussion

The degraded pastures in the Amazon, although highly altered, can be valuable for human use and provide important ecosystem services, such as watershed protection, biodiversity niches, soil fertility recovery by improved fallows (Szott et al., 1991a, 1991b), and sinks for carbon (Fearnside and Guimaraes, 1996; Silver et al., 2000).

The Inga-mahogany–pasture system described is a promising approach to reintroducing mahogany to deforested and degraded lands in the Amazon and upgrading their economic value as well as biological productivity. Our results confirm that mahogany can be sustainably produced by smallholder farmers in association with food crops and pasture species using a range of INM and IPM strategies. However, owing to the severe nutrient mining and soil degradation prevalent on many degraded pastures, it will be necessary for farmers to have access to modest inputs to prime the system to facilitate the efficient functioning of biological nutrient and pest management strategies. For example,

TABLE 21.3

Aboveground Biomass and Nutrient Stocks in 9-Year-Old Mahogany–Pasture System Established on Degraded near Manaus, Brazil

Species	Biomass (t ha^{-1})	N (kg ha^{-1})	P (kg ha^{-1})	K (kg ha^{-1})	Ca (kg ha^{-1})	Mg (kg ha^{-1})
Schizolobium amazonicum (Paricá)	5.86	23.3	3.4	17.9	25.6	2.0
Swietenia macrophylla (Mahogany)	5.23	16.4	3.3	14.3	31.5	2.2
Brachiaria humidicola/brizantha	4.19	46.1	4.19	45.3	19.04	8.6
Desmodium ovulifolium	4.49	67.3	4.5	32.1	35.7	11.0
Invasives[a]	2.25	35.8	2.8	37.0	19.5	5.0
Gliricidia sepium (live fence)	10.48	73.13	5.43	47.5	41.5	7.3
Totals	32.5	262.03	23.62	194.1	172.8	36.1

Nutrient values are weighted averages, calculated as nutrient level x dry weight of trunk + branch + leaves, multiplied by aboveground biomass contribution species^{-1} ha^{-1}.

[a] Predominantly *Rollandra fruticosa* and *Borreria verticillata*.

Rueda et al. (2003) found that in the western Amazon of Brazil, more intensive beef production by judicious fertilization of grass-legume pastures and greater stocking density was the preferable strategy for owners of cattle systems to improve economic returns under current conditions. Ranchers in the state of Acre have also stopped burning pastures to improve pasture palatability and instead use solar-powered electric fencing to facilitate rotational grazing of grass-legume (*Pueraria phaseoloides*) pastures with marked improvement in productivity and sustainability of the pastures.

In 2001, Brazil banned the further exploitation and export of mahogany because of fears that the remaining populations are rapidly becoming endangered. In its native range, mahogany normally occurs at a density of one to three trees per hectare. In our trials we were able to establish approximately 30 mahogany trees per hectare with clean, straight stems, using a combination of INM and IPM techniques that minimized stem damage and mortality by *H. grandella*.

The cultural preferences for beef cattle systems in Brazil will continue for the foreseeable future. The demand and price for Brazilian beef has increased dramatically in recent years because Brazil has largely eliminated foot-and-mouth disease in its beef herd, and consumers are increasingly wary of "mad cow" disease (BSE) in European and North American herds. Our mahogany–pasture system can be applied to supply two of the major commodities (mahogany and beef) via intensively managed systems established on already deforested or degraded pasturelands. Given that these systems also sequester carbon over long rotations, farmers could be provided with payments for carbon sequestration to offset the installation and maintenance costs of the system in the early years.

References

Ackerman, I.L., McCallie, E.L., and Fernandes, E.C.M., Inga and insects: The potential for management in agroforestry, In: *The Utilization of the Genus Inga (Fabacae)*, Pennington, T.A., Ed., Royal Botanic Gardens/International Centre for Research on Agroforelstry, Kew, UK/ Nairobi (1998).

Ae, N. et al., Phosphorus uptake by pigeon pea and its role in cropping systems of the Indian subcontinent, *Science*, **248**, 477–480 (1990).

Altieri, M.A., *Agroecology: The Science of Sustainable Agriculture*, Westview Press, Boulder, CO (1995).

Andow, D.A., Vegetational diversity and arthropod population response, *Annu. Rev. Entomol.*, **36**, 561–586 (1991).

Arevalo, L.A. et al., The effect of cattle grazing on soil physical and chemical properties in a silvopastoral system in the Peruvian Amazon, *Agroforest. Syst.*, **40**, 109–124 (1998).

Barros, E. et al., Development of the soil macrofauna community under silvopastoral and agrosilvicultural systems in Amazonia, *Pedobiologia*, **47**, 273–280 (2003).

Beets, W.C., *Multiple Cropping and Tropical Farming Systems*, Westview Press, Boulder, CO (1982).

Broughton, W.J., Effect of various covers on soil fertility under *Hevea brasiliensis* and on growth of the tree, *Agro-Ecosystems*, **3**, 147–170 (1977).

Carrasco, F., El "defoliador del pacae" — *Automolis inexpectata* Rothschild (Lepidoptera Arctiidae) en el Departamento del Cuzco, *Revista Peruana de Entomología*, **14**, 140–142 (1971).

Chang, J.H., Tropical agriculture: Crop diversity and crop yields, *Econ. Geogr.*, **53**, 241–254 (1977).

Clawson, D.L., Harvest security and intraspecific diversity in traditional tropical agriculture, *Econ. Bot.*, **39**, 56–67 (1985).

Cochrane, T.T. and Sanchez, P.A., Land resources, soils and their management in the Amazon region: A state of knowledge report, In: *Amazonia: Agriculture and Land Use Research*, Hecht, S.B., Ed., International Center for Tropical Agriculture, Cali, Colombia, 137–209 (1982).

Cullota, E., Exploring biodiversity's benefits, *Science*, **273**, 1045–1046 (1996).

Davidson, E.A. et al., Nitrogen and phosphorus limitation of biomass growth in a tropical secondary forest, *Ecol. Appl.*, **14**, S150–S163 (2004).

Denevan, W.M. and Padoch, C., Introduction: The Bora agroforestry project, In: *Swidden-Fallow Agroforestry in the Peruvian Amazon, Adv. Econ. Bot.*, 5. Denevan, W.M. and Padoch, C., Eds., 1–7 (1987).

Eden, M.J. et al., Effects of forest clearance and burning on soil properties in northern Roraima, Brazil, *For. Ecol. Manag.* **38**, 283–290 (1991).

Fearnside, P.M., Amazonian deforestation and global warming: Carbon stocks in vegetation replacing Brazil's Amazon forest, *For. Ecol. Manag.*, **80**, 21–34 (1996).

Fearnside, P.M. and Guimaraes, W.M., Carbon uptake by secondary forests in Brazilian Amazonia, *For. Ecol. Manag.*, **80**, 35–46 (1996).

Feldpausch, T.R. et al., Carbon and nutrient accumulation in secondary forests regenerating on pastures in central Amazonia, *Ecol. Appl.*, **14**, S164–S176 (2004).

Fernandes, E.C.M., Alley Cropping on Acid Soils, PhD dissertation, North Carolina Sate University, Raleigh, NC (1990).

Fernandes, E.C.M., The effects of plant and soil management on nodulation and nitrogen-fixation in the genus Inga, In: *The Utilization of the Genus Inga (Fabacae)*, Pennington, T.A., Ed., Royal Botanic Gardens/International Centre for Research on Agroforestry, Kew, UK/Nairobi (1998).

Fernandes, E.C.M. et al., The impact of selective logging and forest conversion for subsistence agriculture and pastures terrestrial nutrient dynamics in the Amazon, *Cienc. Cult.*, **49**, 34–47 (1997).

Fernandes, E.C.M., Davey, C.B., and Nelson, L., Alley cropping on an Ultisol in the Peruvian Amazon: Mulch, fertilizer and tree root pruning effects, In: *Technologies for Sustainable Agriculture in the Tropics*, ASA Special Publication 56, Ragland, J. and Lal, R., Eds., Agronomy Society of America, Madison, WI, 77–96 (1993a).

Fernandes, E.C.M. et al., Use and potential of domesticated trees for soil improvement, In: *Tropical Trees: The Potential for Domestication*, Leakey, R.R.B. and Newton, A.C., Eds., HMSO, London, 218–230 (1993b).

Fernandes, E.C.M. and Matos, J.C., Agroforestry strategies for alleviating soil chemical constraints to food and fiber production in the Brazilian Amazon, In: *Chemistry of the Amazon: Biodiversity, Natural Products, and Environmental Issues*, Seidl, P.R., Gottlieb, O.R., and Kaplan, M.A.C., Eds., American Chemical Society, Washington, DC, 34–50 (1995).

Fernandes, E.C.M. and Nair, P.K.R., An evaluation of the structure and function of tropical homegardens, *Agric. Syst.*, **21**, 279–310 (1986).

Francis, C.A., *Multiple Cropping Systems*, MacMillan, NY (1986).

Gehring, C. et al., Response of secondary vegetation in eastern Amazonia to relaxed nutrient availability constraints, *Biogeochemistry*, **45**, 223–241 (1999).

Giller, K.E., Beare, M.H., Lavelle, P., Izac, A.-M., and Swift, M.J., Agricultural intensification, soil biodiversity and agroecosystem function, *Appl. Soil Ecol.*, **6**, 3–16 (1997).

Harwood, R.R., *Small Farm Development: Understanding and Improving Farming Systems in the Humid Tropics*, Westview Press, Boulder, CO (1979).

INPE, http://www.inpe.br/Informações_Eventos/amazonia.htm. Instituto Nacional de Pesquisas Espaciais, São José dos Campos, Brazil, 2004.

Keller, M. et al., Ecological research in the large-scale biosphere atmosphere experiment in Amazônia (LBA): Early results, *Ecol. Appl.*, **14**, S3–S16 (2004).

Kerr, W.E. and Posey, D.A., Informações adicionais sobre a agricultura dos Kayapó, *Interciencia*, **9**, 392–400 (1984).

Koptur, S., Experimental evidence for defense of Inga (Mimosoideae) saplings by ants, *Ecology*, **65**, 1787–1793 (1984).

Laurance, W.F., Reflections on the tropical deforestation crisis, *Biol. Conserv.*, **91**, 109–117 (1999).

Lawrence, A., Farmer knowledge and use of Inga species, In: Nitrogen-Fixing Trees for Acid Soils: Proceedings of a workshop held July 3–8, 1994, Turrialba, Costa Rica, Evans D.O. and Szott, L.T., Eds., Nitrogen-Fixing Tree Association, Bangkok, Thailand, 142–151 (1995).

León, J., Central American and West Indian species of Inga, *Ann. Mo. Bot. Gard.*, **53**, 265–369 (1966).

Leston, D., The ant mosaic: Tropical tree crops and the limiting of pests and diseases, *Pest. Art. News Summ.*, **19**, 311–341 (1973).

McCaffery, K.A., Carbon and nutrients in land management strategies for the Brazilian Amazon, Ph.D. dissertation, Cornell University, New York, USA (2003).

McKerrow, A.J., Nutrient stocks in abandoned pastures of the Central Amazon Basin prior to and following cutting and burning, MSc thesis, North Carolina State University, Raleigh, NC (1992).

Nepstad, D., Uhl, C. and Serrao, E.A., Surmounting barriers to forest regeneration in abandoned, highly degraded pastures: A case study from Paragominas, Parà, Brazil, In: *Alternatives to Deforestation: Steps toward Sustainable Use of the Amazon Rain Forest*, Anderson, A.B., Ed., Columbia University Press, New York, 215–229 (1990).

Niell, D.A. and Revello, N., Silvicultural trials of mahogany (*Swietenia macrophylla*) interplanted with two Inga species in Amazonian Ecuador, In: *The Genus Inga: Utilization*, Pennington, T. and Fernandes, E.C.M., Eds., Royal Botanic Gardens, Kew, UK, 141–150 (1998).

Okafor, J.C. and Fernandes, E.C.M., Compound farms of southeastern Nigeria: A predominant agroforestry homegarden system with crops and small livestock, *Agroforest. Syst.*, **5**, 153–168 (1987).

Paddoch, C. and de Jong, W., The house gardens of Santa Rosa: Diversity and variability in an Amazonian agricultural system, *Econ. Bot.*, **45**, 166–175 (1991).

Parrotta, J.A., Turnbull, J.W., and Jones, N., Introduction: Catalyzing native forest regeneration on degraded tropical lands, *For. Ecol. Manag.*, **99**, 1 (1997).

Patiño Valera, R., Genetic resources of *Swietenia* and *Cedrela* in the Neotropics: Proposals for coordinated action, *FAO Forest Genet. Resour.*, **25**, 20–32 (1997).

Pennington, T.D., *The genus Inga: Botany*, Royal Botanic Gardens, Kew, UK (1997).

Pennington, T.D. and Fernandes, E.C.M., Eds., *The genus Inga: Utilization*, Royal Botanic Gardens, Kew, UK (1998).

Peoples, M.B. and Herridge, D.F., Nitrogen fixation by tropical legumes, *Adv. Agron.*, **44**, 155–223 (1990).

Perfecto, I.R. et al., Shade coffee: A disappearing refuge for biodiversity, *BioScience*, **46**, 598–608 (1996).

Posey, D.A., A preliminary report on diversified management of tropical forest by the Kayapó Indians of the Brazilian Amazon, *Adv. Econ. Bot.*, **1**, 112–126 (1984).

Posey, D.A., Native and indigenous guidelines for new Amazonian development strategies: Understanding biological diversity through ethnoecology, In: *Change in the Amazon Basin*, Vol. 1, Hemming, J., Ed., Manchester University Press, Manchester, UK, 156–181 (1985).

Power, A.G., Arthropod diversity in forest patches and agroecosystems in tropical landscapes, In: *Forest Patches in Tropical Landscapes*, Schelhas, J., and Greenberg, R., Eds., Island Press, Washington, DC, 91–110 (1996).

Pretty, J., *Regenerating Agriculture*, Joseph Henry Press, Washington, DC (1995).

Rueda, B., Nutrient Dynamics and Productivity Potentials of Pasture-Based Cattle Systems in the Western Amazon of Brazil, PhD dissertation, Cornell University, Ithaca, NY (2002).

Rueda, B. et al., Production and economic potentials of cattle in pasture-based systems of the western Amazon region of Brazil, *J. Anim. Sci.*, **81**, 2923–2937 (2003).

Sanchez, P.A., Linking climate change research with food security and poverty reduction in the tropics, *Agric. Ecosyst. Environ.*, **82**, 371–383 (2000).

Sanchez, P.A. et al., Soil fertility replenishment in Africa: An investment in natural resource capital, In: *Replenishing Soil Fertility in Africa*, Special Publication No. 51, Buresh, R.J., Sanchez, P.A., and Calhoun, F., Eds., Soil Science Society of America and International Centre for Research on Agroforestry, Madison, WI, 1–46 (1997).

Schroth, G. et al., Subsoil accumulation of mineral nitrogen under polyculture and monoculture plantations, fallow and primary forest in a ferralitic Amazonian upland soil, *Agric. Ecosyst. Environ.*, **75**, 109–120 (1999).

Serrão, A.E. et al., Soil alterations in perennial pasture and agroforestry systems in the Brazilian Amazon, In: *Soil Management: Experimental Basis for Sustainability and Environmental Quality*, Lal, R. and Stewart, B.A., Eds., CRC Press, Boca Raton, FL, 85–104 (1995).

Silver, W.L., Brown, S., and Lugo, A.E., Effects of changes in biodiversity on ecosystem function in tropical forests, *Conserv. Biol.*, **10**, 17–24 (1996).

Silver, W.L., Ostertag, R., and Lugo, A.E., The potential for carbon sequestration through reforestation of abandoned tropical agricultural and pasture lands, *Restor. Ecol.*, **8**, 394–407 (2000).

Smyth, T.J. and Cravo, M.S., Aluminum and calcium constraints to continuous crop production in a Brazilian Amazon Oxisol, *Agron. J.*, **84**, 843–850 (1992).

Snapp, S.S., Mafongoya, P.L., and Waddington, S., Organic matter technologies for integrated nutrient management in smallholder cropping systems of southern Africa, *Agric. Ecosyst. Environ.*, **71**, 185–200 (1998).

Subler, S. and Uhl, C., Japanese agroforestry in Amazonia: A case study in Tomé Açu Brazil, In: *Alternatives to Deforestation: Steps toward Sustainable Use of the Amazon Rainforest*, Anderson, A.B., Ed., Columbia University Press, New York, 152–166 (1990).

Szott, L.T., Fernandes, E.C.M., and Sanchez, P.A., Soil–plant interactions in agroforestry systems, *For. Ecol. Manag.*, **45**, 127–152 (1991a).

Szott, L.T., Palm, C.A., and Sanchez, P.A., Agroforestry in acid soils of the humid tropics, *Adv. Agron.*, **45**, 275–301 (1991b).

Tapia-Coral, S.C. et al., Carbon and nutrient stocks in the litter layer of agroforestry systems in central Amazonia, Brazil, *Agroforest. Syst.*, **65**, 33–42 (2004).

Thrupp, L.A., *Cultivating Diversity: Agrobiodiversity and Food Security*, World Resources Institute, Washington, DC (1998).

Thurston, D., *Slash/Mulch Systems: Sustainable Methods for Tropical Agriculture*, Westview Press, Boulder, CO (1997).

Tilman, D., Wedin, D., and Telford, A.D., Productivity and sustainability influenced by biodiversity in grassland ecosystems, *Nature*, **379**, 718–720 (1996).

Turner, M.G., Gardner, R.H., and O'Neill, R.V., Ecological dynamics at broad scales: Ecosystems and landscapes, *Bioscience*, S29–S35 (1995).

Uhl, C., Buschbacher, R., and Serrão, E.A.S., Abandoned pastures in eastern Amazonia I. Patterns of plant succession, *J. Ecol.*, **76**, 663–681 (1988).

Verissimo, A. et al., Extraction of a high-value natural resource in Amazonia: The case of mahogany, *For. Ecol. Manag.*, **72**, 39–60 (1995).

22

Direct-Seeded Tropical Soil Systems with Permanent Soil Cover: Learning from Brazilian Experience

Lucien Séguy,[1] **Serge Bouzinac**[1] **and Olivier Husson**[2]
[1]*CIRAD, Goiânia, Brazil*
[2]*CIRAD/GSDM, Madagascar*

CONTENTS

Almost one quarter of the land in Brazil is covered by a 200-million-hectare ecosystem known as the *cerrados,* of which at least 50 million hectares can potentially be used for intensive mechanized agriculture (EMBRAPA, 1998). A large proportion of this arable land is in the vast humid tropical zone that lies to the south and west of the Amazon forest. It has been proposed that intensive, high-input cultivation of this "reservoir" of land could produce, without supplemental irrigation, more than 150 million tons of grain, 9 million tons of meat, and more than 300 million m^3 of wood annually while retaining 20% of the area for environmental conservation reasons (Goedert, 1989). The *cerrados* thus constitute a vast, as yet little-exploited reserve available for meeting food needs in the twenty-first century. However, this vast ecosystem suitable for perennial crops, annual food, and industrial crops as well as livestock production should be exploited in a sustainable manner, without degrading it.

Cultivation of the *cerrados* as a humid tropical savannah started at the end of the 1970s when farmers migrated from the southern states of Brazil and rapidly colonized the central parts of the region and then the more humid west. Mechanized agriculture introduced after land clearing established cropping systems based at first on rain-fed rice and extensive grazing (Brachiaria spp.). These were superseded by the industrial cultivation of soybeans as a major crop for export. Farming methods imported from temperate or subtropical situations, based on disk-plowing and monocropping, soon turned out to be disastrous, given high precipitation from 1500 to more than 2000 mm distributed over only 7 months. Soil erosion across whole landscapes ensued (Derpsch et al., 1991), causing first a gradual decrease and then a rapid continuous fall in soil productivity in spite of the increased use of chemical inputs (inorganic fertilizer, pesticides), resulting in some spectacular bankruptcies (Séguy et al., 1996).

CIRAD and its various partners have been operating on the pioneer fronts in the central northern Mato Grosso where more than 1.3 million hectares are cultivated today. The aim was to create the basis for sustainable agriculture in order to help settle this area more sustainably and thus to reduce the pressure on hitherto uncolonized zones. Operations were first conducted in the savannah area from 1983 to 1994, and then in the forest zone as well, to protect the environment in preparation for the anticipated arrival of mechanized operators. As the economic context was unstable and the physical environment very fragile, the sustainable management of soil resources at least cost was the major research objective, being concerned, at the same time, with a reduction of economic risk.

Starting from the widespread soybean monoculture, which was proving to be disastrous for the physical environment, efforts to reduce such damage and economic risk led to gradual development of diversified direct-seeded, mulch-based cropping systems. When direct seeding is combined with permanent soil cover, we refer to such systems as DSPSC. The initial research effort has supported very substantial and rapid expansion of direct seeding in the *cerrados* from the early 1990s onward. It has been enhanced by the work of John Landers and farmer partners in southern Gioás state in the early 1980s, and more recently with dissemination carried out by the Association for Direct Seeding in the Cerrados (APDC) and the regional *Clubs des Amis de la Terre* (CATs). More than 5 million hectares were covered by the new practices within 10 years (FBDP, 2001).

The experience gained in Brazil under these very aggressive climatic conditions, the understanding acquired of how soil systems function, and the identification of universal agronomic principles have all been used to develop concepts for designing sustainable cropping practices in various environments, ranging from tropical, hot/humid climates to subtropical, high-altitude climates and semiarid areas. All designs had to meet the same requirements: cropping systems should be sound and reproducible from an agronomic point of view, technically practicable with conservation of soil fertility as a prime condition, efficient in the use of available water, economically profitable, and more stable than the systems currently in use.

Along with sustainable productivity, these soil systems were planned to increase farmers' value added through diversification (introducing crop rotation) and the quality of products. Varietal improvement programmes were also set up, especially for high-technology rain-fed rice and for cotton. The cropping systems were all developed together with farmers, having both practicability and profitability as major concerns. The knowledge and know-how acquired in this process now allows CIRAD to propose, on a worldwide basis, a range of cropping systems and practices that are adapted to various soil, climate and socioeconomic situations.

22.1 The Rain Forest as a Model for Agricultural Systems

During the research and development phase, the design of direct-seeding techniques benefited from an understanding of the functioning of the major ecosystem that prevails in the region: the existing forest ecosystem. This is naturally efficient and sustainable even under aggressive climatic conditions. The key features of natural forest ecosystems are:

- Total physical protection of the soil by permanent soil cover.
- Differentiated horizons where the top horizon (5 cm) supports intense biological activity that beneficially affects soil structure, nutrient availability, and organic matter dynamics, and supplies most plants' nutrients (Stark and Jordan, 1978).
- Very high levels of primary biomass productivity by making the optimal use of soil resources, even ones chemically very depleted and with acidic substrates, and by optimizing the use of climatic resources, with plants growing for as much time as the climate allows, thereby maximizing photon interception.
- Retention of most of the systems' nutrients in the biomass rather than in the soil by introducing and supporting plants with deep rooting systems. These minimize nutrient loss through their recycling action and sustain the soil–crop system in a relatively closed manner.

These last two factors imply a fast turnover of organic matter and nutrients that is discussed in the next section.

The challenge was to mimic such ecosystem functioning, adapting it to cropped ecosystems. The approach adopted was to introduce elements of cropping practices sequentially, aiming to transform the functioning of the cropped soil systems so that, starting from degraded soils, it would be possible to recover gradually the original mode of ecosystem functioning that can be found under natural forest cover, while at the same time developing higher-yielding, cost-effective, diversified, sustainable, and "clean" farming systems.

22.2 Two Basic Concepts for Soil System Management: Plants as Nutrient Pumps, and the Multifunctionality of Cover Crops

This reconversion required several steps, reflected in the evolution of cropping systems in Mato Grosso as discussed below in the sections on the evolution of crop productivity and soil functioning, and the impacts of direct seeding with permanent soil cover. These systems have been gradually reconstructed with direct-seeding practices through the insertion of supplementary biomass-generating crops, which are intercropped before or after the commercial crops or in relay. The use of cover crops is discussed in more general terms, on a worldwide scale, in Chapter 30. Here, we consider cover crops in the specific transformation of agroecosystems in Brazil.

Cover crops can play a key role in farming systems by mimicking the natural forest ecosystem. They produce biomass during periods when no commercial crops are grown, utilizing available resources through their strong and well-developed root systems. By recycling a major share of the nutrients that would otherwise be leached away, these plants prevent the soil system from letting scarce crop resources seep or erode away. They can therefore be referred to as "nutrient pumps" (Séguy et al., 1998a, 1998b). The additional biomass that they produce is enough to keep the soil permanently covered even under humid tropical conditions with intense rainfall. Implementation of this nutrient-pump concept was an essential first step toward the introduction of sustainable management of grain cropping systems in these regions.

The performance of these systems was then improved through the concept of "multiple function," which can be applied to additional cover crops (Séguy et al., 2001a). Cover crops, in addition to their primary performance as nutrient pumps, can fulfill other agronomic and ecological functions. By supplementing the action of commercial crops, they enhance the efficiency of the whole system (see Figure 22.1). DSPSC functions occur both above- and belowground.

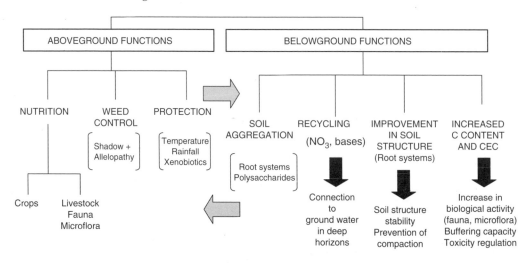

FIGURE 22.1
The multiple functions and effects of soil system processes.

22.2.1 Aboveground

22.2.1.1 Total and Permanent Protection of the Soil Surface

This protects against harsh climatic conditions and other stresses. The mulch layer provides a water- and temperature-regulating shield, a protective screen for fauna and against pesticide compounds, and a buffer to avoid soil compaction under the weight of heavy equipment or animals.

22.2.1.2 Increasing Water Supplies Available for Crops

The mulch that these crops produce reduces runoff and direct evaporation from the soil. The extent of these effects will vary according to the crop residue type and quantity (Scopel et al., 1998).

22.2.1.3 Nutritional Improvements for the Main Crop

Nutrients are returned to the system via mulch mineralization, regulated by the C:N ratio and lignin contents of the aboveground and root parts of the crops. There are also nutritional benefits for livestock when livestock production is supported by the forage produced by cover crops, and for soil fauna and microflora, recuperating biodiversity.

22.2.1.4 Weed Control

Shade and allelopathic effects are utilized for this purpose, an example being the control of *Cyperus rotondus* by using a sorghum cover crop (Séguy et al., 1999).

22.2.2 Belowground

22.2.2.1 Aggregating the Soil

Within the top few centimeters, well-developed root systems provide support for the soil, stabilize it, and avoid compaction.

22.2.2.2 Restructuring Soil through the Aggregation Activities of Crop Root Systems

These systems enhance soil porosity which boosts infiltration and aeration. It also ensures quick drying of the soil profile (rapid drainage of excess water) and high water retention capacity (microporosity). The soil becomes highly resistant to compaction from the movement of heavy machinery and animals, which is reduced with direct seeding anyway. Soil porosity is efficiently maintained via many galleries left in the soil by decomposing roots. Structural stability is enhanced by the production of highly efficient substances for soil aggregation, such as polysaccharides secreted by the roots and arbuscular endomycorrhiza (Doss et al., 1989). *Eleusine coracana*, *Brachiaria ruziziensis*, *B. decumbens* and *B. humidicola* are ideal species in this respect because their roots are highly sheathed in a protective microaggregate sleeve.

22.2.2.3 Tapping Deep Groundwater

There is usually water available in the horizon below that normally exploited by commercial crops, as happens in forest ecosystems during the dry season. The capacity to tap deep reserves of water enables green biomass production during the dry season. It enhances continuous carbon supplies in the deep soil layers because of root production, and maintains sustained biological activity throughout most of the year.

22.2.2.4   Recycling of Nutrients Leached to Deep Soil Horizons

This keeps the soil–crop system relatively closed. Nutrients are drawn up to the surface through very strong root systems that grow down to deep horizons and have a high nutrient and organic molecule-capturing potential. Especially potassium, calcium, and magnesium, along with minerals such as silicon and aluminum, which are critical for the soil mineral composition, are recycled this way (Lucas et al., 1993).

22.2.2.5   Fertility Mobilizing Capacity

Nutrients are extracted through aggressive root systems even in very poor and acid soils, and crops are then provided access to these nutrients via dry matter mineralization. The grass species Eleusine and Brachiaria, for example, fix nitrogen in their rhizospheres by the activity of nonsymbiotic bacteria and of arbuscular mycorrhizas that are capable of mobilizing insoluble phosphorus molecules (Doss et al., 1989). Also, legume species such as Cajanus, Crotalaria, and Stylosanthes symbiotically fix airborne free nitrogen. Such grass and legume crops can be mixed to obtain multiple-function nutrient pumps.

22.2.2.6   Development of High Levels of Biological Activity

The high biomass input to soil systems from DSPSC, derived from both aboveground and belowground plant functioning, provides ideal cropping environments. The soil is protected with very little cultivation, thus favoring the development and activity of soil fauna and microflora. Their activity promotes the quality of nutrients recovered in the system and enhances soil porosity.

22.2.2.7   Using Detoxification Potential for Bioremediation Against Polluting Pesticide Compounds

A mixed sorghum + Crotalaria cover has been found to be highly efficient for recycling the compound Sulfentrazone, for example (Séguy and Maeda, unpublished data). Some cover crops (Cassia, Brachiaria, and Stylosanthes species) can mitigate aluminum toxicity problems or excessive salinity. Various organic acids released during cover-crop biomass mineralization have a high neutralizing and complex-forming potential (Miyazawa et al., 2000). Thus, the soil system, when infused with biological diversity, has its own means for maintaining productive functioning.

22.3   Selection and Use of Plant Species to Enhance Soil System Performance

Selected multiple-function, nutrient-pump crops are usually planted at the onset of the rainy season. They are either cut and dried before regeneration or killed back with herbicide to form a mulch layer for the commercial crops, or they are left on the field after harvesting at the end of the rainy season when they have been utilized by farmers as an attractive added-value crop. These nutrient pumps are chosen on the basis of their ability to tap available runoff water at the beginning of the rains and deep groundwater after the rainy season ends. Often they must function under extremely variable rainfall conditions. High biomass production both at the beginning and after the end of the rainy season is always the main goal (Séguy et al., 1996).

At the end of the rainy season, if rainfall conditions are suitable, additional high biomass-producing species can be intercropped to tap deep groundwater. Their sowing is staggered according to the period of the rainy season and associated risks; maize is sown earlier than sorghum or millet, for instance. Species with very deep rooting systems that continue to produce biomass throughout the dry season are of particular interest, especially if they can provide forage, e.g., Brachiaria, Stylosanthes, and Cajanus, so that livestock grazing them generate supplementary income for farmers (Séguy et al., 2003a, 2003b).

Perennial species that produce runners and rhizomes (Arachis, Stylosanthes, and Pueraria legume species, and Cynodon, Paspalum, Stenotaphrum, and Pennisetum grass species) can also be used as nutrient-pump cover crops. They form living perennial forage covers whose growth can be controlled with very low dosage, nonpolluting herbicide treatments to keep them from competing with the commercial crops. They recover full vegetative growth after the commercial crop is harvested and can be grazed during the dry season (Séguy et al., 2001a).

All perennial species used as living cover exclude annual weeds, thus simplifying the job for farmers, who only have to manage the living cover and the commercial crop which together constitute the cropping system. Intercropped nutrient pumps, which become functional at the end of the rainy season and continue during the dry season, e.g., living perennial covers, can produce abundant biomass throughout the year when well managed within cropping systems. During the dry season, which is cool in the *cerrados*, organic matter mineralization is minimal. High biomass production on the surface and in deep horizons enables maximal carbon accumulation and powerful recycling of leached base compounds and nitrates (Séguy et al., 2001a).

Nutrient-pump species can be planted in cropping systems, depending on the objectives, either by broadcasting under the cover of the commercial crop or by direct seeding for pure or mixed crops. While serving multiple agronomic and ecological functions, these plants must meet certain technical and economic criteria to facilitate their cost-effective and large-scale adoption and duplication by farmers (Séguy et al., 1996). They must be *user-friendly*, being easy to sow and harvest and amenable to good technical control in cropping systems. They must also have *high value-added*, giving bumper forage and grain crops for livestock feed during the dry season and contributing, if possible, to human consumption, e.g., complements to wheat flour, or ingredients for beer brewing or alcohol making.

22.4 Review of Alternative Crop Species

All cover crops do not fulfill these different functions with the same degree of efficiency. Table 22.1 summarizes properties of the main multifunctional cover crops used in DSPSC systems in Brazil, but increasingly in other countries as well. Figure 22.2 shows the evolutionary process for designing cover crop use for biological soil improvement, with a progressive shift from selection of species used as dead mulch, to plants used as living covers and associations of different species that maximize the efficiency of these multifunctional cover crops.

The most suitable systems will be, ultimately, those that through different crop rotations best address the constraints and production objectives of farmers in a given region. Various options are often identified and tested by farmers to expand upon the initial choices. By applying the nutrient-pump concept and by exploiting the multiple functions of cover crops, a number of DSPSC systems have been developed under diverse ecological conditions, being open to expanded functions and additional species capable of efficiently fulfilling these functions.

TABLE 22.1

Properties of the Main Multifunctional Cover Crops Used in DSPSC Systems in Brazil

	Millet	Sorghum	Eleusine coracana	Maize, Millet or Sorghum Associated with Brachiaria ruziziensis or Stylosanthes guyanensis	Cynodon dactylon or Tifton 85	Arachis pintoi cv. Amarillo
Rooting speed	Fast: 2–3 cm day^{-1}	Fast: 2–3 cm day^{-1}	Very fast: 3–5 cm day^{-1}	Fast	Fast	Fast
Roots biomass (at 90 days)	Medium	High	Very high	Very high (Brachiaria)	Very high (rhizomes + stolons)	Medium (stolons)
Root C:N ratio	C:N 41	C:N 60	C:N 51	C:N 35–38	NA	NA
Soil structure improvement	Medium	High	Extremely high	Very high	Very high	Very high
C injection in the profile	Medium (90 days)	High (90–100 days)	Very high (90–100 days)	High (90–100 days) to very high (150–210 days)	High (continuous)	High (continuous)
Speed of decomposition	Fast	Slow	Medium	Medium	Slow	Fast
N immobilization	Low	High	Medium	Medium	Medium	Very low
C:N ratio	C:N: 22–27	C:N: 41–49	C:N: 35	C:N: 77	C:N: NA	C:N: NA
Recommended N application (kg N ha^{-1})	10–15	30	15–20	15–20	20–25	0
Acidity neutralization	NA	NA	High	High		High
Forage quality	Good	Good	Excellent	Excellent	Excellent	Excellent
Ability to control dicotyledons	Medium	High	High	Very high	Very high	Very high
Ability to control grasses	Medium	Very high	High	Very high	Very high	Very high
Ability to control vegetal pests (e.g., Cyperus rotundus)	Low	Very high		Very high	Very high	Very high
Risk of crop infestation after desiccation	Medium (grains)	High (grains + regrowth)	High (grains)	Very low to nil	Very high	Very high
Desiccation of cover crop before sowing	Easy (glyphosate, 2,4-D)	Easy (glyphosate)	Easy (glyphosate, 2,4-D)	Easy (glyphosate)	Easy (paraquat)	Easy (diquat)
Herbicide needs during cropping period	Medium to high	Low to very low	Medium	Low to nil	Very low	Very low
Nutrients very well recycled			K, Ca, Mg			Mg, Zn, Cu, P, K

Source: Adapted from Séguy et al. (2001a).

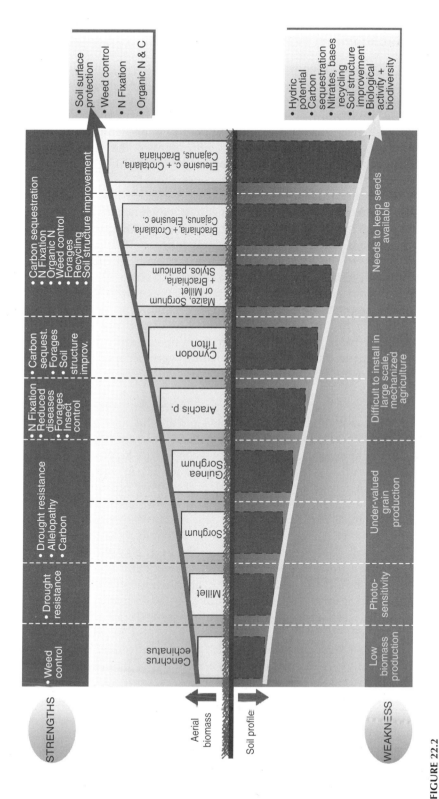

FIGURE 22.2

A comparison of cover-crop species for use in sustainable soil systems. *Source:* Adapted from Séguy et al. (2001c).

22.5   Development of Cropping Systems for the *Cerrados* in Brazil: Adaptation of Direct Planting Systems to Humid Tropical Climates

The evolution of cropping systems in the *cerrados* and in the humid forest ecology of central northern Mato Grosso between 1986 and 2000 is shown in Figure 22.3, along with indications of their consequences for biomass production and for nutrient and water use. This evolution in cropping systems is closely related to the understanding of soil functioning that was developed during the same period. In an iterative process, a better understanding of how soil–plant systems function allowed the introduction of new practices. These in return provided useful knowledge on potential changes in the soil systems made possible by different cropping practices.

By the mid-1980s, crop rotations were introduced (rice or maize in rotation with soybean) to get away from monocropping and to reduce its negative effects, especially on pest and disease incidence and on soil fertility. Direct seeding on crop residues, as practiced in the southern part of Brazil and in temperate or subtropical situations, was introduced to reduce erosion. However, in such a hot, wet climate, mineralization is extremely fast, and the biomass produced from a commercial crop by itself is not sufficient to keep the soil covered permanently and to control major weeds. The 6–8 ton ha^{-1} of biomass that the crop produces is less than is consumed in ongoing processes of mineralization, even for residues with high cellulose and lignin contents. Also, because of their rather shallow rooting (less than 80 cm), commercial crops do not improve soil porosity. Significant nutrient losses through leaching were occurring all year long, but were especially high at the start of the rainy season when a peak of mineralization is observed.

Systems that alternated two annual crops in succession were introduced at the beginning of the 1990s, with direct seeding in year 1 and another annual crop the following year. This alternation increased cropping intensity and biomass production as the first crop was the main commercial crop while the second crop, referred to locally as the *safrinha*, also played a role as a biological nutrient pump. In such systems, biomass production is greatly increased by a longer period of growth, possible because water in deep horizons (down to 2.5 m with plants such as sorghum or pearl millet) was being tapped to produce biomass during part of the dry season. With a total annual biomass production of 18–22 ton ha^{-1}, crop residues were sufficient to cover the soil permanently and control major weeds. Deep rooting of the *safrinha* crop also allows the recycling of nutrients that have been leached, and it improves soil structure at depths that facilitate better rooting of the subsequent commercial crop, enhancing its yield.

In the mid-1990s, this system was further improved by adding forage production to the *safrinha*. Forage species were selected for their ability to grow during the entire dry season, tapping water from very deep horizons (below 2.5 m) and for their forage quality. Brachiaria sp. and *Eleusine coracana* proved to be the most effective. With such systems, biomass production can reach 26–32 ton ha^{-1}. As most of this biomass is returned to the soil, organic matter content of the soil rapidly increases, improving soil structure and also fertility, given the recycling of nutrients that is occurring. This evolution is shown in Figure 22.3.

As the third millennium begins, the quest toward more biology-based agriculture is advancing, assisted by the development of various biological products for pest and disease control. Direct-seeding practices by themselves improve the health status of crops as better-fed and well-watered plants are less susceptible to diseases and pests. However, supplementing this protection is a growing number of biological products becoming

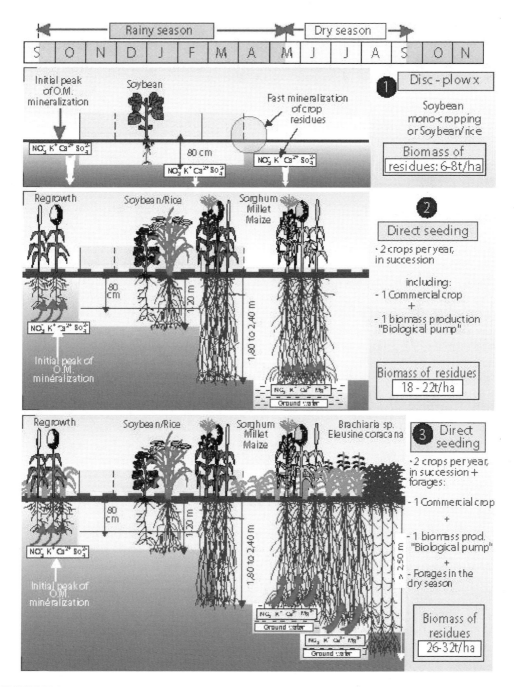

FIGURE 22.3

Evolution of cropping systems, biomass of residues, and water and nutrient use within *cerrados* and tropical humid forest ecologies, central northern Mato Grosso, 1986–2000. *Source:* From Séguy, L. et al., in *II World Congress on Conservation Agriculture, Iguassu Falls 11–15 August*, Federacéo Brasileira de Plantio Direto na Palha, Ponta Grosso, Paraná, Brazil, 2003a, 310–321.

available and affordable, such as *Bacillus thuringensis*, *Metharizium anisoplie*, *Beauveria bassiana*, Derris sp., and Azadiracta (neem) extracts, which are now used against insects as insecticides or repellents. Trichoderma can be used effectively against various fungi (Chapter 34). Elicitors made from enzymes, vitamins, and micronutrients are being found to increase natural resistance of plants against insects. Wood distillate can be used to stimulate the immune system of plants. Also, liquid humus can advantageously replace mineral fertilizers, at least in part. In Brazil, 6 l of liquid humus are giving fertilization effects equivalent to 90 kg N, producing healthier plants with better grain filling. In addition, enzymes can be used to speed up the mineralization of biomass with a high C:N ratio that is decomposing slowly, or they can activate plants' metabolism, increasing their chlorophyll synthesis, sugar production, etc. Such additional biological inputs can enhance DSPSC crop–soil management.

22.6   The Evolution of Crop Productivity in Mato Grosso, Brazil

Soybean is currently the main crop produced in the central northern region of Mato Grosso. Its productivity has been increased with the gradual improvement of DSPSC systems from 1700 to 2000 kg ha^{-1} in 1986 to more than 4500 kg ha^{-1} under experimental conditions since 2000. In the past 5 years, DSPSC systems have become increasingly well managed through the accumulation of knowledge, refining of practices, and improvement of varieties. The results of the research conducted by CIRAD show that soybean yields are closely correlated with the quantity and quality of the grasses used as either dead or live cover. For maize, sorghum, and millet, *Brachiaria ruziziensis* and *Eleusine coracana* grown as mulch are the most beneficial. As living cover, the advantages of *Cynodon dactilon* are becoming more evident.

With very small applications of inorganic fertilizer (40 kg P$_2$O$_5$ + 40 kg K$_2$O ha^{-1}), showing some benefit nutrient supplementation of the soil's own biological production capacity, soybean productivity has increased each year in the best DSPSC systems. Differences have ranged from 12 to 15% in the first year to 45 to 52% in the fifth year. The average annual increase in yield under DSPSC has been more than 700 kg ha^{-1} over a 5-year period.

22.6.1   Soybean

With a low level of fertilization, soybean productivity in the best DSPSC systems from the third year onwards has ranged from 3100 kg ha^{-1} with short-cycle soy crops to over 3500 kg ha^{-1} with intermediate-length cycles. Whatever the soybean cycle, the average yield over a 5-year period is higher in the best DSPSC systems with little fertilization (40 kg P$_2$O$_5$ + 40 kg K$_2$O ha^{-1}) than in the monocropping + disk-plowing system with twice as much fertilization, and it is close to that achieved in the same system with heavy fertilization (160 kg P$_2$O$_5$ + 110 kg K$_2$O ha^{-1}).

With medium fertilizer application (80 kg P$_2$O$_5$ + 80 kg K$_2$O ha^{-1}), the most common practice in the region, high-potential intermediate-cycle soybean cultivars display increased productivity under the best DSPSC systems, producing over a 5-year period 16 to 40% more than the monoculture systems. Yields exceed 4300 kg ha^{-1} from the third year of DSPSC cultivation. The short-cycle varieties have lower potential and a smaller annual gain in yield with DSPSC, 516 kg ha^{-1} in comparison with 934 kg ha^{-1} for the intermediate-cycle cultivars. Over a 5-year period, gains in yields by the best DSPSC systems have ranged from 23 to more than 43%.

These results show the economic as well as agronomic advantages of increasing soil productivity through organobiological pathways under DSPSC. More is produced with much less inorganic fertilizer. This leads to considering as the finest cultivars those that perform best under DSPSC systems, as the heavy use of chemical inputs leads to suboptimal conditions. DSPSC cultivation optimizes the genotype–environment interactions as the environment is very strongly modified by farmers' soil management procedures.

22.6.2 Rice

The potential productivity of rain-fed rice, as observed in controlled trials in the region, has increased from 1800 to 2000 kg ha^{-1} in 1986 to more than 8000 kg ha^{-1} in 2000, with a record yield of 8500 kg ha^{-1} in large-scale cropping at Campo Novo dos Parecis in 1998–1999. This has been accompanied by an improvement in grain quality, which is now evaluated as being as good, and sometimes better, than that of the best irrigated varieties (Séguy et al., 1998b).

As with soybean, rice productivity in these systems is closely correlated with the quantity and quality of the biomass produced by combinations of grasses. The plants with the most powerful root systems for restructuring the soil are: *Eleusine coracana*; maize, sorghum, or millet intercropped with *Brachiaria ruziziensis*; and deep-rooting, nitrogen-fixing legumes such as Crotalaria sp., *Cajanus cajan*, and *Stylosanthes guianensis*.

The creation and selection of rain-fed rice varieties and hybrids is performed for and under the best DSPSC conditions. Their resistance to water shortage and their stable resistance to diseases will make them excellent for use in rain-fed DSPSC systems in the humid tropics, under sprinkler irrigation in areas with little or no rainfall, and in lowland areas and rice perimeters with poor control of water when the facilities are degraded.

22.6.3 Overall Production

After 15 years of research, the best DSPSC systems can thus now produce the following amounts in 1 year: either 4500 kg ha^{-1} of soybean or more than 6000 kg ha^{-1} of rice in the wet season, followed in the dry season by 1500 to 3000 kg ha^{-1} of maize, millet, or sorghum (*Eleusine coracana*). In addition, there can be also 65 to 90 kg ha^{-1} of meat or 3000 to 4500 kg ha^{-1} of cotton in rotation with the preceding grain + pasture systems (Séguy et al., 2001a, 2001b).

Since 2000, changes in soil-system functioning, made possible by direct seeding on permanent soil cover, have allowed the introduction into the humid environment of the *cerrados* a crop that is usually confined to semiarid areas: cotton. Thanks to good soil structure, even at deep horizons, cotton can now be grown during the dry season, often on a sorghum mulch. Within 4 years, such a system has been extended to 300,000 ha in the Amazonian forest ecosystem, with top-level performance and profit, approaching the highest cotton yields in the world, up to 4.8 ton ha^{-1}.

22.7 Soil Functioning and the Impacts of Direct Seeding with Permanent Soil Cover in Mato Grosso

DSPSC systems reproduce within a cropped ecosystem the functioning of a forest ecosystem. The global functioning of such cropped ecosystems, with their multiple interactions, was represented in Figure 1.1 in the first chapter of this volume. It indicates how such systems can be made sustainable because the recursive functioning of this ecosystem, with high biomass production above and below the soil surface, reinforces the processes of soil protection, development of biological activity, and improvement of soil structure. All of these processes improve soil fertility in a broad way. Table 22.2, based on a literature review and our own measurements, summarizes the performance and

TABLE 22.2

Indicators of Soil System Functioning for Tropical Rain Forest and the Best DSPSC Systems

	Forest Ecosystems	Best DSPSC Systems
Litter biomass	8.4 ton ha^{-1}	10–15 ton ha^{-1}
Speed of litter decomposition	50% of dry weight in 37 days (rainy season) 50% dry weight in 216 days (dry season)	50% of dry weight in 30 days (rice, maize)
Root biomass	~5 ton ha^{-1} 60% in horizon 0–20 cm 20% in horizon 20–40 cm	5–7 ton ha^{-1}
Organic matter (0–20 cm)	18 ton ha^{-1} C in litter + roots 55 ton ha^{-1} humus, of which 44 ton ha^{-1} are closely linked to minerals	14–20 ton ha^{-1} in litter + roots >40–50 ton ha^{-1} humus
Porosity	Macropores mainly (0.1–100 μm) MWD between 4 and 5	Macropores mainly (0.1100 μm) Good soil structure over 2 m deep in the profile (grass roots) MWD between 4 and 5
Water use by plants	>1.7 m in dry season	>2 m in dry season (cotton, sorghum, millet, sunflower, forages)

functioning of a tropical forest ecosystem in comparison to the best DSPSC systems. For most functions, the best DSPSC systems are equivalent to a sustainable forest ecosystem.

Figure 22.4 shows an example of changes occurring in soil organic matter content (OM), cation exchange capacity (CEC), and base saturation level (V) induced by DSPSC systems on a ferralitic soil in the *cerrados*. Whereas disk-plowing practices led to a fast decrease in both OM and CEC, direct seeding has made it possible to increase these parameters rapidly to a higher level than in the original ecosystem of savannah, mainly through biological paths.

22.8 Development of Direct-Seeded Systems for Semiarid Tropics

Using the same concepts and principles, DSPSC management systems were developed for contrasting environments, such as the semiarid tropics in Madagascar and Tunisia. This helped advance our learning process and gave us some feedback on soil functioning in terms of the universal processes now identified. In all situations, rotations are based on the same model of forest ecosystem functioning: promote maximum aboveground and belowground biomass production, protect (untilled) soil, increase available water reserves throughout the year, and recover biodiversity through the use of rational rotations and nutrient pumps.

22.8.1 Southwestern Madagascar

On tropical ferruginous soils in southwestern Madagascar, some basic principles for cultivation with direct seeding techniques emerged, especially for the growing of strictly rain-fed crops on sandy alfisols (*sols ferrugineux tropicaux*), with over 70% sand, less than 20% clay, and 2% organic matter.

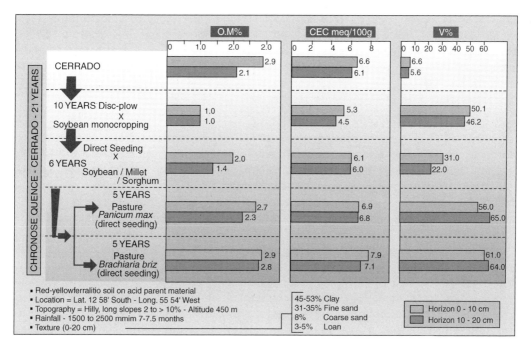

FIGURE 22.4

Changes in organic matter content (OM in %), CEC (meq/100 g) and base saturation level (V in %) in relation to *cerrado* cropping systems in central north Brazil with humid tropical climate. *Source:* From Séguy, L., Bouzinac, S., Maronezzi, A.C., Un dossier du semis direct: Systèmes de culture et dynamique de la matière organique, 2001; Séguy, L. et al., A safrinha de algodão: Opção de cultura arriscada ou alternativa lucrativa dos sistemas de plantio direto nos trópicos úmidos, 2001; Séguy, L. et al., Dossier séquestration du carbone: Et se on avait sous-estimé la potentiel de séquestration du carbone pour le semis direct? Quelle consequences pour la fertilité des sols et la production?, 2001.

22.8.1.1 Early Sowing

Immediately after the first useful rain, sowing is carried out on soil covered with a mulch. In such ecosystems, it is essential to use all available water and to benefit from the high mineralization of nutrients at the beginning of the rainy season. Early sowing is made possible by direct-seeding techniques when mulch has been prepared in the dry season. It is also advisable to keep fragile alfisols permanently covered with a mulch, from the first year, to prevent their hardening and to increase their water reserve (Rollin and Razafintsalama, 2001).

22.8.1.2 Restoring Soil Fertility

To achieve a rapid rise in soil fertility, two solutions are possible: use of mineral fertilizer, which proved efficient, or a process known as soil smoldering, a low-cost but labor-intensive technique, giving a maize yield of over 2.5 ton ha^{-1} without mineral fertilization. This latter method, discussed in greater detail in section 23.2.2.3 of the next chapter, must however be used with caution on soil with low organic matter content. It is recommended only for farmers who already have sufficient experience in direct-seeding practices (Charpentier et al., 2001).

In the first year, cultivation of crop associations that produce significant biomass and can improve soil structure is recommended. Without such an intervention to enhance soil fertility, the best approach is to start direct seeding by associating sorghum with annual long-cycle legumes, such as *Dolichos lab lab, Vigna umbellata,* or nonclimbing *Vigna unguiculata.* These are well-adapted to semiarid conditions, have deep rooting so as to capture water from deep horizons, and are able to sustain their growth during the dry season. These plants improve soil structure, fix nitrogen, recycle nutrients, and permanently cover the soil.

22.8.1.3 Crop Rotations and Associations

Once the soil has been improved by such cultivation, various crops can be grown. However, crop rotation remains a key strategy for maintaining soil structure and fertility, whereas monocropping leads to yield decrease. In the southwest of Madagascar, cultivation of cotton in rotation with an association of cereals (maize, sorghum, or millet) + legume (*Vigna unguiculata* or *Dolichos lab lab*) gives sustainable cotton yields that are two to three times higher than those from the usual practice of cotton monocropping with plowing.

Once soils have been improved by sorghum cultivation in association with legumes, requiring 2 to 3 years of cultivation for the poorest soils such as found around Menabe in the west of Madagascar, any of a number of crops such as rice, groundnut, maize, and cotton can be cultivated. However, production of significant biomass for at least 1 or 2 years is needed to enhance and maintain soil fertility. An association of maize or sorghum with Dolichos or Mucuna can produce 15 to 20 ton of dry matter per hectare. This biomass will be sustained through the dry period, and the increased mulch will be able to control weeds during the following cropping season. Part of the new biomass can also be used for animal feed. The dramatic improvement in yield can be attributed, in part, to the prevention of soil hardening, due to the introduction of mulch, and to the consequent reduction of runoff and improvement of water-use efficiency (Figure 22.5).

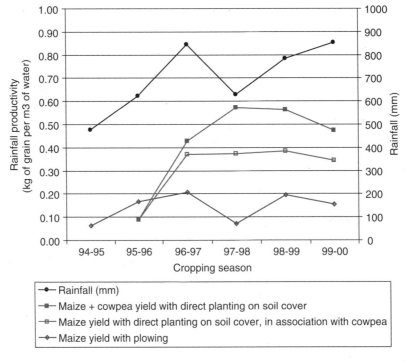

FIGURE 22.5
Effects of soil and crop management under different rainfall conditions over time

22.8.2 Tunisia

On brown clay–lime soils in Tunisia in the Mediterranean Basin, low levels of winter rainfall (400 to 500 mm yr^{-1}) lead to low and irregular yields of wheat, barley, and pasture (rangeland for sheep grazing). DSPSC systems are currently being developed on the basis of the "opportunity farming" principle, whereby all heavy rainfalls (above 40 to 50 mm) are exploited to produce supplementary biomass with nutrient pumps adapted to the specific soil-climate conditions, either before the winter cereal crops are planted or under the cover of these crops in early spring. This permits biomass production to be extended as long as possible into the dry season, both aboveground and belowground, where soil is restructured to promote efficient water infiltration.

22.9 Development of Direct-Seeding Systems for Temperate Areas

In large cereal-cropping regions of central France, on fertile soils (brown clay–lime soils in the Loire valley, Berry, and Cher), the same DSPSC strategies have been applied:

- High biomass production in winter (oats, oats + vetch, mixed species) for direct seeding of spring–summer crops to serve as nutrient pumps (maize, barley, sunflower, sorghum).

- Abundant biomass production in summer within rape/wheat rotations, using mixtures of temperate and tropical species as diversified nutrient pumps.

The systems developed have shown a considerable impact on soil biological activity over time (Figure 22.6). This variability is usually low in temperate areas. The impact on soil physicochemical characteristics is also considerable, even within a few years, as seen in Table 22.3.

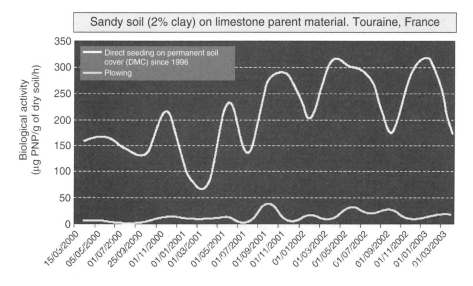

FIGURE 22.6
Soil biological activity according to different cropping practices: fields from J.C. and A. Quillet and J.B. Habert, farmers in Montlouis S/Loire. Analyses made by C Bourguignon, Laboratoire d'Analyses Biologiques des Sols, Marey sur Tille, France. *Source:* From Séguy, L., Bouzinac, S., Maronezzi, A.C., Un dossier du semis direct: Systèmes de culture et dynamique de la matière organique, 2001.

TABLE 22.3

Physiochemical Properties of a Sandy Soil in Touraine, France, and Crop Performance as a Function of Alternative Soil System Management, 1996–2002

Plowing with Rotation, 1996–2002			Direct Seeding with Rotation, 1996–2002		
Crops grown and yield performance (t ha^{-1})			Crops grown and yield performance (t ha^{-1})		
1996	Wheat	3.5	1996	Sunflower[a]	1.0
1997	Barley	3.2	1997	Barley[a]	3.4
1998	Sunflower	1.0	1998	Sunflower[b]	1.5
1999	Wheat	3.2	1999	Oats[b]	5.0
2000	Triticale	3.1	2000	Barley[b]	4.5
2001	Alfalfa	1.5–2.0	2001	Sorghum[b]	6.5
2002	Alfalfa	1.3–3.0	2002	Barley[b]	4.7
Physico-chemical characteristics of top 0–25 cm horizon[c] with DSPSC (1996 and 2003)					
Organic matter content (%)			1.0		2.4
CEC (meq 100 g^{-1})			2.6		4.1
Phosphorus (Joret-Hebert) (% P_2O_5)			0.023		0.953
Assimilable potassium (% K_2O)			0.93		0.124
Total nitrogen (%)			0.64		1.01
C/N ratio			8.6		13.6
Crusting index			2.64		1.76

Fields from J.C. and A. Quillet and J.B. Habert, farmers in Montlouis S/Loire. Analyses made by C. Bourguinon, Laboratoire d'Analyses Biologique des Sols, Marey sur Tille, France.

[a] Simplified cropping techniques.

[b] Direct seeding on permanent soil cover.

[c] Calcareous material, 2–5% clay, 85–90% sand.

Source: Séguy et al. (2001a).

22.10 Discussion

The concepts supporting the formulation of direct-seeded cropping systems based on mulch (DSPSC), in addition to the principles described above, could become a basis for the global development of ecosystem-friendly, sustainable agriculture managed according to a "biology-driven" farming model. In all large tropical, subtropical, and temperate ecoregions, irrespective of the type of agriculture practiced, DSPSC systems provide complete erosion control and are much more profitable than conventional cropping systems with tillage, due to the large cost savings in labor, farm machinery, and fuel with DSPSC systems.

These results highlight that DSPSC systems are more productive, more stable, and environmentally cleaner, with an increasing share of organic fertilization to enhance the soil production potential. Agriculture based on the concepts of nutrient pumping and the multiple functions of cover crops can not only recycle nutrients that will benefit commercial crops but can also act as a CO_2 sink via high biomass inputs into the system. DSPSC systems can quickly have beneficial impacts on the biological quality of soils and water. Such positive environmental impacts of DSPSC systems can induce government and society at large to support farmers utilizing these cropping systems because of their participation in reducing the "greenhouse effect" and for preserving the landscape, rural infrastructure, and wildlife.

Acknowledgments

The applied research reported in this chapter began with a CIRAD team (Séguy and Bouzinac) working with various Brazilian research and development partners together with farmers of the region, including a pioneer of zero-tillage, Munefume Matsubara. Partners included the EMBRAPA rice and beans program (CNPAF) and the Mato Grosso state research center (EMPAER-MT) from 1986 to 1989; then also RHODIA, a Brazilian subsidiary of Rhône Poulenc, and COOPERLUCAS, a cooperative in Lucas do Rio Verde, from 1990 to 1995; and more recently, the Prefecture of SINOP, the MAEDA group, COODETEC, and the AGRONORTE private research enterprise from 1995 to 2002.

References

Charpentier H. et al., *Projet de diffusion de systèmes de gestion agrobiologique des sols et des systèmes cultivés è Madagascar: Rapport de campagne 2000/2001 et synthèse des 3 années du projet TAFA*, Tany sy Fampandrosoana (TAFA), Antananarivo, http://agroecologie.cirad.fr/pdf/charp.pdf (2001).

Derpsch, R. et al., *Controle da erosão no Paraná, Brasil: Sistemas de cobertura de solo, plantio direto e preparo conservacionista do solo*, German Agency for Development Cooperation (GTZ)/Instituto Agronomico do Paraná (IAPAR), Eschborn, Germany/Londrina, Brazil (1991).

Doss, D.D., Bagyaraj, D.J., and Syamasundar, J., Morphological and histochemical changes in the roots of finger millet *Eleusine coracana* colonized by VA mycorrhiza, *Proceedings of the Indian National Science Academy*, New Delhi, 291–293 (1989).

EMBRAPA, *Embrapa cerrados e a região dos cerrados: Informações básicas e dados estatísticos*, EMBRAPA Cerrados (CPAC), Brasilia, Brazil (1998).

FBDP, Expansao da Area Cultivada em Plantio Direto de 1972/73 a 2000/2001, *Federaçao Brasileira de Plantio Direto na Palha*, Ponta Grossa, Paraná, Brazil, www.febrapdp.org.br (2001).

Goedert, W.J., Regiâo dos cerrados: Potencial agricola e politica para seu desnvolvimento, *Pesquisa Agropecuària Brasileira*, **24**, 1–17 (1989).

Lucas, Y. et al., The relation between biological activity of the rain forest and mineral composition of soils, *Science*, **260**, 521–523 (1993).

Miyazawa, M., Pavan, M.A., and Franchini, J.C., *Neutralização da acidez do perfil de solo por resíduos vegetais, Informações Agronômicas da POTAFOS, 92*. Potash and Phosphate Institute, Piracicaba SP, Brazil (2000).

Rollin, D. and Razafintsalama, H., Conception de nouveaux systèmes de culture pluviaux dans le sud ouest malgache: Les possibilités apportées par les systèmes avec semis direct et couverture végétale, In: *Sociétés paysannes transitions agraires et dynamiques écologiques dans le sud ouest de Madagascar*, Razanaka, S. et al., Eds., Centre National de Recherche sur L'Environnement (CNRE-IRD)/Institut de Recherche pour le Développement, Antananarivo/Montpellier, 281–292 (2001).

Scopel, E. et al., Quantifying and modelling the effects of a light crop residue on the water balance: An application to rainfed maize in western mexico, *Proceedings of the XVI World Congress of Soil Science*. CIRAD, Montpellier, France (1998).

Séguy, L. et al., La maîtrise de Cyperus rotundus par le semis direct en culture cotonnière au Brésil, *Agriculture et Développement, 21*, 87–97 (1999)

Séguy, L. et al., Brésil: Semis direct du cotonnier en grande culture motorisée, *Agriculture et Développement*, **17**, 3–23 (1998a).

Séguy, L., Bouzinac, S., and Maronezzi, A.C., *Les plus récents progrès technologiques réalisés sur la culture du riz pluvial de haute productivité et à qualité de grain supérieure, en systèmes de semis direct: Ecologies des forêts et cerrados du Centre Nord de l'état du Mato Grosso*, CIRAD internal document. CIRAD, Montpellier, France (1998b).

Séguy, L., Bouzinac, S., and Maronezzi, A.C., *Un dossier du semis direct: Systèmes de culture et dynamique de la matière organique*, Internal document and CD-ROM. CIRAD, Montpellier, France (2001a).

Séguy, L. et al., *A safrinha de algodão: Opção de cultura arriscada ou alternativa lucrativa dos sistemas de plantio direto nos trópicos úmidos*, Boletim técnico 37. Cooperativa Central Agropecuaria de Desenvolvimento Tecnologico e Economico (COODETEC), Cascavel PR, Brazil (2001b).

Séguy, L. et al., *Dossier séquestration du carbone: Et se on avait sous-estimé la potentiel de séquestration du carbone pour le semis direct? Quelle consequences pour la fertilité des sols et la production?*, CIRAD document. CIRAD, Montpellier, France (2001c).

Séguy, L. et al., The success of no-tillage with cover crops for savannah regions. From destructive agriculture with soil tillage to sustainable agriculture with direct seeding mulch based systems: 20 years of research of CIRAD and its Brazilian partners in the Cerrados region in Brazil, *II World Congress on Conservation Agriculture, Iguassu Falls 11–15 August 2003*, Federacéo Brasileira de Plantio Direto na Palha, Ponta Grosso, Paraná, Brazil, 310–321 (2003a).

Séguy, L. et al., New concepts for sustainable management of cultivated soils through direct seeding mulch based cropping systems: The CIRAD experience, partnership and networks, *II. World Congress on Conservation Agriculture, Iguassu Falls, 11–15 August 2003*, Federacéo Brasileira de Plantio Direto na Palha, Ponta Grosso, Paraná, Brazil, 284–295 (2003b).

Séguy, L. et al., L'agriculture brésilienne des fronts pionniers, *Agriculture et Développement*, **12**, 2–61 (1996).

Stark, N.M. and Jordan, C.F., Nutrient retention by the root mat of an Amazonian rain forest, *Ecology*, **59**, 434–437 (1978).

23

Restoration of Acid Soil Systems through Agroecological Management

Olivier Husson,[1] Lucien Séguy,[2] Roger Michellon[1] and Stéphane Boulakia[3]
[1]*CIRAD, Antananarivo, Madagascar*
[2]*CIRAD, Goiânia, Brazil*
[3]*CIRAD, Phnom Penh, Cambodia*

CONTENTS

Acid soils are widespread, especially in the tropics. Among the most acidic soils are acid sulfate soils, which are estimated to cover over 24 million ha in the world. The pH values of acid sulfate horizons measured in the field commonly range from 3.2 to 3.8 (Dent, 1986), but they can be even lower. In the Plain of Reeds, Vietnam, Verburg (1994) measured the pH of soil solution to be as low as 2.75 at a depth of 40 cm on typic sulfaquepts, while water pH in the irrigation/drainage canals could be as low as 2.2 at the beginning of the rainy season.

In addition, acid ferrallitic soils, which cover tens of millions ha worldwide, present similar problems when cultivated. Their acidity is harmful to plants, impairing the absorption of nutrients, especially of calcium and phosphorus (Sen, 1988). However, the induced effects of acidity, especially aluminum solubility and the resulting toxicity, are likely to contribute even more to poor plant growth on these soils.

Plants may tolerate large concentrations of H^+ ions as long as the concentrations of other cations are also large and the concentration of toxic polyvalent cations is small (Rorison, 1973). When pH values are less than 3.5, H^+ and Fe^{3+} ions may inhibit plant growth. However, the principal hazard to plant health and vigor is likely to be soluble aluminum (Dent, 1986). Exchangeable aluminum in the soil is inversely correlated with pH and rapidly increases whenever $pH_{(KCl)}$ drops below 4.5 (Pionke and Corey, 1967).

The reasons for the injurious effects of aluminum and the possible mechanisms for crop plants' susceptibility to (or tolerance of) it are not yet clear. Achieving a set of physiological adaptations that can act in a coordinated fashion to provide protection against aluminum stress requires an integrated approach (Taylor, 1991). Aluminum has been shown to:

1. Interfere with cell division in plant roots,
2. Fix phosphorus in less-available forms in the soil and in or on plant roots,
3. Decrease root respiration,
4. Interfere with certain enzymes governing the deposition of polysaccharides in cell walls,
5. Increase cell walls' rigidity (by cross-linking pectins), and
6. Interfere with plants' uptake, transport, and use of several elements (Ca, Mg, P, K) and water (Foy et al., 1978).

Al^{3+} can be toxic in concentrations as low as 0.04 to 0.08 mol m^{-3} (1 to 2 ppm), however, there is great variation of tolerance from one species to another and within particular species (Dent, 1986).

Acid soils can be managed in ways that allow for acceptable crop production by using various "tools" that overcome the effects of acidity and aluminum toxicity. However, sometimes the usual interventions for improving soil fertility and dealing with acidity constraints, chemical fertilization and liming, will not permit crop production on these soils, even with very high levels of application. Up to 100 t ha^{-1} of lime has been applied on some acid soils with no improvement as up to 3% pyrite can be found in these soils, each mole of pyrite producing 4 mol of sulphuric acid when oxidized. To cope with such constraints, a special set of practices needs to be developed and applied, including appropriate fertilization and varietal selection, although the most important adaptation is altered water management. Such management practices have been developed by CIRAD teams and their partners for various situations in a number of countries.

23.1 Reclamation of Acid Sulfate Soils in Vietnam

In waterlogged soils that are both rich in organic matter and flushed by dissolved sulfates, as in tidal swamps with mangroves, there is a serious accumulation of pyrite (FeS_2). Acid sulfate soils result when, with drainage, this pyrite is oxidized into sulfuric acid. This is the case in the Plain of Reeds, a low swampy area in the north of the Mekong river delta in Vietnam, where 400,000 ha of acid sulfate soils are found.

In the mid-1980s, a reclamation program was begun whereby a network of primary and secondary canals was dug. Encouraged by the government, local farmers and migrants were responsible for building tertiary canals and for the reclamation of the fields.

Farmers saw a unique opportunity to increase their land area. They first settled on the less or only moderately acid sulfate soils, which were less difficult to reclaim. However, by 1990, only severely acid sulfate soils, covering an estimated area of 150,000 ha, remained uncultivated.

These soils posed tremendous problems, and the first attempts to reclaim them failed. Oxidation of pyritic material in those soils led not only to strong acidification but also to solubilization of aluminum, which rapidly reached toxic levels. Under reduced conditions, plants suffer from ferrous iron present in high concentrations and from hydrogen sulfide, carbon dioxide, and organic acids produced in these organic matter-rich soils. In addition, acid sulfate soils are generally low in available phosphorus and other nutrients and have low base status. They also present physical and biological problems, being poorly structured, with adverse conditions for the growth of microorganisms. Furthermore, these soils are extremely variable over space and time and are highly permeable, making water management difficult.

In this context, the Institute for Agricultural Science of Vietnam and the Fund for Development Cooperation of Belgium formulated a project (IAS/FOS) that set as its main priority the development, with farmer collaboration, of techniques for land reclamation of these severely acid sulfate soils.

23.1.1 Understanding Soil Genesis Processes and Soil Variability

A major characteristic of the Plain of Reeds is tremendous variability of soil characteristics within very short distances, which can be explained by processes of soil genesis. Two major types of soil were identified in the area, reflecting differences in microelevation and water regime:

- In the highest, drier parts (above 85 cm above mean sea level), potential oxidation of pyrite led to formation of jarosite and to acidification of the soil, but also to a "ripening" of the soil through irreversible loss of water and changes in its structure, forming "typic sulfaquepts" (USDA classification). The vegetation is mainly composed of *Aeschymum rugosum* grass.
- In the lower parts (below 75 cm above mean sea level), where water remains most of the time, the pyrite was partly oxidized, leading to soil acidification but under conditions of redox potential not allowing formation of the characteristic pale yellow jarosite. These acid soils have not ripened as typic sulfaquepts, are very rich in organic matter and ferrous iron, and are covered by *Eleocharis dulcis* reeds.

In between these categories, an intermediate soil type is observed. The high soil variability is explained by the strong influence of the water regime on soil genesis, reflecting differences in elevation as horizontal hydraulic conductivity is low (Husson, 1998). This appreciation of soil genesis (and thus of soil characteristics) in relation to microelevations and water regimes facilitated development of differentiated water management strategies for each soil type, easily identified from the natural vegetation, according to the characteristics of each and its constraints for selected crops.

23.1.2 Adapting Water Management and Agricultural Practices to Specific Soil and Water Conditions

The Plain of Reeds experiences annual flooding of the Mekong river during which time the whole plain is submerged for 5 to 9 months with up to 2.5 m of water. Floating rice cultivation, which has unreliable yields, has progressively disappeared, being replaced by rice-growing starting at the end of the flooding period (December to February according to field elevation), taking advantage of the rise in pH linked to submersion, given that the reduction of ferric iron to ferrous iron consumes protons.

23.1.2.1 Typic Sulfaquepts

On the more-elevated typic sulfaquepts, the main problems are linked to strong oxidation in the dry season, leading to acidification and aluminum toxicity. Thus, the time window during which growing conditions are favorable is very short since water recession is rapid for these higher elevations, and irrigation is hardly possible during the first years of cultivation due to very high soil permeability.

To enlarge this time window, farmers in the Plain of Reeds developed the technique of broadcasting pregerminated rice seeds in the water, before the floodwater has completely receded. The sowing date is calculated so that young plants will not have to stay underwater for more than 2 weeks; otherwise, plant mortality will be too high to achieve sufficient plant density. Frequent irrigation, even though water does not stay long in the field after floodwater recession, allows farmers to extend their time window for cultivation of rice varieties that are short-cycle and aluminum-tolerant.

Land preparation, done underwater, aims at rapidly creating a plow pan to improve water control (which takes 2 to 3 years on these soils) and to remove as much fresh organic matter (grasses or reeds remnants) as possible to prevent H_2S and methane toxicity at the beginning of the rice cultivation. With enough time and cultivation, improved water control allows a delay in sowing to obtain a higher plant density while avoiding acidification at the end of the plant cycle by irrigation (Husson, 1998).

With such practices and optimized water management, as well as application of some phosphorus fertilizer at sowing (60 kg P ha^{-1} of thermophosphate), yields over 4 t ha^{-1} of rice can be attained from the first year of cultivation. A poorly solubilizable form of phosphorus is used since a more soluble form such as diammonium phosphate (DAP) leads to the development of algae that suffocate the young rice plants. Thermophosphate, applied as a fine powder, creates high surfaces of microenvironment favoring the development of microorganisms and creating better conditions for plant growth.

23.1.2.2 Hydraquentic Sulfaquepts

Flood water recession on the lower hydraquent sulfaquepts is much slower than on the higher typic sulfaquepts. These soils remain submerged late into the dry season, with the tidal movements of water in the canal system contributing to keeping them submerged most of the time. This results in very slow organic matter mineralization and consequently in high organic matter content, over 15% in all horizons and more than 22% in the A horizon, with possible development into peat soils.

This organic matter helps maintain a very low redox potential and produces numerous toxicities. In these soils, H_2S, FeS, CH_4 (linked to anaerobic decomposition of organic matter), and ferrous iron are present in high quantities. H_2S can affect the plant root system by suffocating it at concentrations as low as 2×10^6 mol m^{-3} (Sylla, 1994). It can also lead to a physiological disease known as *Akiochi* and to deficiency in silicon and bases.

Again, the time window during which cropping conditions are favorable is extremely short in this situation because of excess water and very low redox potential, further reduced by the high organic matter content. In these fields also, sowing of pregerminated seeds is needed to extend the time window for cultivation. However, a high mortality of young plants is observed, even when sowing time has been delayed as long as possible. Roots turn black, and seeds and young leaves are covered by a black coating and rapidly die.

This physiological impairment can be explained by reduction processes and by the activity of sulfate-reducing and ferric iron-reducing bacteria, which gradually colonize seeds and the rice rhizosphere, producing toxic H_2S and FeS (Jacq et al., 1993; Husson, 1998). These substances are oxidized before ferrous iron by the oxygen that is pumped by rice plants into the root zone. In the presence of these compounds, rice roots are unable to maintain on their surface the crust of ferric hydroxides that prevents excessive ferrous iron uptake, thus iron toxicity can develop (Hanhart and Ni, 1993). This leads to plant weakness, low absorption of nutrients, and development of sensitivity to various diseases, including *Helminthosporium* and *Pyricularia oryza*.

A simple practice allows reclamation of these soils, reducing these toxicities and dramatically increasing plant density and rice yield. Pumping the water from fields before sowing, even if it comes back within 24 h (because of the high hydraulic conductivity and the permeability of the newly made dikes), creates a slight oxidation. This is sufficient to reduce significantly the populations of sulfate- and ferric iron-reducing bacteria and their production of toxic substances.

Regularly repeating the operation of pumping water from fields is recommended to prevent development of these bacteria during the plant growth cycle. Together with the application of thermophosphate (60 P ha^{-1}) and planting aluminum-tolerant varieties, such water management practices achieve a profitable rice yield from the first year of cultivation, over 3.5 t ha^{-1}. Yields gradually increase with time and with improvement of water control. This process occurs more slowly on soils that are poorly structured and high in organic matter compared with typic sulfaquepts.

Based on this understanding of soil genesis and soil characteristics and of their implications for rice cultivation, certain adapted cropping practices, especially water management, have been developed in Vietnam and have allowed an amazingly fast reclamation of these very difficult soils. Within 5 years (1992–1997), 120,000 ha of severely acid sulfate soils have been reclaimed by farmers in the Plain of Reeds, and they are now producing over 400,000 tons of rice annually.

23.2 Sustainable Cultivation of Acid Ferrallitic Soils in Madagascar, Brazil, Gabon, and Vietnam

Upland acid ferrallitic soils have very different origins and characteristics, often originating from aluminum-rich and acidic parent material. They have been influenced most evidently by the processes of leaching (lixiviation) of base salts, which are greater in humid climates and build up particular soil conditions. Traditional practices such as slash-and-burn and plowing on these soils have led to erosion, loss of organic matter, and further base lixiviation. Having low pH (pH_{KCl} from 3.5 to 4.5) makes exchangeable aluminum concentration very high. When base levels are very low, the soils' exchange complex can be saturated by aluminum, up to 100% in Northern Vietnam. The main problems for cultivating these soils are the risks of aluminum toxicity, low CEC, low base level, very low nutrient availability (especially phosphorus), and compaction.

23.2.1 Management of Acid Ferrallitic Soil Systems through Direct Seeding on Permanent Soil Cover

As discussed in the preceding chapter, CIRAD and its partners have been developing techniques of direct seeding on permanent soil cover (DSPSC) in various situations, especially for acid ferrallitic soils in countries such as Brazil, Madagascar, Gabon, and Vietnam. These techniques are based on an understanding of soil system functioning inspired by the example of tropical forests in which the soil is permanently covered, with nutrients being recycled by deep root systems and with vigorous biological activity.

One characteristic of these systems is the change in organic matter dynamics and in plant nutrition, induced by cropping practices that aim to maximize biomass production and restore this biomass to the soil. The development of biological activity favored by these practices allows soil structure improvement and stabilization and also solubilization of nutrients such as phosphorus (Chapter 13).

The most essential step in managing these systems is to initiate the process of "pumping" nutrients from lower soil horizons. This requires producing enough biomass in the first year to start the virtuous cycle of soil improvement and plant production, not easy on acid ferrallitic soils with very low fertility.

23.2.2   Initial Biomass Production to "Start the Pump"

Three main approaches can be proposed in order to achieve sufficient biomass production in the first year and to introduce direct seeding systems.

23.2.2.1   Use of Plant Species Able to Grow under Adverse Conditions

The first option is to introduce plants that can grow on low-fertility, often compacted soils. Several species and varieties have been identified and are now widely used for this purpose, especially in Brazil. All these plants are characterized by their strong root systems. Although research is still needed to understand the exact processes occurring, it is suspected that these plants work in association with soil microorganisms which are supported through root exudation.

Annual grasses like sorghum (*Eleusine coracana*) and perennial grasses such as Brachiaria sp., presently growing on over 80 million ha in Brazil, are not only excellent forages for animals; they have remarkable ability to restore soil porosity, to mobilize nutrients that are not accessible to most other plants, and to achieve high biomass production within short periods of time. The shoots of *Brachiaria ruziziensis* can produce over 15 t of dry matter ha^{-1} within 3 months, while *Eleusine coracana* can grow over 4 t ha^{-1} of roots in this period. Among the legumes, *Stylosanthes guianensis* also has the ability to perform well on such poor soils, although its growth is slower than that of selected grasses.

The return of this high biomass production to the soil as litter provides the physical basis for processes of progressive soil-fertility recovery that are strengthened by direct seeding techniques. Table 23.1 gives an idea of the capacity for recycling of micronutrients and macronutrients by two different cover crops or associations in Brazilian forest ecosystems.

When fertilizers and lime are available and affordable to farmers, these mineral materials can be used to increase this initial biomass production since these plants are able to take advantage of the additional fertilization and improvement in pH. The fast-growing plants have the ability to stimulate soil biological activity, due to their roots providing a source of carbon for microorganisms, oxidizing the soil, and improving soil structure. In Vietnam, we found that the microbial populations in degraded ferrallitic soil were restored after 4 years of rice cultivation to the same level as a 10-year-old forest ecosystem simply by having had 2 years of *Brachiaria ruziziensis* growth on it (Husson et al., 2003).

23.2.2.2   Liming and Fertilization

When farmers have no land available for growing cover crops that could improve their crop land during the first year(s), direct-seeding techniques can be applied either with or

TABLE 23.1

Macro- and Micronutrient Content Recycled by Cover Crops (Biological Pumps) at Harvest on Oxidized Ferrallitic Soils in a Forest Ecosystem, Brazil

	N	P	K	Ca	Mg	S	C	C/N(%)	Zn	Cu	Fe	Mn	B
	Macronutrients (kg ha^{-1})								Micronutrients (g ha^{-1})				
Shoots[a]													
Eleusine coracana (cv 5352)	65	2.5	145	60	17	8	2275	35	115	34	915	205	12
Sorghum + *Brachiaria ruziziensis*	104	4.0	120	29	15	5	3830	37	132	63	1912	293	51
Roots[b]													
Eleusine coracana (cv 5352)	44	2.0	6.4	12.8	2	3.6	2240	51	94	52	23592	138	135
Sorghum + *Brachiaria ruziziensis*	52	2.4	24.8	12.8	4	2.8	2000	38	104	46	7532	114	57

[a] Shoot biomass production (dry matter): *E. coracana* 5 t ha^{-1}, Sorghum + *B. ruziziensis*. 8 t ha^{-1}.

[b] Root biomass production (dry matter): *E. coracana* 4 t ha; Sorghum + *B. ruziziensis* 4 t ha^{-1}.

Source: From Séguy, L. et al. *Un dossier du semis direct: Systèmes de culture et dynamique de la matière organique* (2001).

without fertilizer application and liming. For fast recovery of soil fertility, however, liming, fertilizer application, and the correction of any micronutrient deficiencies are required. In Brazil, formulas have been developed to calculate the requirements for liming and fertilizer application based on soil analysis. Such fertilization accelerates the production of grains and biomass, which can be used for soil improvement by following the principles of direct seeding on permanent soil cover.

When fertilizers are not available or not affordable by resource-poor farmers, as is often the case in Madagascar, direct-seeding techniques allow progressive improvement of soil fertility, as shown in Figure 23.1 with data from the Madagascar highlands. Providing only manure for soil nutrient supplementation, maize yields progressively increased with direct seeding, whereas tillage led to stable (but low) or decreasing yields. Within 4 years of direct seeding with use of manure only, maize yields were higher than with plowing and with the addition of recommended mineral fertilization plus liming. Yields were even as high as with plowing and heavy fertilization. This shows that it is possible, with very limited means (manure only), to restore the fertility of soil systems and that agroecological practices can advantageously replace the use of costly mineral fertilization, or enhance its effects when applied.

23.2.2.3 Soil Smoldering as a Means of Rapid Fertility Improvement

Figure 23.1 shows the strong, rapid improvement of yield when using a technique known as soil smoldering. The data are from an intervention begun in 1996 and tracks the first and second years of cultivation thereafter. Originally practiced by Cameroonian farmers, this technique has been adapted in Madagascar for different types of soil,

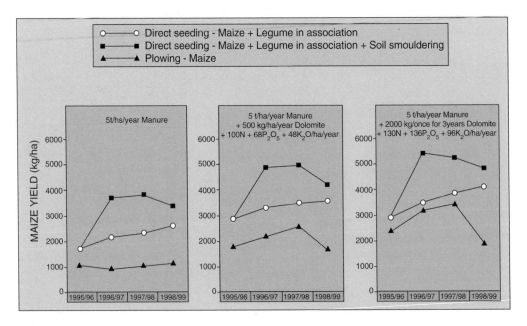

FIGURE 23.1
Trends in maize management as a function of soil and crop management, on acid ferralitic soils, Andranomanelatra, Madagascar highlands, 1995/96–1998/99, with soil smouldering in 1996/97. *Source:* Michellon et al., *Amélioration de la Fertilité par Ęcobuage: Influence de la Fréquence et de l'Intensité de la Combustion selon le Type de Sol de Tanety,* Centre de Coopération Internationale en Recherche Agronomique pour le Développement (CIRAD), Tany sy Fampandrosoana (TAFA), and Centre de Recherche Appliquée au Développement Rural (FOFIFA), Antananarivo, Madagascar, 2002a. With permission.

with studies done on the impact of the frequency of burning (Michellon et al., 2002a) and the burning material used (Michellon et al., 2002b).

Soil smoldering consists of a slow burning of organic material in the soil at a relatively low temperature (250 to 350°C), compared to temperatures of over 1,000 degrees, which are reached in slash-and-burn systems. This avoids nitrogen volatilization and benefits from dramatic changes in soil chemical, physical, and biological features. Trenches are dug (20 to 30 cm wide and about 20 cm deep, varying according to topsoil thickness) in parallel, about 1 m apart. The trenches are filled with organic material to be burned — straw to let air circulate better, and rice husks if available, as these have been identified as the best material for burning — and are then covered with 8 to 12 cm of topsoil. The process of soil smoldering consumes organic matter, thus the most organic horizon should be used.

Chimneys are made of straw every 1.5 to 2 m to let air circulate and to speed up the fire. Speed and temperature of combustion are regulated by the depth of the soil placed back on the burning material; with less than 8 cm, combustion will be too fast, and with high temperatures, there are losses of nutrients by volatilization. If the soil added is more than 12 cm deep, incomplete combustion can lead to poor decomposition of the buried organic matter and to development of toxicity if the soil smoldering operation is not repeated. Optimal burning temperature has been determined to be 250 to 350°C. This method involves 1.5 to 2 days of combustion, fairly similar to charcoal production processes.

The direct effects of soil smoldering as described here are to raise pH (by 1 unit in most situations) and consequently to increase phosphorus availability, raise CEC, and improve soil structure (Michellon et al., 2002a, 2002b). It can also activate organic matter otherwise inaccessible to plants, with a strong effect on crop yields as seen in Figure 23.1. Sometimes it allows farmers to grow crops on soils that have been regarded as not suitable for production, e.g., maize in Northern Vietnam. The increase in yield attributable to soil smoldering is significantly higher than the increase produced by adding ashes from the same quantity of burning material (Michellon et al., 2002a, 2002b). The production of black carbon in the process, with the associated effects described in Chapter 37, may help to explain the rise in yield.

As the process involves burning of organic matter (0.5 to 1%), it should be used sparingly and should not be carried out every year. Frequency of application should be adapted to soil conditions (Michellon et al., 2002a, 2002b), and cropping practices should aim at restoring higher organic matter levels to the soil, as done with direct-seeding techniques. This soil-smoldering methodology can be regarded as an initial booster for the first years. Its effects will decrease with time and will vary with soil type, lasting longer on recent volcanic soils than on degraded ferrallitic soils (Michellon et al., 2002a, 2002b). Although biological organisms in the soil are killed by the fire, 5 to 10 cm around the trenches, fast recovery is observed, and within 3 months, biological activity is usually higher than before the burning because living conditions for microorganisms have been improved in and around the trenches. The organisms that survived between the trenches can recolonize the field rapidly (Boyer, CIRAD, Thailand, unpublished results).

23.2.2.4 Changes in Soil Characteristics Induced by Direct Seeding and Soil Smoldering

Direct seeding on permanent soil cover practices are effective in restoring the fertility of acid soil because they change the functioning of the soil system, especially the dynamics of organic matter and biological activity, which in turn influence various soil characteristics. Figure 23.2 shows examples of such changes in organic matter, CEC, and base saturation level (V) for an acid ferrallitic soil in the Madagascar highlands in response to different soil and crop management practices. It shows that within 5 years,

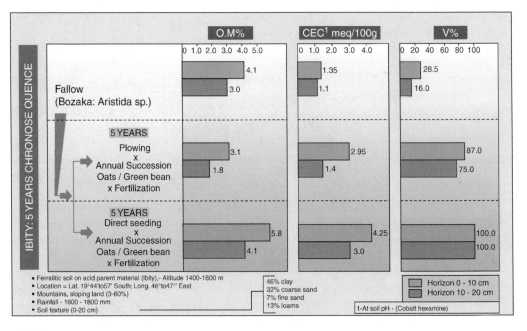

FIGURE 23.2

Changes in organic matter content (OM in %), CEC (meq 100 g^{-1}), and base saturation level (V in %), in relation to cropping system, Ibity, Madagascar Highlands, sub-tropical climate. *Source:* Séguy, et al., *Un dossier du semis direct: Systèmes de culture et dynamique de la matière organique*, Centre de Coopération Internationale en Recherche Agronomique pour le Développement (CIRAD), Agronorte Pesquisas, Groupe Maeda, Tany sy Fampandrosoana, Madagascar (TAFA), Centre de Recherche Appliquée au Développement Rural (FOFIFA) and Association Nationale d'Actions Environementales (ANAE), 2001. With permission.

direct seeding produces a significant increase in organic matter, whereas plowing leads to significant losses in soil organic matter content. CEC is also usually increased in this situation as a consequence of applying mineral fertilizer and manure. The value for base saturation level (V) can reach 100% after being only 28.5% in the original soil.

Figure 23.3, presenting data from Gabon, shows the impact of different cropping systems and practices as well as fertilization on the evolution of V in an acid soil:

- Even with a medium level of fertilization, the base saturation level decreases in systems with annual plowing of the soil for maize cultivation, even when soybean is directly sown in maize straw (thus, one plowing for two crops).
- With direct seeding, systems with crop rotation/association that include maize, sorghum, soybean, and millet increase production with a medium level of fertilization. However, this is not sufficient for a system where maize is cultivated with direct seeding on Calopogonium.

We see that some direct-seeding systems are more efficient than others in improving soil fertility. The plants used as biological nutrient pumps in the different systems largely explain the respective performances of these systems in restoring soil fertility. Plants such as millet, sorghum, Eleusine, and Brachiaria sp. are excellent in this respect.

Figure 23.4 shows, for the Madagascar highlands, an example of the changes in organic matter content, pH, K, Ca, CEC, V, and P that can be achieved in a ferrallitic soil under direct seeding of maize into Desmodium as a living cover crop. It shows for various parameters how this cropping system can lead to very significant increases within 5 years,

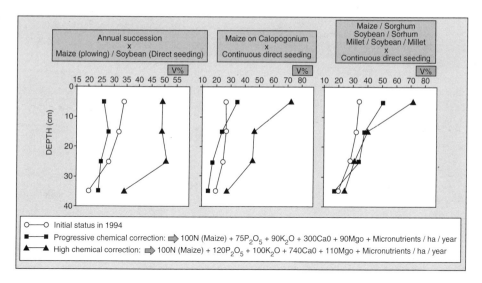

FIGURE 23.3

Evolution over 3 years of exchangeable base saturation levels (V in %) as a function of crop and soil management, in humid tropical climate, Boumango, Gabon, with yellow ferralitic soils on acidic parent material (pH KCl: 4.05, organic matter: 3.7%, P Olsen-Dabin: 72.7 ppm, and CEC 2.87 meq 100 g^{-1} in 0–10 cm horizon) *Source:* Boulakia and Delafond, unpublished results. With permission.

especially at a soil depth between 0 and 10 cm, which provides the main support for increases in soil fertility. The figure also shows some remarkable changes resulting from soil smoldering: an increase in pH of over 1 unit, five times more available phosphorus than under natural vegetation, four times higher CEC, and 100% saturation in bases.

All these changes have a strong impact on soil fertility with a reduction of soil toxicities which explains the strong rise in yields (5.5 to 6.5 t ha^{-1} of maize with fertilization level of 2: 5 t ha^{-1} manure, 500 kg dolomite ha^{-1}, and 20 N, 20 P and 48 K). We note particularly:

- The increase in soil organic matter content. As a consequence, exchangeable aluminum content is reduced since it is correlates negatively with organic matter content (Pionke and Corey, 1967). Also, short-chain organic acids (oxalic, tartaric, and citric) produced by organic matter decomposition probably have a role in aluminum reduction (fixation of aluminum) as does the fast turnover of organic matter.

- The increase in pH due to the recycling of bases by biological pumps, which consequently reduces aluminum solubility and toxicity. With soil smoldering, an immediate increase of 1 unit in pH has a dramatic influence on aluminum toxicity.

- Bases are recycled, and CEC is progressively increased, as base saturation levels (V) reach 100%, improving nutrient availability for plants.

- As soil structure is improved, so is water infiltration and biological activity.

Direct-seeding techniques are particularly interesting for acid soils. They are now applied in the Brazilian *cerrados* on 8 million ha, and they are being extended in other countries. Crops such as maize, which are not suitable for these soils with conventional soil management techniques (unless very high levels of fertilization are maintained), can now be grown by farmers with very limited resources with soil smoldering methods. Farmers in the Lake Alaotra region of Madagascar were interested in the technique of

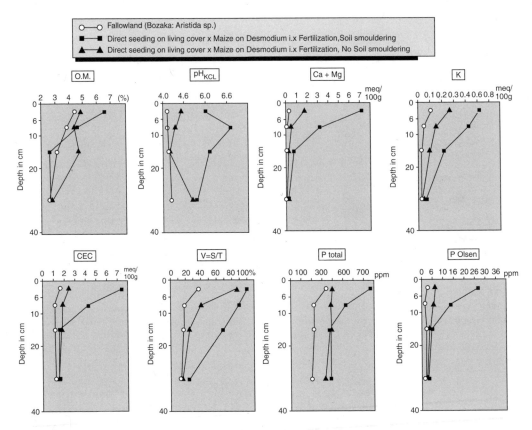

FIGURE 23.4

Impacts of soil and crop management on chemical and physical soil characteristics after 5 years of continuous cultivation, on acid ferrallitic soils of very low fertility, Ibity, Madagascar Highlands. *Source:* Séguy, et al., *Un dossier du semis direct: Systèmes de culture et dynamique de la matière organique*, Centre de Coopération Internationale en Recherche Agronomique pour le Développement (CIRAD), Agronorte Pesquisas, Groupe Maeda, Tany sy Fampandrosoana, Madagascar (TAFA), Centre de Recherche Appliquée au Développement Rural (FOFIFA) and Association Nationale d'Actions Environementales (ANAE), 2001. With permission.

soil smoldering but were limited by labor availability, since digging trenches manually with a simple *angady* (hoe) as initially proposed takes 90 man-days ha^{-1}. To overcome this limitation, they have modified the technique by digging the trenches with a plow pulled by oxen. This dramatically reduces their working time to 4 days ha^{-1}.

23.3   Discussion

These two examples of restoring productivity on acid soils with very different characteristics (lowland acid sulfate soils vs. upland acid ferrallitic soils), using very different strategies and sets of practices, show that knowledge-based management, with an understanding of soil systems' genesis, characteristics, and functioning, allows profitable cultivation and rapid (and we expect, sustainable) improvement of soil fertility.

In the case of acid sulfate soils, reducing strong toxicities is made possible by careful water management and adoption of appropriate cropping practices (strategic fertilization, suitable varieties, adjustments in the cropping period). These alterations lead to changes in the populations of microorganisms, especially to a reduction of sulfate- and iron-reducing bacteria, which produce harmful toxins. As a consequence, yields over 3.5 t ha^{-1} of rice are obtained from the first year of reclamation and will further increase with time and cultivation, while costs decrease as less labor and irrigation are needed with soil improvement. Thus, rather than employ costly conventional reclamation methods, with several years of investment required and negative economic returns before a gain is obtained on these soils otherwise unsuitable for cultivation, land reclamation becomes a relatively inexpensive and an immediately profitable operation.

Acid ferrallitic soil fertility can be improved by direct seeding on permanent soil cover, completely modifying the dynamics of organic matter mineralization and nutrient availability, with results superior to traditional practices of slash-and-burn and conventional tillage. These practices allow farmers to raise soil organic matter content and make progressive changes in biological activity, pH, CEC, bases, nutrient availability, and soil structure. This alleviates toxicities that are linked to low pH and allows the development of favorable cropping conditions.

These changes in soil fertility can be gradual or they can be speeded up by the use of manure, mineral fertilizers and soil smoldering as boosters to initiate biomass production. A growing set of options is being developed to provide different management systems that can be adapted to the various soil characteristics, for various crops, and to farmers' means. Thus, for the poorest farmers, on very degraded soils in Madagascar, where fertilizers are rarely accessible and very expensive, simply associating cassava with Brachiaria sp. leads to a 200 to 400% increase in cassava yield (up to 30 t ha^{-1}), while producing quality forage and improving soil characteristics, which makes rice cultivation possible and profitable with direct seeding from the next season (if herbicides and urea are available). Associating cassava with *Stylosanthes guianensis* also has a positive effect on cassava yield (although less than Brachiaria), and it makes rice cultivation possible without external inputs from the next season.

Simple use of manure with direct-seeding techniques on permanent soil cover will certainly increase soil fertility and yields. For rice or maize cultivation, soil smoldering will boost yields and speed up the process of improving soil systems' fertility through direct-seeding techniques. When fertilizers are accessible, their use will speed up this process by increasing biomass production, which is the "engine" of these systems.

With rice yields ranging from 2.5 to 5 t ha^{-1}, maize yields between 2.5 and 6.5 t ha^{-1}, and soybean yields between 1.5 and 2.5 t ha^{-1}, agroecological crop and soil management with lower costs of production offers very practical ways of increasing farmers' yields and benefits while preserving and even enhancing the natural resource base. Whether farmers attain the high or the low end of new production range possible will depend on their initial soil characteristics, the cropping system employed, and their level of intensification of inputs and management.

References

Dent, D., *Acid Sulphate Soils: A Baseline for Research and Development*, International Land Reclamation Institute, Wageningen, Netherlands (1986).

Foy, C.D., Chaney, R.L., and White, M.C., The physiology of metal toxicity in plants, *Ann. Rev. Plant Physiol.*, **29**, 511–566 (1978).

Hanhart, K. and Ni, D.V., Water management on rice field at Hoa an in the Mekong delta, In: *Selected Papers of the Ho Chi Minh Symposium on Acid Sulphate Soils, Vietnam, March 1992*, Dent, D.L. and Van Mensvoort, M.E.F., Eds., Publication 53, International Land Reclamation Institute, Wageningen, Netherlands, 161–176 (1993).

Husson, O., Spatio-temporal variability of acid sulphate soils in the Plain of Reeds, Vietnam: Impact of soil properties, water management and crop husbandry on the growth and yield of rice in relation to microtopography. Ph.D. thesis, Wageningen Agricultural University, The Netherlands (1998).

Husson, O. et al., *Impacts of Cropping Practices and Direct Seeding on Permanent Vegetal Cover (DSPVC) Techniques on Soil Biological Activity in Northern Vietnam*, Proceedings of the World Congress on Conservation Agriculture, Iguassu, Brazil, August, 460–465 (2003).

Jacq, V.A., Ottow, J.C.G., and Prade, K., Les risques de toxicité ferreuse et sulfureuse en rizière inondée: Symptomologie, écologie et prévention, In: *Bas-fonds et Riziculture. Agence de Coopération Culturelle et Technique*, Raunet, M., Ed., Centre de Coopération Internationale en Recherche Agronomique pour le Développement (CIRAD), Institut Français de Recherche Scientifique pour le Développement en Coopération (ORSTOM), and Centre de Recherche Appliquée au Développement Rural (FOFIFA), Montpellier, France, 283–303 (1993).

Michellon, R. et.al., Amélioration de la fertilité par ęcobuage: Influence de la fréquence et de l'intensité de la combustion selon le type de sol de tanety. Centre de Coopération Internationale en Recherche Agronomique pour le Développement (CIRAD), Tany sy Fampandrosoana (TAFA), and Centre de Recherche Appliquée au Développement Rural (FOFIFA), Antananarivo, Madagascar (2002a).

Michellon, R. et al., Amélioration de la fertilité par ęcobuage: Influence de la nature du combustible selon le type de sol de tanety. Centre de Coopération Internationale en Recherche Agronomique pour le Développement (CIRAD), Tany sy Fampandrosoana (TAFA), and Centre de Recherche Appliquée au Développement Rural (FOFIFA), Antananarivo, Madagascar (2002b).

Pionke, H.B. and Corey, R.B., Relations between acidic aluminum and soil pH, clay and organic matter, *Soil Sci. Soc. Am. Proc.*, **31**, 749–752 (1967).

Rorison, I.H., The effects of extreme acidity on the uptake and physiology of plants, In: *Acid Sulphate Soils. 1:18*, Dost, H., Ed., International Land Reclamation Institute, Wageningen, Netherlands, 223–254 (1973).

Séguy, L., Bouzinac, S., and Maronezzi, A.C., Un dossier du semis direct: Systèmes de culture et dynamique de la matière organique. Centre de Coopération Internationale en Recherche Agronomique pour le Développement (CIRAD), Agronorte Pesquisas, Groupe Maeda, Tany sy Fampandrosoana, Madagascar (TAFA), Centre de Recherche Appliquée au Développement Rural (FOFIFA) and Association Nationale d'Actions Environementales (ANAE). http://agroecologie.cirad.fr/index.php?rubrique = rapports&langue = fr (2001).

Sen, L.G., Influence of various water management and agronomic packages on the chemical changes and on the growth of rice in acid sulphate soils. Ph.D. thesis, Wageningen Agricultural University, Wageningen, The Netherlands (1988).

Sylla, M., Soil salinity and acidity: Spatial variability and effects on rice production in West Africa's mangrove zone. Ph.D. thesis, Wageningen Agricultural University, Wageningen, The Netherlands (1994).

Taylor, G.J., Current views of the aluminum stress response: Physiological base of tolerance, *Curr. Top. Plant Biochem. Physiol.*, **10**, 57–93 (1991).

Verburg, P., Morphology and genesis of soils and evaluation of the side-effects of a new canal in an acid sulphate soil area, Plain of Reeds, Vietnam. *IAS/FOS FSR Project Paper*. Wageningen Agricultural University, Wageningen, The Netherlands (1994).

24

Conservation Agriculture and Its Applications in South Asia

Peter Hobbs,[1] Raj Gupta[2] and Craig Meisner[3]
[1]*Department of Crop and Soil Sciences, Cornell University, Ithaca, New York, USA*
[2]*Rice–Wheat Consortium for the Indo-Gangetic Plains, New Delhi, India*
[3]*International Center for Soil Fertility and Agricultural Development (IFDC), Dhaka, Bangladesh*

CONTENTS

Although agriculture is an essential occupation needed to feed the world's population, it often has negative environmental impacts when practiced without regard to the condition of the soils that it depends on. Intensive modern food production systems are often accompanied by numerous adverse impacts on soil systems: loss of soil organic matter (SOM), erosion by wind and water, reduced soil biological diversity, physical degradation, poor nutrient-use efficiency, groundwater pollution, declining water tables, salinization and waterlogging, greenhouse gas emissions, with accelerating effects on global warming, air pollution, loss of biodiversity, and decline in factor productivity.

The challenge in the next few decades will be to keep pace with the still-growing demand for food, giving attention to the food security needs of the underfed and undernourished, while adopting technologies that are more efficient at using natural resources and have minimal adverse impacts on the environment. In addition, the profitability of agriculture needs to be increased so that farmers around the world, both better-endowed and resource-poor, can make enough income to improve their livelihoods. This chapter considers a set of innovations grouped under the rubric of conservation agriculture and how they are being adopted by farmers as a way to achieve these goals. It draws on data presently being generated within rice–wheat rotational farming systems in the Indo-Gangetic plains of South Asia as an example of what can be done, and what more needs to be done, to ensure sustainable intensive agricultural production.

24.1 Conservation Agriculture Defined

Conservation agriculture (CA), a term introduced in the 1970s, was adopted by the U.N. Food and Agriculture Organization (FAO) in Rome in the 1990s (FAO CA web site, 2004). This term has often been used interchangeably with other terms such as conservation tillage, no-tillage, zero-tillage, direct seeding/planting, *planto directo*, and *siembra directa*. These share some features with CA, but it should be understood as more than just a particular method of cultivation or crop establishment technique. CA is different from conventional agriculture in that it retains crop residues on the soil's surface as a cover, not incorporating them into the soil by tillage. The essential features of CA are:

- Little or no soil disturbance. The native soil microorganisms and soil fauna and the roots of crops and cover crops take over the tillage function and soil nutrient mobilization and balancing. Mechanical tillage disturbs this process. Therefore, zero-tillage is an essential element of CA.

- Utilization of green manure cover crops (GMCCs), other cover crops, and residues from previous crops for permanent or semipermanent organic soil cover. When cover crops are used, they are cut or otherwise killed so they do not compete with crop plants. The dead-residue biomass functions as mulch, protecting the soil physically from sun, rain, and wind and also feeding soil biota. Mineralization and loss of nutrients are reduced, and more satisfactory levels of SOM are built up and maintained. This also moderates soil temperatures in favor of biological activity that is important in tropical and subtropical areas.

- Use of crop rotation to help control pests, diseases, weeds, and other biotic factors. Well-balanced crop rotations can neutralize many of the possibly negative aspects of no-till, such as pest buildup, as they increase the diversity of favorable insects and organisms that can help maintain checks on the spread and impact of pests and diseases.

- Use of other integrated pest management principles. Similar to IPM, CA enhances biological processes. With CA, IPM practices of crop and pest management are expanded into total land husbandry. Indeed, without the use of IPM practices, sufficient buildup of soil biota for biological tillage would not be possible.

- No burning of crop residues since they are made part of the permanent soil cover. Air pollution is reduced where burning is stopped.

- Efficient use and possible reduction in agrochemicals and a definite reduction in the use of fossil fuels. Use of herbicides such as glyphosate is often increased to handle weed control and cover crop control, but this should be compared with the amount of herbicide and pesticide use of traditional tillage systems. As more experience is gained with rotation systems, it is becoming more common to use biological rather than chemical means of weed and pest control.

The goal of conservation agriculture is agricultural sustainability, conserving, improving, and making more efficient use of natural resources through the integrated management of available soil, water, and biological resources, combined with judicious use of external inputs. CA contributes to environmental conservation as well as to enhancing and sustaining agricultural production. It aims to be a more resource-efficient/resource-effective form of agriculture.

24.2   Reasons for Reducing or Stopping Tillage

Traditionally, tillage has been recommended as a necessary component of agriculture, serving the following functions:

- Incorporating the residues of the previous crop and any amendments (fertilizers, organic, and inorganic) into the soil.
- Preparing a seedbed so that normal seeding equipment can penetrate the soil and place the seed at the proper depth into moist soil.
- Controlling any weeds that have germinated or carried over from the previous crop.
- Helping release nutrients through the mineralization and oxidation of SOM.
- Giving relief from compaction by breaking up any compaction layers.

These functions are increasingly being questioned, however, by farmers who have experimented with zero-tillage and who are finding that with some adaptation of other practices, tillage may not be necessary. The following comparisons can be made between tillage and zero-tillage systems:

- Tillage done by tractors consumes large quantities of fossil fuels that add to costs while also emitting greenhouse gases (mostly CO_2) and contributing to global warming. Animal tillage is also expensive since farmers have to maintain and feed a pair of animals for a year in order to undertake tillage operations. Zero-tillage reduces these costs and emissions.
- Tillage often delays timely planting of crops, with subsequent reductions in yield potential. By reducing turnaround time to a minimum, zero-tillage can mean crops are planted on time and thus increase yields without greater input cost.
- Tillage results in decline of SOM due to increased oxidation over time, leading to soil degradation. Although this SOM mineralization liberates nitrogen and can lead to improved yields over the short term, there is always some mineralization of nutrients and loss by leaching into deeper soil layers. This is particularly significant in the tropics where organic matter reduction is processed more quickly, with low soil carbon levels resulting after only one or two decades of intensive soil tillage. Zero-tillage, on the other hand, especially with permanent soil cover, has been shown to result in a buildup of organic carbon in the surface layers (see Figure 24.1; and Campbell et al., 1996).
- Although tillage does afford some relief from compaction, it is itself a major cause of compaction, especially when repeated passes of a tractor are made to prepare the seedbed or to maintain a clean fallow. Zero-tillage reduces dramatically the number of passes over the land and thus compaction.

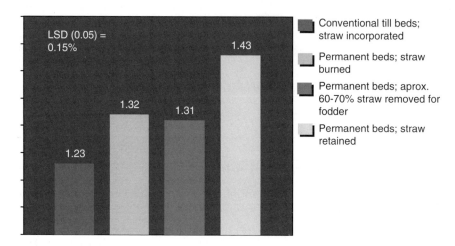

FIGURE 24.1
Percent organic matter in the surface 7 cm in four permanent-bed treatments after 10 years of wheat–maize in Ciano, Ciudad Obregon, Mexico. *Source:* From Sayre, K.D. and Hobbs, P.R., in *Sustainable Agriculture and the Rice–Wheat System*, Marcel Dekker, New York, 2004, 337–355. With permission.

- Use of deep-rooted cover crops and biological agents (earthworms, etc.) can also help to relieve compaction under zero-tillage systems. In some cases, higher bulk densities have been reported under zero-tillage (Gantzer and Blake, 1978), which can lead to yield loss, but this can be corrected by using a permanent soil cover (Sayre and Hobbs, 2004). Even when bulk density is higher in some no-till systems, the infiltration of water is higher.

- Tillage leaves the soil bare, and when this is pulverized excessively and exposed to wind and rain, this leads to soil erosion, especially on sloping land. The loss of topsoil results in significant soil degradation. Runoff and erosion start with raindrop impact on a bare soil surface. This disaggregates the soil into finer particles that clog soil pores and create a surface seal that impedes rapid water infiltration. Most of the rainwater then runs off the land carrying precious topsoil with it. Soil crusting can also occur upon drying and restrict seedling emergence. The permanent crop cover with zero-tillage absorbs the energy of the raindrops and allows more infiltration of water, significantly reducing erosion (Lal, 1989). In Paraguay, topsoil losses of 46.5 ton ha^{-1} have been recorded with conventional tillage on sloping land after heavy rain compared with 0.1 ton ha^{-1} under no-till cultivation (Derpsch and Moriya, 1999).

- Tillage disrupts the root channels left by previous crops, with resulting reduction in water infiltration. It also destroys the channels and nests that are built by soil biota, which results in lower soil microbial populations. In zero-tillage systems, more and greater diversity of earthworms, arthropods (acarina, collembolan, insects, etc.), fungi, and mycorrhiza are found than under conventional tillage (Kemper and Derpsch, 1981; Clapperton, 2003). Zero-tillage thus results in a better balance of microbes and other organisms and thus in healthier soil.

- An added, economic consideration is that tillage results in more wear-and-tear on machinery and higher maintenance costs for tractors than under zero-tillage systems.

24.3   Why Maintain Permanent Soil Cover?

Conservation agriculture is distinguished from conventional agriculture by the presence of permanent soil cover. The benefits of this essentially complement those from zero-tillage as noted above. Evidence of this from Brazil has already been presented in Chapter 22. An example of such cover is shown in Figure 24.2. Main reasons for the use of ground cover include:

- Ground cover increases water infiltration into the soil by negating the energy of raindrops, preventing clogging of soil pores, and allowing percolation of water into the profile. This protects the soil against water and wind erosion.

- Soil mulch reduces water evaporation, conserves moisture, and helps moderate soil temperature, making conditions more hospitable for belowground biota.

- A cover crop helps reduce weed infestation through competition, and also results in weeds being killed at the same time that the cover crop is cut, rolled flat, or killed.

- Cover crops contribute to the accumulation of organic matter in the surface soil horizon.

- Mulch also helps with the recycling of nutrients, especially when GMCC are used, through the association with belowground biological agents and by providing food for microbial populations.

- The presence of the cover crop and minimal soil disturbance leads to an improvement in soil structure and aggregation over time compared with a tilled soil.

- Cover crops and surface mulch help promote biological soil tillage through their rooting but also by support for earthworm, arthropods, and microorganisms belowground. Ground cover promotes an increase in biological diversity both below- and aboveground. The numbers of beneficial insects, for example, has been seen to be higher where there is ground cover and mulch (Jaipal et al., 2002).

FIGURE 24.2
Residue retention distinguishes conservation agriculture from conventional farming systems, which are characterized by leaving the soil bare and unprotected, exposed to climatic agents. The soil cover is not incorporated into the soil by tillage (Photo courtesy of Peter Hobbs).

Little research has been done in South Asia on belowground changes in soil biota and associated changes in soil physical and chemical properties. However, it is likely that the same kinds of results are occurring with CA as have been documented in the U.S.A. Brady and Weil (2002: 891–892) report the results of comparison trials in two different locations:

- In the mid-Atlantic states, six pairs of adjacent fields were evaluated. In one field of each pair, conservation practices (reduced tillage, greater crop diversity, rotation and organic nutrient sources) had been used, while conventional practices (more tillage, less crop diversity, etc.) were used in the other. An average of measurements made for coastal plain and piedmont soils showed CA practices giving greater microbial biomass C (2.5 vs. 1.8% of total organic C), more total organic C (16.1 vs. 11.9 g kg^{-1}), more active organic C (126 vs. 94 mg kg^{-1}), greater aggregate stability (73.5 vs. 62%), and a higher N mineralization rate constant day^{-1} (40 vs. 34.5) (Islam and Weil, 2000).

- In three locations in the state of Nebraska, conservation practices (crop rotations including legumes, crop residues, and manure) were compared with conventional systems (deep plowing and inorganic fertilizers) for continuous maize or maize–soybean rotation. With conservation practices, microbial biomass averaged much higher (928 vs. 590 kg C ha^{-1}) as well as organic C (61.3 vs. 50.4 mg ha^{-1}) and water holding capacity (0.19 vs. 0.16 m^3 m^{-3}); pH was more neutral (6.43 vs. 5.87), and bulk density was less (1.17 vs. 1.30 mg m^{-3}) (Liebig and Doran, 1999).

These are the kinds of changes that we expect are occurring in South Asia with CA.

24.4 Global Use of Conservation Agriculture

CA is gaining popularity among farmers throughout the world. Although it is difficult to get an accurate estimate of the total area covered, Derpsch and Benites (2003) calculate that in 2002, CA covered 72 million hectares (Table 24.1).

One of the main reasons for this tillage revolution has been the greater profitability of CA over conventional systems as a result of lower input costs (less fossil-fuel use and more efficient input use) coupled in most cases with an increase in yield. The multiple benefits listed above add to the economic and time benefits that have initially attracted farmers to this technology, and they result in more sustainable crop production for the future.

TABLE 24.1

Estimated Coverage of Conservation Agriculture Globally, 2002

Country	Area Planted (million ha)	% of Total
United States	22.4	31.1
Brazil	17.3	24.0
Argentina	14.5	20.1
Australia	9.0	12.5
Canada	4.1	5.7
Paraguay	1.3	1.8
South Asia	1.0	1.4
Rest of the World	2.4	3.3
Total	72.0	100.0

24.5 Current Farming Practices in South Asia

The major cropping system in the Indo-Gangetic plains (IGP) of South Asia, covering much of north India, southern Nepal, and major parts of Pakistan and Bangladesh, is alternating rice–wheat, with rice grown in the wet, humid monsoon season, and wheat in the dry, cool winter. Tillage has been and still is promoted as an essential component of management of these two crops in South Asia. For rice, the soils are plowed, flooded, and then puddled (plowed when wet). This is done to reduce the percolation of water and promote ponding; the standing water helps control weeds. Rice seedlings from separately raised seedbeds are transplanted into the softened soil in the main rice field.

The puddling of rice fields degrades the soil physical properties and probably has significant negative impacts on the biological population. The land requires repeated plowings (more on heavier, finer textured soils) after the rice is harvested to bury the rice residues and to obtain a fine seedbed suitable for planting the next crop, usually wheat. This plowing is costly, consumes large quantities of fossil fuels, emits large quantities of greenhouse gases, and delays the planting of wheat, whose yield is affected by delayed crop establishment. The poor physical condition of the soil leads to poor crop stands and to waterlogging after irrigation, with aeration stress and yellowing of the young wheat plants. All these factors take their toll on yield potential, natural resource use efficiency, and environmental quality.

These standard practices are now being replaced by new practices focused on more ecologically sound management of plants, soil, water, and nutrients, supporting beneficial soil biological processes. The whole concept and practice of CA has not been adopted by all farmers, but the main elements of zero-tillage and maintaining residue cover on the soil are gaining wide acceptance. More comprehensive management strategies are still evolving.

24.6 Zero-Tillage Wheat after Rice Harvest

Zero-tillage was initially introduced into Pakistan in the Indo-Gangetic plains for wheat cultivation in 1983. It had been tried in the Indian Punjab in the 1970s but without any substantial adoption because of the nonavailability of suitable equipment. A drill imported into Pakistan from Aitchison Industries (New Zealand) changed the attractiveness of zero-till. This equipment worked very well following manual harvesting of rice since the anchored residue (stubble) did not pose a problem; there were no loose residues to entangle the fixed inverted-T openers. A similar drill was imported into India in 1988 to Pantnagar University, and following these original equipment imports, local manufacturers started making drills at much reduced cost, thus more affordable to farmers.

By 2003, as much as 1.5 million hectares of wheat were being planted this way in the IGP. There are now 20,000 drills made by 68 manufacturers available to farmers. Zero-tilled wheat saves about 50 l of diesel ha^{-1}, or 75 million liters annually, worth US\$37 million. It also reduces CO_2 emissions by 2 million tons. Farmers have adopted zero-tilled wheat quickly after overcoming their mindsets regarding tillage and after experimenting with the technology in their own fields. They realized that they could get more yield by earlier planting, and the lower cost of production adds to their incomes. There are still some in the Pakistan extension service who remain cautious about using zero-till, but with time, these staff also accept that the technology works well.

Zero-till offered an attractive solution to the problem of late planting. Late planting can be caused by late harvest of the previous rice crop, or by the extensive tillage that farmers must do to convert their physically degraded, puddled rice soil into a suitable seedbed tilth for wheat (Hobbs and Gupta, 2003). Data from the South Asian region show a 1 to 1.5% loss in yield potential for every day's delay after the optimum seeding date of November 15. In Pakistan, the majority of the rice grown in the Punjab is a photosensitive, high-quality *basmati* variety that matures in November. Farmers usually make six to eight passes with a soil tillage implement (disk harrow and/or nine-tine cultivator) to prepare the seedbed for wheat. This takes 20 to 30 days and uses 50 l ha^{-1} of diesel to accomplish. The large quantity of rice residue left in combine-harvested fields (50 to 60% of the farms are now harvested this way in Pakistan and NW India) creates a problem for planting the traditional winter-season (*rabi*) wheat, so the majority of farmers burn the loose residue, and instead of drilling wheat, they then sow by broadcast. This creates air pollution problems and poorer plant stands than when seed is drilled.

The approach that worked best for accelerating adoption of zero-till wheat was to supply zero-till drills to innovative farmers and let them experiment in their own fields. As they became more confident about the outcome, they became the best extension channels for the technology to spread to other farmers. Field days and farmer visits were organized to advertise the technology to other farmers.

The main reason given for adoption of zero-tillage since its introduction was the extra yield obtained by planting closer to the optimal sowing time and the cost savings in land preparation and planting (Khan and Hashmi, 2004). Farmers welcome higher yield at less cost. Over time, farmers have realized other environmental and resource-use efficiency benefits. These other benefits have been described in farmer surveys and also by monitoring farmers' fields (zero-till vs. conventional paired plots) over time. Emerging assessments include:

1. Higher yields with zero-till compared with normal tilled wheat. A comparison of the yield of zero-tillage wheat and farmer practice averaged over six farmer sites with paired plots showed a 41% increase (3.67 vs. 2.60 ton ha^{-1}) for zero-tilled plots mainly due to an average 24 days earlier planting in zero-till (Hobbs and Gupta, 2003). Similarly, surveys conducted in Pakistan to assess the impact of zero-tillage wheat showed a 13, 16 and 18% increase in zero-till compared with farmer practice in 1991–1992, 1995–1996, and 2000–2001, respectively (Khan and Hashmi, 2004). These data showed an increase in zero-till yields over time, assumed to be the result of increased skill with the zero-tillage drill. In another survey, by the same authors in 2001–2002, a 14% increase in yield with zero-till was obtained. In India, similar data showed a 400 kg ha^{-1} increase, with 5.2 and 4.8 ton ha^{-1} yields for zero-till and farmer practice, respectively (Singh et al., 2002). In these permanent plots, yields increased for both zero-tillage and farmer practice with time to 5.4 and 5.1 ton ha^{-1}, respectively. Improved weed control was suggested as the main reason for this increase.

2. Less irrigation was needed with zero-till wheat, especially in the first irrigation application. In many cases farmers planted directly after their rice harvest with zero-till, not applying any preplanting irrigation that is usually needed for conventional tillage. Water savings have been calculated as 10 to 18% using zero-till over conventional tillage (RWC, 2004). There is faster movement of water over the soil surface of zero-till fields compared with plowed ones. Savings in water-pumping costs, including fossil-fuel use, can add to this benefit. There was also less waterlogging after irrigation and thus less yellowing of young wheat seedlings.

3. Fewer weeds germinated in the zero-till plots because the winter crop weed seeds were exposed less in zero-tilled plots than in tilled ones. If a farmer followed good weed management in zero-tilled plots, using the same herbicide as in conventional plots and preventing the weeds from setting seeds, herbicides were not needed after 4 to 5 years of zero-tillage (Malik et al., 2004).

4. Farmers observed less lodging in the zero-tilled plots, with improved rooting apparently the main cause of this.

5. By leaving the anchored rice stubble in the field, populations of beneficial insects increased and controlled pests as their habitat was not destroyed by burning or by plowing under the stubble. This could be why there have been no reported outbreaks of rice stemborer in fields using zero-till wheat (Jaipal et al., 2002).

6. Fertilizer efficiency has been increased with zero-till since basal fertilizer can be placed more precisely in a band with the seed drill. In conventional cultivation, the fertilizer is often broadcast and then plowed under because seed drills have difficulty functioning in the loose incorporated stubbles.

7. Plant stands were better and early growth was more vigorous because of the more uniform depth of planting by drill.

8. Environmental benefits have included less air pollution (less burning, although many farmers still do a partial burn) and less greenhouse gas emissions (mostly CO_2) through less use of diesel (Grace et al., 2003).

It took 15 years in Pakistan and 10 years in India to reach significant adoption of the zero-tillage component of CA with wheat. This was mainly because of mindset problems and the fear of failure; CA contravened the conventional wisdom built up over many years about the benefits of tillage. Farmers had to experiment with no-till to convince themselves that it worked. Today in South Asia, there is spreading adoption, and farmers are financially better off. A main factor that facilitated extensive adoption was the willingness and availability of local artisans to build the necessary drills. Without such equipment, success would not have been possible.

24.7   Permanent-Bed Planting in Rice–Wheat Systems

An improvement upon zero-tillage can be brought about if crops are grown on the ridges of a ridge-and-furrow planting configuration (Sayre and Hobbs, 2004). This is, however, profitable only if farmers shift to a permanent bed system, i.e., zero-tilled crops on raised beds. After the wheat harvest, rice can be either direct-sown on the beds or transplanted onto wet beds. There are many advantages to the bed system:

1. Water savings are in the order of 30 to 40% using beds compared with wheat grown on flat soil surfaces. Since water will become increasingly a major limiting resource in the future, this is a critical benefit. In NW Mexico, in the Yaqui valley where CIMMYT undertakes its wheat research in the winter, 90% of the farmers are now growing wheat and other crops on beds because of the limitations of irrigation water (Aquino, 1998).

2. Controlled traffic in the furrows can prevent compaction in the beds. This can be a significant benefit over time, especially in zero-tilled fields. It also results in more compaction at the bottom of the furrow, which restricts vertical water flow (seepage) and helps wet the adjacent beds through better horizontal flow.

3. It offers farmers an alternative to herbicide use for weed control since most weeds can be controlled by mechanical cultivation by driving down the furrows. Herbicide application is also much easier with beds, since nozzles can be prepared to apply the herbicide uniformly over the bed and by using the bed as a guide for the applicator.

4. Similarly, this system offers an opportunity to band basal and top-dress fertilizer applications, instead of broadcasting them, which improves efficiency.

5. The beds permit more diversity in cropping systems during the summer season, as better drainage can be maintained for crops on beds during the wet monsoon period.

6. In addition, permanent beds offer all the other benefits listed above for zero-tillage.

Local manufacturers in India and Pakistan are now making suitable bed formers-planters for practice of this system. There are still some design flaws, which need to be corrected, but many farmers have had success in their experiments with this system.

24.8   Conservation Agriculture in Rice–Wheat Systems

Zero-tillage is just one of the three pillars of CA. The others are permanent ground cover and crop rotation. Work is now being done to promote more use of ground cover in the wheat cycle. First, farmers are encouraged to stop burning their loose residues left after combine-harvesting. A lack of seeding implements that allow planting into loose residue is one of the main limitations at the moment, but local engineers and manufacturers are working together to solve this issue. The following are some promising approaches:

1. Manage the crop residues better during combine-harvesting; essentially build a mechanism in the combine to chop and evenly spread the straw on the field after it leaves the combine. The presently-used inverted-T opener for seeding is not affected by anchored straw and will clog less if the straw is cut into small pieces and spread evenly.

2. Evaluate other systems of planting besides the inverted-T opener. Cutting disks, trash removers, and other systems used in Europe and the U.S. are being tried in South Asia, but the costs of these are still high, and the equipment is heavy and requires larger tractors.

3. A strip-till system where a rotary blade cuts the residue and forms a narrow strip for planting seed and fertilizer is also being tried. There is a Chinese seeder that comes as an attachment for two-wheel tractors that is easily adapted for this purpose. These implements are a common power source in eastern regions of the IGP.

4. In Australia, equipment called a "Happy Seeder," which cuts and picks up the loose straw ahead of the planter, sows the seed without tillage into the soil cleared of straw. The straw that was cut is then blown out behind the straw chopper, being distributed evenly on the ground to form residue mulch. Pakistan engineers are working on a prototype.

Equipment being used in other parts of the world such as South America should be obtained and tested in South Asia. Eventually, prototypes have to be made locally and then mass-produced, fitting the lower horsepower tractors found in Asia. Heavy equipment based on disk-type openers has been tried and is not as popular as the inverted-T opener and other options being promoted by the Rice–Wheat Consortium. Equipment designs used in South America by small-scale farmers may be more suitable to the power systems in Asia.

The cultivation and use of cover crops has been tried little yet since rice residue leaves considerable biomass. There is also the time constraint between the rice harvest and optimal wheat planting, as noted above. However, non-photosensitive, earlier-maturing rice varieties are now being grown in the Indian Punjab with harvest starting at the end of September. These situations could benefit from the use of a cover crop before planting wheat.

24.9　Issues for the Rice–Wheat Farming System

The rice–wheat system is a double-cropping system, and in order to obtain the full benefits of CA, the rice cycle should also use zero-till. This will also be a hard mindset for farmers to change. Growing direct-seeded rice without puddling soils has been tried many times in the region, and data suggest that wheat yields after nonpuddled rice are higher than after puddled rice (Hobbs et al., 2002). However, usually weed growth during the rice cycle has limited the success of this approach. There are a number of issues to be resolved before zero-tilled rice will be acceptable. Many of these are already being researched in a participatory mode with farmers and other stakeholders by RWC partners:

1. Management of weeds with rice. Herbicides are probably the simplest way to attain adequate weed control in the early transition years of CA adoption. Hopefully, as weed problems subside over time, these will no longer be needed. The main objective should be to prevent weeds from setting seeds and to use integrated approaches to achieve this. The use of stale seedbeds, where fields are allowed to germinate weeds, which are then killed or removed before rice seeding, is one approach. Other integrated methods include use of more competitive rice varieties and selective hand weeding. Many herbicidal products are available for transplanted rice based on many years of testing by rice agronomists. Glyphosate is much less toxic than many herbicides presently used for rice cultivation with rapid breakdown on reaching the soil. Persistent use of this herbicide could lead to development of weed resistance, however, as has already occurred with other rice herbicides. With aggressive weed management, the problem should diminish over time.

2. Development of a suitable planter for sowing seed into loose residues and placing seeds more precisely at favorable planting depths (rice should be shallow-planted).

3. Use of cover crops and green manures between the wheat harvest and rice planting. The GMCC would provide biomass for ground cover and help control weeds. The cover crop could be killed by herbicides, followed by the planting of zero-till rice into the mulch, with no incorporation. Any weeds that germinate along with the cover crop would also be killed. There need to be economic assessments for cover crops to see whether the costs involved are greater than the benefits. Data from Parvin et al. (2004) show that no-till cotton following a wheat cover crop is profitable. In the rice–wheat system, the costs of the seed, planting, nutrients, irrigation, and herbicide would all need to be considered.

4. Selection of more vigorous and competitive aerobic rice varieties. Most rice grown in South Asia has been selected under puddled, transplanted conditions; but for CA, varieties are needed that grow under more aerobic conditions, are more competitive with weeds, and withstand the different insect and disease attacks found under more upland conditions. These are being developed by various agencies, international and national, and should become available for testing under this land management system

5. Use of crop rotations to help handle weed and other biotic problems. It would be easier in the wet monsoon season if these crops were planted on beds since waterlogging restricts their use on flat soil surfaces. The discussion of polycropping in Chapter 40 provides some promising data and explanations from China as to how intercropping can be practiced within rice–wheat rotations to improve crop performance.

6. Understanding and using the benefits that can accrue from adopting the system of rice intensification (SRI). The increased rice yields achieved with SRI methods, reported in Chapter 28, derive in large part from the deeper, healthier root systems and increased soil biological activity that these methods induce (Uphoff, 2003). These are two factors associated with successful CA. Not keeping soils flooded during the rice segment of the rice–wheat rotation would provide more suitable soil conditions for growing wheat afterwards, so there could be indirect benefits for that part of the farming system while achieving greater and lower-cost production from the rice cycle. The SRI changes in rice-growing would reduce the requirements for water supply, which is becoming a constraint in some parts of the IGP where the RWS predominates at present. It remains to be seen how SRI practices will respond to no-tillage, mulch, and nonpuddling of rice soils. It is possible that SRI and conservation agriculture could both benefit from combining these strategies for plant, soil, water, and nutrient management.

24.10 Discussion

CA is known to enhance soil biological activity, but research on this topic is difficult to do and poorly understood. Future research should assess and quantify the impacts of CA on biological populations including species identification and function, how microbial populations check negative outbreaks of pests and diseases, consequences of soil biota for soil physical properties, and their role in nutrient recycling and mineralization. Included in such research would be assessments of the effect of fertilizers, herbicides, and pesticides on the viability of the soil biota to give guidance on how these processes can best be enhanced. In the rice–wheat systems of South Asia, introduction of conservation agriculture for entire regions is just beginning. Data are thus not available on the beneficial effects that it will have, if any, on large-scale biological activity. Methodologies are also rudimentary and need to be introduced and refined for such measurements to be taken more easily and accurately.

A question often asked is: how long will it take to obtain the benefits of CA? Also: how sustainable will the benefits be? One of the leaders in the development of conservation agriculture, Rolf Derpsch, has commented:

> In our project in Paraguay we showed that we can regain average productivity on extremely degraded soil in about 3 years when converting to zero tillage and green

manure crops (pigeon pea the first year, mucuna the second year) + fertilization, when about 12 ton ha^{-1} year^{-1} of dry matter (crops + GMCC) are produced. Of course this has been possible in a high rainfall area (1600 mm/year) (Florentin et al., 2001).

There is very little literature on this topic, and for the rice–wheat system, it may be necessary for all, or almost all, farmers to use CA practices before major benefits will be seen. It should be assumed for now that time will be required to obtain full benefits and that there may be a transition period of variable length before farmers see significant positive effects. This should be documented and assessed as more data become available on CA.

Much of present-day agriculture that relies completely on tillage is not sustainable over the long term in more intensive production systems because of significant soil degradation, contributions to global warming, and inefficient use of natural resources. CA has the potential to help reduce the negative effects of intensive production agriculture: it reduces soil degradation; helps build up SOM in the surface layers; improves soil physical and biological properties; reduces the use of fossil fuels; and improves the efficiency of inputs. The main impediments to accelerated adoption are mindsets that favor the status quo on tillage and the lack of access to suitable equipment for planting into the permanent soil cover.

Experimentation by farmers with this resource-efficient technology together with scientific research will determine the best set of practices for each ecoregion. As reported above, at least 72 million hectares of crops were grown around the world with these methods as of 2002, and this number has been growing each year. Other areas have adopted some aspects of this technology, usually zero-tillage, but permanent soil cover needs to be integrated into farming systems to obtain additional benefits. This is the case in the rice–wheat areas of South Asia where farmers are obtaining higher wheat yields at less cost by adopting zero-tillage.

In the next decade, farmers growing rice and wheat in rotation need to find means to change the way that rice is grown by adopting a planting system, such as SRI, not reliant on puddling soils since this practice negates many of the positive effects obtained with CA in the wheat cycle. Changes in practice will depend, in part, on having suitable equipment and finding ways to control weeds and convince the farming community that this new set of practices is preferable. CA has the potential to benefit farmers in other agroecological and cropping-system zones of Asia. Stakeholders in South Asia need to accept that present practices within their intensive food production systems are not sustainable and that a more environmentally friendly and efficient agriculture is needed to produce the extra food needed for the expanding population, doing this in a sustainable way for the benefit of future generations.

References

Aquino, P., *The Adoption of Bed Planting of Wheat in the Yaqui Valley, Sonora, Mexico*, CIMMYT Wheat Special Report 17a. CIMMYT, Mexico, DF (1998).

Brady, N.C. and Weil, R.R., *The Nature and Properties of Soils*, 13th ed., Prentice-Hall, Upper Saddle River, NJ (2002).

Campbell, C.A. et al., Long-term effects of tillage and crop rotations on soil organic C and total N in a clay soil in southwestern Saskatchewan, *Can. J. Soil Sci.*, **76**, 395–401 (1996).

Clapperton, M.J., Increasing soil biodiversity through conservation agriculture: Managing the soil as a habitat, *Proceedings of the Second World Congress on Conservation Agriculture: Producing in*

Harmony with Nature, Iguassu Falls, Parana-Brazil, August 11–15. Food and Agriculture Organization, Rome (2003), (published on CD).

Derpsch, R. and Benites, J.R., Situation of conservation agriculture in the world, In: *Proceedings of the Second World Congress on Conservation Agriculture: Producing in Harmony with Nature, Iguassu Falls, Paraná, Brazil, August 11–15.* Food and Agriculture Organization, Rome (2003), (CD).

Derpsch, R. and Moriya, K., Implications of soil preparation as compared to no-tillage on the sustainability of crop production: Experiences from South America, In: *Management of Tropical Agro-Ecosystems and the Beneficial Soil Biota*, Reddy, M.V., Ed., Science Publishers, Enfield, NH, 49–65 (1999).

Florentin, M.A. et al., Abonos verde y rotación de cultivos en siembra direto: Pequeños propriedades. In: *Proyecto Conservación de Suelos*, MAG-GTZ, DEAG, San Lorenzo, Paraguay (2001).

Gantzer, C.J. and Blake, G.R., Physical characteristics of a La Seur clay loam following no-till and conventional tillage, *Agron. J.*, **70**, 853–857 (1978).

Grace, P.R. et al., The long-term sustainability of the tropical and subtropical rice and wheat system: An environmental perspective, In: *Improving the Productivity and Sustainability of Rice–Wheat Systems: Issues and Impacts*, Ladha, J.K. et al., Eds., ASA Special Publication 65, Agronomy Society of America, Madison, WI, 27–44 (2003).

Hobbs, P.R. and Gupta, R.K., Resource-conserving technologies for wheat in rice–wheat systems, In: *Improving the Productivity and Sustainability of Rice–Wheat Systems: Issues and Impact*, Ladha, J.K. et al., Eds., ASA Special Publication 65, Agronomy Society of America, Madison, WI, 149–171 (2003).

Hobbs, P.R. et al., Direct seeding and reduced tillage options in the rice–wheat systems of the Indo-Gangetic Plains of South Asia, In: *Direct Seeding — Research Strategies and Opportunities: Proceedings of the International Workshop on Direct Seeding in Asian Rice Systems, Strategic Research Issues and Opportunities, 25–28 January 2000, Bangkok, Thailand*, Pandey, S., Mortimer, M., Wade, L., Tuong, T.P., Lopez, K., and Hardy, B., Eds., International Rice Research Institute, Los Baños, Philippines, 201–215 (2002).

Islam, K.R. and Weil, R.R., Soil quality indicator properties in Mid-Atlantic soils as influenced by conservation management, *J. Soil Water Cons.*, **55**, 69–78 (2000).

Jaipal, S. et al., Species diversity and population density of macro-fauna of rice–wheat cropping habitat in semi-arid subtropical northwest India in relation to modified tillage practices of wheat sowing, In: *Herbicide-Resistance Management and Zero-Tillage in the Rice–Wheat Cropping System: Proceedings of an International Workshop*, Malik, R.K. et al., Eds., CCSHA University, Hissar, India, 166–171 (2002).

Kemper, B. and Derpsch, R., Results of studies made in 1978 and 1979 to control erosion by cover crops and no-tillage techniques in Paraná, Brazil, *Soil Till. Res.*, **1**, 253–267 (1981).

Khan, M.A. and Hashmi, N.I., Impact of no-tillage farming on wheat production and resource conservation in the rice–wheat zone of Punjab, Pakistan, In: *Sustainable Agriculture and the Rice–Wheat System*, Lal, R. et. al., Eds., Marcel Dekker, New York, 219–228 (2004).

Lal, R., Conservation tillage for sustainable agriculture: Tropical versus temperate environments, *Adv. Agron.*, **42**, 85–197 (1989).

Liebig, M.A. and Doran, J.W., Impact of organic production practices on soil quality indicators, *J. Environ. Qual.*, **28**, 1601–1609 (1999).

Malik, R.K. et al., No-tillage farming in the rice–wheat cropping system in India, In: *Sustainable Agriculture and the Rice–Wheat System*, Lal, R., Hobbs, P.R., Uphoff, N., and Hansen, D.O., Eds., Marcel Dekker, New York, 133–146 (2004).

Parvin, D., Dabney, S., and Cummings, S., *No-till cotton yield response to a wheat cover crop in Mississippi*, Online Crop Management doi:10.1094/CM-2004-0416-01-RS, http://www.plantmanagementnetwork.org/pub/cm/research/2004/cover/ (2004).

RWC, *RWC Highlights for 2003–2004*, see RWC web site (2004) listed below.

Sayre, K.D. and Hobbs, P.R., The raised-bed system of cultivation for irrigated production conditions, In: *Sustainable Agriculture and the Rice–Wheat System*, Lal, R. et al., Eds., Marcel Dekker, New York, 337–355 (2004).

Singh, S. et al., Long term effect of zero-tillage sowing technique on weed flora and productivity of wheat in rice–wheat cropping zones of the Indo-Gangetic Plains, In: *Proceedings of International Workshop on Herbicide Resistance Management and Zero-Tillage in Rice–Wheat Cropping System. 4–6 March*, Malik, R.K. et al., Eds., Haryana Agricultural University, Hisar, India, 155–158 (2002).

Uphoff, N., Higher yields with fewer external inputs? The System of Rice Intensification and potential contributions to agricultural sustainability, *Int. J. Agric. Sustain.*, **1**, 38–50 (2003).

Websites

FAO Conservation Agriculture: http://www.fao.org/ag/AGS/agse/main.htm
Rolf Derpsch: http://www.rolf-derpsch.com/
Conservation Technology Information Center: http://www.ctic.purdue.edu/CTIC/CTIC.html
Rice–Wheat Consortium (RWC) web site: http://www.rwc-prism.cgiar.org/rwc
SRI web site: http://ciifad.cornell.edu/sri/

25

Managing Soil Fertility on Small Family Farms in African Drylands

Michael Mortimore

Drylands Research, Milborne Port, Somerset, UK

CONTENTS

Crisis narratives about soil management on small family farms in Africa have influenced policy formation for many decades. From the 1930s until the 1970s, wind or water erosion from newly exposed agricultural land or from grazing land attracted much attention. The links with human activities were visible in soil wash, gully formation, and flood waters heavily loaded with sediment, especially in areas recently cleared of natural vegetation. The erosion narrative motivated some colonial and postcolonial departments of agriculture to introduce stringent soil and water conservation policies.

To measure the impact of climate and agricultural practices, the Universal Soil Loss Equation, developed under North American conditions and designed for use there, was applied to Africa. Plot measurements, assumed to be representative, were projected to catchment or regional levels, with some estimates claiming that more than $5-15\,t\,ha^{-1}\,yr^{-1}$ of topsoil was being removed from large areas. On desert margins, sand dunes were said to be expanding, and scenarios of desert advance were boldly quantified. Stebbing's prediction (1935, 1953) of the southern advance of the Sahara was resurrected in many scientific, policy, donor, and NGO reports following the drought cycles of the 1970s and 1980s (Mortimore, 1989). UNEP's *World Atlas of Desertification* (UNEP, 1992) estimated that 24% of Africa was affected by wind erosion and 16% by water erosion in 1990.

In the 1980s and 1990s, however, concern with soil erosion was overtaken by a soil degradation narrative. Nutrient loss through "soil mining" was attributed to a failure fully to replace, either by fallowing or by chemical amendments, the nutrients being removed in crops. Development agency reports frequently cited estimates of rapid depletion of the major nutrients commonly provided in chemical fertilizers, nitrogen (N), phosphorus (P), and potassium (K) (Stoorvogel and Smaling, 1990). By one estimate, sub-Saharan Africa during the period 1982–1984 was losing 22 kg N, 2.5 kg P_2O_5, and 15 kg of K_2O ha^{-1} yr^{-1} (Stoorvogel et al., 1993). A subsequent policy study claimed that 86% of sub-Saharan African countries are losing more than 30 kg N ha^{-1} yr^{-1}, with net combined NPK losses put at 60 to 100 kg ha^{-1} yr^{-1}, and increasing (Henao and Banaante, 1999).

Such estimates went into institutional statements. The World Bank (2003) cited losses of more than 30 kg of N, P, and K ha^{-1} yr^{-1} in all but three African countries. Projections of Stoorvogel and Smaling's figures were used to support claims that 660 to 700 kg N, 75 to 100 kg P and 450 kg K ha^{-1} were lost during the previous 30 years (Sanchez et al., 1997; Gruhn et al., 2000). In policy discussions, direct linkages have been made between "mining natural capital," poverty, and economic underdevelopment (Pinstrup-Andersen and Pandya-Lorch, 2001; World Bank, 2003).

Both the soil erosion and soil degradation narratives have been subjects of criticism on methodological grounds, however (Stocking, 1996; de Ridder et al., 2004; Faerge and Magid, 2004), and the originators of the nutrient-loss scenarios have themselves warned against uncritical use of such estimates (Smaling et al., 1999). Nutrient-depletion scenarios, we have argued elsewhere, are not fully consistent with longer-term data available on the performance of certain farming systems in the region (Mortimore and Harris, 2005). However, the two narratives continue to frame much thinking and proposed remedial actions for soil systems in Africa, as seen in a new international initiative called TerrAfrica recently announced by the Global Environmental Facility (GEF, 2004).

Both scientists and policy-makers tend to attribute losses from erosion or "soil mining" to the practices of small farmers and livestock producers. African drylands, which range across arid, semiarid and dry subhumid agroecological zones, are variously defined. They are generally characterized as having <1000 mm average (and quite variable) annual rainfall; growing seasons of 75 to 180 days each year that are often short and not very predictable; high average temperatures and evaporation; and soils of low natural fertility (Jones and Wild, 1975; Pieri, 1989).

The area affected by soil degradation has been put at 332 million hectares, or 25% of the African continent's surface (Oldeman and Hakkeling, 1990; UNEP, 1990). The causes for this are said to be overgrazing, inappropriate agricultural practices, over-exploitation of the land, and deforestation. Within a Malthusian framework of population-environment relations, African drylands are seen as vulnerable to degradation because of their rapidly growing and sometimes very dense populations that obstruct sustainable ecosystem management.

Previous findings in the Machakos district of Kenya have challenged such a framework (Tiffen et al., 1994). Except for its subhumid hilltops, Machakos qualifies as a dryland area, with low inherent fertility. Yet over a 60-year period, natural resource degradation there was largely brought under control, and significant increases in the value of agricultural output were achieved by a suite of farming practices that managed plants, soil, water, and nutrients differently and congruently. The Machakos and some other African experiences suggest that current narratives on soil fertility management in Africa are in need of review (Tiffen and Mortimore, 2002).

This chapter expands upon the Machakos findings by comparing them with some dryland regions in the West African Sahel. In general, African drylands are diverse rather than homogeneous, both in terms of their agroecological systems (Raynaut, 1997) and in the ways that soil nutrients are managed (Hilhorst and Toulmin, 2000; Scoones, 2001). This restricts the scope for grand theories. However, the long-term, 40- to 60-year performance of some farming systems that have been under stress from growing and dense populations and high environmental risk offers some counterintuitive success stories that justify investment in drylands (Reij and Steeds, 2003; GM-CCD, 2004). These center on smallholders' achievements in restoring or sustaining soil fertility in the face of numerous constraints. They raise questions about equating soil health with simple indicators of erosion or nutrient status. This chapter goes further to suggest that accounting for disparities between predictions and performance, and finding better directions for policy, can be furthered by paying more attention to the maintenance and enhancement of soil biological activity.

The review here begins with a discussion of experience in the former Machakos district of Kenya. Then, two systems in West Africa are examined, the Kano Close-Settled Zone (KCSZ) in Nigeria, and the Diourbel region in the "groundnut basin" of central Senegal. Our interest in West African experience was prompted by the way that Kenyan farmers in East Africa had challenged and contradicted, dramatically, the conventional wisdom about soil system dynamics and capabilities.

25.1 Restoring Degraded and Eroded Soils in Machakos, Kenya

Between the 1930s and the 1990s, the area formerly delineated as the district of Machakos saw its population increase from 250,000 to 1.5 million. This sixfold increase in 60 years was among the fastest in East Africa. Cultivated land expanded from <20% of the total area in the old Ukamba Reserve in the 1930s to >80% in some locations 50 years later. Through migration, the mostly Akamba population was redistributed from the old centers of settlement on subhumid hilltops downslope onto semiarid plains in the east and south.

By the beginning of the 1930s, after two decades of colonial administration, extensive shifting cultivation and expanding cattle herding had reduced most dry woodland areas to degraded scrub; the hills were mostly deforested; and exposed slopes suffered from sheet erosion and gullies (Barnes, 1937). "It is not too much to say," wrote the Kenyan government's Hall Commission in 1929, "that a desert has already been created. Every phase of misuse of land is vividly and poignantly displayed" (Maher, 1937).

Official alarm at the rapid environmental destruction led to attempts by the Department of Agriculture to restrict the use of all woodlands, banning the grazing of stock and imposing a livestock quota system. From the 1940s on, it forced farmers to adopt prescribed soil conservation measures. The Machakos soils are deeply weathered Ferralsols and Luvisols, with strongly developed horizons (except where bedrock is exposed on steep slopes). "The greater part of the Ukamba Reserve has lost [its] topsoil through erosion [and rivers are] carrying tons of the soil of the Reserve towards the Indian Ocean," said a government report (Maher, 1937). Only leached and infertile B-horizon soils were left for farmers' use; 56% of uncultivated land was considered to be eroded.

Yet by the 1990s, this landscape had undergone a notable transformation, and not primarily due to government prescriptions or interventions. Outside of the forest reserves, it is true, little woodland remained. However, instead of becoming a wasteland, the farmed landscape had evolved, through decades of small private investments, into a well-conserved system with terraces and field drains, fodder-grassed banks and woodlots, roof-top water catchments and small dams, hedgerows, enclosed pastures, reclaimed barren land, and cultivated fields studded with citrus, mango, pit-planted bananas and other economic trees (Figure 25.1a,b). Food security had improved, and on higher land, coffee producers had successfully entered the export market, while vegetables were being grown for urban markets. That the soils disparaged by colonial authorities could be made very productive is quite evident.

This landscape of conservation had been extended downslope to encompass progressively more arid and risky environments. Meanwhile, the energetic pursuit of

FIGURE 25.1

(a) A Machakos landscape during the dry season 1937. Note exposed soil on cultivated slopes, degraded natural vegetation, and scarcity of trees and permanent farmsteads. Kalama Hills from Kivandini, June or July, 1937. Map sheet 162/2, GR 3104E, 98302N; Brg 145°. (Photo: R.O. Barnes, used with permission). (b) The same landscape in the dry season 1991. Note terraced, cultivated fields, densely scattered farmsteads, and increased numbers of on-farm economic trees. Kalama Hills from Kivandini, September 1991. Map sheet 162/2, GR 3104E, 98302N; Brg 145°. (Photo: M. Mortimore).

education — at considerable private cost — had opened up employment opportunities in the cities for enterprising Akambas. Incomes became more diversified in rural areas, and funds were brought back to support both consumption and investment in improving farms as well as houses.

At the root of this transformation, reported in detail in Tiffen et al. (1994), was a revolution in soil management. Coercive colonial schemes of soil conservation, which had been accepted by farmers with reluctance, failed in most areas as the introduced structures fell into dilapidation in the run-up to national independence in 1962. This can be seen from old photographs. From the mid-1960s on, a remarkable conversion to conservation values occurred, as farmers discovered for themselves that terraces, especially those of a certain design called *fanya juu*, could enhance soil moisture and hence crop yields. (To make *fanya juu* terraces, the soil is thrown upslope, creating a useful ditch at the top of the field for collecting water and growing water-tolerant plants, rather than downhill as was done to establish terraces in the government-recommended manner.) While government and donor projects invested heavily in promoting terrace construction, even at the peak of this activity (under the Machakos Integrated Development Project, 1978–1988), air photographic evidence shows that twice as many kilometers of terraces were built through private investment as with project assistance.

Concurrently with an expansion of farming to every corner of the district (even in areas having <400 mm average annual rainfall), common-access grazing gradually disappeared as livestock production became based on private pastures, usually enclosed. By the 1990s, a need to support as many animals as possible on reduced grazing resources was motivating livestock owners to reseed grasses and make drainage investments. On many farms, pasture had been so reduced that cattle or small ruminants now depend mainly on crop residues, hedge plants, cut grass from terrace banks or field borders, and tree browse. The cattle corral (*boma*) became a pivotal institution in such landscapes, cycling nutrients between private pasture and farmland.

As intensification became imperative for Machakos farmers, conservation gained social acceptance as best practice. Any family without the means — animals, land, labor — to intensify its production in ways that capitalized on natural resource improvement and maintenance became socially disadvantaged. Inequalities in nutrient ownership and in the enjoyment of corresponding agricultural benefits became aligned with social class differentiation (Rocheleau et al., 1995; Murton, 1999). Households with diminished agricultural opportunities must expend effort to obtain off-farm or out-of-district income. Demand for land has increased with population growth and the investment ambitions of migrant workers. Holdings have become ever smaller through subdivision, encouraging intensified resource management.

In the district as a whole, despite or because of this pressure on the land, the value of agricultural production of crops and livestock increased greatly between 1930 and 1987. In per capita terms, and at constant 1957 prices, the value of agricultural output went up by 3.5 times, and in per-area terms, it increased tenfold per km^2 (Tiffen et al., 1994: 93–95). Although famines had been severe and frequent after droughts up until the 1960s, available evidence indicates that food security has markedly improved since then.

The revolution in soil management, which made these gains in productivity possible, seems contradicted by what little soil fertility data there are. We have found no representative time-series data available on soil fertility in the district. However, Table 25.1, using a spatial analog method, suggests that most of the chemical properties of soils there do deteriorate after prolonged cultivation. The lighter color and the presence of sharp angular rock fragments in both fallow and cultivated soils are consistent with the fear expressed in the 1930s that topsoils had been largely washed away.

TABLE 25.1

Chemical Properties of Kilungu Soils, Machakos District, 1990

Property	Uncultivated[a]	Fallow/Grazing[b]	Cultivated[c]
Soil pH (water)	5.5	5.4	5.0
Carbon (%)	2.49	1.25	0.74
Nitrogen (%)	0.35	0.18	0.11
Phosphorus (ppm)	23.0	14.0	13.0
Potassium (me%)	0.56	0.4	0.29
Calcium (me%)	8.7	2.4	1.1
Magnesium (me%)	3.4	1.4	0.9

Note: These values are averages of three samples from each of three sites, each representing a land use class.

[a] Protected (sacred) forest >60 years.

[b] Fallowed >20 years since cultivation.

[c] Cultivated annually for 40 to 60 years, with little fertilization, and (at two sites) with no terraces or other land improvement.

Source: J.P. Mbuvi, adapted from Tiffen, M. et al. *More People, Less Erosion: Environmental Recovery in Kenya*. J. Wiley, Chichester, 1994. With permission.

How can such measured declines in nutrient availability be reconciled with the evidently productive performance of the farming systems in the district, matched by the visible evidence of private investments in land and housing improvements? Or with the impressive levels of biomass production achieved on the farms where rainfall is satisfactory? To find answers, one needs to consider certain simple soil system management practices of Machakos small farmers: (1) improved field moisture regimes on terraces; (2) manuring and composting; (3) tree protection and planting; and (4) biodiversity conservation. These change the soil environments for crop and livestock production, below- and aboveground.

25.1.1 Improved Field Moisture Regimes

Soil and water conservation was initially promoted in Kenya in response to evident and extensive soil loss. There is a large literature focused on physical parameters of this and on engineering solutions (Karanja and Tefera, 1990). In retrospect, it is surprising that so little attention was paid to the economic (yield) benefits of conservation measures taken by farmers in Machakos, apart from a handful of studies (Figueiredo, 1986; Lindgren, 1988; Fones-Sundell, 1989). In fact, it was the moisture-conserving rather than the soil-conserving effects of terracing that eventually convinced a majority of local farmers to invest their own resources in individual or group conservation activities.

Where rainfall is erratic and intensive, separated by dry spells, and on steep slopes that otherwise had rapid runoff and poor infiltration, terraces have a profound impact on soil moisture regimes. By smoothing variability in the hydrology of soil systems, terraces not only improve the delivery of water to crops but help maintain a favorable biotic environment for organisms whose impact on fertility in Machakos has been little investigated. Decomposition and incorporation of plant residues and manure as well as root development must certainly be affected. It is known that terrace environments are highly variable at the microscale. Farmers know the physical and biological properties of every corner of their holdings and adapt their practice to such variations. Where there is excessive rainfall, on the other hand, coordinated field drains help to regulate soil water-logging and reduce flood damage to field crops.

25.1.2 Livestock, Manuring, and Composting

The low values obtained for chemical indicators of soil fertility when measured on fallow/grazing land in Machakos are the result of the removal of nutrients through grazing for conversion into manure or compost via the *boma* system. Such fallows are not really regenerative; they would be better characterized as a residual category of land use that suffers from chronic underinvestment.

Where farmers lack animals and cannot afford to buy manure or compost, the results for their cultivated land are those seen in Table 25.1, where the sites had received little or no fertilization for prolonged periods. On the other hand, where farmers keep animals and maintain a *boma* — or can afford to supplement their ever-insufficient supply of manure or compost with small doses of mineral fertilizer — the yield potential becomes a function of inputs. Fertility supplementation is an opportunistic matter, however. Manuring takes place immediately before planting or even afterwards, on a stand-by-stand pattern according to availability. Microdoses of inorganic fertilizer, if affordable, are added after the individual plant has shown viability under given rainfall conditions. Fertilization is perceived to be directed to the crop rather than to the soil, however. These activities cannot easily be monitored by soil sampling because nutrient cycling is quick and efficient, at least where soil organisms are abundantly present to break the organic materials down.

25.1.3 Tree Protection and Planting

The creation of government forest reservations, regulation of cutting, and other forms of intrusive management were intended to counter deforestation. To head off an anticipated fuelwood crisis, it was assumed that farmers had to be persuaded to plant and protect trees wherever possible, particularly for firewood. While men were indeed often reluctant to embark on producing timber on woodlots, despite the various incentives and promotions by the forestry department, women in Machakos adopted fruit trees enthusiastically as a means of adding value to their small farms, tapping the growing market for citrus, pawpaw, and bananas. These could grow well if assisted by pits and channeled water runoff from roads or open surfaces.

Data collected in the 1980s suggest a negative correlation between size of holding and the number of farm trees per hectare (Gielen, cited in Tiffen et al., 1994: 221). This means that through indigenous agroforestry, farmers achieved the highest productivity on the smallest farms. There is no doubt that income gains were the chief incentive for planting and protecting such trees. However, ecological gains are also significant.

The beneficial impacts of certain trees on crop yield are well recognized by African farmers, as are possible negative effects, such as shading. Trees use different soil horizons from field crops and help to stabilize surface materials. The landscape-scale impact of tree planting is evident in any photograph of Machakos farmland during the growing season (see Figure 25.1b). Multipurpose, value-adding and seasonally complementary management of trees on and around small fields underlines the fact that the commercial-industrial concept of field is both scientifically and economically inappropriate under these conditions. This theme is taken up again when discussing the West African cases below.

25.1.4 Biodiversity Conservation

Agrobiodiversity in tropical farming systems is promoted by microvariability in the growing environments on a farm holding or even a single terrace (Brookfield, 2001). These variations interact with the effects of rainfall fluctuation to determine the performance of cultivars. Hence, farmers enthusiastically added the "improved" composite maize promoted by the government to the repertoire of "local" cultivars that they planted (hybrid maize being generally unsuitable for dry conditions). However, it failed to achieve the dominance expected. Why? The complexity of local farming systems is not merely a matter of tradition. It is the product of successful adaptive practice in response to new opportunities, e.g., coffee production on hillsides, and expansion of horticultural production. It also reflects new stresses generated by land scarcity and growing domestic consumption. Soil fertility management appropriate for such varied and changing conditions is unlikely to be reducible to certain levels of nitrogen, phosphorus, and potassium. The overall performance of soil systems depends on more than a few nutrients, however important these may be.

25.2 Sustaining Long-Term Intensification in the Kano Close-Settled Zone of Nigeria

Soils around Kano City have been under rotational cultivation for many centuries and were considered to be highly productive by early European travellers (Barth, 1857). Around the time of colonial occupation, a third of the farmland (at most) was under short grass fallow (Gowers, 1913). Soil fertility was maintained through organic soil amendments: manuring, dry composting, and residue incorporation. Legumes (cowpea) and grains (millet and sorghum) were intercropped or rotated. When a railway was opened in 1912, the production of groundnuts, also a legume, was inserted into the cropping system to serve distant market demands. The indigenous farming system was widely regarded as exemplary, and the colonial rulers did not interfere with it. The ferruginous tropical soils, derived from aeolian parent materials, are sandy, well-drained, responsive to organic fertilization, and easy to cultivate with hand technologies.

Until road transport became well-developed in the 1950s, the Close-Settled Zone around Kano City supplied it with most of its grain. During six decades of groundnut exports, from 1912 to 1975, farmers in the zone exploited their comparative advantage for this crop. Already by the 1960s, farming households often bought rather than produced their food grain (Mortimore, 1993). However, by 1975 the combined effects of drought and rosette disease on groundnuts undermined such a strategy. Further, as the urban population grew sixfold from the 1960s to the 1990s, local market demands for food, labor, and services were increased. Kano's central position in regional trading networks facilitated diversification of economic activity into off-farm sectors.

Early "soil mining" estimates were used to promote the use of inorganic fertilizers in northern Nigeria in the 1960s (Watson, 1964). In the early 1980s, fertilizer applications peaked under a subsidy regime sponsored by a World Bank-supported agricultural development program. The impact of these efforts cannot be quantified, but in the long term it appears marginal. Removal of subsidies under structural adjustment after 1986 (Mustapha and Meagher, 2000) has not threatened the continuity of the organically-based system.

Accurate accounting for the volume of nutrients actually leaving the system is difficult, e.g., a large fraction of grain output is consumed at home. A rough calculation suggests that human wastes at a population density of 2.23 per hectare and a mean dry weight of $100 \text{ g person}^{-1} \text{ day}^{-1}$ (based on Burkitt, 1983: 44) would produce $0.8 \text{ t ha}^{-1} \text{ yr}^{-1}$ — twice as much as a reasonable estimate for animal manure produced. It is not known how much of this finds its way back to the fields, but we note that many villagers lack latrines.

By the end of the 20th century, with a population of over 5 million and average densities exceeding 300 persons km^{-2} in central districts (Tiffen, 2001), the KCSZ continued to be characterized by intensive agriculture (Mortimore, 1993; Harris, 1998). On farms within about 35 km of Kano City, where residents may exceed 220 persons km^{-2}, cultivable area averages only 0.5 ha person^{-1}.

Investments by small producers continue, despite the ever-diminishing size of agricultural holdings and persistent poverty among less successful families (Hill, 1972). The cultivated fraction, all of it in permanent manured fields, was already 88% of the land area in 1981. Change is not driven only by markets, however. There are close connections among population growth, agricultural intensification, and livestock management. The higher the density of the farming population, the more labor is available. Tumbau farmers living 30 km from Kano invest 2.5 to 5 times more family labor per hectare in weeding and harvesting than do farmers in Dagaceri, where the rural population density is only one-fifth as high (Mortimore and Adams, 1999: 114). Livestock densities, as measured in standard tropical livestock units, also correlate strongly with human population densities (Hendy, 1977; Bourn and Wint, 1994; de Leeuw et al., 1995), even though rangeland has long since disappeared from the landscape. The Kano experience permits the Machakos findings to be both validated and extended in a West African context.

25.2.1 Improved Field Moisture Regimes

No soil and water conservation structures have been installed in the KCSZ to exert an effect on hydrological regimes comparable in scale to those of Machakos. At first sight, farmers are thus more dependent on nature. For early germination and strong crop growth, it is critical to catch the planting rain at the beginning of the season, normally in May or June, although in drought years not until July. Early rain events may be widely separated, so timely planting allows crop roots to follow the field moisture as it percolates downwards. A day's delay may be too late for some crops. Later on, however, as rainfall increases, gently sloping or flat fields are prone to temporary flooding, and ridges are constructed either before or after row planting to control runoff, retain field water, and protect roots from saturation. Residual moisture after the end of the rains, normally early October, is crucial for the growth of late-maturing sorghum and cowpea.

25.2.2 Livestock, Manuring, and Composting

Kano farmers have demonstrated a clear preference for cycling nutrients through animals, as compared with zero-tillage or other nonanimal-based systems. This is because animals have multiple uses (draft, milk, breeding, fattening) and also synergies with crop and tree production as animals convert crop residues and foliage into manure. Also, animals accumulate significant value, especially small ruminants, as investments for poor people. Incomes from animals benefit women as well as men, although market prices for meat tended downward in the 1990s (Ariyo et al., 2001). It is unlikely that the value of these assets will ever be outweighed by efficiency gains from technical improvements in nutrient management, such as residue incorporation. Animals are fed with crop residues, browse and weeds, using a labor-intensive, cut-and-carry system during the growing season. However, fodder markets have emerged, especially for draft animals and for fattening cattle for market. A key question therefore is: what constraints limit the growth of the animal population, and how can these be relaxed?

Kano farmers produce enough organic material to fertilize all cultivated land every year or every few years, at rates ranging from 3.1 to 5.2 t ha^{-1} and averaging around 4 t ha^{-1}

TABLE 25.2

Association between Fertilization Regime and Relevant Soil Properties in the Kano Close-Settled Zone (measured in top 15 cm)

Fertilization Regime	Clay (%)	Organic C (%)	Total N (%)
(1) Uncultivated grazing land	3.8	0.60	0.025
(2) Intermittent fallow	2.0	0.27	0.025
(3) Inorganic only, ox-plowed	1.2	0.26	0.042
(4) Organic only, hand cultivation	2.0	0.42	0.003
(5) Organic only (cattle manure), ox-plowed	2.4	0.44	0.028
(6) Organic only (cattle), pasture/cultivation rotation	1.7	0.69	0.027
(7) Organic + inorganic, hand cultivation	2.0	0.70	0.032

Source: From Yusuf, M. Soil assessment and indigenous soil management strategies in the semi-arid north-eastern Nigeria. Ph.D. thesis, Department of Geography, Bayero University, Kano, Nigeria, 2001. (Table 6.1). With permission.

(Harris and Yusuf, 2001). The major nutrient input is in a low-quality, dry compost of manure, ash, and crop residues. However, this is never abundant enough to meet most technical recommendations.

A range of fertilization strategies is employed (Yusuf, 2001). Table 25.2 illustrates some associations between fertilization regimes and clay fraction, organic carbon, and total nitrogen. Composite samples were taken from four plots per regime; because no quantification of inputs was possible, the data are more illustrative than definitive. They show the capacity of organic fertilization under known local practice to restore organic carbon under intensive cropping (mainly sorghum, millet, cowpeas, and groundnuts). The fertility trajectory has followed a U-shaped curve, descending from an uncultivated control through degraded fallows and solely inorganic fertilization, then rising again with cattle manure and integrated fertilization. Total nitrogen peaks under inorganic fertilization, but its biological properties are poor, whereas total nitrogen also improves with the organic treatments.

Fertility management practices vary widely, reflecting differing access to factors of production and individual circumstances, with consequent influence on nutrient balances (Harris, 1998). At 59 representative sites in the KCSZ, sampled in 1977 and again in 1990, the chemical properties of cultivated topsoils (measured in terms of C, N, Ca, Mg) did not change significantly on average, although potassium declined, and phosphorus was not measured (Mortimore, 1993).

25.2.3   Tree Protection and Planting

Mature trees growing in the farmed parklands of the KCSZ have mean densities of 10 to 12 per hectare west of the city and 7 to 8 per hectare in an area 35 km east of the city. Regeneration is vigorous although variable between species. Planting as well as protection are practiced to increase tree populations (Cline-Cole et al., 1990). As in Machakos, trees support more diverse and resilient ecosystems and interact with field crops, in ways that can be positive or negative for yields, although generally adding value to total farm output. This accounts for the observation that during the period 1972 to 1985, when there were two major drought cycles and the sale of wood for fuel became a common income strategy, the density of farm trees was sustained, and in some areas even increased around Kano.

Woodfuel crises have dominated discussions of tree management to the exclusion of their ecological and biological functions in farmed parkland landscapes. This has reinforced the preoccupation of government forestry departments with tree protection measures, based on the belief that farmers are agents of deforestation, requiring the imposition of centralized control (Cline-Cole, 1997). However, the enormous variety of medicinal, food, fodder, and construction benefits derived from West African trees, catalogued many years ago (Dalziel, 1937), undergirds a local conservation ethic that is strongly positive for biodiversity (Mortimore and Turner, 2005).

25.2.4   Biodiversity Conservation

As in Machakos, agrobiodiversity is a resource for supporting livelihoods that is maintained at the community level. It is a kind of bank to draw on when conditions change, and it often supports an emergency food reserve. Soil and plant management practices thus need to be fitted into encompassing livelihood strategies. In one village (Tumbau), it has been documented that 76 different cultivars are being maintained, including 12 landraces of pearl millet, 22 of sorghum, 9 of cowpea, 5 of groundnuts and 4 of cassava (Yusuf, 1996, cited in Mortimore and Adams, 1999).

The sustainability of tree populations, it has been often shown, depends on the value of tree products, and this is greater for trees that are alive rather than dead. In addition to trees, every type of herbaceous growth is put to use in the KCSZ. Besides the careful storage of crop residues, for later use as fodder, construction materials, or fuel, volunteer shrubs and weeds are systematically collected for feeding to livestock. Field boundaries are used to grow thatching grass (*Andropogon gayanus*) and other plants. Tree browse is cut for animals, and roots are left in the ground after the crop harvest, adding organic material to the soil. Field trash is first grazed by small ruminants, with the remainder gathered into heaps and burned.

The diversified biomass produced and managed is as relevant for the well-being of rural households undertaking integrated ecosystem management as is their economic yield for market or consumption. An attempt has been made to quantify the productivity of total plant biomass on farmlands using this system (Mortimore et al., 1999). The dry matter produced is at least equal to that produced in a natural savanna woodland, according to a rangeland model based on rainfall (Breman and de Wit, 1983). The same has been found in a less productive, more extensive system under lower rainfall described elsewhere (Harris, 1999). Although these findings are based on small samples, they call into question the assumption, implied by "nutrient mining" scenarios, that farmers are degrading their soil's productive potentials.

25.2.5   Managing Biological Threats

Biological agents are not always favorable. KCSZ farmers face a number of threats common to the Sahel:

- *Parasites*: The annual weed *Striga hermonthica* can reduce yields by 10 to 100%. Pending the development of resistant varieties, farmers know that practical control is best achieved through soil management (long fallows or rotations). Infestation is a function, most directly, of lowered soil organic matter, but in broader terms, it reflects land scarcity (repeated cropping), greater population density, and poverty. Striga control is discussed in Chapter 40.

- *Viruses*: Rosette disease can wipe out an entire groundnut crop over a large area and can persist in a region throughout the year if certain wetland vegetation is available for its vector, *Aphis craccivora*. This occurred in Kano for years after the disease first struck in 1975, when irrigation schemes provided it with refuge during the dry season.

- *Pests*: The grasshopper *Oedaleus senegalensis* can cause as much damage to millet as can a drought, which deprives the herbivorous insect of alternative food and keeps its population in check. This connects the productivity of a cropping system with the grasslands located elsewhere within the local land-cover pattern. Spatial interactions occur on a regional scale with the devastation of locusts, invasive again in the Sahel at the time of writing (2004). Migrating seed-eating rodents, such as the jerboa, spread rapidly among farms and villages, and even across an international border in the 1980s.

The Kano experience suggests that the local knowledge of African farmers prioritizes the management of soil biology, which is but one facet of a wider biological realm in which productivity is determined and livelihoods are supported. In such a circumstance, it is anomalous that so much of the advice given by professionals has dramatized physical or chemical constraints and solutions.

25.3 Adapting Soil System Management to Policy Failure in Diourbel Region, Senegal

Elements of the Machakos and Kano experience are discernible in farming systems elsewhere in the Sahel. In both the Kenyan and Nigerian cases, government policy over the long term tended to be noninterventionist, leaving Akamba farmers to find their own solutions to the challenges of demographic growth and migration, market changes, and economic diversification, while Hausa farmers in Kano adjusted a centuries-old farming system to an oil boom (1970–1978), structural adjustment from 1986 on, and drought in the 1970s and 1980s (Mustapha and Meagher, 2000).

Senegal, on the other hand, experienced strong discontinuity. An agrarian policy based on a dominant role for the state — which supplied credit, seed, subsidized inputs, and technical advice; controlled markets and prices; and promoted the consumption of subsidized rice — failed financially in 1984 under the conjunction of drought and policy changes. Small producers had been reduced almost to the status of sharecroppers in an ex-French grand design to exploit the country's comparative advantage in groundnut production, sustained by high global prices. The structural adjustment policies that were put in place implemented market deregulation, the ending of subsidies, and withdrawal of government services (Gaye, 2000). As groundnut prices and profits stagnated in the 1980s and 1990s, farmers who were now free to sell what and where they wished diverted more of their output to buoyant local markets. Rural investment, formerly dependent on a government credit system, fell dramatically, as preference continued for private investment opportunities in the national capital, Dakar, over those offered in agriculture.

Diourbel region is the center of groundnut production, with rural population densities of 46 to 151 persons km^{-2} in 1988 (Barry et al., 2000). Even though it is not as densely populated as the KCSZ, >90% of the Diourbel area is under either cultivation or grass fallow. Most remaining areas of woodland and wet depressions (bas-fonds) were eliminated between 1978 and 1999 (Ba et al., 2000). Such changes resulted from an interaction between declining rainfall (1960s–1990s) and agricultural expansion, in particular for groundnut cultivation. Using a spatial analog method, it has been established that during 40 years, soils under cultivation every year may have lost 24% of the soil organic carbon found in soil under savanna woodland (Elberling et al., 2003).

Local soil fertility management strategies recognize two types of field: *champs de case*, which are situated close to the house, receive organic manure and waste, and are planted earlier and more thoroughly weeded; and *champs de brousse*, which are farther away. The latter were formerly fallowed periodically and later fertilized with subsidized inorganic amendments, until rising prices of fertilizers made this unaffordable after 1984 (Garin et al., 1999; Badiane et al., 2000). These types of cultivation are distributed in a landscape mosaic of permanent fields, formed in holdings that are fragmented among owners. Thus no regular pattern of "aureoles" around the village — intensive infields with extensive outfields — is discernible. The *champs de case* are used primarily for cereal cultivation and form the basis of the subsistence sector. The *champs de brousse* are still the foundation of groundnut production, using rotations with millet or cowpea or short fallows.

Both modes of management face a crisis. Given the density of the animal and human populations, there is not enough organic material to fertilize more than about 20% of land (the *champs de case*), and fallows are now too short to replace the inorganic amendments on which (with the exception of nitrogen) the *champs de brousse* had become dependent. Consequently, while the *champs de case* more or less maintain their fertility, that of the *champs de brousse* is declining. Also, farm trees are more numerous on the *champs de case* while some *champs de brousse* are reported to be losing trees.

Long-term agricultural intensification, promoted by the state in the interest of export expansion, created a vulnerable farming system dependent on state capitalization. Rural people have for years been diverting a large part of their profits out of agriculture and into urban investments in major cities. Yet in the long run, adaptive strategies were evident outside the groundnut sector. Between 1960–1961 and 1992–1993, while mean annual rainfall declined, millet yields per hectare rose by about 20%. Yields per mm of rainfall doubled, and yields per agricultural worker rose by about 25% after reaching a low point in 1980–1981 (Faye et al., 2000). Market demand for citrus was exploited wherever soil moisture is sufficient, and *Hibiscus sabdariffa*, which commands a niche market, is popular as a minor crop on rainfed farmlands.

While the numbers of cattle, donkeys, and horses in the region fluctuated around mean values that changed little over the 30-year period, small ruminants roughly doubled in number. Fattening animals has become the most profitable enterprise in farming, based on the intensive use and exchange of crop residues (Faye and Fall, 2000). It has become very popular with poor people, particularly women. Increasing the integration of crops and livestock in the farming system is critical for maintaining soil fertility, through manuring or composting (Badiane et al., 2000). Without subsidized fertilizers, only such a strategy for intensification can extend the area of *champs de case*.

In such a situation, the responses of farmers to agrarian crisis, and their strategies for soil fertility management, have tended towards a reintegration of crop and livestock production, in a slow process of spontaneous recapitalization. Despite its obvious constraints, this direction appears to be more sustainable than the colonial export-agricultural model, in large part because it sustains or builds up soil capabilities. Benefits attainable from improved field moisture, manuring, trees, and agrobiodiversity were mostly neglected under the groundnut-led agrarian policy. The Diourbel region, therefore, suggests some scope for closer integration of biological factors into farming systems, against the background of a retreat from a credit-based, subsidized and external-input-dependent system which has proved to be unsustainable. It will be interesting to see what if any impact a return to bumper groundnut crops with the recent improvement in rainfall (2002–2003) will have.

A third West African case, which has been studied in some detail, that of the Maradi department in Niger, is not reported here. It shows evidence of a significant transition in soil system management, even in dry areas, moving from extractive practices to more intensive management in conjunction with animal husbandry and increased organic fertilization, as documented in Mortimore et al. (2001).

25.4 Discussion

Dryland farmers in West Africa are doing better than expected given the many productivity constraints under which they operate. The soil erosion and "nutrient-mining" scenarios, although verifiable at particular sites, do not provide an accurate assessment of the overall performance of soil systems in the three areas considered here. Machakos, Kano, and Diourbel represent a range of terrains, population densities, agroecological conditions, and rapidity of change. To assess soil fertility management only in terms of physical erosion or chemical nutrient balances — data which present a simple picture — can be misleading because farmers have to work their way through a more complex set of factors.

Observations at field and livelihood level show that farmers' practices reflect a holistic framework. A number of these practices recognize, explicitly or implicitly, biological dimensions of soil system productivity. Biological factors include more than soil microbial activity. Farmers attempt to capitalize on biological dynamics across a wide front, including the control of parasites, viruses, and pests, for the most part without access to chemical interventions.

Rather than perpetuate the short-term diagnostic-prescriptive framework adhered to by many development professionals, policies and public-sector practices should take account of factors that determine soil fertility in the longer term. Where possible, they should build upon local knowledge and practice, aiming to make local management effective rather than necessarily to transform it.

Incentives for local people to invest more in soil management should be created. Subsidized inputs and costly technologies are not a tenable option given stagnant or declining global prices for agricultural exports, unfair international trading relationships, resistance to larger allocations for development assistance, and recessions in national economies. On the other hand, opportunities to supply domestic food commodity markets are continuously increasing, utilizing local means of production more efficiently and sustainably.

In this context, farmers own strategies need to be understood better by those who advise policy-makers. However, there is a serious deficit in terms of knowledge. Not enough is yet known about the roles of biological agents in sustaining soil fertility under physical and economic constraints, such as those experienced in African drylands. For these farmers, agriculture may hinge preeminently on managing these biological processes better.

Acknowledgments

The research was funded by the UK Department for International Development, the UK Economic and Social Research Council, and the Rockefeller Foundation under the terms of several projects (1990–2001).

References

Ariyo, J., Voh, J., and Ahmed, B., *Long-Term Change in Food Provisioning and Marketing in the Kano Region*, Working Paper 34. Drylands Research, Crewkerne (2001).

Ba, M. et al., *Région de Diourbel: Cartographie des changements d'occupation-utilisation du sol dans la zone agricole du Sénégal Occidental*, Working Paper 21. Drylands Research, Crewkerne (2000).

Badiane, A., Khouma, M., and Sène, M., *Région de Diourbel: Gestion des sols*, Working Paper 15. Drylands Research, Crewkerne (2000).

Barnes, R.O., *Soil Erosion, Ukamba Reserve, Report to the Department of Agriculture DC/MKS/10a/29/1.* Kenya National Archives, Nairobi (1937).

Barry, A., et al., *Région de Diourbel: Les aspects démographiques*, Working Paper 13. Drylands Research, Crewkerne (2000).

Barth, H., *Travels and Discoveries in North and Central Africa*, Longman Green, London (1857).

Bourn, D. and Wint, W., Livestock and agricultural intensification in Sub-Saharan Africa, Network Paper 37a, Pastoral Development Network, Overseas Development Institute, London (1994).

Breman, H. and de Wit, C., Rangeland productivity and exploitation in the Sahel, *Science*, **221**, 1341–1387 (1983).

Brookfield, H., *Exploring Agrodiversity*, Columbia University Press, New York (2001).

Burkitt, D., *Don't Forget Fibre in Your Diet*, Martin Dunitz, London (1983).

Cline-Cole, R., Promoting (anti)social forestry in northern Nigeria?, *Rev. Afr. Polit. Econ.*, **74**, 515–536 (1997).

Cline-Cole, R. et al., *Wood Fuel in Kano*, United Nations University Press, Tokyo (1990).

Dalziel, M., *The Useful Plants of West Tropical Africa*, Crown Agents for the Colonies, London (1937).

de Leeuw, P.N., Reynolds, I., and Roy, B., Nutrient transfers in West African agricultural systems, In: *Livestock and Sustainable Nutrient Cycling in Mixed Farming Systems in Sub-Saharan Africa*, Vol. II: Technical Papers, Powell, J.M., Fernandez-Rivera, S., Williams, T.O. and Renard, C., Eds., International Livestock Centre for Africa, Addis Ababa, 371–392 (1995).

de Ridder, N. et al., Revisiting a 'cure against land hunger': Soil fertility management and farming systems dynamics in the West African Sahel, *Agric. Syst.*, **80**, 109–131 (2004).

Elberling, B., Touré, A., and Rasmussen, K., Changes in soil organic matter following groundnut-millet cropping at three locations in semi-arid Senegal, West Africa, *Agric. Ecosyst. Environ.*, **96**, 37–47 (2003).

Faerge, J. and Magid, J., Evaluating NUTMON nutrient balancing in Sub-Saharan Africa, *Nutr. Cycling Agroecosyst.*, **69**, 101–110 (2004).

Faye, A. and Fall, A., *Région de Diourbel: Diversification des revenus et son incidence sur l'investissement agricole*, Working Paper 22. Drylands Research, Crewkerne (2000).

Faye, A., Fall, A., and Coulibaly, D., *Région de Diourbel: Evolution de la production agricole*, Working Paper 16. Drylands Research, Crewkerne (2000).

Figueiredo, P., *The Yield of Food Crops on Terraced and Non-terraced Land: A Field Survey of Kenya*, Working Paper 35. Swedish University of Agricultural Sciences, Uppsala (1986).

Fones-Sundell, M., *Land Degradation in Sub-Saharan Africa: Farmer/Government/Donor Perspectives*, Issue Paper 9. Swedish University of Agricultural Sciences, Uppsala (1989).

Garin, P., Guigou, B., and Lericollais, A., Les pratiques paysannes dans le Sine, In: *Paysans Sereer: Dynamiques Agraires et Mobilité au Sénégal*, Lericollais, A., Ed., Editions Institut de Recherches en Développement, Paris, 211–298 (1999).

Gaye, M., *Région de Diourbel: Politiques nationales affectant l'investissement chez les petits exploitants*, Working Paper 12. Drylands Research, Crewkerne (2000).

GEF, *TerrAfrica draft business plan, 2005–6*, Global Environmental Facility, World Bank, Washington, DC (2004).

GM-CCD, *Why Invest in Drylands? A Study Carried out for the Global Mechanism of the Convention to Combat Desertification*, Global Mechanism of the Convention to Combat Desertification, Rome, Rome (2004).

Gowers, W., *Kano Province Annual Report for 1913*, Nigerian National Archives, Kaduna (1913).

Gruhn, P., Goletti, F., and Yudelman, M., *Integrated Nutrient Management, Soil Fertility, and Sustainable Agriculture: Current Issues and Future Challenges*, Agriculture and the Environment Discussion Paper 32. International Food Policy Research Institute, Washington (2000).

Harris, F., Farm-level assessment of the nutrient balance in northern Nigeria, *Agric. Ecosyst. Environ.*, **71**, 201–214 (1998).

Harris, F., Nutrient management strategies of small-holder farmers in a short-fallow farming system in north-east Nigeria, *Geogr. J.*, **165**, 275–285 (1999).

Harris, F. and Yusuf, M., Manure management by smallholder farmers in the Kano Close-Settled Zone, Nigeria, *Exp. Agric.*, **37**, 319–332 (2001).

Henao, J. and Banaante, C., *Estimating Rates of Nutrient Depletion in Soils of Agricultural Lands of Africa*, International Fertilizer Development Center, Mussel Shoals, AL (1999).

Hendy, C., *Animal Production in Kano State and the Requirements for Further Study in the Kano Close-Settled Zone*, Land Resources Report 21. Land Resources Division of the Overseas Development Administration, Tolworth (1977).

Hilhorst, T. and Toulmin, C., *Integrated Soil Fertility Management*, Policy and Best Practice Document 7. Development Cooperation, Netherlands Ministry of Foreign Affairs, The Hague (2000).

Hill, P., *Rural Hausa: A Village and a Setting*, Cambridge University Press, Cambridge (1972).

Jones, M. and Wild, A., *Soils of the West African Savanna: The Maintenance and Improvement of their Fertility*, Commonwealth Agricultural Bureau, Harpenden (1975).

Karanja, G. and Tefera, F., *Soil and Water Conservation in Kenya: Bibliography with Annotations*, Publication 90/1. Department of Agricultural Engineering, University of Nairobi, Nairobi (1990).

Lindgren, B., *Machakos Report 1988: Economic Evaluation of a Soil Conservation Project in Machakos District, Kenya*, Ministry of Agriculture, Nairobi (1988).

Maher, C., *Soil Erosion and Land Utilisation in the Ukamba Reserve (Machakos)*, Report to the Department of Agriculture. Rhodes House Library, Oxford (1937).

Mortimore, M., *Adapting to Drought, Farmers, Famines and Desertification in West Africa*, Cambridge University Press, Cambridge (1989).

Mortimore, M., The intensification of peri-urban agriculture: The Kano Close-Settled Zone, 1964–86, In: *Population Growth and Agricultural Change in Africa*, Turner, B., Kates, R., and Hyden, G., Eds., University Press of Florida, Gainesville, FL, 358–400 (1993).

Mortimore, M. and Adams, W., *Working the Sahel: Environment and Society in Northern Nigeria*, Routledge, London (1999).

Mortimore, M. and Harris, F., Do small farmers' achievements contradict the nutrient depletion scenarios for Africa?, *Land Use Policy*, **22**, 43–56 (2005).

Mortimore, M., Harris, F., and Turner, B., Implications of land use change for the production of plant biomass in densely populated Sahelo-Sudanian shrub-grasslands in north-east Nigeria, *Global Ecol. Biogeogr.*, **8**, 243–256 (1999).

Mortimore, M. and Turner, B., Does the Sahelian smallholder's management of woodland, farm trees and rangeland support the hypothesis of human-induced degradation? *J. Arid Environ.*, **63**, 567–595 (2005).

Mortimore, M. et al., *Department of Maradi: Synthesis*, Working Paper 39e. Drylands Research, Crewkerne (2001).

Murton, J., Population growth and poverty in Machakos District, Kenya, *Geogr. J.*, **165**, 37–46 (1999).

Mustapha, A. and Meagher, K., *Agrarian Production, Public Policy and the State in Kano Region, 1900–2000*, Working Paper 35. Drylands Research, Crewkerne (2000).

Oldeman, R. and Hakkeling, R., *World Map of the Status of Human-Induced Soil Degradation: An Explanatory Note*, United Nations Environment Programme, Nairobi (1990).

Pieri, C., *Fertilité des Terres de Savannes: Bilan de Trente Ans de Recherche et de Développement Agricoles au Sud du Sahara*, Ministère de la Coopération, Centre de Coopération Internationale en Recherche Agronomique pour le Développement, Paris (1989).

Pinstrup-Andersen, P. and Pandya-Lorch, R., *The Unfinished Agenda: Perspectives on Overcoming Hunger, Poverty, and Environmental Degradation*, International Food Policy Research Institute, Washington, DC (2001).

Raynaut, C., *Sahels: Diversité et Dynamiques des Relations Sociétés-Nature*, Editions Karthala, Paris (1997).

Reij, C. and Steeds, D., *Success Stories in Africa's Drylands: Supporting Advocates and Answering Sceptics*, Centre for International Cooperation, Vrije Universiteit Amsterdam, Amsterdam (2003).

Rocheleau, D., Benjamin, P., and Diang'a, A., The Ukambani region of Kenya, In: *Regions at Risk: Comparisons of Threatened Environments*, Kasperson, J., Kasperson, R., and Turner, B., Eds., United Nations University Press, Tokyo (1995).

Sanchez, P. et al., *Soil Fertility Replenishment in Africa: An Investment in Natural Resource Capital*, Special Publication 51. Soil Science Society of America, Washington, DC (1997).

Scoones, I., Ed., *Dynamics and Diversity: Soil Fertility and Farming Livelihoods in Africa*, Earthscan, London (2001).

Smaling, E., Oenema, O., and Fresco, L., *Nutrient Disequilibria in Agroecosystems: Concepts and Case Studies*, CABI Publishing, Wallingford (1999).

Stebbing, E., The encroaching Sahara: The threat to the West African colonies, *Geogr. J.*, **88**, 506–524 (1935).

Stebbing, E., *The Creeping Desert in the Sudan and Elsewhere in Africa 15 to 30 Degrees Latitude*, McCorqudale, Khartoum (1953).

Stocking, M., Soil erosion: Breaking new ground, In: *The Lie of the Land*, Leach, M. and Mearns, R., Eds., Heinemann/James Currey, London, 140–154 (1996).

Stoorvogel, J. and Smaling, E., *Assessment of Soil Nutrient Depletion in Sub-Saharan Africa, 1983–2000, Main Report*, Vol. 1. Winand Staring Centre, Wageningen (1990).

Stoorvogel, J., Smaling, E., and Janssen, B., Calculating soil nutrient balances in Africa at different scales: I. Supra-national scale, *Fertil. Res.*, **35**, 227–235 (1993).

Tiffen, M., *Profile of Demographic Change in the Kano-Maradi Region, 1960–2000*, Working Paper 24. Drylands Research, Crewkerne (2001).

Tiffen, M. and Mortimore, M., Questioning desertification in dryland Africa sub-Saharan Africa, *Nat. Resour. Forum*, **26**, 218–233 (2002).

Tiffen, M., Mortimore, M., and Gichuki, F., *More People, Less Erosion: Environmental Recovery in Kenya*, Wiley, Chichester, UK (1994).

UNEP, *World Map of the Status of Human-Induced Soil Degradation*, United Nations Environment Programme, Nairobi (1990).

UNEP, *World Atlas of Desertification*, Arnold, London (1992), for United Nations Environment Programme.

Watson, K., Fertilizers in Northern Nigeria: Current utilization and recommendations for their use, *Afr. Soils*, **9**, 5–20 (1964).

World Bank, *Sustainable development in a dynamic world: Transforming institutions, growth, and quality of life, World Development Report 2003*. World Bank, Washington, DC (2003).

Yusuf, M., Soil assessment and indigenous soil management strategies in the semi-arid northeastern Nigeria. Ph.D. thesis, Department of Geography, Bayero University, Kano, Nigeria (2001).

26

Restoring Soil Fertility in Semi-Arid West Africa: Assessment of an Indigenous Technology

Abdoulaye Mando,[1] Dougbedji Fatondji,[2] Robert Zougmoré,[3] Lijbert Brussaard,[4] Charles L. Bielders[5] and Christopher Martius[6]

[1]International Center for Soil Fertility and Agricultural Development, Lomé, Togo
[2]International Crop Research Institute for the Semi-Arid Tropics, Niamey, Niger
[3]Institute for Environment and Agricultural Research, Ouagadougou, Burkina Faso
[4]Wageningen University, Wageningen, The Netherlands
[5]University of Louvain, Louvain, Belgium
[6]Center for Development Research, Bonn, Germany

CONTENTS

Low soil fertility and surface sealing, leading to severe water loss through runoff and to a drastic decline in vegetation cover, are major Sahelian agricultural constraints (Casenave and Valentin, 1989; Bationo and Mokwunye, 1991). Owing to lack of financial resources it is often not possible for farmers to utilize external inputs to solve these problems. Increasingly, attention has focused on low-cost but effective alternative solutions. Given the region's poverty, new innovations will only get adopted if they are cheap, easily accessible, and minimize the use of external inputs. Further, the chances for adoption will be higher if the proposed technology is based on some improvement of traditional practices.

The demand in the Sahel for suitable agricultural techniques and methods has never before been so great as at present due to continuing population growth and an unprecedented rate of soil degradation (Mando et al., 2001). To solve the problems that confront rural areas in the Sahel, to secure food production, and curtail further soil degradation, a mixture of initiatives is needed to increase the productivity of available arable land and to extend the area of land under cultivation or pasture by rehabilitation of wasteland.

This chapter shows how traditional soil management practices can combat land degradation and improve productivity. Its focus is on on-farm technologies rather than on farming system improvement assessed at a regional level as considered in the preceding chapter. However, in both cases, the aim is to raise and sustain soil productivity by relying mostly on local resources.

It has been well established that soil nutrient deficiencies as well as the lack of adequate soil water supply — resulting from physical degradation (soil sealing) or irregular intra-annual rainfall distribution exacerbating nutrient shortages — are the principal constraints to crop production in arid and semiarid regions (Bationo and Mokwunye, 1991; Mando, 1997; Zougmoré, 2003). Soil system management to improve crop production therefore needs to address both water and nutrient constraints. On degraded land, both of these problems trace back, in part, to deficiencies in the soil biological realm.

This chapter considers the reintroduction of a traditional practice known as the *zaï* technology. This addresses both water and nutrient limitations by stimulating biological activity to rehabilitate wastelands (Mando et al., 2001). The method reported on here requires only limited use of external inputs, which is an important consideration since the limited availability of organic resources in the Sahel is a constraint on any strategy to improve soil productivity (Timothy et al., 1995; Ouédraogo, 2004).

The zaï technique relies largely on biological processes to improve soil productivity. Indeed, the improvement in soil structure following soil faunal activity leads to increased water infiltration and better drainage, lower runoff, and reduced soil resistance to root penetration. The application of organic inputs and/or mineral fertilizers not only enhances soil nutrient availability, but also improves crop nutrient uptake from soil reserves. The application of either organic material or mineral fertilizer alone is less effective than their combined application, so this combination opens up interesting possibilities for achieving sustainable increases in the productivity of now-crusted and agriculturally useless soils in sub-Saharan Africa.

After describing the technology briefly, this chapter examines the effect of zaï practices on soil structure and physical properties, their effects on decomposition and mineralization of soil organic matter, their impact on soil and crop performance, and their potential for restoring floral and faunal life cycles. The chapter also discusses some limitations of the technology and offers some recommendations on how to facilitate its larger-scale implementation.

26.1   Zaï Holes: An Indigenous Technique for Land Rehabilitation

The traditional technique of zaï holes has been used to combat land degradation and restore soil fertility in the Sahel for many years, although not on as wide a scale as could be beneficial (Mando et al., 2001; Fatondji, 2002). The technique is accessible to most farmers given that it relies on locally available labor and materials and needs only small amounts of external inputs (organic amendments and fertilizers). The name appears to be derived from the word *zaïgre*, which in one of the languages of Burkina Faso means "to get up early and prepare one's land" (Roose et al., 1999). In the Hausa region of Niger, the zaï technique is called *tassa*.

With the zaï methodology, desirable physical and chemical properties of the soil are restored by mixing small quantities of organic material, e.g., compost or manure and fertilizer (when available) at a rate of typically a few hundred grams per hole, in small

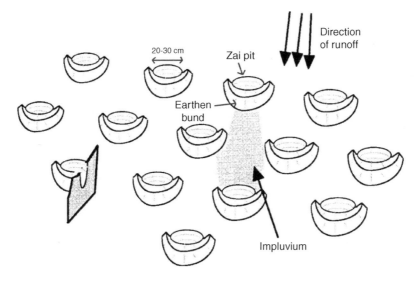

FIGURE 26.1
Spatial placement of zaï holes in a field (http://www.wocat.net; drawn by M. Evéquoz). In lower left of figure, a zaï hole cross-section is shown.

holes 20 to 30 cm in diameter and 10 to 15 cm deep that have been dug into the degraded, crusted soil, with pits dug in alternate rows as shown in Figure 26.1 (Fatondji, 2002).

26.2 Biophysical and Biochemical Processes of Zaï Holes in Restoring Soils

26.2.1 Soil Structure and Soil Physical Properties

In the Sahelian zone where soils are generally poorly endowed with organic matter, crusting is a major problem (Casenave and Valentin, 1989), limiting infiltration and therefore water availability for crop growth. By breaking the soil crust, pit digging facilitates more water infiltration. The pit also serves to harvest runoff water from the bare crusted areas in between the pits. This water-harvesting function is enhanced by using the earth that was dug out of the pit to form a small earthen bund downslope of the pit.

However, due to the aggressiveness of rainfall in the region, the low organic matter content of the degraded soils and the small extent of vegetative cover, a new seal quickly builds up, and the earthen bund can be carried away by runoff after a few showers. This will hamper further water infiltration and sustain rapid runoff (Mando et al., 2001). Adding some organic amendments into the pit reinforces the physical processes of rehabilitation through biological processes. Organic matter attracts termites, which have significant effects on soil structure. As discussed in Chapter 11, they open up large and numerous macropores throughout the entire soil profile as a result of their nesting and foraging activities, which facilitate water infiltration.

The size of the macropores and their density per unit of soil surface depend on termite species, their population size, and the quality of the organic material at stake. The most common termites that have the most significant ecological role in the Sahel are Odontotermes and Macrotermes. A micromorphological study by Mando (1997) indicated that termites excavate irregular-shaped macropores, i.e., channels and chambers, throughout the entire soil profile. In addition, termites contribute to the aggregation of

TABLE 26.1

Soil Resistance to Cone Penetration, within and between Zaï Pits

	Resistance to Cone Penetration (MPa)	
Treatment	In the Pits	Outside the Pits
Control plots	>1	>1
Zaï pitting + compost (1.5 t ha^{-1})	0.4 a	0.9 a
Zaï pitting + straw of *Loudetia togoensis* (1.5 t ha^{-1})	0.2 b	0.6 a

Note: Treatments followed by the same letter are not significantly different ($P = 0.05$).
Source: From Zombré N.P. et al., in *La Jachère en Afrique Tropicale, de la Jachère Naturelle à la Jachère Améliorée: Le Point des Connaissances*, John Libbey Eurotext, Paris, 2000, 771–777. With permission.

the soil through their building structures, which sustain the voids. Termite voids loosen the soil and reduce soil resistance to root penetration that could be a major constraint to crop growth in hardening and crusted soils (Table 26.1).

Greater reduction of the soil resistance is obtained from the addition of undecomposed organic material such as straw than with material under decomposition such as compost as termites are more active in straw than in compost (Ouédraogo, 2004). Moreover, the soil fauna-mediated voids increase water infiltration and drainage due to improved hydraulic conductivity. This was demonstrated by earlier findings in Niger, by Chase and Boudouresque (1987), who established that 100 l h^{-1} of rainfall could be drained by a single termite-made void in 30 min. Ambouta et al. (2000) have reported an annual mean runoff coefficient about six to ten times lower on zaï plots than on control plots. As a result of increased infiltration, there is an increase in water availability for crop growth. Also, drainage is often increased in the zaï system (Roose et al., 1999). This reduces the likelihood that waterlogging will occur as a result of water-harvesting interventions (Figure 26.1).

26.2.2 Decomposition and Nutrient Release

In semiarid and arid ecosystems, soil moisture is generally a limiting factor to decomposition processes. The zaï system accelerates decomposition under semiarid conditions by improving soil water storage and drainage. Soil fauna-mediated voids within zaï holes improve air circulation and the soil microbiological activity that enhances decomposition. An on-farm study in Niger conducted in 1999 and 2000 at Damari (a site with high termite activity) and Kakassi (a site where termites are quasi-absent) compared the decomposition of millet straw and cattle manure with soil-surface broadcast application, using litterbags of 2 mm mesh size in the zaï holes (Fatondji, 2002). This revealed faster litter decay in the zaï pit compared with the soil surface at Kakassi, indicating that the water harvested in the pits created better conditions for litter decay compared with soil surface exposure (Table 26.2). At Damari, the influence of termites on litter disappearance overshadowed that of zaï holes, with greater litter disappearance on the soil surface than in the zaï holes. The control of termite activity in selected plots of the experiment (through pesticide application) showed that termites played a key role in the decomposition at that location (Figure 26.2).

These results confirmed findings by Mando and Brussaard (1999), which indicated that in the Sahel, termites have a tremendous effect on the decomposition of organic matter. A study in Burkina Faso by Ouédraogo (2004) confirmed the major contribution of soil fauna as well as the effect of pit construction on greater decomposition. Coarse and fine

TABLE 26.2

Litter Disappearance (decomposition rate constant *k*) as Affected by Environment, Mode of Application (whether in Zaï Hole or on surface applied around planting hills), and Amendment Quality

Sites	Zaï Pit		Flat Surface Applied	
	Cattle Manure	Millet Straw	Cattle Manure	Millet Straw
Damari	0.0212	0.0010	0.0328	0.0121
Kakassi	0.0169	0.0129	0.0077	0.0089
SED (±)	0.0026 (between sites),			
	0.0029 (within sites)			

Source: Adapted from Fatondji, D., Ph.D. thesis, Center for Development Research, University of Bonn. Cuvillier Verlag, Gottingen, Germany, 2002. With permission.

mesh litterbags were filled with different types of organic matter (manure, maize straw, *Andropogon* grass straw) and were either buried or placed on the soil surface during the year 2000 rainy season. When termites were excluded by using fine-mesh litterbags, up to 90% of organic inputs remained undecomposed after one cropping season. The role of termites appeared greater in maize straw (C:N 59; lignin:N 0.21) >manure (C:N 40; lignin:N 0.52) >grass straw (C:N 153; lignin:N 0.41). The main role that termites play in decomposition and, therefore, in nutrient release from organic inputs, is comminution, i.e., reduction of litter and other organic matter to small pieces, which enhances soil microbial activity and the turnover of organic material (Mando and Brussaard, 1999).

Litter decomposition is controlled not only by soil fauna and soil moisture but also by the quality of the organic inputs, e.g., their N concentration, C:N ratio, and lignin:N ratio. The decay coefficient of manure (C:N = 20), for example, was twice the coefficient for millet straw (C:N = 50) (Fatondji, 2002). Similar results were obtained by Ouédraogo (2004). Of relevance for planning remediation programs, the addition of fertilizer to organic inputs was shown to be effective in boosting microbial activity. However, decomposition rates were increased mainly when low-quality organic inputs were applied (Ouédraogo, 2004).

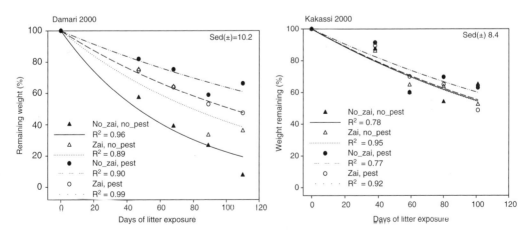

FIGURE 26.2

Effect on organic matter decomposition of (i) suppressing termites and other soil organisms through pesticide application and (ii) method of cattle manure application (in zaï pits vs. on surface) at Damari and Kakassi, Niger, in rainy season 2000. Pest = pesticide application; No zai = surface application. Sed = standard error of difference between means. *Source:* Fatondji (2002). With permission.

Decomposition rates of various organic amendments in the zaï hole determine how much of the nutrients will be available to the crop, and whether nutrient release matches crop demand. In the study reported above, Fatondji (2002) documented that the rate of nutrient release from organic inputs was higher in the zaï holes, and that over 80% of nutrients were released within 4 months.

26.3   Results Associated with Zaï Hole Interventions

26.3.1   Zaï Effects on Crop Performance

Owing to its ability to improve water status in the soil, to increase decomposition and nutrient release, and to reduce soil resistance to root penetration, the zaï system has a great impact on crop performance under semiarid conditions. The zaï hole technique leads to improved nutrient uptake and use-efficiency by plants (Fatondji, 2002). All studies in the region indicate that crop performance in zaï holes depends on the quality and the nature of the applied amendments, suggesting an important role of nutrients in sustaining production.

The experiments managed by Kaboré (1995) compared the following treatments during two cropping seasons:

- Control (no zaï hole, no inputs)
- Pits alone
- Pits + neem tree leaves
- Pits + compost at 3 t ha^{-1}
- Pits + NPK fertilizer (10–20–10 kg ha^{-1})
- Pits + compost (3 t ha^{-1}) + NPK fertilizer (10–20–10 kg ha^{-1})

The plots were all sown with sorghum. Table 26.3 presents the yield data. Control plots and pits alone resulted in similar grain and biomass production, suggesting that under the semiarid conditions, a lack of water was not the main limiting factor as there were also nutrient constraints. The addition of neem leaves slightly increased yield and biomass production, but not significantly. The application of compost made some improvement in crop nutrient uptake, but the effect was short-lived. The large differences in yields between 1992 and 1993, as reported by Kaboré, were due to nutrient shortages in 1993 as no amendments were made in that year.

Fatondji (2002) reported that the total macronutrients added per hectare through manure were 41 kg N, 19 kg P, and 20 kg K, respectively. At harvest, the nitrogen uptake by millet was twice the amount applied and potassium uptake was four times the amount applied. This suggests that substantial amounts of additional nutrients were taken from the soil stocks to meet plant demand, and further, that the zaï technology contributed to this soil nutrient acquisition. The improved physical and biological conditions of the soil in the zaï holes may increase the decomposition of native soil organic matter, and therefore the availability of nutrients endogenous to the soil system. Furthermore, the improved soil conditions in zaï holes enhance the growth of rooting systems that are better able to explore the soil for nutrients than are the roots developed in crusted, compacted soils. The soil nutrient mining effect of zaï holes is worsened where heavy rains induce leaching losses (Fatondji, 2002).

TABLE 26.3

Sorghum Grain and Biomass Production (kg ha^{-1}) on Deep, Brown Eutropept Soil, 1992 and 1993, at Taonsogo, Burkina Faso

Treatment	1992			1993		
	kg ha^{-1}	± S.D.	Test[a]	kg ha^{-1}	± S.D.	Test[a]
Grain production						
Control	150	± 154	a	3	± 0.6	a
Pit	200	± 63	a	13	± 4.2	a
Pit + neem leaves	395	± 151	ab	24	± 7.3	a
Pit + compost[b]	654	± 145	abc	123	± 82.5	a
Pit + mineral fertilizer	1383	± 236	bc	667	± 256.3	b
Pit + compost + mineral fertilizer	1704	± 305	bc	924	± 346.8	b
Biomass production						
Control	946	± 529	a	167	± 75	a
Pit	1329	± 549	a	292	± 49	a
Pit + neem leaves	1990	± 207	ab	875	± 172	ab
Pit + compost[b]	2843	± 945	abc	1417	± 511	bc
Pit + mineral fertilizer	4839	± 1105	bc	2375	± 706	bcd
Pit + compost + mineral fertilizer	5333	± 1490	bc	3250	± 857	cd

[a] Tukey-Kramer test. Treatments followed by the same letter are not significantly different from each other ($P < 0.05$).

[b] Compost: 3 t ha^{-1} of a mixture of dry manure, straw, and various crop residues composted during 3 dry months.

Source: Adapted from Roose, E. et al., *Arid Soil Res. Rehabil.*, 13, 343–355, 1999. With permission.

The addition of fertilizer greatly improved the production of grain and straw, especially when fertilizer was combined with compost, as seen in Table 26.3. After reviewing the research on zaï holes in Mali, Burkina Faso, and Niger, Mando et al. (2001) reported results that confirm the trends observed in Table 26.3. Neither the alleviation of water constraints nor the addition of organic resources alone was able to boost crop productivity. The combined use of organic inputs and fertilizer enhances nutrient-use efficiency, reduces leaching as a result of immobilization, improves water-use efficiency, and increases decomposition.

Mechanisms involved in fertilizer and organic input interaction in the soil include the capture of fertilizer nutrients by the soil microbial community whose activities are boosted by the organic input. This appears to improve synchrony between the supply and demand of crop nutrients and to reduce nutrient losses to the environment. The alleviation of nutrient limitations on organic amendment decomposition following the addition of organic fertilizer, mainly when low quality organic inputs are utilized, is another mechanism of significance.

26.3.2 Zaï Effects on Vegetation Rehabilitation

All studies throughout the subregion have indicated that zaï practices lead to the establishment of diverse woody and herbaceous vegetation on formerly bare soil. Kaboré (1995) and Zombré et al. (2000) reported the reestablishment on formerly bare soil of over 20 herbaceous species and 15 woody species following two consecutive years of zaï in the central part of Burkina Faso. The herbaceous seeds are either seeds that survived in the soil or were transported by wind or runoff and trapped in the microcatchments.

Most of the woody seeds are brought in through the addition of manure as it contains many seeds given the diets and foraging practices of livestock in the Sahel. Roose et al. (1999) identified the seeds of 13 different woody species in manure in Ouahigouya in northern Burkina Faso. The germination of seeds, which have passed through animal guts, is much easier because the acidic conditions in the gut weaken the integument of the seeds that could have impeded germination. The zaï technique could become an efficient technique of reforestation because of the high germination potential of seeds brought in manure and the improved moisture condition in the zaï holes. Furthermore, the voids made by termites in the pits facilitate root development that improves soil structure over a larger area.

26.4   Discussion

Zaï holes are efficient in improving soil physical properties through physical and biological processes, and therefore in improving crop yield, biomass production, and water and nutrient use efficiencies. However, implementation of zaï technology should be carried out only under certain conditions. Roose et al. (1999) concluded from studying a transect from the forest to the semiarid zone in Burkina Faso that the optimum climatic condition for introducing zaï holes is between the isohyets of 300 and 800 mm rainfall. Zaï holes are unlikely to be sufficient to alleviate drought constraints under rainfall conditions lower than 300 mm and may provoke leaching and waterlogging and subsequent yield loss if rainfall is more than 800 mm.

The zaï technique at present is still labor-intensive. About 60 working days, based on an average of $5 \, h \, d^{-1}$, are needed to dig 1 ha of zaï holes. However, since zaï holes can be dug during the dry season, when much labor is unemployed and opportunity costs are low, labor limitations are generally of minor concern. An attempt to overcome these constraints was made by scientists at INERA in Burkina Faso. Based on their experiments, they have recommended a so-called "mechanical zaï" that could alleviate the work required of farmers (Barro et al., 2001). Holes can be made mechanically with animal-drawn tools such as the Dent IR12 for sandy soils or the Dent RS8 for other types of soils. This reduces by more than 90% the amount of time required for making the pits as it takes only 11 to $22 \, h \, ha^{-1}$ to construct these pits with oxen that are well-fed with crop residues.

The availability of organic inputs in the Sahel is often a major constraint to the adoption of the zaï practice, imposing a strong need to promote biomass production and management through, for example, a purposeful integration of food crop production and livestock. There is a need to promote the integrated use of organic inputs and mineral fertilizers in the holes to avoid further nutrient depletion and to ensure sustainable productivity and better economic returns to farmers. Indeed, the nutrient content in most compost or manure is too low to ensure sustainable crop production. However, manure or compost addition can mitigate possible environmental problems such as soil acidification linked to the sole use of fertilizer and can increase fertilizer use-efficiency, resulting in better yield and higher biomass production. Appropriate policies are needed to facilitate the accessibility of inputs as fertilizers are presently rarely available or affordable to most farmers in the region. Moreover, participatory learning and action research will be needed to develop sustainable technological options that combine organic inputs and the judicious use of mineral fertilizer tailored to farmers' conditions.

References

Ambouta, K., Moussa, I.B., and Ousmane, S.D., Réhabilitation de la jachère dégradée par les techniques de paillage et de zaï au Sahel, In: *La Jachère en Afrique Tropicale, de la Jachère Naturelle à la Jachère Améliorée: Le Point des Connaissances*, vol. 1, Floret, C. and Pontanier, R., Eds., John Libbey Eurotext, Paris, 751–759 (2000).

Barro, A., Zougmoré, R., and Ouédraogo-Zigani, P., Réalisation du zaï mécanique en traction animale pour la réhabilitation des terres encroûtées. Fiche technique réalisée par INERA. Programme GRN/SP, Koudougou, Burkina Faso (2001).

Bationo, A. and Mokwunye, A.U., Role of manure and crop residue in alleviating soil fertility constraints to crop production, with special reference to the Sahelian and Sudanian zones of West Africa, *Fertil. Res.*, **29**, 117–125 (1991).

Casenave, A. and Valentin, C., Les états de surface de la zone sahélienne: Influence sur l'infiltration. Collections didactiques, Editions de l'ORSTOM, Paris (1989).

Chase, R. and Boudouresque, E., Methods to stimulate plant regrowth on bare Sahelian forest soils in the region of Niamey, Niger, *Agric. Ecosyst. Environ.*, **18**, 211–221 (1987).

Fatondji, D., Organic amendment decomposition, nutrient release and nutrient uptake by millet (*Pennisetum glaucum*) in a traditional land rehabilitation technique (zaï) in the Sahel. Ph.D. thesis, Center for Development Research, University of Bonn. Cuvillier Verlag, Gottingen, Germany (2002).

Kaboré, V., Amélioration de la production végétale des sols degradés (Zipillés) au Burkina Faso par la technique des poquets (zaï). Ph.D. thesis, Ecole Polytechnique de Lausanne, Switzerland (1995).

Mando, A., The role of termite and mulch in the rehabilitation of crusted Sahelian soils. Ph.D. thesis, Wageningen Agricultural University (1997).

Mando, A. and Brussaard, L., Contribution of termites to the breakdown of straw under Sahelian conditions, *Biol. Fertil. Soils*, **29**, 332–334 (1999).

Mando, A. et al., Réhabilitation des sols dégradés dans les zones semi-arides de l'Afrique subsaharienne, In: *La Jachère en Afrique Tropicale, de la Jachère Naturelle à la Jachère Améliorée: Le Point des Connaissances*, vol. 2, Floret, C. and Pontanier, R., Eds., John Libbey Eurotext, Paris, 311–339 (2001).

Ouédraogo, E., Soil quality improvement for crop production in semi-arid West Africa. Ph.D. thesis, Wageningen University, The Netherlands (2004).

Roose, E., Kaboré, V., and Guenat, C., Zaï practice: A West African traditional rehabilitation system for semiarid degraded lands: A case study in Burkina Faso, *Arid Soil Res. Rehabil.*, **13**, 343–355 (1999).

Timothy, O. et al., Manure availability in relation to sustainable food crop production in semi-arid West Africa: Evidence from Niger, *Quart. J. Int. Agric.*, **34**, 248–258 (1995).

Zombré, N.P., Mando, A., and Ilboudo, J.B., Impact des techniques de conservation des eaux et des sols sur la restauration des jachères très dégradées au Burkina Faso, In: *La Jachère en Afrique Tropicale, de la Jachère Naturelle à la Jachère Améliorée: Le Point des Connaissances*, vol. 1, Floret, C. and Pontanier, R., Eds., John Libbey Eurotext, Paris, 771–777 (2000).

Zougmoré, R., Integrated water and nutrient management for sorghum production in semi-arid Burkina Faso. Ph.D. thesis, Wageningen University, The Netherlands (2003).

27

Leguminous Biological Nitrogen Fixation in Sustainable Tropical Agroecosystems

Robert M. Boddey, Bruno J.R. Alves and Segundo Urquiaga

EMBRAPA Agrobiologia Center, Seropédica, Rio de Janeiro, Brazil

CONTENTS

The world's increasing human population and the demand for the elimination of poverty and hunger lead to ever-increasing pressures on the planet's diverse ecosystems. The richer societies of the industrialized world want to see the remaining areas of tropical forests and other natural ecosystems preserved with their biodiversity intact. In the Third World, the model of land use for agriculture has been either the intensification of slash-and-burn systems, with decreasing fallow periods and resulting decline in soil fertility, or the continuous clearing of new land, followed by intensive exploitation and then abandonment (Boddey et al., 2003).

The result is that in various regions of Latin America, Africa, and South and Southeast Asia, there are large swathes of degraded lands and pastures that have extremely low biological diversity and are unsuitable for productive agriculture. In many of these degraded areas farmers eke out a living on small holdings, where nutrient exports through production and losses by burning, erosion, and leaching exceed any organic or inorganic nutrient inputs. This leads to ever-decreasing crop yields due to nutrient mining (Batiano et al., 1998; Sanchez, 2004). In Chapter 12, the biophysical processes whereby biological nitrogen fixation (BNF) occurs in plant-microbial associations were discussed. The most reliable of these is in the root nodules of legumes. This chapter focuses on the contribution that the use of such species within cropping systems can make to sustainability through BNF.

27.1 Grain Legumes in Agricultural Systems

There is general consensus that legumes within a cropping system "improve" the soil as a consequence of the input of biologically fixed nitrogen. However, in the case of some grain legumes, where the concentration of nitrogen as protein in the grain and the harvest index (HI) of the plant are both high, it is possible that more nitrogen is exported from the agricultural system in grain than is obtained from BNF. An example of this possibility is the agriculturally very important legume, soybean. Total world production of soybean is almost 190 million tons (Mt) (2003 data, www.fao.org/faostat/). While this is far higher than any other grain legume, it is still far behind the annual production of wheat, rice, or maize, which ranges from 555 to 640 Mt. However, total protein harvested annually in soybeans is approximately 80 Mt, compared with 83, 38, and 54 Mt for these three major cereal crops, respectively.

In Brazil, most soybeans are grown on Oxisols and Ultisols, which are acidic and especially deficient in phosphorus and nitrogen. While farmers need to apply lime, phosphorus, potassium, and micronutrient fertilizers for best results, no applications of nitrogen fertilizer are required. Investigations have shown that application of nitrogen fertilizer to Brazilian soybeans does not increase grain yields, but only substitutes for the nitrogen otherwise derived from BNF (Alves et al., 2003). Recently, similar results were reported from Argentina with its local soybean varieties (Gutiérrez-Boem et al., 2004).

When no nitrogen fertilizer is added, crop nitrogen can only be obtained from the mineralization of soil organic matter (SOM) or derived from the air through bacterial processes of BNF. If more nitrogen is exported as grain than is supplied by BNF, this implies that SOM reserves will become depleted, with the consequence of increasing the susceptibility of the soil to erosion, lowering its nutrient storage capacity, and eventual detrimental effects for soil physical properties. Possibly nitrogen can be provided through other processes in the soil and in plants. An example of the contribution of endophytes is given in Table 28.3 of the following chapter.

27.2 Soybean-Based Crop Rotations

Brazil is already the world's largest exporter of soybeans. It is steadily increasing its production so that in a few years it will probably overtake the U.S. as the world's largest producer (2004 production estimates for Brazil and the U.S., respectively, were 52 and 66 Mt). In Brazil, soybean is usually grown in rotation or in sequence with other crops. In the southern region, where cropping is year-round, the rotations frequently include oats, wheat, or green manure crops in the winter months and mainly maize as an alternative to soybean in the summer. In the tropical savannah *cerrado* region, soybean is usually followed by maize, sorghum, or millet, but crop growth ceases for 4 to 5 months owing to the harsh dry season in this region.

In order to assess the long-term viability of these rotation systems, a team of Embrapa-Agrobiologia researchers has quantified the BNF inputs and the nitrogen outputs in harvested products at several sites in both the southern and *cerrado* regions. More than half of the soybean-based rotations today are managed under zero-tillage (ZT), similar to that discussed in Chapter 22. This system leaves crop residues on the soil surface throughout the year in order to protect it against erosion and conserve soil water; the planting process itself causes very little soil disturbance.

Under the conventional system, turning over of the soil stimulates the mineralization of SOM. Studies quantifying BNF have shown that under conventional tillage (CT), the mineral nitrogen in NH_4^+ and NO_3^- forms produced from mineralization of SOM inhibits early nodulation of the soybean, and total BNF contributions are thus lower in this system than under ZT (see Table 27.1; also Alves et al., 2002). Under both tillage systems, the proportion of nitrogen derived from BNF was high, but our studies revealed that there is

TABLE 27.1

Grain Yield, N Derived from BNF (Ndfa) and N Balance[a] in a Soybean-Based Crop Rotation Planted under Conventional and No-Tillage in Paraná, Brazil

Crop	Crop Conditions/History	Grain Yield (t ha^{-1})	Ndfa (%)	N Balance (kg N ha^{-1})
Maize	No-till; after oats (harvest 1997/1998)	4.3	—	16.2[c]
	Conventional; after oats (harvest 1997/1998)	4.9	—	2.5[c]
	No-till; after oats (harvest 1997/1998)	5.9	80.9	−6.9[d,*]
Soybean	Conventional; after oats (harvest 1997/1998)	5.4	74.1	−24.3[d]
	No-till; after soybean (harvest 1998)	2.5	—	−16.7[c]
Wheat	Conventional; after soybean (harvest 1998)	2.3	—	−15.4[c]
Lupins	No-till; after oats (harvest 1998)	9.4[b]	74.4*	202.5[e]
	Conventional; after soybean (harvest 1998)	11.2[b]	68.8	216.5[e]
Maize	No-till; after lupins (harvest 1998/1999)	9.3**	—	−28.1[c]
	Conventional; after lupins (harvest 1998/1999)	7.8	—	−46.3[c]

*, **, denote the difference between means significant at $p < 0.05$ and 0.01, respectively.

[a] Partial total N balance = Total N exported in grain — BNF input — N fertilizer added. No estimates of gaseous or leaching N losses were made.

[b] Difference between N added as fertilizer and that exported in grain.

[c] Difference between total grain N and N fixed by plant (roots included).

[d] Total shoot dry matter.

[e] Total BNF contribution (^{15}N natural abundance technique).

Source: Adapted from Zotarelli (2000).

rarely more nitrogen put into the soil/plant system than is exported in the grain. This nitrogen balance is always more negative under CT than ZT.

These estimates of nitrogen balance may be underestimated because recent work from Australia suggests that considerable nitrogen may be deposited in the soil as root exudates and senescent roots (McNeill et al., 1997; Khan et al., 2003). However, on the other hand, these simple balances do not take into account various nitrogen losses via leaching or in gaseous forms, so firm conclusions are not yet possible.

In Brazil, nitrogen fertilizer is almost never used with oats or other winter green-manure crops, and the inputs of nitrogen fertilizer for wheat and maize are modest, typically 50 and 100 kg N ha^{-1}, respectively, and rarely exceeding by more than 10 or 20 kg N ha^{-1} the nitrogen exported in grain (Table 27.1). Hence, the most favorable scenario is that in most rotations under ZT, SOM levels are being maintained, while under CT, SOM levels are being lowered. Since CT stimulates SOM decomposition, the leaching losses of all mobile nutrients are generally higher.

On sloping land, particularly on sandier soils, conventional tillage twice a year can lead also to severe erosion (de Maria, 1999), eventually making cropping there completely unviable. The inclusion of a leguminous green manure in a soybean-based rotation can lead to a considerable potential positive nitrogen balance (Table 27.1). Under CT, much of this nitrogen may be lost when crop residues with a low C:N ratio rapidly mineralize after incorporation. However, under ZT, where crop residues remain undisturbed on the soil, surface decomposition is slower, and the benefit to a subsequent maize crop can be considerable, also favoring SOM accumulation.

A recent evaluation of a long-term experiment in southern Brazil by Sisti et al. (2004) showed no difference in SOM stocks between CT and ZT under continuous wheat/soybean cultivation. However, where vetch was included in a 2- or 3-year rotation before maize (with soybean as the summer crop in one or two of the years, respectively), after 13 years of cropping, approximately 10 more tons of soil carbon had accumulated under these rotations (at 0 to 100 cm depth) compared with a continuous sequence of soybean/wheat. Where vetch was included in the rotation under CT, SOM stocks were still 17 t ha^{-1} lower than under ZT.

While there appear to be no other studies in Brazil that compare in the same experiment the effect of introducing a leguminous green-manure, those studies where such legumes were present in the rotation showed definite gains in SOM under ZT (Bayer et al., 2000a, 2000b; Diekow et al., 2005). Conversely, studies where there was no legume in the rotation, or only soybean, have showed no accumulation of SOM under ZT (de Maria, 1999; Machado and Silva, 2001; Freixo et al., 2002).

If an agricultural system is continuously exporting more nutrients than are being added, with intensive tillage and erosion visually apparent, the negative results are immediately evident. However, in many cases this phenomenon is more gradual, and changes in the stocks of SOM can only be evaluated after several years, typically 5 to 10. The above-mentioned studies on nitrogen balance and changes in SOM stocks show that nutrient-balance studies can be an accurate short-term indicator of the sustainability of nutrient and SOM stocks in agroecosystems.

27.3   Legumes in Smallholder Agricultural Systems

Many traditional cropping systems used by resource-poor peasant farmers involve the use of legumes, either as monocrops, intercrops (e.g., maize and beans), in relay cropping/rotation, or as leguminous leys. Attempts to increase agricultural productivity and family welfare in poor communities in many parts of the developing world, especially sub-Saharan Africa, continue to focus on the use of legumes of various types to add nitrogen to agricultural systems (Keatinge et al., 2001; Sanchez and Jama, 2002).

Mucuna (velvet bean, usually *Mucuna pruriens*), often used as a crop for short-term fallows, has been widely adopted, with great benefit in Latin America and Indonesia (Sanchez and Salinas, 1981; Van Noordwijk et al., 1995) as well as in West Africa (Carsky et al., 2001; Anthofer and Kroschel, 2002; see also Chapter 30). Mucuna is most usually integrated into maize-cropping systems, sometimes being planted when the maize is at the early reproductive stage of growth (tasseling) so that by the time of harvest there is complete soil cover. Mucuna has been shown to be able to provide considerable quantities of nitrogen from BNF in a variety of regions (50 to 200 kg N ha^{-1}), and there are many reports of increased yields of maize with its nitrogen inputs (Carsky et al., 2001).

In field-station trials, responses to inoculation with Rhizobium have been rare, so such inoculation has not been generally recommended. However, in 7 out of 34 sites in southern Benin where no nodules were observed on the roots of Mucuna (Sanginga et al., 1996), and on 15 farms in this same region, Houngnandan et al. (2000) reported that Rhizobium inoculation increased dry matter yields by 28% on average. In general, incorporating Mucuna residues into the soil (through tillage) has shown greater increases in maize yields than leaving the Mucuna on the soil as a mulch. Possibly the nitrogen derived from the mulch is more susceptible to immobilization owing to its slow release. For farmers who want short-term benefits from higher maize yield, this is a disadvantage; however, for the longer-term benefit of building up SOM, slow release would offer some advantage.

Other herbaceous plants that can be used for such fallows include kudzu (*Pueraria phaseoloides*), sunnhemp (*Crotalaria juncea*), and various forage legumes, such as forage groundnut (*Arachis pintoi*) and Stylosanthes spp. (Giller, 2001; Tian et al., 2001; Sanchez and Jama, 2002). Some of these legumes can have other beneficial effects on soil fertility and crop management. There is evidence that Mucuna mobilizes soil phosphorus and that phosphorus applications for a subsequent maize crop can be reduced (LeMare et al., 1987; Vanlauwe et al., 2000). Several of these legumes can form a complete ground cover and hence reduce labor costs for weed control. Both Crotalaria spp. and Mucuna have been cited for their reduction in the population of certain root-pathogenic nematodes. Several authors (e.g., Carsky et al., 2001; Giller, 2001) have pointed out that it is often these properties, rather than their capacity to acquire nitrogen from N_2 fixation, that lead farmers to include these legumes in their cropping systems. This range of microbial services parallels that reported for associative diazotrophic bacteria which contribute BNF to nonleguminous plants (Dobbelaere et al., 2003).

Recently, considerable attention has been given to the adoption of agroforestry systems and the introduction of deep-rooting woody legumes into farming systems (Sanchez and Jama, 2002). The deep roots of these trees can recover nutrients leached below the rooting depth of annual crops. Thus if their prunings are applied to the field, the crops benefit not only from the fixed nitrogen but also from other nutrients recovered from lower soil horizons.

In a recent study at our field station near Rio de Janeiro, the consequences of using a fallow compared with Mucuna or groundnut followed by maize were studied for their impact on gross nitrogen balance (Okito et al., 2004). N_2 fixation in the three green-manure treatments (the fallow contained a high proportion of the legume *Indigofera hirsuta*) was quantified using the ^{15}N natural abundance technique. This showed inputs of approximately 30, 40, and 60 kg ha^{-1} for the fallow, groundnut, and Mucuna, respectively (Table 27.2). Average yields of the two varieties of maize in the experiment were 2.06, 2.36, and 2.80 t ha^{-1} after fallow, groundnut, and Mucuna, respectively. To the farmer, planting maize after Mucuna would earn him \$255 ha^{-1} at current currency conversion rates; doing the same after groundnut would net \$221. However, with the latter rotation, approximately 1 t ha^{-1} of groundnut, worth an additional \$247, was harvested from that treatment. Not only would this increase farmers' gross annual income per hectare, but their fields would produce income twice during the year.

However, an examination of the crude nitrogen balance data presented in Table 27.2 shows that in a groundnut-maize sequence, the biological nitrogen fixation input is 40 to 50 kg N ha^{-1} less than the nitrogen that is exported as grain, while in the Mucuna-maize sequence, there is a small positive balance of $+7$ to $+17$ kg N ha^{-1}. We acknowledge that the data used to construct the nitrogen balance in Table 27.2 were extremely simple and unavoidably incomplete. Also, there was no estimate of nitrogen losses or examination

TABLE 27.2

Simple Nitrogen Balance for the Sequence of Groundnut, Velvet Bean, or Natural Fallow Followed by Two Maize Varieties (Sol de Manhã and Caiana Sobralha), Calculated from Export of Nitrogen in Grains Minus Input of Nitrogen from BNF in Aerial Tissue

Green Manure	Maize Variety	N Derived from BNF	N Exported in Legume Grain	N Exported in Maize Grain	Overall N Balance
Groundnut	Caiana Sob.	40.9	49.4	34.1	−42.6
Velvet bean	Caiana Sob.	59.6	0.0	42.6	+17.0
Fallow	Caiana Sob.	30.9	0.0	32.2	−1.3
Groundnut	Sol de Manhã	40.9	49.4	42.7	−51.2
Velvet bean	Sol de Manhã	59.6	0.0	51.8	+7.8
Fallow	Sol de Manhã	30.9	0.0	33.6	−2.7

Source: Adapted from Okito, A. et al., *Pesquisa Agropecuária Brasileira*, 2004. With permission.

of the balance of other nutrients. However, it appears that the groundnut-maize system as now practiced may not be sustainable over the long run, and that nitrogen reserves from SOM are being reduced by this cropping system, which overexploits the soil system.

27.4 Discussion

This chapter has considered the potential importance of legumes in agroecosystems of the tropics as a contribution toward their sustainability, particularly with regard to maintaining the balance of nitrogen in soil systems. Experience with mechanized, high-input agriculture in Brazil has pointed to conclusions that are also applicable in general terms to resource-poor farmers in other regions. As also discussed in other chapters, minimum soil disturbance and year-round soil cover by live crops or residues are essential to minimize soil erosion and to preserve SOM. Zero-tillage and planted fallows and pastures can preserve soil integrity and SOM. In run-down or degraded agricultural systems, to increase SOM, maintaining a positive nitrogen balance for the cropping system is essential. Crops such as soybean, with a very high nitrogen HI, often leave little nitrogen for SOM buildup. Other grain crops, such as groundnut and pigeon pea, which have much lower nitrogen HI, add more to this balance. This makes these crops more desirable environmentally and for long-term economic productivity. The inclusion of green-manure legumes can make significant contributions to building up SOM, but this will only occur if soil physical disturbance is minimal.

Growing herbaceous or woody legumes can be attractive because they increase yields and farm income. Their capacity to reduce farmers' risks represents an additional incentive for adoption. When considering interventions in farming systems, it is important that at least a simple nutrient balance of the system (or whole property) be considered. If legumes are to be introduced to provide increased nitrogen in the system, as is usually desirable, one part of the nutrient balance calculation should estimate the contributions that their N_2 fixation can make. When yields are improved, however, export from the farming system of nutrients in the agricultural products is also increased. Failure to compensate for these increased nutrient outputs, by having increased inputs from one source or many, can lead to losses of SOM and soil nutrient reserves that need to be set against short-term gains for the farmer. Any deficits that are not made up can have disastrous consequences in the long term because of erosion and soil system degradation.

Acknowledgments

The authors thank Dra. Maria Cristina P. Neves for her critical reading of the text and suggestions for modifications.

References

Alves, B.J.R., Boddey, R.M., and Urquiaga, S., The success of BNF in soybean in Brazil, *Plant Soil*, **252**, 1–9 (2003).

Alves, B.J.R. et al., Soybean benefit to a subsequent wheat cropping system under zero tillage, In: *Nuclear Techniques in Integrated Plant Nutrient, Water and Soil Management*, IAEA-CSP-11/P, 87–93 (2002).

Anthofer, J. and Kroschel, J., Partial macronutrient balances of mucuna/maize rotations in the forest savannah transitional zone of Ghana, In: *Integrated Plant Nutrient Management in Sub-Saharan Africa: From Concept to Practice*, Vanlauwe, B. et al., Eds., CABI, Wallingford, UK, 87–96 (2002).

Batiano, A., Compo, F. and Koala, S., Research on nutrient flows and balances in West Africa: State-of-the-art, *Agric. Ecosyst. Environ.*, **71**, 19–35 (1998).

Bayer, C. et al., Effect of no-till cropping systems on soil organic matter in a sandy clay loam Acrisol from Southern Brazil monitored by electron spin resonance and nuclear magnetic resonance, *Soil Till. Res.*, **53**, 95–104 (2000a).

Bayer, C. et al., Organic matter storage in a sandy clay loam Acrisol affected by tillage and cropping systems in southern Brazil, *Soil Till. Res.*, **54**, 101–109 (2000b).

Boddey, R.M. et al., Brazilian agriculture: The transition to sustainability, *J. Crop Prod.*, **9**, 593–621 (2003).

Carsky, R.J., Becker, M., and Hauser, S., *Mucuna* cover crop fallow systems: Potential and limitations, In: *Sustaining Soil Fertility in West Africa*. Soil Science Society of America, Madison, WI, 111–135 (2001).

Diekow, J. et al., Soil C and N stocks as affected by cropping systems and nitrogen fertilisation in a southern Brazil Acrisol managed under no-tillage for 17 years, *Soil Till. Res.*, **81**, 87–95 (2005).

Dobbelaere, S., Vanderleyden, J., and Okon, Y., Plant growth-promoting effects of diazotrophs in the rhizosphere, *Crit. Rev. Plant Sci.*, **22**, 107–149 (2003).

Freixo, A.A. et al., Soil organic carbon and fractions of a Rhodic Ferrasol under the influence of tillage and crop rotation systems in southern Brazil, *Soil Till. Res.*, **64**, 221–230 (2002).

Giller, K.E., *Nitrogen Fixation in Tropical Cropping Systems*, CABI, Wallingford, UK (2001).

Gutiérrez-Boem, F.H. et al., Late season nitrogen fertilization of soybeans: Effects on leaf senescence, yield and environment, *Nutr. Cycl. Agroecosys.*, **68**, 109–115 (2004).

Houngnandan, P. et al., Response of *Mucuna pruriens* to symbiotic nitrogen fixation in farmers' fields in the derived savanna of Benin, *Sustaining Soil Fertility in West Africa*. Soil Science Society of America, Madison, WI, 111–135 (2000).

Keatinge, J.D.H. et al., Sustaining soil fertility in West Africa in the face of rapidly increasing pressure for agricultural intensification, *Sustaining Soil Fertility in West Africa*. Soil Science Society of America, Madison, WI, 1–22 (2001).

Khan, D.F. et al., Effects of below-ground nitrogen on N balances of field-grown favabean, chickpea and barley, *Aust. J. Agric. Res.*, **54**, 333–340 (2003).

LeMare, P.H., Pereira, J., and Goedert, W.J., Effects of green manure on isotopically exchangeable phosphate in a dark-red latosol in Brazil, *J. Soil Sci.*, **38**, 199–209 (1987).

Machado, P.L.O. de A. and Silva, C.A., Soil management under no-tillage systems in the tropics with special reference to Brazil, *Nutr. Cycl. Agroecosys.*, **61**, 119–130 (2001).

de Maria, I.C., Erosão e terraços em plantio direto, *Viçosa Boletim Informativo da Sociedade Brasiliera de Ciência do Solo*, **1**, 17–21 (1999).

McNeill, A.M., Zhu, C., and Fillery, I.R.P., Use of *in situ* ^{15}N-labelling to estimate the total below-ground nitrogen of pastures legumes in intact soil–plant systems, *Aust. J. Agric. Res.*, **48**, 295–304 (1997).

Okito, A. et al., N$_2$ fixation by groundnut and velvet bean and residual benefit to a subsequent maize crop, *Pesquisa Agropecuária Brasileira*, **39**, 1183–1190 (2004).

Sanchez, P.A., Soil fertility and hunger in Africa, *Science*, **295**, 2019–2020 (2004).

Sanchez, P.A. and Jama, B.A., Soil fertility replenishment takes off in East and Southern Africa, In: *Integrated Plant Nutrient Management in Sub-Saharan Africa: From Concept to Practice*, Vanlauwe, B. et al., Eds., CABI, Wallingford, UK, 23–45 (2002).

Sanchez, P.A. and Salinas, J.G., Low-input technology for managing Oxisols and Ultisols in tropical America, *Adv. Agron.*, **34**, 279–406 (1981).

Sanginga, N. et al., Evaluation of symbiotic properties and nitrogen contribution of mucuna to maize grown in the derived savanna of West Africa, *Plant Soil*, **179**, 119–129 (1996).

Sisti, C.P.J. et al., Change in carbon and nitrogen stocks in soil under 13 years of conventional or zero tillage in southern Brazil, *Soil Till. Res.*, **76**, 39–58 (2004).

Tian, G. et al., *Pueraria* cover crop fallow systems: Benefits and applicability, In: *Sustaining Soil Fertility in West Africa*. Soil Science Society of America, Madison, WI, 137–155 (2001).

Van Noordwijk, M. et al., Nitrogen supply from rotational or spatially zoned inclusion of Leguminosae for sustainable maize production on an acid soil in Indonesia, In: *Plant-Soil Interactions at Low pH*, Date, R.A., Ed., Kluwer, Dordrecht, The Netherlands, 779–784 (1995).

Vanlauwe, B. et al., Utilization of rock phosphate by crops on a representative toposequence in the northern Guinea savanna zone of Nigeria: Response by *Mucuna pruriens*, *Lablab purpureus*, and maize, *Soil Biol. Biochem.*, **32**, 2063–2077 (2000).

Zotarelli, L., Balanço de nitrogénio na rotação de culturas em sistemas de plantio direto e convencional na regiaño de Londrina, PR. MSc thesis, Universidade Federal Rural do Rio de Janeiro, Itaguaí, Rio de Janeiro (2000).

28

Soil Biological Contributions to the System of Rice Intensification

Robert Randriamiharisoa,[1] **Joeli Barison**[2] **and Norman Uphoff**[3]
[1]*University of Antananarivo, Antananarivo, Madagascar*
[2]*Chemonics International, Antananarivo, Madagascar*
[3]*Cornell University, Ithaca, New York, USA*

CONTENTS

The System of Rice Intensification (SRI) synthesized 20 years ago in Madagascar by Fr. Henri de Laulanié, S.J., has been gaining acceptance around the world because of the remarkable increases in factor productivity that its methods are making possible. Irrigated rice yields are being raised by 50–100%, and often more, without requiring the use of purchased inputs and without changing varieties. Such inputs and such changes were the foundation of the Green Revolution. The higher yields with SRI practices are achieved with any and all rice varieties while using less water and with little or no use of mineral fertilizer. Because the resulting plant phenotypes are more resistant to damage by pests and diseases, there is usually little or no need for chemical crop protection. All this has been quite unexpected.

FIGURE 28.1
On left, roots of three conventionally grown rice plants, transplanted at 28 days vs. the root of a single SRI plant, on the right, transplanted at 8 days. Both are the same Madagascar variety (2787). *Source:* From Barison (2002: Figure 4).

While the mechanisms that make these benefits attainable are not yet fully understood, it appears that SRI effects are the result of greater root growth and of increases in soil biological activity that are induced by different cultural practices. This inference is consistent with recent research, discussed in Chapter 15, that has identified genetic mechanisms affecting better plant performance and health that are associated with certain changes in cultural practices and with a reduction in inorganic fertilization (Kumar et al., 2004). Analyses of the expression of certain genes in leaf tissue cells has shown a positive correlation between alternative plant, soil and nutrient management, on the one hand, and the switching on or off of certain genes that are known to contribute to plant senescence or to biochemical processes that protect plant health, on the other. The larger, healthier root systems that are observed in the trials with certain alternative practices are consistent with the production by soil microbes of phytohormones (Chapter 14) that could be contributing to the greater root health and vigor seen with SRI (Figure 28.1).

Laulanié spent two decades working with farmers in Madagascar to improve their rice yields before he assembled the set of practices that constitute SRI (Laulanié, 1993). He recommended to farmers that, instead of buying new seeds or applying fertilizer, they change the way they manage their rice plants, soil, water and nutrients. Farmers were also encouraged to add compost or any other organic matter to their fields, but not all undertook this practice. The recommended SRI practices changed the E in the interaction of genetics and environment (G × E), creating more favorable growing environments for rice plants and particularly for their roots.

28.1   Inducing Phenotypical Changes

SRI changes in the management practices for plants, soil, water and nutrients have produced, in all rice genotypes used with SRI methods thus far, rather different and more productive phenotypes as described in Table 28.1. Because scientific evaluation of SRI began only about 5 years ago, much of what is known about it is based on farmer or NGO reports rather than on scientific measurement and documentation, a distinction made in

TABLE 28.1

Phenotypical Differences in Rice Plants and Grains Associated with SRI Practices

Measured Differences

Tillering: 30–50 tillers plant^{-1} are common with SRI methods, and twice this number or more can be attained with good use of the methods and with improved soil.

Root Growth: see data in text and Figure 28.1.

Grain Filling: positive correlation between number of panicles and size of panicles, contrary to the typical rice phenotype grown with continuous flooding, for which the correlation is negative, as discussed in text; usually reduced % of unfilled grains.

Grain Weight: a 5–15% increase is common, contributing to higher yield; this can occur without an increase in grain size, which represents greater grain density.

Reduced Time for Maturation: often 1–2 weeks less time for same variety to ripen when SRI methods are used.

Grain Quality: 10–15% higher outturn of milled rice from SRI paddy, due to fewer unfilled grains (less chaff) and fewer broken grains during milling (less shattering), reflecting higher grain density. A Sichuan Agricultural University evaluation in China found a 16% increase in outturn of milled rice, and also less chalkiness.

Reported Differences

Resistance to Pests and Diseases: widely reported by farmers in different countries, saying that pests and diseases do not cause enough damage to warrant use of biocides.

Resistance to Drought, Wind and Rain Damage, and Cold Temperatures: larger root systems appear to account for the resistance to drought and cold as well as to cyclone and typhoon damage as observed in India, China and Sri Lanka (see Figure 28.3); lodging is rare with SRI plants despite larger panicles, even under extreme weather conditions.

the table. The main requirements for getting SRI results are good water control, being able to apply minimum amounts of water as needed, and initially more labor and more careful management of the crop. After a few years, when SRI practices have been mastered, farmers find that SRI can even become labor saving (Anthofer, 2004; Li et al., 2005; Sinha and Talati, 2005; Uphoff et al., 2005). The use of organic amendments is a variable factor, not necessarily required for higher yields with SRI, but it usually improves them enough so that the extra effort involved is profitable.

Although SRI results sound too good to be true, they have been observed now in 22 countries, from China to Cuba and from Philippines to Peru. By raising the productivity of land, labor, capital and water all at the same time, SRI goes against the law of diminishing returns and the apparent necessity of making tradeoffs among factors of production. This apparent impossibility kept us from accepting SRI for several years after we first learned about it from Association Tefy Saina, the NGO that promotes SRI in Madagascar. However, those farmers who took up SRI under Tefy Saina's tutelage in the peripheral zone around Ranomafana National Park averaged over 8 ton ha^{-1} for three consecutive years (and more), on soils where they had previously produced only 2 ton ha^{-1} (Uphoff, 1999). They were not relying on inorganic fertilizer, and the extent to which they were adding organic materials varied widely. It became clear to us that the system warranted broader interest and thorough assessment.

A few of the farmers around Ranomafana attained yields in the range of 12–16 ton ha^{-1} with SRI methods, considered to be beyond the biological maximum (Khush and Peng, 1996). In fact, these yields were achieved on soils that have been evaluated for the Soil Science Department of North Carolina State University as some of the poorest in Africa, in chemical and physical terms: very acid with pH 3.8–5.0; serious Fe toxicity and Al saturation problems; low to very low cation exchange capacity (CEC) in all horizons;

and seriously deficiencies in available phosphorus (P) (Johnson, 1994, 2002). This analysis concluded that, given their deficient parent material, these soils would remain unproductive unless large amounts of chemical fertilizer were applied. However, by using SRI methods, farmers were able to increase their yields without using such amendments, just adding compost to their soil, and many did not even do this.

The unexpected results produced by SRI methods interested us in the biological dimensions of soil fertility. The first author, a natural scientist, undertook a series of detailed research projects to construct explanations for the anomalous benefits with SRI through studies by a succession of University of Antananarivo students, among them the second author. The third author, a social scientist by training, sought to get SRI evaluated in other countries and searched the literatures in different disciplines for possible explanations, also assembling results from a growing number of countries where SRI trials were carried out which gave perspective and insight into this novel method of crop management (Uphoff, 2003).

28.2   Improved Crop Performance with SRI Practices

The System of Rice Intensification changes certain practices for growing irrigated rice that farmers have used for centuries or longer (Table 28.2). These changes have the effect of inducing observable, measurable changes in the resulting rice plants. The most notable feature of SRI rice if one examines the entire plant is its much larger root system. Tests of root-pulling resistance (RPR) have shown that conventionally grown clumps of 4 to 6 rice plants required, on average, 22.00, 35.00 and 20.67 kg plant^{-1} at the stages of panicle initiation, anthesis, and maturity, respectively. In contrast, for *single* SRI plants at these respective stages, RPR was 55.19, 77.67 and 53.00 kg plant^{-1} (Barison, 2002: Table 14). This represents a 10-fold difference in the amount of force per plant. The evident difference in root growth can be seen in Figure 28.1.

With their larger root systems, SRI rice plants tiller more profusely and are less subject to senescence and to pest or disease damage. With SRI, there is a positive correlation between the number of panicles per plant and the size of panicles (number of grains). This is contrary to the negative relationship that is reported in most of the scientific literature, for rice plants grown conventionally with flooding and close spacing (Ying et al., 1998; Sheehy et al., 2004).

SRI plants perform differently from plants of the same variety cultivated with usual practices, among other things, maturing up to 15 days sooner (Uprety, 2005). Standard practices, besides creating anaerobic soil conditions that compromise root growth and function, crowd plants close together, both within hills and between hills.

- The roots of rice plants grown in flooded fields remain near the surface, about 75% in the top 6 cm (Kirk and Solivas, 1997). This enables the plants to capture dissolved O_2 in the irrigation water, with root growth into lower soil levels limited by the hypoxic conditions. Root length density (cm cm^{-3}) measured at soil depths of 30–50 cm was 0.53 for SRI rice with compost added; 0.45 for SRI rice without compost, 0.27 with "modern" methods using NPK plus urea, 0.22 for modern methods without fertilization, and 0.19 with conventional farmer practice, the latter three treatments having continuous flooding (Barison, 2002: Table 13).

- When the rhizosphere soil is kept continuously saturated, most of the root system will degenerate by the time of flowering. Controlled comparison studies have

TABLE 28.2

SRI Methods Contrasted with Conventional Practices

Conventional Practices	SRI Methods
Mature seedlings, transplanted at 3–4 weeks of age	*Very young seedlings*, 8–12 days old, are transplanted, at most 15 days old, i.e., before start of fourth phyllochron (see below)
3–4 or more seedlings per hill are transplanted in clumps by plunging them deeply into a flooded (hypoxic) soil environment	*Single seedlings* are transplanted, very shallow (1–2 cm) into a muddy, unflooded field after uprooting them gently from a garden-like, unflooded nursery
Large plant populations are established in rows, with a seeding rate of 50–100 kg ha^{-1}	*Sparse plant populations*, widely spaced in a square pattern at least 25 × 25 cm, with a seeding rate of 5–10 kg ha^{-1}
Soil saturation with paddies kept flooded throughout the growing cycle	*Soil aeration* is maintained during the vegetative growth period, avoiding continuous soil saturation; after panicle initiation, shallow flooding (1–2 cm); alternate wetting and drying throughout cycle is preferable in some soils
Weeds are controlled by flooding and also by hand weeding and/or *herbicides*	*Weeds are controlled with a rotary* hoe that aerates the soil as it eliminates weeds; weeding 3–4 times before canopy closes
Chemical fertilizer is applied, providing up to 100–150 kg ha^{-1} N; soil organic matter is optional; little/no attention to soil biota	*Compost* is recommended, as much as possible to build up soil organic matter, to support larger populations of soil biota

shown 78% of rice roots degenerated when grown in continuously flooded soil compared with almost no degeneration of roots in well-drained, aerated soil (Kar et al., 1974).

- Under hypoxic soil conditions, root cortex cells disintegrate to form aerenchyma, air pockets that permit the diffusion of oxygen from the shoot into the root. Kirk and Bouldin (1991) report that this disintegration is "often almost total (and) must surely impair the ability of the older part of the plant to take up nutrients and convey them to the stele." In the latter part of the growing cycle, they write, "the main body of the root system is largely degraded and seems unlikely to be very active in nutrient uptake" (pp. 197–198). This would account for the easier uprooting of conventionally grown plants.

- Under SRI soil conditions, in contrast, roots do not atrophy, maintaining more ability to function throughout the growth cycle. Researchers at Nanjing Agricultural University compared the oxygenation ability of SRI and conventional rice roots by measuring the α-naphthylamide (α-NA) levels for the same variety (Wuxianggeng 9) at different stages of plant growth. During the effective tillering, jointing and heading stages, α-NA levels were 1.9–2.3 times higher in SRI roots; at maturity when grain filling is being completed, the level of α-NA was 2.9 times higher (Wang et al., 2002).

- This same evaluation also found that plant metabolic processes were different in SRI plants. In SRI leaves, the contents of soluble sugar, nonprotein N, MDA and proline were consistently higher compared to those of conventional rice: 12 to 54% at the jointing stage, and 23 to 104% at the heading stage (Wang et al., 2002). Further, with SRI practices, the accumulation of carbohydrates, N and dry matter was greater in the vegetative organs, and the absolute partitioning rate of stored matter from the vegetative organs to grains was remarkably higher, more than

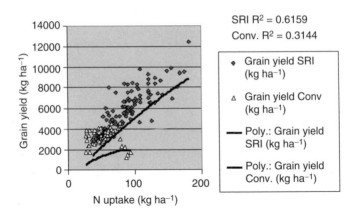

FIGURE 28.2

Relationship between N uptake and grain yield in rice plants grown with SRI vs. conventional methods, analyzed with QUEFTS modeling, sampled from 108 farms in Madagascar. *Source:* Barison (2002: Figure 9).

>   3 times greater from the leaves, and 1.4 to 1.7 times greater from stems and sheaths. Total translocation from leaves, stems and sheaths was found to be greater by 66.9% (Wang et al., 2002).

These measurements of physiological processes mirror the visible differences in plant size and structure, all representing different phenotypical results from the same genetic foundation.

Rice plants' ability to take up nutrients and convert them into grain is different when they are grown under SRI conditions. An agronomic analysis comparing SRI and conventional rice plants grown on the same farms and by the same farmers in Madagascar (N = 108), using the QUEFTS analytical model to assess internal nutrient efficiency, found that SRI plants were producing about twice the yield as conventional ones with the same amounts of N (Figure 28.2).

The same relationship was seen for P and for K (Barison, 2002). This reflects the phenotypical differences associated with different management of plants, soil, water and nutrients, evoking productive potentials existing in the rice genome when encountering favorable environments.

28.3   Possible Explanations for Induced Phenotypical Differences

Attention initially focused on characteristics of the plants, seeking physiological explanations endogenous to the plants themselves. It is well known that lower plant density will contribute to larger root systems and canopies. However, having a reduced plant population will give *lower yield* unless the fewer plants compensate with more and larger panicles and with higher grain weight. Such increases are observed when SRI practices are used together.

28.3.1   Aboveground Effects

Root systems support canopy growth by providing nutrients and water, while canopies in turn produce photosynthate, some part of which is shared with root systems for their nutrition. Positive feedback between root growth and shoot growth is easy to understand as a larger root system can support a larger canopy, and vice versa. Larger roots provide more water and nutrients to the canopy; a bigger canopy produces more photosynthate that benefits the roots. There are also phytohormone effects, as shoots release auxins which enhance root initiation, while roots produce cytokinins that stimulate the development of shoots (Oborny, 2004). These interactions occur within the plant. Less obvious but possibly more important are the effects of synergy between the plant and soil biota, discussed in Section 28.4. We need to refer to such interaction in this section because these internal and external processes are interrelated.

For years, agronomists have recognized what they call "the border effect" — the larger, more robust growth of plants on the edges of a field which are more exposed to sunlight and air circulation. With wider spacing of plants, SRI creates this effect throughout whole fields, as its plants are uniformly more vigorous, not just on the borders. Measurements made at the Indonesian Rice Research Institute at Sukamandi evaluated the effects of wider SRI spacing on photosynthesis within leaves at different levels of the canopy. With conventional close spacing, lower leaves did not receive enough solar radiation to carry out photosynthesis, so they were taking from rather than contributing to the plant's total production of photosynthate (Dr. Anischan Gani, personal communication). Also, as seen in Section 28.3.4 below, the growth of rice plant tillers and roots is slowed when seedlings are transplanted after the start of their fourth phyllochron of growth. This adds to the possible physiological explanations for SRI's observed phenotypical differences.

28.3.2 Root System Effects

Conventional practices of plant, soil, water and nutrient management have belowground effects as well. When seedlings are not transplanted quickly and carefully soon after removal from their nursery, their roots become desiccated, and having been raised under flooded conditions they are hypoxic. If in addition the roots are traumatized by rough handling during transplanting, it is understandable why conventionally transplanted seedlings commonly require 7 to 14 days to recover from the shock of relocation (Kirk and Solivas, 1997). This diminishes the tillering that can be accomplished before panicle initiation. Such loss reduces the potential for grain formation and sink size for filling.

One possible additional factor inhibiting yield, adding to the effects of anaerobic soil conditions, could be the effects that chemical fertilizers and biocides have on soil biota. This has not yet been studied scientifically, however, so it remains only a hypothesis at present. It could explain why SRI yields are often higher on farmers' fields than on experiment stations. The first SRI trial at IRRI headquarters at Los Baños gave a yield of only 1.44 ton ha^{-1} (Rickman, 2003) at the same time that other SRI yields in the Philippines were reaching 6 to 12 ton ha^{-1} (ATI, 2002). Usually farmers have difficulty replicating research-ers' results, but with SRI, the opposite has more often been true. SRI methods perform better where soils have not been heavily and continuously fertilized with also large applica-tions of agrochemicals. But these relationships remain to be evaluated systematically.

The share of the photosynthate that is sent to the roots and exuded into the rhizosphere (carbohydrates, amino acids, vitamins, enzymes and other compounds) supports micro-bial populations that provide services to plants and are in turn substrate for higher-level fauna in the soil food web (Neumann and Römheld, 2001; Dakora and Phillips, 2002; Wardle, 2002; also Chapters 5 and 7). Various estimates suggest that as much as 40 to 60% of plants' photosynthate is transported via the phloem to the root system. Most is used for root metabolism, generating energy needed for acquiring and transporting nutrients and water. However, 20 to 40% of this carbon and other compounds are exuded into the rhizosphere (Brimecombe et al., 2001).

Little research has been done on exudation from rice roots, except recently as this affects methane production or has allelopathic effects, perhaps because continuously flooded root systems are stunted and degenerating so that there is little exudation to consider. Even plants that have normally growing roots in well-drained rather than submerged soil present difficulties for measurement of exudates.

28.3.3 Soil Chemistry Effects

One possible explanation for SRI performance, particularly relevant for tropical soils that are acidic and have high Fe and Al concentrations, can be found in soil chemistry. It has been axiomatic that flooded soil conditions are preferable for rice performance (Sanchez, 1976; De Datta, 1981), even though the resulting reduced soil conditions increase methane production and solubilize Fe and Mn. When soils high in Fe are kept flooded, ferro-hydroxide is formed which is not beneficial for plant growth. Alternate flooding and draining of such fields, as done with SRI, can reduce Fe toxicity and Al saturation thereby diminishing this particular constraint on plant growth (Dobermann, 2004). However, this explanation for SRI performance can at best be partial because over 1,500 comparison trials evaluating SRI across all 22 districts of Andhra Pradesh state in India have found positive responses to its methods on all kinds of soils, with an SRI yield advantage of 2.5 ton ha^{-1} (Uphoff et al., 2005). Further, alternate wetting and drying (AWD) as a strategy for water management can make a positive contribution to SRI yields by promoting both biological N fixation and P solubilization as discussed in Section 28.4. Various complementary effects are probably at work above- and belowground.

Another chemical effect relating to plant physiology associated with soil and water management is the finding of Kronzucker et al. (1999) that rice yields can be higher by 40 to 60% when plants take up their nitrogen from the soil in both ammonium (NH_4) and nitrate (NO_3) forms, rather than absorbing their N entirely as ammonium. This finding was unexpected because NH_4 should require less energy from the plant for its utilization; yet a substantial yield premium is observed from there being a diversity of N forms in the soil. In flooded rice production, most of the N taken up will be as ammonium, whereas in soil that is alternatively wetted and dried, NO_3 also becomes available. This could be contributing to "the SRI effect," augmented by what is known in the literature as "the Birch effect," i.e., the flush of N mineralization that occurs with the rewetting of dry soil (Birch, 1958). Another soil chemistry factor could be effects on silicon uptake, which is blocked by high soil acidity. SRI soil and water management practices could reduce soil acidity by organic matter buffering and reduced generation of H^+, so that Si uptake is enhanced (Mark Laing, personal communication). This would explain the stiffer stalks and leaves of SRI plants that give reduced lodging (Figure 28.3) and more resistance to pest attack. These possible explanations warrant detailed investigation.

28.3.4 Phyllochron Regulation of Growth

An additional explanation in terms of plant physiology derives from understanding phyllochrons, a periodicity in the formation of tillers and roots that is found with rice as well as other gramineae (grass) species. Phyllochrons were first analyzed in the 1920s and 1930s by the Japanese scientist T. Katayama who carefully analyzed the growth patterns of rice, wheat and barley. Unfortunately, his results were not published until after World War II (Katayama, 1951) and then only in Japanese. For some reason, wheat scientists (see special issue of *Crop Science*, 35:1, 1995) and forage scientists are better acquainted with phyllochrons than are rice scientists. Phyllochrons are similar to degree-days in their effect on growth, but they provide a more sophisticated physiological understanding of relationships among plant growth, temperature, shading, etc. (Nemoto et al., 1995).

FIGURE 28.3
Experimental plots supervised by researchers from Tamil Nadu Agricultural University, India, after heavy rainstrom. Plot in foreground cultivated with standard methods; plot behind in center of picture, with SRI methods, with other factors (soil, variety) constant. Picture courtesy of Dr. T. M. Thiyagarajan.

Laulanié discovered empirically that rice seedlings transplanted before their 15th day, i.e., about the beginning of their fourth phyllochron of growth, grow larger and are more productive than ones that are transplanted later than this from their nurseries, especially if these are flooded. This effect, discussed in Stoop et al. (2002), is clearly seen in Figure 28.1, although it may be attributable also to the production of phytohormones by soil bacteria and fungi.

These possible explanations, still to be regarded as hypotheses, all have some basis in the literature. Scientific examinations of SRI have not advanced far enough to have assessed their separate and joint effects, confirming some and disconfirming others. There are, however, in addition, a number of other explanations, some mentioned already, that derive from an understanding of the biological dimensions of soil systems. These deserve consideration not just because they facilitate the production of more rice with fewer external inputs, but because of how they can inform us about those systems.

28.4 Soil Biological Processes

28.4.1 Biological N Fixation

That biological nitrogen fixation (BNF) occurs in the root nodules of leguminous plant species is well-known and much-studied, as discussed in Chapter 12. But BNF can also occur in, on and around the roots of all gramineae (grass) species, including rice (Döbereiner, 1987; Boddy et al., 1995). Bacteria classified as diazotrophs synthesize the enzyme nitrogenase which enables them to convert N_2 into the ammonium form (NH_4) that plants can take up. Such associative BNF is very difficult to measure, however, being a more diffuse process than the BNF that occurs in the nodules of leguminous species.

BNF in rice has been evaluated for many years by scientists working with the International Rice Research Institute (IRRI) (Ladha and Reddy, 1999; Ladha et al., 2000). However, despite extensive research findings confirming the existence and ubiquity of these processes, it has not been demonstrated to scientists' satisfaction that there are significant nutritional benefits to rice plants from BNF. In the case of SRI, however, there must be considerable BNF occurring in rice root zones because an average yield of 8 ton ha^{-1} requires larger supplies of N than Barison's soil analysis found in Ranomafana soils, 4.3% N-Kjeldahl (Joelibarison, 1998).

Controlled experiments by Magdoff and Bouldin (1970) have documented that BNF increases by orders of magnitude when aerobic and anaerobic bulk soils are mixed, rather than having aerobic and anerobic soils segregated. While these researchers did not establish the mechanisms for this effect, they suggested that bacterial activity appeared greatest at the interface between aerobic and anaerobic soils. SRI water management practices, which alternately flood and drain the soil create such heterogeneous soil conditions. Moreover, use of a 'rotating hoe' which churns up the surface soil to remove weeds would not only aerate the top 5 to 10 cm but it would also mix anaerobic and aerobic soils together. Unfortunately, we have not yet been able to study the soil microbiological processes and populations associated with SRI water management and weeding.

We do have some data relating to endophytic BNF that can be carried out by bacteria living inside the roots and other plant tissues (see Chapter 12; also Reinhold-Hurek and Hurek, 1998; Kannaiyan et al., 1999; Feng et al., 2005). In the village of Anjomakely, 18 km south of Antananarivo on the high plateau ($\sim$1200 m elevation), a set of factorial trials (N = 240) was done on farmers' fields using Fisher block design with randomized sets of treatments. These assessed the separate and combined effects of five SRI practices (age of seedlings, number of plants hill^{-1}, spacing, water management, and fertilization) on two kinds of soil (clay-loam and sandy loam). The same traditional variety (*riz rouge*) was used for all the trials. Since there were no significant differences between spacing trials — both spacings, 25 × 25 cm vs. 30 × 30 cm, being within the recommended SRI range — there were actually six replications for each of the combinations of treatments evaluated. Soil tests of the plots showed pH$_{H_2O}$ 6.5; N 1.2–1.5%; C 22.8–24.0%; C:N ratio 16–19:1; P$_2$O$_5$ 0.12–0.18%; and K$_2$O 2.8–3.2 meq/100 g (Andriankaja, 2001).

Of interest here is the correlation of endophytic microbes with tillering and with yield according to different combinations of plant, soil and water management practices interacting with (a) no fertilization, (b) the addition of a recommended dose of 6–22–16 NPK chemical fertilizer (150 kg ha^{-1}), or (c) application of compost made from rice straw, manure and leguminous plant materials (2 ton ha^{-1}).

The bacterium Azospirillum was studied because it is a well-documented N-fixer and could be counted reliably with laboratory facilities in Madagascar. Other organisms can be also involved in N-fixation and other plant-supportive and -protective services, so Azospirillum was not considered to be solely responsible for the plant effects observed. Rather it is regarded as an indicator species, representing overall changes in microbial populations and activity in, on, and around the plant roots and in soil–plant interaction generally. The results of these trials are shown in Table 28.3. Yields of over 10 ton ha^{-1} with SRI methods on clay soil with compost have been measured elsewhere in replicated trials (Randriamiharisoa, 2002), so these results are not idiosyncratic.

We had expected to find differences in the populations of Azospirillum in the rhizosphere, but this was not the case as soil samples from the plots with different treatments all had Azospirillum counts $\sim$25 × 10^3 ml^{-1} of rhizosphere soil. The effect of SRI practices on the soil and plants was seen in the roots themselves, which were sampled from the different plots and prepared for microscopic counting. SRI methods alone, inducing larger root systems through wider spacing and aerobic soil conditions, gave a three-fold increase in yield even without soil amendments. Adding inorganic nutrients (NPK) to clay soil increased yield by 50% over that from unfertilized SRI plots. On the other hand, compost amendments increased SRI yield even more, by an additional 17%. On loamy soil, the increase from SRI practices without any fertilization compared to conventional practices was relatively small, just 16%. However, when compost was added,

TABLE 28.3

Endophytic Azospirillum Populations, Tillering, and Rice Yield Associated with Alternative Cultivation Practices and Nutrient Amendments

	Azospirillum Count in Roots (10^3 ml^{-1})	Tillers Plant^{-1}	Yield (ton ha^{-1})
Clay soil			
Conventional cultivation with no nutrient amendments	65	17	1.8
SRI cultivation with no nutrient amendments	1100	45	6.1
SRI cultivation with NPK amendments	450	68	9.0
SRI cultivation with compost amendments	1400	78	10.5
Loam soil			
SRI cultivation with no nutrient amendments	75	32	2.1
SRI cultivation with compost amendments	2000	47	6.6

Source: From Andriankaja, A.H., Mise en evidence des opportunités de développement de la riziculture par adoption du SRI, et evaluation de la fixation biologique du l'azote. Mémoire de fin d'etudes, École Supérieure des Sciences Agronomiques. University of Antananarivo, Madagascar (2001).

providing more energy sources for microbes and nutrients for the crop, both plant tillering and microorganism populations were much greater, giving a tripling of yield.

The absolute numbers in such evaluations will vary depending on soil, temperature, crop variety and other differences. One can expect large differences among clay, loam and sandy soils due to their respective particle structures, the availability of oxygen therein, and resulting biological dynamics. This pattern of response has been seen several times in replicated trials, and often in farmers' fields, so it deserves serious consideration. BNF in the soil has been difficult to promote by inoculation or soil amendments, and even more difficult to measure (Chapter 12). However, significant BNF must be occurring as a result of the SRI changes in cultural practices since the plants achieve large increases in yield in the absence of corresponding organic or inorganic soil amendments. The additional N going into the plants and grain has to come from somewhere.

There can be also other benefits from microbial activity in plant roots as seen from the research on rhizobia in rice roots reported in Chapter 8 (also Doebbelaere et al., 2003). Table 8.1 in that chapter showed much larger populations of rhizobia in the roots of rice plants growing in soils that are well-drained rather than continuously flooded. So BNF is not the only or necessarily the most important contribution made by these diazotrophic organisms.

28.4.2 Phosphorus Solubilization

The other macronutrient that has a major effect on yield is phosphorus. As reported above, the soils in Madagascar where we first evaluated SRI are particularly deficient in available P. Of 10 paddy soil cores sampled around Ranomafana National Park, two had <3 mg P kg^{-1} soil, and three had <4 mg P kg^{-1} in all horizons (Olsen tests used), the other five sample all tested <6 mg P kg^{-1} in all horizons (Johnson, 2002). Such levels are less than half of what is usually considered necessary for an acceptable crop yield; 10 mg of available P kg^{-1} is commonly cited as a threshold below which yields will be unacceptable. Barison's nutrient analysis of soils around Ranomafana found 8 mg available P kg^{-1} (Olsen-P), only

80% of the usual minimum; 21 mg available P kg^{-1} is usually considered necessary for a 6–7 ton ha^{-1} grain yield (Joelibarison, 1998). Yet somehow, Ranomafana farmers averaged 8 ton ha^{-1} using SRI methods, with no P amendments, and some yields were much higher. As they continued practicing SRI, most farmers achieved increasing yields rather than declining ones as would be expected on soil with such limited P availability.

This result could be at least partly explained by the P solubilization by microorganisms. Probably 90% or more of the P in most soils is "unavailable," i.e., complexed in molecules such as inositol phosphates (Turner et al., 2000; see also Table 35.2 for data on this from India). Phosphate ions can be made available by aerobic bacteria that solubilize and absorb P, for their own purposes, from sources not available to plants. When the soil is flooded, these organisms lyse (burst, die) under anaerobic conditions and release their P into the soil solution (Turner and Haygarth, 2001). With SRI water management, soils are alternately saturated and aerated, so this could accelerate P solubilization. This is the simplest explanation for how SRI farmers could get such high yields from soils that are so severely P-constrained according to standard assessments.

This is a subject area where research is growing. A review by Gyaneshwar et al. (2002) describes in detail the processes for such mobilization of P. Researchers face the same kinds of problems with measurement of P, especially *in situ*, as do researchers studying BNF. But SRI suggests that such processes must be going on even if it is difficult to measure them precisely.

28.4.3   Other Biological Processes and Support

28.4.3.1   Mycorrhizal Fungi

Most terrestrial plants depend at least in part on benign "infections" of their roots by mycorrhizal fungi as discussed in Chapter 9. Since fungi are aerobic organisms, they will not survive or grow well in continuously flooded soil (Ilag et al., 1987), and reduction in mycorrhizal fungi populations adversely affects yields (Ellis, 1998). Growing rice in submerged soil systems greatly reduces or even eliminates the contribution that fungi can make to irrigated rice plants' performance.

It is likely that SRI practices enhance the nutritional status of rice at least in part through mycorrhizal associations, although we have no direct evidence of this yet. It is known that even under the unfavorable soil environment of flooding, mycorrhizal inoculations of irrigated rice can increase its yield by 10% (Solaiman and Hirata, 1997). Presumably even more benefit would come from mycorrhizal fungi under more aerobic soil conditions.

28.4.3.2   Protozoan Cycling

Protozoan cycling of N is a soil process well documented in the microbiology literature (Bonkowski, 2004); however, little attention has been paid to it agronomically. As rhizobacteria feed on plant roots' exudation and as their populations increase on rhizoplanes (root surfaces), they are "grazed" by protozoa, which are orders of magnitude larger, much like cattle graze on grass. Protozoa are in turn "grazed" by nematodes, which are even higher in the soil food web (Chapter 5). Because protozoa have a lower C:N ratio than the bacteria that they consume, the "excess" N that they ingest is excreted into the rhizosphere, where it is conveniently accessible to plant roots. By such processes, starting with the photosynthetic fixing of CO_2 from the atmosphere into energy sources that are shared with soil biota through root exudation, plants can indirectly enhance the availability of N in their root zones.

Since SRI increases both the canopies and root systems of rice plants, there should be more exudates available to support larger populations of rhizobacteria, which in turn sustain larger populations of protozoa. The increased turnover of organisms creates a growing supply of N in the root zone through a positive-feedback loop. Nobody knows how much this process contributes to SRI performance; however, a change from anaerobic to mixed aerobic-anaerobic soil conditions, with plants having larger root systems and canopies, should increase "N harvesting" in the soil. By some estimates, belowground net primary productivity is at least as great as NPP aboveground (Eissenstat and Yanai, 2002).

28.4.3.3 Phytohormone Production

Another possible contribution to the improved performance of SRI plants, particularly through their root systems, could be from the production by soil bacteria and fungi of phytohormones, i.e., auxins, cytokinins, gibberellins, ethylene, and abscisic acid. What is known about these has been summarized in Chapter 14. Unfortunately, it takes very sophisticated equipment to make measurements, and the results of studies are often contradictory or inconsistent. Very small amounts, and even small changes in those amounts, can have different and sometimes even opposite effects, since the effects are very sensitive to the *timing* of phytohormone influence during plants' growth cycles. Some will regard much of this research as inconclusive because results have been so varied, yet researchers who work with phytohormones are convinced that there are causal relations, only very sensitive and complex ones (Frankenberger and Arshad, 1995).

In our efforts to come up with explanations for SRI performance, the contribution of phytohormones by soil microorganisms is a leading candidate because aerobic bacteria and fungi would be inhibited or reduced in soil that is kept continuously flooded. The growth of larger rice plant roots with SRI management, sometimes massive as seen in Figure 28.1, is probably not due simply to endogenous physiological processes. Some stimulation of root and shoot growth by aerobic (and some anaerobic) soil organisms seems likely to be part of the explanation.

28.4.4 Other Services and Benefits

In addition to increasing the supply of available nutrients and synthesizing phytohormones and vitamins, soil microorganisms make other contribution to plant growth and protection. These have been discussed in reviews by Whipps (2001) and Doebbelaere et al. (2003) of the extensive recent literature, for example. There is considerable evidence that soil organisms can improve the uptake of available nutrients. This has been documented for N, P and K, although the mechanisms are not all agreed upon. Probably other nutrients are also accessed in greater amounts or more continuously through microbial intermediation, but this has not been studied.

Soil organisms can enhance drought-tolerance and oxidative-stress resistance. Their contribution to biocontrol, preventing the deleterious effects of various pathogenic microorganisms, is another complicated research domain. This involves synthesis of antibiotics and/or fungicidal compounds, competition for nutrients, and the induction of systemic resistance (ISR) to pathogens. It is well known that microorganisms contribute to soil aggregation and improve soil structure, but they also increase roots' adherence to soil so that plants can benefit more from their contact with soil particles. One of the organisms most often shown to have such effects is Azospirillum, although there are dozens of others.

An interesting finding is that often the presence of multiple organisms has additive and/or synergistic effects. This means that much of the research done on single species, sterilizing the soil to create gnotobiotic conditions, produces only partially valid or maybe even invalid results. This could be one reason for the frequent finding in laboratories that soil biota do not fix enough N or solubilize enough P to make a significant contribution to plant nutrition or do not have a significant effect on plant protection. These effects need to be evaluated *in situ*.

28.5 Discussion

The study of soil biology, as seen from earlier chapters, has been immense even though it is not well integrated into the main body of agronomic literature and practice. SRI challenges scientists to come up with adequate and properly measured explanations of how and why certain changes in the growing conditions for rice are enhancing yields so remarkably, utilizing fewer external inputs rather than more. This is occurring too often, and in too many countries, to be dismissed as an artifact of measurement or a result of wishful thinking.

Probably the frame of reference should be soil ecology rather than conventional soil biology, so that various ensembles of organisms are considered in terms of what they can contribute together. This will include the study of more than certain combinations of microorganisms but also of the interactions that range from bacteria and fungi, even viruses, to earthworms and other soil fauna that change the physical as well as chemical characteristics of soil systems. Such changes make these environments more hospitable for billions and billions of soil organisms. SRI suggests some new possibilities for achieving higher and sustainable production in ways that are environmentally friendly and accessible to resource-limited farmers.

Acknowledgments

The second and third authors wish to acknowledge not only the contributions made to this chapter by the first author, Prof. Robert Randriamiharisoa, who passed away in August 2004, but also his contributions to a scientific understanding of SRI. Without his interest, efforts and insights, the knowledge that we and others have about this methodology and what it can tell us about soil systems would not have advanced as far or as fast.

References

Andriankaja, A.H., Mise en evidence des opportunités de développement de la riziculture par adoption du SRI, et evaluation de la fixation biologique du l'azote. Mémoire de fin. École Supérieure des Sciences Agronomiques. University of Antananarivo, Madagascar (2001).

Anthofer, J., An evaluation of the System of Rice Intensification in Cambodia. Report to the German Agency for Development Cooperation (GTZ), Phnom Penh, Cambodia (2004).

ATI, System of Rice Intensification (SRI) Field Day Report, October 23, 2002, Agricultural Training Institute, Department of Agriculture, Cotabato, Mindanao, Philippines (2002).

Barison, J., Nutrient-use efficiency and nutrient uptake in conventional and intensive (SRI) rice cultivation systems in Madagascar. Master's thesis, Crop and Soil Sciences Department, Cornell University, Ithaca (2002).

Birch, H.F., The effect of soil drying on humus decomposition and nitrogen, *Plant Soil*, **10**, 9–31 (1958).

Boddy, R.M. et al., Biological nitrogen fixation associated with sugar cane and rice: Contributions and prospects for improvement, *Plant Soil*, **174**, 195–209 (1995).

Bonkowski, M., Protozoa and plant growth: The microbial loop in soil revisited, *New Phytol.*, **162**, 616–631 (2004).

Brimecombe, M.J., De Leij, F.A., and Lynch, J.M., The effect of root exudates on rhizosphere microbial populations, In: *The Rhizosphere: Biochemistry and Organic Substances at the Soil–Plant Interface*, Pinton, R. et al., Eds., Marcel Dekker, New York, 95–140 (2001).

Dakora, F.D. and Phillips, D.A., Root exudates as mediators of mineral acquisition in low-nutrient environments, *Plant Soil*, **245**, 35–47 (2002).

De Datta, S.K., *Principles and Practices of Rice Production*, Wiley, New York (1981).

Döbereiner, J., *Nitrogen-Fixing Bacteria in Non-Leguminous Crop Plants*, Springer, Berlin (1987).

Dobermann, A., A critical assessment of the system of rice intensification (SRI), *Agric. Syst.*, **79**, 261–281 (2004).

Doebbelaere, S., Vanderleyden, J., and Okon, Y., Plant growth-promoting effects of diazotrophs in the rhizosphere, *Crit. Rev. Plant Sci.*, **22**, 107–149 (2003).

Eissenstat, D.M. and Yanai, R.D., Root life span, efficiency and turnover, In: *Plant Roots: The Hidden Half*, 3rd ed., Waisel, Y. et al., Eds., Marcel Dekker, New York, 221–238 (2002).

Ellis, J.F., Post-flood syndrome and vesicular–arbuscular mycorrhizal fungi, *J. Prod. Agric.*, **11**, 200–204 (1998).

Feng, C., et al., Ascending migration of endophytic rhizobia, from roots to leaves, inside rice plants and assessment of benefits to rice growth physiology, *Appl. Envir. Microbiol.* **71**, 7271–7278 (2005).

Frankenberger, W.T. and Arshad, M., *Phytohormones in Soils: Microbial Production and Functions*, Marcel Dekker, New York (1995).

Gyaneshwar, F. et al., Role of soil microorganisms in improving P nutrition of plants, *Plant Soil*, **245**, 83–93 (2002).

Ilag, L.L. et al., Changes in the population of infective endomycorrhizal fungi in a rice-based cropping system, *Plant Soil*, **103**, 67–73 (1987).

Joelibarison, Perspective de developpement de la region de Ranomafana: Les mechanisms physiologiques du riz sur de bas-fonds: Case du SRI. Mémoire de fin d'etudes. Ecole Supérieure des Sciences Agronomique, University of Antananarivo, Madagascar (1998).

Johnson, B.K., Soil survey, In: *Final Report for the Agricultural Development Component of the Ranomafana National Park Project in Madagascar*. Soil Science Department, North Carolina State University, Raleigh, NC, 5–12 (1994).

Johnson, B.K., Soil characterization and reconnaissance survey of the Ranomafana National Park Area, Southeastern Madagascar. Ph.D. thesis, Soil Science Department, North Carolina State University, Raleigh, NC (2002).

Kannaiyan, S. et al., The xylem of rice (*Oryza sativa*) is colonized by *Azorhizobium caulinodans*, *Proceedings of the Royal Society: Biological Sciences*, **267**, 103–107 (1999).

Kar, S. et al., Nature and growth pattern of rice root system under submerged and unsaturated conditions, *Il Riso (Italy)*, **23**, 173–179 (1974).

Katayama, T., *Ine mugi no bungetsu kenkyu* (Studies on tillering in rice, wheat and barley), Yokendo Publishing, Tokyo (1951).

Khush, G.S. and Peng, S., Breaking the yield frontier of rice, In: *Increasing Yield Potential in Wheat: Breaking the Barriers: Proceedings of a workshop held in Ciudad Obregon, Sonora, Mexico*. CIMMYT, Mexico, D.F., 36–51 (1996).

Kirk, G.J.D. and Bouldin, D.R., Speculations on the operation of the rice root system in relation to nutrient uptake, In: *Simulation and Systems Analysis for Rice Production*, Penning de Vries, F.W.T. et al., Eds., Pudoc, Wageningen, 195–203 (1991).

Kirk, G.J.D. and Solivas, J.L., On the extent to which root properties and transport through the soil limit nitrogen uptake by lowland rice, *Eur. J. Soil Sci.*, **48**, 613–621 (1997).

Kronzucker, H.J. et al., Nitrate–ammonium synergism in rice: A subcellular flux analysis, *Plant Physiol.*, **119**, 1041–1045 (1999).

Kumar, V. et al., An alternative agricultural system is defined by a distinct expression profile of select gene transcripts and proteins, *Proc. Natl. Acad. Sci. USA*, **101**, 10535–10540 (2004).

Ladha, J.K., de Bruijn, F.J., and Malik, K.A., Eds., *Opportunities for Biological Fixation in Rice and Other Non-Legumes*, Kluwer/International Rice Research Institute, Dordrecht, Netherlands/Los Baños, Philippines (2000).

Ladha, J.K. and Reddy, P.M., Eds., The quest for nitrogen fixation in rice, In: *Proceedings of a Workshop, 9–12 August, 1999, Los Baños, Laguna, Philippines*, International Rice Research Institute, Los Baños, Philippines (1999).

Laulanié, H., Le système de riziculture intensive malgache, *Tropicultura* (Brussels), 11, 110–114 (1993).

Li, X., Xu, X., and Li, H., A socio-economic assessment of the system of rice intensification (SRI): A case study from Xiusheng Village, Jianyang County, Sichuan Province, College of Humanities and Rural Development, China Agricultural University, Beijing (2005).

Magdoff, F.R. and Bouldin, D.R., Nitrogen fixation in submerged soil–sand–energy material media and the aerobic–anaerobic interface, *Plant Soil*, 33, 49–61 (1970).

Nemoto, K., Morita, S., and Baba, T., Shoot and root development in rice related to the phyllochron, *Crop Sci.*, 35, 24–29 (1995).

Neumann, G. and Römheld, V., The release of root exudates as affected by the plant's physiological status, In: *The Rhizosphere: Biochemistry and Organic Substances at the Soil–Plant Interface*, Pinton, R. et al., Eds., Marcel Dekker, New York, 41–93 (2001).

Oborny, B., External and internal control of plant development, *Complexity*, 9, 22–27 (2004).

Randriamiharisoa, R.P., Research results on biological nitrogen fixation with the System of Rice Intensification, In: *Assessments of the System of Rice Intensification: Proceedings of an International Conference*, Sanya, China, April 1–4, Uphoff, N. et al., Eds., CIIFAD, Ithaca, NY, 148–157 (2002).

Reinhold-Hurek, B. and Hurek, T., Life in grasses: Diazotrophic endophytes, *Trends Microbiol.*, 6, 139–144 (1998).

Rickman, J.F., Preliminary results: Rice production and the system of rice intensification (SRI), International Rice Research Institute, Los Baños, Philippines (2003).

Sanchez, P.A., *Properties and Management of Soils in the Tropics*, Wiley, New York (1976).

Sheehy, J. et al., Fantastic yields in the system of rice intensification: Fact or fallacy?, *Field Crops Res.*, 87, 131–154 (2004).

Sinha, S.K. and Talati, J., Impact of the System of Rice Intensification (SRI) on Rice Yields: Results of a New Sample Study in Purulia District, India, Paper for IWMI-Tata Fourth Annual Partners Meeting, February 24–26. Anand, Gujarat, India (2005).

Solaiman, M.Z. and Hirata, H., Responses of directly seeded wetland rice to arbuscular mycorrhizal fungi inoculation, *J. Plant Nutr.*, 20, 1479–1487 (1997).

Stoop, W., Uphoff, N., and Kassam, A., A review of agricultural research issues raised by the system of rice intensification (SRI) from Madagascar: Opportunities for improving farming systems for resource-poor farmers, *Agric. Syst.*, 71, 249–274 (2002).

Turner, B.L. and Haygarth, P.M., Phosphorus solubilization in rewetted soils, *Nature*, 411, 258 (2001).

Turner, B.L. et al., Inositol phosphates in the environment, *Phil. Trans. R. Soc., Lond. Ser. B*, 357, 449–469 (2000).

Uphoff, N., Agroecological implications of the system of rice intensification (SRI) in Madagascar, *Environ. Dev. Sustainability*, 1, 297–313 (1999).

Uphoff, N., Higher yields with fewer external inputs? The system of rice intensification and potential contributions to agricultural sustainability, *Intl. J. Agric. Sustainability*, 1, 38–50 (2003).

Uphoff, N., Satyanarayana, A., and Thiyagarajan, T.M., Prospects for rice sector improvement with the system of rice intensification, considering evidence from India, Paper presented to 16th International Rice Conference, Bali, Indonesia, Sept. 11–14, Agency for Agricultural Research and Development, Jakarta, Indonesia, and International Rice Research Institute, Los Baños, Philippines (2005).

Uprety, R., Performance of SRI in Morang District, 2004. Seasonal Report, District Agricultural Development Office, Biratnagar, Morang, Nepal (2005).

Wang, S. et al., Physiological characteristics and high-yield techniques with SRI rice, In: *Assessments of the System of Rice Intensification: Proceedings of an International Conference*, Sanya, China, April 1–4, Uphoff, N. et al., Eds., CIIFAD, Ithaca, NY, 116–124 (2002).

Wardle, D.A., *Communities and Ecosystems: Linking the Aboveground and Belowground Components*, Princeton University Press, Princeton, NJ (2002).

Whipps, J.M., Microbial interactions and biocontrol in the rhizosphere, *J. Exp. Bot.*, 52, 487–511 (2001).

Ying, J. et al., Comparison of high-yield rice in tropical and subtropical environments: I: Determinants of grain and dry matter yields, *Field Crops Res.*, 57, 71–84 (1998).

29

Contributions of Managed Fallows to Soil Fertility Recovery

Erika Styger and Erick C.M. Fernandes

The World Bank, Washington, DC, USA

CONTENTS

Fallow management and improvement is as old as agriculture itself. Fallows are currently still intrinsic parts of many tropical farming systems. The development of agricultural systems and the spatial dynamics of land use are strongly correlated with the evolution of fallows patterns. Indeed, some classifications of farming systems have been based on fallow characteristics, e.g., Boserup (1965) and Ruthenberg (1980). While fallow is commonly referred to a resting period for agricultural land between two cropping cycles during which soil fertility is restored, it has more roles than just fertility restoration.

Fallows functions include weed control and the interruption of pest and disease cycles. They provide cash income in times of immediate need and help to balance food supply (Styger et al., 1999). They produce wood, fibers, and medicinal plants for households and can serve as pastures for livestock. For resource-poor farmers with constraints on their labor, inputs, and access to new techniques, fallows are economically often a good option for optimizing agricultural production, especially when noncrop products can be harvested. Fallows can also be important reservoirs of above- and belowground biodiversity. A wealth of indigenous knowledge is associated with the diversity and richness of these fallow systems.

The oldest form of fallow management was shifting-cultivation systems where very long fallow periods (>20 years) alternated with short cropping periods (1 to 2 years). A multitude of fallow systems has developed across the tropics over many centuries, diverging from this primordial model as farmers have developed diverse strategies for intensified fallow use.

Two major pathways of fallow intensification were identified at the conference on indigenous strategies for intensification of shifting cultivation in Southeast Asia in 1997 (Cairns, 2004b). The first strategy has as its main objective, increasing fallows' economic productivity. Fallow lengths stay the same or even lengthen as farmers add value by introducing economic perennial species and take appropriate steps to enhance soil fertility. These systems include interstitial tree-based improved fallows, perennial–annual crop rotations, and what are called complex agroforests. These fallows can be characterized as more *productive* fallows. The second strategy aims to improve the biological efficiency of a particular fallow system, seeking to achieve greater benefits within the same or shorter time frame. These include shrub-based accelerated fallows and short-term herbaceous fallows. They can be referred to as more *effective* fallows. Biophysical and economic benefits can accrue from either strategy (Cairns, 2004a).

Given the wide range of fallow systems, dynamic and continually changing, the operational definition of managed fallows covers a spectrum of practices, from growing viney legumes husbanded as dry-season fallows for a few months, to long-term complex agroforests, which capitalize on multiple synergies, tapping opportunities at various heights and depths above- and belowground. For improvement and optimization of farming systems, the management of fallow and cropping cycles needs to address continually the dual aims of production and soil fertility enhancement. For this reason, Brookfield (2004) characterizes such systems as *farmer-guided ecological change* rather than as fallow improvement, since the objectives include biological diversity and quality and quantity of livelihoods, rather than just soil improvement or combating soil degradation.

29.1 Fallowing for Soil Fertility Improvement

Keeping in mind this broader concept of fallow and crop management, this chapter concentrates on the roles and functions of fallows for sustainable soil fertility improvement. Across all continents, poor farmers in the tropics face similar dynamics and constraints. As populations grow and pressure on land area increases, fallow periods are reduced, and traditional fallow management is often not sufficient to restore soil fertility, leading over time to soil and vegetation degradation and to noticeable declines in crop yields. Given the fact that future food production in the world will have to be achieved with less land and water per capita, and given the expectations that poverty will be reduced, efforts for sustainable intensified agriculture should give priority to developing knowledge and skills for optimizing ecological and agricultural environmental conditions. Associated priority should be attached to getting more efficient labor use, capital, and external inputs, addressed later.

The agronomic performance of agroecosystems can be enhanced by managing more efficiently the biological cycles and interactions among the components that determine crop productivity. Improved fallow management as an entry point has the potential to capture nutrients within the system and make them plant-available, to reduce weed pressure, to restore soil organic matter (SOM), litter layers, and biological activity in the surface soil, and to rehabilitate soil micro- and macroorganisms that were reduced during the cultivation phase. These concepts of improved fallow management can be extended to the rehabilitation of degraded and abandoned land. However, what is actually happening within a fallow cycle and how do fallows work?

29.2   Soil Fertility Restoration during a Fallow Cycle

At the start of a fallow period, the fallow vegetation grows rapidly, from new seedlings and from the root systems already in place from previous crop and fallow periods. Nutrients are taken up from the surface soil and subsoil and are stored in the vegetation. They are in part returned to the soil through litter and rain wash from the aboveground vegetation. With intense biotic activity at the soil surface, litter decays rapidly and is transformed partly into SOM. At the same time, erosion is minimized, as is leaching, due to ground cover and rooting mass. Humus and litter layers increase until they reach an equilibrium between build-up and rate of oxidation (Nye and Greenland, 1960).

29.2.1   Soil Chemical Improvements

Buildup and maintenance of SOM is critical to soil productivity and generally corresponds to nutrient buildup. SOM increases the cation exchange capacity (CEC) of the surface soil which is especially important in kaolinitic soils. Of special importance, increasing SOM can reduce phosphorus fixation in soils with high iron and aluminum oxide content. In West African Alfisols, SOM accounts for 80% of CEC, and available P, K, Mg, Ca and CEC are highly correlated with SOM levels (Agboola, 1994). The big opportunity for management in such soil systems is that SOM is a renewable resource, whose level can be replenished by additions of organic inputs (Fernandes et al., 1997).

29.2.2   Soil Physical Improvements

Beneficial soil physical changes occur as well. The fine surface roots of the fallow vegetation mold the soil into soft and porous granules or crumbs, worms deposit their casts on the soil surface, and drainage channels are created. Such a soil surface permits rapid infiltration of water and resists erosion unless completely unprotected. SOM, especially in sandy soils, has a profound impact on soil structure. Physical properties such as hydraulic conductivity, bulk density, total porosity, and aggregate stability all decrease as SOM falls, as a result of the duration and intensity of cropping (Nye and Greenland, 1960; Agboola, 1994).

29.2.3   Soil Biological Improvements

Biomass production, litter fall, and SOM influence microclimates and the substrates for soil fauna. Litter fall under fallows has been shown to increase microarthropod and earthworm population in Southwestern Nigeria (Salako and Tian, 2001). A cover crop within an agroforestry system in Amazonia has been seen to exert a favorable effect on the soil fauna, presumably by keeping the soil moist and shaded and by providing litter as a substrate (Barros et al., 2003). In a comparison of a 15-year fallow and a 3-year potato field in the high tropical Andes, labile carbon and nitrogen soil pools under fallow increased significantly; microbial biomass nearly doubled; the microbial community was much more diverse; and the rate of plant material decomposition was much faster under fallow (Sarmiento and Bottner, 2002). Furthermore, belowground organic matter quality, SOM, fine and coarse roots, and heterotrophic biota have been shown to have a suppressive effect on phytoparasitic nematodes.

In the sudanian zone of Senegal, it has been shown that woody species can control harmful nematodes better then in herbaceous fallows, since coarse roots play a key role in maintaining nematode diversity, with nonparasitic species competing with those that are parasitic (Manley et al., 2000). SOM decline also exacerbates the problems with Striga infestation. In northern Ghana, in fields closer to the homestead, where organic matter inputs had increased SOM to 2.54% compared with 1.42% in the more distant fields, the close-by soils had 40% higher CEC and microbial biomass was four times greater. Striga infestation was found to be inversely correlated with the total nitrogen content and microbial biomass in the soil (Sauerborn et al., 2003).

29.2.4   Time Dynamics of Nutrient Accumulation and Soil Fertility Restoration

In the early stages of a fallow when biomass is increasing and nutrient uptake is rapid, there may actually be a net loss of nutrients from the topsoil. It is only later in the fallow development, when litter fall greatly exceeds the increase of nutrient uptake into biomass, that the amount of nutrients in the topsoil may be increased and restored. Looking at nutrient stocks during the fallow cycle in an Ultisol in the Peruvian Amazon, total stocks of phosphorus and potassium under three types of fallow, *Inga edulis*, *Desmodium ovalifolium*, and natural fallow, were, respectively, about 40, 80, and 12% greater at 53 months than initial values; however, calcium and magnesium stocks were reduced by 25 to 40% in the three fallow types. While nitrogen was restored quickly, carbon restoration was estimated to take at least 8 to 10 years. Since calcium and magnesium were incompletely restored, there is a concern about the sustainability of short-term, fallow-based systems on such acidic, infertile soils. Either fallow species with deep root systems should be selected for more efficient uptake of these elements, or inorganic inputs have to be applied to provide the key limiting elements (Szott and Palm, 1996).

In natural fallows in eastern Madagascar, nutrients regenerated rapidly in the fallow vegetation, attaining already within one year 36 to 57% of the previous phytomass pools, whereas topsoil nutrient concentrations started to increase only after 3 to 5 years of fallow (Brand and Pfund, 1998). In a study looking at carbon and nutrient accumulation in secondary forests regenerating on pastures in Central Amazonia, Feldpausch et al. (2004) found that in a 14-year-old forest, most of the carbon and nitrogen had been stored within the soil, while the increased phosphorus, potassium, magnesium, and calcium resided more within the vegetation. These findings have important implications for management. When nutrients are accumulated in the vegetation rather than in the soil, there is a danger of nutrient stock depletion when biomass is exported or burned rather than being carefully recycled within the plot.

29.3   Fallow Systems

Two major fallow systems are generally distinguished: (1) fallows composed of natural vegetation and (2) fallows that consist of deliberately planted, introduced species.

29.3.1 Natural Vegetation Fallows

The most common approach to fallow management is the opportunistic use of the germplasm available *in situ*. In systems with long fallow periods and in early cycles after deforestation, natural fallow establishment is characterized by regenerating trees either coppiced from old plants or growing from seeds. If fallow periods shorten and cropping frequencies increase, tree seedlings do not survive disturbances from cultivation and fire use, tree stumps die, and tree seedbanks in the soil become progressively depleted. Thus, trees are replaced by shrubs that are either indigenous or exotic naturalized species. Then, with further intensity of cropping/fallowing (meaning reduction in the length of fallow period), shrubby fallows are replaced by herbaceous fallows that have lower soil-restoring abilities and are often difficult to control, which may result in abandonment of the land for agriculture.

Trees often depend upon bats or birds for seed dispersal whereas some shrubby species, grasses, and forbs have wind-dispersed seeds and are able to colonize new areas more rapidly. Also, when nutrients released during burning and are liable to rapid loss through leaching or runoff, those species that can establish themselves quickly will steadily increase in density and cover (Uhl et al., 1981).

These transitions in vegetation composition can occur very rapidly. A fallow study in the rainforest region of eastern Madagascar documented what happens when fallow periods are significantly reduced. Within the past 30 years, the length of fallow periods declined from 8 to 15 years to currently 3 to 5 years. These current fallow periods are much too short to maintain soil and vegetation productivity, and the fertility of soil systems is being lost. A very rapid increase of upland degradation and transition from primary forest to abandoned grasslands is occurring within periods of 20 to 40 years, five to ten times faster than previously reported (Styger, 2004).

29.3.1.1 Natural Tree Fallows

Natural trees establish in young fallows only if the previous fallow periods were long enough to regenerate soil system fertility or if pioneer tree species can regenerate rapidly from existing seed pools after forest disturbance. If shrubs, forbs, or grasses come to dominate the early fallow phases, shade-tolerant trees establish only gradually under the canopy of other species, and shade-intolerant species lose out. This considerably slows the speed of successions into mixed forest vegetation and can mean that such mixture never materializes (Uhl et al., 1981).

Farmers pursue various management techniques in natural tree fallows: selective weeding, protecting favored species during the cropping cycle, and enrichment planting with favored species. The selective retention of desirable species gradually alters the forest fallows with the production of desirable and economically valuable species to farmers. The ability of tree fallows to regenerate, referred to as their resilience, can vary considerably. In Thailand, for instance, farmers are able to maintain a productive system of upland rice production by depending on the regrowth of the pioneer species *Macaranga denticulata* (Euphorbiaceae) within 7-year fallow/cropping cycles (Yimyam et al., 2003). In eastern Madagascar, on the other hand, the pioneer tree *Trema orientalis* (Ulmaceae) is able to colonize and dominate only the first fallow cycle after deforestation. In the second fallow cycle, *Trema* is displaced from the agroecosystem, and shrubby fallows establish themselves instead (Styger, 2004).

29.3.1.2 Natural Shrubby Fallows

Progressive loss of forest cover and increased land-use intensity favors the expansion of pioneer shrubs at the expense of trees. In many locations in the tropics, exotic, naturalized, and invasive shrubs have replaced the indigenous vegetation. In many cases they have proved themselves successful in restoring soil fertility and in delaying further transition to herbaceous fallows. A prime example is *Chromolaeana odorata* (Asteraceae) which once introduced into Asia and Africa spread aggressively, altering the natural fallow vegetation so that in many parts of the world it is considered a noxious weed.

However, in Southeast Asia, farmers learned to appreciate the ability of *C. odorata* to colonize young fallows rapidly, developing dense, almost monospecific thickets that protect the soil. Very importantly, this plant can shade out *Imperata cylindrica* and other light-demanding gramineous weeds. *C. odorata*'s rapid biomass accumulation and copious leaf litter appears to accelerate nutrient cycling and increase SOM. Although not a nitrogen fixer, Chromolaena plays a critical role in nutrient conservation by aggressively scavenging labile nutrients from the soil nutrient pool. *C. odorata* plants have been shown to have higher nitrogen, phosphorus, and calcium contents compared with indigenous fallow (Roder et al., 2004).

Another species from the Asteracaeae family with similar properties is *Tithonia diversifolia*. Both species have nematocidal properties, reducing disease problems in the subsequent cropping phase. Indeed, farmers use Tithonia juice extracts as an ingredient in botanical pesticides. Exotic Asteraceae are seldom subject to insect herbivory. Most importantly, these fallows require little labor, establish spontaneously when seed is available, require no special management, and are easily cleared (Cairns, 2004a).

In eastern Madagascar, the second fallow cycle after initial deforestation is most often dominated by the indigenous shrub species *Psiadia altissima* (Asteraceae). However, this species is in turn quickly outcompeted in subsequent fallows by *Rubus moluccanus* (Rosaceae) and/or *Lantana camara* (Verbenaceae), two exotic invasive species. *Rubus* is the most aggressive among all the shrubby fallow species and quickly forms a thick stand. In eastern Madagascar, Rubus accumulates the highest nutrient stocks in short-term fallows up to 5 years, while Trema tree fallows accumulate higher nutrient stocks only beyond 5 years. Rubus also has very prolific litter production, and its root biomass constitutes 35% of total biomass compared with 15 to 20% for Psiadia and Trema. This enables the species to recover quickly and be competitive after the frequent disturbances by burning and cropping. However, Rubus is not robust enough to withstand repetitive slashing and burning, so after 3 to 4 fallow cycles, it gives way to ferns and to *I. cylindrica* (Poaceae) (Styger, 2004).

These exotic and naturalized fallows can be labeled "improved fallows" if they restore soil fertility more rapidly than would the indigenous vegetation and at minimum labor cost to farmers. These fallow plants, however, have some negative qualities as their rapid dispersal and colonization make it difficult to control their expansion. Rubus is a spiny plant, and this makes its plant residues difficult to manage. Farmers use fire as the most practical solution for field preparation. Chromolaena, despite its desirable qualities as a fallow species, has shown allopathic effects on tree growth, and because both species suppress the regeneration of desirable woody species, they represent a threat to native biodiversity. Also, because neither species is palatable to livestock, their expansion reduces grazing (Roder et al., 2004; Styger, 2004).

29.3.1.3 *Natural Herbaceous Fallows*

Continuation of high frequency of land use, with shorter and shorter fallow times, means that natural fallows will eventually be dominated by herbaceous plants, forbs, and grasses. *I. cylindrica* is the most widespread and aggressive species in Asia and Africa, with some spread also in Latin America, although for unclear reasons, it has proved to be less aggressive there. In Madagascar, Imperata and the ferns *Pteridium aquilinum*

(Dennstaedtiaceae) and *Sticherus flagellaris* (Gleicheniaceae) replace Rubus as soil fertility levels decline through successive cycles of cropping and fallow. Once herbaceous fallows are established, farmers cease upland rice cultivation and may plant root crops for one or two more seasons before the land is completely abandoned. In Madagascar, unlike in Asia, Imperata is not the end stage of succession; it is eventually replaced by Aristida grasses (Poaceae). These grasses are nutrient-poor, lack good soil coverage, and thus favor soil erosion. They are burned periodically for low-productivity cattle grazing (Styger, 2004).

29.3.2 Planted Fallows

Under intensified rotations, natural fallows show considerable limitations characterized by soil and vegetation degradation and by species-replacement from woody to herbaceous species. The herbaceous species are seldom able to produce large amounts of good-quality biomass and to restore soil fertility quickly. A desirable step in agricultural intensification is, therefore, to plant fallow species that improve soil systems more efficiently than do the natural fallows.

Exotic, leguminous tree, shrub, or herbaceous species are often selected and planted in fallows and intercrops from a few months up to 3 to 5 years (Fernandes et al., 1994). Since the 1980s, research on improved planted fallows has increased, and adoption of new techniques is taking place in Latin America, Africa, and Southeast Asia. The main species used are legumes of the genus Sesbania, Tephrosia, Crotalaria, Cajanus, Leucaena, Indigofera, Mimosa, and herbaceous species Centrosema, Pueraria, and Mucuna (Sanchez, 1999). Results with several of these species used in fallow are discussed in Chapter 19.

Planted fallows are adopted where labor and technologies are available and where land has become a limiting factor. They are also favored when land tenure is secure and when markets exist for the variety of products that can be grown (or grown better) with improved soil quality. Technical issues that arise are selection of the best germplasm (species/varieties), cropping techniques such as optimal planting dates, plant densities, establishment methods, harvesting techniques, below- and above-ground residue management, and ease of seedbed preparation for the first crop after fallow. These new techniques demand access to germplasm, specific skills, and knowledge, and more labor than do natural fallows, and this is often a limitation for the adoption of these techniques. Furthermore, research and extension services need to support the process of innovation and adaptation in collaboration with the farmers since fallow systems need to be continually evaluated and often modified in light of local conditions that are often changing biophysically but also socioeconomically, e.g., affecting labor availability.

29.3.2.1 *Planted Tree and Shrub Fallows*

Planted tree fallows are often used when fallow periods are longer then 5 years and when additional products such as wood or fruits can be produced as an added benefit of the short-fallow restoration. Shrub species are often used in fallows from 6 months to 3 years. In general, the higher the biomass production, the higher are the nutrient stocks established and the better the soil fertility improvement, resulting in higher crop yields. This has been shown in various experiments, many reported in Chapter 19.

In one evaluation not reported in Chapter 19, the maize yields resulting after 2- and 3-year fallows with *Sesbania sesban* in eastern Zambia were 5.0 and 6.0 t ha^{-1}, respectively, compared with 4.9 and 4.3 t ha^{-1} from continuously cropped maize with fertilizer

(112 kg N ha^{-1}), and 1.2 and 1.9 t ha^{-1} without fertilizer. The total yield over four cropping seasons following a 2-year fallow was 12.8 t ha^{-1} compared with 7.6 t ha^{-1} for six seasons of continuous unfertilized maize. In addition, 15 and 21 t ha^{-1} of fuelwood were harvested after the 2 and 3-year fallows, respectively, a large benefit to households in the area (Kwesiga et al., 1999).

In western Kenya, a species screening of 22 shrubby and herbaceous legumes in a 2-year fallow showed 30 to 100% increases in maize yield compared with the natural fallow (Niang et al., 2002). In Uganda, with a one-season fallow of *Crotalaria ochroleuca*, which was intercropped and then mulched with maize or beans, yields decreased during the intercropping season by an average of 30%; but then the yields in the first season after mulching increased by an average of 40%, with a best response being 68%. In the second season, there was still an increase of greater than 20%. In fields where Crotalaria was planted, it was also seen that water infiltration increased and bulk density of the soil decreased (Fischler et al., 1999).

29.3.2.2 Planted Herbaceous Legumes

Herbaceous legumes or cover crops are often relay-planted during the cropping cycle, and once the crop is harvested, the soil surface is covered quickly and nutrient losses are minimized. Cover crops and green manures are discussed in more detail in the next chapter. An ideal cover crop is:

- Easy to establish
- With few external inputs and costs
- Fast-growing
- Adapted to a broad range of soil conditions
- Has few pests and diseases
- Provides good soil cover and erosion control
- Improves soil fertility
- Fixes nitrogen (Hairiah, 2004)

In addition to soil fertility restoration, some cover crops also produce grains for human consumption or fodder for livestock. Although such benefits are appealing, there is a trade-off as the grain production and harvest will affect the amount of nutrient accumulation in the soil. If the harvest index increases, a species intended for fallow enrichment may become really a legume crop and part of a permanent crop rotation.

29.3.2.3 Natural vs. Planted Fallows

It is not always the case that planted fallows outperform natural fallows. An evaluation of 2-year leguminous shrub fallows, comparing this with Chromolaena as a natural fallow in northern Laos, showed that leguminous shrubs produced 4 to 6 times higher aboveground biomass (20 t to 30 t ha^{-1}) than did Chromolaena (5 t ha^{-1}). However, the superior biomass did not have any significant impacts on rice production or soil parameters (Roder and Maniphone, 1998). A planted Tithonia fallow in western Kenya did not differ in the quantity of aboveground biomass produced compared with a natural weed fallow, although soils were improved more efficiently, indicating that some important rhizosphere processes were taking place (George et al., 2002).

Species difference can be remarkable. In the Peruvian Amazon, a comparison between single species fallows showed that *I. edulis* and *D. ovalifolium* produced the highest nutrient stocks after 53 months, followed by the natural fallow, which outperformed the other planted species, Centrosema, Stylosanthes, Cajanus, and Pueraria (Szott and Palm, 1996). Fallow improvement is thus not only a function of the amounts of inputs invested, but depends on a host of interacting factors, including the quality, timing, and location of inputs (above- or belowground), soil biological dynamics, and the effect of these differences on SOM and nutrient availability patterns (Fernandes et al., 1997). The interactions among these factors remain in many instances not well-known.

29.3.2.4 Supplementary Inorganic Inputs

Fallowing alone may not be sufficient to achieve a productive and sustainable system. This can be the case where soil systems have experienced serious depletion due to overexploitation and erosion or in soils with low pH, low CEC, high aluminum content, or specific nutrient deficiencies. In addition to organic inputs, raising pH through liming and targeted use of inorganic fertilizers can often jump-start the system, stimulating soil microbial activities and increasing the efficiency for soil restoration. A combination of organic and targeted inorganic inputs has often been reported to have synergistic effects that exceed the effect of applying a single kind of amendment with the same level of nutrients. Combinations can provide a solid basis for sustained system improvements.

An experiment in eastern Madagascar compared the productivity and effect on soil fertility of traditional slash-and-burn techniques with a crop and soil management system that mulched the slashed fallow biomass and recycled crop residues. The trials included plots with no added phosphorus (M0) and others with either 40 kg P ha^{-1} (M40) or 80 kg P ha^{-1} (M80), applied through a locally available guano-phosphate on a rotation of upland rice, beans, ginger, and *C. grahamiana* fallow. The aboveground nutrient stocks of the Crotalaria fallows in M0, M40 and M80 were, after 1 year, 2–3, 3.5–4, and 5–6 times higher, respectively, compared with natural fallow. Given the nutrient stock accumulation curve for natural fallows, it would take a natural fallow 5 to 30 years, depending on the fallow species and elements, to achieve the same nutrient stocks that the Crotalaria M80 treatment achieved in 1 year.

Yield comparisons showed that M80 yields were 200, 340, 155, and 180% higher for rice, beans, ginger, and *Crotalaria*, respectively, than for M0. Net monetary returns per hectare, subtracting the guano cost, were about 50% higher for M80 compared with M0 (US$ 2760 vs. 1860) (Styger, 2004).

29.4 Limitations on Fallow Management

Fallow systems face a fairly inexorable limitation. With increased cropping frequency, i.e., shorter fallow periods, natural vegetation will change in its species composition from woody to herbaceous species, and it will, over time, become degraded as the loss in biodiversity aboveground parallels that associated with soil degradation belowground. In overexploited systems, more nutrients will be exported than are replenished, SOM is likely to decrease, and the environment for soil biota will become less favorable.

Fallowing faces some serious limitations if one wants to build up nutrient pools within short time-periods. The time needed for nutrient restoration varies greatly across elements; nitrogen, for instance, can be restored in less than 2 years, whereas in the same soil, calcium needs 15 to 20 years (Szott et al., 1999). In young fallows, soil nutrient pools may decrease temporarily and may accumulate in fallow biomass until returned through litter back to the soil. Management interventions at that stage need to manage fallow biomass carefully, minimizing its export and avoiding burning it.

Planted leguminous fallows often have higher biomass production and nutrient stocks compared with natural fallow, and this often translates into higher crop yields as seen above. However, this is not always the case. Careful testing of new fallow species and fine-tuning of management techniques needs to be carried out to optimize fallow functions. Exotic species have to be carefully chosen or else avoided, since some have the potential to become weeds outside their native locations. Yield increases for subsequent crops will be highest in the first crop season, but then will decline, sometimes rapidly, in the second and third crop. Additional inputs that correct soil pH, build up SOM and CEC, increase nutrient availability, and complement the limiting elements are welcome so that yield increases can be maintained longer. The change to planted fallows occurs when agricultural systems intensify, which demands more labor. Research and extension support are important for making best use of the available resources. Often the lack of well-suited germplasm can be a limiting factor for the adoption of new fallow species.

29.5 Discussion

The sustainable intensification of fallow and cropping systems should aim to provide continuous soil cover, maintain SOM, preserve nutrients within the systems, and keep nutrient cycling as efficient as possible, building on species diversity in fallows and in crops, with each contributing according to its specific comparative advantages to the overall increase of the system's productivity.

The selection of fallow species is critical for optimizing fallow functions. The species selected should be able to take up nutrients from poor soils, should rapidly accumulate biomass, produce high quality litter, and have balanced nutrient profiles in their biomass. Species with deeper rooting depth are better able to capture nutrients from subsoil. Those that suppress weeds efficiently and break cycles of pest and diseases make different but important contributions. Mixing species can increase fallow efficiency as the integration of deep-rooting woody plants is complemented by association with shallow but densely rooted, rapidly colonizing cover crops. It is important also to include species that form symbioses with mycorrhizal fungi and favor other beneficial soil microbes or macrofauna.

Optimizing organic matter recycling is important for successful fallow regeneration. Organic matter produced within the system, fallow biomass, weeds, and crop residues should be recycled as efficiently as possible and should not be exported or burned. As much as possible, other inputs such as homestead residues, livestock manure, and biomass transfer should be used if available to augment the buildup of organic matter. The maintenance of SOM, especially in tropical ecosystems, is dependent on the continual input of organic materials. SOM losses and gains depend on soil type, climate, and crop and fallow management practices, on length of fallowing and cropping, tillage, residue management, etc. The proportion of fresh material converted to soil humus will probably be between 10 and 20%. For tropical forests, these numbers can be closer to 10 than 20%. Root material that is amenable to mineralization contributes to soil humus between 20 and 50% of annual root growth (Nye and Greenland, 1960).

Combining organic matter recycling with inorganic nutrient amendments opens up some new possibilities. On infertile acid soils, phosphorus and calcium are often needed to prime biological processes such as nutrient cycling and nitrogen fixation. Mulches and the addition of organic matter to the soil can keep phosphorus fertilizer from becoming bound to aluminum and other ions in acid soils, thus making it more available for plants (Schlather, 1998).

The maintenance of biodiversity within soil systems and in the vegetative biomass helps to make intensification of farming systems more sustainable. This is important for ecological balance and to address product diversification better within the fallow and cropping cycles. Landscape management should strive for mosaics of different land-use forms where a range and variety of cropping and fallow systems are operated with short- to long-term fallows fulfilling various economic and ecological functions. If a broad range of farming systems is available, farmers can choose, adapt, and use various niches within the landscape that are as yet not used optimally.

More attention should be paid when developing these systems to the management and enhancement of environments for soil micro- and macrofauna. At this moment, it is obvious that many landscapes are degrading. Low yields are easily attributed to measurable deficiencies in available soil chemicals, and physical degradation is easy to see. What is not so easily evident is the loss of soil biota and of their contributions to a productive soil system. The current problem for agriculture is not so much that land *per se* is not available, but that fertile land for agricultural use is becoming scarce. Intensification is an almost unavoidable response. If improved management strategies with short-term fallows can be implemented, this can buy time for marginal land, which is degraded or not currently suitable for agriculture, to be rehabilitated for future production through longer-term restorative measures.

Published reports indicate that farmers in some locations have been quick to take up managed fallow technologies. For example, in Central America and West Africa, farmers have rapidly adopted the herbaceous legume Mucuna in rotation with maize (Buckles and Triomphe, 1999; Tarawali et al., 1999). In southern Africa, widespread adoption of *S. sesban* fallows has been reported by Kwesiga et al. (1999). While the large-scale adoption of single-species fallows is having positive results, this gives some cause for concern. If thousands of farmers convert landscapes to monospecific fallows, it is possible that pests or pathogens could emerge that decimate their fallow and/or cropping systems. The example of the widespread promotion and subsequent collapse of *Leucaena leucocephala*-based systems in Southeast Asia because of the psyllid pest in the 1980s underscores the need to maintain both crop and fallow species diversity at the landscape scale.

The regeneration of woody species supplemented by enrichment plantings can lead to stable agroforestry systems that fulfill ecological functions, such as watershed protection or the establishment of biodiversity corridors, while being economically productive. We expect that the emerging science and practice of *payments for environmental services* will eventually result in many farmers using fallows not for soil fertility restoration and low-value staple crops but for higher-value ecosystem services and associated payments.

For example, the World Bank's BioCarbon Fund is already funding the retention on or reforestation of degraded lands with native trees and shrubs to sequester carbon (U.S.$ 2–5 t^{-1} CO_2 equivalent). The private sector is also paying land-users to improve the hydrological cycle with tree-based systems rather than continue pasture and annual crops (Fernandes, 2005). Fallow management is always location-specific, needing to suit and be adapted to soil, slope, and socioeconomic factors. Natural and managed fallows, which protect landscape integrity and farmland productivity as well as watershed service functions, will be increasingly important to both rural and urban communities as they adapt to climate change and the increasing frequency and severity of extreme weather events.

References

Agboola, A.A., A recipe for continuous arable crop production in the forest zone of western Nigeria, In: *Alternatives to Slash-and-Burn Agriculture: Proceedings of 15th International Soil Science Congress, Acapulco, Mexico*, Sanchez, P.A. and Van Houten, H., Eds., International Centre for Research in Agroforesry and International Society of Soil Sciences, Nairobi, 106, (1994).

Barros, E. et al., Development of the soil macrofauna community under silvopastoral and agrosilvicultural systems in amazonia, *Pedobiologia*, **47**, 273–280 (2003).

Boserup, E., *The Conditions of Agricultural Growth*, Earthscan Publications, London (1965).

Brand, J. and Pfund, J.L., Site and watershed-level assessment of nutrient dynamics under shifting cultivation in Eastern Madagascar, *Agric. Ecosyst. Environ.*, **71**, 169–183 (1998).

Brookfield, H., Working with plants, and for them: Indigenous fallow management in perspective, In: *Voices from the Forest: Farmer Solutions towards Improved Fallow Husbandry in Southeast Asia*, Cairns, M., Ed., International Centre for Research in Agroforestry, Southeast Asia Program, Bogor, Indonesia, 8–14 (2004).

Buckles, D. and Triomphe, B., Adoption of mucuna in the farming systems of northern Honduras, *Agroforest. Syst.*, **47**, 67–91 (1999).

Cairns, M., Conceptualizing indigenous approaches to fallow management: Road map to this volume, In: *Voices from the Forest: Farmer Solutions towards Improved Fallow Husbandry in Southeast Asia*, Cairns, M., Ed., Johns Hopkins University Press, Baltimore, MD, 15–32 (2004a).

Cairns, M., Ed., *Voices from the Forest: Farmer Solutions towards Improved Fallow Husbandry in Southeast Asia*, Johns Hopkins University Press, Baltimore, MD (2004b).

Feldpausch, T.R. et al., Carbon and nutrient accumulation in secondary forests regenerating on pastures in Central Amazonia, *Ecol. Appl.*, **14**, S164–S176 (2004).

Fernandes, E.C.M., Integrated water management to enhance watershed functions and to capture payments for environmental services, In: *Shaping the Future of Water for Agriculture: A Sourcebook for Investment in Agricultural Water Management*, Dinar, A. and Dargouth, S., Eds., World Bank, Washington, DC, 226–230 (2005).

Fernandes, E.C.M. et al., Use and potential of domesticated trees for soil improvement, In: *Tropical Trees: The Potential for Domestication and the Rebuilding of Forest Resources*, Leakey, R.R.B. and Newton, A.D., Eds., HMSO, London, 137–147 (1994).

Fernandes, E.C.M. et al., Management control of soil organic matter dynamics in tropical land-use systems, *Geoderma*, **79**, 49–67 (1997).

Fischler, M., Wortmann, C.S., and Feil, B., Crotalaria (*C. ochroleuca* g don) as green manure in maize-bean cropping systems in Uganda, *Field Crops Res.*, **61**, 97–107 (1999).

George, T.S. et al., Utilisation of soil organic P by agroforestry and crop species in the field, Western Kenya, *Plant Soil*, **246**, 53–63 (2002).

Hairiah, K., Introduction to part IV: Herbaceous legume fallows, In: *Voices from the Forest: Farmer Solutions towards Improved Fallow Husbandry in Southeast Asia*, Cairns, M., Ed., Johns Hopkins University Press, Baltimore, MD (2004).

Kwesiga, F.R. et al., *Sesbania sesban* improved fallows in eastern Zambia: Their inception, development and farmer enthusiasm, *Agroforest. Syst.*, **47**, 49–66 (1999).

Manley, R.J. et al., Relationships between abiotic and biotic soil properties during fallow periods in sudanian zone of Senegal, *Appl. Soil Ecol.*, **14**, 89–101 (2000).

Niang, A.I. et al., Species screening for short-term planted fallows in the highlands of Western Kenya, *Agroforest. Syst.*, **56**, 145–154 (2002).

Nye, P.H. and Greenland, D.J., *The Soil under Shifting Cultivation*, Commonwealth Agricultural Bureau, Farnham Royal, UK (1960).

Roder, W. and Maniphone, S., Shrubby legumes for fallow improvement in Northern Laos: Establishment, fallow biomass, weeds, rice yield, and soil properties, *Agroforest. Syst.*, **39**, 291–303 (1998).

Roder, W. et al., Fallow improvement in upland rice systems with *Chromolaena odorata*, In: *Voices from the Forest: Farmer Solutions towards Improved Fallow Husbandry in Southeast Asia*, Cairns, M., Ed., Johns Hopkins University Press, Baltimore, MD, 134–143 (2004).

Ruthenberg, H., *Farming Systems in the Tropics*, 3rd ed., Clarendon Press, Oxford, UK (1980).

Salako, F. and Tian, G., Litter and biomass production from planted and natural fallows on a degraded soil in Southwestern Nigeria, *Agroforest. Syst.*, **51**, 239–251 (2001).

Sanchez, P.A., Improved fallows come of age in the tropics, *Agroforest. Syst.*, **47**, 3–12 (1999).

Sarmiento, L. and Bottner, P., Carbon and nitrogen dynamics in two soils with different fallow times in the high tropical Andes: Indications for fertility restoration, *Appl. Soil Ecol.*, **19**, 79–89 (2002).

Sauerborn, J., Kranz, B., and Mercer-Quarshie, H., Organic amendments mitigate heterotrophic weed infestation in savannah agriculture, *Appl. Soil Ecol.*, **23**, 181–186 (2003).

Schlather, K.J., The dynamics and cycling of phosphorus in mulched and unmulched bean production systems indigenous to the humid tropics of Central America. Ph.D. dissertation, Department of Crop and Soil Sciences, Cornell University, Ithaca, New York (1998).

Styger, E., Fire-less alternatives to slash-and-burn agriculture (*tavy*) in the rainforest region of Madagascar. Ph.D. dissertation, Department of Crop and Soil Sciences, Cornell University, Ithaca, New York (2004).

Styger, E. et al., Indigenous fruit trees of Madagascar: Potential components of agroforestry systems to improve human nutrition and restore biological diversity, *Agroforest. Syst.*, **46**, 289–310 (1999).

Szott, L.T. and Palm, C.A., Nutrient stocks in managed and natural humid tropical fallows, *Plant Soil*, **186**, 293–309 (1996).

Szott, L.T., Palm, C.A., and Buresh, R.J., Ecosystem fertility and fallow function in the humid and subhumid tropics, *Agroforest. Syst.*, **47**, 163–196 (1999).

Tarawali, G. et al., Adoption of improved fallows in West Africa: Lessons from mucuna and stylo case studies, *Agroforest. Syst.*, **47**, 93–122 (1999).

Uhl, C. et al., Early plant succession after cutting and burning in the upper Rio Negro region of the Amazon basin, *J. Ecol.*, **69**, 631–649 (1981).

Yimyam, N., Rerkasem, K., and Rerkasem, B., Fallow enrichment with pada (*Macaranga denticulata* (bl.) muell. Arg.) trees in rotational shifting cultivation in Northern Thailand, *Agroforest. Syst.*, **57**, 79–86 (2003).

30

Green Manure/Cover Crops for Recuperating Soils and Maintaining Soil Fertility in the Tropics

Roland Bunch

World Neighbors, Oklahoma City, Oklahoma, USA

CONTENTS

Some 30 years ago, a number of agronomists around the world independently realized that chemical fertilizers had become a tremendous financial drain for smallholders. At the same time, they realized that composting takes more labor than will be invested in most extensive subsistence crops; and animal manure, at least among the poor, is usually unavailable or too limited to maintain soil fertility levels. Furthermore, there had been no significant adoption of conventional green-manuring practices by smallholders. So a search began for new, less expensive ways that smallholders could maintain or increase the fertility of their soils. This research focused on agroforestry systems (Chapters 19, 20 and 21) and in what are now referred to as green manure/cover crops (GMCCs). These systems have achieved, to a remarkable degree, their goal of offering farmers widely applicable systems that can maintain or increase soil fertility in the tropics where rainfall is adequate, at little or no cost.

30.1 Defining Green Manure/Cover Crops

The term "green manure/cover crops" does not refer to traditional green manures, which were leguminous plants grown as a monocrop to be cut down at flowering stage and incorporated into the soil before growing a main crop. Many attempts to introduce this method for providing organic matter to farms in the tropics have shown that this practice is not attractive to smallholders there.

The GMCCs discussed here are not necessarily leguminous species. Together with colleagues Milton Flores and Gabino Lopez, the author has compiled an inventory of 150 GMCC systems around the world, not yet published. We found that 9% of these systems do not involve legumes at all. Over 60% of the systems documented involve intercropping or relaying the legumes into traditional crops or planting them under tree crops. But almost all GMCCs are applied to the soil surface, usually *in situ*, rather than being incorporated into the soil. The plants are almost always applied to the soil after maturation, because the plants' seeds are valuable. Farmers want to eat the seeds, to sell them, or to feed them to their animals; and they almost always want to save at least some of the seeds for future planting. Thus, while the earlier green-manure concept of using plant material to fertilize the soil is still operative, most of the rest of the practices of traditional green manuring have been found to be inappropriate for smallholder farming systems and priorities. Thus, the definition used here for a green manure/cover crop is a species of plant, often but not always leguminous, whether a tree, bush, vine or crawling plant, that is used by farmers for multiple purposes, at least one of which is maintaining or improving soil fertility and/or controlling weeds.

While the database cited above includes some 150 GMCC systems, we know that many other systems have been observed around the world but not documented sufficiently for analysis and evaluation. Extrapolation from the various lists of GMCCs known to be in use suggests that there may be more than 500 GMCC systems in use in the tropical world today. Furthermore, many of these systems have spread from farmer to farmer fairly rapidly. The Mucuna–maize system developed in Mesoamerica, for instance, has spread spontaneously, i.e., without outside promotion, among some 20,000 farmers in four nations in a period of just 55 years. Smallholder farmers have developed more than 60% of these 150 systems. This shows how appropriate these systems can be for such farmers, and how interested smallholders themselves are in finding and adopting alternatives to chemical fertilizer.

30.2   Why GMCCs Work: Dynamics of Soil Nitrogen and Phosphorus

The two nutrients that most often limit crop growth in tropical soils are nitrogen and phosphorus. Therefore, a technology that will greatly increase the amount of N and P that plants can access will dramatically raise crop production in the majority of cases, assuming that the other factors of plant growth, such as the availability of water and sunshine, are also favorable.

Virtually all N in the soil is within the soil organic matter, either dead or alive, or as chemical fertilizer. The amount of N that must usually be available in the soil for good plant nutrition is about 120 kg of N crop^{-1} (Martin and Leonard, 1967). Many GMCC species such as Mucuna, jackbean (*Canavalia ensiformis*), alfalfa (*Medicago sativa*) and tarwi (*Lupinus mutabilis*) can fix much more than 120 kg ha^{-1} crop^{-1}. Some common legumes such as cowpeas (*Vigna unguiculata*), mungbean (*V. radiata*) and pigeon peas (*Cajanus cajan*) produce closer to about 80 kg ha^{-1}, and therefore, depending on local productivity levels, these may need supplementation with small amounts of inorganic fertilizer (Kilham, 1994).

The access that crops have to P in the soil is far more important than the total quantity of P in the soil. Why? Normally only 1–2% of all the P in a given soil is available to crops at any one time, and in poorer, often acidic tropical soils, the percentage available can be 0.3% or less. Large amounts of unavailable P are physically or chemically inaccessible (Chapter 14). If farmers can double or triple the percentage of P that is accessible to their crops, this will result in much more accessible P than if they add even a large amount of P to the soil, most of which will immediately become complexed and inaccessible to the crop.

The percentage of P accessible to plants in the same type of soil can easily vary by a factor of three or more (Cardoso, 2002). One of the most effective ways of making P more accessible to plants is by minimizing its contact with the soil, which will adsorb or otherwise immobilize it. In conventional agricultural practice, it is recommended that P fertilizer be "banded" — applied in a band so that much of the P is in contact with the rest of the inorganic fertilizer, not the soil. However, P accessibility is equally well served if the P is applied as part of some form of organic matter. P applied to the soil surface in an organic form is often 3 to 10 times more accessible to plants than the P already in the soil, or that which is applied as inorganic fertilizer.

There are other issues that need to be taken into account, of course. Synchronization of the applications of N over time with the crops' need for N will be addressed below. Whereas the above strategy for application of P will supply crops with more than enough P in the short term, over the long term it will be mining the P in the soil, unless biological processes mobilize significant amounts of P from otherwise unavailable sources in the soil. Some other kinds of organic amendments might be used to deal with this problem, e.g., purchased organic matter rich in P such as animal manure, coffee pulp, or sugarcane bagasse.

The easiest solution in many cases will be to apply replacement quantities of P in the form of inorganic fertilizer. By using GMCCs together with the application of small amounts of inorganic fertilizer (often one-quarter to one-third of the amounts normally recommended), farmers can sustainably maintain the fertility of their soils, producing grain yields around 3 ton ha^{-1}, above most current smallholder production levels. If farmers' productivity is still below 3 ton ha^{-1}, they may find that using more organic fertilization relative to inorganic forms can raise this.

30.3 Benefits of Using GMCCs

30.3.1 Increases in Yield

The increases in crop yields achieved by GMCCs vary widely, depending on a number of factors, from soil quality and pH to the species and practices used. Two of the most intensively studied systems, the Mucuna–maize system in Honduras, Guatemala and Mexico, and the tarwi (lupine)–potato system in Peru and Bolivia, provide fairly typical results. In Honduras, average maize yields throughout the country stand at about 850 kg ha^{-1}. However, average yields of maize after 10 years of using Mucuna remain around 2.5 ton ha^{-1}, with no use of chemical fertilizer, and 3.2 ton ha^{-1} when farmers also add about 70 kg of urea ha^{-1} yr^{-1} (Flores and Estrada, 1992). In Peru, experiments have showed at least a doubling of potato yields the very first year after the tarwi was plowed under, and often yields were tripled. In general, based on my own experience with some 40 such systems, smallholders who start with yields less than 1.5 ton ha^{-1} can expect, at a minimum, to double their crop yields within a few years.

30.3.2 Sustainability of Yield

Farmers using the Mucuna–maize system in northern Honduras are achieving, on average, slightly better yields after 40 years of using the system than do farmers who have only used the system for 10 years (Bunch and Kadar, 2004). None of the farmers studied was applying any inorganic P. In the half dozen systems that I have visited and assessed in which large numbers of farmers have used GMCCs for 15 years or more with no inorganic fertilizer, farmers have reported that they have observed no decline in their productivity.

30.3.3 Reduced Cost of Production

The net costs of GMCC systems vary tremendously, depending especially on the GMCC species used. With the Mucuna–maize system in northern Honduras, Flores and Estrada (1992) found that since this system has eliminated the need for soil preparation and has drastically reduced the need for weeding or fertilization, the total costs per ton of maize produced (including the cost of growing the Mucuna) were 29% lower than those for nearby tractor-based conventional farmers. This means that the Mucuna was improving the soil at the same time that it reduced total production costs. Fertilization of the land with *Mucuna*, rather than entailing added costs for farmers, was in fact saving farmers money. Such reduction in cost represents a net benefit.

This advantage of GMCCs is more pronounced where the species used controls major weed problems or allows a transition to zero-tillage. The net advantage of GMCCs will be enhanced even more when they also produce food for human consumption. Depending on agronomic and economic levels attained, GMCCs can increase smallholder yields, often dramatically, either with or without small added amounts of inorganic fertilizer, at very low or even reduced cost.

These examples suggest that GMCCs are not so much a substitute for inorganic fertilizers as a potential complement to them. In fact, the use of GMCCs often improves the effect of inorganic fertilizers — by supplying micronutrients not provided in NPK fertilizer, making fertilizer nutrients more soluble, maintaining soil moisture, reducing soil crusting and maintaining good soil structure — thereby making inorganic fertilizers more economically attractive. Even so, when the use of GMCCs entails some increased costs of production, inorganic fertilizers may still become unattractive by comparison.

30.3.4 Agronomic Advantages

30.3.4.1 Increased Organic Matter and Nutrient Recycling

GMCCs are capable of adding up to 50 ton ha^{-1} (green weight) of organic matter (OM) to the soil during each application. This OM has, in turn, a whole series of positive effects on the soil, such as recycling and pumping nutrients up to the soil surface and improving the soil's water-holding capacity, nutrient content, cation exchange capacity, nutrient balance, numbers of macro- and microorganisms, soil friability and pH.

30.3.4.2 Nitrogen Fixation

GMCCs can add significant quantities of N to farming systems. Many, if not most, of the widely used leguminous GMCCs are capable of fixing more than 75 kg N ha^{-1}, while a few GMCC species fix much more: velvet bean can fix 140 kg N ha^{-1} crop^{-1}, jackbean up to 240 kg-N, and tarwi (*Lupinus mutabilis*), fava beans (*Vicia faba*) and *Sesbania rostrata* are capable of fixing 400 kg N ha^{-1} or more (NAS, 1979; Tisdale et al., 1993). Even at just 140 kg ha^{-1}, this means that, even with considerable volatilization, farmers can add to their systems quantities of N that would cost them at least US\$ 75 ha^{-1} for equivalent inorganic fertilizer. This addition of both N and OM can increase soil fertility so significantly that GMCC programs in much of Latin America now speak of "soil restoration" and "soil recuperation," not just fertility supplementation.

30.3.4.3 Weed Control and Reduced Use of Agrochemicals

As already mentioned, GMCCs can also be an important factor in reducing weed control costs. By controlling weeds, they often decrease the labor requirements of various farming systems.

GMCCs significantly reduce the use (and expense) of agrochemicals, often cutting chemical fertilizer use by 60%, and usually reducing and in many cases eliminating the use of herbicides. Specific species can also substitute for other chemical uses: the velvet bean is a wide-spectrum nematicide, and sunnhemp (*Crotalaria ochroleuca*) can control grain storage pests. As reported in Chapter 15 with tomatoes, increased yield and delayed senescence as well as suppression of weeds and pathogens have been reported when using hairy vetch mulch, rather than black plastic mulch. If similar results can be achieved with other legume cover crops, they could be recommended to reduce the need for herbicides and other chemical biocides.

30.3.4.4 Soil Cover

The soil cover provided by most GMCCs can be very important for soil conservation, an importance that has generally been greatly underestimated. The kinetic energy of raindrops falling from the sky is far greater than that of an equal amount of water running down a hillside. Since the amount of erosion caused is closely related to the water's kinetic energy, good soil cover (plus an increased infiltration rate due to increased soil OM) can virtually eliminate water erosion. Smallholder experience confirms this fact. A study for the International Development Research Centre (IDRC) has shown that farmers monocropping maize on 35% slopes with a 2000 + mm rainfall in northern Honduras are actually increasing the productivity of their soil year by year, in effect experiencing "negative erosion," with the only soil conservation practices being their maintaining velvet bean on their fields for 10 months each year (Buckles et al., 1998).

30.3.4.5 Improved Soil Moisture

GMCC soil cover plus the increased infiltration and water-holding capacity of the soil brought about by the OM increases crops' resistance to drought. In one experiment in southern Honduras carried out during a drought season, maize fertilized with inorganic fertilizer died 1 month into the drought; maize fertilized with animal manure died about 2 weeks later; maize fertilized with jackbean still managed to produce a small harvest (Bunch and Kadar, 2004).

30.3.4.6 Zero Tillage

The experience of tens of thousands of farmers in Brazil, Paraguay, Argentina and Honduras shows us that after 1 to 4 years of heavy applications of OM from GMCCs, farmers can move to zero-till systems that retain very high levels of productivity with greatly reduced costs (Chapter 22; also Chapter 24). Farmers in northern Honduras, using velvet beans and no inorganic fertilizers, are maintaining yields of nonrotated, zero-tilled maize of over 2.5 ton ha^{-1}, and achieve yields of maize of 4 ton ha^{-1} with very small applications of urea. In Brazil, farmers using rotations and medium applications of inorganic fertilizer along with GMCCs, regularly harvest 7–8 ton ha^{-1} of maize without having tilled the soil in over 10 years (Monegat, 1991; Bunch, 1994).

30.4 Economic Advantages

30.4.1 No Transportation Costs

GMCC additions of OM and N, unlike for inorganic soil amendments, entail no transportation costs since they are produced in the field and are already well distributed.

30.4.2 Negligible Cash Investment

GMCCs require no capital outlay once the farmer has purchased his or her first few handfuls of seed. There are no other materials or inputs that need to be procured.

30.4.3 Competitiveness with Mechanized Production

Weeding and plowing are the two heavy operations that have previously provided a major advantage to those farmers capable of mechanizing their agriculture. Since GMCCs can often eliminate the need for both of these operations, they can give nonmechanized and/or hillside farmers a better chance of competing with their wealthier, mechanized competitors. In an age of falling trade barriers, this fact alone could justify the use of GMCCs among the world's resource-poor farmers, who need to minimize their costs of production in a globalized market.

30.4.4 Additional Benefits

Many GMCCs produce other benefits such as food, feed or salable commodities that improve farm households' nutritional status and incomes. To be sure, any of these uses will to some extent reduce the total amount of OM and nutrients being recycled onto the soil, thereby making them somewhat less valuable as soil amendments. Still, the farming system as a whole will usually be much more productive and efficient.

When the above advantages are compared with those of composting, in most cases GMCCs will be more attractive. The major exception would be where farmers are growing very high-value crops and/or have access to a very limited amount of land (less than 0.5 ha, for instance), so that the opportunity costs of growing a GMCC is a deterrent.

Each of the above advantages needs to be analyzed and assessed when choosing which GMCCs to recommend. It is rare that farmers will be primarily attracted by GMCCs' ability to increase soil fertility. Farmers are most often motivated by GMCCs' potential to add to their food supplies, usually the highest priority among the above advantages, or their ability to control weeds, given the arduousness of weeding by hand or the cost of using chemicals. Generally speaking, GMCCs should probably be promoted to farmers on the basis of these other advantages, referring to them as a food crop or a "green herbicide" rather than presenting them primarily as a strategy for soil fertility enhancement.

30.5 Disadvantages of GMCCs

In spite of these various advantages, GMCCs are sometimes difficult to introduce to farmers. There are some notable disadvantages to using GMCCs, and unless these are overcome, GMCCs will not be successfully or sustainably introduced.

30.5.1 Possible Opportunity Cost of Land

Farmers may be reluctant to expend effort on something that only fertilizes their soil where they could plant either a subsistence or cash crop. Unless the GMCC also produces food, the land used to grow GMCCs must have no obvious opportunity cost. The fact that traditional green manure systems typically used land that had an opportunity cost is probably the main reason why these have had very little uptake among smallholders. The systems discussed here are integrated in time and/or space with other crops so as to minimize opportunity costs.

30.5.2 Sometimes Slow Results

The improvement of soil systems is often a long-term process, with results not immediately evident to farmers. Usually, significant improvement in productivity does not occur until after the first GMCC crop has been applied to the soil, which means that visible results, i.e., "recognizable success," are not apparent until well into the second cropping cycle. Delayed appearance of results, which when they come may not be evidently attributable to GMCCs, complicates their adoption. This is another reason why it is usually better to promote GMCCs for some reason other than soil fertility. If farmers are not aware of the value of soil OM, one can in the first year also make a heavy application of animal manure so that the farmers will better understand the benefits of OM.

30.5.3 Dry Season Problems

Often GMCCs must produce their OM at the end of the wet season or must continue to grow during the dry season. Grazing animals, wild animals, termites, agricultural burning or bush fires and several other problems may destroy the biomass before farmers can utilize it in the following rainy season. In fact, in very hot climates where there is no shade, most of the nitrogen and OM will be burned off if the biomass is lying on top of the soil for 6 months or more. This problem can be avoided by having shade, using GMCC species that grow well into or completely through the dry season, or growing GMCCs whose biomass is utilized during the same rainy season.

30.5.4 Difficult Growing Conditions

Extremely low or irregular rainfall, extremes in soil pH, severe drainage problems, or a combination of these problems that are all too common on the farms of resource-poor farmers will reduce the growth of GMCCs, thereby reducing their impact. Through the years, ways have been found to overcome many such problems, often by using GMCC species that are resistant to certain problems (assessed in Table 30.1). But such solutions are often achieved at the cost of reduced biomass production, fewer additional benefits, or a lesser number of niches in which the GMCCs can fit within local farming systems.

30.5.5 Synchronization

GMCCs will boost farmers' productivity only when the provision of these nutrients, especially N, is reasonably well synchronized with those periods when crops most need the nutrients. In some GMCC systems, this synchronization is either impossible or very difficult to achieve. In such cases, natural foliar sprays can be used, e.g., solutions of crushed *Gliricidia sepium* leaves or of cattle urine or small amounts of inorganic fertilizer, to supply nutrients at critical times when the crops' needs are greater than the amount of nutrients that the GMCC is supplying.

30.6 Factors Affecting the Adoption of GMCCs

There has been a great deal of discussion about the levels of adoption or abandonment of GMCC systems by smallholders in different countries. From the survey referred to in section 30.1, we have identified at least 90 GMCC systems around the world that have been developed by smallholders themselves and that are operating successfully. We infer from this that smallholders will adopt and maintain such systems if these meet perceived needs. Many traditional GMCC systems have been given up in recent decades as Green Revolution technologies have been promoted, as traditionally produced and consumed foods have become unfashionable, as often-subsidized inorganic fertilizer has become more widely available, and as extension personnel have disparaged farmers for doing things not considered as "modern." This gradual process of abandonment has apparently affected systems like the Vigna–maize systems that once extended from Mexico through Nicaragua, and the scarlet runner bean–maize system that once reached from the northern United States down through Bolivia.

Nevertheless, many GMCC systems have spread widely and quickly, right up to the present time. One of the major examples that seemed to show that GMCC systems were being disadopted by farmers was the Mucuna–maize system in northern Honduras (Neill and Lee, 2000). However, a recent study that examined this system throughout Mexico, Guatemala and Honduras has found that it is spreading, in western Honduras, in the Guatemalan Peten, and in eastern Mexico as much as it is receding in other parts of the region (Bunch and Kadar, 2004). This varying experience brings up the question: what makes some GMCC systems more popular than others? Most of the programs that have successfully introduced sustainable GMCC systems have conformed to the following rules of thumb (Bunch, 1995).

30.6.1   Cost

GMCCs should be low-cost, and in particular, the land used must entail no opportunity cost. This rule sounds very limiting, but there are many ways and places where GMCCs can be introduced without increasing farmers' direct or indirect costs or with compensating benefits. Of course, if the GMCC produces a valued food, it can be grown in any way that fits into the farming system like any other equally valued crop, and its benefits offset the costs of growing the GMCC. Examples of ways to avoid opportunity costs include the following.

1. The GMCC can be grown intercropped with another food, such as jackbean with maize or cassava, or perennial groundnut (*Arachis pintoi*) with coffee. This is presently the most popular niche for introduced GMCC systems.

2. The GMCC can be grown on wasteland for the first year or two of a fallow, making it an improved fallow, e.g., in Vietnam, broadcasting *Tephrosia candida* seeds into the first-year fallow reduces the normal 5-year fallow to just 1 or 2 years.

3. The GMCC can be grown during the dry season, either relayed into normal rainy season crops, such as the cowpea/maize and Lablab/maize systems in northern Thailand; planted after the normal crops, as in the ricebean (*Vigna umbellate*)–rice system used in Vietnam; or intercropped with the usual crop and then allowed to continue growing through the dry season, e.g., the sweet clover (*Melilotus albus*)–maize system in Oaxaca state of Mexico.

4. The GMCC can be grown under fruit or forest trees or almost any perennial crop. In this case, one needs highly shade-resistant GMCCs like jackbeans, perennial groundnut or *Centrosema pubescens*.

5. Other small, occasional niches can be found, such as during periods of frost (when tarwi does very well), in extremely acid soils (velvet bean or buckwheat), or during very short periods of time (*Sesbania rostrata*).

30.6.2   Out-of-Pocket Expense

GMCCs should require little or no cash expenditure. This means that farmers must be able to produce their own seed year after year, and the GMCCs must have no disease or insect problems that significantly slow down their biomass production. Also, there should be no need to use inoculants. In the event that an insect or disease does become a major problem, it is probably best to discard the affected GMCC and use some alternative species.

30.6.3   Labor Requirements

GMCCs should require little or no additional labor. This means that except where animal traction or tractors are available, GMCCs will have to be applied to the soil surface rather than being incorporated into the soil. It also means that the intercropping of GMCCs is particularly advantageous because the resulting reduction in expenditures for weed control can offset the labor that must go into planting and cutting down the GMCC. Farmers' need to minimize their labor requirements makes GMCCs' ability to move farmers to a zero-till system an important consideration. Farmers can often be motivated to plant GMCCs by the prospect of no longer having to plow or hoe their fields.

30.6.4   Fit with Existing Farming Systems

GMCCs will be seen, at least for the first few years, as much less important than food or cash crops. Thus the GMCCs will have to be adjusted to fit into the already-existing farming system, rather than the other way around.

30.6.5   Non-Soil Benefits

The GMCC chosen should provide at least one major benefit other than improving the soil. A worldwide survey of introduced GMCC systems found a very high correlation between those systems that had lasted long after the introducing organization had left the area and systems that could produce definite benefits other than soil improvement alone. Thus, whenever possible, the GMCC species proposed should be one that can be eaten, fed to animals, and/or provide some other benefits for which a strong felt need exists among farmers.

30.6.6   Appropriate Species

The GMCC species used should fit the available biophysical niche(s) as well as possible. In general, desirable GMCC species establish themselves easily and grow vigorously under local conditions. They are able to cover weeds quickly and either to fix nitrogen or to concentrate phosphorus in significant amounts. They should be resistant to insects, diseases, grazing animals, bush fires, droughts, or any other problem they may have to face within the particular system. They should also have multiple uses and should produce viable seeds in sufficient quantities for future planting. If the GMCCs are to be intercropped, they should withstand shade and fit in with the cycle of the main crop(s).

Great care should be used not to introduce GMCCs into new areas where they might become pests. The most dangerous species in this respect seem to be creeping perennials. Common kudzu (*Pueraria lobata*) is famous as a potential pest, having long ago "worn out its welcome" in the southern United States. Tropical kudzu (*Pueraria phaseoloides*) has also brought some complaints from farmers, as have the perennial groundnut and perennial soybean.

We have learned, while trying to apply these rules in different situations around the world, that finding acceptable, widely adoptable systems for (or preferably, with) farmers requires a great deal of flexibility and creativity. No textbook or computer program can yet prescribe which technology should be used in each circumstance. The experience and perspective of local farmers should be sought and respected in working out prospective GMCC innovations, and reliable evaluations need to await a number of years of actual field results.

GMCCs have become very useful for large-scale farmers who have as much as 100,000 ha in Brazil. Among smallholders, GMCCs tend to be most useful for farmers who have access to between 0.5 and 10 ha. Farmers who have more than 10 ha per household can still use shifting agriculture in ways that do not destroy their soils; it can be difficult for a GMCC technology to compete with such a system in returns to labor. For farmers under 0.5 ha per household, the use of the land is often (although not always) so intense that there may be no niches for which the opportunity cost is low. Small-scale paddy farmers tend to fall in this category. In such cases, farmers are often better off making compost or buying soil amendments, organic or inorganic.

Among the remaining smallholders — most of whom have between 0.5 and 10 ha — niches for GMCCs can generally be found. Generally, the most successful approach is first to observe the local farming systems, looking for the above-mentioned niches. In the absence of such possibilities, one can try growing the GMCCs during the drier seasons or as "fallow improvers" so that farmers can begin planting again within a year or two instead of waiting 4 or 5 years or more. Another important niche for the use of GMCCs is in the recuperation of lands taken over by Imperata grass (*Imperata cylindrica*).

30.7 GMCC Species Most Used

Some 41 different species are used in the 150 documented systems referred to above. The Brazilian research institute EPAGRI reports the use of over 60 species of GMCC in the state of Santa Catarina alone, where there are less than a dozen of the systems on the list. Furthermore, a good number of known GMCC tree species are not included in the inventory. Thus, the total number of GMCC species already in use around the world is probably close to 100. Many of these systems are described in Monegat (1991) and Calegari et al. (1993). The ones most widely used today are the following.

- **Scarlet Runner Bean (*Phaseolus coccineus*).** This legume is grown in parts of all the major highland areas from northern Mexico to southern Bolivia, as well as in limited areas of highland central Africa and Southeast Asia (Van der Maesen and Somaatmadja, 1992). It is mostly grown intercropped with maize and produces an edible, tasty bean. Many farmers have grown maize and scarlet runner beans on the same land for 20 years straight with no chemical fertilizer or any visible diminution in productivity.
- **Pigeon Peas (*Cajanus cajan*), Common Beans (*Phaseolus vulgaris*), Soybeans (*Glycine max*) and Oats (*Avena* spp.).** These are more widely grown than any other GMCC species because they have major importance as commercial and subsistence crops. It is not known to what extent they are grown by smallholders at least partly because they help fertilize the soil and/or control weeds.
- **Velvet Bean (*Mucuna*).** This is probably the GMCC species that has been most heavily promoted by development programs. In Mesoamerica, Brazil and West Africa, this species has been successfully introduced to more farmers than any other, even though the Mucuna bean is not usually consumed by humans (Flores et al., 2002). Its wide spread is probably a result of professionals' lack of attention to other GMCCs that are food-producing; its very aggressive weed control behavior, which is especially important in West Africa where it is grown largely to control Imperata grass; and possibly the fact that it effectively controls nematodes and several plant diseases, including *Phytophthora* and *Rhizoctonia* root rots.
- ***Vigna* spp.** This very popular, often overlooked genus of GMCCs includes mung-beans, green beans, cowpeas and the ricebean. These species are all tasty, easily grown and drought-resistant.
- **Jackbean (*Canavalia ensiformis*).** This is the second most widely used introduced GMCC. Along with its cousin, the swordbean (*Canavalia gladiatus*), this species is very useful because it is capable of growing well under the worst of conditions. Jackbean is extremely resistant to drought, very low pH, insects and diseases. Thus it can often be introduced during the dry season, in very marginal environments where crops will not grow, and for recuperating wastelands. It is capable of fixing up to 240 kg N ha^{-1}, withstands heavy pruning, and can be intercropped with anything from maize, cassava and sorghum to tomatoes and chili peppers. Table 30.1 summarizes the capabilities and uses of these and some other common GMCC species.

TABLE 30.1

Characteristics of the Most Commonly Used Green Manures/Cover Crops

Common Name	Scientific Name	Resistance to Shade	Resistance to Poor Soil	Resistance to Drought	N-Fixation	Erect or Climbing	Annual or Perennial	Eaten by Humans	Control of Weeds	Other Uses
Velvet bean	*Mucuna* spp.	3	3	3	140 kg ha^{-1}	Climbing	Both	Only with processing	4	Medicine, animal feed
Jackbean	*Canavalia ensiformis*	4	4	4	240 kg ha^{-1}	Both	Perennial	Only tender pods	3	None
Lablab bean	*Lablab purpureum* or *Dolichos lablab*	3	1	4	130 kg ha^{-1}	Both	Perennial	Yes, pods, green or dry seeds	3	Animal feed, especially dry season
Cowpea	*Vigna unguiculata*	3	3	Some vars. 4	80 kg ha^{-1}	Both	Annual	Yes, pods and seeds	3	None
Rice bean	*Vigna umbellate*	3	3	3	80?	Both	Perennial	Yes, very good taste	2	None
Mung bean	*Vigna radiata*	3	2	2?	80?	Both	Annual	Yes	2	Bean sprouts, poultry feed
Pigeon pea	*Cajanus cajan*	3	3	4	70 +	Erect	Perennial	Yes	2	Animal feed
Tephrosia	*Tephrosia vogelii* or *T. Candida*	2	4	4	?	Erect	Perennial	No!	2	Insecticide
Sunn hemp	*Crotalaria ochroleuca*	3	3	3	?	Erect	Annual	No	2	Insecticide for stored grains

Key: 4 = extremely good, 3 = good, 2 = fair, 1 = poor.

30.8 Discussion

This chapter has covered only the best-known and most widely used species. This is an area of agronomic practice where knowledge is still developing, and researchers have only recently started to evaluate species and farming systems performance. Probably in the next decade, more species and more benefits, as well as possible disadvantages, will become better known.

The fact that GMCCs are already widely grown in the tropics by small farmers — perhaps 100 species are used in an estimated 500 + systems — shows that resource-poor farmers have already been searching for alternatives to dependence on inorganic fertilizer and have in many parts of the world found and adopted beneficial alternatives. Nevertheless, a tremendous amount of work must be done to find additional species and systems, as well as to learn about and disseminate those that are already known.

The potential of GMCCs as well as the need for them is evident. Deteriorating shifting agriculture systems are destroying soils, reducing the area under forests, and even contaminating the air. At the same time, hundreds of millions of people now depend on inorganic fertilizers, the price of which is already prohibitive for tens of millions of smallholders. The chance of such fertilizer becoming less costly in the future is small. Moreover, increased use is likely to have some adverse effects on water quality through accumulation of nitrates. The desirability of having alternative methods for enhancing soil fertility is obvious.

Evaluations by persons who have worked with GMCCs suggest that around 60% of the world's smallholder farmers could benefit from the use of GMCCs. If productive, reliable and profitable alternatives to inorganic fertilization do not become available, so that these farmers can continue to produce food at least at present levels and if possible higher levels, hunger and malnutrition will become more serious scourges than they are at present.

References

Buckles, D., Triomphe, B., and Sain, G., *Cover Crops in Hillside Agriculture: Farmer Innovation with Mucuna*, International Development Research Centre (IDRC)/International Maize and Wheat Development Center (CIMMYT), Ottawa/Mexico City (1998).

Bunch, R., EPAGRI's work in the State of Santa Catarina, Brazil: Major new possibilities for resource-poor farmers, unpublished report (1994).

Bunch, R., *The Use of Green Manures by Villager Farmers: What We Have Learned to Date*, 2nd ed., Technical Report No. 3. Cover Crops International Clearinghouse (CIDICCO), Tegucigalpa, Honduras (1995).

Bunch, R. and Kadar, A., The adoption, adaptation and disadoption of Mucuna systems in Mesoamerica, *ILEIA Newslett.*, **20**, 16–18 (2004).

Calegari, A. et al., *Adubação Verde no Sul do Brasil*. Assesoria e Servicos a Projets em Agricultura Alternativa, Rio de Janiero (1993).

Cardoso, I., *Phosphorus in Agroforestry Systems: A Contribution to Sustainable Agriculture in the Zona de Mata of Minas Gerais, Brazil*, Wageningen University, Wageningen, The Netherlands (2002).

Flores, M. and Estrada, N., Estudio de caso: La utilizacion del frijol abono (*Mucuna* spp.) como alternativa viable para el sostenimiento productivo de los sistemas agricolas del Litoral Atlantico, Paper presented to Center for Development Studies, Free University of Amsterdam (1992).

Flores, M. et al., Eds., *Food and Feed from Mucuna: Current Uses and the Way Forward: Proceedings of an International Workshop*, CIDICCO, Tegucigalpa, Honduras (2002).

Kilham, K., *Soil Ecology*, Cambridge University Press, Cambridge, UK (1994).

Martin, J. and Leonard, W., *Principles of Field Crop Production*, 2nd ed., Macmillan, Toronto (1967).

Monegat, C., *Plantas de Cobertura del Suelo: Características y Manejo en Pequeñas Propiedades*, CIDICCO, Tegucigalpa, Honduras (1991).

NAS, *Tropical Legumes: Resources for the Future*, National Academy of Sciences, Washington, DC (1979).

Neill, S. and Lee, D.R., Examining the adoption and disadoption of sustainable agriculture: The case of cover crops in Northern Honduras, *Econ. Dev. Cult. Change*, **49**, 793–820 (2000).

Tisdale, S.L., Nelson, W.L., and Beaton, J.D., *Soil Fertility and Fertilizers*, 5th ed., Macmillan, New York (1993).

van der Maesen, L. and Somaatmadja, S., Eds., *Plant Resources of South-East Asia, 1: Pulses*, Plant Resources of Southeast Asia (PROSEA), Bogor, Indonesia (1992).

31

Compost and Vermicompost as Amendments Promoting Soil Health

Allison L.H. Jack[1] and Janice E. Thies[2]

[1]*Department of Plant Pathology Cornell University, Ithaca, New York, USA*
[2]*Department of Crop and Soil Sciences, Cornell University, Ithaca, New York, USA*

CONTENTS

Before the advent of modern industrialized agriculture, farmers relied almost entirely on raw and composted animal manures and agricultural residues as soil fertility amendments. Now, post-Green Revolution, scientists are shifting their focus away from agricultural systems that rely on synthetic fertilizers and pesticides, toward systems that incorporate the use of composted organic materials. A growing understanding of the complex ecological mechanisms producing the observed soil and plant growth benefits of traditional soil amendment methods prompts this. Practitioners, researchers and entrepreneurs in both developed and developing nations are experimenting with novel composting technologies, using a wide variety of organic materials, and reporting positive results (Bailey and Lazarovits, 2003; Arancon et al., 2003; Bhadoria and Prakash, 2003).

The current emphasis on composting as a means to stabilize manure comes from increasing public concern over nutrient run-off and the eutrophication of aquatic ecosystems associated with the over-application of raw animal manures to soils. Composting animal manures and plant residues can increase their bulk density, kill weed seeds, and decrease the levels of human and plant pathogens. However, the composting process can lead to an overall loss of nutrients compared to the starting material (Sommer, 2001). This potential drawback could be outweighed by the increased benefits to plants and the environment, which are beginning to be quantified and understood. For example, a recent field study comparing the effects of fresh vs. thermally composted swine manure on the growth and yield of maize showed a 10% increase in yield, and up to a 15% increase in aboveground biomass for plants receiving the composted manure treatment (Loecke et al., 2004). In this study, treatments were normalized for total nitrogen (N) content, so the increased yield could not be explained by a difference in the overall quantity of N applied, although differences in the chemical form of N might have affected plant availability. Soil physical and biological factors, including the microbial communities that the composted material benefited, were probably responsible for the increased yield in this case.

This chapter presents an overview of the current understanding of how both traditional compost and the recently popularized vermicompost affect plant growth and overall plant function. Experience from three countries in different parts of the world is reviewed to illustrate the beneficial use of compost in sustainable agricultural systems.

31.1   Definitions of Traditional Compost and Vermicompost

There are two major categories of compost: those that incorporate a biologically driven heating phase and those that do not. The first category will be referred to here as traditional compost, while vermicomposting will be used as the primary example of the latter category (Figure 31.1).

Traditional compost is the stabilized product of the decomposition of plant and animal residues at high temperatures (40–70°C) by the activity of thermophilic (heat-loving) microorganisms. The traditional composting process involves an initial stage conducted at moderate temperatures (10–40°C), during which labile organic matter is rapidly consumed by mesophilic microorganisms, followed by a stage when thermophilic

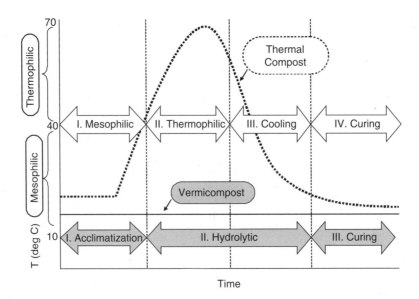

Time

FIGURE 31.1

Theoretical time vs. temperature curves for thermogenic (thermal) compost and vermicompost. Arrows represent major phases in the composting process. Phases for thermal compost are adapted from Chefetz et al. (1996) and those for vermicompost are adapted from Benitez et al. (2000).

microorganisms drive the temperatures up to 60°C at which point lipids, proteins and complex carbohydrates are consumed and broken down. During the final curing stage, when the material cools down, mesophilic organisms are able to recolonize and break down the remaining recalcitrant organic matter (Chefetz et al., 1996).

The vermicomposting process has been described as "biooxidation and stabilization of organic material involving the joint actions of earthworms and (mesophilic) microorganisms" (Aira et al., 2002). In contrast to traditional compost, vermicompost never heats up much above ambient temperatures. Many feedstocks can be used directly in vermicomposting systems; however, animal manures can drive temperatures into the thermophilic range during decomposition, which can kill compost earthworms. Many practitioners combine the two techniques, initially doing a partial precomposting at high temperatures followed by a finishing stage of vermicomposting (Frederickson et al., 1997). The phases in vermicomposting involve a period where worms acclimatize to the substrate that they are placed in. There is a hydrolytic phase in which readily degradable organic matter is broken down, and then a curing phase where the more recalcitrant organic matter is broken down (Benitez et al., 2000). More information on the vermicomposting process is provided in Aranda et al. (1999).

Microbes living in traditional compost effectively go through a selection event during the heating phase where the material is populated by only specially adapted thermophilic bacteria, most of which are as of yet uncultured (Dees and Ghiorse, 2001). The microbial community in finished traditional compost is derived from microbes that are facultative thermophiles and can survive in mesophilic temperatures, surviving the hot phase by forming spores or recolonizing during the mesophilic curing stage. Vermicompost, on the other hand, maintains a wide diversity of organisms throughout the entire process, including saprophytic bacteria and fungi, protozoa, nematodes and microarthropods. There is evidence that the compost worm *Eisenia fetida* has unique indigenous gut-associated microflora (Toyota and Kimura, 2000) which could be contributing to the microbial community in cured vermicompost.

Earthworm fecal pellets, or casts, are different from traditional compost in that they are covered with a mucus layer generated by the worms' intestinal tract. This layer provides a readily available source of carbon for soil microbes and leads to a flush of microbial activity in freshly deposited casts. The effects of earthworm-derived polysaccharides on soil microbial populations are reviewed in Brown et al. (2000). Polysaccharides of plant, bacterial, fungal origin have been shown to enhance soil aggregate stability (Oades, 1984). Several recent studies have investigated the effect that polysaccharides of earthworm origin may have on the physical aspects of soil as well as the documented effects on soil microbiota (Albiach et al., 2001; Ge et al., 2001).

The microbial communities in traditional compost and vermicompost are thought to be distinct because of the differences between the two processes just described. However, many of the studies carried out so far have used composts made from different feedstocks (raw organic materials). This makes it difficult to determine whether the resulting microbial community differences are due to the respective composting processes or are an artifact of the starting materials. A culture-based study using Biolog plates (Biolog Inc., Hayward, CA, USA) found that microbial populations in vermicomposts were able to metabolize a greater number of single C substrates than microbial populations in traditional composts made from different feedstocks (Atiyeh et al., 2000b). This implies a greater metabolic diversity in the culturable bacterial communities in vermicompost.

It should be pointed out, however, that culturable soil microbial communities represent only a small fraction of the total communities present, so other research techniques are needed to access a greater proportion of the total microbial community (Hill et al., 2000). Several separate studies using molecular techniques to track microbial communities have shown that traditional composts and vermicomposts have strikingly different taxonomic groups present; however, the composts used in these studies were prepared from different types of feedstocks (Alfreider et al., 2002; Verkhovtseva et al., 2002; Schloss et al., 2003).

The entire field of soil microbiology is rapidly expanding as new techniques provide greater insight into the structure and function of soil microbial communities. Our understanding of compost microbiology should increase greatly as researchers use more of these techniques to assess the differences between traditional and vermicompost and between composts made from different feedstocks, relating this information to their overall effects on soil and plant function.

31.2 Review of Research Findings

31.2.1 Historical Overview

The oldest recorded mention of the use of manure in agriculture is on clay tablets from the Akkadian Empire of the Mesopotamian Valley nearly 4500 years ago, and both the Greeks and the Romans wrote about compost (Rodale, 1960). Sir Albert Howard is widely considered the father of modern scientific composting through his pioneering research and international work in the early 1900s in India where he helped develop the Indore method of compost production (Howard, 1943). Rodale, champion of organic farming and founder of the Rodale Institute in Kutztown, PA, USA, continued Howard's work and published several classic, still relevant texts on composting in the USA (Rodale, 1945, 1960).

Traditional compost has been widely used as a potting mix amendment in the horticulture, turf, landscaping and nursery industries for decades (Roe, 1998), but it is rarely used on field crops in industrialized nations. Very little recent scientific information is available on the large-scale field application of composts, although some research is beginning to appear on this topic, most notably in developing nations such as India, as discussed later in section 31.3.

Although the 1960 edition of Rodale's book of composting contains an entire chapter on home and farm production of earthworm compost, vermicomposting on a commercial scale is a more recent practice. Large-scale vermicompost production was pioneered by Clive Edwards who worked on developing the continuous flow-through reactor design at the Rothamsted Experimental Station in the 1970s, and the industry has been growing ever since. Edwards has contributed greatly to the scientific literature on vermicompost production and use (Edwards et al., 1984; Edwards and Neuhauser, 1988) as well as to general earthworm ecology (Edwards, 2004).

In the United States, vermicompost is most commonly used as a potting mix amendment for cultivating vegetables, ornamentals and seedling starts. It is primarily marketed to home gardeners in small, highly packaged quantities through horticulture catalogues and retailers. Prices run up to 10 times those for traditional compost, which leads to the image of vermicompost as a "luxury" soil amendment. However, in developing nations, both traditional compost and vermicompost are applied to field soil at rates of several t ha^{-1} in an effort to decrease dependence on synthetic fertilizers and pesticides and to replace organic matter lost through years of intensive conventional farming.

31.2.2 Plant Growth Promotion: Integrated Nutrient Management

When traditional composts and vermicomposts have been compared directly with respect to their chemical characteristics, vermicompost has been shown to be a potentially better growth medium amendment than traditional compost based on such analyses (Vinceslas-Akpa and Loquet, 1997; Chaoui et al., 2003). In one study, woody plant materials were composted using both thermal composting and vermicomposting, and the chemical composition of the organic matter in the resulting composts was compared using NMR spectroscopy (Vinceslas-Akpa and Loquet, 1997). The organic matter in the vermicompost was more humified and had undergone more lignolysis compared to the traditional compost.

This result indicates that vermicompost is a more stable form of organic matter than the thermal compost, and thus less likely to immobilize N when used as a growth medium amendment. In another study, vermicompost was shown to be a slower-release fertilizer than traditional compost, giving an extended release of N over time, which reduces the risk of nutrient leaching in container-based systems (Chaoui et al., 2003), although it is not known if the composts used in this study were made from the same type of feedstock. In the same study, vermicompost also had a lower risk of causing salinity stress in plants when compared to traditional compost.

Some of the more interesting effects of both traditional and vermicompost on plants are nonnutrient related and point to the possibility that using composts may be a sustainable way to increase biological soil health. Experiments have shown that vermicompost in both solid and liquid forms promotes plant growth (Buckerfield et al., 1999) and enhances seedling germination (Ayanlaja et al., 2001). With all mineral nutrients held equal, the addition of as little as 10–20% vermicompost to the growth medium resulted in significant increases in plant biomass and yield in greenhouse tomatoes (Atiyeh et al., 1999) and marigolds (Atiyeh et al., 2002b). More compost is not necessarily better in this case, as plants grown in 100% vermicompost were significantly smaller than the controls (Atiyeh et al., 2000a).

Subsequent field studies carried out by the same research group found that 5–10 t ha^{-1} applications of vermicompost significantly increased marketable yields of pepper, tomato and strawberry when compared to synthetic fertilizers (Arancon et al., 2003). Plant growth promotion depended on the type of feedstock used; vermicomposted cow manure consistently gave the highest yields compared with food-waste and paper mill-waste vermicomposts (Arancon et al., 2003). A subsequent greenhouse comparison of traditional vs. vermicomposts found an inconsistent growth-promoting effect compared to a Metro-Mix control (Atiyeh et al., 2000b). However, the same study did find a significant growth-enhancing effect when 20% vermicompost was added to soil and used to grow raspberry plants. Vermicomposted pig manure solids gave the highest plant biomass when compared to vermicomposted food waste and to traditionally composted biosolids, leaf waste, yard waste, bark, and chicken manure (Atiyeh et al., 2000b). No clear conclusions about the plant growth-promoting effects of different composting processes (thermal and vermicompost) can be made until composts made from the same feedstock are compared.

Humic acids extracted from vermicompost have been shown to promote plant growth (Atiyeh et al., 2002a) as well as physiological changes in plant roots including a greater number of sites of lateral root emergence and greater total root area (Canellas et al., 2002). One of the ways that humic acids are thought to enhance plant growth is by binding plant-growth hormones present in the soil. The contributions of phytohormones to plant growth and health have been discussed in more detail in Chapter 14. Auxin (indole acetic acid, or IAA) has been extracted from traditional compost (Garcia Martinez et al., 2002) and has been detected in humic acid complexes in vermicompost (Canellas et al., 2002). Auxin is present in the soil solution partly due to the action of certain groups of rhizobacteria referred to as plant growth-promoting rhizobacteria (PGPR). These are soil bacteria that aggressively colonize plant roots and have been shown to promote plant growth and suppress plant disease. PGPR that produce auxin-like compounds have been isolated from traditional composts (De Brito Alvarez et al., 1995) and are thought to contribute to the increased plant growth observed with compost amendments.

In addition to bacteria that secrete plant-growth hormones, PGPR include associative nitrogen-fixing bacteria and phosphate-solubilizing bacteria, considered in Chapters 12 and 13. Associative nitrogen-fixing bacteria such as Azospirillum spp. do not enter the plant root or form root nodules as in the classic Rhizobium–legume symbiosis, but they fix atmospheric N_2 into cellular amino acids, which are subsequently mineralized into plant-available N forms (ammonium and nitrate) upon cell death. Phosphate-solubilizing bacteria convert mineral forms of P into soluble, plant-available forms as discussed in Chapter 13. Although research in this field is expanding, there has been limited commercialization of PGPR as biofertilizers. Finding strains that will perform consistently in different plant cultivars and under different field conditions is an ongoing challenge, but there is great potential for the use of PGPR to help reduce dependence on synthetic fertilizers (Vessey, 2003).

Other microorganisms that are associated with the plant-growth promotion effect of composts are the arbuscular mycorrhizal fungi (AMF). As discussed in Chapter 9, AMF colonize plant roots and form a symbiotic relationship that facilitates plant nutrient uptake in exchange for photosynthetically-derived carbon (C). AMF colonization of rice (Kale et al., 1992) and sorghum (Cavender et al., 2003) was found to increase significantly with vermicompost applications, and AMF colonization has also been correlated with traditional compost applications (Tarkalson et al., 1998), although this relationship has not been studied in depth. In the case of sorghum, plant growth was actually found to decrease with vermicompost applications, even though there was an increase in AMF colonization. This could be due to the plants not being nutrient-limited during the experiment, so the AMF colonization was more parasitic than beneficial (Cavender et al., 2003).

Both composts and biofertilizers can be used in Integrated Nutrient Management (INM) schemes, where lower rates of synthetic fertilizers are applied in conjunction with use of organic or biological amendments. In a recent field study in India, researchers found that providing rice with 50% of its N requirement from synthetic fertilizer and 50% from vermicompost, with the addition of *Azospirillum lipoferum* and *Bacillus megaterium* var. *phosphaticum* as biofertilizers, could increase yields up to 15% compared with the control (receiving 100% of the recommended dose of synthetic fertilizer N) (Jeyabal and Kuppuswamy, 2001). Since organic amendments and biofertilizers were not tested separately, it is impossible to draw conclusions on how much of the observed effect was due to the vermicompost or the manure treatments alone. Further evidence of the benefits from INM is found in a greenhouse study from Venezuela where vermicomposted cow manure and coffee pulp composed up to 45% of the potting medium for papaya plants. Plant growth in treatments with a mixture of intermediate amounts of synthetic N fertilizer and vermicompost was higher than in treatments with the full rates of one or the other source of nutrients (Acevedo and Pire, 2004).

31.2.3   Plant Disease Suppression

Traditional composts have widely documented plant disease-suppressive properties and are a common component of potting media used in greenhouse production systems (Hoitink and Fahy, 1986; Hoitink and Kuter, 1986; De Ceuster and Hoitink, 1999). Liquid preparations of compost, known as compost teas, have also been shown to prevent various plant diseases (Weltzien, 1992; Scheuerell and Mahaffee, 2002). Less widely known are the disease-suppressive qualities of vermicompost. The addition of vermicompost to growth media has resulted in the significant suppression of the following plant diseases: damping off (Pythium, Rhizoctonia) (Chaoui et al., 2002); wilts (Verticillium) (Chaoui et al., 2002), Fusarium (Szczech et al., 1993); root rot (Phytophthera) (Szczech et al., 1993; Szczech and Smolinska, 2001); club root (Plasmodiophora) (Szczech et al., 1993; Nakamura, 1996); white rot (Sclerotium) (Pereira et al., 1996); and the sugar beet cyst nematode (*Heterodera schachtii*) (Szczech et al., 1993).

A recent study showed vermicompost tea to be as effective in controlling bacterial canker (*Clavibacter michiganensis* spp. *michiganensis*) in tomatoes as several commercially available biocontrol agents, giving a 63% reduction in disease for inoculated seedlings (Utkhede and Koch, 2004). Field trials have also shown a reduction in plant damage between conventionally fertilized rice and rice fertilized with vermicompost at equivalent NPK levels. The plants receiving the vermicompost treatment had significantly less damage due to the brown plant hopper (*Nilapavata lugens* (Stal) (Homoptera: Delpecidae)) and sheath blight caused by the fungus *Rhizoctonia solani* compared with plants in synthetic fertilizer treatment, both with and without chemical pest control (Bhadoria et al., 2003).

Disease suppression has been seen to vary depending on the type of feedstock. In one trial, biosolids vermicompost was less suppressive toward Phytophthera than vermicompost made from sheep, horse or cow manure (Szczech and Smolinska, 2001). The observed suppression was thought to be biological in nature because heat-sterilized vermicompost was not found to be disease-suppressive (Szczech, 1999). In addition to their role in plant growth promotion, PGPR isolated from traditional compost have been shown to inhibit fungal pathogens *in vitro* (De Brito Alvarez et al., 1995).

There are several ways in which beneficial microbes such as PGPR can inhibit plant pathogens. They can outcompete the pathogen for a nutrient source; they can directly parasitize the pathogen; or they can stimulate the plant to make physiological changes that will decrease its susceptibility to infection through a process known as Induced Systemic Resistance (ISR) (Pieterse et al., 2003). One of the physiological changes induced in plants by PGPR is an increased production of antioxidants. Scientists are just beginning to make the connection between soil health, plant health and human health by documenting cases of higher levels of antioxidants in foods grown without synthetic pesticides (Carbonaro et al., 2002).

Along with the hormone-producing and nutrient-mineralizing strains of PGPR, many PGPR have been shown to prevent a wide variety of plant diseases in greenhouse and field trials. Many practitioners who are committed to limiting chemical pesticide use are interested in using PGPR and other biocontrol agents as alternatives, as discussed in Chapter 32. Even with the great potential documented in the scientific literature, very few plant disease-suppressing PGPR are available commercially to growers (Nelson, 2004; USEPA, 2004).

There is evidence that certain composts naturally contain many species of PGPR (De Brito Alvarez et al., 1995), as well as plant-growth hormones such as auxin (Canellas et al., 2002; Garcia Martinez et al., 2002). Owing to the complex microbial ecology of composts, the presence of many strains of PGPR is not necessarily an indicator that the compost as a whole will be suppressive to plant diseases (McKellar and Nelson, 2003). With increasing knowledge of compost microbiology, the use of compost could potentially be a low-cost, sustainable way to inoculate agricultural soils with beneficial bacteria that can biologically enhance plant growth.

31.3   Some International Experience with Composting and Vermicomposting

31.3.1   Cuban Experience

After the Cuban revolution of 1959 and the subsequent US embargo, Cuban agriculture relied almost entirely on imports for maintaining its food and agricultural systems. At the time, over half of the daily caloric intake of Cubans was obtained through imported foods. To support the Cuban economy, the Soviet Union bought Cuban sugar at above world market prices, and it provided synthetic fertilizers and pesticides, farm equipment and fuel to support "modernization" of Cuban agriculture. The collapse of the USSR at the end of the 1980s and a strengthening of the US embargo in 1992 led to a food security crisis in Cuba. Historical circumstance thus created a nationwide experiment in more biologically based agriculture, the results of which have been well documented (Funes et al., 2000; Warwick, 2001).

The Cuban government responded to the crisis by enlisting the support of researchers and practitioners to shift the national agricultural system toward low-input, organic, small-farm and urban garden plots with an emphasis on civic participation. Traditional compost and vermicompost have played important roles in this shift toward sustainable agriculture along with regional small-scale production and use of various biocontrol agents against both microbial and insect pests (Altieri, 1999). Already by 1993, Cuba had 197 large-scale vermicomposting centers that produced 93,000 t vermicompost year^{-1} from a mixture of cow manure, sugar cane press mud, coffee pulp, plantain waste and municipal garbage feedstocks (Gersper et al., 1993). The Ministry of Agriculture estimates that about 600,000 metric tons of vermicompost were produced in urban areas of Cuba in 2002, while approximately 4.4 million t of other organic soil amendments were used (Nodals, 2004). The Ministry also stated that agronomic studies have documented yield increases of up to 40% in various crops when vermicompost is substituted for synthetic fertilizers (Vanasselt and Bourne, 2000), although none of these reports is accessible in the USA. Chapters 32 and 33 report experience of Cuban researchers with the use of bacterial and fungal inoculants in their own and other countries.

31.3.2   Indian Experience

There is a large sustainable agriculture movement in India that is seeking to combine traditional knowledge with modern science. Understanding and using beneficial soil organisms is part of the overall effort to reduce dependence on synthetic fertilizers and pesticides (Sinha, 1997; see also Chapter 35). Many government and civil society groups are promoting vermicomposting both as a system of waste management and as a valuable soil amendment. Recent reports in the *Indian Journal of Agronomy* have been providing evidence that vermicompost applications to field crops can significantly reduce synthetic fertilizer inputs.

Researchers have found some synergistic effects from satisfying only 50–75% of the crop's recommended NPK need with synthetic fertilizer while providing the remainder through vermicompost. A combination of synthetic and organic fertilizers has, in many cases, resulted in higher yields than from either nutrient source used alone. Yields from the integrated use of synthetic and organic fertilizers (traditional compost, farmyard manure or vermicompost) that are comparable to or greater than those achieved with 100% recommended NPK fertilizer applications have been documented for rice (Jeyabal and Kuppuswamy, 2001; Bhadoria and Prakash, 2003; Bhadoria et al., 2003), sunflower

(Dayal and Agarwal, 1998), guinea grass (George and Pillai, 2000), forage oats (Jayanthi et al., 2002), maize (Nanjappa et al., 2001), and wheat (Ranwa and Singh, 1999).

An economic analysis of the INM system has shown little benefit to small farmers if vermicompost had to be purchased off-site from a commercial producer (George and Pillai, 2000). In light of this finding, several Indian nongovernmental organizations such as Morarka Rural Research Foundation (Morarka, 2004) and the Bharatiya Agro Industries Foundations Institute for Rural Development have responded by training over a million farmers in on-site vermicompost production (BAIF, 2004). Pune, India, the location for the Bhawalkar Ecological Research Institute (BERI), is considered by some to be the vermicompost capital of the world. BERI promotes the *in situ* use of deep-burrowing earthworms for processing organic wastes in agricultural fields, as well as vermicompost-ing toilets and other urban waste-management systems (BERI, 2004; Bhawalkar, 2004).

In addition to being used with field crops, vermicompost is being used in various tree crops, including agroforestry systems. One case study describes the success of one farmer who dug a small basin around each coconut palm tree and applied 5 kg of active vermicompost as an initial inoculum, a thin layer of cow manure, and a layer of plant debris as mulch. The mulch is replenished periodically, and vermicompost is continually produced under each tree (OFAI, 2004). In teak production, vermicompost used in combination with the ring basin irrigation method has led to increased growth of young teak trees through nutrient additions and improved moisture retention (Koppad and Rao, 2004). Indian researchers and practitioners are also enriching vermicompost with different species of PGPR, including nitrogen-fixing and phosphate-solubilizing bacteria, with vermicompost serving as a substrate for producing and applying of biofertilizers (Kumar and Singh, 2001). Emerging local systems for producing and distributing vermicompost and biofertilizers in India are discussed in Chapter 45.

31.3.3 Australian Experience

Australia is home to one of the larger vermicompost equipment companies, VermiTech, which manufactures large-scale, continuous flow-through, reactor-type vermicomposting systems. Many of Australia's agricultural soils are low in organic matter and have poor water infiltration and water-holding capacities. Trials that VermiTech has carried out with cooperating farmers on the effects of applying vermicompost to field soils have showed potential for $3-5\,t\,ha^{-1}$ application of vermicompost to significantly decrease soil compaction, measured as depth to 300 psi with a penetrometer (Patten, personal communication). Another field trial showed the potential of the same application of vermicompost to increase pH, cation exchange capacity (CEC), and soil organic matter (SOM), but so far, no peer-reviewed studies have been published (P. Patten, personal communication).

Vermicompost has the potential to become a useful soil amendment in the Australian and New Zealand wine industries. Field trials carried out by Australia's Commonwealth Scientific and Industrial Research Organization (CSIRO) with vermicomposts made from mixtures of grape marc waste, animal manure and other agricultural wastes have shown significant increases in yield without decreases in fruit quality (Buckerfield and Webster, 1998). In one study, vermicompost was used as mulch under the vines, applied about two inches deep and covered with a thick layer of straw. When vermicompost was applied alone, there was no significant increase in yield; however, when the vermicompost applied was covered with straw, there was a 56% increase in yield. The straw is thought to protect the microorganisms in the vermicompost from UV radiation and desiccation. The following fruit quality factors were measured: Brix (% sugar content), pH and titratable acidity, none of which was adversely affected by vermicompost treatments (Buckerfield and Webster, 1998). The effects of the undervine vermicompost and straw combination were still measurable two years after the original application, which enhanced the economic return for the grower (Buckerfield and Webster, 2000). Papers in peer-reviewed journals describing these trials are forthcoming.

Even with several successful field trials, grower adoption of this practice is relatively low, however, owing to a lack of consistent quality-assured product and high cost (K. Webster, personal communication). There has been some grower adoption of vermicompost amendments in the US wine industry. One well-established worm farm in Sonoma Valley, CA, charges \$375 yd^{-3} for dairy-manure vermicompost and recommends that it be used in very small amounts, just one small cup when planting new vines (Anonymous, 2003). Outreach and extension efforts as well as market prices and the availability of quality vermicompost will determine adoption rates and the ways in which vermicompost is used by practitioners.

31.4 Discussion

Composts play an essential role in any type of sustainable agricultural system, effectively recycling municipal, industrial or agricultural wastes and residues, and returning organic matter to the soil ecosystem. Adding organic matter to soils can relieve compaction and increase aggregate stability, thus greatly improving soil structure, which has favorable effects on soil biota. Composting processes, both traditional and vermicomposting, can reduce the number of human and plant pathogens present in many types of organic residues before they are applied to agricultural soils. The complex microbial communities in high-quality cured compost have the ability to increase plant growth and decrease the incidence of plant disease, reducing the need for synthetic fertilizers and pesticides. Compost applications, both in research and in practice, are demonstrating that they can increase soil health and therefore the overall sustainability of food-producing systems.

The future of compost use in sustainable agriculture will depend on collaborative efforts among researchers, entrepreneurs and practitioners. Groups in the United States such as the US Composting Council, the Cornell Waste Management Institute, and JG Press, the publishers of BioCycle magazine and Compost Science & Utilization, are providing channels for communication among these groups. Two US Department of Agriculture programs, Associated Technology Transfer to Rural Areas (ATTRA) and Sustainable Agriculture Research and Education (SARE), have compiled and disseminated extensive information on composting.

Several international conferences have contributed to the growing scientific base of knowledge such as the International Symposium on Composting and Compost Utilization in Columbus, Ohio, held in 2002, and the First International Soil and Compost Eco-biology (SoilACE) Conference convened in Leon, Spain, in 2004. A number of international NGOs are carrying out compost research and extension on a large scale. Even with so much international interest in compost science and use, much of the information available on producing and using compost is difficult to access. It is scattered throughout numerous trade journals, scientific journals from widely varying disciplines, diverse Internet sites, or is passed on entirely by word of mouth.

Books published recently in Europe (de Bertoldi et al., 1996; Insam et al., 2002) are helping to assemble and disseminate available scientific and practical information on compost production and use. Increased attention by governments, NGOs and agribusinesses to the role of compost in sustainable agriculture will contribute to the development of new and better compost technologies. More mechanistic (laboratory) and empirical (field) studies on the production and use of composts will be mutually reinforcing. Knowledge of actual practices and their effects can help keep laboratory experimentation relevant, and a mechanistic understanding of the beneficial effects that composts have on soils and plants should help drive innovative new practices.

References

Acevedo, I.C. and Pire, R., Effects of vermicompost as substrate amendment on the growth of papaya (*Carica papaya* L.), *Interciencia*, **29**, 274–279 (2004).

Aira, M. et al., How earthworm density affects microbial biomass and activity in pig manure, *Eur. J. Soil Biol.*, **38**, 7–10 (2002).

Albiach, R. et al., Organic matter components and aggregate stability after the application of different amendments to a horticultural soil, *Bioresour. Technol.*, **76**, 125–129 (2001).

Alfreider, A. et al., Microbial community dynamics during composting of organic matter as determined by 16S ribosomal DNA analysis, *Compost Sci. Util.*, **10**, 303–312 (2002).

Altieri, M.A., The ecological role of biodiversity in agroecosystems, *Agric. Ecosyst. Environ.*, **74**, 19–31 (1999).

Anonymous, Terra Squirma: Worm farm sets out to revolutionize agriculture with its magical vermicompost, *North Bay Biz*, Sonoma, Marin, Napa (2003), 32–37.

Arancon, N.Q. et al., Effects of vermicomposts on growth and marketable fruits of field-grown tomatoes, peppers and strawberries, *Pedobiologia*, **47**, 731–735 (2003).

Aranda, E. et al., Vermicomposting in the tropics, In: *Earthworm Management in Tropical Agroecosystems*, Hendrix, P., Ed., CABI Publishing, New York, 253–287 (1999).

Atiyeh, R.M. et al., Growth of tomato plants in horticultural potting media amended with vermicompost, *Pedobiologia*, **43**, 724–728 (1999).

Atiyeh, R.M. et al., Influence of earthworm processed pig manure on the growth and yield of greenhouse tomatoes, *Bioresour. Technol.*, **75**, 175–180 (2000a).

Atiyeh, R.M. et al., Effects of vermicomposts and composts on plant growth in horticultural container media and soil, *Pedobiologia*, **44**, 579–590 (2000b).

Atiyeh, R.M. et al., The influence of humic acids derived from earthworm-processed organic wastes on plant growth, *Bioresour. Technol.*, **84**, 7–14 (2002a).

Atiyeh, R.M. et al., The influence of earthworm-processed pig manure on the growth and productivity of marigolds, *Bioresour. Technol.*, **81**, 103–108 (2002b).

Ayanlaja, S.A. et al., Leachate from earthworm castings breaks seed dormancy and preferentially promotes radicle growth in jute, *Hortscience*, **36**, 143–144 (2001).

BAIF, Bharatiya Agro Industries Foundation Institute for Rural Development, http://www.baif.com, accessed on: June 2, [Online] (2004).

Bailey, K.L. and Lazarovits, G., Suppressing soil-borne diseases with residue management and organic amendments, *Soil Till. Res.*, **72**, 169–180 (2003).

Benitez, E. et al., Isolation by isoelectric focusing of humic-urease complexes from earthworm (*Eisenia fetida*)-processed sewage sludges, *Biol. Fertil. Soils*, **31**, 489–493 (2000).

BERI, Bhawalkar Ecology Research Institute, http://www.members.tripod.com/eco_logic/index.htm, accessed on: June 2, [Online] (2004).

Bhadoria, P.B.S. and Prakash, Y.S., Relative influence of organic manures in combination with chemical fertilizer in improving rice productivity of lateritic soil, *J. Sustain. Agric.*, **23**, 77–87 (2003).

Bhadoria, P.B.S. et al., Relative efficacy of organic manures on rice production in lateritic soil, *Soil Use Manage.*, **19**, 80–82 (2003).

Bhawalkar, U., *Vermiculture Ecotechnology*, Bhawalkar Earthworm Research Institute, Pune, India (2004).

Brown, G.G., Barois, I., and Lavelle, P., Regulation of soil organic matter dynamics and microbial activity in the drilosphere and the role of interactions with other edaphic functional domains, *Eur. J. Soil Biol.*, **36**, 177–198 (2000).

Buckerfield, J.C. and Webster, K.A., Worm-worked waste boosts grape yields: Prospects for vermicompost use in vineyards, *Aust. NZ Wine Ind. J.*, **13**, 73–76 (1998).

Buckerfield, J.C. and Webster, K.A., Vermicompost — more than a mulch? Vineyard trials to evaluate worm-worked wastes, *The Australian Grapegrower and Winemaker Annual Technical Issue*, 160–166 (2000).

Buckerfield, J.C. et al., Vermicompost in solid and liquid forms as a plant-growth promoter, *Pedobiologia*, **43**, 753–759 (1999).

Canellas, L.P. et al., Humic acids isolated from earthworm compost enhance root elongation, lateral root emergence, and plasma membrane H^+-ATOase activity in maize roots, *Plant Physiol.*, **130**, 1951–1957 (2002).

Carbonaro, M. et al., Modulation of antioxidant compounds in organic vs conventional fruit (Peach, *Prunus persica* L., and pear, *Pyrus communis* L.), *J. Agric. Food Chem.*, **50**, 5458–5462 (2002).

Cavender, N.D., Atiyeh, R.M., and Knee, M., Vermicompost stimulates mycorrhizal colonization of roots of *Sorghum bicolor* at the expense of plant growth, *Pedobiologia*, **47**, 85–89 (2003).

Chaoui, H.I., Zibilske, L.M., and Ohno, T., Effects of earthworm casts and compost on soil microbial activity and plant nutrient availability, *Soil Biol. Biochem.*, **35**, 295–302 (2003).

Chaoui, H.I. et al., Suppression of the plant diseases, *Pythium* (damping-off), *Rhizoctonia* (root rot) and *Verticillilum* (wilt) by vermicomposts. *Proceedings of Brighton Crop Protection Conference — Pests and Diseases*, Brighton, UK (2002).

Chefetz, B. et al., Chemical and biological characterization of organic matter during composting of municipal solid waste, *J. Environ. Qual.*, **25**, 776–785 (1996).

Dayal, D. and Agarwal, S.K., Response of sunflower (*Helianthus annuus*) to organic manures and fertilizers, *Indian J. Agron.*, **43**, 469–473 (1998).

de Bertoldi, M. et al., Eds., *The Science of Composting*, Blackie Academic and Professional, London (1996).

De Brito Alvarez, M., Gagne, S., and Antoun, H., Effect of compost on rhizosphere microflora of the tomato and the incidence of plant growth-promoting rhizobacteria, *Appl. Environ. Microbiol.*, **61**, 194–199 (1995).

De Ceuster, T.J.J. and Hoitink, H.A.J., Prospects for composts and biocontrol agents as substitutes for methyl bromide in biological control of plant diseases, *Compost Sci. Util.*, **7**, 6–15 (1999).

Dees, P.M. and Ghiorse, W.C., Microbial diversity in hot synthetic compost as revealed by PCR-amplified rRNA sequences from cultivated isolates and extracted DNA, *FEMS Microbiol. Ecol.*, **35**, 207–216 (2001).

Edwards, C.A., Ed., *Earthworm Ecology*, Lewis Publishers/CRC Press, Boca Raton, FL (2004).

Edwards, C.A. and Neuhauser, E.F., *Earthworms in Waste and Environmental Management*, SPB Publishers, The Hague, Netherlands (1988).

Edwards, C.A. et al., The use of earthworms for composting farm wastes, In: *Composting of Agricultural and Other Wastes*, Gasser, J.K.R., Ed., Elsevier, Essex, UK, 229–253 (1984).

Frederickson, J. et al., Combining vermiculture with traditional green waste composting systems, *Soil Biol. Biochem.*, **29**, 725–730 (1997).

Funes, F. et al., *Sustainable Agriculture and Resistance: Transforming Food Production in Cuba.* Food First Books, ACTAF (Cuban Association of Agricultural and Forestry Technicians) and CEAS (Center for the Study of Sustainable Agriculture, Agrarian University of Havana), San Francisco, CA (2000).

Garcia Martinez, I. et al., Extraction of auxin-like substances from compost, *Crop Res.*, **24**, 323–327 (2002).

Ge, F. et al., Water stability of earthworm casts in manure- and inorganic-fertilizer amended agroecosystems influenced by age and depth, *Pedobiologia*, **45**, 12–26 (2001).

George, S. and Pillai, G.R., Effect of vermicompost on yield and economics of guinea grass (*Panicum maximum*) grown as an intercrop in coconut (*Cocos nucifera*) gardens, *Indian J. Agron.*, **45**, 693–697 (2000).

Gersper, P.L., Rodriguez-Barbosa, C.S., and Orlands, L., Soil conservation in Cuba: A key to the new model for agriculture, *Agric. Hum. Values*, **10**, 16–23 (1993).

Hill, G.T. et al., Methods for assessing the composition and diversity of soil microbial communities, *Appl. Soil Ecol.*, **15**, 25–36 (2000).

Hoitink, H.A.J. and Fahy, P.C., Basis for the control of soilborne plant pathogens with composts, *Annu. Rev. Phytopathol.*, **24**, 93–114 (1986).

Hoitink, H.A.J. and Kuter, G.A., Effects of composts in growth media on soilborne pathogens, In: *The Role of Organic Matter in Modern Agriculture*, Avnimelich, Y., Ed., Nijhoff Publishers, Boston, 289–306 (1986).

Howard, S.A., *An Agricultural Testament*, Oxford University Press, New York (1943).

Insam, H., Riddech, N., and Klammer, S., Eds., *Microbiology of Composting*, Springer, Berlin (2002).

Jayanthi, C. et al., Integrated nutrient management in forage oat (*Avena sativa*), *Indian J. Agron.*, **47**, 130–133 (2002).

Jeyabal, A. and Kuppuswamy, G., Recycling of organic wastes for production of vermicompost and its response in rice–legume cropping system and fertility, *Eur. J. Agron.*, **15**, 153–170 (2001).

Kale, R.D. et al., Influence of vermicompost application on the available macronutrients and selected microbial-populations in a paddy field, *Soil Biol. Biochem.*, **24**, 1317–1320 (1992).

Koppad, A.G. and Rao, R.V., Influence of *in situ* moisture conservation methods and fertilisers on early growth of teak (*Tectona grandis*), *J. Trop. Forest Sci.*, **16**, 218–231 (2004).

Kumar, V. and Singh, K.P., Enriching vermicompost by nitrogen fixing and phosphate solubilizing bacteria, *Bioresour. Technol.*, **76**, 173–175 (2001).

Loecke, T.D. et al., Corn growth responses to composted and fresh solid swine manures, *Crop Sci.*, **44**, 177–184 (2004).

McKellar, M.E. and Nelson, E.B., Compost-induced suppression of *Pythium* damping-off is mediated by fatty-acid-metabolizing seed-colonizing microbial communities, *Appl. Environ. Microbiol.*, **69**, 452–460 (2003).

Morarka, http://www.morarkango.com, accessed on: June 2, [Online] (2004).

Nakamura, Y., Interactions between earthworms and microorganisms in biological control of plant pathogens, *Farming Japan*, **30**, 37–43 (1996).

Nanjappa, H.V., Ramachandrappa, B.K., and Mallikarjuna, B.O., Effect of integrated nutrient management on yield and nutrient balance in maize (*Zea mays*), *Indian J. Agron.*, **46**, 698–701 (2001).

Nelson, E.B., Biological control of oomycetes and fungal pathogens, In: *Encyclopedia of Plant and Crop Science*, Goodman, R.M., Ed., Marcel Dekker, New York, 137–140 (2004).

Nodals, A.R., La agricultura urbana en Cuba: Impactos economicos, sociales y productivos, *Revista Bimestre Cubana, de la Sociedad Economica de Amigos de Pais*, **20**, 103–124 (2004).

Oades, J.M., Soil organic-matter and structural stability: Mechanisms and implications for management, *Plant Soil*, **76**, 319–337 (1984).

OFAI, Organic Farming Association of India: Case Study. C. Dutt, http://www.ofai.org/case/ancas.htm, accessed on: June 2, [Online] (2004).

Pereira, J.C.R. et al., Control of *Sclerotium cepivorum* by the use of vermicompost, solarization, *Trichoderma harzianum* and *Bacillus subtilis*, *Summa Phytopathol.*, **22**, 228–234 (1996).

Pieterse, C.M.J. et al., Induced systemic resistance by plant growth-promoting rhizobacteria, *Symbiosis*, **35**, 39–54 (2003).

Ranwa, R.S. and Singh, K.P., Effect of integrated nutrient management with vermicompost on productivity of wheat (*Triticum aestivum*), *Indian J. Agron.*, **44**, 554–559 (1999).

Rodale, J.I., *Pay Dirt: Farming and Gardening with Composts*, Devin-Adair, New York (1945).

Rodale, J.I., *The Complete Book of Composting*, Rodale Books, Emmaus, PA (1960).

Roe, N.E., Compost utilization for vegetable and fruit crops, *Hortscience*, **33**, 934–937 (1998).

Scheuerell, S. and Mahaffee, W., Compost tea: Principles and prospects for plant disease control, *Compost Sci. Util.*, **10**, 313–338 (2002).

Schloss, P.D. et al., Tracking temporal changes of bacterial community fingerprints during the initial stages of composting, *FEMS Microbiol. Ecol.*, **46**, 1–9 (2003).

Sinha, R.K., Embarking on the second green revolution for sustainable agriculture in India: A judicious mix of traditional wisdom and modern knowledge in ecological farming, *J. Agric. Environ. Ethic*, **10**, 183–197 (1997).

Sommer, S.G., Effect of composting on nutrient loss and nitrogen availability of cattle deep litter, *Eur. J. Agron.*, **14**, 123–133 (2001).

Szczech, M.M., Suppressiveness of vermicompost against *Fusarium* wilt of tomato, *J. Phytopathol.*, **147**, 155–161 (1999).

Szczech, M. and Smolinska, U., Comparison of suppressiveness of vermicompost produced from animal manures and sewage sludge against *Phytophthora nicotinae* Breda de Haan var. *nicotinae*, *J. Phytopathol.*, **149**, 77–82 (2001).

Szczech, M. et al., Suppressive effect of a commercial earthworm compost on some root infecting pathogens of cabbage and tomato, *Biol. Agric. Hortic.*, **10**, 47–52 (1993).

Tarkalson, D.D. et al., Mycorrhizal colonization and nutrient uptake of dry bean in manure and compost manure treated subsoil and untreated topsoil and subsoil, *J. Plant Nutr.*, **21**, 1867–1878 (1998).

Toyota, K. and Kimura, M., Microbial community indigenous to the earthworm *Eisenia foetida*, *Biol. Fertil. Soils*, **31**, 187–190 (2000).

USEPA, *Biopesticide Active Ingredient Fact Sheets*, http://www.epa.gov/pesticides/biopesticides/ingredients/index.htm, accessed on: January 10, [Online] (2004).

Utkhede, R. and Koch, C., Biological treatments to control bacterial canker of greenhouse tomatoes, *Biocontrol*, **49**, 305–313 (2004).

Vanasselt, W. and Bourne, J., Cuba's agricultural revolution: A return to oxen and organics, *World Resources 2000–2001: People and Ecosystems, the Fraying Web of Life*. Elsevier Science, Oxford, UK (2000), 389.

Verkhovtseva, N.V. et al., Comparative investigation of vermicompost microbial communities, In: *Microbiology of Composting*, Klammer, S., Ed., Springer, Berlin, 99–111 (2002).

Vessey, J.K., Plant growth promoting rhizobacteria as biofertilizers, *Plant Soil*, **255**, 571–586 (2003).

Vinceslas-Akpa, M. and Loquet, M., Organic matter transformations in lignocellulosic waste products composted or vermicomposted (*Eisenia fetida andrei*): Chemical analysis and ^{13}C PMAS NMR spectroscopy, *Soil Biol. Biochem.*, **29**, 751–758 (1997).

Warwick, H., Cuba's organic revolution, *Forum Appl. Res. Public Policy*, **16**, 54–58 (2001).

Weltzien, H.C., Biocontrol of foliar plant fungal diseases with compost extracts, In: *Microbial Ecology of Leaves*, Hirano, S.S., Ed., Springer, New York, 431–449 (1992).

32

Practical Applications of Bacterial Biofertilizers and Biostimulators

Rafael Martinez Viera and Bernardo Dibut Alvarez

Institute for Basic Research in Tropical Agriculture (INIFAT), Havana, Cuba

CONTENTS

Achieving further increments in agricultural productivity with a reduction in agrochemical use for economic or environmental reasons will need a new generation of technologies. Among these are a growing number of biofertilizers and biostimulators (Burdman et al., 2000; Bauer, 2001). Advances in soil microbiology and in applied agricultural biotechnology, including bioengineering, are allowing the exploitation of selected soil microorganisms that have high potential for fixing atmospheric nitrogen, solubilizing fixed phosphorus in the soil, and synthesizing biochemical substances that, when interacting with plants, stimulate the plants' metabolism. Worldwide over the last half century, many potentials of different microorganisms for having beneficial effects on crop productivity have been demonstrated in diverse cultivation systems. However, lack of technical and economic support has impeded the practical applications of these bacteria and their general use for large areas of agricultural production (Sasson, 2000).

The long-term sustainability of agricultural systems will depend on more effective handling of the internal resources of agroecosystems. From the results reported below, we believe that biofertilizers are likely to become vital components of these systems in the future, offering economically attractive and ecologically acceptable ways to reduce external inputs while improving the quantity and quality of internal resources.

Biofertilizers are manufactured products that contain certain microorganisms which normally inhabit the soil but in relatively small populations. When their populations are increased by means of artificial inoculation of the soil, they are able, by their biological activity, to provide some often important shares of the nutrients that plants need for their development, as well as a variety of hormones or other plant growth promoters/regulators that benefit crops. Such bioproducts supply or mobilize nutrients with minimum reliance on nonrenewable resources.

Biocompounds elicit microbial processes that are usually quick and can be applied in small agricultural units to solve specific local problems. Today, they are becoming more attractive and more acceptable to farmers, in part because subsidies for the purchase of inorganic fertilizers are decreasing almost everywhere. This is a favorable moment for the development of both basic and applied research to evaluate and promote the use of these microorganisms. This needs to include the education of farmers on the benefits of these biocompounds and on how to apply them properly. Unfortunately, in most countries there are few specialists with much knowledge and interest in the practical aspects of using biofertilizers. For their use to expand significantly, more specialized personnel will be needed who know how to work with these materials, who are able to respond to the modern conditions of agricultural production, and who are attentive to the growing concerns for sustainability and environmental protection.

The biofertilizers most widely utilized in many countries are ones based on Rhizobium and Bradyrhizobium, bacteria that establish symbiotic fixation of atmospheric nitrogen in leguminous plants. The mechanisms for this fixation have been reviewed in Chapter 12. Because such biofertilizers are well known and often applied, in this chapter we consider results obtained from the application of selected strains of *Azotobacter chroococcum* and *Azospirillum brasiliense*. These are fixers of atmospheric dinitrogen and producers of plant-growth stimulators, by themselves or in mixed applications. We also consider the effects of *Pseudomonas fluorescens* and *Bacillus megatherium* var. *phosphaticum* as soil phosphorus solubilizers. These are not the only bacteria that can be used for such purposes, but they are ones we currently know the most about and have been able to use successfully at field scale.

32.1 Capitalizing on Biological Processes in Different Settings

Biofertilizers are an appropriate biotechnology because their use is technically feasible in countries with high or low scientific-technical levels, providing tangible benefits to farmers and being environmentally safe as well as socioeconomically and culturally acceptable (Izquierdo et al., 1995). Due regard must be shown, however, for differences in the way that different microorganisms perform, separately and collectively, under varying soil and climatic conditions with a variety of crops and cultivars. Chapter 45 describes village-based production of biofertilizers in India.

32.1.1 Differences between Tropical and Temperate Agroecosystems

There are often disparities among the results achieved with the application of biofertilizers in temperate and tropical countries. Most published results to date have been obtained in temperate regions, and in many cases, the effects of inoculation have not been very satisfactory. An atmosphere of distrust has been created toward biocompounds so that their use in agricultural production has not become very widespread. However, few generalizations are tenable when taking into account the diversity of plant–soil–microbial interactions that exist in soil systems. These are affected by climatic variation and by the biological, physical, and chemical differences associated therewith.

Some research has shown, for example, that the excretions of plant roots in tropical regions have higher concentrations and greater diversity as sugars and organic acids as well as various hormones, enzymes, and other biological compounds (Dibut, 2000). It thus should not be surprising that different results occur in temperate regions where there is less luminous energy available. Under tropical conditions, carbon fixation in plant canopies reaches up to $20 \, t \, ha^{-1} \, yr^{-1}$ compared to $12 \, t \, ha^{-1} \, yr^{-1}$ in temperate regions (Debinstein, 1970). What goes on in canopies has direct implications for what will go on in the soil, and vice versa.

32.1.1.1 Phyllosphere Processes

In the leaf zone of tropical plants, referred to as the phyllosphere, there are high populations of diazotrophic bacteria, specifically on leaf surfaces (phylloplane), as noted in Chapter 2. The canopy acts as a support center for nutrient production and water storage for these bacteria, which process nitrogen compounds that are then distributed throughout the plant (Bhat et al., 1971). As long as 50 years ago, Ruinen established the presence of diazotrophic bacteria in the phyllosphere, isolating them on 192 out of 198 leaves collected from trees, shrubs, epiphytes, and seacoast vegetation in Indonesia (Ruinen, 1974).

This has been confirmed for cocoa and coffee leaves in Surinam and on sugarcane leaves in Cuba and Brazil (Döbereiner, 1983; Martínez Viera, 1986). In Egypt, Abd-el-Malek (1971) isolated diazotrophic bacteria on 56 of 58 leaves collected from trees, field crops, vegetables, ornamentals, and aquatic plants. The presence of the nitrogenase enzyme on inner sugarcane leaves in Brazil has been demonstrated by Olivares et al. (1997). The positive effects of foliar applications of selected bacteria on citrus and mango plantations in Cuba and Mexico are shown in section 32.3. While foliar sprays have application in temperate agriculture, there are reasons to expect they can give more benefit under tropical conditions given the biodiversity present.

32.1.1.2 Rhizosphere Processes

Not all of the strains of any particular microbial species will behave in the same way in their response to the root exudates of a certain plant species. Neglect of this fact could account for some of the inconsistency of results that have been obtained to date in different countries with the application of biofertilizers. It is necessary to carefully select those strains of bacteria that have highest effectiveness for each plant species, not expecting that a single strain can be used for all cultures. This is what we have been learning how to do in our laboratories and field trials.

The root exudates of each plant species have their own composition and will attract (or repel) organisms with different intensity. This attraction (or repulsion), known as chemotaxy, will vary among microbial species or even across the strains of a single species. Organisms respond, positively or negatively, with different speed and intensity along a chemical gradient from most chemical sources, by either moving toward them or away by whatever means of locomotion are available. It has been seen, for example, that plants with C4 paths of photosynthesis have associations principally with *Azospirillum lipoferum*, while C3 plants are more associated with *A. brasiliense*. The differences are attributable to chemotaxic effects (Döbereiner, 1983).

We have examined how different strains of *Azotobacter chroococcum* respond to onion root exudates by evaluating them in a chemotaxism chamber (Dibut, 2000). The INIFAT-17 and INIFAT-3 strains of *A. chroococcum* were found to be approximately 3000 times more attracted by the root exudates of onion than was the INIFAT-10 strain. Subsequently, the ARA test of acetylene reduction indicated that INIFAT-17 is much more efficient in nitrogen fixation than INIFAT-3, so the first strain was selected for commercial use with onions. We had confidence that abundant populations of this strain would settle in the rhizosphere of the target plants, and also that these plants would receive N benefits from inoculation.

Using this same analytical process, the strain INIFAT-12 has been selected for use with tomato and other vegetables, and INIFAT-9 has been chosen for gramineaceous species and banana. The fact of such dramatic intra-specific differences challenges researchers to study plant–microbial interactions more closely with practical benefits in mind.

32.1.1.3   Nutrient Mobilization and Phytohormone Production

Phosphorus is known to be an essential element for the growth of plants. However, the availability of soil phosphorus is widely restricted, especially in the tropics, by the complexing of this element with soil cations and by its adsorption on soil particle surfaces. These reactions lead to rather low efficiency in plants' use of phosphorus fertilizer, especially in calcareous soils or in soils high in iron and aluminum oxides. The efficiency of P fertilizer applied on soil generally ranges between 10 and 25% (Paul and Clark, 1989), so the quantity used by plants is only a small part of the total P present in the soil, with a large proportion in nonavailable forms, as discussed in Chapter 13.

Many microorganisms, including bacteria, fungi, and actinomycetes, are capable of solubilizing this unavailable inorganic P by their production of organic and inorganic acids, which react with the insoluble forms of P and transform them into soluble forms. The organisms involved in these processes are often present in low populations in the soils, existing exist mainly in the rhizosphere of plants.

We should also remember that associative nitrogen-fixing bacteria and soil-P solubilizers synthesize significant amounts of phytohormones, various active substances that stimulate plant growth reviewed in Chapter 14. These substances are assimilated by plants through their roots and are taken up at successive stages of growth. Some of these substances stimulate the development of the roots or of the whole plant; others induce flowering or reduce floral abortion; still others facilitate the formation and maturation of fruit. These interactions can support precocious development in the more vigorous plants and also increments of yield, as seen from the results reported in Section 32.3. The effects of phytohormones are hard to disentangle from nutrient mobilization effects, but practically speaking this is not necessary.

32.1.1.4   Biochemical Synthesis

Synthesis of these various substances is high in some strains isolated from tropical soils. Table 32.1 shows the concentrations of active substances that we found were synthesized by just one commercial strain of *A. chroococcum* isolated from a Cuban soil (Dibut, 2000). This table documents that some tropical organisms are veritable biochemical "factories." Our analysis has also shown some interesting differences between the synthesizing activity of this highly active strain, INIFAT-12, in comparison with another strain of Azotobacter that was isolated from a Russian soil under temperate climatic conditions (Fadeiev, 1986). The most evident disparity was regarding different cytokinins, a category of plant-growth regulators discussed in Chapter 14. Both strains produced isopentenil adenosine, isopentinil adenine, and trans zeatin; but the Cuban strain produced zeatin monophosphate, zeatin ribose, and another, unidentified cytokinin (Dibut, 2000), whereas the Russian strain did not. Instead, it produced other cytokinins, cis zeatin, and dihydro zeatin. This underscores the specificity and diversity that one must deal with when working at the microbial–plant interface.

TABLE 32.1

Production of Plant-Growth-Promoting Substances, Vitamins, and Different Amino Acids
Synthesized by *A. chroococcum* Strain INIFAT-12

Categories of Plant-Growth Regulators	Activity (μg L^{-1})	Vitamins	Concentration (μg L^{-1})	Amino Acids	Concentration (nmol mL^{-1})
Auxins (IAA)	14.47	Thiamine	5.7	Aspartic acid	71.05
Gibberellins (A3G)	30.20	Riboflavin	44.0	Serine	61.65
Cytokinins (Kinetin)	32.50	Pyridoxine	18.0	Glycine	127.35
		Folic acid	3.5	Valine	38.70
				Isoleucine	20.05
				Glutamic acid	82.15
				Ornithine	0.83
				Lysine	9.40
				Arginine	4.45
				Threonine	58.80
				Leucine	35.95
				Phenylalanine	65.55
				Proline	60.60
				Total concentration	728.90

32.1.2 Effects of Combination

It has been shown that increases of soluble soil P by the action of solubilizer organisms can increase the efficiency of N fixation by bacteria (Subba Rao, 1996). Such results indicate that there can be beneficial associative effects from combining such P-solubilizing microorganisms with nitrogen fixers in the soil and in the roots by producing and applying mixed inoculants (Subba Rao, 1996). Parr et al. (1994) and Higa (1995) have suggested that the advantages from applying beneficial microorganisms in soil will be greatest when the inoculum used has sufficient density and variety of these organisms. However, not enough research has yet been carried out to know how much density and variety will be optimal. This will surely vary for crop species and soil types.

Benefits have been demonstrated from an association of Rhizobium and Azotobacter, with increments in N-fixation efficiency achieved from Rhizobium in the legume to which it is applied and also an increase in agricultural yield attributable to the action of active substances synthesized by the Azotobacter (Peoples and Craswell, 1992). Similarly, the association between *Bradyrhizobium japonicum* and *A. chroococcum* has been demonstrated to increase the nodulation and yield of soybeans (Singh and Subba Rao, 1979; Katayama et al., 1996), as has that of Azospirillum and Rhizobia for different leguminous species (Burdman et al., 2000). The propensity of modern science to focus on one organism or one function at a time, considering other organisms and functions as interferences in efforts to identify and measure "true" effects, can impede our understanding of processes that display ecological and symbiotic dynamics. While combinatory approaches complicate the task of scientists, they are showing practical results.

32.2 Producing Biofertilizers

Present information indicates compatibility among the most important N-fixing bacteria, symbiotics and associatives, and between them and P-solubilizing bacteria. Making mixed bioproducts with several of these microorganisms combined together is proving to be feasible. In Cuba, bioproducts have been developed containing *A. chroococcum* and *P. fluorescens*; Rhizobium spp. and *P. fluorescens*; as well as *A. chroococcum* and *A. brasiliense* (Martínez Viera et al., 2004); and also *A. chroococcum* and *B. megatherium* (Dibut and Martínez, 2003). All of these have demonstrated high effectiveness when applied under agricultural conditions. These combinations are amenable to large-scale production of the respective bioproducts in a single process, which reduces costs of production compared to their having to be produced separately.

To apply this strategy effectively in agricultural practice, it is necessary to begin with diverse collections of bacterial species that have been isolated from different soils and regions and that have the characteristics desired for biofertilizers or biostimulators. Various assessments then need to be carried out with these collections, particularly chemotaxis studies. Tests are then carried out to assess N fixation (by acetylene reduction) and P solubilization, followed by screening in trays to determine the stimulation capacity of the different strains. Through such investigations, the most effective strains can be selected for a certain plant species and certain soil conditions (Martínez Viera et al., 2004).

New culture media have been developed that allow fast bacterial multiplication and increase the synthesis of active substances. Industrial fermentation technologies have had to be developed for different microbial species. To scale-up production for large-scale use, factories with industrial fermenters have been established that can produce 500–1000 l of desired biocompounds within 20–24 h. These industrial-scale operations require steam and air fountains as well as other auxiliary media. Smaller artisanal operations can process materials in smaller laboratories with only air fountains. These produce 80 l of output within 24 h. Powdered inoculants are produced under laboratory conditions that use different organic substrates and have capacities of approximately 300 kg weekly.

Simple or mixed biofertilizers have been developed that contain populations up to 10^{14} cells ml^{-1} of each desired species. This material can be applied at a rate of 2 l ha^{-1} through an irrigation system or by other means of dispersion, either hand or machine sprayers at field level or by airplane over larger surfaces. With such concepts and methods, farmers are able to make substantial savings on fertilization costs and to shorten the growing cycle of crops due to the action of bacterially-produced substances (Subba Rao, 1996; Dibut and Martínez Viera, 2003; Martínez Viera et al., 2004). These practices have reduced environmental contamination while increasing farm productivity and profits, as seen in the following section.

32.3   Results of Biofertilization

Because of its particular national circumstances during the 1990s, Cuban researchers have taken a special interest in the development of biofertilization and biostimulation, seeking to compensate for the reduced availability and higher costs of continuing inorganic fertilization and crop protection, especially materials derived from petroleum. The knowledge and applications that have been developed in Cuba have been tested in other countries, and collaboration with colleagues around the world has helped to advance our understanding of what is possible and what, so far, is not. The utilization of bacterial organisms to enhance crop performance is beneficial not just in terms of yield, but also in terms of speed of maturation and resistance to pests and diseases.

In Cuba, in the production of vegetables generally, the effect of inoculation with *A. chroococcum* has been to reduce the need for nitrogen fertilizer by 40%. An example for tomato production has been documented by Martínez Viera and Dibut (1996). First, the active substances synthesized by bacteria accelerate the development of the plants as seedlings. This reduces the period between sowing of seedlings and their transplanting, saving 7–10 days and thereby saving also water, energy, pesticides, and labor, while shortening the total cycle of cultivation. The number of flowers and fruits per plant is higher, and fructification is earlier in the treated fields. An increment of yield is obtained, and the quality of the fruits is superior, with average weight and diameters significantly higher for the treated plants. Figure 32.1 shows some of these benefits.

Table 32.2 shows some results obtained with the application of *Azotobacter chroococcum* on different crops in Turkey (Dibut and Martínez, 2003). It shows the increments in yield attained from seven crops to which biofertilizers were applied. The increases ranged from 26% for tomatoes to 45% for cotton, with an average of 34%, while nitrogen fertilizer was reduced by 30%. The average weight of fruit went up by 36% with inoculation treatments.

In the case of the roots and tubers, the nitrogen-fixing activity and the effects of active substances synthesized by the bacteria stimulate photosynthesis and reduce the respiration of plants (Martínez Viera and Dibut, 1996). This allows more photosynthate storage which is the basic requirement for formation of roots and tubers, built up by reserve material. Figure 32.2 shows the effects of the application of *A. chroococcum* on roots (cassava) and tubers (potato).

When determining the methods of biofertilizer application, it is necessary to consider what is known about the status of the phyllosphere, which is important for making effective foliar applications. Table 32.3 presents the results of aerial applications of *A. croococcum* biofertilization on 100 ha of citrus groves in Cuba, with an addition of only 50% of the N fertilizer generally recommended. On another 100 ha, there was 100% N

FIGURE 32.1
Effect of *A. chroococcum* on the development of tomato plants, with bunches of 12 plants grown from seed. Control is on left; plants in the center and on right had liquid or solid inoculation.

TABLE 32.2

Effect of the Application of *A. chroococcum* on Different Crops under Farm Conditions in Izmir Region, Turkey

Culture	Treatment	Yield (t ha^{-1})	Average Weight of Fruit (g)	Increase of Yield (%)
Tomato	Control	36.43	189.28	—
	Inoculated	45.87	255.97	26
Pepper	Control	18.92	17.32	—
	Inoculated	24.93	21.95	30
Eggplant	Control	91.2	207.55	—
	Inoculated	127.18	261.93	39
Cotton	Control	4.09	5.78	—
	Inoculated	5.95	8.52	45
Maize	Control	8.32	200.67	—
	Inoculated	11.15	290.55	34
Soy bean	Control	2.82	—	—
	Inoculated	3.62	—	28
Sunflower	Control	4.12	—	—
	Inoculated	5.7	—	38

fertilization without any biofertilization, and on still another 100 ha, only 50% of the recommended N fertilizer was applied (Martínez Viera et al., 2001). The substitution of biofertilizer for half of the N application produced 10% more oranges and one-third more grapefruit.

With aerial applications of *A. chroococcum* biofertilizer on mango orchards in Mexico, there was an increment of 28% in the total number of fruits harvested and 20% more total weight of fruits, as seen in Table 32.4 (Martínez Viera et al., 2001). From an economic perspective, an important additional effect of biofertilization was the acceleration of fructification; 38% of the production could be picked in the first harvest of the biofertilized trees, while from the control trees, it was only 22%. The second

(a)   (b)

FIGURE 32.2
(a) Comparison of cassava plants inoculated with *A. chroococcuum*. Left: Inoculated plant; right: control.
(b) Comparison of potato plants inoculated with *A. chroococcum*. Left: Control plants; right: inoculated plants.

TABLE 32.3

Effect of Foliar Applications of *A. chroococcum* on Grapefruit and Orange Yields in Cuba

Treatment	Yield (t ha^{-1})
Grapefruit	
50% N + Azotobacter	73.00
100% N	66.50
50% N	57.90
Orange	
50% N + Azotobacter	48.00
100% N	36.25
50% N	27.60

Source: From Martínez Viera, R., et al., in *Latinoamericano de las Ciencias del Suelo,* Sociedad Latinoamericana de la Ciencia del Suelo, Varadero, Cuba, 2001, 265–274. With permission.

picking was lower from the biofertilized trees, because fruit had been more heavily removed in the first picking and the trees were recuperating. The biofertilized trees then gave a third picking 44% higher, extending the harvesting season. The value of the first and third pickings was higher because demand-supply relations were more favorable and give a higher price in the market.

Table 32.5 presents the results obtained with the application to rice of a mixed biofertilizer based on *A. chroococcum* and *Azospirillum brasilense,* applied in combination with the phosphorus-solubilizing fungus *Penicillium bilaii* in rice fields in the Tolima department of Colombia. Application of N fertilizer was reduced by 40% on the biofertilized fields while the control fields received the recommended inorganic fertilization (100%) (Martínez Viera et al., 2001). The mixture of bacteria is able to contribute, by means of biological fixation, 40% of the reduced nitrogen fertilizer, and even to increase the yield by 4–9% through the bacterial provision to plants of phytohormones and other active substances. When combined with a phosphorus solubilizer, the yield was increased by 15–18%.

TABLE 32.4

Effect of Foliar Applications of *A. chroococcum* on Mango Crop in Sinaloa State, Mexico

Collection	No. of Fruits		Weight of Crop (kg)	
	Control	Biofertilizer	Control	Biofertilizer
1st	328	716	201.8	349.2
2nd	796	540	310.4	218.8
3rd	348	630	210.6	314.6
Total	1472	1886	730.8	882.6

Source: From Martínez Viera, R., et al., in *Latinoamericano de las Ciencias del Suelo,* Sociedad Latinoamericana de la Ciencia del Suelo, Varadero, Cuba, 2001, 265–274. With permission.

TABLE 32.5

Effect of Applications of a Mixed Biofertilizer, with *Azospirillum Chroococcum* and *A. Brasiliense* in Combination with *Penicillium Bilaii*, on Rice Fields, Tolima Department, Colombia

Rice Fields	Treatment	Grains/Panicle	Empty Grains (%)	Yield (t ha^{-1})	Increase (%)
S. Lorenzo (4 ha)	Control	145	6	7.81	—
	Mixed	150	4	8.12	4
	Mixed + *P. bilaii*	153	4	8.96	15
La Pilar (3 ha)	Control	133	6	6.14	—
	Mixed	148	4	6.74	9
	Mixed + *P. bilaii*	155	4	7.32	18

32.4 Discussion

It is becoming evident that bacterial biofertilizers and biostimulators can be produced and applied with high efficacy and economic benefit in a variety of agricultural systems. But it is necessary to select appropriate strains for the respective plant species and to use effective culture medium and fermentation techniques in order to obtain higher bacterial biomass and higher concentration of active substances. Further, we now know that it is possible to increase the effectiveness of these biocompounds by formulating and applying mixed preparations based on microorganisms that perform and contribute different biological functions.

The application of these bioproducts is possible in a wide variety of environmental conditions, because it is not necessary to establish the bacteria permanently in the soil. One needs only to establish them temporarily in the rhizosphere for 3–4 months, during which time the microorganisms are able to act in association with root secretions. It has been reported in the literature, and is confirmed by our experience, that acid soils are not very conducive for *Azotobacter* spp, for example. However, we have obtained significant yield results in such soils with application of this bacteria isolated from a neutral soil (Table 32.5), even though 3 months after application the inoculated bacterial population was very scarce.

The results obtained have important economic benefits for producers, not only because of the yield increases obtained from biofertilizer applications, but also because such treatments support earlier maturation, which enables farmers to get their product to market when there is little competition and prices are highest. Over time, this advantage will diminish as other producers follow suit. However, by then we expect that the production of biofertilizers will have expanded in scale and the cost of these materials will be less than at present. This is a dynamic area where experience and new products are both increasing rapidly. Scientists and producers need to be prepared for unforeseen disadvantages that could arise in the future from such methods. However, so far, the balance is heavily weighted in favor of advantages, and the disadvantages of chemical fertilization are likely to become greater and more obvious in the years ahead.

References

Abd-el-Malek, Y., Free living nitrogen-fixing bacteria in an Egyptian soil and their possible contribution to soil fertility, *Plant Soil*, 423–442 (1971), special volume.

Bauer, T., *Microorganismos Fijadores de Nitrógeno*, (http://11www.microbiologia com.ar/suelo/rhizobium.html) (2001).

Bhat, J.B., Limayev, E.S., and Vasantharajam, B.L., *Ecology of the Leaf-Surface Microorganisms*, Indian Academy of Sciences, New Delhi (1971).

Burdman, S., Hamaoui, B., and Okon, Y., *Improvement of Legume Crop Yields by Co-Inoculation with Azospirillum and Rhizobium*, Center for Agricultural Biotechnology, Jerusalem (2000).

Debinstein, J., *A Tropical Rain Forest*, Wiley, New York (1970).

Dibut, B., Obtención de un bioestimulador del crecimiento y el rendimiento vegetal para el beneficio de la cebolla (*Allium cepa* L.). Ph.D. Thesis, Comisión Nacional de Grados Científicos, Ministerio de Educación Superior, Havana (2000).

Dibut, B. and Martínez Viera, R., Biofertilizantes y bioestimuladores: Métodos de inoculación, In: *Manual de Agricultura Orgánica Sostenible*. UN Food and Agriculture Organization, Havana, 17–22 (2003).

Döbereiner, J., Dinitrogen fixation in rhizosphere and phyllosphere associations, In: *Encyclopedia of Plant Physiology*. Springer, Berlin, 332–350 (1983).

Fadeiev, N.I., *Microbial Plant-Growth Substances*. USSR Academy of Sciences, Moscow (1986) (in Russian).

Higa, T., Effective microorganisms: Their role in Kusei nature farming and sustainable agriculture, In: *Proceedings of the Third International Conference on Kusei Nature Farming*. US Department of Agriculture, Washington, DC, 161–184 (1995).

Izquierdo, J., Ciampi, L., and de García, E., *Biotecnología Apropiable: Racionalidad de su Desarrollo y Aplicación en América Latina y el Caribe*, UN Food and Agriculture Organization, Santiago, Chile (1995).

Katayama, A., Rao, T.P., and Loto, O., Root system development of components in intercropping, In: *Dynamic of Roots and Nitrogen in Cropping Systems of the Semi-Arid Tropics*. ICRISAT, Patancheru, India, 619–623 (1996).

Martínez Viera, R., *Ciclo Biológico del Nitrógeno en el Suelo. Ed*, Cientifico-Técnica, Havana (1986).

Martínez Viera, R. and Dibut, B., Los biofertilizantes como pilares Básicos de la Agricultura Sostenible, In: *Memorias del Taller Internacional sobre Gestión Medioambiental de Desarrollo Rural*. INIFAT, Havana, 62–81 (1996).

Martínez Viera, R. et al., Trascendencia internacional de los biofertilizantes cubanos, In: *Memorias del V Congreso Latinoamericano de las Ciencias del Suelo*. Sociedad Latinoamericana de la Ciencia del Suelo, Varadero, Cuba, 265–274 (2001).

Martínez Viera, R. et al., Reducción de la fertilización nitrogenada en distintos cultivos económicos mediante la aplicación de biofertilizantes, *Memorias del Congreso Trópico 2004*. Academia de Ciencias de Cuba, Havana, 146–161 (2004).

Olivares, F.L. et al., Injection of mottled stripe disease in susceptible and resistant sugar-cane varieties by the endophytic diazotroph Herbaspirillum, *New Phytol.*, **135**, 723–737 (1997).

Parr, J.F., Hornick, S.B., and Kaufman, D.D., *Use of Microbial Inoculants and Organic Fertilizers in Agricultural Production*, Food and Fertilizer Technology Center, Taipei, Taiwan (1994).

Paul, E.A. and Clark, F.E., Phosphorus transformation in soil, In: *Microbiology and Biochemistry*, Paul, E.A. and Clark, F.E., Eds., Academic Press, New York, 222–232 (1989).

Peoples, M.B. and Craswell, E.T., Biological nitrogen fixation: Investments, expectations and actual contributions to agriculture, *Plant Soil*, **141**, 13–39 (1992).

Ruinen, J., Foliar association in higher plants, In: *The Biology of Nitrogen Fixation*, Quispel, A., Ed., North Holland, Amsterdam, 210–246 (1974).

Sasson, A., La contribución de las biotecnologías a la alimentación, *Biotecnología Aplicada*, **17**, 2–6 (2000).

Singh, S.C. and Subba Rao, N.S., Associative effect of *Azospirillum brasiliense* and *Rhizobium japonicum* on nodulation and yield of soybean, *Plant Soil*, **53**, 387–392 (1979).

Subba Rao, N.S., Interaction of nitrogen-fixing microorganisms with other soil microorganisms, In: *Biological Nitrogen Fixation*, Marcel Dekker, New York, 37–63 (1996).

33

Inoculation and Management of Mycorrhizal Fungi within Tropical Agroecosystems

Ramon Rivera and Felix Fernandez
National Institute of Agricultural Sciences (INCA), Havana, Cuba

CONTENTS

The importance of mycorrhizal symbiosis for plants is widely recognized by the international scientific community which appreciates that this increases plants' capacity for water and nutrient absorption, produces a degree of tolerance for root diseases and pests, and at the same time improves some of the soil's physical properties. The literature supporting these conclusions has been reviewed in Chapter 9. These changes support plants' adaptation to stressful conditions, so promoting the colonization of crop roots by mycorrhizal fungi is considered a means for enhancing agricultural production, particularly under marginal soil conditions. The use of this production practice is limited, however, by the current paucity of effective mycorrhizal products tailored for different production systems and by a lack of research and application programs that proceed with a systems perspective. The broader utilization of mycorrhizal fungi is also apparently constrained by an assumption that they should be used mainly for crops growing on marginal soils. In fact, we have found that the benefits of inoculation are not thus limited.

In the early 1990s, given certain constraints that the country was experiencing, a full-scale research program was started in Cuba with the objective of establishing a solid scientific understanding of how mycorrhizal symbiosis can be managed as a positive element of agroecosystems. With such knowledge, we have sought to develop mycorrhizal products to be applied at low doses, evaluating them on a productive scale, not only in their own country but also in others in the Latin American region. This chapter summarizes the results of this research.

33.1 Basic Knowledge Needed for the Effective Management of Mycorrhizal Symbiosis

The Cuban research program was based on three main premises: (i) that efficient inoculation of plants with arbuscular mycorrhizal fungi (AMF) species is technically and economically feasible, (ii) that the edaphic environment is very important for determining and selecting which strains will be most efficient for use in inoculation, and (iii) that external nutrient amendments will influence the effectiveness of fungi–plant symbiosis. Experiments were undertaken under glasshouse, microplot, and field conditions, with a wide variety of crops, soils, and AMF species. In general, positive effects on yield have been recorded for all crops and on all soils studied, from Acrisols with low fertility to calcareous Cambisols and Vertisols having high fertility (FAO, 1989). A summary of the program results has been presented in Rivera and Fernández (2003). The main results were the following.

33.1.1 Arbuscular Mycorrhizal Fungi Strain-Crop Specificity

In our research program we have found that for specific edaphic conditions, the most effective AMF strain, the one that produces the highest yields from different crops, was the same for each of the crops studied. This does not mean that all crops had the same inoculation response or the same mycorrhizal dependency, or that all plant species were equally associated with the different AMF strains studied. Rather our finding is that a most "efficient" AMF strain could be identified for all the crops evaluated under a specific set of soil conditions. This we call "low efficient strain-crop specificity" (Rivera and Fernández, 2003). It contrasts with the highly specific relationships that have been observed between Rhizobia strains and legume species.

Table 33.1 summarizes the results of experiments on the effects of inoculation on a diverse set of crops grown in the same soil (calcareous Cambisols), including results for some AMF species that were not studied for all crops. The interaction between different

TABLE 33.1

Effectiveness of Inoculated Strains with Different Crops in Calcareous Cambisols in Cuba, 1994–1998 (each value is average for 3 years of results)

	EI%							
Strains	**Potato**	**Cassava**	**Sweet Potato**	**Malanga**	**Pepper**	**Cucumber**	**Tomato**	**Banana**
Glomus intraradices	43.9a	48.8a	397.6a	110.0a	77.7a	74.2a	148.5a	68.0a
G. fasciculatum	31.2ab	27.4bc	319.5b	6.6bc	38.2b	44.5b	28.3c	56.3a
G. mosseae	24.7bc	1.1d	186.5c	20.0b	37.6b	9.4c	92.1b	10.5cd
G. clarum	18.0bc	38.0a	7.3d	3.3bc	26.1c	18.7c	23.3d	29.2bc
G. occultum	5.4c	29.8bc	3.6d	18.3b	—	—	—	17.9cd
A. scrobiculata	1.8d	20.2c	0.0d	−10.0c	—	—	—	45.2ab
G. spurcum	—	—	—	—	73.1a	71.2a	130.0a	—
C_V (%)	12.8	7.1	6.9	8.6	6.35	11.3	6.35	12.6

Numbers in the columns with the same letters are not significantly different ($p < 0.05$) from each other, Duncan's test. EI% (Effectiveness Index) = ((AMF yield − control yield)/Control yield) × 100. C_V: coefficient of variation (%) of the original analysis of variance (ANOVA).

Source: From Rivera R. and Fernández-K. S., *El manejo efectivo de la simbiosis micorrízica, una vía hacia la agricultura sostenible: Estudio de caso El Caribe*, National Institute of Agricultural Sciences (INCA), Havana, 2003. With permission.

crops and AMF species exhibited a highly variable Effectiveness Index (EI %). However, the results associated with *Glomus intrarradices* under this particular soil condition showed consistently good performance for all the crops evaluated. At the same time, *G. fasciculatum* presented usually adequate results, so both fungi can be considered "efficient" AMF species for this kind of soil situation. We note that *G. spurcum*, although used with just three crops, also exhibited very good behavior, similar to *G. intraradices*, so possibly it can also be selected as an "efficient" AMF species after further studies.

This pattern of response (low efficient strain-crop specificity), could simplify the management of AMF symbiosis because recommendations of an efficient AMF strain to be used for inoculation with a particular soil would not to be limited to certain crops. Low efficient strain-crop specificity has also been reported by Siqueira and Franco (1988) and Sieverding (1991).

33.1.2 Arbuscular Mycorrhizal Fungi Strain-Soil Specificity

Table 33.2 shows strain recommendations derived from studies carried out to assess the performance of mycorrhizally infected coffee seedlings across a range of different soils. Most of the AMF strains evaluated came from preexisting collections, but some were native AMF isolations from the studied soils, with *Glomus* sp. 1 and sp. 2 proving to be the most effective among the latter. However, while these latter *Glomus* strains were found to be effective in some soils, they were not generally superior to the AMF strains already better known.

Soil type, with associated fertility expressed here in terms of interchangeable Ca^{2+} and Mg^{2+}, affects the mycorrhizal behavior of different AMF strains. In soils of low fertility, we found that strains such as *Acaulospora scrobiculata* and *G. clarum* gave better performance,

TABLE 33.2

Efficient AMF Strains per Type of Soil for Coffee Seedling Production in Cuba during 1990–1998 Period and Interchangeable $Ca^{2+} + Mg^{2+}$ Soil Contents

Soils	Soil Content Ca^{2+} + Mg^{2+} (cmol kg^{-1})	AMF Strains Recommended
Haplic Acrisol (Rhodic Kandiudult)	2.8	*Glomus clarum, G.* sp. 1, *Acaulospora scrobiculata*
Distric Nitisol (Rhodic Kandiudalf)	4.0	*G. clarum, G. intrarradices, A. scrobiculata*
Humic–Cromic and Cromic Cambisol, Humic Eutrudept and Humic Eutrustept	6.7–9.4	*G. clarum* and *G.* sp. 2
Cromic Luvisol (Chromic Hapludalf)	8.7	*G. fasciculatum, G. mosseae*
Gleyic Cambisol (Aquic Haplustalf)	10.4	*G. intrarradices, G. mosseae, G. fasciculatum*
Éutric Ferralsol (Rhodic Eutrustox)	12.0–15.0	*G. fasciculatum, G. intrarradices*
Éutric and Humic–Eutric Cambisol (Humic Eutrustept)	16.8–39.9	*G. fasciculatum*

Sources: From Fernández, F., Manejo de las asociaciones micorrízicas arbusculares sobre la producción de posturas de cafeto (*C. arabica*, L. var. Catuaí) en algunos tipos de suelos. Ph.D. Thesis, National Institute of Agricultural Sciences (INCA), Havana, 1999. Sánchez, C., Uso y manejo de los hongos micorrizógenos y abonos verdes en la producción de posturas de cafeto en algunos suelos del macizo Guamuhaya, Ph.D. Thesis, National Institute of Agricultural Sciences (INCA), Havana, 2001. Joao, J.P., Efectividad de la inoculación de cepas de HMA en la producción de posturas de cafeto sobre suelos Ferralítico Rojo compactado y Ferralítico Rojo Lixiviado de montaña, Master's Thesis, National Institute of Agricultural Sciences (INCA), Havana, 2002. With permission.

while in medium- and high-fertility soils, *G. fasciculatum* and *G. intraradices* had more effect.

Since the most efficient strains appear to perform well in a rather wide range of soils, it should be possible to achieve good results on a wide spectrum of soils by working with just three or four efficient AMF species. According to our experience, inoculation must be conducted using selected single-strain inoculants. This will avoid possible competition between the recommended high-infectivity strains. Some of these may not be very efficient under a certain soil condition, and their success in root infection will prevent more efficient strains from getting established. Our assessment and application of these results has enabled us to identify and recommend certain strains most appropriate for given agro-ecosystems, having looked for low efficient strain-crop specificity and adequate strain effectiveness per group of soils (Rivera and Fernández, 2003).

33.1.3 Nutrient Delivery and Mycorrhizal Effectiveness

A large number of experiments were conducted both on microplots and under field conditions to assess the effects of practices for delivering nutrients on mycorrhizal effectiveness. It was already known that the amount of chemical fertilizer applied to the soil can affect mycorrhizal symbiosis. This was confirmed in our experiments as shown in Figure 33.1. Symbiosis effectiveness was found to be increased where a low fertilizer rate was used with efficient AMF-inoculated crops that were being grown under low or medium nutrient availability conditions. Such conditions led to higher mycorrhizal colonization and yields.

There is presumably an optimal fertilizer rate associated with efficient AMF presence that gives a maximum symbiotic efficiency under certain soil conditions. Such an optimal rate, compatible with an efficient AMF strain for mycorrhizal plants, would assure higher yields and lower costs, by reducing application of chemical fertilizer, compared with

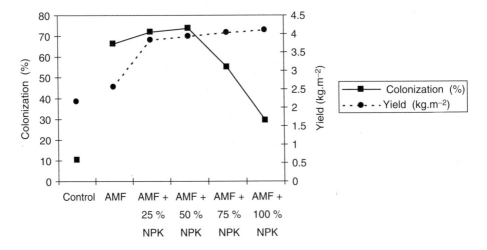

FIGURE 33.1

Effect of fertilization (%NPK) on mycorrhizal colonization and sweet potato yield in efficient AMF strain-inoculated plants on Calcareous Cambisols microplot. *Source:* From Ruiz, L., Efectividad de las asociaciones micorrízicas en especies vegetales de raíces y tubérculos en suelos Pardos con carbonatos y Ferralíticos Rojos de la región central de Cuba, Ph.D. thesis, National Institute of Agricultural Sciences (INCA), Havana, 2001. With permission.

uninoculated plants that received 100% of the recommended NPK dose. This saving would reflect the greater soil nutrient and fertilizer absorption capacity of mycorrhizally inoculated plants.

Application of fertilizer nutrients at higher-than-optimal rates decreases mycorrhizal colonization and inhibits this in the presence of 100% NPK doses, with no yield decrease in the latter situation. That there is no decline indicates that plants are meeting their nutritional requirements but not through AMF activity. Mycorrhizal symbiosis is a mechanism that allows plants to meet their nutritional needs, to obtain up to their full potential yield by depending on existing available soil or substrate nutrients. Similar results had been obtained with organic nutrient supply. Mycorrhized plants not only present a higher capacity for absorption of soil nutrients and fertilizers, but also have more tolerance for water deficits and other stresses, as discussed in Chapter 9.

For best crop yields and maximum symbiotic effectiveness, it is necessary to complement efficient AMF strain inoculation with low quantities of nutrients from organic and/or mineral sources. Appropriate management of nutrient availability makes the recommended efficient AMF strains also effective. This suggests that the purposeful use of mycorrhizal symbiosis should be beneficial not only for marginal soil conditions, but also in better-endowed, intensive production systems, as can be seen from the evaluations reported in Table 33.4 below.

33.1.4   Co-Inoculation with Rhizosphere Bacteria

The joint application, or co-inoculation, of AM fungi together with bacteria from the Rhizobium, Bradyrhyzobium, Azotobacter, Azospirillum, and Burkholderia genera has generally shown positive results. Superior yields are obtained compared to those from applying only mycorrhizal inoculants, always taking into account that appropriate Rhizobacterium-crop specificity must be observed, i.e., only Rhizobacteria should be used for which there is experimental evidence of positive crop response. Co-inoculation effects are a consequence of positive relations established among these genera within the mycorrhizal plant rhizosphere (Fitter and Garbaye, 1994; Höflich et al., 1994; Gryndler, 2000). These favorable results suggest that bacterial-association mechanisms are complementary with mycorrhizal symbiosis. They indicate further that the application of AMF with other biofertilizers can be recommended as part of modern agricultural production.

33.1.5   Mycorrhization and Crop Sequences

Some experiments have been completed evaluating how AMF inoculations can be best used within different cropping systems, but general recommendations are not yet ready. We can report the following results, however. In two crop sequences — potato–sweet potato–cassava–sweet potato on Calcareous Cambisols (Ruiz, 2001) and soybean–corn–sweet potato and soybean–sunflower–sorghum on Eutric Ferralsols (Riera, 2003) — it was found that by inoculating with an efficient strain every two crops, mycorrhizal functioning and yields are similar to those obtained when inoculation was carried out for every crop. On the other hand, when inoculation was carried out every three crops, mycorrhizal functioning and yields were significantly lower than when inoculation was carried out with every crop. Evaluation of soil AMF spore dynamics showed that inoculation raised levels in any of the sequences, depending on the inoculation frequency, the crop concerned, and the preceding crop (Riera, 2003). These results are also indicative of the low efficient strain-crop specificity discussed above.

33.2 Development of Mycorrhizal Inoculants

A mycorrhizal inoculant in solid form, known as EcoMic®, has been developed in Cuba that can colonize roots quite effectively. Its excellent adhesive properties improve the retention of propagules on the seed surface and make it appropriate for seed-dressing technology (Fernández et al., 1999). EcoMic® inoculant has a high content of spores and extraradical mycelium per gram of substrate. In recent years, extraradical mycelium has come to be seen as offering the most effective propagules, especially for Glomus strains. According to Jasper et al. (1989), extraradical mycelium has certain mechanisms that allow it to remain infective in the substrate so long as its physical structure is well preserved. On the other hand, Bago et al. (2000) have pointed out that the cytoplasmic content of external hyphae, the most plentiful ones in the fungal net, function differently from branching absorption structures (BAS) or from absorptive hyphae, remaining intact for long periods, even years, so perhaps this contributes to high EcoMic® performance.

Table 33.3 presents information on the main mycorrhizal characteristics of certified inoculants produced in 5-kg pots comparing "traditional" production methods with those of EcoMic®. The differences between the two methods of inoculant development appear to relate to variations both in substrates and environmental conditions, particularly pH. During production supported by EcoMic® (with pH over 7), not only did more external mycelia emerge, but spores were also more numerous. Several authors have reported significant increases in the production of extraradical mycelium when AMF were grown in media over 7.0 pH (Abbott and Robson, 1985; Porter et al., 1987; van Aarle et al., 2002). Nevertheless, the factors that determine extramatrical mycelia architecture of AM fungi remain one of the most puzzling research topics in this domain.

The commercial product is obtained by inoculating the host plant *Brachiaria decumbens* with a small amount of certified inoculant on a specific substrate in beds of $3-6 \text{ m}^{-3}$. When there is a greater volume of production than this, the quality of the product is somewhat reduced. However, it is possible to maintain EcoMic® quality advantages in

TABLE 33.3

Colonization of Roots, Growth of External Mycelium, Sporulation, and Arbuscule Frequency by *Glomus fasciculatum* Inoculation in Association with *Sorghum vulgare* in Two Different Substrates

Substrate	Colonization (%)		External Mycelium (mg g^{-1} substrate)		Sporulation (spores g^{-1})		Arbuscules (%)	
	60d	120d	60d	120d	60d	120d	60d	120d
Ecomic®	42.1c	65.6a	34.3b	56.8a	120b	245a	67.4a	42.9b
Traditional Mixture	53.4b	68.3a	15.2d	23.9c	43c	86c	62.9a	36.6b
SE	1.4***		0.38***		2.6***		2.1***	

Comparison of EcoMic® with traditional inoculation mixture of soil:sand:organic manure (ratio 3:3:1) after 60 and 120 days of plant growth in 5-kg pots. Values are the means for eight observations. Numbers in the columns with the same letters are not significantly different ($p < 0.001$), Duncan's test. ***Significant < 0.001.
Source: Authors' data.

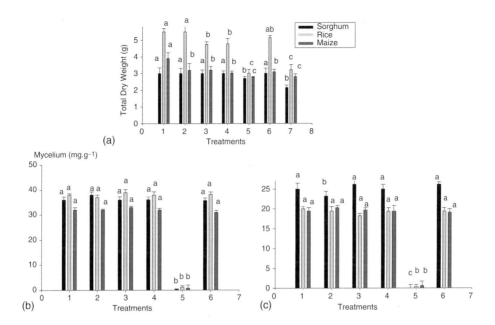

FIGURE 33.2

Effects on dressed seed of sorghum (*Sorghum vulgare*), rice (*Oriza sativa*), and maize (*Zea mays*) with different EcoMic® doses according to (a) total dry seed weight (in g), (b) external mycelium at 30 days of inoculation, and (c) mycorrhizal colonization (in %). *Source:* Authors' data. Doses of EcoMic®, ratio of inoculum weight-to-dressed seed weight: 1 = 1:1; 2 = 1:2; 3 = 1:5; 4 = 1:10; 5 = 1:1 dose of inoculant in soil:sand:organic matter to dressed seed; 6 = traditional inoculation process: 5 g plant^{-1}; 7 = Uninoculated control.

relation with traditional commercial inoculants as seen in Table 33.3. Commercial EcoMic® can commonly lead to between 20 and 30 spores g^{-1} with substantial mycelia production.

The success of this technology is probably due to the greater amount of external mycelium present in the commercial product. According to our general understanding, 50% of Glomus spore populations are not likely to be viable or will have very long dormancy periods, and another 25% will have long dormancy periods, so only approximately 25% will become active propagules. This makes it impossible for a seed-dressed technology that starts only with spores to have the desired effect of mycorrhization. From our experience, we consider the mycelium of Glomus to be the most important fungal propagules for starting the colonization process.

The dry inoculant, packed and stored at 25°C, is effective for up to 1 year, although we have had some instances of 2 years of viability. The drying process and keeping product humidity low, less than 5%, are very important for sustaining long periods of viability.

EcoMic® has been found to be effective in seed dressing when applied in low doses of even just 10% of seed weight (Figure 33.2). Differences were not observed between the traditional inoculant used for seed treatment and seeds dressed with different EcoMic® percentages by weight. This makes it evident that, independent of plant species, significant agrobiological effects can be achieved, even with very low doses of inoculant (10–20%).

In general, the seed-dressing dose of EcoMic® that showed the most total dry matter production was accompanied by high values of root colonization (%) and of external mycelium growth. On the other hand, dressing seeds with the traditional inoculation medium, at even 100% of seed weight, did not show positive effects, indicating that seed dressing is not a viable strategy for the traditional method. That these experiments confirmed the productivity of seed dressing with very low inoculant doses of EcoMic® (10%) showed that this technology is feasible for use under current agricultural conditions.

33.3 The Use of Mycorrhizal Symbiosis on a Large Scale

An extensive program of validation for EcoMic® biofertilizer has been carried out using the coating of seeds as means of application for many common crops, both for low external-input, small-scale agricultural production as well as with high-input management on larger areas. This has been seen in Cuba and in other countries in the region such as Colombia, Bolivia, and Mexico. Evaluations have shown yield increases ranging between 15 and 40% in both kinds of production systems, whether in Cuba or different countries (Table 33.4 and Table 33.5). The crops benefited have been soybean, sorghum, rice, maize, beans, sunflower, cotton, cassava, wheat, and vegetables, among others. In the case of leguminous crops, efficient AMF strain as EcoMic® has been applied on a commercial scale together with Rhizobial biofertilizers (Chapter 32).

The results of inoculation technology reported in Table 33.4 and Table 33.5 have confirmed that "adequate" strain inoculation can successfully raise yields in a cost-effective way on a scale that is meaningful for large, commercial farmers as well as for small or subsistence ones. The average size of validation areas in Table 33.4 was 125 ha, while that of areas in Table 33.5 was 1.5 ha, showing that positive results are attainable under widely varying conditions of production.

The application of this technology is certainly not limited to small-scale agricultural production. Machines can be used to coat seeds with the inoculant very efficiently, and seed drills can be employed in mechanized production systems to cover large sowing areas. Mechanical coating of seeds has permitted a reduction in EcoMic doses to as little as 6% of seed weight. They have further demonstrated the efficacy of seed inoculation for

TABLE 33.4

Results of EcoMic® Validation Campaigns in Larger, High-Input Agricultural Operations, with Different Crops and Soils Types; Doses = 6–10% Seed Weight, Applied by Coating

Crop — Country	Soils	Área (ha)	Yield with AMF (ton ha^{-1})	Yield with Control (ton ha^{-1})	Increase (%)
Rice — Colombia	Eutric Fluvisol	16	4.8	2.70	77.7
Cotton — Bolivia	Eutric Luvisol	94	0.94	0.69	38.0
Maize — Cuba	Eutric Ferrasol	16	2.84	2.34	21.3
Maize — Cuba	Eutric Ferrasol	16	5.41	3.04	77.9
Maize — Bolivia	Eutric Luvisol	150	2.92	2.16	35.1
Maize — Bolivia	Eutric Luvisol	150	3.12	2.51	24.3
Wheat — Bolivia	Eutric Luvisol	50	3.19	2.75	16.0
Wheat — Bolivia	Eutric Luvisol	50	3.12	1.82	71.4
Soybean — Bolivia	Eutric Luvisol	150	2.73	1.94	40.7
Soybean — Bolivia	Eutric Luvisol	750	2.93	2.32	26.3
Bean — Bolivia	Eutric Luvisol	11	1.71	0.96	78.1
Sunflower — Bolivia	Eutric Luvisol	40	1.23	0.86	43.1

Eutric Fluvisol: (Typic Haplustalf); Eutric Ferralsol (Rhodic Eutrustox); Eutric Regosol (Eutric Usthorthent).
Source: From INCA, *Efecto de las aplicaciones del biofertilizante Ecomic (HMA) en cultivos de interés económico, durante el periodo 1990–1998*, Research Report, National Institute of Agricultural Sciences (INCA), Havana, 1999, Pejeira, L. et al., *Informe final sobre campaña del Biofert-Bol (EcoMic®) en Sta Cruz de la Sierra, Bolivia*, National Institute of Agricultural Sciences (INCA), Havana, 1998. With permission.

TABLE 33.5

Results of EcoMic® Validation Campaigns in Smaller, Low-Input Agricultural Operations, with Different Crops and Soil Types; Doses = 10% Seed Weight, Applied by Hand Coating

Crop — Country	Soils	Area (ha)	Yield with AMF (ton ha^{-1})	Yield with Control (ton ha^{-1})	Increase (%)
Rice — Cuba	Petroferric Gleysol	1.0	6.80	4.60	47.8
Rice — Colombia	Eutric Fluvisol	2.0	2.15	1.30	65.3
Rice — Colombia	Eutric Fluvisol	2.0	2.40	1.40	71.4
Cotton — Colombia	Molic Gleysol	1.5	2.60	2.20	18.2
Cotton — Colombia	Molic Gleysol	1.5	2.50	1.90	31.5
Maize — Colombia	Eutric Fluvisol	1.0	2.96	1.64	80.5
Maize — Colombia	Eutric Fluvisol	1.0	2.59	1.45	78.6
Bean — Colombia	Eutric Fluvisol	1.0	0.50	0.29	72.4
Bean — Cuba	Eutric Ferralsol	1.0	1.01	0.69	46.4
Bean — Cuba	Eutric Ferralsol	2.5	1.20	0.96	25.0
Soybean — Cuba	Eutric Ferralsol	0.5	2.63	1.50	75.3
Maize — Cuba	Eutric Ferralsol	3.0	2.83	2.44	16.0
Groundnut — Cuba	Eutric Ferralsol	1.2	1.12	1.00	12.0

Petroferric Gleysol (Kandic Plinthaquult); Molic Gleysol (Typic Endoaquoll).

Sources: From INCA, *Efecto de las aplicaciones del biofertilizante Ecomic (HMA) en cultivos de interés económico, durante el periodo 1990–1998,* Research Report, National Institute of Agricultural Sciences (INCA), Havana, 1999. Sosa, J., *Informe final de resultados de validación del biofertilizante Biomonte (Ecomic®) en municipios del Departamento de Córdoba, Colombia,* Fundación San Isidro, National Institute of Agricultural Sciences (INCA), Havana, 1999. Calderón, A., Corbera, J., and Medina, N., El papel de los biofertilizantes en la actividad de extension agrícola del INCA: Sa crecimiento en los últimos tres años, Research Report, National Institute of Agricultural Sciences (INCA), Havana, 2002. With permission.

different soil groups and low strain-crop specificity, which makes the innovation broadly applicable. The effectiveness of AMF in nutrient delivery has been validated, along with the effectiveness of seed coating, also showing the benefits of co-inoculation with other biofertilizers as well as EcoMic® biofertilizer.

33.4 Discussion

Mycorrhizal fungi are as old as plants, as these biological kingdoms have co-evolved together over a period of 400 million years. Their joint management can make certain fungi species more effective components of agroecosystems offering benefits to plant crops that can be economically exploited. This symbiosis enables plants to enhance their water and nutrient absorption process, enabling them to adapt to nutritional and water-stress conditions. This symbiotic relationship is a general one, whose practical use should not be limited to marginal economic or environmental conditions.

Mycorrhizal symbiosis is closely related with plant development in general, much as Rhizobia are linked with leguminous plants generally (Chapter 12) and with cereals as seen in Chapter 8. Our expanding knowledge of these relationships is now permitting more effective incorporation of microbial populations into operational agricultural systems. Efficiently mycorrhized plant models are establishing and developing new agricultural practices that can help farmers to achieve system optimization. These systems are valid not only under low-input conditions, but also for more capital-intensive agriculture, obtaining higher yields with an increase in soil life while reducing soil burdens from excessive fertilizer applications, at the same time mitigating the negative effects of drought or other abiotic stresses.

It is important that AMF strain inoculation is not thought of as just the application of another input in agriculture. Rather it is a matter of adopting a different concept of agriculture, mobilizing and channeling endogenous biological processes and capabilities. Agriculture with an agroecological and conservationist orientation, protecting the environment and the soil, will guarantee our ability to meet food security needs while responding appropriately to the economic, environmental, and social expectations for 21st century agricultural production.

References

Abbott, L.K. and Robson, A.D., The effect of soil pH on the formation of VA mycorrhizas by two species of Glomus, *Aust. J. Soil Res.*, **23**, 253–261 (1985).

Bago, B., Azcón-Aguilar, C., and Shachar-Hill, Y., El micelio externo de la micorriza arbuscular como puente simbiótico entre la raíz y su entorno, In: *Ecología, Fisiología y Biotecnología de la Micorriza Arbuscular*, Alarcón, A. and Ferrera-Cerrato, R., Eds., Colegio de Postgraduados/Mundi Prensa, Montecillos/México, DF, 78–92 (2000).

Calderón, A., Corbera, J., and Medina, N., El papel de los biofertilizantes en la actividad de extension agrícola del INCA: Sa crecimiento en los últimos tres años, Research Report, National Institute of Agricultural Sciences (INCA), Havana, 2002.

FAO, *Soil Map of the World*, revised legend. Reprint of World Soil Resources Report 60, U.N. Food and Agriculture Organization, Rome (1989).

Fernández, F., Manejo de las asociaciones micorrízicas arbusculares sobre la producción de posturas de cafeto (*C. arabica*, L. var. Catuaí) en algunos tipos de suelos, Ph.D. Thesis, National Institute of Agricultural Sciences (INCA), Havana (1999).

Fernández, F., Rivera, R., and Noval, B., Metodología de recubrimiento de semillas con inoculo micorrizógeno, Intellectual Property Cuban Office (OCPI), Havana (1999), No. 22641.

Fitter, A.H. and Garbaye, J., Interactions between mycorrhizal fungi and other soil organisms, *Plant Soil*, **159**, 123–132 (1994).

Gryndler, M., Interactions of arbuscular mycorrhizal fungi with other soil organisms, In: *Arbuscular Mycorrhizas: Physiology and Function*, Kapulnik, Y. and Douds, D.D., Eds., Kluwer Academic Publishers, The Netherlands, 239–262 (2000).

Hoflich, G. et al., Plant growth stimulation by inoculation with symbiotic and associative rhizosphere microorganism, *Experientia*, **50**, 897–905 (1994).

INCA, *Efecto de las aplicaciones del biofertilizante Ecomic (HMA) en cultivos de interés económico, durante el periodo 1990–1998*, Research Report. National Institute of Agricultural Sciences (INCA), Havana (1999).

Jasper, D.A., Abbott, L.K., and Robson, A.D., Hyphae of vesicular arbuscular mycorrhizal fungus maintain infectivity in dry soil, except when the soil is disturbed, *New Phytol.*, **112**, 101–107 (1989).

Joao, J.P., Efectividad de la inoculación de cepas de HMA en la producción de posturas de cafeto sobre suelos Ferralítico Rojo compactado y Ferralítico Rojo Lixiviado de montaña, Master's Thesis, National Institute of Agricultural Sciences (INCA), Havana (2002).

Pijeira, L., Lara, D., and Mederos, J.D., Informe final sobre campaña de validación del Biofert-Bol (EcoMic®) en Sta Cruz de la Sierra, Bolivia, National Institute of Agricultural Sciences (INCA), Havana (1998).

Porter, W.M., Robson, A.D., and Abbott, L.K., Field survey of the distribution of vesicular–arbuscular mycorrhizal fungi in relation to soil pH, *J. Appl. Ecol.*, **24**, 659–662 (1987).

Riera, M., Manejo de la biofertilización con hongos micorrízicos arbusculares y rizobacterias en secuencias de cultivos sobre suelo Ferralítico Rojo, Ph.D. Thesis, National Institute of Agricultural Sciences (INCA), Havana (2003).

Rivera, R. and Fernández, K.S., El manejo efectivo de la simbiosis micorrízica, una vía hacia la agricultura sostenible: Estudio de caso El Caribe, National Institute of Agricultural Sciences (INCA), Havana (2003), (http://169.158.24.166/texts/pd/959-16/250/959-7023-24-5.pdf)

Ruiz, L., Efectividad de las asociaciones micorrízicas en especies vegetales de raíces y tubérculos en suelos Pardos con carbonatos y Ferralíticos Rojos de la región central de Cuba, Ph.D. Thesis, National Institute of Agricultural Sciences (INCA), Havana (2001).

Sánchez, C., Uso y manejo de los hongos micorrizógenos y abonos verdes en la producción de posturas de cafeto en algunos suelos del macizo Guamuhaya, Ph.D. Thesis, National Institute of Agricultural Sciences (INCA), Havana (2001).

Sieverding, E., *Vesicular Arbuscular Mycorrhiza in Tropical Agrosystems*, Deutsche Gesellschaft für technische Zusammenarbeit (GTZ), Eschborn, Germany (1991).

Siqueira, J.O. and Franco, A.A., Biotecnología do solo: Fundamentos e perspectiva, Ministerio de Educação (MEC)/Associacao Brasileira de Educacao Agricola Superior (ABEAS)/Escola Superior de Agricultura Lavras (ESAL)/Fundação de Apoio ao Ensino, Pesquisa e Extensão (FAEPE), Brasília (1988).

Sosa, J., *Informe final de resultados de validación del biofertilizante Biomonte (Ecomic$^{®}$) en municipios del Departamento de Córdoba, Colombia, Fundación San Isidro*. National Institute of Agricultural Sciences (INCA), Havana (1999).

van Aarle, I.M., Olsson, P.A., and Söderström, B., Arbuscular mycorrhizal fungi respond to the substrate pH of their extraradical mycelium by altered growth and root colonization, *New Phytol.*, **155**, 173–182 (2002).

34

Trichoderma: An Ally in the Quest for Soil System Sustainability

Brendon Neumann and Mark Laing

Department of Plant Pathology, University of KwaZulu-Natal, Pietermaritzburg, South Africa

CONTENTS

The Trichoderma species of fungus was first established in 1794 by the Dutch scientist C.H. Persoon, but it remained poorly characterized for many years. In 1932, R. Weindling made the first detailed descriptions of Trichoderma as a parasite of other soil fungi and concluded that, under certain conditions, Trichoderma might be used for the biological control of fungal diseases. Apart from some further work by Weindling (1934), however, Trichoderma as a biocontrol agent received little further attention until the late 1970s. This gap in biocontrol research is typically explained by the intervening successes of chemical pesticides. Today, however, with hundreds of studies having being conducted on various aspects of Trichoderma, this fungus has enjoyed commercial success as a soil inoculant and seed treatment of agricultural crops, with numerous commercial products being registered around the world. Here we offer a summary overview of this remarkable organism; for a more extensive review of Trichoderma, see Harman et al. (2004).

34.1   Characterization of Trichoderma

34.1.1   Taxonomy

The genus Trichoderma falls within the Hyphomycetes, while the sexual stage, when found, is classified as part of the Ascomycetes, usually in the genus Hypocrea. Ever since the genus was established over 200 years ago, there has been much confusion surrounding its taxonomy, however. Of the four original species proposed for this genus, only one still remains (Cook and Baker, 1983). Rifai (1969) revised the genus and created nine species aggregates, of which the most important are *Trichoderma viride* Pers.: Fr., *T. hamatum* (Bon.) Bainier, *T. harzianum* Rifai, *T. koningii* Oud., and *T. polysporum* (Link: Fr.) Rifai.

34.1.2   Activity

The primary activity for which Trichoderma is known is that of biological control of root pathogens. Other activities within the biocontrol field include the control of fruit pathogens such as Botrytis, as well as some recent work on the control of nematodes (Spiegel and Chet, 1998; Sharon et al., 2004). Trichoderma's ability to control plant pathogens can significantly improve plant growth in infected soils. However, it is interesting that plants in apparently healthy soils, uninfected with root pathogens, often demonstrate a positive growth response after being treated with Trichoderma. This has led to the identification of Trichoderma as a growth stimulant as well.

A third activity of Trichoderma allied to growth stimulation is that of environmental buffering. Through the promotion of healthy root development, Trichoderma helps plants to cope with environmental stresses such as drought, waterlogging, and nutritional stress. A final area in which Trichoderma has been used is in the commercial production of enzymes, specifically cellulases. This is thus a very versatile organism. Given this book's focus, this chapter will focus on Trichoderma activities that take place in the soil and contribute to the sustainability of soil systems.

34.2   Biocontrol

Examples of the successful use of Trichoderma spp. as biocontrol agents to reduce or prevent damages to plant roots are numerous and can be found for almost all the major root pathogens that affect all major crops. Some isolates of Trichoderma spp. clearly have the ability to function as effective controls of a wide range of soil pathogens. Table 34.1 lists some of the most important biocontrol activities.

Trichoderma biocontrol activity has also been reported for pigeon pea wilt (*Fusarium udum*) (Biswas and Das, 1999), tomato wilt disease caused by *F. oxysporum* f. *lycopersici* (Sarhan et al., 1999), potato late blight (Arora, 2000), Rhizoctonia black scurf of potatoes (Haggag and Nofal, 2000), stem canker of tomato caused by *Phytophthora parasitica* (Besoain et al., 2001), damping off of tomato caused by *Pythium aphanidermatum* (Gnanavel and Jayaraj, 2003), and many other diseases. The levels of control are often as good or better than those achieved by conventional chemical fungicides, thus making this organism an effective and sustainable alternative to synthetic products for disease control.

TABLE 34.1

Examples of Biocontrol by Trichoderma

Crop	Pathogen	Isolate	Outcome	Reference
Beans, tomato	*Sclerotium rolfsii*; *Rhozoctonia solani*	*T. harzianum* Drenching in greenhouse	Control	Elad et al. (1980)
Beans, tomato, cotton	*Sclerotium rolfsii*; *Rhozoctonia solani*	*T. harzianum* Drenching in field	Control	Elad et al. (1980)
Cotton	Seedling disease complex	*T. virens* Seed treatment + metalaxyl	Equal to fungicides	Howell et al. (1997)
Wheat	Fusarium foot rot	*T. atroviride*, *T. longibrachi-atum*; *T. harzianum*	Increased emergence; winter survival; number of heads and yield increased, lower disease incidence and severity	Roberti et al. (2001)
Maize	Stalk rot (*Pythium aphanidermatum*; *Fusarium graminearum* [*Gibberella zeae*])	*T. viride* and Pseudomonas sp.	Control	Jie et al. (1999)
Maize	Banded leaf and sheath blight (*Rhizoctonia solani* f. sp. *sasakii*)	*T. harzianum*	49.2% control vs. 52.1% by carbendazim	Meena et al. (2003)

34.3 Growth Stimulation

The ability of Trichoderma spp. to promote plant growth when there is apparent absence of pathogens that it could be controlling makes it a more interesting ally for healthier and more productive soil systems. Chang et al. (1986) found that when *T. harzianum* was added to either steamed or raw soil, it reduced the time to flowering for periwinkle (*Vinca minor*) and increased the number of blooms/plant in chrysanthemums. It was also found to increase the heights and weights of other plants. Windham et al. (1986) studied the effects of Trichoderma on tomato and tobacco seedling growth in autoclaved soil. Their work showed an increased rate of seedling emergence and considerable increases in root and shoot dry weights after 8 weeks (213 to 275% in tomato, and 259 to 318% in tobacco). Windham and co-authors concluded that Trichoderma spp. produce a growth-regulating factor that increases seedling emergence as well as shoot and root dry weights.

Ousley et al. (1993, 1994) worked with several isolates of *T. harzianum* and *T. viride* to assess the ability of the isolates to promote growth in a wide range of host plants. Ousley et al. (1993) found that the plant growth responses in lettuce treated with the different Trichoderma isolates resulted from a balance between inhibition and growth promotion as discussed for phytohormones in Chapter 14. These authors conducted similar trials on *Tagetes patula*, petunia and verbena plants to establish the feasibility of using Trichoderma in the promotion of bedding plant growth (Ousley et al., 1994). In their study it was found that, depending on the concentration, each strain was able to increase either the number of flowers, the weight of flowers, shoot wet weights, or shoot dry weights, but seldom all four parameters at the same time. Plant growth promotion by Trichoderma spp. thus appears to be concentration- as well as host-dependent.

Kleifeld and Chet (1992) obtained more consistent growth stimulation responses in experiments with a strain of *T. harzianum* on bean, radish, tomato, pepper, and cucumber

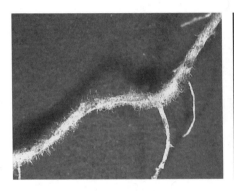
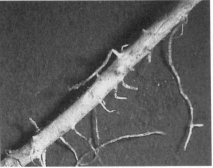

FIGURE 34.1
Section of root from a Trichoderma-treated maize plant (left) showing significantly more tiny root hairs from main root than are seen on the untreated-plant main root (right).

plants. Typical responses included increases in seedling emergence, plant height, leaf area, and dry weight. These authors also observed that there was presence of Trichoderma inside the roots of treated plants. This led them to suggest that Trichoderma may function as a mycorrhizal organism. Trials conducted on maize (Neumann, unpublished) have resulted in noticeable changes in root structure, with more lateral roots and root hairs in Trichoderma-treated plants compared with untreated ones (Figure 34.1). These results are similar to those reported by Harman (2000).

34.4   Environmental Buffering

Through the development of healthier and larger root systems, as already discussed, Trichoderma is able to buffer plants against environmental stresses such as drought, waterlogging, and nutritional stress. In fact, plant growth responses measured in terms of increased yield are often most pronounced with Trichoderma when the plants are under some form of stress.

Neumann and Laing (2002) showed that lettuce plants were not significantly influenced by changes in growth-medium moisture content in the absence of a pathogen (Pythium), and concurrently, at optimum soil moisture conditions, Pythium inoculation did not have a significant effect on lettuce yield. However, when the pathogen was added to plant soil in over- or under-watered conditions, the resulting yield loss was significant. In all cases, the addition of Trichoderma as a biocontrol agent (Eco-T®) resulted in an improvement in lettuce yield. This showed Trichoderma's ability to overcome the combined negative effects of poor soil moisture management and Pythium infection. The results indicated that the organism's impact was on neutralizing the adverse environmental conditions as well as on the fungal disease.

In field trials, Neumann (unpublished) observed significant differences in transplant survival of cabbage seedlings treated with Trichoderma in the greenhouse at 4 weeks prior to transplanting, compared with untreated seedlings. These observations could be explained by the fact that treated seedlings developed much larger, healthier root systems during their 4 weeks of growth in the greenhouse prior to transplanting. When transplanted at a 34°C daytime air temperature, the Trichoderma-treated seedlings were better able to cope with the transplanting shock (Figure 34.2).

FIGURE 34.2

Effect of Trichoderma treatment (left) on the survival of cabbage seedlings, transplanted under extreme temperature and drought stress conditions.

34.5   Modes of Action

Considering that the survival and proliferation of Trichoderma in the soil environment is largely dependent on the existence of large, healthy root systems, it is suggested that, as a genus, Trichoderma has evolved several mechanisms for ensuring and promoting the growth of such roots. Weindling (1932) described in detail the first observations of Trichoderma as a parasite of other soil fungi, especially *Rhizoctonia solani*. In this work, he described how Trichoderma hyphae physically coiled around Rhizoctonia hyphae and could destroy colonies of this fungus. He noted that it was conceivable that Trichoderma might act merely as a competitor for food, while incidentally killing other fungi, although observations strongly suggested that substances from the host hyphae were used as nutrients. It was further noted that it had not been determined whether enzymes or toxins were involved in the parasitic action, although in subsequent work, Weindling proposed that the parasitic activity of Trichoderma was made possible by what he termed "a lethal principle." In the last 70 years, a number of mechanisms used by Trichoderma spp. to achieve control of soil pathogens and enhance plant growth have been identified (Tronsmo and Hjeljord, 1998). These are summarized in Table 34.2.

TABLE 34.2

Proposed Mechanisms of Biocontrol and Growth Stimulation by Trichoderma

Biocontrol	Growth Stimulation
Mycoparasitism	Control of sublethal pathogens
Antibiosis	Environmental buffering (against pH, drought, waterlogging, cold, heat)
Competition (for nutrients or space)	Solubilization of sparingly soluble minerals such as P and Mo
Induction of systemic acquired resistance	Production of plant growth hormones, stimulating root production
Inactivation of pathogens' enzymes	Degradation of allelochemicals, and buffering against soil toxins
Siderophore production (sequestering iron needed by pathogens)	

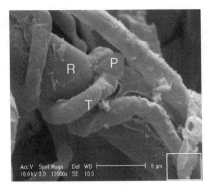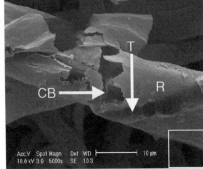

FIGURE 34.3
Trichoderma hyphae (T) coiling around Rhizoctonia (R), penetrating it (P) with resulting breakdown of *Rhizoctonia* hyphal walls (CB).

34.5.1 Mycoparasitism

Mycoparasitism is the best documented mechanism of control by Trichoderma. This is accomplished by Trichoderma through the physical mechanism of coiling hyphae around the pathogen, penetrating and degrading it, as first observed and reported by Weindling (1932). This physical parasitism can easily be observed with the use of scanning electron microscopy (Figure 34.3). A number of enzymes have been identified that are involved in the process of mycoparasitism including cellulase, β-1,3-glucanase, and chitinase.

34.5.2 Antibiosis

Several antibiotics such as gliovirin and gliotoxin have been shown to be part of the biocontrol activity by Trichoderma spp. Antibiotic-deficient mutants have been shown to lose their biocontrol capacity, confirming the importance of antibiotic substances in hyperparasitism. Benhamou and Chet (1997) proposed the following scheme for the interaction of mycoparasitism, enzyme production, and antibiosis in the biocontrol of Pythium by *T. harzianum*:

1. Recognition of Pythium by *T. harzianum* and attachment of latter's hyphae.
2. Production of β-1,3-glucanases to weaken the host cell wall accompanied by the slight production of cellulase to facilitate penetration by Trichoderma hyphae.
3. Production of antibiotic substances to deregulate host cell metabolism.
4. Host cell invasion and increased production of cellulases resulting in the breakdown of pathogen cells.

34.5.3 Acquired Systemic Resistance

Yedidia et al. (1999) showed that plants treated with Trichoderma produced higher levels of enzymes such as peroxidase and chitinase associated with increased plant resistance to pathogens, while Howell et al. (2000) showed that biocontrol activity was highly correlated with the induction of terpenoid synthesis in cotton roots by Trichoderma spp. These responses have been shown to be systemic through split-root experiments in which Trichoderma is only applied to half of the root system. The induction of resistance responses in the remaining untreated roots demonstrates the systemic nature of this effect.

34.5.4 Inactivation of Pathogen Enzymes

Kapat et al. (1998) studied the effect of *T. harzianum* strains on the activity of hydrolytic enzymes produced by *Botrytis cinerea*. These enzymes are responsible for the breakdown of plant cell material and are thus essential in the pathogens' ability to cause disease. Protease enzymes produced by Trichoderma are found to break down the enzymes produced by Botrytis, resulting in a reduction in disease severity. The importance of this mechanism was highlighted by the fact that the addition of protease inhibitors, which block the functioning of protease enzymes, nullified the effectiveness of Trichoderma in the control of *B. cinerea*.

34.5.5 Siderophore Production

Many fungal and bacterial biocontrol agents release siderophores, such as pseudobactines and pyoverdins (Bakker et al., 1990). Trichoderma spp. have been shown to be prolific producers of siderophores (Casale, 1995). These compounds compete very efficiently for iron in the soil to the extent that plant pathogens such as Fusaria are successfully suppressed. The impact of siderophores is strongly influenced by soil pH, as they are effective in neutral to alkaline soil in which iron is not readily available (Duijff et al., 1995). With such a wide array of functional modes of action, Trichoderma-based biocontrol products are unlikely to suffer resistance build-up by pathogens, making them a sustainable alternative to traditional chemical fungicides.

34.6 Mechanisms Apparently Involved

34.6.1 Control of Sub-Lethal Pathogens

It has been suggested that the increase in plant growth obtained through inoculation with Trichoderma can be attributed to the control of sub-lethal pathogens (Kleifeld and Chet, 1992). These pathogens live on fine roots and root hairs and in so doing, reduce the growth of the host plant without causing visible symptoms of disease. However, Trichoderma has been shown to promote plant growth in sterile soil, suggesting that plant growth stimulation involves mechanisms other than the control of lethal or sub-lethal pathogens. This mode of action may operate with Trichoderma, but at present it is not clearly demonstrable.

34.6.2 Solubilization of Sparingly-Soluble Minerals

Altomare et al. (1999) investigated the ability of *T. harzianum* to solubilize certain insoluble or sparingly-soluble minerals by various mechanisms such as acidification of the medium, production of chelating metabolites, and redox activity. These authors concluded that medium acidification did not play a major role in mineral solubilization as the pH of cultures never fell below 5.0. However, chelation and reduction were both found to be positively involved in the process. These researchers concluded that the ability of Trichoderma to solubilize minerals such as MnO_2, metallic zinc, and rock phosphate explains, at least in part, the ability of Trichoderma strains to increase plant growth.

34.6.3 Plant Growth Hormone Production

The early work of Kampert and Strzelczyk (1975a, 1975b) on this topic showed that isolates of Trichoderma produced detectable amounts of auxin- and gibberellin-like substances. Reddy and Reddy (1987) studied the synthesis of the auxin IAA by seed-borne fungi of maize and found that of the 27 isolates screened, *T. viride* synthesized the most IAA, which had a positive impact upon root production of the colonized plants. Neumann (unpublished) has shown that seed treatment of maize and wheat can result in more than doubling of the root surface area, and a quantum increase in root hair production (seen in Figure 34.1). These are apparent responses to the presence of auxins, which were reviewed in Chapter 14.

34.6.4 Degradation of Allelochemicals, and Buffering against Soil Toxins

In view of the importance of allelochemicals (Chapter 16), it is worth considering the suggestion that Trichoderma could play a key role in degrading these organic molecules before they stunt the growth of host plants, as this would diminish the root zone that is available for colonization by Trichoderma (Laing, unpublished). Similarly, Trichoderma may be buffering against the toxic effects of aluminum and manganese toxicity in acid soils, typical of tropical regions. These and other aspects of Trichoderma's influence on plants and other soil organisms warrant extensive and intensive investigation given the demonstrable effects that this remarkable fungus can have on crop and soil system performance.

34.7 Discussion

In order to survive, Trichoderma needs healthy plants and healthy root systems. Members of this genus have evolved numerous mechanisms for promoting the sustainable development and growth of healthy plant root systems in the soil environment. Their multiple roles in biocontrol and growth stimulation are now well documented. However, their ability to buffer plants against adverse environmental stress is a relatively recent discovery that could be exploited more profitably in sustainable, low-input agriculture. Our research and that of others has shown that Trichoderma should be considered as a powerful ally in our quest for soil sustainability.

References

Altomare, C. et al., Solubilization of phosphates and micronutrients by the plant-growth-promoting and biocontrol fungus *Trichoderma harzianum* Rifai 1295-22, *Appl. Environ. Microb.*, **65**, 2926–2933 (1999).

Arora, R.K., *Biocontrol of potato late blight, Potato Global Research and Development: Proceedings of the Global Conference on Potato, New Delhi, India, December*, Vol. 1, 620–623 (2000).

Bakker, P.A.H.M., van Peer, R., and Schippers, B., Specificity of siderophores and siderophore receptors and biocontrol by Pseudomonas spp, In: *Biological Control of Soilborne Plant Pathogens*, Hornby, D., Ed., CAB International, Wallingford, UK, 131–142 (1990).

Benhamou, N. and Chet, I., Cellular and molecular mechanisms involved in the interaction between *Trichoderma harzianum* and *Pythium ultimum*, *Appl. Environ. Microbiol.*, **63**, 2095–2099 (1997).

Besoain, X. et al., Biological control of *Phytophthora parasitica* in greenhouse tomatoes using *Trichoderma harzianum*, *Bull. OILB/SROP*, **24**, 103–107 (2001).

Biswas, K.K. and Das, N.D., Biological control of pigeonpea wilt caused by *Fusarium udum* with Trichoderma spp., *Ann. Plant Protect. Sci.*, **7**, 46–50 (1999).

Casale, W.L., Biological control of Phytophthora root rot. *Proceedings of the California Avocado Society and University of California-Riverside Avocado Research Symposium*, 51–53 (1995).

Chang, Y.-C. et al., Increased growth of plants in the presence of biological conrol agent *Trichoderma harzianum*, *Plant Dis.*, **70**, 145–148 (1986).

Cook, R.J. and Baker, K.F., *The Nature and Practice of Biological Control of Plant Pathogens*, American Phytopathological Society, St. Paul, MN (1983).

Duijff, B.J. et al., Influence of pH on suppression of Fusarium wilt of carnation by *Pseudomonas fluorescens* WCS417, *J. Phytopathol.*, **143**, 217–222 (1995).

Elad, Y., Chet, I., and Katan, J., *Trichoderma harzianum*: A biocontrol agent effective against *Sclerotium rolfsii* and *Rhizoctonia solani*, *Phytopathology*, **70**, 119–121 (1980).

Gnanavel, I. and Jayaraj, J., Integration of soil solarization and *Trichoderma viride* Pers.: Fr. to control damping-off of tomato (*Lycopersicum esculentum* Mill.), *J. Biol. Control*, **17**, 99–101 (2003).

Haggag, W.M. and Nofal, M.A., Application of formulated biocontrol fungi against Rhizoctonia black scurf disease of potato, *Arab. Univ. J. Agric. Sci.*, **8**, 319–334 (2000).

Harman, G.E., Myths and dogmas of biocontrol: Changes in perceptions derived from research on *Trichoderma harzianum* T-22, *Plant Dis.*, **84**, 377–393 (2000).

Harman, G.E. et al., *Trichoderma* species: Opportunistic, avirulent plant symbionts, *Microbiology*, **2**, 43–56 (2004).

Howell, C.R. et al., Field control of cotton seedling diseases with *Trichoderma virens* in combination with fungicide seed treatments, *J. Cotton Sci.*, **1**, 15–20 (1997).

Howell, C.R. et al., Induction of terpenoid synthesis in cotton roots and control of *Rhizoctonia solani* by seed treatment with *Trichoderma virens*, *Phytopathology*, **90**, 248–252 (2000).

Jie, C. et al., Infection mechanisms and biocontrol of major corn fungal diseases in Northern China, *Res. Prog. Plant Protect. Plant Nutr.*, 78–84 (1999).

Kampert, A. and Strzelczyk, E., Synthesis of auxins by fungi isolated from the roots of pine Seedlings (*Pinus silvestries* L.) and soil, *Acta Microbiol. Polon. Ser. B*, **7**, 223–230 (1975a).

Kampert, M. and Strzelczyk, E., Influence of soil microorganisms on growth of pine seedlings (*Pinus silvestris* L.) and microbial colonization of the roots of these seedlings, *Polish J. Soil Sci.*, **8**, 59–66 (1975b).

Kapat, A., Zimand, G., and Elad, Y., Effect of two isolates of *Trichoderma harzianum* on the activity of hydrolytic enzymes produced by *Botrytis cinerea*, *Physiol. Mol. Plant Pathol.*, **52**, 127–137 (1998).

Kleifeld, O. and Chet, I., *Trichoderma harzianum*: Interaction with plants and effect on growth response, *Plant Soil*, **144**, 267–272 (1992).

Meena, R.L., Rathore, R.S., and Mathur, K., Efficacy of biocontrol agents against *Rhizoctonia solani* f.sp. *sasakii* causing banded leaf and sheath blight of maize, *J. Mycol. Plant Path.*, **33**, 310–312 (2003).

Neumann, B.J. and Laing, M.L., Soil moisture and root zone pH as tools for enhancing biocontrol of *Pythium* by *Trichoderma*, *Bull. OILB/SROP*, **25**, 89–92 (2002).

Ousley, M.A., Lynch, J.M., and Whipps, J.M., Effect of *Trichoderma* on plant growth: A balance between inhibition and growth promotion, *Microb. Ecol.*, **26**, 277–285 (1993).

Ousley, M.A., Lynch, J.M., and Whipps, J.M., Potential of *Trichoderma* spp. as consistent plant growth stimulators, *Biol. Fert. Soils*, **17**, 85–90 (1994).

Reddy, V.K. and Reddy, S.M., Synthesis of IAA by some seed-borne fungi of maize (*Zea mays* L.), *Natl Acad. Sci., India, Sci. Lett.*, **10**, 267–269 (1987).

Rifai, M.A., A revision of the genus Trichoderma The Commonwealth Mycological Institute, *Mycological Papers*, **116**, 56 (1969).

Roberti, R. et al., Biological control of wheat foot rot by antagonistic fungi and their modes of action, *Bull. OILB/SROP*, **24**, 13–16 (2001).

Sarhan, M.M. et al., Application of *Trichoderma harzianum* as biocontroller against tomato wilt disease caused by *Fusarium oxysporom f. lycopersici*, *Egypt. J. Microbiol.*, **34**, 347–376 (1999).

Sharon, E. et al., Biocontrol of plant parasitic nematodes by *Trichoderma harzianum*, *Bull. OILB/SROP*, **27**, 247–249 (2004).

Spiegel, Y. and Chet, I., Evaluation of *Trichoderma* spp. as a biocontrol agent against soilborne fungi and planto-parasitic nematodes in Israel, *Integr. Pest Manag. Rev.*, **3**, 169–175 (1998).

Tronsmo, A. and Hjeljord, L.G., Biological control with Trichoderma species, In: *Plant-Microbe Interactions and Biological Control*, Boland, G.J. and Kuhkendall, L., Eds., Marcel Dekker, New York, 111–126 (1998).

Weindling, R., *Trichoderma lignorum* as a parasite of other soil fungi, *Phytopathology*, **22**, 837–845 (1932).

Weindling, R., Studies on a lethal principle effective in the parasitic action of *Trichoderma lignorum* on *Rhizoctonia solani* and other soil fungi, *Phytopathology*, **24**, 1153–1179 (1934).

Windham, M.T., Elad, Y., and Baker, R., A mechanism for increased plant growth induced by *Trichoderma* spp., *Phytopathology*, **76**, 508–521 (1986).

Yedidia, I., Benhamou, N., and Chet, I., Induction of defense responses in cucumber plants (*Cucumis sativus* L.) by the biocontrol agent *Trichoderma harzianum*, *Appl. Environ. Microb.*, **65**, 1061–1070 (1999).

35

Evaluation of Crop Production Systems Based on Locally Available Biological Inputs

O.P. Rupela,[1] C.L.L. Gowda,[1] S.P. Wani[1] and Hameeda Bee[2]

[1]*International Crops Research Institute for the Semi-Arid Tropics (ICRISAT), Patancheru, India*
[2]*Department of Microbiology, Osmania University, Hyderabad, Andhra Pradesh, India*

CONTENTS

Crop production systems that require chemical fertilizers, pesticides, machinery for tillage, and irrigation water are expensive. In countries such as India, they have started to undermine the water security of future generations, contributing to soil and water pollution particularly when synthetic pesticides are not used properly. It is true that agriculture as practiced 100 years ago without modern inputs had lower productivity than present systems of production. However, many premodern practices, such as the use of organic manures to enhance soil fertility and of herbal extracts to protect crops, can be made more efficient by the scientific knowledge that has been gained over the past century, making crop production more sustainable while still achieving high productivity.

This is becoming more evident from the published literature on practices such as the use of organic manures and biopesticides (e.g., Carpenter-Boggs et al., 2000; Stockdale et al., 2001; Kough, 2003) and experience with conservation tillage (discussed in Chapters 22 and 24). This chapter reports the results from an ongoing, long-term experiment started at ICRISAT in June 1999 on a rainfed Vertisol at Patancheru, Andhra Pradesh, India. It examines the possibility of achieving high yields using low-cost inputs, plant biomass in particular, that are available within the vicinity of the farm or that could be produced *in situ*. The field trials utilized biological approaches reported in the published literature and from traditional knowledge.

While some of these methods require considerable labor, more than many large farmers might be able or willing to invest, they could be relevant to a large number of small and marginal farm households in the semiarid tropics that have family labor available but very little cash. The methods reviewed here are proving to be profitable in terms of their returns to labor as well as to the other factors of production.

35.1 Designing Crop Production Systems for Sustainability

Production practices, such as putting on crop residues or other biomass as surface mulch, using compost and green manures, intercropping of legumes in cropping systems, and biocontrol of insect pests and diseases, all help to enhance yields and sustain soil fertility and health (e.g., Willey, 1990; Reganold et al., 1993; Fettell and Gill, 1995; van Keulen, 1995; Mäder et al., 2002; Delate and Cambardella, 2004). Appropriate use of such biologically-based approaches has been reported to enhance soil microorganisms and macrofauna (e.g., Kukreja et al., 1991; Fatondji, 2002), thereby enhancing microbial transformations of different nutrients from bound to available form. These various approaches can be combined into an integrated soil–plant–animal cropping system for attaining sustainable high yields. Such a system, which is depicted in Figure 35.1, has been tested since 1999 and is explained below.

While a variety of crops and practices are known to be able to contribute to farming system success, it is not known to what extent they can be used jointly in ways that are sufficiently productive and profitable, as well as sustainable, to improve the lives of farmers. It is not necessary that any system be advantageous for all farmers, since no single

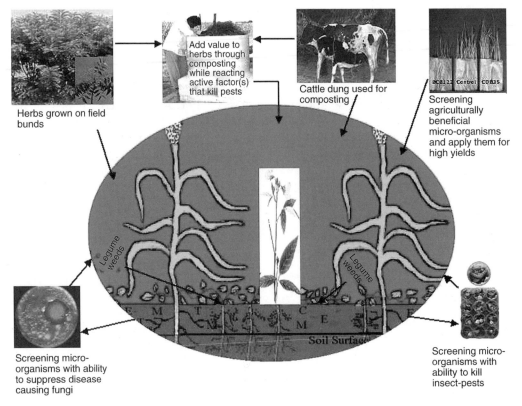

FIGURE 35.1
Elements of a biologically-based, integrated soil–plant–animal cropping system.

farming system should be expected to be optimal for everyone. Our effort was to design a crop production system that could be particularly beneficial for small landholdings. It drew on existing knowledge that:

- Legume and nonlegume crops can improve soil fertility when grown as intercrops (as examined further in Chapter 39).
- Crop residues produced *in situ* can improve the soil's physical and biological properties when retained as surface mulch, without tillage.
- Selected weeds can promote crop growth when grown under the main crop, i.e., not all weeds are deleterious.
- Where relevant or required, some amount of external inputs, preferably low cost, can be applied to the soil or crop on an as-needed basis to good effect.
- Certain soil microorganisms have beneficial traits, e.g., biological nitrogen fixation, plant growth promotion, or antagonism to disease-causing soil organisms (fungi, nematodes) or to insect pests. These can be effectively applied either as soil inoculants or sprayed on plants.
- Certain plant extracts sprayed on crops in a timely way, according to traditional knowledge, can protect crops from many if not all insect pests.
- Compost can be more than a source of nutrients for the soil, being also a soil-building substance and a source of beneficial microorganisms (Chapter 31).

As seen from our results, these practices are indeed quite compatible with one another, and as discussed in Chapter 17, cattle should be regarded as an important component of such systems. In the system that we designed and tested, only the grain produced is exported from the system. Crop stover is retained as surface mulch. Where stover is needed for economic purposes, e.g., as cattle feed, an equivalent quantity of biomass having no such economic value is returned to the field, i.e., foliage or loppings from shrubs or trees grown on field bunds or from outside the farm. The system is understood to function as a single entity, within which all of the functions in the soil, among plants, and at the soil–plant interface are highly interactive for producing yield.

Such a system is relevant to millions of small and marginal farmers in developing countries of the humid, subhumid, and semiarid tropics. About three-quarters of farmers in India have either small holdings (0.4 to 1.4 ha) or marginal holdings (<0.4 ha). They have little scope to benefit from technologies or implements designed for larger farms. This does not mean, however, that these small holdings are less productive. Actually, on a per-hectare basis they usually outperform larger farms, even by orders of magnitude (Feder, 1985; Rosset, 1999). Larger farms operate extensively rather than intensively and amass their higher total returns from their size of operation rather than from greater factor productivity or efficiency. The model presented in Figure 35.1 assumes that small and marginal farmers can and will mobilize family labor, their major asset, to undertake intensive crop and animal management if this is productive and profitable enough, i.e., if they can get higher returns per hour or per day of labor invested.

35.2   Design of the Long-Term Experiment

To examine whether yields comparable to conventional agriculture can be attained using the kinds of strategies and inputs reviewed in the preceding section, a multiyear experiment was designed to compare and evaluate four different systems of crop husbandry (T1 to T4). Since it was assumed that very small farmers would own few animals and therefore would not have enough manure, the use of other organic matter was planned for. However, the systems being tested would benefit from the addition and incorporation of animal production and the use of animal wastes, whatever the availability. The results reported can quite certainly be improved upon to the extent that animals are incorporated into the farming system. We did not want our findings to be limited to a better-case scenario.

The major objective of the experiment was to learn whether plant biomass, added to three of the four systems evaluated, could be used profitably as a source of crop nutrients instead of being burned, which is common practice in South Asia (Sidhu et al., 1988). Details of these four systems are given in Table 35.1. Note that T3 is the treatment most similar to conventional current cropping systems, i.e., relying for its nutrient inputs on inorganic fertilizers, while T1 and T2 represent low-cost systems where crop nutrients are provided from biomass inputs, in addition to what can be mobilized from the soil through biotic activity. T4 is a combination of conventional and alternative systems as it receives the same organic inputs that are provided for T2 plus the T3 chemical fertilizer applications.

The experiment is being conducted on a 1.5 m deep Vertisol, with pH in the top 15 cm ranging from 8 to 8.2 and with electrical conductivity 0.16 to 0.22 dSm^{-1}. The area is fully rainfed, with annual mean rainfall at Patancheru of 783 mm. This allows two crops to be grown in a year, either as intercrops (in all years) or as sequential crops, with a probability of success in 6 of 10 years, given the possibility that the rains can fail. To be certain of some production, given the variability in timing of rainfall, second crops have to be sown as intercrops during the rainy season, in June or July. In each year of the first 6 years of the experiment, different crops were grown, as seen in footnote to Table 35.1, but they were always the same across all four treatments. The experiment is providing an excellent field site for testing the overall hypothesis that treatments receiving high biomass as a source of nutrients — and that consequently exhibit high soil biodiversity and support higher levels of biological activity (both intervening variables being tested in our experiment) — will produce good agronomic results.

Rather than conduct the experiment on a large number of small replicated plots, the design was to use larger plots, 0.2 ha for each treatment, with a total area of 1.02 ha including noncropped area. This design has permitted observation of the effects of using biopesticides (bacteria in particular) for insect–pest management on fields of normal size and under conditions matching those of farmers' fields. We have monitored Helicoverpa pod borer, the major pest in the area, and also two of its natural enemies as well. This approach to evaluation of field-scale treatments is not new (Guthery, 1987; Guldin and Heath, 2001). It seems acceptable and appropriate for our purposes of evaluation since small replicated plots could not control for and assess so well the effects of above- and belowground biotic relationships.

Each of the treatments, T1 to T4, has 30 plots, each 9×7.5 m, laid out in six strips with five plots. Observations for yield and some other parameters have been made and analyzed for all plots. For those observations that are more costly, such as soil properties, samples are drawn from all the plots and are pooled strip-wise (and depth-wise where relevant) before analysis. There are thus 30 data points (internal replications) for parameters such as yield in our evaluation, with six data points (based on internal replications) for the different soil properties.

The concepts of sustainable agriculture expressed in Figure 35.1 apply to the first two of the four treatments, T1 and T2, in this ongoing experiment. They receive plant biomass as their major source of crop nutrients and depend on herbal extracts and agriculturally

TABLE 35.1

Treatments Used in a Continuing Long-Term Experiment at ICRISAT, Patancheru, India, June 1999 to December 2004[a]

Treatments	T1	T2	T3	T4
Inputs	Low-cost system I, based on rice straw	Low-cost system II, based on farm waste	Conventional agriculture	Conventional agriculture + T2 biomass
Land preparation and intercultivation	None	None	Conventional (bullock plow)	Conventional (bullock plow)
Sowing	Bullock-drawn drill	Bullock-drawn drill	Bullock-drawn drill	Bullock-drawn drill
Microbial inoculants	Added	Added	None	None
Biomass (first 3 years only)	10 t ha^{-1} yr^{-1} with rice straw as surface mulch	10 t ha^{-1} yr^{-1} with farm waste, stubble and hedgerow foliage as surface mulch	None	10 t ha^{-1} yr^{-1} with farm waste, stubble and hedgerow foliage incorporated
Compost	1.5–1.7 t ha^{-1} yr^{-1}	1.5–1.7 t ha^{-1} yr^{-1}	1.8 t ha^{-1} in years 2, 4, 6	1.8 t ha^{-1} in year 2, 4, 6
Fertilizer (N)	None	None	80 kg N ha^{-1} in 2 split doses yr^{-1}	80 kg N ha^{-1} in 2 split doses yr^{-1}
Fertilizer (P)	20 kg ha^{-1} as rock phosphate	20 kg ha^{-1} as rock phosphate	20 kg ha^{-1} as single super phosphate (SSP)	20 kg ha^{-1} as single super phosphate (SSP)
Plant protection	Biopesticides	Biopesticides	Chemical pesticides	Chemical pesticides
Weeding	Manual, weeds retained	Manual, weeds retained	Manual, weeds discarded	Manual, weeds discarded

a Same crops were grown in all plots each year: Crop rotations for all four treatments were: *Year 1.* Pigeon pea-chick pea sequential (June 1999 to May 2000); *Year 2.* Sorghum/pigeon pea intercrop (June 2000 to May 2001); *Year 3.* Cowpea/cotton intercrop (June 2001 to May 2002); *Year 4.* Maize/pigeon pea intercrop (June 2002 to May 2003); *Year 5.* Cowpea/cotton intercrop (June 2003 to May 2004); *Year 6.* Maize/pigeon pea intercrop (June 2004 to May 2005); pigeon pea not yet harvested when chapter was written.

beneficial microorganisms as soil inoculants and biopesticides. Both are cultivated with minimum tillage, where only the sowing is done with bullock-drawn implements. For the first 3 years, T1 received 10 t ha^{-1} of rice straw and T2 was given the same quantity of farm waste (crop stubble, leftovers after cattle have eaten, and tree leaves). Both treatments received these applications as surface mulch soon after sowing.

The conventional agriculture treatment, T3, received: 80 kg N and 20 kg P ha^{-1} yr^{-1}; regular tillage (land preparation, sowing, and intercultivation to remove weeds with a bullock-drawn tropicultor); chemical pesticides for managing pests; manual weeding; and 1.8 t ha^{-1} compost in alternate years. The T4 plots had the same inputs used for conventional agriculture, but in addition, they received 10 t ha^{-1} yr^{-1} of biomass (for the first 3 years only) similar to the T2 plots. This biomass has been incorporated into the T4 plots rather than left as surface mulch. From year 4, no further biomass from external sources has been added to any of the four treatments, except compost at rates shown in Table 35.1. The uneconomic parts of plants, e.g., leaves and stem stover, have all been retained on plots in treatments T1, T2, and T4. From year 5, loppings of *Gliricidia* grown on the plot bunds have been added during the crop growth period in equal quantities two to three times a year to all four treatment plots.

As depicted in Figure 35.1, the foliage of *Gliricidia sepium* and neem (*Azadirachta indica*) has been composted in separate tanks, and the wash from this (50 l ha^{-1} at least five times per season) has been sprayed on plants in T1 and T2 to protect crops from insect pests. The wash from neem, a known biopesticide, and from *Gliricidia* has been found to contain siderophore-producing bacteria (O.P. Rupela, unpublished study). These microbes have also been reported as promoting plant growth (Kloepper et al., 1980).

Certain bacterial preparations, e.g., EB35 and CDB35, which degrade cellulose, solubilize phosphorus, promote plant growth, and suppress disease-causing fungi (H. Bee, unpublished studies), have been applied as sand-coat inoculants and sown along with seeds in T1 and T2. A certain bacterium (*Bacillus subtilis* strain BCB 19) and also a selected fungus (*Metarrhizium anisoplliae*), both ICRISAT research products, have shown the ability under laboratory conditions to kill young larvae of *Helicoverpa armigera*, a major pest of cotton and legumes in the region. These preparations have been used as biopesticides in T1 and T2 only, along with other low-cost materials of traditional knowledge. Earthworms plus cattle dung (applied as 1% dung slurry in water to soak into the biomass as a food for earthworms) are important ingredients for composting in the tank shown in Figure 35.1.

The experiment completed its first 5 years in May 2004, so we are able to report and discuss here all the variables, including yield, with particular attention to soil biological factors. The work is ongoing, so there are also some data from the sixth year. As this is an ongoing evaluation, more long-term results will become available.

35.3 Crop Growth and Yield

The high variability in precipitation that farmers in this region have to cope with can be seen from the annual rainfall totals (in mm) for the different years: 580 (year 1), 1473 (year 2), 688 (year 3), 628 (year 4), 926 (year 5), and 610 (year 6). The different crops grown in the last 6 years (soybean, pigeon pea, maize, sorghum, cowpea, and cotton) all emerged well, including those in T1 and T2, which had to emerge through about 10 cm of biomass applied as surface mulch. The incidence of collar rot, caused by Sclerotium, was expected to increase on T1 and T2 in the presence of biomass, but this problem has been virtually nonexistent (<5% mortality of seedlings), at par with or even marginally lower than in T3.

Except in year 1, when T1 and T2 yields were 35 to 62% lower as the transition was made to biological production methods, as discussed further below, the yields of the different crops in T1 and T2 over the first 5 years, produced with lower cash cost, have been on a par with T3 or at most 14% lower. The reasonably high yields of pigeon pea in year 2 and of cotton in year 3 for both T1 and T2 were associated with the effective management of Helicoverpa by using biopesticides. Conversely, the low yields in T1 and T2 from pigeon pea in 2002 (year 4) and cotton in 2003 (year 5) were associated with poor success in managing insect pests mostly other than Helicoverpa. Detailed information and data on crop yields in the different years are given in Rupela et al. (2005). Annual productivity of T1 and T2 — the combined yield of legumes + nonlegumes, e.g., the mass of cowpea grains and seed cotton (lint + seeds) in year 5 — was high in all 5 years except year 1 (Figure 35.2).

Most significant for farmers, the net income from crops in each year except year 1, which was essentially a year of learning, has been higher — even much higher — in T1 and T2 than T3. The differential has ranged between 1.3 and 4.6 times (Figure 35.2), showing that in economic terms, the low-input strategy is proving to be much more profitable. In this calculation, each input was costed (except the cost of biomass and labor). Biomass was assumed to be available with little or no opportunity cost, having been saved from burning and being handled by family labor. Labor is not a free resource, of course, but it is the one most available to poor households, who are primarily constrained in terms of their land area and cash. Thus, labor was not considered to be the resource from which economic returns had to be maximized.

It should be noted that in year 3, there was a substantial loss (US$156 ha^{-1}) from growing cotton in T3 in contrast to a substantial net income gained from the cotton crop on the sustainable agriculture plots — US$210 ha^{-1} from T1 and US$140 ha^{-1} from T2

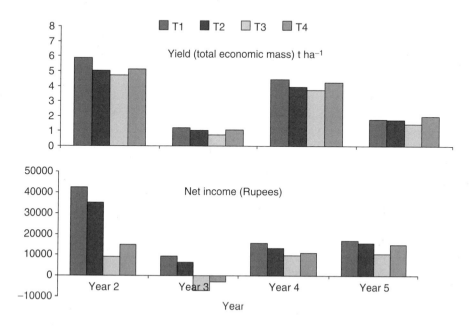

FIGURE 35.2

Yield and net income (in rupees) over years 2 through 5 from the four different systems of crop production (T1 to T4) in long-term experiment at ICRISAT, Patancheru, India. Income was calculated by putting a common price across all treatments for each item (both inputs and outputs). Per-day labor was priced @ Rs. 75 per day for both farmers and family members. (1 US$ = ~Rs. 45)

(Figure 35.2). The low-input strategy of T1 and T2 has, therefore, in some years performed much better agronomically than the more costly conventional cropping system. This makes the economic advantages even greater.

35.4   Soil Properties and Nutrient Balances

Every year in April/May, for all treatments, soil samples from three depths (0 to 15, 15 to 30, and 30 to 60) were collected from each plot before sowing crops, using a 40 mm diameter soil core. The samples from a set of five plots were pooled as indicated previously and were analyzed for total and available nitrogen, total and available phosphorus, total potassium, and organic carbon. Methods of analysis for the different parameters were the same as described by Okalebo et al. (1993). Soil bulk density measured in April 2002 (at the end of year 3) was similar across the treatments and ranged from 1.19 to 1.36 at the different depths. Electrical conductivity and pH were measured. Data for total nitrogen, total and available phosphorus for the first 4 years are given in Table 35.2 as means from the three depths for which measurements were made.

It was important to note that at the same time that T1 and T2 produced yields comparable to T3 — without receiving any chemical fertilizer amendments — they actually showed increases (rather than decreases) in their concentrations of soil nutrients compared with T3. In years 3 and 4, there were increases of 11 to 34% in total nitrogen and 11 to 16% in total phosphorus in T1 and T2, relatively more than in T3. However, it was noted that the mean nitrogen and phosphorus in all four treatments, after improving up to year 3, was reduced in year 4 (Table 35.2). The reasons for this reduction are still being considered.

Soil biological properties, presented in Table 35.3, were assessed only once, close to the time of crop harvest in year 5, using soil depths of 0 to 10 and 10 to 20 cm. The methods used were the same as those in Jenkinson (1988) for microbial biomass carbon and microbial biomass nitrogen; in Anderson and Domsch (1978) for microbial biomass carbon; in Casida et al. (1964) for soil dehydrogenase activity; and in Eivazi and Tabatabai (1977) for acid and alkaline phosphatases.

Of the different parameters measured to assess the biological activity in soil samples from the four different systems of crop husbandry, more activity was noted in T1, T2, and T4 compared with T3. Soil respiration was more by 17 to 27% than in T3; microbial biomass carbon was 28 to 29% higher; microbial biomass nitrogen was 23 to 28% more; and acid and alkaline phosphatases were 5 to 13% higher. While these different parameters are reported as point-in-time measurements of microbial activity under laboratory conditions, they depict treatment differences.

In this experiment, 79 to 109 kg N ha^{-1} were noted to be associated with microbial biomass in the top 20 cm profile, which is more than usually reported for such soils, and this needs further examination. Wani et al. (2003) reported 42 kg N ha^{-1} in the top 60 cm profile of plots using traditional methods of cropping, compared with 86 kg N ha^{-1} in plots using an improved system of cropping. The microbially-bound nitrogen is likely to be mineralized for use by plants when microorganisms die naturally or due to unfavorable factors, such as soil drying or application of chemical pesticides to soils.

The overall results on the different soil biological parameters strongly suggest that the soils from plots T1 and T2 were consistently more active microbiologically than those of T3 (Table 35.3). While the total bacterial populations were not that different across all four treatments, 5.3 to 5.7 ($\log_{10}$ g^{-1} soil), the population of Pseudomonas spp. was about 10

TABLE 35.2

Total Nitrogen (mg kg^{-1} soil), Total and Available Phosphorus (mg kg^{-1} soil) in Top 60 cm profile (mean of three depths: 0 to 15, 15 to 30 and 30 to 60 cm), Field BW3, ICRISAT, Patancheru, AP, India

Treatment	Total N					Total P					Available P				
	Year 1	Year 2	Year 3	Year 4	Mean	Year 1	Year 2	Year 3	Year 4	Mean	Year 1	Year 2	Year 3	Year 4	Mean
T1	452 (.80)	569 (21.1)	690 (30.1)	492 (17.5)	553	175 (6.3)	231 (7.0)	253 (15.7)	194 (9.0)	213	1.2 (0/08)	1.7 (0.34)	2.1 (0.31)	0.7 (0.24)	1.4
T2	458 (12.6)	643 (16.2)	681 (30.9)	489 (32.4)	575	189 (7.2)	257 (10.5)	263 (10.9)	213 (11.3)	230	0.7 (0.02)	1.3 (0.26)	1.7 (0.33)	0.6 (0.24)	1.1
T3	506 (22.1)	651 (73.4)	514 (12.3)	440 (17.9)	528	204 (3.9)	263 (49.2)	227 (3.3)	175 (8.1)	222	1.00 (0.13)	1.4 (0.34)	2.0 (0.29)	0.4 (0.11)	1.2
T4	500 (10.5)	588 (49.3)	586 (61.9)	429 (13.4)	526	244 (23.7)	213 (21.1)	232 (3.9)	177 (1.8)	218	0.5 (0.09)	1.6 (0.34)	2.4 (0.47)	0.3 (0.10)	1.2
Mean	489	613	618	462		203	247	244	189		0.8	1.5	2.0	0.5	

Data in parentheses are ± SE.

TABLE 35.3

Biological Properties of Soils with Different Cropping System Treatments Assessed in Top 20 cm Profile, Field BW3, ICRISAT, Patancheru, Close to Harvest in Year 5

Properties	T1	T2	T3	T4	Mean
Soil respiration (kg C ha^{-1}10 d^{-1})	330 (19.5)	360 (18.6)	283 (14.3)	436 (25.9)	352
Microbial biomass C (kg C ha^{-1})	1550 (110.3)	1535 (120.1)	1202 (66.8)	1510 (104.1)	1449
Microbial biomass N (kg N ha^{-1})	97 (6.7)	109 (8.9)	79 (4.0)	98 (7.5)	96
Organic carbon (t C ha^{-1})	23 (1.5)	20 (1.1)	17 (0.9)	22 (1.1)	20
Acid phosphatase (μg p-NP g^{-1} h^{-1})[a]	310 (38.8)	332 (32.5)	294 (36.0)	357 (39.8)	323
Alkaline phosphatase (μg p-NP g^{-1} h^{-1})[a]	937 (103.2)	1008 (111.3)	890 (114.8)	1011 (113.1)	962
Dehydrogenase (μg TPFg^{-1} 24h^{-1})[b]	133 (28.0)	137 (29.2)	130 (23.8)	142 (27.7)	136
Bacterial population (log$_{10}$ g^{-1} soil)	5.6	5.6	5.3	5.7	5.6
Pseudomonas spp populations (log$_{10}$ g^{-1} soil)	4.1	4.6	3.3	3.2	3.8

Numbers in brackets are ±SE

[a] *p*-NP = para nitro phenol

[b] TPF = triphenylformazan

times more in T1 and T2 than in T3 and T4 (4.1 to 4.6 vs. 3.2 to 3.3 log$_{10}$ g^{-1} soil). Several soil isolates of this species are suppressive to disease-causing fungi and nematodes, and this trait can therefore be regarded as an indicator of soil health. The measured differences are likely to be due to the inoculant bacteria that were added at sowing of the T1 and T2 crops each year.

It should be noted that less than 10% of microorganisms that live in the soil can be cultured in laboratory media (Ward et al., 1990). Some researchers think that this number is less than 5 or even 1%. One cannot say the exact number since the denominator is unknown, which is indicative of how little we know yet about the earth's microbiota. This fact suggests, in any case, that soil respiration and microbial biomass carbon and nitrogen are going to be more reliable parameters of soil biological activity, reflecting the total microbial community, than are counts of microbial population using laboratory media.

A balance sheet of nitrogen and phosphorus, the two macronutrients considered most critical for crop production, was prepared for all four treatments. For this purpose, all the materials added to the different treatments plots, e.g., crop residues, compost, and those removed (e.g., grain), were fully accounted for. Figure 35.3 shows the amounts of total nitrogen and phosphorus added and removed, and the balance for the first 5 years across the four different crop husbandry systems. T1 and T2, which received plant biomass, compost, and microorganisms as their major sources of crop nutrients, ended up receiving substantially more nitrogen (27 to 52%) and phosphorus (50 to 58%) than was added to T3 (604 kg N ha^{-1} and 111 kg P ha^{-1}, largely as chemical fertilizers). Of course, T4, having both sources, received the largest quantities of nitrogen (1232 kg ha^{-1}) and phosphorus (193 kg ha^{-1}). It is therefore not surprising that T1, T2, and T4 resulted in having a much larger balance of nitrogen (2.5 to 10 times) and phosphorus (12 to 13 times) than was measured for T3 (55 kg N ha^{-1} and 5 kg P ha^{-1}).

This does not mean, however, that the crops in the low-cost systems, T1 and T2, had access to more nitrogen and phosphorus than those in T3, the conventional system. Nutrients when added as biomass are not in a readily available form for crops and need to be mineralized by microbial activity. Also, since the biomass was added as surface mulch, microbial activity at the soil surface might not be sufficient for its decomposition. It is

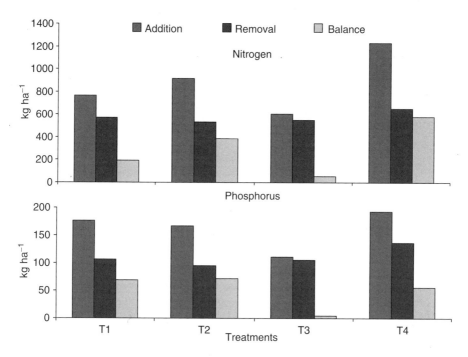

FIGURE 35.3

Nutrient (N and P) balance of the four different systems of crop production (T1 to T4) after 5 years, in long-term experiments, ICRISAT, Patancheru, India.

widely accepted that only a proportion of the nitrogen applied as biomass to the soil through soil incorporation is recovered by the crop (Schomberg et al., 1994; Thönnissen et al., 2000). According to T. J. Rego, ICRISAT (unpublished data), under Patancheru conditions this proportion would be less than 10% in year 1.

This helps to explain the lower yield obtained in year 1 in T1 and T2 (lower by 35 to 62%) than that produced by T3, which received chemical fertilizer. The longitudinal yield data suggest, however, that in subsequent years, microorganisms, whether in the soil or applied externally, were able to decompose the biomass sufficiently so that the released nutrients could readily meet crop demand, when T1 and T2 yields were on a par with or very close to those from T3.

If T1 and T2 received substantially more nitrogen and phosphorus and their removal was similar to that in T3 (Figure 35.3), then the soil systems of T1 and T2 should have substantially higher amounts of nitrogen and phosphorus. This was observed, at least in the measurements up to the end of year 4 (data for subsequent years are yet to be analyzed). The top 15 cm soil profile for T1 and T2 had 30 to 41% more nitrogen (an additional 355 to 483 kg ha^{-1}) and 0.2 to 17% more phosphorus (an additional 2 to 129 kg ha^{-1}) compared with the level of nitrogen and phosphorus for T3 (1192 kg N ha^{-1} and 716 kg P ha^{-1}). The amounts of nitrogen and phosphorus in the biomass still remaining as surface mulch on T1 and T2 from recent additions are not accounted for in this analysis. Much of the biomass applied at sowing had largely, except for thick plant stems, disintegrated by the end of the rainy season each year, suggesting that all the leafy materials added at sowing time were decomposed during the rains, particularly in a normal to good rainfall year.

35.5   Discussion

From the data collected during the first 5 years of the long-term experiment presented here, it is apparent that the two crop husbandry systems, T1 and T2, which received locally available, low-cost and eco-friendly materials such as biomass and compost, along with agriculturally-beneficial microorganisms, were able to produce yields that match those from the T3 system that relies on purchased inputs, e.g., chemical fertilizers and pesticides, and that also continued conventional tillage practices. Labor was the major input in T1 and T2. While this has opportunity costs for small and marginal farmers, these producers have relatively more access to labor than to cash, so their binding constraint is land and capital rather than labor.

Inputs of the agriculturally-beneficial microorganisms used in this study are not yet widely available, although efforts are beginning in India not just to produce them in large commercial operations but also at village level by villagers, as discussed in Chapter 45 (see also Bhattacharyya and Dwivedi, 2004).

In the second year, 20 mm of rain was received in the first week of January 2001, about 10 days before pigeon pea was to be harvested. For a conventional system (T3), this rain meant less strenuous tillage effort for the bullocks after harvest. For the no-till systems (T1 and T2), it was an opportunity to harvest more. Pigeon pea, particularly the non-determinate cultivars, has a tendency to regrow after harvest if soil moisture is conducive. Since such regrowth was noticed, it was decided to harvest by picking pods rather than by the normal method of cutting plants close to the ground. This resulted in 0.69 to 0.77 t ha^{-1} additional pigeon pea harvest, about 25% of total yield. The no-till system gives farmers more flexibility for using opportunities given by nature.

Sowing crops when there is surface mulch is a potential hindrance to adoption of the concept of sustainable agriculture represented in Figure 35.1. Sowing in the long-term experiment described here was done using a bullock-drawn implement. Manual sowing is an option, but both have high labor requirements. Before using the bullock-drawn implement for sowing, we had to rake off the biomass (largely crop stems) from the soil surface and spread it again soon after sowing. A machine punch-planter, which is able to sow crops through surface mulch, has recently become available in India and will be used and evaluated in the future. This machinery will reduce labor requirements substantially.

Earthworms are widely accepted as having a beneficial influence on soil structure and chemistry that promotes plant growth. We have recorded the presence of large numbers of siderophore-producing bacteria (1.2×10^4 to 4.5×10^6 ml^{-1}) in the wash of compost that was made from neem and Gliricidia foliage using earthworms (O.P. Rupela, unpublished study). It is likely that other agriculturally-beneficial microorganisms, such as ones able to suppress disease-causing fungi, are present in certain compost used by organic farmers (Rupela et al., 2003). If locally available earthworms that feed aggressively on biomass placed on the soil surface can be identified and introduced in large numbers in the future, this will obviate the need to spray compost wash on the crop, reducing further the labor requirement for such biological management of the crop and soil systems.

It was apparent that plant biomass was the engine of crop productivity in T1 and T2, mediated by biological processes that enhance soil fertility. It is generally argued that biomass is required to feed cattle in South Asia, and therefore is not available for application to the soil to enhance crop production as has been done in T1 and T2. Being able to apply the levels of biomass used in T1 and T2 over time will require special efforts from any farmers who want to utilize this biologically-based cropping system. However, there are many ways in which biomass supply can be augmented for a system such as this.

In the long-term experiment, 4.5 t of biomass (containing 103 kg N and 6.7 kg P ha^{-1}) was available annually from year 5 on from the fast-growing Gliricidia grown on bunds (190 m long × 1.5 m wide, separating the four treatments) and on the boundary (218 m long) around the 1.02 ha field. Some crops, such as pigeon pea, which drop their leaves, can contribute biomass and nutrients directly to the soil system. In this experiment, 22 kg N and 2 kg P in year 2 were assessed to be added through the 3.1 t ha^{-1} of fallen leaves of pigeon pea when this was grown as the economic crop.

Fallen leaves and loppings of tree branches on-farm are another source of biomass, and many non-arable areas within the farming community could produce more biomass cheaply from fast-growing shrubs and trees introduced on wasteland, not displacing any agricultural production, provided that there is sufficient rainfall. It is important to note that deep-rooted shrubs and trees are an important biological tool that can acquire nutrients for crops, extracting them from lower layers of the soil and providing them on the surface layer in the form of fallen leaves, thus improving soil fertility; alternatively, these can be used as surface mulch or applied after composting.

A number of leguminous species offer opportunities to enhance biomass availability as cover crops or green manures, as discussed in Chapter 30. Farmers practicing alternative agriculture need to appreciate the value of biomass and to develop multiple practices and technologies that can harness this source of nutrients for crop production. Producing yields on a par with or higher than their neighbors without incurring the cash costs of chemical fertilizers and pesticides offers farmers a significant incentive for change.

A recent study by Delate and Cambardella (2004) has reported yields and differentials similar to those we report here, for the production of maize and soybeans in Iowa, USA, using organic (nonchemical) vs. conventional farming practices over a 3-year period converting from conventional to organic production. The study reported here from India likewise suggests that biological approaches to crop production can sustain soil systems profitably for farmers, provided they have sufficient labor and its opportunity costs are not too high.

Making alternative agriculture systems more productive than conventional agriculture will be essential for their spread, although we must remember not to consider yield alone, a physical measure of success that ignores economic considerations. Costs of production per unit of output need to be assessed, including water-use efficiency. This was not considered in our trials because water provision was beyond our control in a purely rainfed system. However, rainwater harvesting was better in the low-cost systems (T1 and T2) than from the conventional system (T3), as seen from the reduced runoff (Rupela et al., 2005).

The scientific underpinnings for more biologically-based systems have been built up by researchers and practitioners over the past 50 years while Green Revolution technologies were receiving all the public attention and most of the public financial support. Many more studies are needed to be certain of the net value of alternative production systems, for different cropping patterns, on different soil types, and in different climatic regimes. Moreover, one cannot expect to evaluate the effects of biologically-based systems in a single year or two. Longitudinal evaluations are necessary to track the dynamic changes, positive and/or negative, in the many factors that operate in soil systems. This is why this particular long-term experiment was undertaken.

Overall, the biological approaches reported here — use of plant biomass as surface mulch, agriculturally beneficial microorganisms, and other practices — have enhanced soil biological and chemical properties of a rainfed Vertisol in the semiarid tropical environment in southern India. Yields were comparable to the conventional system of crop production that used standard agrochemical inputs. In the crop husbandry systems receiving biological inputs only, depending on the crops grown that year, stover yield ranging from 6.6 to 11.6 t ha^{-1} and grain yield ranging from 4 to 5.9 t ha^{-1} was harvested annually when there was ≥ 628 mm of rainfall. There is, however, the need to evaluate such systems in other locations for soil and climatic differences, so that we can better understand the many interfaces between biotic and abiotic subsystems as they respond to anthropogenic interventions in pursuit of human livelihoods and sustenance.

References

Anderson, J.P.E. and Domsch, K.H., A physiological method for the quantitative measurement of microbial biomass in soils, *Soil Biol. Biochem.*, **10**, 215–221 (1978).

Bhattacharyya, P. and Dwivedi, V., *Proceedings of National Conference on Quality Control*, National Biofertilizer Development Centre, Ghaziabad, Uttar Pradesh, India (2004).

Carpenter-Boggs, L., Reganold, J.P., and Kennedy, A.C., Effects of biodynamic preparations on compost development, *Biol. Agric. Hort.*, **17**, 313–328 (2000).

Casida, L.E., Klein, D.A., and Santoro, T., Soil dehydrogenase activity, *Soil. Sci.*, **98**, 371–376 (1964).

Delate, K. and Cambardella, C.A., Organic production: Agroecosystem performance during transition to certified organic grain production, *Agron. J.*, **96**, 1288–1298 (2004).

Eivazi, F. and Tabatabai, M.A., Phosphatases in soil, *Soil Biol. Biochem.*, **9**, 167–172 (1977).

Fatondji, D., Organic amendment decomposition, nutrient release and nutrient uptake by millet (*Pennisetum* glaucum (L.) R. Br.) in a traditional land rehabilitation technique (Zaï) in the Sahel (www.zef.de/download/zefc_ecologydevelopment/ecol_dev_1_text.pdf) (2002).

Feder, G., The relationship between farm size and farm productivity, *J. Dev. Econ.*, **18**, 297–313 (1985).

Fettell, N.A. and Gill, H.S., Long-term effects of tillage, stubble, and nitrogen management on properties of a red-brown earth, *Aust. J. Exp. Agric.*, **35**, 923–928 (1995).

Guldin, J.M. and Heath, G., *Underplanting Shortleaf Pine Seedlings Beneath a Residual Hardwood Stand in the Ouachita Mountains: Results after Seven Growing Seasons*, U.S. Department of Agriculture, Forest Service Research Note SRS-90, Washington, DC (2001).

Guthery, F.S., Guidelines for preparing and reviewing manuscripts based on field experiments with unreplicated treatments, *Wildlife Soc. Bull.*, **15**, 306 (1987).

Jenkinson, D.S., The determination of microbial biomass carbon and nitrogen in soil, In: *Advances in Nitrogen Cycling in Agricultural Ecosystems*, Wilson, J.R., Ed., CAB International, Wallingford, UK, 368–386 (1988).

Kloepper, J.W. et al., Enhanced plant growth by siderophores produced by plant growth-promoting rhizobacteria, *Nature*, **286**, 885–886 (1980).

Kough, J., The safety of *Bacillus thuringiensis* for human consumption, In: *Bacillus thuringiensis: A Cornerstone of Modern Agriculture*, Metz, M., Ed., Food Products Press, New York, 1–10 (2003).

Kukreja, K. et al., Effect of long-term manorial application of microbial biomass, *J. Ind. Soc. Soil Sci.*, **39**, 685–688 (1991).

Mäder, P. et al., Soil fertility and biodiversity in organic farming, *Science*, **296**, 1694–1697 (2002).

Okalebo, J.R., Gathua, K.W., and Woomer, P.L., *Laboratory Methods of Soil and Plant Analysis: A Working Manual*, Tropical Soil Biology and Fertility Program, Nairobi, Kenya (1993).

Reganold, J.P. et al., Soil quality and financial performance of biodynamic and conventional farms in New Zealand, *Science*, **260**, 344–349 (1993).

Rosset, P.M., The multiple functions and benefits of small farm agriculture in the context of global trade negotiations (http://www.foodfirst.org/pubs/policybs/pb4.pdf) (1999).

Rupela, O.P. et al., A novel method for the identification and enumeration of microorganisms with potential for suppressing fungal plant pathogens, *Biol. Fert. Soils*, **39**, 131–134 (2003).

Rupela, O.P. et al., Lessons from non-chemical input treatments based on scientific and traditional knowledge in a long-term experiment, In: *Agricultural Heritage of Asia: Proceedings of an International Conference, 6–8 December, 2004, Hyderabad*, Nene, Y.L., Ed., Asian Agri-History Foundation, Secunderabad, India (2005).

Schomberg, H.H., Steiner, J.L., and Unger, P.W., Decomposition and nitrogen dynamics of crop residues: Residue quality and water effects, *Soil Sci. Soc. Am. J.*, **58**, 372–381 (1994).

Sidhu, B.S. et al., Sustainability implications of burning rice and wheat straw in Punjab, *Econ. Polit. Weekly*, A163–A168 (1998).

Stockdale, E.A. et al., Agronomic and environmental implications of organic farming systems, *Adv. Agron.*, **70**, 261–326 (2001).

Thönnissen, C. et al., Legume decomposition and nitrogen release when applied as green manures to tropical vegetable production systems, *Agron. J.*, **92**, 253–260 (2000).

van Keulen, H., Sustainability and long-term dynamics of soil organic matter and nutrients under alternative management strategies, In: *Eco-Regional Approaches for Sustainable Land Use and Food Production*, Bouma, J. et al., Eds., Kluwer, Dordrecht, Netherlands, 353–375 (1995).

Wani, S.P. et al., Improved management of Vertisol in the semiarid tropics for increased productivity and soil carbon sequestration, *Soil Use Manage.*, **19**, 217–222 (2003).

Ward, D.M., Weller, R., and Bateson, M.M., 16S rRNA sequences reveal numerous uncultured microorganisms in a natural community, *Nature*, **345**, 63–65 (1990).

Willey, R.W., Resource use in intercropping systems, *Agric. Water Manage.*, **17**, 215–231 (1990).

36

Bio-Char Soil Management on Highly Weathered Soils in the Humid Tropics

Johannes Lehmann[1] and Marco Rondon[2]
[1]*Cornell University, Ithaca, New York, USA*
[2]*International Center for Tropical Agriculture (CIAT), Cali, Colombia*

CONTENTS

Maintaining an appropriate level of soil organic matter and biological cycling of nutrients is crucial to the success of any soil management in the humid tropics. Cover crops, mulches, compost, or manure additions have been used successfully, supplying nutrients to crops, supporting rapid nutrient cycling through microbial biomass, and helping to retain applied mineral fertilizers better (Goyal et al., 1999; Trujillo, 2002). The benefits of such amendments are, however, often short-lived, especially in the tropics, since decomposition rates are high (Jenkinson and Ayanaba, 1977) and the added organic matter is usually mineralized to CO_2 within only a few cropping seasons (Bol et al., 2000). Organic amendments therefore have to be applied each year to sustain soil productivity.

Management of black carbon (C) — increasingly referred to as bio-char — may overcome some of those limitations and provide an additional soil management option. This is a highly aromatic form of organic matter that is present in most soils to varying extents (Schmidt and Noack, 2000; Skjemstad et al., 2002). Interest in and application of biomass-derived black carbon — using incompletely combusted organic matter such as charcoal (Glaser et al., 2002) — was prompted by studies of soils found in the Amazon Basin, referred to as *Terra Preta de Indio* (Lehmann et al., 2003c). These Amazonian Dark Earths are anthropic soils that were created by Amerindian populations between 500 and 2500 years ago. They have maintained high amounts of organic carbon, and their high fertility, even several thousand years after they were abandoned by the indigenous population, contrasts distinctly with the low fertility of the adjacent acid upland soils (Lehmann et al., 2003b).

The reasons for these soils' high fertility are multiple, but the source of the large amounts of organic matter and their high nutrient retention has been attributed to the extraordinarily high proportions of black carbon (Glaser et al., 2001). Such large amounts of black carbon can only originate from incompletely combusted biomass carbon, such as wood from kitchen fires or possibly from in-field burning (Smith, 1980; Hecht, 2003). This chapter considers the beneficial effects of this bio-char soil management system and discusses opportunities for applying such management within a sustainable system that can be called "slash-and-char," as well as within other smallholder agricultural systems.

36.1 Bio-Char Management and Soil Nutrient Availability

Black carbon is found along a continuum of forms of aromatic carbon, from charred organic materials to charcoal, soot, and graphite (Schmidt and Noack, 2000). Biomass-derived black carbon, or bio-char, is produced through burning at 300 to 500°C under partial exclusion of oxygen (Antal and Gronli, 2003). The result is a highly aromatic organic material with carbon concentrations of about 70 to 80% (Lehmann et al., 2002).

Increases in soil fertility attributable to charcoal are known from naturally occurring fires (although the increases have often been attributed to adsorption of phenolics, e.g., Wardle et al., 1998) and from remnants of charcoal hearths (Chidumayo, 1994; Mikan and Abrams, 1995; Young et al., 1996; Oguntunde et al., 2004). Additions of bio-char to soil have shown definite increases in the availability of major cations and phosphorus as well as in total nitrogen concentrations (Glaser et al., 2002; Lehmann et al., 2003a). Both CEC and pH are also frequently increased through such applications, by up to 40% of initial CEC and by one pH unit, respectively (Tryon, 1948; Mikan and Abrams, 1995; Topoliantz et al., 2002). Higher nutrient availability for plants is the result of both the direct nutrient additions by the bio-char and greater nutrient retention (Lehmann et al., 2003a), but it can also be an effect of changes in soil microbial dynamics, discussed in the following section.

Yield increases have frequently been reported that are directly attributable to the addition of bio-char over a control without bio-char (Lehmann et al., 2003a). However, growth depressions have been found in some instances (Mikan and Abrams, 1996). The immediate beneficial effects of bio-char additions for nutrient availability are largely due to higher potassium, phosphorus, and zinc availability, and to a lesser extent, calcium and copper (Lehmann et al., 2003a). Longer-term benefits for nutrient availability include a greater stabilization of organic matter, concurrent slower nutrient release from added organic matter, and better retention of all cations due to a greater cation exchange capacity.

The effect of bio-char on plant productivity depends on the amount added. Progressive growth improvement with greater bio-char applications is seen with comparatively low levels of bio-char; already at very low application rates of 0.4 to 8 t C ha^{-1}, significant improvements in productivity can be observed ranging from 20 to 220% (biomass production equal to 120 to 320% of the control in Figure 36.1). In many cases, nitrogen limitation will be the reason for declining yields at high application rates, as nitrogen availability decreases through immobilization by microbial biomass at high C:N ratios (Lehmann et al., 2003b), although other growth-limiting factors may be responsible as well. With increasing rates of application, plant response at a given site is positive until some maximum is reached, above which growth response is negative, as shown for beans with applications of 31 to 93 t C ha^{-1} (data points 16, 17 and 18 in Figure 36.A1).

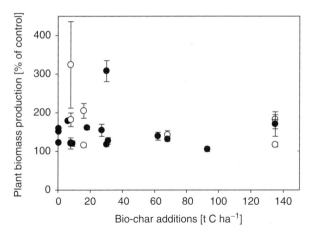

FIGURE 36.1

Plant biomass increase over control as a function of the amount of bio-char applied (means and standard errors). Open symbols = unfertilized; filled symbols = fertilized. Site and crop information are given in Table 36.A1 in annex to this chapter.

The response function is additionally dependent on the properties of the bio-char (Tryon, 1948), soil properties (greater response occurs on nutrient-deficient, sandy soils), concurrent nutrient and organic matter additions, and plant species. Legumes appear to thrive under greater bio-char additions, more than do gramineae species, since they can compensate for limited nitrogen availability by increased biological nitrogen fixation (BNF). Additions of nutrients from using inorganic or organic fertilizers are usually essential for high productivity and increase the positive response of the bio-char amendment. However, the relative effect of the bio-char addition may not be as high as for unfertilized crops (Figure 36.1).

Bio-char soil management, with associated increases in nutrient availability and pH, not only enhances crop yields and decreases risk of crop failure, but also opens new possibilities for cropping, i.e., high-value crops can be produced on sites that would normally not be suitable for production. Nutritious and easily marketable produce can improve cash returns and health among farmers who currently only have access to poor soils. Carrots and beans grown on steep slopes and on soils with a soil reaction of less than pH 5.2 have had yields significantly improved by bio-char additions (Rondon et al., 2004). Improvement of rural livelihoods is possible not only through increased crop yields, but also through increased quality and variety of the crops grown.

36.2 Microbial Cycling of Nutrients in Soils with Bio-Char

Interactions of bio-char with soil microorganisms are complex. On the one hand, soil microbial diversity and population size, as well as population composition and activity, may be affected by the amount and type of bio-char present or added to soil. On the other hand, microorganisms are able to change the amount and properties of bio-char in soil. Both effects will have significant influence on nutrient cycles and nutrient availability to plants.

Some indications exist from soils that are rich in bio-char that microbial community composition, species richness, and diversity change with greater bio-char concentrations (Pietikäinen et al., 2000; Yin et al., 2000; Thies and Suzuki, 2003). Pietikäinen et al. (2000) found a greater bacterial growth rate in layers of charcoal than in the underlying organic horizon in a temperate forest soil. Already small amounts of 7.9 t C ha^{-1} of bio-char in a highly weathered soil in the tropics significantly enhanced microbial growth rates when nutrients were supplied by fertilizer (Steiner et al., 2004). A greater microbial biomass was reported in forest soils in the presence of charcoal by Zackrisson et al. (1996), and higher microbial activity (CO_2 production as well as organic matter decomposition) was found in soils exposed to black carbon aerosols derived from charcoal making (Uvarov, 2000). Apparently, bio-char provides a suitable habitat for a large and diverse group of soil microorganisms.

A higher retention of microorganisms in bio-char soils may be responsible for greater activity and diversity due to a high surface area as well as surface hydrophobicity of both the microorganisms and bio-char. A strong affinity of microbes to bio-char can be expected since the adhesion of microorganisms to solids increases with higher hydrophobicity of the surfaces (Stenström, 1989; Huysman and Verstraete, 1993; Castellanos et al., 1997; Mills, 2003). Activated carbon, which is chemically similar to black carbon or bio-char in soil, has been shown to sorb microorganisms strongly, and this adsorption was seen to increase with higher hydrophobicity (Rivera-Utrilla et al., 2001).

Strong microbial adhesion can be achieved between microorganisms and organic surfaces in the presence of divalent cations and specifically of Ca^{2+} (Rivera-Utrilla et al., 2001; Mills, 2003). Whether the mechanism is electrostatic bondage or increased hydrophobicity is not yet clear. Pore geometry and size distribution has been found definitely to promote the growth and activity of certain microorganisms.

Bio-char is also able to serve as a habitat for extraradical fungal hyphae that sporulate in their micropores due to lower competition from saprophytes, and it can therefore act as an inoculum for arbuscular mycorrhizal fungi (Saito and Marumoto, 2002). Root infection by arbuscular mycorrhizae significantly increased by adding 1 kg m^{-2} of bio-char to alfalfa in a volcanic ash soil that related well with growth of alfalfa ($r = 0.88; P < 0.01; N = 7$) being 40 to 80% greater after the application (Nishio and Okano, 1991; Nishio, 1996). Similarly, mycorrhizal infection increased when bio-char (7 g kg^{-1} soil) was added to soil that was inoculated with spores of *Glomus etunicatum*, improving the yields of onion (Matsubara et al., 1995).

Methods have already been developed to inoculate soils with ectomycorrhizas using carbonized rice husks (Mori and Marjenah, 1994). A more rapid cycling of nutrients in soil organic matter and microbial biomass as well as better colonization of roots by arbuscular mycorrhizal fungi will improve nutrient availability and crop yields by (1) retention of nutrients against leaching in highly weathered soils of the humid tropics that have little cation exchange capacity, and (2) a better access of the plants to fixed phosphorus due to inoculation by mycorrhizae.

The effect of microorganisms on bio-char is difficult to determine considering its long half-life. Bio-char is quite recalcitrant to microbial attack. However, we know that even bio-char must ultimately be broken down (Schmidt and Noack, 2000). Mineralization rates are not clear, and available data show rates of decomposition that are both rapid (Shneour, 1966; Bird et al., 1999) and slow (Shindo, 1991). It appears that a large part of bio-char is mineralized over a short time-scale, and a small part remains in a very stable, highly aromatic form, displaying greater ^{14}C age than the oldest SOM fractions (Pessenda et al., 2001).

The greater amount of cation exchange capacity per unit C found in soils with high amounts of bio-char such as the Amazonian Dark Earths (Sombroek, 1966) may be the result of a greater surface area of the bio-char and a higher charge density per unit surface area. A concomitant adsorption of low-molecular organic matter has been proposed since cation adsorption could be increased by coating of bio-char with manure extracts (Lehmann et al., 2002). In either case, the consequence is a greater cation exchange capacity per unit C found in soil with large amounts of bio-char compared with those with low amounts. Abiotic oxidation was also found to increase abundance of carboxylic acids on bio-char surfaces (Adams et al., 1988), and this may contribute to higher CEC after long periods of time in tropical ecosystems that experience high soil temperatures.

36.3   Biological Nitrogen Fixation in Soils with Bio-Char

Bio-char additions not only affect microbial populations and activity in soil, but also plant–microbe interactions through their effects on nutrient availability and modification of habitat. Rhizobia spp. living in symbiosis with many legume species are able to reduce atmospheric N_2 to organic nitrogen through a series of enzymatic reactions (Giller, 2001). This BNF is regarded as an important opportunity to mitigate nitrogen deficiency in cropping systems worldwide (Chapter 27). BNF significantly decreases, however, if available nitrate concentrations in soils are high, and if available calcium, phosphorus, and micronutrient concentrations are low (Giller, 2001).

Soils with appreciable concentrations of bio-char show the reversed situation as evident from Amazonian Dark Earths. With large bio-char concentrations, available nitrate concentrations are usually low and available calcium, phosphorus, and micronutrient concentrations are high, which is ideal for maximum BNF (Lehmann et al., 2003b). Indeed, BNF by common beans (CIAT BAT477), as determined by ^{15}N dilution, increased from 50 to 72% of total nitrogen uptake with increasing rates of bio-char additions (0, 31, 62, and 93 t C ha^{-1}) to a low-fertility Oxisol (Rondon et al., in preparation).

In addition to changing nutrient availabilities that are conducive to high BNF, inoculation with Rhizobia may be more effective in the presence of bio-char due to the habitat offered by the bio-char. In fact, several studies indicate that bio-char is an excellent support material for Rhizobium inoculants (Pandher et al., 1993; Lal and Mishra, 1998). Consequently, BNF determined by nitrogen difference was found to be 15% higher when bio-char was added to soil at early stages of alfalfa development, and 227% higher when nodule development was greatest (Nishio, 1996). Bio-char additions are, therefore, able to increase the net input of nitrogen into agricultural landscapes.

This does not necessarily mean that the nitrogen nutrition of the legume is improved, especially if large amounts of bio-char are added. Lehmann et al. (2003a) showed that while biomass production and nitrogen uptake of cowpea increased through large amounts of bio-char additions, plants' nitrogen nutrition decreased. With appropriate application rates of bio-char and supplementary nutrient additions, nitrogen input to agricultural systems can be increased without decreasing plant productivity. Such a soil management system may be interesting in the context of mixed legume–cereal intercropping or of agroforestry with woody legumes. Soil nitrogen stocks and eventually nitrogen availability can be increased and be made available to the nonlegume in a rotational system.

36.4   Slash-and-Char as an Alternative to Slash-and-Burn

Building on the evidence that bio-char additions significantly improve soil fertility as seen above, land-use systems have been developed that incorporate this technology. One such technology is a slash-and-char system, conceived as an alternative to slash-and-burn, and building on the socioeconomic and biophysical environment of such shifting-cultivation systems. This alternative entails — similar to slash-and-burn — that biomass from a given area is used as a soil conditioner for that same unit area while at the same time the field is cleared in a rapid and cost-efficient way. Woody biomass is charred on site in simple pits or under grass covers similar to small-scale charcoal production systems. For this effort to be successful, sufficient biomass has to be available, and the system has to utilize tree biomass efficiently for the production of bio-char. In fact, often in a slash-and-burn system, large proportions of the biomass are not burned completely, e.g., large branches, trunks, and roots. Such large woody debris are only partially charred and are usually seen as an obstacle for field preparation. However, in a slash-and-char system, they can become a source of bio-char.

36.4.1 Efficiency of Conversion

Analyses of conversions of woody biomass to bio-char have shown an average recovery of 54% of the initial carbon in the bio-char (Lehmann et al., 2002). This value was largely derived from laboratory experiments and commercial charcoal operations, however. Conversion using earthen pits and mounds will more likely range around 30 to 40%. The production of bio-char in agricultural fields using recently slashed organic matter requires specific skills. However, farmers who practice slash-and-burn are intimately familiar with cutting of biomass and the process of burning, and many farmers regularly produce charcoal for sale in local markets. Therefore, charring organic matter in simple earthen mounds or pits should not be limited by the availability of local knowledge. Improvements in the wood-to-bio-char conversion efficiency are feasible with changes in the geometry of the pits or piles and in management of the air supply during the charring process.

36.4.2 Supply of Material

Amounts of aboveground biomass largely depend on site characteristics, the age of the forest, whether it is a primary or secondary forest, as well as the history of land use (Buschbacher et al., 1988; Silver et al., 2000). A guideline presented in Figure 36.2 shows the amounts of bio-char that can be produced and applied to 1 ha of land as a function of forest age for various regions. Comparing these values with the positive growth and yield responses already at 5 to 10 t C ha^{-1} obtained from several experiments (Figure 36.1), about 4 to 8 years of growth of a secondary forest are needed to produce the bio-char equivalent to improve effectively the productivity of crops grown in a slash-and-char system. These amounts of regrowth necessary for the system can clearly be achieved under humid tropical conditions as seen in Figure 36.2.

These data establish that bio-char management is feasible with the amounts of biomass already available for systems such as slash-and-burn or slash-and-mulch without requiring any biomass transfers. This conclusion confirms the feasibility of creating the amounts of bio-char found in Amazonian Dark Earths with the resources that are available on the same piece of land that will be cropped. Applying a Terra Preta soil management system is therefore feasible under most conditions in the humid tropics. Therefore, slash-and-char does not increase but reduces anthropogenic CO_2 emissions, and bio-char even constitutes a long-term carbon sink in addition to improving soil fertility and production potential.

36.4.3 Duration

Slash-and-char not only increases crop yields after bio-char application, but also increases the number of cropping cycles before crop yields decrease to unacceptably low levels that

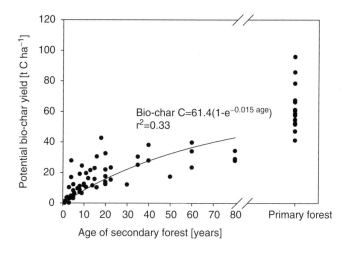

FIGURE 36.2

Potential bio-char yield from the woody biomass of secondary and primary forests in the humid tropics. A typical biomass loss of 69.4% upon bio-char conversion and a C concentration of 75.7% was calculated from data in Lehmann et al. (2002). The maximum of 61.4 t ha^{-1} for the plotted regression was the mean of all primary forest sites. Data on woody biomass volumes were obtained from Toky and Ramakrishnan (1983), Uhl and Jordan (1984), Buschbacher et al. (1988), Saldarriaga et al. (1988), Lugo (1992), Fearnside et al. (1993), Foster Brown et al. (1995), Kauffman et al. (1995), Alves et al. (1997), Camargo et al. (1999), Fearnside et al. (1999), Gehring et al. (1999), Graca et al. (1999), Gerwing and Farias (2000), Hughes et al. (2000), Mackensen et al. (2000), Sorrensen (2000), Johnson et al. (2001), and Feldpausch (2002). Woody biomass for the Alves, Camargo, Gehring and Sorrensen studies was assumed be equal to the average proportion of woody biomass of total biomass calculated from all other cited publications, i.e., 89.7% for older secondary forests, and 92% for primary forests.

make a fallow period necessary. A recent study has shown crop productivity to be low directly after slash-and-burn on a Typic Hapludox in the central Amazon, and even this could not be sustained for more than 1 year after slashing using mineral fertilizer, whereas application of bio-char resulted in low but sustainable grain yields (Steiner et al., 2004).

Continuous cultivation on Amazonian Dark Earths rich in bio-char has been reported to be more than 40 years in some instances (Petersen et al., 2001). On these soils, historically amended with bio-char, recovery of soil conditions is usually rapid, and only very short fallow periods of 1 to 2 years are required (German, 2001). On the other hand, slash-and-burn cycles are typically 2 to 3 years of cropping followed by 10 years of fallow (Nye and Greenland, 1960). In the absence of long-term studies with slash-and-char, its ratio between crop and fallow cycles can only be estimated, but this should be greater than unity.

36.4.4 Sustainability

In addition to higher proportions of crop vs. fallow periods, slash-and-char opens up the possibility for farmers of changing to permanent cropping. Black carbon is usually several thousand years older than nonblack carbon in soils and sediments (Masiello and Druffel, 1998; Schmidt et al., 2002). The polyaromatic structure of black carbon is extremely resistant to microbial attack, and only specialized fungi with unique extracellular enzymes are able to mineralize black carbon, as shown with coal (Willmann and Fakoussa, 1997; Hofrichter et al., 1999).

We can therefore expect that the beneficial effects of a slash-and-char intervention will persist in soil for a significant period of time. In contrast, amendments of mulches, manures, or composts are typically mineralized to carbon dioxide within a time frame of months to years, due to rapid decomposition rates in tropical soil ecosystems (Jenkinson and Ayanaba, 1977). The fraction of bio-char and nonbio-char, which is stabilized inside aggregates and organo-mineral complexes, can currently only be estimated, but it is certainly much higher for bio-char than for manures or mulches.

36.4.5 Labor Requirements

Labor inputs will most likely be greater during land clearing using a slash-and-char technique as woody biomass has to be charred and applied to soil. In many situations the additional labor may be low, however, as secondary burns and piling up or removal of unburned branches are already part of many slash-and-burn practices (Ketterings et al., 1999). This technology therefore has the potential for farmers in many ecoregions of the humid tropics to escape from the vicious cycle of declining productivity and soil degradation as a result of high population pressure and resulting shortened fallow periods.

36.5   Discussion

Bio-char can be obtained from a variety of other sources in addition to slashed woody biomass, which was described above, and bio-char production can be built into several land-use systems. Crop residue from grain processing (e.g., rice husks) and wood residue from lumber processing (e.g., sawdust) can be used to produce bio-char. The production of bio-char from rice husks is a procedure recommended by the Food and Fertilizer Technology Center for the Asian and Pacific Region (FFTC, 2001). Such conversions of crop residues can be done with locally available techniques, such as using residue mounds and firing (FFTC, 2001), simple firing chambers, or more sophisticated furnaces. Such obtained bio-char can be applied to any cropping system.

A better utilization of residues from charcoal production itself provides opportunities for a combination with a bio-char soil management system. Charcoal is still used in many parts of the world as an important source of energy for food preparation (World Energy Council, 2001). Worldwide an estimated 41 million tons (Mt) of charcoal were produced in 2002, with Brazil being the largest producer with 12 Mt (FAO, 2004). Much more charcoal is produced in developing countries (40 Mt in 2002) than in developed countries (1.4 Mt). Africa is the highest producer (21 Mt), followed by South America (14 Mt) and Asia (4 Mt).

During any production process, a significant proportion accumulates as waste that cannot be sold as fuel. The percentage varies, of course, and depends on the charring method as well as the wood type and environmental conditions, such as moisture content and temperature, as well as on postcharring management and packing of the charcoal. An estimate of waste accumulation from simple small-scale production systems of charcoal for barbeques in Colombia was 30 to 40% waste by weight of the produced charcoal. However, a growing local demand for charcoal of smaller sizes (1 to 2 cm) was observed as support material for growing ornamental plants in nurseries. The fraction of waste material that is not suitable for the market, and thus is discarded, lies between 10 and 15%.

Another source of bio-char is the pyrolysis of biomass (from agricultural and forestry sources) to produce energy, where bio-char is a by-product (Day et al., 2005). An additional improvement of the properties of such bio-char provides the conversion of gas emissions from the power plant (or other CO_2-producing industries) into NH_4HCO_3 by using the H_2 that was produced during the combustion process (Lee and Li, 2003). This ammonium bicarbonate can be adsorbed to the bio-char to increase nitrogen levels similar to commercially available nitrogen fertilizers. The resulting relatively low C:N ratios would make the bio-char a source and not a sink of nitrogen. Once refined, this relatively new technology could be used in small communities to provide energy while producing a soil conditioner rich in stable carbon and nitrogen.

Bio-char amendments have so far shown beneficial effects when applied to rice, sorghum, maize, various beans (soybean, common bean, cowpea, mungbean), banana, and vegetables such as carrots. Since bio-char is a valuable commodity, applications can be successfully applied to smaller units of area than entire fields, such as high fertility trenches in contour rows of steep land for vegetables or grain crops, and planting holes for tree crops. Bio-char has been widely applied in tree nurseries and is a recommended amendment (Jaenicke, 1999). It is used for propagation, in some cases due to its ability to adsorb inhibitory substances (Nhut et al., 2001). The particle size of the bio-char appears to play a minor role in its effect on soil fertility and crop production (Lehmann et al., 2003a), which simplifies the application of the technology.

Many questions still remain to be answered regarding the mechanisms governing surface properties of bio-char and how nutrient dynamics are affected by bio-char. The opportunities for carbon sequestration and the reduction of greenhouse gas emissions have not been explored at all, but they are potentially significant. It is clear already, however, that bio-char can significantly improve soil fertility in acid and highly weathered soils, and it has the potential for widespread application under various agroecological situations by mobilizing and improving the complex of chemical, physical, and biological properties of soil systems.

Annex: Site and crop information for Table 36.1

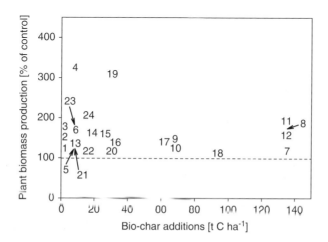

FIGURE 36.A1
Plant biomass increase over control as a function of the amount of bio-char applied, identifying data points in Figure 36.1 of this chapter. Numbers refer to site and crop information given in Table 36.A1.

TABLE 36.A1

Crop Species and Experimental Details of Trials with Added Bio-Char That Are Shown in Figure 36.A1. Trial Numbers (left column) Refer to Numbers in Figure 36.A1

Trial Number[a]	Plant Species	Soil	Amount of Bio-Char Applied[b] (t C ha^{-1})	Origin of the Bio-Char	Type of Experiment (Replicates)	Source
1	Mungbean (*Vigna radiata*)	Nd	0.38[c]	Commercial charcoal	Pots ($N = 3$)	Iswaran et al. (1980)
2	Soybean (*Glycine max*)	Nd	0.38	Commercial charcoal	Pots ($N = 3$)	Iswaran et al. (1980)
3	Pea (*Pisum sativum*)	Nd	0.38	Commercial charcoal	Pots ($N = 3$)	Iswaran et al. (1980)
4, 5	Rice (*Oryza sativa*)	Oxisol	7.9	Commercial charcoal	Field ($N = 5$)	Nehls (2002)
6	Alfalfa (*Medicago sativa*)	Andosol	6.1	Commercial charcoal from bark	Pots ($N = 3$)	Nishio and Okano (1991)
7, 8	Rice (*Oryza sativa*)	Oxisol	68	Commercial charcoal	Lysimeters ($N = 4$)	Lehmann et al. (2003a)
9, 10	Cowpea (*Vigna unguiculata*)	Oxisol	68	Commercial charcoal	Pots ($N = 5$)	Lehmann et al. (2003a)
11, 12	Cowpea (*Vigna unguiculata*)	Oxisol	135	Commercial charcoal	Pots ($N = 5$)	Lehmann et al. (2003a)
13	Soybean (*Glycine max*)	Oxisol	9	Commercial charcoal (*Swinglea glutinosa*)	Pots ($N = 4$)	Rondon et al. (unpublished data)
14	Soybean (*Glycine max*)	Oxisol	18	Commercial charcoal (*Swinglea glutinosa*)	Pots ($N = 4$)	Rondon et al. (unpublished data)
15	Soybean (*Glycine max*)	Oxisol	27	Commercial charcoal (*Swinglea glutinosa*)	Pots ($N = 4$)	Rondon et al. (unpublished data)
16	Bean (*Phaseolus vulgaris*)	Oxisol	31	Commercial charcoal (*Eucalyptus deglupta*)	Pots ($N = 4$)	Rondon et al. (unpublished data)
17	Bean (*Phaseolus vulgaris*)	Oxisol	62	Commercial charcoal (*Eucalyptus deglupta*)	Pots ($N = 4$)	Rondon et al. (unpublished data)
18	Bean (*Phaseolus vulgaris*)	Oxisol	93	Commercial charcoal (*Eucalyptus deglupta*)	Pots ($N = 4$)	Rondon et al. (unpublished data)
19	Carrots (*Phaseolus vulgaris*)	Andosol	30	Commercial charcoal (*Gliricida sepium*)	Field ($N = 4$)	Rondon et al. (2004)
20	Bean (*Phaseolus vulgaris*)	Andosol	30	Commercial charcoal (*Gliricida sepium*)	Field ($N = 4$)	Rondon et al. (2004)
21	Maize (*Zea mays*)	Oxisol	8	Commercial charcoal	Field ($N = 3$)	Rondon et al. (unpublished data)
22	Maize (*Zea mays*)	Oxisol	16	Commercial charcoal	Field ($N = 3$)	Rondon et al. (unpublished data)
23	Native savanna	Oxisol	8	Commercial charcoal	Field ($N = 3$)	Rondon et al. (unpublished data)
24	Native savanna	Oxisol	16	Commercial charcoal	Field ($N = 3$)	Rondon et al. (unpublished data)

[a] First number = unfertilized plants, second number = fertilized plants.

[b] In case of pot experiments, calculated for a depth of 0.1 m.

[c] Assuming 76% C in bio-char (Lehmann et al., 2002).

References

Adams, L.B. et al., An examination of how exposure to humid air can result in changes in the adsorption properties of activated carbons, *Carbon*, **26**, 451–459 (1988).

Alves, D.S. et al., Biomass of primary and secondary vegetation in Rondonia, Western Brazilian Amazon, *Global Change Biol.*, **3**, 451–461 (1997).

Antal, M.J. and Gronli, M., The art, science, and technology of charcoal production, *Ind. Eng. Chem. Res.*, **42**, 1619–1640 (2003).

Bird, M.I. et al., Stability of elemental carbon in a savanna soil, *Global Biogeochem. Cycles*, **13**, 923–932 (1999).

Bol, R. et al., Tracing dung-derived carbon in temperate grassland using ^{13}C natural abundance measurements, *Soil Biol. Biochem.*, **32**, 1337–1343 (2000).

Buschbacher, R., Uhl, C., and Serrao, E.A.S., Abandoned pastures in eastern Amazonia II. Nutrient stocks in the soil and vegetation, *J. Ecol.*, **76**, 682–699 (1988).

Camargo, P.B. et al., Soil carbon dynamics in regrowing forest of eastern Amazonia, *Global Change Biol.*, **5**, 693–702 (1999).

Castellanos, T., Ascencio, F., and Bashan, Y., Cell-surface hydrophobicity and cell-surface charge of *Azospirillum* spp, *FEMS Microbiol. Ecol.*, **24**, 159–172 (1997).

Chidumayo, E.M., Effects of wood carbonization on soil and initial development of seedlings in miombo woodland, Zambia, *Forest Ecol. Manage.*, **70**, 353–357 (1994).

Day, D. et al., Economical CO_2, SO_x, and NO_x capture from fossil-fuel utilization with combined renewable hydrogen production and large-scale carbon sequestration, *Energy*, **30**, 2558–2579 (2005).

FAO, *FAOSTAT Data*, Food and Agriculture Organization of the United Nations, Rome (2004), http://apps.fao.org/default.jsp

Fearnside, P.M., Leal, N., and Fernandes, E.M., Rainforest burning and the global carbon budget: Biomass, combustion efficiency, and charcoal formation in the Brazilian Amazon, *J. Geophys. Res.*, **98**, 16733–16743 (1993).

Fearnside, P.M. et al., Tropical forest burning in Brazilian Amazonia: Measurement of biomass loading, burning efficiency and charcoal formation at Altamira, Para, *Forest Ecol. Manage.*, **123**, 65–79 (1999).

Feldpausch, T.R., Carbon and nutrient accumulation, forest structure, and leaf area in secondary forests regenerating from degraded pastures in central Amazonia, Brazil. M.Sc. thesis, Cornell University, Ithaca, New York (2002).

FFTC, *Application of Rice Husk Charcoal*, FFTC Leaflet for Agriculture 2001 no. 4. Food and Fertilizer Technology Center, Taipei (2001).

Foster Brown, I. et al., Uncertainty in the biomass of Amazonian forests: An example from Rondonia, Brazil, *Forest Ecol. Manage.*, **75**, 175–189 (1995).

Gehring, C. et al., Response of secondary vegetation in eastern Amazonia to relaxed nutrient availability constraints, *Biogeochemistry*, **45**, 223–241 (1999).

German, L., The dynamics of Terra Preta: An integrated study of human–environmental interaction in a nutrient-poor Amazonian ecosystem. Ph.D. Thesis, University of Georgia, USA (2001).

Gerwing, J.J. and Farias, D.L., Integrating liana abundance and forest stature into an estimate of total aboveground biomass for an Eastern Amazonian forest, *J. Trop. Ecol.*, **16**, 327–335 (2000).

Giller, K.E., *Nitrogen Fixation in Tropical Cropping Systems*, 2nd ed., CAB International, Wallingford (2001).

Glaser, B. et al., The Terra Preta phenomenon: A model for sustainable agriculture in the humid tropics, *Naturwissenschaften*, **88**, 37–41 (2001).

Glaser, B., Lehmann, J., and Zech, W., Ameliorating physical and chemical properties of highly weathered soils in the tropics with charcoal: A review, *Biol. Fertil. Soils*, **35**, 219–230 (2002).

Goyal, S. et al., Influence of inorganic fertilizers and organic amendments on soil organic matter and soil microbial properties under tropical conditions, *Biol. Fertil. Soils*, **29**, 196–200 (1999).

Graca, P.M.L., Fearnside, P.M., and Cerri, C.C., Burning of Amazonian forest in Ariquemes, Rondonia, Brazil: Biomass, charcoal formation and burning efficiency, *Forest Ecol. Manage.*, **120**, 179–191 (1999).

Hecht, S., Indigenous soil management and the creation of Amazonian dark earths: Implications of kayapo practices, In: *Amazonian Dark Earths: Origin, Properties, Management*, Lehmann, J., et al., Ed., Kluwer, Dordrecht, 355–371 (2003).

Hofrichter, M. et al., Degradation of lignite (low-rank coal) by lignolytic basidiomycetes and their peroxidase system, *Appl. Microbiol. Biotechnol.*, **52**, 78–84 (1999).

Hughes, R.F., Kauffman, J.B., and Cummings, D.L., Fire in the Brazilian Amazon: 3. Dynamics of biomass, C, and nutrient pools in regenerating forests, *Oecologia*, **124**, 574–588 (2000).

Huysman, F. and Verstraete, W., Effect of cell surface characteristics on the adhesion of bacteria to soil particles, *Biol. Fertil. Soils*, **16**, 21–26 (1993).

Iswaran, V., Jauhri, K.S., and Sen, A., Effect of charcoal, coal and peat on the yield of moong, soybean and pea, *Soil Biol. Biochem.*, **12**, 191–192 (1980).

Jaenicke, H., *Good Tree Nursing Practices: Practical Guidelines for Research Nurseries*, International Centre for Research in Agroforestry, Nairobi (1999).

Jenkinson, D.S. and Ayanaba, A., Decomposition of carbon-14 labeled plant material under tropical conditions, *Soil Sci. Soc. Am. J.*, **41**, 912–915 (1977).

Johnson, C.M. et al., Carbon and nutrient storage in primary and secondary forests in eastern Amazônia, *Forest Ecol. Manage.*, **147**, 245–252 (2001).

Kauffman, J.B. et al., Fire in the Brazilian Amazon: 1. Biomass, nutrient pools, and losses in slashed primary forests, *Oecologia*, **104**, 397–408 (1995).

Ketterings, Q.M. et al., Farmers' perspectives on slash-and-burn as a land clearing method for small-scale rubber producers in Sepunggur, Jambi Province, Sumatra, Indonesia, *Forest Ecol. Manage.*, **120**, 157–169 (1999).

Lal, J.K. and Mishra, B., Flyash as a carrier for Rhizobium inoculant, *J. Res* (Birsa Agric. Univ.), 10, 191–192 (1998).

Lee, J.W. and Li, R., Integration of fossil energy systems with CO_2 sequestration through NH_4HCO_3 production, *Energy Convers. Manage.*, **44**, 1535–1546 (2003).

Lehmann, J. et al., Slash-and-char: A feasible alternative for soil fertility management in the central Amazon? *17th World Congress of Soil Science, Bangkok, Thailand*. CD-ROM Paper no. 449, 1–12 (2002).

Lehmann, J. et al., Nutrient availability and leaching in an archaeological Anthrosol and a Ferralsol of the Central Amazon basin: Fertilizer, manure and charcoal amendments, *Plant Soil*, **249**, 343–357 (2003a).

Lehmann, J. et al., Soil fertility and production potential, In: *Amazonian Dark Earths: Origin, Properties, Management*, Lehmann, J., Ed., Kluwer, Dordrecht, 105–124 (2003b).

Lehmann, J. et al., Eds., *Amazonian Dark Earths: Origin, Properties, Management*, Kluwer, Dordrecht (2003c).

Lugo, A.E., Comparison of tropical tree plantations with secondary forests of similar age, *Ecol. Monogr.*, **62**, 1–41 (1992).

Mackensen, J. et al., Site parameters, species composition, phytomass structure and element stores of a terra-firme forest in East-Amazonia, Brazil, *Plant Ecol.*, **151**, 101–119 (2000).

Masiello, C.A. and Druffel, E.R.M., Black carbon in deep-sea sediments, *Science*, **280**, 1911–1913 (1998).

Matsubara, Y.I., Harada, T., and Yakuwa, T., Effect of inoculation density of VAM fungal spores and addition of carbonized material to bed soil on growth of Welsh onion seedlings, *J. Jpn. Soc. Hortic. Sci.*, **64**, 549–554 (1995).

Mikan, C.J. and Abrams, M.D., Altered forest composition and soil properties of historic charcoal hearths in southeastern Pennsylvania, *Can. J. Forest Res.*, **25**, 687–696 (1995).

Mikan, C.J. and Abrams, M.D., Mechanisms inhibiting the forest development of historic charcoal hearths in southeastern Pennsylvania, *Can. J. Forest Res.*, **25**, 687–696 (1996).

Mills, A.L., Keeping in touch: Microbial life on soil particle surfaces, *Adv. Agron.*, **78**, 1–43 (2003).

Mori, S. and Marjenah, A., Effect of charcoal rice husks on the growth of Dipterocarpaceae seedlings in East Kalimantan with special reference to ectomycorrhizal formation, *J. Jpn. Forest Soc.*, **76**, 462–464 (1994).

Nehls, T., Fertility improvement of a Terra Firme Oxisol in central Amazonia by charcoal applications. M.Sc. thesis, University of Bayreuth, Germany (2002).

Nhut, D.T. et al., Effects of activated charcoal, explant size, explant position and sucrose concentration on plant and shoot regeneration of *Lilium longiflorum* via young stem culture, *Plant Growth Regul.*, **33**, 59–65 (2001).

Nishio, M., *Microbial Fertilizers in Japan*, FFTC Extension Bulletin. Food and Fertilizer Technology Center, Taipei (1996).

Nishio, M. and Okano, S., Stimulation of the growth of alfalfa and infection of mycorrhizal fungi by the application of charcoal, *Bull. Natl. Grassland Res. Inst.*, **45**, 61–71 (1991).

Nye, P.H. and Greenland, D.J. *The Soil under Shifting Cultivation*, Commonwealth Bureau of Soils Technological Communication, **51** (1960).

Oguntunde, P.G. et al., Effects of charcoal production on maize yield, chemical properties and texture of soil, *Biol. Fertil. Soil*, **39**, 295–299 (2004).

Pandher, M.S. et al., Studies on growth and survival of Rhizobium isolates in different carriers, *Indian J. Ecol.*, **20**, 141–146 (1993).

Pessenda, L.C.R., Gouveia, S.E.M., and Aravena, R., Radiocarbon dating of total soil organic matter and humin fraction and its comparison with ^{14}C ages of fossil charcoal, *Radiocarbon*, **43**, 595–601 (2001).

Petersen, J., Neves, E.G., and Heckenberger, M.J., Gift from the past: Terra preta and prehistoric amerindian occupation in Amazonia, In: *Unknown Amazonia*, McEwan, C., Ed., British Museum, London, 86–105 (2001).

Pietikäinen, J., Kiikkilä, O., and Fritze, H., Charcoal as a habitat for microbes and its effects on the microbial community of the underlying humus, *Oikos*, **89**, 231–242 (2000).

Rivera-Utrilla, J. et al., Activated carbon surface modifications by adsorption of bacteria and their effect on aqueous lead adsorption, *J. Chem. Technol. Biotechnol.*, **76**, 1209–1215 (2001).

Rondon, M., Ramirez, A., and Hurtado, M., Charcoal additions to high fertility ditches enhance yields and quality of cash crops in Andean hillsides of Colombia. *CIAT Annual Report 2004*, Cali, Colombia (2004).

Saito, M. and Marumoto, T., Inoculation with arbuscular mycorrhizal fungi: The status quo in Japan and the future prospects, *Plant Soil*, **244**, 273–279 (2002).

Saldarriaga, J.G. et al., Long-term chronosequence of forest succession in the Upper Rio Negro of Colombia and Venezuela, *J. Ecol.*, **76**, 938–958 (1988).

Schmidt, M.W.I. and Noack, A.G., Black carbon in soils and sediments: Analysis, distribution, implications, and current challenges, *Global Biogeochem. Cycles*, **14**, 777–794 (2000).

Schmidt, M.W.I., Skjemstad, J.O., and Jäger, C., Carbon isotope geochemistry and nanomorphology of soil black carbon: Black chernozemic soils in central Europe originate from ancient biomass burning, *Global Biogeochem. Cycles*, **16**, 70 (2002).

Shindo, H., Elementary composition, humus composition, and decomposition in soil of charred grassland plants, *Soil Sci. Plant Nutr.*, **37**, 651–657 (1991).

Shneour, E.A., Oxidation of graphitic carbon in certain soils, *Science*, **151**, 991–992 (1966).

Silver, W.I., Ostertag, R., and Lugo, A.E., The potential for carbon sequestration through reforestation of abandoned tropical agricultural and pasture land, *Ecol. Restor.*, **8**, 394–407 (2000).

Skjemstad, J.O. et al., Charcoal carbon in U.S. agricultural soils, *Soil Sci. Soc. Am. J.*, **66**, 1249–1255 (2002).

Smith, N.J.H., Anthrosols and human carrying capacity in Amazonia, *Ann. Assoc. Am. Geogr.*, **70**, 553–566 (1980).

Sombroek, W., *Amazon Soils: A Reconnaissance of Soils of the Brazilian Amazon Region*, Centre for Agricultural Publications and Documentation, Wageningen (1966).

Sorrensen, C.L., Linking smallholder land use and fire activity: Examining biomass burning in the Brazilian Lower Amazon, *Forest Ecol. Manage.*, **128**, 11–25 (2000).

Steiner, C. et al., Microbial response to charcoal amendments of highly weathered soil and Amazonian dark earths in central Amazonia: Preliminary results, In: *Amazonian Dark Earths: Explorations in Time and Space*, Glaser, B. and Woods, W.I., Eds., Springer, Heidelberg, 95–212 (2004).

Stenström, T.A., Bacterial hydrophobicity: An overall parameter for the measurement of adhesion potential to soil particles, *Appl. Environ. Microbiol.*, **55**, 142–147 (1989).

Thies, J. and Suzuki, K., Amazonian dark earths: Biological measurements, In: *Amazonian Dark Earths: Origin, Properties, Management*, Lehmann, J. et al., Eds., Kluwer, Dordrecht, 287–332 (2003).

Toky, O.P. and Ramakrishnan, P.S., Secondary succession following slash and burn agriculture in northeastern India, *J. Ecol.*, **71**, 735–745 (1983).

Topoliantz, S. et al., Effect of organic manure and endogeic earthworm *Pontoscolex corethrurus* (Oligochaeta: Glossoscolecidae) on soil fertility and bean production, *Biol. Fertil. Soils*, **36**, 313–319 (2002).

Trujillo, L., Fluxos de nutrientes em solo de pastagem abandonada sob adubacao organica e mineral na Amazonia central. M.Sc. thesis, INPA and University of Amazonas, Brazil (2002).

Tryon, E.H., Effect of charcoal on certain physical, chemical, and biological properties of forest soils, *Ecol. Monogr.*, **18**, 81–115 (1948).

Uhl, C. and Jordan, C.F., Succession and nutrient dynamics following forest cutting and burning in Amazonia, *Ecology*, **65**, 1476–1490 (1984).

Uvarov, A.U., Effects of smoke emissions from a charcoal kiln on the functioning of forest soil systems: A microcosm study, *Environ. Monit. Assess.*, **60**, 337–357 (2000).

Wardle, D.A., Zackrisson, O., and Nilsson, M.C., The charcoal effect in boreal forests: Mechanisms and ecological consequences, *Oecologia*, **115**, 419–426 (1998).

Willmann, G. and Fakoussa, R.M., Extracellular oxidative enzymes of coal-attacking fungi, *Fuel Process. Technol.*, **52**, 27–41 (1997).

World Energy Council, 19th World Energy Council Survey of Energy Resources (2001), http://www.worldenergy.org.

Yin, B. et al., Bacterial functional redundancy along a soil reclamation gradient, *Appl. Environ. Microbiol.*, **66**, 4361–4365 (2000).

Young, M.J., Johnson, J.E., and Abrams, M.D., Vegetative and edaphic characteristics on relic charcoal hearths in the Appalachian mountains, *Vegetatio*, **125**, 43–50 (1996).

Zackrisson, O., Nilsson, M.C., and Wardle, D.A., Key ecological function of charcoal from wildfire in the boreal forest, *Oecologia*, **77**, 10–19 (1996).

37

Improving Phosphorus Fertility in Tropical Soils through Biological Interventions

Astrid Oberson,[1] Else K. Bünemann,[2] Dennis K. Friesen,[3] Idupulapati M. Rao,[4] Paul C. Smithson,[5] Benjamin L. Turner[6] and Emmanuel Frossard[1]

[1]*Institute of Plant Sciences, Swiss Federal Institute of Technology, Zurich, Switzerland*

[2]*School of Earth and Environmental Sciences, University of Adelaide, Adelaide, Australia*

[3]*International Center for the Improvement of Maize and Wheat (CIMMYT) and International Fertilizer Development Center (IFDC), Addis Ababa, Ethiopia*

[4]*Tropical Soil Biology and Fertility Institute, International Center for Tropical Agriculture (CIAT), Cali, Colombia*

[5]*Berea College, Berea, Kentucky, USA*

[6]*Smithsonian Tropical Research Institute, Panama City, Republic of Panama*

CONTENTS

Low availability of phosphorus is a major constraint on agricultural productivity in highly weathered tropical soils. Such soils have a significant capacity to sorb large amounts of phosphorus, taking them out of the soil solution. This limits the availability of inorganic phosphorus for plants, whether it is already contained in the soil or added as fertilizer. Further, some tropical soils contain only small amounts of total phosphorus, with a relatively large proportion of this present in organic forms. This makes biological processes vitally important for enhancing phosphorus availability to crops in tropical soils, especially those that are receiving organic amendments as their major nutrient source (Nziguheba and Bünemann, 2005; Oberson and Joner, 2005).

Soil microbes play a central role in enhancing phosphorus availability to plants because they mediate the turnover of phosphorus contained in organic amendments and in soil organic matter. At the same time, the incorporation of phosphorus into microbial cells prevents its strong sorption to soil constituents, thereby maintaining it in a form that can be released subsequently into the soil solution following microbial turnover. Microbial processes are driven by the availability of decomposable organic carbon, which highlights the importance of sustaining and improving soil organic matter concentrations if large populations of microbes are to be active in the soil.

Organic amendments such as manures and plant residues are a major source of organic carbon to the soil. Plants also promote microbial populations and subsequent turnover by exuding organic carbon from their roots (Merckx et al., 1985). Some plants are well-adapted to low phosphorus availability owing to their rooting pattern and root characteristics (Rao et al., 1996), their associations with mycorrhizal fungi (Sieverding, 1991), and/or their ability to take up soil phosphorus from recalcitrant compounds (George et al., 2002b). When the selection and breeding of germplasm is adapted for these traits, plants can further enhance the biological cycling of soil phosphorus.

Biological processes that enhance soil phosphorus availability are discussed in detail in Chapter 13, detailing the turnover of phosphorus through the microbial biomass. Such processes can be difficult to assess because they are subject to subtle and complex interactions with various other processes. It is also difficult to synchronize nutrient release by microbial processes with plant demand, an issue that requires further study. Consequently, it is difficult to make broad predictions or generalizations about the impact that biological improvements will have on soil nutrient availability. However, there is sufficient evidence of positive effects in field trials from a variety of agroecosystems that such practices merit serious consideration, even if the scientific basis for understanding the whole process is incomplete.

In this chapter we report on cases where improvements in soil phosphorus availability and turnover were achieved by enhancing soil biological activity. We focus on examples from Colombia and Kenya that combined organic matter management with the use of adapted germplasm and strategic inputs of low doses of phosphorus fertilizers to raise crop productivity and enhance soil fertility. Several additional examples of similar improvements using different biological interventions are considered briefly in the third section.

37.1   Introduced Pastures and Cropping Systems in Eastern Colombia

The eastern plains of Colombia include 17 million ha of land dominated by acidic, nutrient-poor Oxisols with a well-defined dry season of 3 to 4 months. The herbaceous native savanna vegetation is of low nutritional value and is used traditionally for extensive beef production. The introduction into pastures of exotic tropical grasses that are adapted to acidic soils, mostly Brachiaria spp., either alone or in conjunction with tropical forage legumes, can increase the productivity of the grazing animals by between 10- and 15-fold (Lascano and Estrada, 1989).

After cultivars adapted to acidic soils were identified by testing, new alternatives for agricultural production were introduced into the area. A major soil chemical constraint on agricultural productivity on the Colombian savannas is the low concentration of total and plant-available phosphorus (Friesen et al., 1997). To understand and improve the efficiency of phosphorus use and cycling in production systems in this region, a series of field evaluations was undertaken in the eastern plains of Colombia to assess the interactions among soil phosphorus status (including organic phosphorus), the introduction of pastures or different cropping systems, and microbial phosphorus turnover (Rao et al., 2004). Data from these studies are presented in Table 37.1.

37.1.1 Measuring and Explaining Differences in Phosphorus Availability

Improved pastures with new grass species yielded a remarkable increase in beef production with only modest inputs of mineral phosphate fertilizer. This indicated that the efficiency of fertilization was greater than expected on these soils that otherwise strongly sorbed phosphorus. To understand the effect of introduced pastures on phosphorus cycling and availability, phosphorus budgets were estimated, and soil phosphorus status was characterized for soils from (a) unfertilized native savanna pastures, or (b) fertilized, introduced pastures. The latter were either (b1) grass-only (*Brachiaria decumbens* cv. Basilisk) or (b2) a grass–legume mixture (*B. decumbens* with *Pueraria phaseoloides*, commonly known as kudzu) (Oberson et al., 1999).

The activity of soil phosphatase enzymes and the amount of phosphorus in the soil microbial biomass were found to be greater in grass–legume than in grass-only or native savanna pastures. Seasonal analysis of soil phosphorus indicated that the grass–legume system maintained greater concentrations of organic and plant-available phosphorus with less temporal variation than in the two other systems. Analysis of the phosphorus associated with humic and fulvic acids by solution ^{31}P nuclear magnetic resonance spectroscopy revealed in the grass–legume soils greater reserves of phosphate diesters (Guggenberger et al., 1996), which are usually assumed to contribute substantially to plant nutrition (see Chapter 13). Thus the improvement in soil phosphorus availability, as determined by several methods, was clearly greater in grass–legume than in grass-only pastures (Oberson et al., 1999). What could be the reasons for this?

The improved phosphorus availability in grass–legume pastures was not due to differences in fertilizer inputs or exports from the system; alternatively, it could be attributed to changes in the overall biological activity in the soil–plant system caused by the presence of legumes in the vegetation cover. Total carbon and organic phosphorus concentrations (Oberson et al., 1999) as well as macrofaunal activity (Decaens et al., 1994) were all found to be significantly greater in grass–legume soils. Roots from legumes decomposed faster than the roots from Brachiaria grasses (Gijsman et al., 1997a), and annual root production, in terms of both biomass and length, was measured to be significantly greater in the introduced pastures (grass-only and grass–legume) compared with native pasture (Rao et al., 2001).

Mean aboveground litter production in the introduced pastures during the wet season was about $40 \, g \, m^{-2}$ month^{-1} (Thomas et al., 1993). Greater pasture productivity with legumes, associated with greater inputs of plant litter and excrement from the grazing animals in grass–legume pastures, probably provided more consistent organic phosphorus inputs and greater phosphorus cycling and availability, which was in agreement with the results of a simulation of the nitrogen cycle in grass–legume vs. grass-only pastures (Thomas, 1992).

The increased activity of macrofauna in the introduced pastures interacted positively with the increased phosphorus cycling, since total phosphorus was markedly higher in earthworm casts than in the surrounding soil (Jiménez et al., 2003). Increases in microbial phosphorus and plant-available phosphate in the casts were even more pronounced. This suggested that greater concentrations of labile organic phosphorus in pasture soils were linked to a greater abundance of earthworms. Levels of labile organic phosphorus,

TABLE 37.1

Characteristics of Colombian Oxisols Evaluated

Location and Grassland Type	Soil Type and Texture	Depth (cm)	pH	Organic C (g kg⁻¹)	Total P (mg kg⁻¹)	Organic P (mg kg⁻¹)	Microbial P[a] (mg kg⁻¹)	Plant-Available P (mg kg⁻¹)		Reference
								(Bray-II)	Resin-Extractable P	
Introduced pastures replacing native savanna	Oxisol: 39% clay, 19% sand	0–10								Oberson et al. (1999)
Native savanna			4.8	23.5	179	64	5.2	1.3	3.5	
Grass-only			4.85	23.2	228	68	5.9	1.4	4.0	
Grass–legume			4.96	24.9	226	77	7.3	2.2	5.2	
Rice–pasture rotations and rice monocrop replacing native savanna	Oxisol: 40% clay, 30% sand	0–10								
Native savanna			4.7	26.5	178	85	3.9	5.7	2.1	Gijsman et al. (1997b)
Rice–grass–legume			4.8	31.7	209	81	5.2	7.8	3.0	A. Oberson (unpublished)
Rice-only			4.7	29.2	221	95	4.6	6.4	3.1	
Rice–grass–legume–rice			4.6	27.5	250	101	5.2	21.2	8.3	
Rice–grass–rice			4.6	30.7	255	95	5.4	23.5	9.4	
Continuous rice			4.8	26.8	307	107	3.8	32.2	11.0	
Bulk soil and earthworm casts from native savanna and grass–legume pasture[b]										
Native savanna	Oxisol: 39% clay, 19% sand									Jiménez et al. (2003)
Soil		0–15	5.1	21	179	73	2.5	1.0	0.8	
Casts		Surface	5.4	41	267	108	4.0	6.3	7.8	
Grass–legume										
Soil		0–15	5.2	21	194	62	4.1	2.5	1.4	
Casts		Surface	5.8	52	396	136	10.9	11.0	12.8	
Introduced pastures and rice replacing native savanna	Oxisol: silty clay	0–10								Oberson et al. (2001)
Savanna			4.9	26.1	216	82	5.4	1.4	2.6	
Grass–legume			4.8	28.4	272	98	6.6	3.4	4.8	
Continuous rice			4.7	24.7	354	98	2.6	15.6	14.3	

[a] All values present microbial phosphorus extracted after soil fumigation (i.e., no conversion factors (k_P) applied).

[b] Same site as in Oberson et al. (1999).

including phosphate diesters such as DNA, were found to be enriched in earthworm casts (Guggenberger et al., 1996), which could in turn be linked to the greater phosphorus availability in pasture soils.

Earthworms, which play a fundamental role in promoting soil nutrient availability (Jiménez and Decaens, 2004), can be greatly influenced by changes in land use (Decaens et al., 1994). Since earthworm species differ in their nutritional behavior and inhabit different soil layers, the composition of the earthworm community will affect the impact that these organisms have on phosphorus availability. For instance, there is greater phosphate availability in earthworm casts of the geophagous earthworm *Pontoscolex corethrurus* compared with the bulk soil of an Oxisol, which can be attributed to a combination of selective ingestion of small particles and the partial mineralization of organic phosphorus (Chapuis-Lardy et al., 1998).

The endogeic earthworm, *Polypheretima elongate*, on the other hand, has been found to increase phosphorus availability by digestive and microbial processes during gut transit (Brossard et al., 1996). However, in both cases, total phosphorus content did not differ between earthworm casts and bulk soil. Still different, the anecic species, *Martiodrilus carimaguensis*, increases available and total phosphorus content in its casts compared with that in the bulk soil, presumably through the ingestion of phosphorus from plant residues or other organic sources (Jiménez et al., 2003). Earthworms play critical roles within soil food webs and in improving soil physical structure, as discussed in Chapters 5 and 11.

37.1.2   Assessing Phosphorus Turnover

The beneficial effect that tropical pastures have on phosphorus cycling through the microbial biomass was evident when rice–pasture rotations replaced native savanna (Gijsman et al., 1997b). Microbial phosphorus, when measured, was lower under monocropped rice and savanna soils than in crop–pasture rotations. In the latter, the activity of the microbial biomass increased in parallel with the overall improvement in soil fertility.

The relationship between microbial phosphorus cycling and the dynamics of extractable organic phosphorus fractions has been studied by Oberson et al. (2001), Bühler et al. (2002). Grass–legume pastures with greater soil organic phosphorus concentrations were associated with greater soil biological activity than under native savanna or soils cropped continuously with rice. Isotopic labeling indicated a rapid microbial phosphorus turnover that was greatest in grass–legume soils. Phosphorus mineralization rates appeared to be greater in grass–legume than in continuous-rice soils, suggesting that the former had a greater potential to render organic phosphorus available to plants (Oberson et al., 2001).

Studies using radio-labeled phosphate have documented the rapid turnover of phosphorus through both the inorganic and organic pools. In one example, a significant proportion of radio-labeled phosphate was recovered in organic phosphorus fractions 2 weeks after labeling, demonstrating that the dynamics of organic phosphorus are important when soil phosphate availability is limited (Bühler et al., 2002). Isotopic labeling has also shown that phosphorus flowed rapidly through, but did not accumulate in, organic phosphorus pools, at least in the short term (Rao et al., 2004). In an experiment with crop rotation and ley pasture, chemical fractionation of soil phosphorus indicated that applied phosphorus fertilizer moved preferentially into labile phosphate pools and then slowly, via biomass production and microbes, into organic phosphorus pools (Friesen et al., 1997).

37.1.3   Implications of Experience from Colombia

Overall, the results obtained on low-phosphorus, acidic soils in the eastern plains of Colombia have demonstrated that the combined use of improved germplasm and modest, strategic inputs of fertilizer resulted in significantly increased performance of both pastoral and arable production systems, provided that application to arable systems is combined with appropriate tillage and crop rotations (Phiri et al., 2003). The increase in system performance, whether in terms of livestock or crop yields, was clearly related to enhanced soil biological activity and microbial phosphorus turnover.

However, the widespread adoption of agropastoral systems by farmers requires some investment in germplasm and fertilizers as well as knowledge of pasture and crop management techniques. These investments, which occur in conjunction with an intensification and diversification of the overall farm operations, are likely to be made only when remunerative markets for the products are accessible. In the Colombian llanos, local institutions and the Colombian government have restricted their activities and investments, mainly because of social insecurity (Guimaráes et al., 2004). Even though the infrastructure is insufficient there, official efforts continue to make this region more productive and its residents better off.

37.2 Managed Short-Term Leguminous Fallows in Western Kenya

Soils in the West Kenyan highlands are predominantly fine-textured, phosphorus-sorbing Oxisols and Alfisols. The rural population there, with 500 to 1200 people km^{-2}, is one of the densest in the world. In the traditional system of shifting cultivation, cropping periods of 1 to 4 years duration were alternated with fallow periods of up to 15 years. This allowed restoration of soil organic matter by above- and belowground biomass production of woody secondary vegetation. However, the duration of fallow periods has now declined to only one to two growing seasons, i.e., one half to one year, due to the high population density and scarcity of arable land. The biomass production of short, weedy fallows is usually insufficient to replenish soil organic matter lost during the cropping period. Many farmers do not fallow their land at all, or only infrequently, with the poorest farmers who have the least amount of land doing the least fallowing if any. This leads to the further impoverishment of both their soils and themselves.

Depletion of soil organic matter leads to a corresponding depletion of nutrients, in part, because mineral fertilizers are too expensive for the rural poor. To address the need to enhance soil system fertility among poor farmers, the World Agroforestry Centre (ICRAF) has developed crop rotations that include more productive fallows, often involving planted legumes, which restore soil fertility and organic matter (Sanchez, 1999). Results of this research program are reported from Zambia and some other countries in Chapter 19 under the rubric of "fertilizer trees." Here, we examine one application of this concept with particular interest in phosphorus mobilization.

Leguminous fallow plants can fix significant amounts of nitrogen from the atmosphere, although the proportion of fixed nitrogen that builds up in plant biomass as well as the total amount of fixed nitrogen will vary according to the species. Gathumbi et al. (2002) reported that in the absence of phosphorus and potassium limitation, the proportions of total plant nitrogen that were fixed from atmospheric nitrogen ranged between 35 and 83%, with the amount of fixed nitrogen being between 8 and 140 kg ha^{-1}. Of the seven leguminous species studied, *Crotalaria grahamiana* yielded the highest values of fixed nitrogen. The leguminous fallow tree *Sesbania sesban* is particularly effective in obtaining nitrogen from the subsoil (Hartemink et al., 1996).

However, despite its overall importance, nitrogen may not be the most limiting nutrient in many situations. On many soils in western Kenya, phosphorus availability is the main constraint, as seen from the fact that nitrogen fertilization without the simultaneous addition of phosphorus often fails to increase maize yields (Jama et al., 1997). Even in the absence of external phosphorus inputs, annual maize yields were significantly higher in maize–legume fallow rotations than in continuous maize or maize-natural fallow rotations (Niang et al., 2002; Smestad et al., 2002). This suggests that phosphorus availability was enhanced by alternation with a legume fallow. Secondary effects of fallows on maize growth, e.g., modified soil moisture conditions, may also play a role.

37.2.1 Planted Fallow Alternatives in the Crop Rotation

We consider here data on the effects of a leguminous fallow on soil phosphorus dynamics from a case study that lasted 5.5 years (Bünemann, 2003) (Table 37.2). It investigated the relationships among organic matter inputs made through the fallow plant *C. grahamiana*, soil microbial activity, phosphorus availability, and maize crop performance (Bünemann et al., 2004b). The field experiment, established on an Oxisol, included three maize-based rotations (continuous maize, maize–Crotalaria fallow, and maize-natural fallow rotation) with two levels of phosphorus fertilization (0 and 50 kg P ha^{-1}, applied as triple superphosphate). The maize in each rotation was grown during the long rainy season lasting from March to August. The fallows were maintained in the rotations during the short rainy season lasting from September to February, while in the continuous maize system a second maize crop was planted during the short rains.

Seasonal grain yield in the continuous maize system without phosphorus fertilization was low, varying between 0.1 and 1.5 t ha^{-1}. Yields during the short rains are usually poor because the rain is unreliable, and there is little or no break between long-rain harvest and short-rain planting. This increases pest pressure, which in turn spreads the maize streak virus that reduces yields. In all rotations, maize production was roughly doubled by phosphorus fertilization, indicating significant phosphorus deficiency at this site. Maize grain yields reported from the same region by other authors have averaged about 1 t ha^{-1} without and 2 t ha^{-1} with phosphorus fertilization (Maroko et al., 1999; Nziguheba et al., 2000; Kwabiah et al., 2003). This low productivity can be due to a variety of factors, such as limited water availability, deficiency in nutrients other than phosphorus, and weed, disease, and pest problems.

Cumulative maize production over 5.5 years did not differ significantly among the three rotations, indicating that the maize yield forgone during the fallow season was compensated for by higher postfallow yields (Figure 37.1). During the first fallow season, the Crotalaria fallow was more productive than the natural fallow, recycling 5.3 t dry matter ha^{-1} compared with 3.2 t ha^{-1} for the natural fallow. The subsequent maize yield was doubled after the Crotalaria fallow at both levels of phosphorus fertilization. However, Crotalaria growth decreased during the course of the experiment due to pest problems, such as infestation with planthopper and subsequent colonization of stem lesions by saprophytic fungi (Girma, 2002). As a consequence, with longer duration, the introduced fallow was no longer beneficial for the subsequent maize crop compared with continuous maize, and at the end of the field experiment, the cumulative maize yield was similar between the three rotations (Figure 37.1).

37.2.2 Soil Microbial Involvement in Crop–Fallow Systems

In terms of changes in the soil system, both fallow types reversed the trend of soil organic matter depletion observed under continuous maize. The highest concentrations of soil organic matter and microbial nutrients were found in the maize–Crotalaria rotation, while plant-available soil phosphate concentrations were similar in all three rotations, being

TABLE 37.2

Characteristics of Kenyan Oxisols Evaluated

Location and Grassland Type	Soil Type and Texture	Depth (cm)	pH	Organic C (g kg^{-1})	Total P (mg kg^{-1})	Organic P (mg kg^{-1})	Microbial P[a] (mg kg^{-1})	Plant-Available P (Resin-P) (mg kg^{-1})	Reference
Maize cropping systems, zero phosphorus treatment									
Continuous maize	Oxisol:	0–15	5.0	24.0	720	271	3.4	1.8	Bünemann (2003), Bünemann et al. (2004b)
Weed fallow	39% clay,		5.0	25.8	703	267	5.2	1.7	
Crotalaria fallow	37% sand		5.0	25.6	721	282	6.2	1.7	
Maize cropping systems, fertilized treatment (50 kg P ha^{-1} year^{-1})									Same
Continuous maize	Oxisol:	0–15	4.8	23.7	838	257	3.5	6.9	
Natural fallow	39% clay,		5.1	25.2	837	278	5.4	6.7	
Crotalaria fallow	37% sand		5.0	26.3	829	289	6.6	6.3	

[a] All values present microbial phosphorus extracted after soil fumigation (i.e., no conversion factors (k_P) applied).

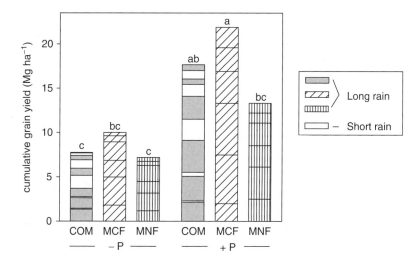

FIGURE 37.1

Cumulative maize grain yields during 5.5 years of field experimentation. COM, continuous maize; MCF, maize–crotalaria fallow, MNF, maize-natural fallow. *Source:* Authors' data.

significantly increased by phosphorus fertilization (Bünemann et al., 2004b) (Table 37.2). The composition of the microbial community, as indicated by phospholipid fatty acid analysis, differed somewhat between soils under continuous maize and in maize–Crotalaria rotation (Bünemann et al., 2004a). In particular, soils with greater microbial biomass concentrations contained a greater relative abundance of fungi and Gram-negative bacteria.

An incubation study with soils from this field experiment indicated that microorganisms took up phosphorus from stable pools in the soil following the addition of *C. grahamiana* residues labeled with radioactive phosphate (Bünemann et al., 2004c). Some incorporation of labeled phosphate into stable soil organic phosphorus supported the shift toward organic phosphorus observed in fallowed vs. monocropped soils. However, during the 9-week incubation period, a large proportion of phosphorus from the residues remained in the microbial biomass. This prevented sorption of phosphorus to the soil matrix, but further work is needed to understand how microbial immobilization and remineralization cycles can be managed in order to render microbial phosphorus more available to field crops grown after fallow periods. Environmental alternations, such as drying and rewetting of soils, may solubilize phosphorus immobilized in the soil microbial biomass (Chapter 13).

The shift toward organic and microbial nutrients in maize–fallow rotations was also observed by Smestad et al. (2002), who measured increases in phosphorus contained in the soil microbial biomass, particulate soil organic matter, and stable soil organic phosphorus. The activation of native soil phosphorus by microbial uptake in response to organic matter addition to highly weathered soils has also been reported by Guppy (2003), but cannot be expected if total soil phosphorus contents are very low (Compaoré et al., 2003).

37.2.3 Effects of Fallows on the Overall System

Studies in southern Cameroon have observed no beneficial effect of Calliandra tree fallows on the yield of the main crops (Nolte et al., 2005). This indicated that short-term fallow plants, which provide an immediate benefit to the farmer, e.g., through the production of firewood, even if readily adopted can lead to substantial nutrient removal from the system (Nolte et al., 2003). Managed fallows may therefore lead to more negative phosphorus balances than those reported generally for sub-Saharan Africa (Smaling et al., 1997) if there are no corresponding phosphorus inputs (Bünemann et al., 2004b).

Problems arising from pests when fallow systems lack biodiversity were observed in other experiments as well as on farms (Drechsel et al., 1996; Smestad et al., 2002). Of particular concern is an increase in pathogenic nematodes (Desaeger and Rao, 2000). These examples demonstrate the need to test carefully the various biophysical effects of inserting particular fallow plants into cropping systems. They also confirm the value of participatory approaches to improve adoption by farmers. This does not diminish the value of systematic screening efforts that assess which of many leguminous species can, from a soil fertility standpoint, best serve as managed fallows (Niang et al., 2002).

Besides leguminous fallows, a nonleguminous shrub in the family Asteraceae, *Tithonia diversifolia*, has been widely tested as a fallow and green manure plant (Jama et al., 2000). It can improve phosphorus availability to crops through the acquisition of stable soil phosphorus, including organic phosphorus (George et al., 2002a), with subsequent rapid release of this phosphorus following residue amendment (Kwabiah et al., 2003). Being a nonlegume, it makes no net nutrient input of any element except carbon, although it may transfer nutrients from hedgerows or field boundaries onto arable areas (Jama et al., 2000).

37.2.4 Implications of Experience from Western Kenya

Short-term fallows have the potential to reverse soil organic matter depletion and to increase phosphorus held in the soil microbial biomass and in more stable soil organic phosphorus pools. They can mobilize phosphorus held in recalcitrant soil phosphorus pools through the effect of plant roots (George et al., 2002b) and enhanced microbial activity (Bünemann et al., 2004c). They may also reduce the phosphorus sorption capacity of soils during decomposition (Nziguheba et al., 1998).

It is noteworthy that these effects, which indicate enhanced soil organic phosphorus turnover, are not necessarily measurable as an increase in plant-available phosphate (next section). All these processes increase the availability of soil phosphorus, but must be complemented by balanced addition of mineral or organic fertilizer to prevent nutrient depletion of the soil (Smithson and Giller, 2002). Furthermore, attention has to be paid to the diversity of planted fallow species in order to avoid pest problems.

37.3 Some Further Examples of Improved Soil Phosphorus Availability by Enhancing Soil Biological Activity

The two experiences considered above focused on combining organic matter management with the use of selected germplasm and strategic inputs of low doses of phosphorus fertilizer. However, there are some good examples of similar improvements made by enhancing soil phosphorus availability through other kinds of biological interventions.

Planted tree or shrub fallows of *Calliandra calothyrsus*, *Indigofera constricta*, and *T. diversifolia* increased the amounts of carbon, nitrogen, and phosphorus in the sand-sized soil organic matter in a volcanic-ash soil of the hillsides in southwestern Colombia (Phiri et al., 2001). Similarly, tree crops improved soil phosphorus availability to perennial crops in central Amazonian Oxisols (Lehmann et al., 2001). In the Amazon research reported in the preceding chapter, enhanced phosphorus cycling through the soil microbial biomass and between plants and soil meant that the need for phosphorus fertilizer application could be reduced to only as much as would replenish the phosphorus exported in harvested crops.

Agroforestry systems in Brazil have improved phosphorus availability in a shaded coffee cultivation system compared with the conventional unshaded system, apparently by promoting the turnover of organic phosphorus (Cardoso et al., 2003). In agroforestry production systems established on sandy savanna soils in Venezuela, the continued addition of animal manure improved soil physical and chemical characteristics and enhanced soil microbial activity (Lopez-Hernandez et al., 2004). Enhanced earthworm abundance was related to greater enzyme activities and microbial biomass. The concentrations of both inorganic and organic phosphorus in the soil were increased. Leguminous cover crops grown in the interspaces of coconut plantations have increased plant-available phosphorus and the rates of biochemical processes in a sandy clay–loam soil in a humid tropical region of India (Dinesh et al., 2004).

Soil microbiological and biochemical parameters were sensitive indicators of soil quality under managed fallow systems with herbaceous or shrubby legumes established on sites with varying degrees of soil degradation in southwestern Nigeria (Wick et al., 1998). In particular, changes in alkaline phosphatase activity can highlight interactions between organic phosphorus dynamics and overall soil fertility in tropical agroecosystems.

Greater soil microbial activity by itself may be beneficial only if the system is modified. Fire-free alternatives to smallholders' slash-and-burn agriculture have been evaluated in the eastern Amazon by Denich et al. (2004). In this region, a 2-year cropping period is alternated with 3 to 7 years of fallow, during which time a woody secondary-forest vegetation regenerates by resprouting from the roots. Traditional burning of fallow vegetation causes most of the carbon, nitrogen, and phosphorus stocks in the aboveground biomass to be lost by volatilization or ash-particle transfer (Sommer et al., 2004). In contrast, the use of mechanized chop-and-mulch technology for land preparation avoids the nutrient losses caused by burning, yet crops grown in the mulch layer do not immediately benefit. This occurs because the availability of nutrients, especially phosphorus, is reduced by microbial immobilization (Bünemann et al., 1998). This probably explains, at least in some cases, the lower yields in mulched compared with burned fields (Kato et al., 1999). With moderate quantities of NPK fertilizer, the first two crops (rice and cowpea) performed equally well in burned and mulched areas. Only the last crop of the sequence, cassava, did not require fertilizer to yield as well in mulched as in burned areas.

The change from burning to mulching certainly stimulates soil biological activity and preserves soil organic matter and associated nutrients. To be successful, however, there may need to be other modifications of the system. These can include selection of cultivars adapted to the more acidic surface soils in mulched as compared with burned areas, and inversion of the cropping sequence by planting first the crop with the lowest nutrient requirements. This cascade of required changes poses a challenge to farmers and researchers. The mulch system, compared with traditional burning, requires additional labor and changes in the cropping system as well as certain agronomic changes. However, it offers opportunities such as more flexible planting dates and an extended cropping phase that allows a maximization of benefits from the slowly decomposing mulch layer (Denich et al., 2004).

37.4 Assessing and Improving Phosphorus Fertility in Tropical Soils

Assessing the success of biological approaches to improving phosphorus availability in tropical soils remains a challenge. In current agricultural practice, phosphorus availability in soils is determined by measuring the size of a phosphate pool, for example, by extraction with bicarbonate (e.g., Olsen et al., 1954), by dilute mineral acid (e.g., Bray and Kurtz, 1945), or by anion exchange resins (Amer et al., 1955). Such "snapshot" measurements of phosphate availability are related empirically to observed crop growth using field experiments, and these relationships are then used to determine fertilizer requirements.

Such methods adapted and calibrated for the specific soil under study can provide relevant information on the amount of phosphate that is available to the plant in a tropical soil (Bühler et al., 2003). However, they do not provide information on the turnover of (organic) phosphorus, even though this is closely related to the actual phosphorus availability in tropical soils (Tiessen and Shang, 1998). In fact, a low extractable phosphate concentration may conceal a rapid rate of organic phosphorus turnover that contributes significantly to the supply of phosphorus to plants. This is apparent when conventionally managed farmland is first brought under organic cultivation (Oberson and Frossard, 2005).

The dependence of phosphorus fertility on organic matter turnover leads to feedback mechanisms that confound further the assessment of phosphorus availability in tropical soils by chemical tests. For example, if it is organic phosphorus turnover that regulates the availability of phosphate, then a decrease in the rate of turnover may place a nutrient limitation on organic matter decomposition. This, in turn, will constrain the turnover of organic phosphorus (Tiessen and Shang, 1998).

Such complexity means that phosphorus availability in tropical cropping systems based on organic matter inputs cannot be assessed accurately by conventional soil testing procedures. These were developed in industrialized countries of the temperate zone where mineral fertilizer is readily accessible to farmers, and the soils have generally less capacity to sorb phosphorus compared with those in many tropical regions. In temperate soils, a single measurement of extractable phosphate can yield results with some relevance to crop uptake, whereas in tropical soils there is a clear need also to assess continually the rates of phosphorus turnover (Organic Phosphorus Workshop, 2005). Unfortunately, there are currently no straightforward methods for assessing organic phosphorus turnover in tropical soils.

37.5 Discussion

Productivity of many tropical agricultural systems must be urgently increased to meet the needs of a growing and often malnourished world population. Phosphorus is a crop-nutrient element that is fundamental in any attempt to achieve this objective because crop production in the tropics is so often limited by the availability of soil phosphorus. For most farmers in tropical regions, mineral phosphate fertilizer is an expensive and often inaccessible means of improving phosphorus fertility. The case studies from the eastern plains of Colombia and western Kenya show that biological interventions are available that are relatively simple and low cost and can improve phosphorus fertility in tropical agroecosystems. The common feature of these interventions is the input of organic matter into the soil, mainly through the residues from leguminous pasture or fallow plants that are adapted to low-phosphorus tropical soils.

The use of well-adapted germplasm together with low doses of phosphorus fertilizers can enhance system productivity, in turn adding to the supply of organic residues above- and belowground. Phosphorus acquisition varies according to germplasm much as symbiotic nitrogen fixation does in legumes, with the quality as well as the amount of residues returned to the soil being important. Since organic matter boosts soil microbial activity, the microbiologically-driven processes in soil phosphorus dynamics are enhanced, and the microbial phosphorus pool is increased when greater amounts of organic matter are made available. Phosphorus recycles more efficiently among plants, microorganisms, and organic forms of phosphorus in the soil where it is protected from strong sorption in highly weathered tropical soils. Enhanced turnover can increase phosphorus availability for crops.

This said, the widespread adoption of biological interventions by farmers requires investments in germplasm, fertilizers, and knowledge of pasture and/or crop management techniques. Such investments, which must occur alongside an intensification and diversification of the overall farm operations, are made only when farmers perceive and expect tangible benefits from self-supply and when favorable markets for their produce are accessible. Finally, sustainable agricultural systems require that phosphorus inputs and outputs be in balance over the long term, in order to avoid an excessive depletion of soil phosphorus stocks.

References

Amer, F. et al., Characterisation of soil phosphorus by anion exchange resin adsorption and 32P equilibration, *Plant Soil*, **6**, 391–408 (1955).

Bray, R.H. and Kurtz, L.T., Determination of total, organic and available forms of phosphorus in soils, *Soil Sci.*, **59**, 39–45 (1945).

Brossard, M., Lavelle, P., and Laurent, J.Y., Digestion of a vertisol by the endogeic earthworm *Polypheretima elongata*, Megascolecidae, increases soil phosphate extractability, *Eur. J. Soil Biol.*, **32**, 107–111 (1996).

Bühler, S. et al., Sequential phosphorus extraction of a P-33-labeled oxisol under contrasting agricultural systems, *Soil Sci. Soc. Am. J.*, **66**, 868–877 (2002).

Bühler, S. et al., Isotope methods for assessing plant available phosphorus in acid tropical soils, *Eur. J. Soil Sci.*, **54**, 605–616 (2003).

Bünemann, E., Phosphorus dynamics in a Ferralsol under maize–fallow rotations: The role of the soil microbial biomass. Ph.D. Thesis, Swiss Federal Institute of Technology, Zurich (2003) (http://e-collection.ethbib.ethz.ch/cgi-bin/show.pl?type = diss&nr = 15207).

Bünemann, E. et al., Fertilizer response of maize and cowpea under conditions of fire-free land preparation in NE Pará, In: *Proceedings of the Third SHIFT-Workshop, Manaus, March 15–19, 1998*, Lieberei, R., Voß, K., and Bianchi, H., Eds., BMBF/GKSS, Hamburg, 157–159 (1998).

Bünemann, E. et al., Microbial community composition and substrate use in a highly weathered soil as affected by crop rotation and P fertilization, *Soil Biol. Biochem.*, **36**, 889–901 (2004a).

Bünemann, E. et al., Maize productivity and nutrient dynamics in maize–fallow rotations in western Kenya, *Plant Soil*, **264**, 195–208 (2004b).

Bünemann, E. et al., Phosphorus dynamics in a highly weathered soil as revealed by isotopic labeling techniques, *Soil Sci. Soc. Am. J.*, **68**, 1645–1655 (2004c).

Cardoso, I.M. et al., Analysis of phosphorus by (PNMR)-P-31 in Oxisols under agroforestry and conventional coffee systems in Brazil, *Geoderma*, **112**, 51–70 (2003).

Chapuis-Lardy, L. et al., Phosphorus transformations in a ferralsol through ingestion by *Pontoscolex corethrurus*, a geophagous earthworm, *Eur. J. Soil Biol.*, **34**, 61–67 (1998).

Compaoré, E. et al., Influence of land-use management on isotopically exchangeable phosphate in soils from Burkina Faso, *Commun. Soil Sci. Plant Anal.*, **34**, 201–223 (2003).

Decaens, T. et al., Impact of land management on soil macrofauna in the Oriental Llanos of Colombia, *Eur. J. Soil Biol.*, **30**, 157–168 (1994).

Denich, M. et al., Mechanized land preparation in forest-based fallow systems: The experience from Eastern Amazonia, *Agroforest. Syst.*, **61**, 91–106 (2004).

Desaeger, J. and Rao, M.R., Parasitic nematode populations in natural fallows and improved cover crops and their effects on subsequent crops in Kenya, *Field Crops Res.*, **65**, 41–56 (2000).

Dinesh, R. et al., Long-term influence of leguminous cover crops on the biochemical properties of a sandy clay loam Fluventic Sulfaquent in a humid tropical region of India, *Soil Till. Res.*, **77**, 69–77 (2004).

Drechsel, P., Steiner, K.G., and Hagedorn, F., A review on the potential of improved fallows and green manure in Rwanda, *Agroforest. Syst.*, **33**, 109–136 (1996).

Friesen, D.K. et al., Phosphorus acquisition and cycling in crop and pasture systems in low fertility tropical soils, *Plant Soil*, **196**, 289–294 (1997).

Gathumbi, S.M., Cadisch, G., and Giller, K.E., N-15 natural abundance as a tool for assessing N-2-fixation of herbaceous, shrub and tree legumes in improved fallows, *Soil Biol. Biochem.*, **34**, 1059–1071 (2002).

George, T.S. et al., Changes in phosphorus concentrations and pH in the rhizosphere of some agroforestry and crop species, *Plant Soil*, **246**, 65–73 (2002a).

George, T.S. et al., Utilisation of soil organic P by agroforestry and crop species in the field, western Kenya, *Plant Soil*, **246**, 53–63 (2002b).

Gijsman, A.J., Alcaron, H.F., and Thomas, R.T., Root decomposition in tropical grasses and legumes, as affected by soil texture and season, *Soil Biol. Biochem.*, **29**, 1443–1450 (1997a).

Gijsman, A.J. et al., Nutrient cycling through microbial biomass under rice–pasture rotations replacing native savanna, *Soil Biol. Biochem.*, **29**, 1433–1441 (1997b).

Girma, H., Insect pest problems and their management in planted fallow crop rotation in western Kenya. Ph.D. Thesis, Kenyatta University, Nairobi, Kenya (2002).

Guggenberger, G. et al., Assessing the organic phosphorus status of an Oxisol under tropical pastures following native savanna using 31P NMR spectroscopy, *Biol. Fertil. Soils*, **23**, 332–339 (1996).

Guimaráes, E.P. et al., Research on agropastoral systems: What we have learned and what we should do, In: *Agropastoral Systems for the Tropical Savannas of Latin America*, Guimaráes, E.P. et al., Eds., CIAT/EMBRAPA, Cali, Colombia/Brazil, 326–336 (2004).

Guppy, C.N., Phosphorus and organic matter interactions in highly weathered soils. Ph.D. Thesis, University of Queensland, Australia (2003).

Hartemink, A.E. et al., Soil nitrate and water dynamics in sesbania fallows, weed fallows and maize, *Soil Sci. Soc. Am. J.*, **60**, 568–574 (1996).

Jama, B., Swinkels, R.A., and Buresh, R.J., Agronomic and economic evaluation of organic and inorganic sources of phosphorus in Western Kenya, *Agron. J.*, **89**, 597–604 (1997).

Jama, B. et al., *Tithonia diversifolia* as a green manure for soil fertility improvement in western Kenya: A review, *Agroforest. Syst.*, **49**, 201–221 (2000).

Jiménez, J.J. and Decaens, T., The impact of soil organisms on soil functioning under neotropical pastures: A case study of a tropical anecic earthworm species, *Agric. Ecosyst. Environ.*, **103**, 329–342 (2004).

Jiménez, J.J. et al., Phosphorus fractions and dynamics in surface earthworm casts under native and improved grasslands in a Colombian savanna oxisol, *Soil Biol. Biochem.*, **35**, 715–727 (2003).

Kato, M.S.A. et al., Fire-free alternatives to slash-and-burn for shifting cultivation in the eastern Amazon region: The role of fertilizers, *Field Crops Res.*, **62**, 225–237 (1999).

Kwabiah, A.B. et al., Phosphorus availability and maize response to organic and inorganic fertilizer inputs in a short term study in western Kenya, *Agric. Ecosyst. Environ.*, **95**, 49–59 (2003).

Lascano, C. and Estrada, J., Long-term productivity of legume-based and pure grass pastures in the eastern plains of Colombia, *Proceedings of the XVI International Grassland Conference, Nice, France*, 1179–1180 (1989).

Lehmann, J. et al., Phosphorus management for perennial crops in central Amazonian upland soils, *Plant Soil*, **237**, 309–319 (2001).

Lopez-Hernandez, D. et al., Changes in soil properties and earthworm populations induced by long-term organic fertilization of a sandy soil in the Venezuelan Amazonia, *Soil Sci.*, **169**, 188–194 (2004).

Maroko, J.B., Buresh, R.J., and Smithson, P.C., Soil phosphorus fractions in unfertilized fallow–maize systems on two tropical soils, *Soil Sci. Soc. Am. J.*, **63**, 320–326 (1999).

Merckx, R., den Hartog, A., and Van Veen, J.A., Turnover of root-derived material and related microbial biomass formation in soils of different texture, *Soil Biol. Biochem.*, **17**, 565–569 (1985).

Niang, A.I. et al., Species screening for short-term planted fallows in the highlands of western Kenya, *Agroforest. Syst.*, **56**, 145–154 (2002).

Nolte, C. et al., Effects of Calliandra planting pattern on biomass production and nutrient accumulation in planted fallows of southern Cameroon, *Forest Ecol. Manage.*, **179**, 535–545 (2003).

Nolte, C. et al., Groundnut, maize and cassava yields in mixed-food crops fields after Calliandra tree fallow in southern Cameroon, *Exp. Agric.*, **41**, 21–37 (2005).

Nziguheba, G. and Bünemann, E., Organic phosphorus dynamics in tropical agroecosystems, In: *Organic Phosphorus in the Environment*, Turner, B.L., Frossard, E., and Baldwin, D.S., Eds., CAB International, Wallingford, UK, 243–268 (2005).

Nziguheba, G. et al., Organic residues affect phosphorus availability and maize yields in a Nitisol of western Kenya, *Biol. Fertil. Soils*, **32**, 328–339 (2000).

Nziguheba, G. et al., Soil phosphorus fractions and adsorption as affected by organic and inorganic sources, *Plant Soil*, **198**, 159–168 (1998).

Oberson, A. and Frossard, E., Phosphorus management for organic agriculture, In: *Phosphorus: Agriculture and the Environment*, Sims, J.T. and Sharpley, A.N., Eds., ASA, CSSA and SSSA, Madison, WI, 761–779 (2005).

Oberson, A. and Joner, E.J., Microbial turnover of phosphorus in soil, In: *Organic Phosphorus in the Environment*, Turner, B.L., Frossard, E., and Baldwin, D.S., Eds., CAB International, Wallingford, UK, 133–164 (2005).

Oberson, A. et al., Phosphorus status and cycling in native savanna and improved pastures on an acid low-P Colombian Oxisol, *Nutr. Cycl. Agroecosyst.*, **55**, 77–88 (1999).

Oberson, A. et al., Phosphorus transformations in an Oxisol under contrasting land-use systems: The role of the soil microbial biomass, *Plant Soil*, **237**, 197–210 (2001).

Olsen, S.R. et al., *Estimation of Available Phosphorus in Soils by Extraction with Sodium Bicarbonate*, United States Department of Agriculture, Washington, DC (1954).

Organic Phosphorus Workshop, Synthesis and recommendations for future research, In: *Organic Phosphorus in the Environment*, Turner, B.L., Frosssard, E., and Baldwin, D.S., Eds., CAB International, Wallingford, UK, 377–380 (2005).

Phiri, S. et al., Changes in soil organic matter and phosphorus fractions under planted fallows and a crop rotation system on a Colombian volcanic-ash soil, *Plant Soil*, **231**, 211–223 (2001).

Phiri, S. et al., Constructing an arable layer through chisel tillage and agropastoral systems in tropical savanna soils of the llanos of Colombia, *J. Sustain. Agric.*, **23**, 5–29 (2003).

Rao, I.M., Plazas, C., and Ricaurte, J., Root turnover and nutrient cycling in native and introduced pastures in tropical savannas, In: *Plant Nutrition: Food Security and Sustainability of Agro-Ecosystems through Basic and Applied Research*, Horst, W.J. et al., Eds., Kluwer Academic Publishers, Dordrecht, Netherlands, 976–977 (2001).

Rao, I.M. et al., Adaptive attributes of tropical forage species to acid soils I. Differences in shoot and root growth responses to varying phosphorus supply and soil type, *J. Plant Nutr.*, **19**, 323–352 (1996).

Rao, I.M. et al., Soil phosphorus dynamics, acquisition and cycling in crop–pasture–fallow systems in low fertility tropical soils of Latin America, In: *Modelling Nutrient Management in Tropical Cropping Systems*, Delve, R.J. and Probert, M.E., Eds., Australian Center for International Agricultural Research (ACIAR), Canberra, Australia, 126–134 (2004).

Sanchez, P.A., Improved fallows come of age in the tropics, *Agroforest. Syst.*, **47**, 3–12 (1999).

Sieverding, E., *Vesicular–Arbuscular Mycorrhiza Management in Tropical Agrosystems*. Deutsche Gesellschaft für Technische Zusammenarbeit, Eschborn, Germany (1991).

Smaling, E.M.A., Nandwa, S.M., and Janssen, B.H., Soil fertility in Africa is at stake, In: *Replenishing Soil Fertility in Africa*, Buresh, R.J., Sanchez, P.A., and Calhoun, F., Eds., Soil Science Society of America and American Society of Agronomy, Madison, WI, 47–61 (1997).

Smestad, T.B., Tiessen, H., and Buresh, R.J., Short fallows of *Tithonia diversifolia* and *Crotalaria grahamiana* for soil fertility improvement in western Kenya, *Agroforest. Syst.*, **55**, 181–194 (2002).

Smithson, P.C. and Giller, K.E., Appropriate farm management practices for alleviating N and P deficiencies in low-nutrient soils of the tropics, *Plant Soil*, **245**, 169–180 (2002).

Sommer, R. et al., Nutrient balance of shifting cultivation by burning or mulching in the Eastern Amazon: Evidence for subsoil nutrient accumulation, *Nutr. Cycl. Agroecosyst.*, **68**, 257–271 (2004).

Thomas, R.J., The role of the legume in the nitrogen cycle of productive and sustainable pastures, *Grass Forage Sci.*, **47**, 133–142 (1992).

Thomas, R.J. et al., Nutrient cycling via forage litter in tropical grass/legume pastures. In: *XVII International Grassland Congress*, Hodgson J, Ed., NZGA, TGSA, NZSAP, ASAP-Qld. and NZIAS, Palmerston North, New Zealand, 508–509 (1993).

Tiessen, H. and Shang, C., Organic matter turnover in tropical land use systems, In: *Carbon and Nutrient Dynamics in Natural and Agricultural Tropical Ecosystems*, Bergström, L. and Kirchmann, H., Eds., CAB International, Wallingford, UK, 1–14 (1998).

Wick, B., Kühne, R.F., and Vlek, P.L.G., Soil microbiological parameters as indicators of soil quality under improved fallow management systems in south-western Nigeria, *Plant Soil*, **202**, 97–107 (1998).

38

Profile Modification as a Means of Soil Improvement: Promoting Root Health through Deep Tillage

Nico Labuschagne[1] and Deon Joubert[2]

[1]*Department of Microbiology and Plant Pathology, University of Pretoria, Pretoria, South Africa*
[2]*Unifrutti SA Pty Ltd., Kirkwood, Eastern Cape, South Africa*

CONTENTS

This chapter is based on a case study in South Africa of the relationship between soil compaction and root disease of citrus, and conversely, of the beneficial effects that deep tillage can have to improve root development and suppress root rot caused by *Phytophthora nicotianae*. It focuses on soil profile modification (SPM) as a means of improving root growth and root health and of suppressing soil-borne pathogens. It does not undertake a review of the various methods and mechanisms of SPM which have previously been reviewed by Eck and Ungerer (1985). We focus on this case because we have good and extensive data to examine within the larger context of soil profile modification.

Deep tillage as a method for this is undertaken to alleviate specific physical and consequently biological problems that create suboptimal soil conditions, such as those resulting from compaction. It does not contradict the value of zero-tillage practices discussed in Chapters 22 and 24. Whenever soil aeration has been reduced or interrupted, restoring it can have very positive effects on the diversity and activity of soil biotic communities. Deep tillage is thus reviewed here as a once-off operation expected to improve soil aeration and drainage, as well as to reduce populations of certain soil-borne plant pathogens, to enhance the overall health of soil systems.

Although SPM is a physical intervention, it has definite effects on the biological components of soil systems as well. Any physical alteration of soils, whether by mechanical or biological means, will have a cascading effect on a complex set of soil factors well beyond those that are classified as physical. It might seem obvious that the microbial realm is influenced by the primary and secondary effects of physical modifications, but we want to underscore the fact that soil profile modification — by intervening in the physical structure and multiple relationships within soil systems — improves the performance of crops by enhancing the health and vigor of the root system. The largest effects of SPM, the responses of root systems, are largely unseen. What is visible are the reflections of subsurface changes which become evident in increased crop yield.

38.1   Soil Profile Modification as an Intervention

Profile modification is conceived and intended to alleviate specific problems associated with suboptimal soil conditions such as those created by compaction from tillage, with or without chemical or physical amendments, at depths greater than the disturbance created by ordinary plowing. There are various kinds of SPM, including deep plowing, subsoiling, vertical mulching, disrupting and mixing of the profile, trenching, and the installation of subsurface barriers. In the South African citrus example discussed below, the main method of SPM was deep tillage or ripping to a depth of 0.75–1.0 m to alleviate compaction.

As underscored by Eck and Ungerer (1985), the overall objective of profile modification is to increase crop production, primarily by providing more favorable zones for root growth, proliferation, and activity. There are situations where SPM is inappropriate and ineffective, however, where soil systems will not respond positively to such manipulation.

- An example was reported by Fritton et al. (1983) in their study of the modification of an Erie channery silt loam soil in Pennsylvania, USA, in which a shallow fragipan horizon was limiting crop production. In this particular soil, it was concluded, mechanical loosening would have only a short-term effect. More beneficial would be the addition of 2% organic matter because this could delay the deterioration of the soil that was occurring with rapid increase of bulk density when no organic matter was added.

- Similarly, Robertson et al. (1977) concluded that in Michigan, USA, deep tillage would have little effect under average conditions and was not likely to improve crop yields except on problem soils. Deep tillage had the greatest effect on crop yields when there were artificially produced compacted zones immediately below the plow layer.

In their review of SPM, Eck and Ungerer (1985) gave a systematic overview of responses to SPM within a range of soils that included plowpan soils, fragipan soils, duripan soils, claypan soils, slowly permeable clay soils, soils with high clay horizons, and poorly drained soils. They gave numerous examples showing that positive responses to deep tillage were only obtained where there were specific problems to be alleviated.

The problems that may be alleviated by SPM have been classified by Ungerer (1979) into three categories:

1. Problems caused by soil layers that restrict root growth and water movement (plowpans, fragipans, duripans, claypans, high clay horizons), with resulting:

(a) Reduced root penetration

(b) Poor infiltration of water

(c) Poor water storage and distribution

(d) Restricted drainage

(e) Less than desirable leaching

2. Problems caused by undesirable substances on or near the surface (salts, toxic materials, radioactive fallout) can also result in:

(a) Poor plant growth

(b) Poor leaching

(c) Poor soil–water relations

(d) Plants with high concentrations of undesirable elements.

3. Problems caused by coarse-textured materials at the surface or to great depths (sandy soils), having the following effects:

(a) Excessive percolation

(b) Low fertility

(c) Excessive leaching

(d) High wind erosiveness

Where there are soil problems that restrict root growth and water infiltration, e.g., compaction, the equipment used to correct the problem must be capable of reaching the problem zone or horizon in the soil. Effectiveness of the operation will depend to a large degree on having a soil water content that is at or near the optimum level for the equipment used. A summary of some profile modification implements, along with their depths of operation, type of action in soils, and soil water conditions for optimum operation, is given in Table 38.1, summarized from Eck and Ungerer (1985). When deep tillage is applied correctly to alleviate soil compaction, the final outcome is improvement of the crop yield.

In terms of changed soil properties, the benefits of deep tillage in compacted or poorly drained soil include decreased soil bulk density and less penetrometer soil strength (PSS) (Voorhees et al., 1975), increased porosity (more macropores $>50\ \mu m$ and fewer micropores $<6\ \mu m$), improved aeration (oxygenation and gas exchange), improved water infiltration and drainage, and increased leaching of salts.

When profile modification was applied to poorly drained soils in California, USA, it resulted in improved drainage and leaching, reduced soil strength, and increased plant growth (Kaddah, 1976). In South Africa, deep tillage has had in some instances a dramatic

TABLE 38.1

Profile Modification Implements

Implement	Operating Depth (m)	Action in Soil	Optimum Soil Water Content
Moldboard plow	0.3 to 1.1	Inversion and some mixing	Mid to upper plastic range
Disk plow	0.3 to 0.8	Some inversion and mixing	Mid to upper plastic range
Slip plow	2.0	Shattering, considerable mixing	Around lower plastic limit
Chisel, subsoiler, or ripper	0.9 to 2.0	Shattering, little mixing	Around lower plastic limit
Trenching machine	1.5	Complete mixing	Mid-plastic range
Barrier installer	0.6	Undercutting, lifting, applying barrier	Around upper plastic limit

Source: From Eck, H.V., Ungerer, P.W., *Adv. Soil Sci.*, **1**, 65–100, 1985. With permission.

effect on the root growth of established citrus trees, whereas little or no benefit has been observed in some other instances, as discussed later in this chapter.

38.2 Soil Compaction

Soil compaction is undoubtedly a major limiting factor for crop growth in many agricultural soils. This is substantiated by the many examples of positive crop response reported after compaction was neutralized by means of deep tillage. Reduced growth and yield due to excessive soil compaction have been reported for many crop species (Zimmerman and Kardos, 1961; Rosenburg, 1964; Voorhees et al., 1975; Smucker and Erickson, 1987; Allmaras et al., 1988; Tu and Tan, 1988). Further, the adverse effects of mechanical impedance on root growth have been well documented (Russel and Goss, 1974; Goss and Russel, 1980). In various studies on a number of crops, researchers have shown with penetrometer probes that the critical penetration resistance at which root growth is impeded lies between 2000 and 2500 kPa (Greacen et al., 1969; Bar-Yosef and Lambert, 1981).

Compacted soil presents unfavorable conditions for root growth because in addition to high resistance to penetration (as reflected in increased penetrometer soil strength), there are small and rigid pore systems (reduced macropores), reduced water drainage and gas exchange (Gupta et al., 1989), and decreased O_2 transport to root surfaces (Grant, 1993). Typically plants growing in compacted soil have impeded root systems (Barley, 1963; Thompson et al., 1987) with deformed and thickened feeder roots (Gerhard et al., 1972; Russel and Goss, 1974; Barley, 1976), resulting in reduced plant vigor and yield (Gerhard et al., 1982; Grimes et al., 1982; McAfee et al., 1989; Rusanov, 1991).

One of the primary causes of root stress in compacted soil is waterlogging, which results in anoxia and accumulation of toxic substances in and around the roots (Allmaras et al., 1988). Even a short period of anoxia can cause severe root damage (Greenwood, 1969). Furthermore, prolonged wet conditions prevailing in compacted soil after irrigation or rain reduce plant respiration (Lambers, 1988) and increase root exudation (Kuan and Erwin, 1980; Smucker and Erickson, 1987) to compensate for the adverse conditions.

38.3 Relationship between Soil Compaction and Root Diseases

In addition to the direct detrimental effect of soil compaction on root growth and vitality, another important consequence is an increase in root diseases. This can be due to plant roots becoming predisposed by stress or to conditions in the soil and around the root to succumb to pathogens. Root diseases associated with soil compaction have, for example, reportedly led to abandonment of conservation tillage systems in some areas (Schmitthenner and van Doren, 1985). A correlation between soil compaction and increased disease incidence has been reported for many types of soil-borne fungi, including various Fusarium spp., *Thielaviopsis basicola*, and various Phytophthora and Pythium spp. (Burke et al., 1972; Raghavan et al., 1982; Bhatti and Kraft, 1992).

Wet conditions and increased root exudation in irrigated, compacted soil are known to stimulate the germination of zoospores and chlamydospores of pathogenic fungi (Sterne et al., 1977a, 1977b; Kaun and Erwin, 1980). Increased mechanical resistance of soil is known to stimulate root exudation (Barber and Gunn, 1974). This, in turn, promotes chemotactic attraction of Phytophthora zoospores toward plant roots (Kew and Zentmeyer, 1973), particularly to the area behind the root tip where most infections by Phytophthora occur (Rovira and Davey, 1974; Kuan and Erwin, 1980). Enhanced host–pathogen contact can therefore be expected if the rate of root growth is reduced and exudation is stimulated by an increase in soil bulk density (Huisman, 1982; Allmaras et al., 1988).

Soil compaction with its associated waterlogged conditions after rain or irrigation is particularly conducive to zoosporic pathogens, such as Phytophthora and Pythium, primarily because zoospore release and movement are both enhanced under such conditions (Duniway, 1976; Duniway, 1979). A number of reports, for example, indicated a significant increase in the incidence of Phytophthora root rot of soybean associated with soil compaction (Fulton et al., 1961; Gray and Pope, 1986; Moots et al., 1988).

Aggravation of Phytophthora root rot by soil which is saturated for prolonged periods of time is a common phenomenon on a wide variety of plants (Wilcox and Mircetich, 1979; Zentmeyer, 1980). With respect specifically to citrus, Joubert (1993) and Joubert and Labuschagne (1998) have reported on the relationship between soil compaction and Phytophthora root rot. By separating the direct effects of increased soil bulk density from the indirect effects of increased water content, they demonstrated that root rot can be increased by soil compaction alone. However, it is further aggravated by high levels of moisture in the soil.

38.4 Soil Profile Modification in Citrus Orchards in South Africa

That soil compaction is a limiting factor in citrus tree performance is well established (Ford, 1959; Mikhael and El-Zeftawi, 1979). In Florida, USA, the suitability of soils for citrus growth is classified based on rooting depth in the soil (De la Rosa and Carlisle, 1978). In South Africa, a number of studies have been conducted to determine the prevalence of soil compaction in citrus orchards and the possible effect of deep tillage as a means of alleviating compaction. A survey was conducted in the Sundays River Valley citrus production area in the Eastern Cape province by Van Huyssteen (1987), who concluded that restricted root depth was a major limiting factor in citrus orchards. A linear correlation was found between root depth and the yield of navel oranges on rough lemon rootstocks in the Sundays River Valley. Navel and Valencia orange trees on rough lemon rootstock that were uprooted because of their declining productivity commonly showed shallow root systems.

In a study of soil factors affecting tree growth and root development in a citrus orchard in South Africa, Nel and Bennie (1984) found a decrease in root and tree growth of Valencia orange trees on rough lemon rootstock when soil penetration resistance exceeded values of 500 kPa. In their analysis of the influence of selected soil properties on citrus tree and root growth in well-drained soil in the absence of poorly permeable soil layers, they report that under these conditions, tree and root growth were restricted wherever the long-term air-filled porosity of the soil was almost continuously less than 15%. They also showed that unconfined penetrometer resistance and air capacity (at field water capacity) correlated well with tree growth and root development. Tree and root growth decreased with an increase in penetrometer resistance above 0.5 kg m^{-2} for small trees on soils with a soil layer that had resistance more than 2.5 kg m^{-2}.

38.4.1   Plowing as a Stimulus for Root Growth and Soil Biota

When considering soil profile modification, and in particular deep tillage/plowing as a means of alleviating soil compaction, it is necessary to make a distinction between preplanting soil preparation and interventions applied in established orchards. When applying deep tillage to an existing orchard, a number of additional factors come into play, such as the effects of root pruning.

A study of the effect of deep plowing on performance of an established navel orange orchard was conducted by Abercrombie and Hoffman (1996) at the Addo Research Station in the Eastern Cape. The relatively low-yielding orchard was approximately 32 years old and was planted on an Oakleaf soil with an A horizon of 15–20% clay, 12–15% silt, and 15–20% fine sand, and a B horizon of 20–25% clay, 15–20% silt, and 25–75% fine sand. These soils were classified according to available soil depth, defined as the depth at which soil penetration resistance (SPR) as measured by a penetrometer exceeded 2200 kPa.

Both a low SPR and a high SPR site were included in the study. Soil compaction occurred at depths of 20 to 30 cm. Deep plowing was conducted by means of an adapted wing-delve plow to a depth of 80 cm at a distance of 1 m from the tree stems on one side of tree rows, followed with deep plowing on the other side 16 months later. Unplowed treatments had SPR values of >2 kPa at a soil depth of 20 cm. In the deep-plowed treatments, SPR values were significantly reduced up to 3 years after treatment.

Root regeneration was assessed in 1 m deep soil profile pits that were dug in the rooting zone 18 months after the treatments had been applied. Deep-plowed soil showed strong root regrowth at 20–80 cm soil depth compared to fewer roots in the unplowed soil. Furthermore, in the unplowed control plots, approximately 80% of the roots occurred in the 0–50 cm soil depth zone. In the 20–30-cm zone where roots were restricted by soil compaction, feeder roots were absent, whereas feeder roots occurred up to depths of 80 cm in deep-plowed soils assessed 18 months after treatment. Prolific regrowth of roots was observed at the points where roots were pruned, and the roots appeared to be healthy.

38.4.2   Results from Deep Plowing

A significant increase in yield and fruit size was recorded over five seasons in the deep-plowed plots. Abercrombie and Hoffman (1996) reported that deep plowing increased cumulative yields and large fruit mass by 44 and 52%, respectively, over three seasons of their trial. They concluded that deep plowing in a mature navel orange orchard could increase fruit yield if there is shallow soil compaction and the trees are relatively low yielding. Notably, these increases were not achieved with increased application of fertilizer. These researchers recommended that plowing be done to a soil depth of 80 cm on compacted soil to establish a favorable soil-rooting volume for regenerated roots. They also recommended that a proper soil compaction survey be conducted before considering deep plowing of existing citrus orchards.

Apart from the direct effect of soil compaction on the growth and yield of citrus trees, resulting in greater soil aeration, the interaction between soil compaction and root diseases and soil-borne pathogens should be considered. Lutz et al. (1986) reported a high incidence of root rot in citrus orchards where impervious soil layers exist.

The effect of soil compaction on Phytophthora root rot of citrus was studied by the authors of this chapter in greenhouse experiments with rough lemon and Troyer citrange citrus seedlings (Joubert, 1993; Joubert and Labuschagne, 1998). Seven-month-old seedlings were planted in virgin Oakleaf sandy loam soil (38% fine sand, 15% clay). Increasing soil bulk densities (SBDs) of 1400, 1500, 1600, and 1700 kg m^{-3} were artificially created by means of compaction. Importantly, a constant soil moisture content of 13–18%

(and air-filled porosity higher than 20%) was maintained by subjecting each pot to suction under vacuum to remove excess moisture. For each SBD, half of the seedlings were inoculated with *Phytophthora nicotianae*, whereas the other half remained uninoculated. After 5 months, the experiment was terminated and assessments were made.

Growth parameters were measured, root rot was assessed, primary and feeder roots were weighed separately, and total length and mean thickness of feeder roots were determined. Shoot growth was not significantly affected by *P. nicotianae* at any SBD over the experimental period, but feeder root loss incurred as a result of infection increased, on average, from 17.6% at SBD of 1400 kg m^{-3} to 22.7, 35.1, and 35.6% at SBDs of 1500, 1600, and 1700 kg m^{-3}, respectively.

38.4.3 Impacts on Roots

These findings demonstrated that the effect of *P. nicotianae* on citrus roots could be aggravated by soil compaction even without the involvement of waterlogging and deoxygenation. Even at the highest SBD of 1700 kg m^{-3}, the soil used in this study still had an air-filled porosity of 20%, which is well above the critical minimum level of 10–15% for citrus root growth (Patt et al., 1966). The effects on root growth and Phytophthora root rot observed in this study were evidently due mainly to increased SBD and to the consequent increase in mechanical resistance of the soil to root penetration.

In addition to the effects of increased SBD on root growth and root rot-related feeder root loss, Joubert and Labuschagne (1998) observed a thickening and stunting of citrus feeder roots in compacted soil. This could be ascribed to shortening and thickening of the cortical cells (Wilson and Robards, 1978). For optimal root growth of citrus, it was recommended that a SBD of less than 1400 kg m^{-3}, or PSS not exceeding 500 kPa, be maintained in citrus orchard soils.

In subsequent experiments, Joubert (1993) studied the relationship between soil compaction, soil moisture, and Phytophthora root rot on rough lemon citrus seedlings. An Oakleaf sandy loam soil was used, and the treatments included an uncompacted soil (SBD of 1.4 ton m^{-3}) and a compacted soil (SBD 1.7 ton m^{-3}). At each SBD, the soil was subjected to three different soil moisture levels (SML), i.e., low moisture (7–11% m/m), intermediate moisture (12–16% m/m), and high moisture level (17–28% m/m). Furthermore, at each SBD and SML, the plants were either inoculated with *P. nicotianae* or left uninoculated. Results of the study showed that at low SBD (1.4 ton m^{-3}), Phytophthora root rot-related feeder root loss was not significant in the dry and intermediate SML, but there was 61% root loss in the wet treatment. At high SBD (1.7 ton m^{-3}), 15.6% root loss was recorded under dry conditions, 34.2% root loss under intermediate SML, and 91.9% under wet conditions.

Greater attraction of Phytophthora zoospores to roots grown in compacted soil was demonstrated by means of the modified dialysis membrane techniques described by Botha et al. (1990). As in the previous study, stunting and thickening of the feeder roots was observed in compacted soil, with few or no lateral branching of the roots. Light- and electron microscopical observations revealed deformed cortex cells with few intercellular spaces present in roots grown in compacted soil.

In a greenhouse study of the effects of soil compaction on a selection of six citrus rootstocks, Joubert (1993) demonstrated that Troyer citrange, Carrizo citrange, Swingle citrumelo, and Express mandarin were the most sensitive to compaction, whereas rough lemon and Volckameriana lemon were most tolerant. These findings indicated that the latter vigorously-growing plants were able to produce the most roots in compacted soil, while the slower-growing rootstocks were inhibited most. Soil compaction caused a decrease of 71–91% in the total lengths of feeder roots in the absence of the pathogen.

In plants inoculated with *P. nicotianae*, root damage was again more severe in compacted soil (SBD of 1.7 ton m^{-3}) compared with uncompacted soil (SBD of 1.4 ton m^{-3}), causing feeder root losses of 33–50% compared to <20%, respectively. In these experiments, as in the others, a constant soil moisture was maintained by means of suction under vacuum.

38.4.4   Other Observations

A number of important field observations have been made by various citrus researchers and growers on the effects of deep plowing in established citrus orchards in South Africa. Most notable was the importance of irrigation management and fertilization in conjunction with deep plowing in existing orchards. In a deep-plowed orchard of Valencia oranges on rough lemon rootstock at Letaba Estates, Northern Province, no positive response was observed. This was ascribed to suboptimal irrigation during the prevailing drought (S.H. Swart, QMS Agriscience, Letsitele, personal communication). In established citrus orchards in the Sundays River Valley and in Swaziland that were subjected to deep tillage, massive root flush occurred upon deep plowing, especially at the point of root pruning (D. Joubert, Dunbrody Estates, Swaziland, unpublished). This corresponds with the observations of Abercrombie and Hoffman (1996).

Howeʋer, the response in terms of tree growth, yield, and root growth was best in young orchards (10 or less years old). In some instances in older orchards, deep plowing caused a set-back in terms of tree growth and yield. Furthermore, observations indicated that trees should not be plowed on both sides at the same time, but the opposite side should only be done a year later. Tree age and condition are apparently important whenever deep-plowing of existing orchards is considered. It is therefore advisable to consider both these latter factors, as well as the soil factors identified by Abercrombie and Hoffman (1996) before SPM is implemented in an existing citrus orchard.

38.4.5   Nutrient Mobilization and Microbial Activity

It is important to keep in mind that improved root growth and yield of citrus in the above-mentioned studies was obtained without application of additional fertilizer. The same holds true for deep tillage with many other crops. The common conclusion that most researchers make is that removal of the physical impedance (soil compaction) results in enhanced root growth and proliferation, and that as a consequence of this, the root system is better able to exploit a larger soil volume, which then supports an increase in plant growth and yield. However, the role of soil microbes in this enhancement of plant growth and yield after deep plowing should also be considered. One can expect a shift in the soil microflora in deep-tilled soil because of increased aeration and a shift in the soil water balance, along with other factors.

A question naturally arises as to what contribution this shift in microbial populations makes to the increase in root and plant growth? There is a growing literature on the plant growth-stimulating effects of plant growth-promoting rhizobacteria (PGPR) with their production of phytohormones among other services to plants (Bloemberg and Lugtenberg, 2001; Bai et al., 2002; Bashan and de-Bashan, 2002; Asghar et al., 2004; also Chapter 14). It is quite likely that a shift from anaerobic to aerobic soil conditions could stimulate populations of PGPR that would contribute to increased plant growth. Further, deep tillage could have an impact on populations of phosphate-solubilizing bacteria (Chapter 13).

Admittedly, it is difficult to study individual factors such as these separately in a complex system such as soil, where a myriad of interrelated factors are at work. Often, seemingly conflicting results are obtained. It has been reported, e.g., that with ryegrass, phosphorus uptake per unit length of root was higher from compacted than from uncompacted soil at bulk densities of 1300–1500 kg m^{-3} (Schierlaw and Alston, 1984). Nevertheless, the response of soil microbes on deep plowing seems to be a promising area to be investigated.

38.5 Discussion

Although studies have shown the importance of soil compaction and its alleviation for the health and productivity of citrus trees in South Africa, the response of established citrus orchards to deep plowing has not yet been thoroughly investigated. There is still need for systematic studies of the long-term effects of deep tillage on trees of different categories, in terms of age and condition, and on different types of soil. Furthermore, the effect of deep tillage on tree-decline complexes that may have different etiologies, and on different populations of pathogens such as the citrus nematode (*Tylenchulus semipenetrans)* and *P. nicotianae*, needs to be studied in depth on a larger scale. As the South African citrus industry has gone through significant restructuring in the last few years, these long-term research projects might now be able to get the attention they deserve.

For better soil system functioning, the beneficial effects of alleviating soil compaction or impervious soil layers by means of deep tillage have been well demonstrated in terms of improved crop condition and yield. This intervention is, however, clearly not a panacea for all soil situations. As with other soil profile modification methods, specific criteria and parameters need to be considered before implementing a soil intervention of this nature.

Acknowledgments

Funding for this work conducted on citrus in South Africa was provided by the South African Citrus Growers Association and the Department of Trade and Industry through the THRIP program managed by the National Research Foundation.

References

Abercrombie, R.A. and Hoffman, J.E., The effect of alleviating soil compaction on yield and fruit size in an established navel orange orchard, *Proc. Int. Soc. Citricult.*, **2**, 979–983 (1996).

Allmaras, R.R., Kraft, J.M., and Miller, D.E., Effects of soil compaction and incorporated crop residue on root health, *Ann. Rev. Phytopathol.*, **26**, 219–243 (1988).

Asghar, H.N., Zahi, Z.A., and Arshad, M., Screening rhizobacteria for improving the growth, yield, and oil content of canola (*Brassica napus* L.), *Aust. J. Agric. Res.*, **55**, 187–194 (2004).

Bai, Y. et al., Isolation of plant growth promoting Bacillus strains from soybean nodules, *Can. J. Microbiol.*, **48**, 230–238 (2002).

Barber, B.A. and Gunn, K.B., The effect of mechanical forces on the exudation of organic substances by the roots of serial plants grown under sterile conditions, *New Phytol.*, **73**, 39–45 (1974).

Barley, K.P., Influence of soil strength on growth of roots, *Soil Sci.*, **96**, 175–180 (1963).

Barley, K.P., Mechanical resistance of the soil in relation to the growth of roots and emerging shoots, *Agrochemica*, **20**, 171–181 (1976).

Bar-Yosef, B. and Lambert, J.R., Corn and cotton root growth in response to soil impedance and water potential, *Soil Sci. Soc. Am. J.*, **45**, 930–935 (1981).

Bashan, Y. and de-Bashan, L.E., Protection of tomato seedlings against infection by *Psuedomonas syringae* pv tomato by using the plant growth-promoting bacterium *Azospirillum brasiliense*, *Appl. Environ. Microbiol.*, **68**, 2637–2643 (2002).

Bhatti, M.A. and Kraft, J.M., Influence of soil bulk density on root rot and wilt of chickpea, *Plant Dis.*, **76**, 960–963 (1992).

Bloemberg, G.V. and Lugtenberg, B.J.J., Molecular basis of plant growth promotion and bio-control by rhizobacteria, *Curr. Opin. Plant Biol.*, **4**, 343–350 (2001).

Botha, T., Wehner, F.C., and Kotze, J.M., Evaluation of new and existing techniques for *in vitro* screening of rootstocks tolerance to *Phytophthora cinnamomi* rands in avocado, *Phytophylactica*, **22**, 335–338 (1990).

Burke, D.W. et al., Counteracting bean root rot by loosening the soil, *Phytopathol.*, **62**, 306–309 (1972).

De la Rosa, D. and Carlisle, V.W., Correlation of productive capacity and estimated yields for selected Florida soils, *Soil Crop Soc. Fla Proc.*, **37**, 134–138 (1978).

Duniway, J.M., Movement of zoospores of *Phytophthora cryptogea* in soil with various textures and matrix potentials, *Phytopathol.*, **66**, 877–882 (1976).

Duniway, J.M., Water relations of molds, *Ann. Rev. Phytopathol.*, **17**, 431–460 (1979).

Eck, H.V. and Ungerer, P.W., Soil profile modification for increasing crop production, *Adv. Soil Sci.*, **1**, 65–100 (1985).

Ford, H.W., Growth and root distribution of orange trees on two different rootstocks as influenced by depth to subsoil clay, *Am. Soc. Hortic. Sci.*, **74**, 313–321 (1959).

Fritton, D.D., Swader, J.N., and Hoddinott, K., Profile modification persistence in a fragipan soil, *Soil Sci.*, **136**, 124–130 (1983).

Fulton, J.M., Mortimore, C.G., and Hildebrand, A.A., Note on the relation of soil bulk density to the incidence of Phytophthora root and stalk rot of soybeans, *Can. J. Soil Sci.*, **41**, 247 (1961).

Gerhard, C.J., Mehta, H.C., and Hinojosa, E., Root growth in a clay soil, *Soil Sci.*, **114**, 37–49 (1972).

Gerhard, C.J., Sexton, P., and Shaw, G., Physical factors influencing soil strength and root growth, *Agron. J.*, **74**, 875–879 (1982).

Goss, H.J. and Russel, R.S., Effects of mechanical impedance on root growth in barley (*Hordeum vulgare* L.), III: Observations on the mechanisms of response, *J. Exp. Bot.*, **121**, 577–578 (1980).

Grant, R.F., Simulation model of soil compaction and root growth, *Plant Soil*, **150**, 1–14 (1993).

Gray, L.E. and Pope, R.A., Influence of soil compaction on soybean, stand, yield and Phytophthora root rot incidence, *Agron. J.*, **78**, 189–191 (1986).

Greacen, E.L., Barley, K.P., and Farrel, D.A., The mechanics of root growth in soils with particular reference to the implications for root distribution, In: *Root Growth*, Whittington, W.J., Ed., Butterworths, London, 256–267 (1969).

Greenwood, D.J., Effect of oxygen distribution in the soil on plant growth, In: *Root Growth*, Whittington, W.J., Ed., Butterworths, London, 202–221 (1969).

Grimes, D.W., Wiley, P.L., and Carlton, A.B., Plum root growth in a variable-strength field soil, *J. Am. Soc. Hortic. Sci.*, **107**, 990–992 (1982).

Gupta, S.C., Sharma, P.P., and De Franchi, S.A., Compaction effects on soil structure, *Adv. Agron.*, **42**, 311–338 (1989).

Huisman, O.C., Interrelations of root growth dynamics to epidemiology of root-invading fungi, *Ann. Rev. Phytopathol.*, **20**, 303–327 (1982).

Joubert, D., Relationship between soil compaction and Phytophthora root rot of citrus. M.Sc. thesis, Department of Microbiology and Plant Pathology, University of Pretoria, South Africa, 1993.

Joubert, D. and Labuschagne, N., Effect of soil compaction on *Phytophthora nicotianae* root rot in Rough lemon and Troyer citrange seedlings, *Afr. Plant Prot.*, **4**, 123–128 (1998).

Kaddah, M.T., Subsoil chiselling and slip plowing effects on soil properties and wheat grown on a stratified fine sandy soil, *Agron. J.*, **68**, 36–39 (1976).

Kew, K.L. and Zentmeyer, G.A., Chemotactic response of zoospores of five species of Phytophthora, *Phytopathol.*, **63**, 1511–1517 (1973).

Kuan, T.L. and Erwin, D.C., Predisposition effect of water saturation of soil on Phytophthora root rot of alfalfa, *Phytopathol.*, **70**, 981–986 (1980).

Lambers, H., Growth, respiration, exudation and symbiotic associations: The fate of carbon translocated to the roots, In: *Root Development and Function*, Gregory, P.J., Lake, J.V., and Rose, D.A., Eds., Society for Experimental Biology, Cambridge University, Cambridge, UK, 125–147 (1988).

Lutz, A.L., Menge, J.A., and O'Connel, N., Citrus root health: Hardpans, claypans and other mechanical impedances, *Citrograph*, **71**, 57–61 (1986).

McAfee, M., Lindström, J., and Johansson, W., Effects of pre-sowing compaction on soil physical properties, soil atmosphere and growth of oats on a clay soil, *J. Soil Sci.*, **40**, 707–717 (1989).

Mikhael, E.H. and El-Zeftawi, B.M., Effect of soil types and rootstocks on root distribution, chemical composition of leaves and yield of Valencia oranges, *Aust. J. Soil Res.*, **17**, 335–342 (1979).

Moots, C.K., Nickell, C.D., and Gray, L.E., Effects of soil compaction on the incidence of *Phytophthora megasperma* f.sp. *glycinea* in soybean, *Plant Dis.*, **72**, 896–900 (1988).

Nel, D.J. and Bennie, A.T.P., Soil factors affecting tree growth and root development in a citrus orchard, *S. Afr. J. Plant Soil*, **1**, 39–47 (1984).

Patt, J., Carmeli, D., and Zafrir, I., Influence of soil physical conditions on root rot development and productivity of citrus trees, *Soil Sci.*, **102**, 82–84 (1966).

Raghavan, G.S.V. et al., Effect of compaction and root rot disease on development and yield of peas, *Can. Agric. Eng.*, **24**, 31–34 (1982).

Robertson, L.S., Erickson, A.E., and Hansen, C.M., Tillage systems for Michigan soils and crops. Pt I: Deep, primary, supplemental and no-till, *Mich. Agric. Exp. Station Bull.*, E-1041 (1977).

Rosenburg, N.J., Response of plants to the physical effect of soil compaction, *Adv. Agron.*, **16**, 181–196 (1964).

Rovira, A.D. and Davey, C.B., Biology of the rhizosphere, In: *The Plant Root and Its Environment*, Carson, E.W., Ed., University Press of Virginia, Charlottesville, VA (1974).

Rusanov, V.A., Effect of wheel and track traffic on the soil and crop growth and yield, *Soil Till. Res.*, **19**, 131–143 (1991).

Russel, R.S. and Goss, M.J., Physical aspects of soil fertility: The response of roots to mechanical impedence, *Neth. J. Agric. Sci.*, **22**, 305–318 (1974).

Schierlaw, J. and Alston, A.M., Effect of soil compaction on root growth and phosphorous, *Plant Soil*, **77**, 15–28 (1984).

Schmitthenner, A.F. and van Doren, D.M., Integrated control of root rot of soybean caused by *Phytophthora megasperma* f.sp. *glycinea*, In: *Ecology and Management of Soil-Borne Plant Pathogens*, Parker, C.A. et al., Eds., American Phytopathology Society, St Paul, MN, 263–266 (1985).

Smucker, A.J.M. and Erickson, A.E., Anaerobic stimulation of root exudates and diseases of peas, *Plant Soil*, **99**, 423–433 (1987).

Sterne, R.E., Zentmeyer, G.A., and Kaufmann, M.R., The effect of matric and osmotic potential of soil on Phytophthora root disease of *Persea indica*, *Phytopathol.*, **67**, 1491–1494 (1977a).

Sterne, R.E., Zentmeyer, G.A., and Kaufmann, M.R., The influence of matric potential, soil texture, and soil amendment on root disease caused by *Phytophthora cinnamomi*, *Phytopathol.*, **67**, 1495–1500 (1977b).

Thompson, P.J., Jansen, I.J., and Hooks, C.L., Penetration resistance and bulk density as parameters for predicting root system performance in mine soils, *Soil Sci. Soc. Am. J.*, **51**, 1288–1293 (1987).

Tu, J.C. and Tan, S.C., Soil compaction effect on photosynthesis, root rot severity and growth of white beans, *Can. J. Soil Sci.*, **68**, 455–459 (1988).

Ungerer, P.W., Effects of deep tillage and profile modification on soil properties, root growth and crop yields in United States and Canada, *Geoderma*, **22**, 275–295 (1979).

Van Huyssteen, L., *Verslag oor besoek aan Sondagsriviervallei in verband met wortelverspreiding en grondvoorbereiding (28 Sept–1 Okt 1987)*, Viticultural and Oenological Research Institute, Stellenbosch, South Africa (1987).

Voorhees, W.B., Farrel, D.A., and Larson, W.E., Soil strength and aeration effects on root elongation, *Soil Sci. Soc. Am. Proc.*, **39**, 948–953 (1975).

Wilcox, W. and Mircetich, S.M., The influence of different levels of soil moisture on Phytophthora root rot and crown rot of Mahaleb cherry rootstock (Abstr.), *Phytopathology*, **69**, 1049 (1979).

Wilson, A.J. and Robards, A.W., The ultrastructural development of mechanically impeded barley roots: Effects on the endodermis and pericycle, *Protoplasma*, **95**, 255–265 (1978).

Zentmeyer, G.A., *Phytophthoa cinnamomi and the Diseases it Causes*, Monograph 10. American Phytopathology Society, St Paul, MN (1980).

Zimmerman, R.P. and Kardos, L.T., Effect of bulk density on root growth, *Soil Sci.*, **91**, 280–288 (1961).

39

Rhizosphere Management as Part of Intercropping and Rice–Wheat Rotation Systems

Liu Xuejun, Li Long and Zhang Fusuo

College of Resources and Environmental Sciences, China Agricultural University, Beijing, China

CONTENTS

Most of the current crop production in China produces high outputs in terms of grain and fiber yields per unit area, but with a corresponding requirement for high inputs, including seeds, irrigation, and various chemicals. At the same time, China is known for the low quality of its outputs, and there is increasing concern for environmental problems. China, with only 7% of the world's arable land area, must support 22% of the world's population. At present, the arable land area in China is decreasing due to the expansion of land uses for nonagricultural purpose, while the population in China continues to increase. For some time to come, China must expect an annual population increase of 10 million while there is an annual decrease of 350,000 ha of cultivated land. To support a larger population with a declining farmland area, China needs to develop production systems that make the most efficient use of the limited land resources, improve soil fertility and productivity, and increase farming profitability.

For the sustainable development of agriculture, efficient exploitation of natural resources is required with minimal negative impacts on the natural resource base. Fertilizer application has been one of the most important means for increasing crop production over the past century, but the negative impacts of fertilizer use on the environment have now become a matter of concern, for several reasons including food quality, water quality, and diminished diversity of flora and fauna. Efforts to improve the productivity of agronomic resources, especially water and soil nutrients, with decreased reliance on the chemical inputs for agricultural production, are spreading around the world.

At present, about 70% of farm production in China comes from modernized cropping systems that utilize improved seeds and fertilizer inputs. Considerable contributions to China's grain production have come from traditional multiple-cropping systems that include rotations, intercropping, and related practices (Liu, 1994; Tong, 1994). Adapting such cropping systems with their efficient use of soil nutrients and limited reliance on chemical inputs is one way to develop the sustainable agriculture while keeping yields high (Zhang and Shen, 1999). This chapter considers two cropping systems — intercropping of various field crops, and rice–wheat rotation. These systems face several nutritional problems and constraints that can be improved through better understanding and management of rhizosphere processes (Zhang et al., 2004).

39.1 Explaining the Benefits of Intercropping

In intercropping systems, two or more crops are grown simultaneously on the same field with their periods of overlap long enough to include their vegetative stage (Gomez and Gomez, 1983). Although this would appear to be disadvantageous to both crops if there is only a fixed amount of nutrients available in the soil, net increases in production result for many if not all crop combinations. That farmers in northwestern China have been using a variety of associations for many years over substantial areas confirms the advantages of intercropping empirically as farmers would not persist with such cropping systems unless beneficial. We were curious to know what could be the scientific explanations for this. Our research has focused on Gansu province and Ningxia autonomous region, where the wheat/maize association covered over 275,000 ha in 1995 when our work began. In addition, there were wheat/soybean, wheat/fava bean, and groundnut/maize intercropping systems widely practiced in the region (Zhang and Li, 2003).

We have evaluated the contributions made to increased yields and nutrient uptake by intercropped wheat arising from belowground interactions compared with those above ground (Li et al., 2001a, 2001b). Previously in maize/soybean strip intercropping, a 26% increase in maize yield was observed along with a 27% yield reduction in soybean border rows when these were located outside eight-row sets of alternating strips in Indiana, USA (West and Griffith, 1992). In a different evaluation in Iowa, USA, it was found that strip intercropping led to 20–24% greater maize yields and 10–15% lower soybean yields in adjacent border rows within a similar intercropping system (Ghaffarzadeh et al., 1994).

In our study, we found a yield advantage for intercropped wheat resulting from both border-row and inner-row effects in both wheat/maize and wheat/soybean intercropping systems (Li et al., 2001a). However, the yield in border rows of intercropped wheat, adjoining either maize or soybean, was definitely higher than in inner rows, being influenced less, if at all, by the other crop. This difference was true for both intercropping systems (Zhang and Li, 2003).

When the total increase of grain yield in the wheat/maize system (74%) was disaggregated, we found that 47% was attributable to aboveground and 27% to belowground interactions. In the wheat/soybean system, the 53% increase in total grain yield could be broken down into 30% for aboveground and 23% from belowground interactions between two species.

- In wheat/maize intercropping, the aboveground and belowground interactions, respectively, increased the nitrogen uptake by wheat by 50 and 59%; while for wheat/soybean intercropping, the additions were 23 and 19%, respectively.

- Similarly for the measured increase in phosphorus uptake, the contributions of aboveground and belowground interactions were 56 and 42%, respectively, for wheat/maize intercropping; and 26 and 28%, respectively, for wheat/soybean intercropping (Zhang et al., 2001).

These comparisons show that interspecific belowground interaction, i.e., rhizosphere effects between intercropped species, play an important role in creating the yield advantage from intercropping. This led us to try to develop explanations for these effects through field investigations.

Interspecific competition is usually expected when two crops are grown together (Van der Meer, 1989). Such competition usually decreases survival, growth, or reproduction of at least one of the species (Crawley, 1997). However, the widespread use of intercropping systems in northwest China shows that combining crop species must offer some advantages. We started our work with field trials in two locations in Gansu province in 1997 and 1998 to see why intercropping could produce more benefits compared to monocropping, enhancing soil fertility as well as yield. Details of these trials have been published in Li et al. (2001a, 2001b).

39.1.1   The Competitive-Recovery Production Principle

To understand the effects of intercropping, we believe it is necessary to look at interactions between plants above- and belowground in time, as well as space. The advantages of one plant species at one point in time can give way subsequently to advantages for the other. Spatially their roots and shoots exploit the resources available more completely and efficiently in different zones in the soil and in the canopy space, improving resource utilization. This is an effect parallel to that discussed in section 39.1.2, where two or more species are able to increase the total amount of available nutrients to their mutual benefit.

39.1.1.1 *Dominant-Species Yield Advantage in Interspecific Competition*

In the wheat/maize or wheat/soybean intercropping systems in north China, there is an overlapping growth period in the field of 70–80 days. This causes intense interspecific interaction between the two intercropped species. These interactions occur mostly at the interface between the two species, where they were closest, with ensuing short-term increases or decreases in growth, development, and possibly yield.

The observed yield advantage in the border row of intercropped wheat with maize or soybean probably derives from differences in interspecific competitiveness. Wheat is seen to be more competitive relative to maize and soybean during the overlapping growth periods in both the wheat/soybean and wheat/maize intercropping systems.

A measure of aggressivity (A) in the interspecies competition in intercropping based on relative yield changes of the two component crops has been proposed by Willey and Rao (1980):

$$A_{ab} = Y_{ia}/(Y_{sa} \times F_a) - Y_{ib}/(Y_{sb} \times F_b) \tag{1}$$

where Y_{ia} and Y_{ib} are yields of crops a and b in intercropping, Y_{sa} and Y_{sb} are yields of crops a and b in sole cropping, with F_a and F_b being the proportion of the respective areas occupied by crop a and crop b in the intercropping system. When A_{ab} is greater than 0, we say that that crop a has greater competitive ability compared with crop b when intercropped.

Li et al. (2001a) calculated the aggressivity of wheat relative to either maize or soybean and found it to be consistently greater: $A_{wm} = 0.26$–1.64 vis-à-vis maize and $A_{ws} = 0.35$–0.95 vis-à-vis soybean. The nutrient competitive ratio, 1.09–7.54 for wheat relative to maize, and 1.2–8.3 for wheat relative to soybean, showed further that wheat has a greater capability of acquiring nutrients compared with soybean and maize. Determining the mechanisms that account for this difference should be helpful for controlling interspecific competition and for gaining more yield advantage from intercropping.

39.1.1.2 *Subordinate-Species Compensation during a Later Growth Stage*

In wheat/maize or wheat/soybean intercropping, up to the time of wheat harvesting, nutrient uptake and growth are reduced in the subordinate species (maize or soybean) due to interspecific competition. After the earlier-maturing species is harvested, however, there is a recovery of nutrient uptake and growth by the later-maturing species. This compensates for the retardation in its earlier growth phase, partly by having developed a larger root system in the meantime.

In a wheat/maize field trial in 1997 conducted at the Baiyun site, the rates of dry matter accumulation in the intercropped maize (10–$20\ \mathrm{g\ m^{-2}\ day^{-1}}$) were significantly lower than those of sole maize (17–$35\ \mathrm{g\ m^{-2}\ day^{-1}}$) during the early stage from 7 May to 3 August. Thereafter, however, the rates of dry matter accumulation for intercropped maize increased substantially, to 59–$70\ \mathrm{g\ m^{-2}\ day^{-1}}$. This was significantly greater, almost double, that in sole maize (23-$52\ \mathrm{g\ m^{-2}\ day^{-1}}$). Nutrient acquisition showed the same trend as growth.

In similar trials in 1998 at the Jingtan site, the disadvantage of the border row of intercropped maize resulting from interspecific competition diminished after the wheat harvest, and it had disappeared by time of maize maturity (Li et al., 2001b). However, this recovery was limited to the treatment that had not received nitrogen and phosphorus fertilizer applications (Li et al., 2001b). This needs further study to understand the mechanisms involved, to understand yield advantages in intercropping where short-season species are associated with long-season ones.

39.1.2   Interspecific Belowground Growth Facilitation by Nutrient Mobilization

The first set of effects discussed above deals with how given volumes of space above- and belowground will be exploited by plant shoots and roots, capturing available radiation, O_2, N_2, water, and nutrients over time. We have found evidence that the crop species in association can enhance the pool of nutrients through their production and exudation of enzymes, hormones, siderophores, and other compounds. Our work has only made a start in examining this domain of plants' interaction with each other and with organisms in the soil.

39.1.2.1   *Improvement in the Fe Nutrition of Groundnut through Maize Intercropping*

Groundnut (*Arachis hypogaea* L.), a major oilseed crop in China, accounts for 30% of the cropped area and 30% of the total oilseed production in the country. However, iron-deficiency chlorosis frequently occurs in groundnuts grown on calcareous soils, especially in northern China. It has been evident for some time that iron-deficiency chlorosis of groundnut is more severe in sole-cropping systems than when this species is intercropped with maize on these soils. This observation has aroused considerable interest in how Fe deficiency can be corrected or prevented. It has now been determined that the improvements in Fe nutrition of groundnut when intercropped with maize are due mainly to rhizosphere interactions between groundnut and maize (Zuo et al., 2000).

The severity of iron-deficiency chlorosis in young groundnut leaves in the intercropping systems was closely related to the distance of the groundnut plants from the maize roots when treatments were assessed during the flowering period for groundnut. In the unrestricted intercropping treatment, where neighboring roots of groundnut and maize intermingled fully, the young leaves of groundnut plants in rows 1–3 from the maize grew without any visible symptoms of iron deficiency, while those in rows 5–10 showed variable degrees of chlorosis.

In the treatment where the groundnut and maize roots were kept physically separated, the groundnut plants in all rows were chlorotic, except that most of the groundnut plants in the first row from maize showed only slight signs of chlorosis (80% of their young leaves were green). In contrast, in the treatment where a 400-mesh nylon mesh separation (nominal aperture 37 μm) was inserted into the ground between the two crop species to prevent direct root contact — and interactions were still possible through mass flow and diffusion — the plants in rows 1 and 2 remained completely green, while those in rows 3–10 were chlorotic to varying extents. In the monocropped control treatment, about 90% of the groundnut plants showed severe Fe deficiency chlorosis in the young leaves.

Groundnut is known to be a plant with strategy I in response to Fe deficiency in the soil, i.e., it lowers the pH in its rhizosphere through exudates to solubilize Fe(III) and reduce this to Fe(II). In calcareous soils, this strategy is less productive, however, and the plant more readily exhibits Fe-deficiency chlorosis. Maize, on the other hand, as a gramineae species can deal with Fe deficiency through a strategy II response, which is more efficient under these soil conditions. Maize plant roots excrete phytosiderophores (non-protein amino acids) into the rhizosphere and these solubilize Fe(III) to form Fe-phytosiderophore complexes. This makes it easier for the maize plant root to acquire iron.

It is likely that when groundnut and maize are grown together, phytosiderphores released from maize roots are mobilizing Fe(III) in the soil, and some of this is benefiting the neighboring groundnut plants which are lacking in Fe nutrition. We have as yet no direct evidence of this, but the biological process is well established (Marschner, 1995), and this is an evident and parsimonious explanation for the observed effects of intercropping maize and groundnut vs. cropping them separately in calcareous soils.

39.1.2.2   Benefits to Maize from Association with Fava Bean

Interspecific root interactions between intercropped fava bean (*Vicia fava*) and maize play an important role in yield advantage and in acquisition of nitrogen and phosphorus within intercropping systems (Li et al., 1999; 2003a). When the roots of the two species are kept completely separate, we have calculated the land-equivalent ratio to be 1.06, and the uptake of nitrogen and phosphorus to be, respectively, 31.8 g N m^{-2} and 3.3 mg P m^{-2}. On the other hand, when the roots of the two species are intermingled, these figures were boosted 14–27%, to a ratio of 1.21 and 38.4 g N, and 4.2 g P m^{-2} (Li et al., 2003a).

In our studies, we have not yet obtained sufficient information to be able to establish the mechanisms for the nitrogen and phosphorus uptake. Possibly there could be increased phosphorus in the soil that is mobilized by fava bean from otherwise unavailable sources, and this could be absorbed by the maize. Furthermore, fava bean, a legume, could be transferring nitrogen to maize. More research needs to be done on parameters such as differences in root distribution and in the ability of fava bean and maize to mobilize soil phosphorus. Also, nitrogen transfer from legumes in intercropping systems should be investigated, along with the possible effects that maize could have on N_2 fixation by fava bean. It is evident that this is an effect to be explained — and exploited.

39.1.2.3   Chickpea Facilitation of Phosphorus Nutrition in Associated Wheat

In pot experiments with no root barrier, neither a nylon-mesh root barrier, nor a solid root barrier, the biomass of wheat was significantly increased when the roots of intercropped wheat and chickpea could intermingle compared with the other two treatments, with either a solid root barrier or a nylon mesh barrier, regardless of whether phosphorus was supplied as phytate or as $FePO_4$.

Where phytate was applied, the root intermingling of wheat and chickpea increased the phosphorus concentration in wheat and the phosphorus uptake by wheat, compared with treatments where roots were separated by nylon mesh or a solid barrier. In addition, phosphorus uptake by chickpea was greater in the treatment with the root barrier and an organic source of phosphorus source, compared with the treatment that had no root barrier. This suggested that the mobilization of organic phosphorus by chickpea probably contributed to the phosphorus uptake by wheat.

In another study, Rao et al. (1999) found that acquisition of phosphorus by the legume plant was markedly greater than that of the grass species, regardless of whether phosphorus was supplied in inorganic or organic form. These results indicated that wheat is less able to use organic phosphorus than inorganic phosphorus, whereas chickpea is able to use both phosphorus sources equally effectively.

When wheat and chickpea were grown together, the following may happen. First, chickpea can mobilize and absorb some organic phosphorus by releasing phosphatase into soil, and this leaves some inorganic phosphorus for wheat. Second, wheat with its greater competitive ability acquires more phosphorus from the root zone of both wheat and chickpea, resulting in phosphorus depletion in the chickpea rhizosphere (Li et al., 2003b). This phosphorus depletion in the chickpea rhizosphere then induces the chickpea roots to release substances that mobilize organic phosphorus, to compensate for having given up available phosphorus to the more aggressive root system of wheat. Further study should focus on differences in phytase and phosphatase production by chickpea and wheat roots.

39.1.2.4   Intercropping to Reduce Nitrate Accumulation in the Soil Profile

In recent years there has been a rapid increase of fertilizer application in China to achieve higher yields, especially in the wheat/maize intercropping systems. The average rate of nitrogen application in China is 180 kg N ha^{-1}. In the irrigated areas of northwest China, application rates of N fertilizers are now 450 kg N ha^{-1} for wheat/maize intercropping on average, which increases the risk of groundwater nitrate pollution. Applying nitrogen up to the economic optimum rate (estimated from yield and nitrogen rate data from individual trials) is associated with fairly small increases in soil nitrate-N after harvest (the mean increase was 4 kg N ha^{-1}). Where the optimum nitrogen rate has been exceeded, however, soil nitrate-N levels are increased to a greater extent (Richards et al., 1996).

Intercropping fava bean with oats or spring wheat is one way to reduce nitrate accumulation in soil profiles (Stuelpnagel, 1993). Intercropping maize with ryegrass is also an effective way for increasing nitrogen uptake under conditions of high nitrogen application (Zhou et. al., 2000). Nitrogen uptake by intercropping of wheat and maize is greater than that by corresponding sole cropping under same nitrogen supply (Li et al., 2001a). Therefore, we would propose wheat/maize intercropping to reduce nitrate accumulation in the soil rather than continue with sole maize and wheat production.

In an irrigated area of Hexi corridor, northwest China, we have seen that the amounts of NO_3^- present in the soil profile after wheat harvest are greater for sole wheat than for intercropped wheat with maize, and also for sole fava bean than for intercropped fava bean with maize with applications of either 200 or 400 kg N ha^{-1}. The decrease is about 0–41% for wheat and 0–31% for fava bean. Amounts of NO_3^- in soil after maize harvest were in this order: sole wheat and fava bean $>$ intercropping wheat and fava bean $>$ maize intercropped with fava bean and wheat. These results confirm that intercropping can decrease the accumulation of nitrate in soil profiles.

39.1.3   Interspecific Competition and Facilitation: Concurrent Effects

How interspecific competition and facilitation occur at the same time has been considered by Geno and Geno (2001). This builds on previous analysis by Van der Meer (1989) which discussed the interaction of competition and facilitation in many intercropping systems. He proposed that their net result can be usefully analyzed in terms of a land equivalent ratio (LER), which is an indicator of intercropping advantage. A LER value of >1 indicates that complementary facilitation is contributing more to the interaction than is competitive interference. A positive LER value can result from low interspecific competition or from strong facilitation effects.

A pot experiment was conducted to investigate the relative strengths of interspecific competition and facilitation in a wheat/soybean association. The pots were separated into two compartments either by a plastic sheet that eliminated root contact and any solute movement, or by a nylon mesh (30 μm) that prevented root contact but permitted solute exchange between the compartments. In addition, there was a control treatment with no separation made in the pot. Phosphorus was added as KH_2PO_4 at rates of 0, 150, and 300 μg P g^{-1} soil. Wheat plants were grown in one compartment, and soybean in the other.

The obvious facilitation by soybean of wheat uptake of phosphorus was observed in the pot with the mesh separation, where competition of wheat for phosphorus was eliminated. In the absence of any separation, biomass and phosphorus uptake of wheat were increased by 25% and by 7% compared with results from plastic-sheet partition. However, biomass and phosphorus uptake of soybean in the nylon-mesh separation treatment were much higher, 150% and 134%, respectively, than when there was plastic-sheet separation. In the pots with the nylon-net separation, more roots of soybean were distributed near the nylon mesh on which wheat formed a dense root mat at the other side. In contrast, soybean roots were distributed more or less evenly throughout the whole compartment in the treatment with plastic-sheet separation.

These measurements and observations suggest that the beneficial effect of soybean growth resulted from a rhizosphere effect of wheat. At the same time, biomass and phosphorus uptake of soybean were significantly reduced when there was intermingling of roots of the two species, indicating that there was an interspecific competition between wheat and soybean for phosphorus when in close proximity. Our study showed the coexistence of interspecific competition and facilitation in the wheat/ soybean association.

39.2   The Rice–Wheat Rotation Cropping System

The rice–wheat cropping system is a long-established, major cereal production system in China and South Asia (Ellis and Wang, 1997; Timsina and Connor, 2001). About 10 million ha are now under rice–wheat rotation in China, and another 13.5 million ha in India, Bangladesh, Nepal, and Pakistan (Lal et al., 2004). We have been interested particularly in the effects of manganese (Mn) deficiency, which is a major constraint for this cropping system, and how to reduce this problem by managing crops, soil, and nutrients differently so as to promote rhizosphere processes that are more favorable to enhanced crop yield. This is important both for making production more profitable and for making these farming systems more sustainable.

39.2.1   Manganese Deficiency in Wheat Induced by Rice–Wheat Rotation

With prolonged reducing conditions in the soil under continuous flooding during the rice-growing period, downward movement of manganese occurs through the soil profile (Lu et al., 1990; Li, 1992), with a gradual decrease in soil Mn concentration, particularly in the surface layer. Manganese deficiency is common in wheat and other upland crops in rice–wheat rotation systems in China (Hu et al., 1981; Yang, 1984; Chu and Li, 1985; Ma et al., 1985; Lu and Zhang, 1997) and also in the Indo-Gangetic plains (Takkar and Nayyar, 1981; Nayyar et al., 1985). When all the soils with wheat Mn deficiency were compared, we found three common aspects: (1) soil pH value is usually >7.0, (2) soil is light or coarse-textured, and (3) continuous rice–wheat rotation is practiced (Liu, 1997; Lu et al., 2004). Rice–wheat rotation is thus a very important factor inducing Mn deficiency in wheat.

39.2.1.1   Manganese Accumulation in the Soil Profile and Effect on Wheat Growth

An integrated study by Lu et al. (2004) using field surveys, field trials and soil column experiments was conducted to determine the relationship among Mn leaching and distribution within soil profiles, paddy rice cultivation, and wheat responses to Mn fertilization. At five field sites surveyed, total Mn and active Mn concentrations in the topsoil layers under rice–wheat rotation were only 42 and 11% of those measured in upland systems, without paddy rice. Total and available Mn increased substantially at lower depths in the soil under rice–wheat rotations, showing significant spatial variability of Mn in the soil profile.

In coarse-textured soil with high pH, manganese leaching was the main pathway for Mn loss, while in clay-textured and acid soils, excessive Mn uptake was the main pathway for this loss (Liu et al., 1999a). When Mn was deficient in the topsoil, the amount of

Mn contained in the subsoil is critical for the growth and Mn nutrition of wheat. Sufficient Mn in the subsoil contributed to reasonable growth and more Mn uptake by wheat, but insufficient Mn in the subsoil resulted in plant Mn deficiency and poor performance in wheat (Lu et al., 2004).

39.2.1.2 Different Responses of Wheat and Rapeseed to Manganese Stress

As mentioned already, Mn deficiency in wheat is a typical nutritional disorder in rice–wheat rotations on light or coarse-textured soils, especially in areas draining into rivers, resulting in the reduction of wheat yield by 30–50% (Nayyar et al., 1985; Lu, 1992). However, under the same field conditions, rapeseed (*Brassica napu* L.) does not develop Mn deficiency. Why?

The differences in response to Mn deficiency by these two crops can be traced to differences in root distribution and in rhizosphere interactions between wheat and rapeseed (Fang et al., 1998; Lu et al., 2002). First, the wheat cultivars (N02 and 942) used in the studies of Fang et al. (1998) and Lu et al. (2002) were ones with high-yielding potential, but ones that are sensitive to Mn deficiency. Rapeseed responded to a pretreatment of Mn deprivation by increasing its Mn uptake by four- to six-fold (Table 39.1). In contrast, the wheat cultivars N02 and 942 showed little response. The tolerant wheat cultivar 80-8 showed a moderate response to Mn deprivation (Table 39.1). After a 2-week pretreatment of Mn deprivation, there was an evident color reduction in the agar sheet along the roots of the wheat genotype 80-8 and a significant increase in the Mn concentration in the shoots, suggesting a definite rhizosphere effect related with the increased root reduction ability or Mn mobilization.

Second, the shallow root system of wheat within the soil profile restricted Mn uptake from the Mn-sufficient subsoil (Liu, 1997). Lu et al. (2002) have reported that root restriction with a nylon net bag could induce wheat Mn deficiency in soils with potential Mn-deficiency. On the other hand, rapeseed can mobilize large amounts of soluble Mn from the soil through acidification and reduction processes in the rhizosphere (Fang et al., 2000). Liu et al. (1999b) reported that deep-rooting wheat genotypes are more tolerant of Mn deficiency, suggesting that the penetration of wheat roots or the downward extension of wheat rhizosphere contributes to overcoming Mn deficiency under rice–wheat rotation.

TABLE 39.1

Mn Concentrations (mg kg^{-1}) in the Shoots of Wheat and Rapeseed Genotypes[a]

Pretreatment in Nutrient Solution	Mn Supply in Agar (mM)	Wheat Genotype			Rapeseed Genotype	
		N02	942	80-8	CY11	CY14
− Mn	1	83	80	104	290	356
+ Mn	1	81	79	75	48	72
Difference		NS	NS	*	**	**

[a] Plants were grown in complete nutrient solution for 2 weeks and then were pretreated either with Mn (+Mn) or without Mn (−Mn) for another 2 weeks. Afterwards, the pretreated plants were transplanted into agar sheets containing 1 mM KMnO$_4$ and other essential nutrient elements, and color indicator for Mn reduction. Plants were harvested after 72 h culture. Wheat genotype 80-8 with root distribution in deep soil is tolerant to Mn deficiency, whereas genotypes 942 and N02 are sensitive to Mn deficiency. Rapeseed genotypes CT11 and CY14 are both highly tolerant to Mn deficiency.

Source: Adapted from Zhang et al., *Plant Soil*, **260**, 89–99 (2004). With permission.

NS, *, and ** represent not significant, significant at 0.05 and 0.001 levels.

We can summarize the main differences in rapeseed and wheat mechanisms to adapt to Mn deficiency as follows. For rapeseed, the main mechanism is to mobilize various Mn oxides from soils through stronger acidification and reducing capacity in the rhizosphere compared with wheat (Fang et al., 2000). The mechanism for wheat, on the other hand, depends mainly on the penetration capacity of roots into Mn-sufficient subsoils (Lu et al., 2002).

39.2.1.3 *Improvement of Wheat Manganese Nutrition in Rice–Wheat Rotation through Rhizosphere Management*

The increasing occurrence of Mn deficiency and decline in wheat grain yield at many sites without Mn application has stimulated studies to develop techniques for efficient management of this problem. Various Mn fertilizers (Mn salts, chelates such as Mn-EDTA, and preparations consisting of mixtures of micronutrients) and various application methods (soil, foliar, and seed application) have been evaluated for their efficiency in correcting Mn deficiency in wheat (Nayyar et al., 1985; 2001; Liu et al., 2001). Most studies have sought to overcome the deficiency of this micronutrient through the application of Mn fertilizers.

However, the possibility of managing the spatial distribution of Mn in the soil profile to improve the Mn nutrition of wheat has seldom been considered. Since there is usually sufficient available Mn in the subsoil under rice–wheat rotation, we can develop some rhizosphere management techniques to utilize this Mn resource. First, deep plowing has proved to be one suitable method to overcome Mn deficiency in wheat, physically bringing soil with higher Mn concentration up to the surface (Liu, 1997). Because deep plowing also makes root penetration into the subsoil easier, this significantly improves the Mn nutrition of wheat shoots when planted after rice compared with a control with no deep plowing (Table 39.2).

Second, selection and planting of wheat genotypes with deeper root systems is an option for surviving in Mn-deficient soils with the rice–wheat rotation. In addition, the growth and Mn nutrition of rice and growth of successive wheat could be improved by directly injecting Mn fertilizer solution into the root zone (rhizosphere) of rice. This would increase the availability of Mn in the soil under nonflooded rice–wheat rotation (Zhang et al., 2004). These various results suggest the possibilities for optimizing Mn management based on a better understanding of the rhizosphere processes involved, so as improve the Mn nutrition of crops in the rice–wheat rotation of China.

TABLE 39.2

Effects of Deep Plowing on Mn Concentration, Mn Uptake, and Grain Yield of Wheat following Rice as Compared to Rice-Straw Mulching and Mn Fertilization in a Sandy Loam Soil

Treatment	Shoot Mn Concentration (mg kg^{-1})	Shoot Mn Uptake (g ha^{-1})	Shoot Dry Matter or Grain Yield (kg ha^{-1})
Shooting Stage			
Control	14 b	18 b	1250 b
Deep plowing	27 a	44 a	1580 a
Harvesting Stage			
Control	11 b	46 b	2160 b
Deep plowing	24 a	148 a	2930 a

Values with different letters in the same column are significantly different at 0.05 level.
Source: Adapted from Lu et al., *Plant Soil*, **261**, 39–46 (2004). With permission.

39.2.2 Application of Nonflooded Mulching Cultivation for Rice–Wheat Rotation

Mulching with plastic film (or plastic mulch, PM) has been imported and developed as a new rice cultivation technique in many regions of China since the 1980s (Zhao and Xiao, 1982; Peng et al., 1999). This practice is employed under nonflooded conditions with limited irrigation, thus creating conditions that are substantially different from both traditional flooded and rain-fed rice cultivation. This new cultivation technique has led to improved water-use efficiency, higher soil temperatures, inhibition of weed growth, and increased yields (Peng et al., 1999; Liang et al., 2000).

Other mulching materials such as wheat or rice straw have subsequently been developed in rice production systems (Fan et al., 2002; Lin et al., 2002). This innovative water-saving technology has been applied in the rice–wheat rotational cropping system in Sichuan province of China since 1999 (Wang, 2001). We have investigated whether both plastic film mulching and wheat straw mulching in rice cultivation have potential to provide a more favorable rhizosphere environment for crops and can achieve similar grain yields as traditional rice–wheat rotation when there is proper nutrient management, especially for nitrogen.

Results from three field experiments (two of which lasted more than 3 years) have demonstrated that there are beneficial interactive effects of nonflooded mulching cultivation and lower nitrogen application rate on crop yield, system productivity, crop nitrogen uptake, and nitrogen cycling (Liu et al., 2003, 2005; Fan et al., 2005a; 2005b). Moreover, these effects were quite consistent and stable, despite year-to-year variations in climatic conditions.

39.2.2.1 Aggregate System Productivity

Nonflooded and mulched cultivation of rice greatly reduced the duration of soil submergence, which makes soil hypoxic. These changes in hydrological status had large effects on crop yield and system productivity under low nitrogen rates, although the effects were relatively small when nitrogen application rates were high. PM resulted in similar or slightly higher rice yield compared with traditional flooding (TF), but use of wheat straw mulching (SM) led to lower rice yield in two long-term field experiments (Liu et al., 2003; Fan et al., 2005a).

Decline in soil temperature at early growing stage seems to be as the main reasons why rice yields were lower with SM, particularly with a lower nitrogen rate. When there was double mulching of wheat straw plus plastic film (SM + PM), thereby diminishing the soil temperature effect, rice yield could achieve the same level as either TF or PM (Liu et al., 2005). In contrast, wheat yields were similar with PM and TF, while SM led to greater yields of wheat — especially when a lower nitrogen rate was applied, e.g., 60 kg N ha^{-1}. System productivity, i.e., total grain yield from rice plus wheat, stayed within a relatively narrow range, however, ranging between 10.2 and 13.2 t ha^{-1}, following the sequence of PM $\geq$ TF $\geq$ SM in spite of nitrogen input levels made to the whole rotation system (Liu et al., 2003).

39.2.2.2 Mulch–Nutrient Interactions

Across several years and experiments, system productivity was, not surprisingly, greater under higher nitrogen input, e.g., 150 kg N ha^{-1} in rice plus 120 kg N ha^{-1} in wheat, than under lower nitrogen input, e.g., 150 kg N ha^{-1} in rice plus 60 kg N ha^{-1} in wheat, for the TF and PM systems. Notably, the highest system productivity with SM was observed with the lower nitrogen input. When SM is used, fertilizer inputs can be reduced from 120 to 60 kg N ha^{-1} for wheat following rice without any significant yield loss (Liu et al., 2003; see Table 39.3). Similar results have been obtained by Fan et al. (2005a). They found that

TABLE 39.3

Net Nutrient Balances in Three Cycles of Rice–Wheat Rotation as Affected by Nonflooded Mulching Cultivation and Nitrogen Rate

Mulching	Nitrogen	Fertilizer Input (1)	Addition from Straw (2)	Uptake by Crops (3)	Net Balance (= 1 + 2 − 3)
Nitrogen (kg N ha^{-1})					
TF	150–60	630	0	560	70
	150–120	810	0	604	206
PM	150–60	630	0	609	21
	150–120	810	0	697	113
SM	150–60	630	75	582	123
	150–120	810	75	595	310
Phosphorus (kg P ha^{-1})					
TF	150–60	197	0	138	59
	150–120	197	0	145	52
PM	150–60	197	0	138	59
	150–120	197	0	153	44
SM	150–60	197	15	135	77
	150–120	197	15	134	78
Potassium (kg K ha^{-1})					
TF	150–60	375	0	612	− 237
	150–120	375	0	632	− 257
PM	150–60	375	0	706	− 331
	150–120	375	0	742	− 367
SM	150–60	375	173	660	− 112
	150–120	375	173	669	− 121

Mulching treatments: TF, traditional flooding; PM, plastic film mulching; SM, wheat-straw mulching. Nitrogen (kg N ha^{-1}): the first value is for rice, and the second is for wheat.
Source: Adapted from Liu et al., *Field Crops Res.*, **83**, 297–311 (2003). With Permission.

using SM in the preceding rice season had a residual positive effect on the succeeding wheat crop, as evidenced by the higher wheat yield with SM than under PM or TF under N0 conditions.

There appears to be an important time effect here. The larger fraction of wheat straw remaining in the soil at the end of the rice crop is likely to decompose during the wheat season and may have beneficial residual effects on nitrogen supply capacity and soil productivity. Eagle et al. (2000) reported that rice yields from N0 plots in which rice straw was retained by incorporation and rolling were greater by the third year of the experiment compared with other treatments, such as burning and removal of rice straw.

A treatment in which straw was applied surpassed a zero-fertilizer treatment in both nitrogen uptake and yield by the second season in a study conducted in the Philippines (Becker et al., 1994). Aulakh et al. (2000) also reported residual effects of green manure on wheat yields in rice–wheat rotations. In general, nonflooded mulching cultivation, especially with PM, showed potential to sustain similar system productivity in rice–wheat rotation under suitable nitrogen-input condition.

39.2.2.3 *Nutrient Balance*

Nutrient balance is usually calculated to evaluate the nutrient-use efficiency in a plant-soil system. Total nutrient balances in the three cycles of rice–wheat rotation are significantly affected by the levels of nitrogen inputs, of course, but also by the mulching practice. The straw mulch (SM) option gave a higher nutrient balance at both the higher nitrogen rate (150–200 kg N ha^{-1}) and at a reduced nitrogen rate (150–60 kg N ha^{-1}). Among the three systems of management, PM led with the lowest net nitrogen balance for having the highest nitrogen uptake and thus the highest output from the soil system. SM had the highest net nitrogen balances due to having had the highest nitrogen inputs, with nitrogen outputs on a par with or marginally lower than TF. Fan et al. (2005a) observed similar phenomena in another study.

Earlier studies have also shown retention of straw on the surface as increasing soil nitrate concentrations, nitrogen uptake, and yield compared to burning the straw (Bacon and Cooper, 1985a; 1985b; Bacon, 1987). This means that SM could be a good option from the viewpoint of nitrogen balance and soil fertility. As a further consideration, ^{15}N balance results from Liu et al. (2005) have showed that fertilizer nitrogen recovery was not affected by nonflooded mulching cultivation within the whole rice–wheat rotation.

Net phosphorus balance was positive due to the much higher phosphorus inputs than phosphorus outputs in spite of different nitrogen application rates. In contrast, net potassium (K) balance was always negative because total K inputs were much lower than total K outputs. Among the three treatments, net P balance was higher with SM than that with TF or PM, mainly due to the P addition from wheat straw (15 kg P ha^{-1}); P outputs (134–153 kg P ha^{-1}) were similar in the three treatments at two nitrogen rates.

Mean K balance was significantly greater with SM (-120 kg K ha^{-1}) than under TF (-247 kg K ha^{-1}) or PM (-349 kg K ha^{-1}). This applied across both nitrogen rates mainly because of the considerable K inputs from wheat straw (173 kg K ha^{-1}). These NPK balance results suggest that the same N, P, and K rates for rice–wheat rotation may not be appropriate for maintaining the sustainability of the whole rice–wheat rotation system with different management of organic and inorganic nutrients.

39.2.3   Soil Properties

After 3 years of practicing nonflooded, mulched cultivation, some selected soil properties were assessed from 0–10 and 10–20 cm soils (Liu et al., 2003). Both SM and PM caused a decline in soil pH at 0–10 cm, compared with TF. Soil organic carbon (OC) and total nitrogen (TN) contents were not affected by either PM or SM in the 0–10 and 10–20 cm soil layers. In contrast, both soil Olsen-P at 0–10 cm and 10–20 cm layers (8.3–12.0 mg kg^{-1} and 2.9–4.1 mg kg^{-1}, respectively) and exchangable K in the 0–10 cm layer (24.1–33.7 mg kg^{-1}) were significantly affected by the nonflooded, mulched cultivation, following the same sequence of SM > TF and PM. Soil-available K in the 10–20 cm layer (17.7–19.3 mg kg^{-1}), however, was not significantly different among the TF, PM, and SM treatments. Fan et al. (2005b) found a similar trend in soil chemical properties in the same field after 5 years of nonflooded, mulched rice–wheat rotation.

In general, most soil properties (except pH) were improved (with SM) or not influenced (with PM) by nonflooded, mulched cultivation in comparison with TF after three cycles of rice–wheat rotation. This suggests that at least in the short term, nonflooded mulching has no negative effects on soil fertility. Further, residual inorganic nitrogen in the soil profile (0–60 cm) after harvest of rice and wheat was not significantly different between TF and PM or SM (Fan et al., 2005a). Such results indicate that PM and SM are not likely to lead to declining soil fertility or to exacerbate related environmental issues, such as nitrate leaching. Indeed, they apparently confer considerable benefit.

39.3 Discussion

Intercropping and rice–wheat rotation are two particular cropping systems that are important in Chinese agriculture (Liu, 1994). Rhizosphere processes are significantly modified by the specific cropping patterns of intercropped or rotated planting, with important impacts on soil nutrient availability in the rhizosphere. There are distinct interspecific rhizosphere interactions in the intercropping systems as well as in the rice–wheat rotation systems. More attention should be paid to rhizosphere processes at an ecosystem level, focusing on the broad interaction processes among plants, soil, microorganisms, and their environment at various scales. Specific mechanisms have been discussed for agriculture generally in Chapter 7.

Managing the rhizosphere ecosystem and regulating rhizosphere processes to promote sustainable development may be an effective approach to enhancing nutrient-resource use efficiency and improving productivity in various cropping systems. These two examples from China reveal that both sustainable productivity and efficient nutrient utilization can be achieved by engaging in appropriate rhizosphere management.

Acknowledgments

This research was funded by the Major State Basic Research Development Program (grant No. G1999011709), the National Natural Science Foundation of China (grant Nos. 30000102 and 30390080), and the 948 Major-Import Programme of the Ministry of Agriculture (grant No. 202003Z53).

References

Aulakh, M.S. et al., Yields and nitrogen dynamics in a rice–wheat system using green manure and inorganic fertilizer, *Soil Sci. Soc. Am. J.*, **64**, 1867–1876 (2000).

Bacon, P.E., Effect of nitrogen fertilization and rice stubble management techniques on soil moisture content, soil nitrogen status, and nitrogen uptake by wheat, *Field Crops Res.*, **17**, 75–90 (1987).

Bacon, P.E. and Cooper, J.L., Effect of rice stubble and fertilizer nitrogen management on wheat growth after rice, *Field Crops Res.*, **10**, 229–239 (1985a).

Bacon, P.E. and Cooper, J.L., Effect of rice stubble and fertilizer nitrogen management techniques on yield of wheat sown after rice, *Field Crops Res.*, **10**, 241–250 (1985b).

Becker, M., Ladha, J.K., and Ottow, J.C.G., Nitrogen losses and lowland rice yield as affected by residue nitrogen release, *Soil Sci. Soc. Am. J.*, **58**, 1660–1665 (1994).

Chu, T.D. and Li, X.P., Wheat manganese deficiency occurred in calcareous soils in Northern China, *Soils Fertil.*, **4** (1985), (in Chinese).

Crawley, M.J., *Plant Ecology*, Blackwell Science, Cambridge, UK (1997).

Eagle, A.J. et al., Rice yield and nitrogen utilization efficiency under alternative straw management practices, *Agron. J.*, **92**, 1096–1103 (2000).

Ellis, E.C. and Wang, S.M., Sustainable traditional agriculture in the Tai Lake Region of China, *Agric. Ecosyst. Environ.*, **61**, 177–193 (1997).

Fan, M.S. et al., System productivity and nitrogen utilization of rice–wheat rotations under non-flooded mulching cultivation and varying nitrogen inputs, *Field Crops Res.*, **91**, 307–318 (2005a).

Fan, M.S. et al., Crop yields, internal nutrient use efficiency, and changes in soil properties in rice–wheat rotations under non-flooded mulching cultivation, *Plant Soil*, **277**, 265–276 (2005b).

Fan, X., Zhang, J., and Wu, P., Water and nitrogen use efficiency of lowland rice in ground covering rice production system in south China, *J. Plant Nutr.*, **25**, 1855–1862 (2002).

Fang, Z., Lu, S.H., and Zhang, F.S., Study on tolerance of different wheat cultivars or lines to manganese deficiency, *J. Plant Nutr. Fertil. Sci.*, **4**, 277–283 (1998), (in Chinese).

Fang, Z. et al., Mechanisms of difference in Mn efficiency between wheat and oilseed rape, *Pedosphere*, **10**, 213–220 (2000).

Geno, L. and Geno, B., *Polyculture Production: Principles, Benefits and Risks of Multiple Cropping Land Management Systems for Australia*, RIRDC Publication No. 01/34. Rural Industries Research and Development Corporation, Barton ACT, Australia (2001).

Ghaffarzadeh, M., Prechac, F.G., and Cruse, R.M., Grain yield response of corn, soybean, and oat grown in a strip intercropping system, *Am. J. Altern. Agric.*, **9**, 171–177 (1994).

Gomez, A.A. and Gomez, K.A., *Multiple Cropping in the Humid Tropics of Asia*, International Development Research Centre, Ottawa, Canada (1983).

Hu, S.N. et al., Study on wheat manganese deficiency with manganese fertilizer experiment on calcareous soils in Wenjiang county, *Turang*, **13**, 100–103 (1981), (in Chinese).

Lal, R. et al., Eds., *Sustainable Agriculture and the International Rice–Wheat System*, Marcel Dekker, New York (2004).

Li, Q.K., Ed., *Paddy Soils in China*, Chinese Scientific Press, Beijing (1992), (in Chinese).

Li, L. et al., Wheat/maize or soybean strip intercropping. II. Recovery or compensation of maize and soybean after wheat harvesting, *Field Crops Res.*, **71**, 173–181 (2001a).

Li, L. et al., Wheat/maize or wheat/soybean strip intercropping. I. Yield advantage and interspecific interactions on nutrients, *Field Crops Res.*, **71**, 123–137 (2001b).

Li, L. et al., Chickpea facilitates phosphorous uptake by intercropped wheat from an organic phosphorus source, *Plant Soil*, **248**, 297–303 (2003b).

Li, L. et al., Interspecific complementary and competitive interaction between intercropped maize and faba bean, *Plant Soil*, **212**, 105–114 (1999).

Li, L. et al., Interspecific facilitation of nutrient uptakes by intercropped maize and faba bean, *Nutr. Cycl. Agroecosys.*, **65**, 61–67 (2003a).

Liang, Y.C. et al., An overview of rice cultivation on plastic film mulched dryland, In: *Studies on Plant Nutrition: Progress and Overview*, China Agricultural University Press, Beijing, 114–127 (2000), (in Chinese).

Lin, S. et al., A successful new approach to save water and increase nitrogen fertilizer efficiency?, In: *Water-Wise Rice Production*, Bouman, B.A.M. et al., Eds., International Rice Research Institute, Los Baños, Philippines, 187–195 (2002).

Liu, X.H., *Tillage Science*, China Agricultural University, Beijing (1994), (in Chinese).

Liu, X.J., Mechanisms and prevention strategies of manganese deficiency in wheat on soils under rice-wheat rotation in Sichuan province, Ph.D. thesis, China Agricultural University, Beijing (1997), (in Chinese).

Liu, X.J. et al., Crop production, nitrogen recovery and water use efficiency in rice–wheat rotation as affected by non-flooded mulching cultivation (NFMC), *Nutr. Cycl. Agroecosys.*, **71**, 281–299 (2005).

Liu, X.J. et al., Effect of water and fertilization on movement of manganese in soils and on its uptake by rice, *Acta Pedologica Sinica*, **36**, 369–375 (1999a), (in Chinese).

Liu, X.J. et al., Effect of Mn application depth on Mn deficiency of two wheat genotypes, *Chin. J. Appl. Ecol.*, **10**, 179–182 (1999b), (in Chinese).

Liu, X.J. et al., Effects of non-flooded mulching cultivation on crop yield, nutrient uptake and nutrient balance in rice–wheat cropping systems, *Field Crops Res.*, **83**, 297–311 (2003).

Liu, X.J. et al., Effects of Mn application periods on Mn nutrition and grain yield of different wheat genotypes, *Southwest. China J. Agric. Sci.*, **14**, 39–43 (2001), (in Chinese).

Lu, S.H., Soil manganese and application of manganese fertilizer under condition of rice-wheat rotation, *J. Sichuan Agric. Univ.*, **10**, 75–79 (1992), (in Chinese).

Lu, S.H. et al., Effect of manganese spatial distribution in soil profile on wheat growth in rice–wheat rotation, *Plant Soil*, **261**, 39–46 (2004).

Lu, S.H., Xu, Y.X., and Hu, S.N., Features of manganese of paddy soil conditions of manganese deficiency on wheat, *Southwest. China J. Agric. Sci.*, **3**, 87–91 (1990), (in Chinese).

Lu, S.H. et al., Effects of root penetration restriction on growth and Mn nutrition of different winter wheat genotypes in paddy soils, *Agric. Sci. China*, **1**, 667–673 (2002).

Lu, S.H. and Zhang, F.S., Review and outlook of ten years research on crop Mn nutrition in soils with upland-paddy rotation, *Soil Bull.*, **12**, 1–7 (1997), (in Chinese).

Ma, G.R., Shi, W.Y., and Yin, X.X., Investigation of manganese deficiency symposium in wheat, *Zhejiang Agric. Sci.*, **2**, 85–87 (1985), (in Chinese).

Marschner, H., *Mineral Nutrition of Higher Plants*, Academic Press, London (1995).

Nayyar, V.K., Arora, C.L., and Kataki, P.K., Management of soil micronutrient deficiencies in the rice–wheat cropping system, In: *The Rice–Wheat Cropping System of South Asia: Efficient Production Management*, Kataki, P.K, Ed., Food Products Press, New York, 87–131 (2001).

Nayyar, V.K., Sadana, U.S., and Takkar, T.N., Methods and rates of application of Mn and its critical levels for wheat following rice on coarse textured soils, *Fertil. Res.*, **8**, 173–178 (1985).

Peng, S. et al., A new rice cultivation technology: Plastic film mulching, *Int. Rice Res. Notes*, **24**, 9–10 (1999).

Rao, I.M. et al., Adaptive attributes of tropical forage species to acid soils. V. Differences in phosphorus acquisition from less available inorganic and organic sources of phosphate, *J. Plant Nutr.*, **22**, 1175–1196 (1999).

Richards, I.R., Wallace, P.A., and Paulson, G.A., Effects of applied nitrogen on soil nitrate–nitrogen content after harvest of winter barley, *Fertil. Res.*, **45**, 61–67 (1996).

Stuelpnagel, R., Intercropping of faba beans (*Vicia faba* L.) with oats or spring wheat, In: *Proceedings of the International Crop Science Congress, July, 1992*. Iowa State University, Ames, Iowa, 14–44 (1993).

Takkar, P.N. and Nayyar, V.K., Preliminary field observation of manganese deficiency in wheat and berseem, *Fertil. News*, **26**(33), 22–23 (1981).

Timsina, J. and Connor, D.J., Productivity and management of rice–wheat cropping systems: Issues and challenges, *Field Crops Res.*, **69**, 93–132 (2001).

Tong, P.Y., Achievements and perspectives of tillage and cropping systems in China, *Cropping Syst. Cultivation*, **77**, 1–5 (1994), (in Chinese).

Van der Meer, J., *The Ecology of Intercropping*, Cambridge University Press, New York (1989).

Wang, J.C., Crop yields and nutrient dynamics impacted by different mulching styles in upland rice/wheat rotation systems, Ph.D. dissertation, China Agricultural University, Beijing (2001), (in Chinese).

West, T.D. and Griffith, D.R., Effect of strip-intercropping of corn and soybean on yield and profitability, *J. Prod. Agric.*, **5**, 107–110 (1992).

Willey, R.W. and Rao, M.R., A competitive ratio for quantifying competition between intercrops, *Exp. Agric.*, **16**, 117–125 (1980).

Yang, Z., Soil manganese deficiency and wheat color-leaf and death of seedling, *Yunnan Agric. Sci. Technol.*, **6**, 23–25 (1984), (in Chinese).

Zhang, F.S. and Li, L., Using competitive and facilitative interactions in cropping systems enhances crop productivity and nutrient-use efficiency, *Plant Soil*, **248**, 305–312 (2003).

Zhang, F.S., Li, L., and Sun, J.H., Contribution of above- and belowground interactions to intercropping, In: *Plant Nutrition: Food Security and Sustainability of Agro-Ecosystems*, Horst, W.J. et al., Eds., Kluwer, Dordrecht, Netherlands, 979–980 (2001).

Zhang, F. and Shen, J., Progress in plant nutrition and rhizosphere research, In: *Research Progress in Plant Protection and Plant Nutrition*, China Agronomy Society, China Agriculture Press, Beijing, 458–469 (1999).

Zhang, F., Shen, J., Li, L., and Liu, X., An overview of rhizosphere processes under major cropping systems in China, *Plant Soil*, **260**, 89–99 (2004).

Zhao, Q. and Xiao, M., Upland rice cultivation technique under plastic film mulching in Northeast region of Japan, *J. Niaoning Agric. Sci.*, **4**, 52–56 (1982), (in Chinese).

Zhou, X.M. et al., Corn yield and fertilizer N recovery in water-table-controlled corn-rye-grass, *Eur. J. Agron.*, **12**, 83–92 (2000).

Zuo, Y.M. et al., Studies on the improvement in iron nutrition of peanut by intercropping with maize on a calcareous soil, *Plant Soil*, **220**, 13–25 (2000).

40

Managing Polycropping to Enhance Soil System Productivity: A Case Study from Africa

Zeyaur Khan,[1] Ahmed Hassanali[1] and John Pickett[2]
[1]*International Centre of Insect Physiology and Ecology, Nairobi, Kenya*
[2]*Biological Chemistry Division, Rothamsted Research, Harpenden, Herts, UK*

CONTENTS

Polycropping is the concurrent use of multiple crops that are beneficial to each other, with the output of each crop becoming the input of another, creating a balance in the soil and environment (Geno and Geno, 2001). Although Geno (1976) considered polycropping as a term referring to mixed farming, Kass (1978) proposed it as the best term to distinguish all multiple-cropping situations from single cropping, i.e., monoculture, denoting an area that is being used for more than one crop at a time.

Traditional farmers all over the world have long favored biodiversity as a way to maintain long-term agricultural productivity. Many indigenous cultures in Mexico and Central America, for example, have traditionally planted their staple crops of maize, beans, and squash all together rather than in separate fields (Tuxill, 2000). This polycropping approach has multiple advantages. The beans fix organic nitrogen, thereby enhancing soil fertility and improving maize growth. The maize plants in turn provide trellises for the bean vines, and the squash plants with their wide, shady leaves help suppress weeds. Polyculture is also a time-honored form of pest control (Finckh and Karpenstein-Machan, 2002; Khan, 2002; Karpenstein-Machan and Finckh, 2002; Thomas, 2002; Mensah and Sequeria, 2004).

Planting different crops together has tended to create more ecological niches for beneficial organisms, such as parasitic wasps or predators, which attack and contain pests (Midega and Khan, 2003). Of course, more diverse plantings may also offer more niches for pests and diseases too, but the likelihood of any one organism breaking out in epidemic levels is greatly reduced, since none are likely to affect all crops equally.

During the 1950s and 1960s, the agricultural Green Revolution brought extremely uniform, high-input varieties to the developing world. In contrast to landraces, these crops do not generally perform well without substantial doses of artificial fertilizer, pesticides, and water. In many areas, grain production increased sharply, but at a substantial cost. The old polycultural landscape yielded to monoculture, where the new regime usually produced only one commodity, instead of a range of foods, medicines, and other plant products. In the staple crops of Asia, Latin America, and Africa, pest and disease problems became more acute, and new pests and diseases emerged swiftly. Moreover, monocropping depletes soils more rapidly, and excessive use of fertilizers, pesticides, and herbicides can cause toxic accumulations in soil.

Therefore, for enhancing soil system productivity for resource-poor farmers of Africa, Asia, and Latin America, who can not afford high-input monoculture agriculture, there is need to develop appropriate, more productive polycropping systems, one such polycropping strategy, for the management of cereal stemborers and the Striga weed in maize-based farming systems has been developed for small- to medium-scale farmers in Africa who practice mixed agriculture. The strategy involving "push–pull" and allelopathic tactics serves as a model for income generation and poverty alleviation through polycropping, while conserving soil system capabilities and biodiversity (Khan and Pickett, 2004).

40.1 Management of Cereal Stemborers and Striga Weed through a Novel Polycropping Strategy

Maize and sorghum are the principal food and cash crops for millions of the poorest people in eastern Africa. Stemborers — *Chilo partellus* (Swinhoe) (Lepidoptera: Crambidae) and *Busseola fusca* Fuller (Lepidoptera: Noctuidae) — and the parasitic Striga weed (either *Striga hermonthica* or *S. asiatica*) are major biotic constraints to increased cereal production in the region. Stemborers can cause yield losses of 20–40% of potential output and are difficult to control largely because of the cryptic and nocturnal habits of the adult moths and the protection provided by the stem of the host crop during immature stages (Ampofo, 1986; Seshu Reddy and Sum, 1992). The main method of stemborer control which is currently recommended to farmers by the Ministry of Agriculture is use of chemical pesticides. However, chemical control of stemborers is costly and impractical for many resource-poor, small-scale farmers.

The parasitic weed Striga threatens the well-being of over 100 million people in Africa as it is already infesting 40% of arable land in the savanna region, causing estimated annual losses of $7–13 billion (M'Boob, 1989; Musselman et al., 1991; Lagoke et al., 1991). Infestation by Striga has resulted in the abandonment of much arable land by farmers in Africa because it impairs cereal yields so much that attempts at cereal production become quite unprofitable.

The problem is more widespread and serious in areas where both soil fertility and rainfall are low. Most of the time it falls to African women to try to weed out Striga, which is a time-consuming and labor-intensive activity. Recommended control methods to reduce Striga infestation include heavy applications of nitrogen fertilizer, crop rotation, use of trap crops, hoeing and hand pulling, herbicide application, and the use of resistant or tolerant crop varieties.

An ingenious method is to apply chemicals to the soil that stimulate early seed germination, which is suicidal for the weed because it needs crop roots as a host for its survival. All these methods, including the most widely practiced hoe weeding, are seriously limited by the reluctance of farmers to utilize them widely or actively, for both biological and socioeconomic reasons (Lagoke et al., 1991).

The recent technology for controlling Striga through switching to Impazyr, a herbicide-resistant mutant maize referred to as IR maize, is still under field testing in four eastern African counties (Kanampiu et al., 2003). There are reports of significant increases in maize yields from the field trials. However, the success of IR maize will depend on how widely it can and will be adopted by resource-poor farmers in Striga-infested areas. Few of the small- and medium-scale farmers in these areas purchase certified seeds every year; most plant their own maize seeds because of frequent crop failures due to drought. Another problem will be that this IR maize does not address the problem of stemborers, which is also a major constraint in cereal farming.

Finding a way to reduce the losses caused by both stemborers and Striga through improved management strategies could significantly increase cereal production and result in better nutrition and purchasing power for many maize and sorghum producers. Several national and international agricultural research centers are searching for ways to improve stemborer and Striga management, but no single method of control had so far provided a solution to both the stemborer and Striga problems.

Efforts by the International Centre for Insect Physiology and Ecology (ICIPE) to control both cereal stemborers and Striga weed in cereal crops (Khan et al., 1997a, 1997b, 2000) have led to a polycropping strategy for maize-based farming systems (www.push-pull. net) that combines *in situ* suppression and elimination of Striga by crop and soil management (Khan et al., 2002). This "push–pull" tactic to control stemborers can be described in technical terms as a stimulodeterrent diversionary method of control (Khan et al., 2000, 2001). The "push" in this situation comes from the use of repellent intercrops to drive stemborers away, while the "pull" comes from trapping them on highly susceptible trap plants (Figure 40.1).

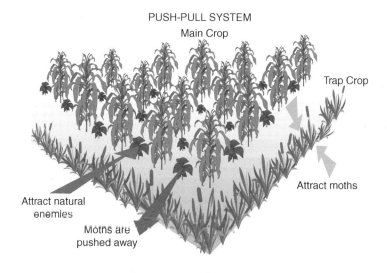

FIGURE 40.1

A diagrammatic presentation of the "push–pull" strategy for insect pest management. (Courtesy of Prof. Johnnie van den Berg, North West University, South Africa.)

A number of plants that attract or repel stemborers and/or inhibit Striga have been identified, e.g., Napier grass (*Pennisetum purpureum*), Sudan grass (*Sorghum vulgare sudanense*), molasses grass (*Melinis minutiflora*), and Silverleaf Desmodium (*Desmodium uncinatum*), a legume that has forage uses. Napier grass and Sudan grass, in particular, have shown potential for use as trap plants, whereas molasses grass and Silverleaf Desmodium repel ovipositing stemborers. Molasses grass, when intercropped with maize, not only reduces infestation of the maize by stemborers, but also increases their being parasitized by the wasp *Cotesia sesamiae*, a natural enemy (Khan et al., 1997b). In addition, Desmodium plants, when intercropped with maize, inhibit Striga.

All four plants are of economic importance to farmers in eastern Africa as livestock fodder and were found to be quite acceptable in initial participatory trials with small-scale farmers needing stemborer and Striga management in Kenya (Table 40.1). During the last 7 years, continuing on-farm trials with more than 2000 farmers in Kenya and Uganda have confirmed that these technologies are effective and have significant impacts on food security and income generation for resource-poor maize farmers.

The polycropping strategy based on the push–pull and Striga suppression–elimination tactics, conducted in 10 districts in Kenya and 3 districts in Uganda, has helped participating farmers to increase their maize yields by an average of 20% in areas where only stemborers are present, and by >50% in areas where both stemborers and Striga are problems (Table 40.2).

TABLE 40.1

Benefits of Push–Pull Strategy as Perceived by Farmers in Seven Districts of Western Kenya, 2002

Benefits	Percent of Farmers Responding from Various Districts in Kenya						
	Trans-Nzoia (*n* = 96)	Suba (*n* = 65)	Rachu-onyo (*n* = 58)	Kisii (*n* = 54)	Vihiga (*n* = 63)	Bungoma (*n* = 70)	Busia (*n* = 82)
Increased maize yield	85	90	88	78	90	86	85
Reduced stemborer damage	80	86	70	67	74	85	80
Reduced Striga infestation	[a]	98	92	83	88	95	95
Increased milk production	73	67	45	63	65	70	60
Reduced use of pesticides to control stemborers	88	[b]	[b]	67	74	63	74
Income generation from sale of forage grass	45	60	79	73	86	74	38
Income generation from sale of forage seed	50	25	48	36	[c]	84	63
Reduced soil erosion	53	47	48	62	60	83	56
Controlled other weeds/ less labor for weeding	38	88	53	41	38	60	64

[a] No Striga present in Trans-Nzoia.

[b] Farmers in Suba and Rachuonyo districts do not use pesticides.

[c] First year of field trials in Vihiga district.

TABLE 40.2

Comparison of Stemborer and Striga Infestation, and Yield of Maize Crop in Push–Pull and Control Maize Fields in 10 Districts of Kenya and 3 Districts of Uganda, 2003

Country	District	No. of Push–Pull Farmers	Stemborer Infestation (%)		Striga Infestation (Striga plants/100 maize plants)		Yield (t ha^{-1})	
			Push–Pull	Control	Push–Pull	Control	Push–Pull	Control
Kenya	Trans-Nzoia	550	7.1	17.2[*]	[a]	[a]	5.5	4.2[*]
	Suba	375	6.1	34.9[**]	12.8	296.0[**]	3.4	1.3[**]
	Kisii	130	7.4	20.0[**]	20.3	166.5[**]	3.8	2.4[*]
	Rachu-onyo	120	5.1	11.8[*]	20.4	257.5[**]	3.3	1.8[*]
	Bungoma	150	12.8	27.7[*]	7.2	368.2[**]	3.8	1.6[*]
	Busia	130	14.0	35.7[*]	11.4	421.3[**]	3.9	2.8[*]
	Vihiga	50	15.7	29.6[*]	356.1	1,076.4[**]	4.1	1.8[**]
	Migori	10	6.2	10.6[ns]	318.0	521.0[*]	4.3	3.2[*]
	Homa Bay	10	6.3	17.7[*]	141.0	431.1[**]	3.4	2.3[*]
	Siaya	10	16.9	32.8[*]	257.3	575.7[*]	3.7	1.8[**]
Uganda	Kap-chorwa	45	12.1	29.4[**]	[a]	[a]	3.9	2.2[*]
	Bugiri	46	26.5	42.5[*]	2.1[b]	3.8[b,d]	2.6	1.2[**]
	Tororo	43	15.4	22.2[ns]	2.0[b]	3.0[b,d]	2.0[*]	1.6[ns, c]

[a] No Striga infestation in Trans-Nzoia (Kenya) and Kapchorwa (Uganda).

[b] In Uganda, Striga infestation was rated on a 1–5 scale.

[c] The yield difference was not expressed due to drought in this district.

[*] Difference significant ($p < 0.05$).

[**] Difference significant ($p < 0.01$).

40.2 How the Push–Pull Polycropping Strategy Works

The integrated polycropping strategy undertakes a holistic approach to understanding and exploiting chemical ecology and agrobiodiversity (Khan et al., 2001; Khan and Pickett, 2004). The plant chemistry responsible for stemborer control involves release of attractant semiochemicals from the trap plants and repellent semiochemicals from the intercrops (Khan et al., 2000). With molasses grass, certain chemicals that are repellent to ovipositing adults also increased parasitism of stemborers (Khan et al., 1997b). The Striga control tactic is based on the use of Desmodium as an intercrop that acts through a combination of mechanisms, including the induction of premature germination of Striga seeds so that the young plants are aborted because they fail to develop and attach onto a maize host's roots (Khan et al., 2002; Tsanuo et al., 2003).

To understand the chemical ecology of the push–pull system, volatile chemicals from trap and repellent plant have been investigated by gas chromatography coupled–electroantennography (GC–EAG) on the antennae of stemborers (Khan et al., 2000). GC peaks consistently associated with EAG activity were tentatively identified by GC coupled–mass spectrometry (GC–MS), and identity was confirmed using authentic samples. Six active compounds in trap plants were identified: octanal, nonanal, naphthalene, 4-allylanisole, eugenol, and linalool. Behavioral tests, employing oviposition on to an artificial substrate treated with individual compounds, demonstrated positive activity for all these compounds. Coupled GC–EAG with volatiles from molasses grass showed a wide range of peaks associated with EAG activity (Figure 40.2).

FIGURE 40.2

A "push–pull" polycropping strategy for controlling stemborers and Striga weed. Desmodium plants, intercropped with maize, repel stemborers away from the maize crop as well as inhibit Striga weed. Napier grass planted on the border around maize acts a trap plant for stemborers as they lay more eggs on Napier grass compared to maize, even though their rate of larval survival on Napier grass is very poor.

A general hypothesis developed during our work on insect pests is that nonhost plants are recognized by colonizing insects through the release of repellent or masking semiochemicals, although it is almost inevitable that compounds also produced by hosts will be present. In this case, the host cereal plants and the nonhost molasses grass would be expected to have a number of volatiles in common as they are both members of the Poaceae family. For molasses grass, five new peaks with EAG activity were identified, in addition to the attractant compounds and others normally produced by members of the Poaceae (Kimani et al., 2000; Khan et al., 2000). These were: (E)-β-ocimene, α-terpinolene, β-caryophyllene, humulene, and (E)-4,8-dimethyl-1,3,7-nonatriene. Ocimene and non-atriene had already been encountered as semiochemicals produced during damage to plants by herbivorous insects (Turlings et al., 1990, 1995).

It was likely that these compounds, being associated with a high level of stemborer colonization and, in some circumstances, acting as foraging cues for parasitoids, would be repellent to ovipositing stemborers. This was subsequently demonstrated in behavioral tests. Investigating legume volatiles, it was shown that Desmodium also produced ocimene and nonatriene, together with large amounts of other sesquiterpenes, including α-cedrene (Khan et al., 2000).

A clear allelopathic mechanism involving Striga suppression by Desmodium was subsequently demonstrated (Khan et al., 2002). This was confirmed by a dramatic reduction in *S. hermonthica* infestation when irrigation water passing through the roots of Desmodium was introduced into pots of maize growing in soil seeded with high levels of *S. hermonthica* (Figure 40.3). Growth of the parasitic weed could be almost completely suppressed, whereas extensive infestation occurred in the control where its water supply was not affected by Desmodium. The allelopathic mechanism was found to involve a chemical inhibition of the development of projections on the hyphae of *S. hermonthica* that enable the fungus to feed on host plant roots.

Work is ongoing to identify the compounds, released from Desmodium roots, that are involved in suppression of the parasite. Three new isoflavanones and a previously known

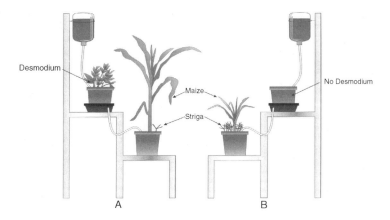

FIGURE 40.3

Diagram of an experiment to investigate the allelochemical mechanism of *Desmodium uncinatum* in suppressing *Striga hermonthica* infestation in maize. Comparison was made between maize plants irrigated by root elutes of *D. uncinatum* (A) with those irrigated by water passing through pots containing only autoclaved soil (B) (from Khan et al., 2002).

isoflavanone (genistein) have so far been isolated and characterized spectroscopically from Desmodium root exudates. One of the new isoflavanones (Uncinanone-B) is a Striga germination stimulant, and another (Uncinanone-C) moderately inhibited radical growth of the germinated Striga seeds. Uncinanone-A has no apparent role as a germination stimulant or as a postgermination growth inhibitor (Tsanuo et al., 2003). Further isolation and assays of other constituents are expected to clarify the precise biochemical traits associated with the effects of Desmodium on Striga and to assess their potential for other biocontrol. The sophisticated mode of action demonstrated here, when fully elucidated, may give more exploitable leads that are beneficial not only in subsistence agriculture, but also supportive for agricultural production in general.

40.3 Benefits of Push–Pull Polycropping System

The "push–pull" polycropping program is a good example of how basic science can be linked with technology transfer, with farmer uptake leading to spontaneous technology transfer between farmers. Because of its multiple benefits shown in Figure 40.4, the program has drawn together and motivated scientists and extension workers from diverse disciplines. The substantial increase in maize yields accompanied by many other beneficial features of this management system have contributed greatly to the high farmer adoption rates.

40.3.1 Food Security

The contribution that push–pull strategies can make to food security is very great. Intercropping or mixed cropping of maize, grasses, and fodder legumes, by increasing crop yields, has enabled farmers in Kenya to improve their families' food security and nutrition. The Striga and stemborer menaces have seriously and negatively affected sorghum farming in arid and semi-arid areas in many parts of eastern, central, and southern Africa, to the extent that many farmers in those areas have not been able to produce adequate food for themselves.

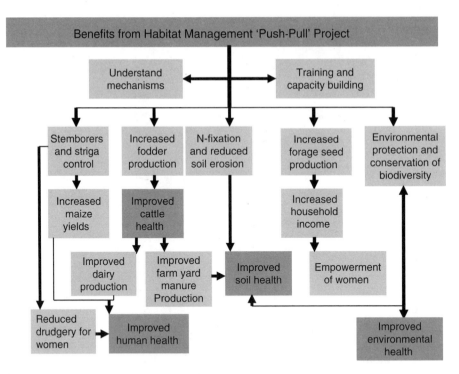

FIGURE 40.4
Benefits from habitat management through "push-pull" strategy.

40.3.2 Dairy and Livestock Production

The push–pull and Striga-suppression/elimination tactics have contributed significantly to increased production of milk and meat by providing more fodder and different crop residues, especially on small farms where competition for land is quite high. For example, the Suba district of Kenya, a milk-deficit region on the shores of Lake Victoria which has mostly indigenous (zebu) livestock, produces only 7 million liters of milk, far short of the estimated annual demand of 13 million liters. A major constraint to keeping improved dairy cattle for milk production has been the unstable availability and seasonality of feed, often of low quality.

A habitat management strategy based on the new polycropping system, integrating crop and fodder production, which has been adopted by more than 350 farmers in this district. This promotes livestock production and improved milk supply now that more and higher-quality feed is available for cattle. The number of improved dairy cattle in the district has increased from four in 1997, when on-farm trials were initiated, to more >500 in June 2004, contributing to a 1 million liter increase in milk production.

40.3.3 Soil Conservation and Fertility

Soil erosion and low fertility are very common problems in eastern Africa. The habitat management strategy just mentioned has become more multifunctional. For example, the cultivation of Napier grass which had been promoted for livestock fodder and soil conservation now has added benefit as a trap plant for stemborer management. Similarly, Desmodium was already being grown for improving soil fertility because it is a legume

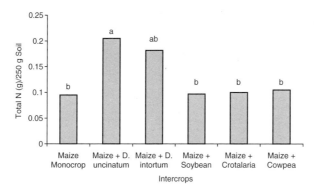

FIGURE 40.5
Total nitrogen fixed in soil with different intercrops (means followed by the same letters are not significantly different).

that is very effective in N-fixation (Figure 40.5) and also for high-quality fodder. Now it is valued also as an effective Striga suppressant.

40.3.4   Enhancing Biodiversity

This innovation is contributing to the promotion and conservation of biodiversity because the habitat management approach embodies maintenance of species diversity. A recent study with the University of Haifa in Israel has demonstrated that the numbers of beneficial soil arthropods in maize-Desmodium fields were significantly higher than the numbers in fields of monocropped maize (Table 40.3). The destruction of biodiversity is linked to the expansion of crop monoculture at the expense of diverse vegetation. Similarly, in a field trial conducted in 2001, both diversity and population of natural enemies on maize in a push–pull system were significantly more than on maize monocrop (Figure 40.6). This biodiversity has direct agronomic and economic benefits.

40.3.5   Protecting Fragile Environments

Existing evidence indicates that higher crop yields and improved livestock production, resulting from habitat management strategies, has the potential of supporting many rural households under existing socioeconomic and agroecological conditions. This can reduce pressure for human migration into environments needing and designated for protection. Moreover, farmers using such strategies have less reason to use synthetic pesticides that affect many flora and fauna.

TABLE 40.3

Soil Microarthropods in Push–Pull Experiment and in Maize Monocrop

Treatment	Collembola	All Acari	Microarthropods
Maize monocrop	8.2	22.5	37.1
Maize and Desmodium	52.5*	104.1*	170.5*

* Difference significant ($p < 0.01$).

Source: Authors' data in collaboration with Professor M. Broza, University of Haifa, Israel.

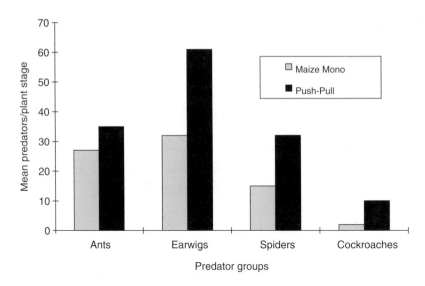

FIGURE 40.6
Comparison of stemborer predators on maize plants in push–pull and monocrop-control fields, Lambwe, Kenya (means followed by different letters are not significantly different ($p < 0.05$)).

40.3.6 Income Generation and Gender Empowerment

The push–pull strategies have shown promise of not only significantly enhancing farm incomes but also of empowering rural women, who can earn more from the sale of farm grain surpluses, fodder, and Desmodium seed. In addition, improved rural productivity and quality of life is expected to make rural life more attractive for youth and to help stem rural–urban migration.

40.4 The Economics of Polycropping Strategy

The economic gains from polycropping generally depend on the balance between lowered costs of controlling insect pests and weeds and the possibly higher costs of maintaining polycropped fields, along with any decreases (or increases) in yield of the main crop due to greater plant competition. As seen in the preceding chapter, having a variety of plants with complementary rooting systems makes net gains possible from polycropping. In East Africa, water is often a limiting factor, so interplant competition can result in lower yield from one or more plant species.

Net profit can be increased if companion plants favorably change the balance between farmers' income and costs. Economic data assessing financial returns, not just agronomic data on biological effects, are needed for making decisions on the use of polycropping for insect pest and weed control. Benefit:cost ratios that compare total measurable economic benefits divided by total expenditures on production assess the economic efficiency of a technology. If the benefits are no greater than costs, a practice is economically marginal, though possibly consideration of some uncounted benefits (or costs) would tilt decisions in favor of (or against) the practice in question. Where costs exceed benefits, there are incentives to reject it.

Commonly, benefit:cost ratios $>1.2:1$ are considered worthwhile, as there are usually some costs of adoption or making a change. Farmers need also to consider risk factors, whether the net benefit is assured or only intermittent. Because they cannot afford to have net losses in any season, a favorable benefit:cost ratio that is uncertain is unlikely to be attractive. In East Africa and in dryland areas generally (Chapter 25), farmers face high climatic risks as well as market failures and distortions in output and input markets.

Benefit:cost ratios for the push–pull crop management strategy were analyzed for several years in two districts of Kenya. The new push–pull technology has consistently posted benefit:cost ratios greater than 2, indicating that the technology is very beneficial, rather reliably giving farmers returns for their investments in both Trans-Nzoia and Suba districts in Kenya. As important, though less quantifiable is the reduction in risk that this cropping strategy provides as the build-up of more fertile, water-retaining soil systems should give added protection to the crop in periods of water shortage and stress.

40.5 Discussion

The polycropping approach described here is now expanding into various parts of Kenya, Uganda, and Tanzania for stemborer and Striga control. A pilot program has been initiated in southern Africa, addressing stemborer control and soil erosion in the arid and semi-arid areas of the Limpopo province of South Africa. However, each region has, in addition to varying climatic conditions and use of alternative cultivars, some differences in crops that must be taken into account. Whereas maize is the main crop in the farming systems in Kenya and Uganda, sorghum, pearl millet, and maize are planted in southern Africa.

Pest management options in this region are affected by low rainfall, the limited extent to which cattle are kept, and the fact that the cattle are largely free-grazing. However, wherever these approaches are adapted to the specific needs of local farming practices and communities, it is essential that the scientific basis for the modified systems should be clarified and explained by appropriate research. Such work lays the foundation for still wider application of these principles and serves as a model for the management of other pests in Africa and beyond. Issues of pest and disease control within a more general context of biologically-framed cropping systems are considered in the following chapter, introducing Part IV of this book.

References

Ampofo, J.K.O., Maize stalk borer (Lepidoptera: Pyralidae) damage and plant resistance, *Environ. Entomol.*, **15**, 1124–1129 (1986).

Finckh, M.R. and Karpenstein-Machan, M., Intercropping for pest management, In: *Encyclopedia of Pest Management*, Pimentel, D., Ed., Marcel Dekker, New York, 423–425 (2002).

Geno, L., Ecological agriculture: The environmentally appropriate alternative, *Alternatives*, **5**, 53–63 (1976).

Geno, L. and Geno, B., *Polyculture Production: Principles, Benefits and Risks of Multiple Cropping Land Management Systems for Australia*, RIRDC Publication No. 01/34. Rural Industries Research and Development Corporation, Barton, ACT, Australia (2001).

Kanampiu, F., Friessen, D., and Gressel, D., A new approach to Striga control, *Pestic. Outlook*, **14**, 51–53 (2003).

Karpenstein-Machan, M. and Finckh, M., Crop diversity for pest management, In: *Encyclopedia of Pest Management*, Pimentel, D., Ed., Marcel Dekker, New York, 162–165 (2002).

Kass, D.L., *Polyculture Cropping Systems: Review and Analysis*, Cornell International Agriculture Bulletin 32. New York State College of Agriculture and Life Sciences, Cornell University, Ithaca, New York (1978).

Khan, Z.R., Cover crops, In: *Encyclopedia of Pest Management*, Pimentel, D., Ed., Marcel Dekker, New York, 155–158 (2002).

Khan, Z.R. and Pickett, J.A., The 'push–pull' strategy for stemborer management: A case study in exploiting biodiversity and chemical ecology, In: *Ecological Engineering for Pest Management: Advances in Habitat Manipulations for Arthropods*, Gurr, G., Waratten, S.D., and Altieri, M.A., Eds., CSIRO and CABI Publishing, Wallingford, UK, 155–164 (2004).

Khan, Z.R. et al., Utilisation of wild gramineous plants for the management of cereal stemborers in Africa, *Insect Sci. Appl.*, **17**, 143–150 (1997a).

Khan, Z.R. et al., Intercropping increases parasitism of pests, *Nature*, **388**, 631–632 (1997b).

Khan, Z.R. et al., Exploiting chemical ecology and species diversity: Stem borer and Striga control for maize and sorghum in Africa, *Pest Manage. Sci.*, **56**, 957–962 (2000).

Khan, Z.R. et al., Habitat management for the control of cereal stem borers in maize in Kenya, *Insect Sci. Appl.*, **21**, 375–380 (2001).

Khan, Z.R. et al., Control of the witchweed, *Striga hermonthica*, by intercropping with *Desmodium* spp., and the mechanism defined as allelopathic, *J. Chem. Ecol.*, **28**, 1871–1885 (2002).

Kimani, S.M., et al., Airborne volatiles from *Melinis minutiflora* P. Beauv., a non-host plant of the spotted stem borer, *J. Essent. Oil Res.*, **12**, 221–224 (2000).

Lagoke, S.T.O., Parkinson, V., and Agunbiade, R.M., Parasitic weeds and control methods in Africa, In: *Combating Striga in Africa, Proceedings of International Workshop, 22–24 August 1988*, Kim, S.K., Ed., International Institute for Tropical Agriculture, Ibadan, Nigeria, 3–14 (1991).

M'Boob, S.S., A regional programme for West and Central Africa, In: *Striga: Improved Management in Africa, Proceedings of FAO/OAU All-African Government Consultation on Striga Control, 20–24 October 1988, Maroua, Cameroon*, Tobson, T.O. and Broad, H.R., Eds., 190–194 (1989).

Mensah, R.K. and Sequeria, R.V., Habitat manipulation for pest management in cotton cropping systems, In: *Ecological Engineering for Pest Management: Advances in Habitat Manipulations for Arthropods*, Gurr, G., Waratten, S.D., and Altieri, M.A., Eds., CSIRO and CABI Publishing, Wallingford, UK, 155–164 (2004).

Midega, C. and Khan, Z.R., Habitat management system and its impact on diversity and abundance of maize stemborer predators in western Kenya, *Insect Sci. Appl.*, **23**, 301–308 (2003).

Musselman, L.J. et al., Recent research on biology of *Striga asiatica*, *S. gesnerioides* and *S. hermonthica*, In: *Combating Striga in Africa: Proceedings, International Workshop, 22–24 August 1988*, Kim, S.K., Ed., International Institute for Tropical Agriculture, Ibadan, Nigeria, 31–41 (1991).

Seshu Reddy, K.V. and Sum, K.O.S., Yield-infestation relationship and determination of economic injury level of the stem borer, *Chilo partellus* (Swinhoe), in three varieties of maize, *Zea mays* L. *Maydica*, **37**, 371–376 (1992).

Thomas, F., Crop covers for weed suppression, In: *Encyclopedia of Pest Management*, Pimentel, D., Ed., Marcel Dekker, New York, 159–161 (2002).

Tsanuo, M.K. et al., Isoflavanones from the allelopathic aqueous root exudates of *Desmodium uncinatum*, *Phytochemistry*, **64**, 265–273 (2003).

Turlings, T.C.J., Tumlinson, J.H., and Lewis, W.J., Exploitation of herbivore-induced plant odors by host-seeking parasitic wasps, *Science*, **250**, 1251–1253 (1990).

Turlings, T.C.J. et al., How caterpillar-damaged plants protect themselves by attracting parasitic wasps, *Proc. Natl Acad. Sci.*, **92**, 4174 (1995).

Tuxill, J., The biodiversity that people made, *World Watch*, 24–35, May–June (2000).

PART IV: RELATED ISSUES

41

Effects of Soil and Plant Management on Crop Pests and Diseases

Alain Ratnadass,[1] **Roger Michellon,**[1] **Richard Randriamanantsoa**[2] **and Lucien Séguy**[3]

[1]*CIRAD, Antsirabe, Madagascar*
[2]*FOFIFA, Antsirabe, Madagascar*
[3]*CIRAD, Goiânia, Brazil*

CONTENTS

An ecosystem perspective on crop and soil management underscores the importance of understanding the multiple interactions among the various flora and fauna that inhabit soil systems and affect the success of any particular cropping system. By definition, pests and pathogens have impacts on crops and other plants. However, conversely, soil and crop management practices, together with the management of nutrients, water and other plants, have demonstrable effects on the many populations of organisms that have parasitic, toxic, or other impacts of significance to farmers.

In this chapter, we consider evidence on this reciprocal relationship, on how soil and crop management practices that capitalize on certain biological dynamics can have desirable and cost-effective impacts on the control of various floral and faunal bio-aggressors. This is or should be part of what has come to be known as integrated pest management (IPM). As a new paradigm for agriculture, IPM shares many principles with the biologically-based systems presented in this book, especially the direct seeding through permanent soil cover (DSPSC) cropping systems discussed in Chapters 22 and 23.

These systems are location-specific and knowledge-intensive with the pros and cons of any particular practice needing to be considered and balanced to achieve the greatest possible net favorable impact on pests and on production. These strategies require a dynamic perspective, taking into account the temporal dimension because some time is invariably required for new biological equilibria to get established after a change is made in cultural practices.

Both time dimension and location-specificity are central themes of emerging biologically-driven agricultural practices. We will report particularly on experience that is being gained in Madagascar, where CIRAD scientists are working with national and local partners to reduce the negative environmental impacts of shifting cultivation of upland rice. We draw also on the experience of CIRAD researchers in other countries and on scientific literature more generally. Because we have most direct experience with the changes associated with DSPSC innovations presented in Chapter 22, we consider especially what has been learned about the impacts on pests and disease from curtailing tillage and from keeping soil surfaces covered.

An example of the importance of location-specificity is our finding that when DSPSC methods were introduced and evaluated in the highlands of Madagascar, attacks by white grubs and black beetles on upland rice were demonstrably reduced after just a few years of this new management (Michellon et al., 1998, 2001; Ramanantsialonina, 1999). This was very encouraging. However, when the same practices were undertaken at lower elevations, the beneficial effects were not observed (Charpentier et al., 2001). These particular pests remain one of the main obstacles to broader success with rainfed rice production and to the adoption of DSPSC.

To develop diagnostic strategies and interventions for dealing with pests and diseases, one needs to appreciate the complexity of interacting factors, both biotic and abiotic, that shape crop production outcomes. Pest control methods should be conceived and developed concomitantly with other crop and soil management techniques, to achieve the best compromises among effectiveness in pest control, cost-efficiency, and environmental impact, recognizing that what constitute optimal combinations of practices may need to change over time.

41.1   Assessing the Effects of Tillage and Its Cessation

Direct seeding in place of conventional tillage makes significant changes in the environment in which plants grow, particularly in the top layer of soil and on the soil surface by not disturbing them. Cessation of tillage has definite impacts on pests and diseases, although not always the same effect nor always desired ones.

When plowing is done, insect pest species that live or pupate in the soil and/or those that live or shelter in crop debris or in weeds before a new crop is planted may be killed or buried to a depth from where they cannot emerge. Others die of the temporary drought created in the upper soil layers, or they get exposed on the surface where they are desiccated or consumed by predators. Hence, the most immediate effect of reducing tillage is to diminish the level of pest control that is achieved through mechanical means. This is particularly important for some general soil pests such as cutworms, wireworms, and slugs (Leake, 2001). For instance, plowing considerably reduces the larval and pupal populations of white grubs, as seen in the case of Heteronychus spp. (Walker, 1968), and *Hoplochelus marginalis* (Vercambre, 1993).

However, some reverse effects can also be observed, because plowing can have adverse impacts on the larval or pupal stages of important predators of pests, ones that help to keep these pests in check. For instance, parasitoids of the cabbage stem weevil (*Ceutorhynchus pallidactylus*) and pollen beetle (*Meligethes aeneus*) were found to be damaged more severely by stubble cultivation with subsequent plowing than by direct seeding (Wamhoff et al., 1999). So cultivation can reduce the predator pressure on pests, thereby contributing to pest prevalence.

Plowing does not necessarily help control pathogens in the soil either. True, it generally buries pathogenic inoculum present on the soil surface and on the stubble of the previous crop. Pathogens that would otherwise attack crops at the base of their stems are removed from their usual entry point, and the new crop can more readily escape infection by fungi, such as Rhizoctonia or Sclerotium (Davet, 1996). However, such temporary dislocation of sclerotes does not necessarily affect their viability, and the next plowing can bring back to the surface a very active inoculum that infects the next crop. In addition, since many fungi have the capacity of penetrating any part of the root system, plowing can result in a more extensive, subsurface distribution of the inoculum by tillage implements (Davet, 1996). The survival of *Sclerotinia sclerotiorum* has been found to be enhanced when the inoculum was buried deeply by plowing (Wamhoff et al., 1999). Thus, tillage may give only short-term control of this pathogen.

While plowing can give mechanical control of some weed species, it can have the reverse effect for some major plant pests. When the tubers of *Cyperus rotundus* are cut and disseminated by plowing during rainy periods, repeated plowing increases the volume of infested soil, and it contributes to a wider distribution of *Striga asiatica* seeds (Andrianaivo et al., 1998). Compared to DSPSC, plowing also increases distribution of Striga seeds via wind and rain water (Andrianaivo et al., 1998). So, it is not necessarily true that tillage reduces pest and disease incidence. The beneficial effects are often short-lived or even subsequently negated.

41.2 Effects of Mulching

41.2.1 Direct Physical Effects

Crop residue left on the soil surface directly supports survival of certain residue-borne pathogens by providing substrates for their growth and by positioning the pathogens at the soil surface where spore release can occur (Kuprinsky et al., 2002). However, the incidence of foliar diseases can be reduced by having a cover-crop mulch, primarily because this prevents the dispersal of pathogen propagules through rain splashing and/or wind-borne processes (Teasdale et al., 2004). Mulches can also suppress the establishment of soil-inhabiting herbivores, such as Colorado potato beetles, by disrupting their emergence and migration behavior (Teasdale et al., 2004).

There are other documented instances and ways in which mulch reduces crop vulnerability to pest or disease loss. When sorghum is directly seeded after wheat, it is much less prone to attacks by *Fusarium moniliforme*, an opportunistic fungus that is favored by water stress and high temperatures (Doupnik and Boosalis, 1980); both of these conditions are minimized by the mulch component of DSPSC. Zero-tillage with its resultant rice-stubble mulch reduces populations of the leafhopper *Amrarasca bigutulla* and the bean fly *Ophiomyia phaseoli*, because these insects have a strong preference for landing on bare soil. The mulch and stubble left on zero-tillage treatments appear to obstruct long-wavelength radiation that these insect pests rely on (Litsinger and Ruhendi, 1984).

In Brazil, DSPSC methods, planting cotton on dead sorghum mulch, have made it possible to reclaim fields so infested with *C. rotundus* weeds that they could not be controlled with conventional farming procedures (Séguy et al., 1999a). The control mechanisms introduced by DSPSC include the shading and physical obstruction caused by mulch, as well as competition and allelopathy, plus better water and mineral nutrition of crops which makes them better able to compete (see section 41.4). Similarly, in Cameroon, Brachiaria mulch has been observed to have positive effects on the main sorghum crop, notably by lowering soil temperature which results in lower *S. hermonthica* incidence (Naudin, 2002). Net effects of plowing are not always predictable, so an empirical frame of mind is needed, with careful attention to effects on predators and beneficials as well as pests.

41.2.2 Indirect Effects through Increased Fauna Abundance and Diversity

Dead plant cover from previous crops left on top of the soil serves as a refuge for an enormous number of invertebrates. While some can be economically important crop pests, many species are beneficial organisms, including nutrient recyclers, pest predators, and parasitoids (Pruett and Guaman, 2003). Certainly, slugs, crickets, plant bugs, leafhoppers, and spittlebugs may be significant pests of alfalfa seedlings in conservation-tillage systems that depend upon the existing cover of vegetation (Grant et al., 1982; Byers et al., 1983). Also, in the regions of Madagascar around Lake Alaotra and Manakara, dramatic damage by black beetles (Heteronychus spp.) has been observed on rice that is cropped on mulch at the beginning of the season, with a significant impact on yield, with damage correlated with mulch thickness (Charpentier et al., 2001). This is a common effect with many mulch-based systems that must be reckoned with.

On the other hand, results from Queensland, Australia, have indicated that long-term reduced or zero-tillage need not lead to increased problems with soil insect pests. Zero-tillage was seen to have the greatest diversity of macrofauna species, while there was no change in the population density of soil herbivores, particularly the three major agricultural pests for emerging seedlings: earwig (*Nala lividipes*), wireworm (*Agrypnus variabilis*), and false wireworm (*Cestrinus trivialis*) (Wilson-Rummenie et al., 1999). The more diverse and continuous availability of food sources that mulch provided in this case improved the survival and activity of predators so that pests could not predominate.

Many farmer fears that mulch will magnify their pest problems are not well-founded. These effects are specific to locations and crops, as well as to kinds of mulch, necessitating *in situ* evaluation. That mulch and zero-tillage generally produce higher crop yields suggests that their net effect is likely to be positive and that pest problems are not exacerbated.

41.3 Effects of Rotations and Crop Associations

One of the reasons why lower numbers of plant-feeding (phytophagous) insects are often found in complex environments such as polycultures, considered in the preceding two chapters, is because countervailing populations of predators and parasitoids can be larger and more effective in such situations, notably because of the more continuous availability of food sources and favorable microhabitats. For example, attacks on maize by the pink borer (*Sesamia calamistis*) were reduced in Reunion when the maize was undersown with birdsfoot trefoil (*Lotus uliginosus*), and when earthworms were added. The cover crop plus earthworms created conditions allowing the development of soil macrofauna that are antagonistic to the pink borer (Boyer et al., 1999).

In Vietnam, CIRAD researchers and colleagues have documented: (1) a rapid decrease in the biodiversity and density of macrofauna associated with conventional systems of rice monocropping that had bare soil compared to the preceding forest; (2) a rapid raise in the biodiversity and density of macrofauna in previously, degraded soil when it was cultivated with a permanent vegetal cover (Brachiaria) associated to the main crop of peanut (Arachis); and (3) the replacement of ants and termites by earthworms under mulch which were not present when the soil was not kept covered (Husson et al., 2003). Some examples provided below do not directly involve soil management; however, as part of DSPSC or IPM strategies, they avoid or minimize the use of synthetic chemicals for pest control that could have adverse effects on soil organisms.

Physical obstruction and visual camouflage are two explanations that can be offered as to why fewer specialized pest insects are found on host plants that grow in diverse backgrounds compared with similar plants being grown in bare soils (Finch and Collier, 2000). Phytophagous insects are more likely to find and remain on host plants growing in dense, nearly-pure stands, whereas a second plant species in the field disrupts the ability of insects to efficiently attack their intended proper host (Asman et al., 2001).

From an aboveground perspective, the more nonhost plants that are removed from a crop area, the greater is the chance that an insect will find a host plant. Bare-soil cultivation that eliminates all plants but the crop ensures that it becomes exposed to the maximum pest-insect attack possible in that particular locality (Collier et al., 2001). There is evidence indicating that in high-trash situations, apterous aphid vectors are unable to identify their host and consequently their colonization is reduced (A'Brook, 1968). Studies on the influence of crop background on aphids and other phytophagous insects on Brussels sprouts have suggested that the maintenance of some weed cover can be useful in integrated control of certain Brassica pests (Smith, 1976).

Polycultures, as a rule, support lower herbivore loads than do monocultures. One possible reason is that specialized herbivores are more likely to find and to remain on pure crop stands that provide them concentrated resources and monotonous physical conditions (Altieri, 1999). The numbers of pest insects found on crop plants can be reduced considerably when the crop is undersown with a living mulch such as clover (Finch and Collier, 2002). Attacks on geranium (Pelargonium) by the weevil *Cratopus humeralis* were reduced when the crop was undersown with birdsfoot trefoil in Reunion (Quilici et al., 1992; Michellon, 1996; Michellon et al., 1996a). Also, the root system of Kikuyu grass seems to reduce the damage done to geranium roots by the white grub *H. marginalis* (Michellon et al., 1996b). Intercropped plants that draw on the same nutrient pool as the desired crop can compensate for the nutrients taken up by giving protection to the crop against its pests.

On the other hand, it is known that volunteer crop plants and weeds can be hosts and reservoirs for many crop diseases or for their insect vectors (Kuprinsky et al., 2002). Some pests sustain themselves on cover crops that thus serve as hosts and favor the build-up of infestation. In Benin, for instance, the cover plant species *Canavalia ensiformis* and *Mucuna pruriens* were found to be good alternate host species for the maize pest *Mussidia nigrivenella* (Schulthess and Setamou, 1999). So in this situation, use of these particular cover crops was disadvantageous.

In Kenya, as discussed in the preceding chapter, a "push pull" or "stimulo-deterrent diversionary" strategy (Miller and Cowles, 1990) has been able to control stemborers affecting maize. This strategy combines the use of trap and repellent fodder plants, so that stemborers are at the same time repelled from the maize crop and attracted to the trap crop. The semiochemicals that mediate this behavior of the pests and parasitoids have been isolated, so the mechanisms are clearly identified (Khan et al., 1997a, 1997b, 2003).

A number of vegetative covers possess allelopathic potential and release chemicals into the soil that inhibit the germination and growth of certain weeds (Weston, 1996). It is reported from Côte d'Ivoire that maize infestation by Striga when undersown with *Pueraria phaseoloides* and *Calopogonium mucunoides* as live cover crops was drastically reduced. There was also some improvement with the sowing of *Cassia rotundifolia* (Charpentier, 1999).

Diseases can thus be avoided through crop selection and the rotation of crops to include some nonhost crops. This is most effective for pathogens that are soil- or residue-borne (Kuprinsky et al., 2002). Diverse cropping rotations contribute to better and more balanced soil fertility for supporting crops because each crop species has different nutritional requirements for optimum growth and development, and each draws on individual nutrients from the soil at different rates (Kuprinsky et al., 2002). This balance has a positive effect on crop resistance to diseases, as discussed in the next section. Although the results of many studies conducted so far highlight the difficulty of predicting exactly how the vegetational diversity introduced through undersowing of live mulches will affect pests and diseases, the general effect is positive.

41.4   Effects of Different Management Practices

Occasionally, no-till practices are associated with an increase in pest and disease severity compared to conventional tillage. However, such differences do not necessarily result in a negative impact on yield. In Mexico, for instance, Kumar and Mihm (2002) found that despite higher damage by Lepidopteran pests, maize production remained higher under no-till than in conventional tillage systems. Also, although white grub presence was reported at Chequèn, Chile, when DSPSC techniques were introduced, no noticeable damage was observed. Particularly the ability of the scarab beetle (Bothynus spp.) to damage plant roots was compensated for by a positive effect on greater soil macroporosity that enhanced the soil's ability to draw organic matter down into lower soil layers, which enhanced crop performance (Crovetto Lamarca, 1999).

Recent research in Brazil has distinguished among different subfamilies of Scarabaeidae. Dynastinae normally feed on organic matter and rarely on roots, while Melolonthinae feed mostly on roots and less on organic matter. Root-feeding species become predominant in soils where biodiversity has been reduced, relative to species that decompose litter and other organic matter and do little damage to roots. The total volume of the holes opened by the latter, notably Bothynus spp., was as much as 10 times greater in no-tillage agroecosystem than in conventional tillage (Brown and Oliveira, 2004). These saprophagous species bury large amounts of plant litter in the soil, significantly increasing P, K, and organic matter in their tunnels compared to adjacent soil. These tunnels, up to 3 cm in diameter and even more than 70 m^{-2}, extend from the surface to >1-m depths, putting them, along with earthworms and termites, in the category of ecosystem engineers (Chapter 11).

There is an indirect positive effect from mulching and the use of cover crops of better crop nutrition from minerals derived from the decomposition of organic matter. Balanced and adequate fertility for any crop reduces plant stress, improves physiological resistance to pest attack, and decreases risk of disease. It also results in induced resistance of plants vis-à-vis pests through nonpreference (antixenosis), tolerance, and compensation mechanisms. These mechanisms derive from the biological dimension of soil systems.

41.4.1 Effects on Microbial Communities

Microbial community management is a key element of cultural practices, though it is sometimes underestimated given a lack of knowledge about causal relationships. For millennia, soil health has been maintained empirically, particularly through applications of organic matter, mainly as manure. Faced with possible crop losses due to parasitic attacks, the first farmers progressively adapted their cropping systems to keep risks at acceptable levels (Altieri, 1999).

Ecosystem health has been defined in terms of an ecosystem's stability and resilience in response to some disturbance or stress. It has been known for some time that certain soils are "disease-suppressive" (Corman et al., 1986). This quality can be viewed as a manifestation of ecosystem stability and health (van Bruggen and Semenov, 2000). Soils with high fertility and high levels of organic matter appear to enhance natural mechanisms for biocontrol of pathogens, as suggested by the fact that in some soils pathogens cause little or no disease, despite an apparently favorable environment for them to grow in. There are many ways in which an antagonist can operate to curb or control pathogens. There can be rapid colonization in advance of pathogen presence to pre-empt space and substrates, or subsequent competition may lead to exclusion from a given ecosystem niche. Antibiotics may be produced, or there may be mycoparasitism or lysis of the pathogen (Altieri, 1999).

In other cases, the suppressiveness is probably due to the activity of soil microbiota since suppressive soils consistently show higher populations of actinomycetes and bacteria than do soils conducive to disease. Additions of organic material increase the general level of microbial activity; and the more microbes there are in the soil, the greater are the chances that some of them will be antagonistic to pathogens (Altieri, 1999).

The rotation of diverse crops provides a heterogeneous food base for microorganisms that offers more ecological niches and encourages microbial diversity. Reduced tillage contributes to this diversity because more heterogeneous residues accumulate on the soil surface over time (Kuprinsky et al., 2002). However, high microbial biomass and activity in soils under organic and integrated farming are not always correlated with high disease suppression. Specific organic amendments, such as mulching with straw and the practice of using lucerne as a break-crop in cereal cultivation, have been seen to influence the inoculum potential of *F. culmorum* and resulting disease outbreak and suppression, for example (Knudsen et al., 1999).

Some microorganisms simply assist crop plants to grow better, so that even if a disease is present, its symptoms are masked or impeded (Altieri, 1999). A positive impact of DSPSC techniques, notably using live mulch of *Arachis pintoi*, has been a lower incidence of fungal and bacterial diseases on rainfed rice and cotton, as reported from the humid tropical zone of north-central Brazil (Séguy et al., 1999b). Possible explanations are that better and more stable regulation of water and mineral plant nutrition under DSPSC may minimize water stress and help the crop plants to resist parasitic aggression.

41.4.2 Interactions with Manure

The application of contaminated manure can have some adverse effects on soil and plant health, so such biological amendments can be counterproductive. In Mali, the frequent use of organic manure, often contaminated with the seeds of parasitic plants, is known to favor their dissemination (Hoffmann et al., 1997). Striga has been found to be concentrated in certain cattle grazing zones and on their itineraries where manure deposition and application sustain the weed populations (Bengaly and Defoer, 1997). In Madagascar, cattle eating Striga plants do not digest the seeds, and thus they contribute to Striga dissemination, either directly through feces or through subsequent applications of manure (Andrianaivo et al., 1998).

Organic manure, particularly cow dung, can be a source of infestation by white grub species, particularly *Heteronychus plebejus* (Rajaonarison and Rakotoarisoa, 1994; Bourguignon, 1997). In addition, certain antagonisms that are enhanced by the addition of organic manure can work against entomopathogenic fungi (see Section 41.4.1). So the application of organic manure, depending on the conditions, can have either positive or negative effects on pests, notably white grubs and Striga. Some adverse effects can be solved by pretreatment of organic manure, for instance, by heating as achieved with composting (Chapter 31). No such problem is foreseen with DSPSC practices, however, since this crop and soil management system relies on organic matter derived from litter decomposition, which has positive effects through better plant nutrition.

Plants whose root systems are well developed can sustain a parasitic load higher than others with less favorable growing conditions. Application of organic matter in a soil often has positive effects on root systems' health status, an indirect effect that supports another discussed in the following section. Manure and compost have been found to reduce attacks of *Rhizoctonia solani* on radish and bean (Voland and Epstein, 1994) and of *Pyrenochaeta lycopersici* and *Phytophthera parasitica* on tomato (Workneh et al., 1993). While the reasons for this are not all certain, the effect is widely seen and often reported by farmers who rely on organic nutrient inputs.

41.4.3   Nutrient Effects

By facilitating the quick absorption of any excess nitrogen, plowing modifies plant physiology, and the absorption of other minerals is slowed down (Séguy et al., 1981, 1989). This is particularly the case with *Pyricularia grisea*, the pathogen responsible for rice blast, which is the most important disease of rainfed rice worldwide (Ou, 1985). Blast is a problem for farmers in Madagascar, and its damage is aggravated by the application of inorganic nitrogen fertilizers. This effect is probably due partly to the injurious effects of ammonium accumulation in the cells of plants treated with high N (Ou, 1985). However, also an abundance of soluble nitrogen, particularly amino acids and amines in plants, may serve as a suitable nutrient for fungus growth. Therefore, to minimize the adverse impact from rice blast infection, moderate doses are usually recommended when applying N-fertilizer. Plants receiving large amounts of nitrogen have less silication of epidermal cells and thus lower resistance to herbivores. The application of nitrogen also reduces hemicellulose and lignin in the cell wall and weakens plants' mechanical resistance to blast (Ou, 1985).

Massive nitrogen applications have multiple consequences for plant physiology and for host population structure, thus on plant receptivity to certain diseases. Agricultural practices that lead to significant discrepancies in nitrogen availability (in terms of quantity, form, and balance with other nutrients) are likely to translate into variations in the amount of disease (Primavesi et al., 1972; Séguy et al., 1981, 1989; Chaboussou, 2004).

Rice resistance to *P. oryzae* in volcanic soils may be due to the greater presence and availability of micronutrients such as Cu and Mn, while susceptibility might be linked to the high content of amino acids in plant tissues and to reducing sugars that sustain pathogen development (Chaboussou, 2004). In Cameroon, it has been found that soil type — through its effect on rice plant nutrition — was a determinant in rice plant resistance to blast (Séguy et al., 1981). In the Lake Alaotra plain of Madagascar, on peat soils recently put under cultivation, nitrogen release during the first year was so much that rice plants had abnormal growth and were destroyed by rice blast (Séguy et al., 1981). The pathways of influence and causation in the domain of plant nutrition are multiple, and effects can be ambiguous, because of countervailing influences. However, studies are showing that many soil factors affecting plants' nutrient access and supply directly affect the damage caused by pests.

41.5 Assessing the Effects of Pesticides

DSPSC systems, in addition to minimizing the effects that mineral fertilizers can have on soil biota, reduce reliance on pesticides, particularly herbicides, to control weeds and kill them. Cover crops are used instead to form a suitable mulch that protects the soil and suppresses weed growth. In some cases, carefully selected insecticides are applied, such as seed treatments, and there can be use of herbicides in cases where the ground cover strategy is not sufficient at first. Experiments are ongoing with different crop rotations that can minimize or end the use of agrochemicals, but DSPSC is not a strictly "organic" system. It combines inputs and practices with the aim of mobilizing biological processes for farmers' benefit. Its attitude toward the use of agrochemicals is therefore empirical and pragmatic.

Herbicide applications can, in fact, increase pest damage to crops by removing weeds as alternate hosts or by driving the pests on to nearby crops. For instance, the larvae of stalk borer (*Papaipema nebris*) typically move from grassy weeds to maize when herbicides are applied, so that certain areas within maize fields become more damaged after weed removal (Stinner and House, 1990). Musick and Suttle (1973) have observed that armyworm adult moths (*Pseudaletia unipunctata*) oviposit on small grain cover crops, such as rye and wheat, in which maize was planted directly, so when herbicides kill these grasses, the larval armyworms feed on maize.

If one has to drill seed directly into a green crop or weed residues, for example, to take advantage of rain in the winter time, a synthetic pyrethrum insecticide may be applied first in a tank mix with glyphosate to prevent insect attacks on the emerging small plants (Pruett and Guaman, 2003). Because of the problems associated with specific pest species in conservation-tillage farming, considerable effort has been directed toward developing tailored insecticide control measures for these systems. In temperate climates, for example, implementation of DSPSC often results in increased slug problems during the first years. In such instances, the application of a molluscicide after crop emergence can be highly beneficial (Leake, 2001). Also, in Madagascar, in areas where black beetle attacks are greater on mulch than on plowed plots, seed treatment is mandatory for upland rice production, although its cost may reduce the attractiveness of DSPSC (Charpentier et al., 2001).

No-till cultivation systems, on the other hand, may buffer the impacts of insecticide on the arthropod assemblage, thus minimizing its effect so it has less impact than in conventional cultivation. With deltamethrin, for instance, there are significant decreases in arthropod abundance in the maize canopy compared with conventional tillage (Badji et al., 2004). On the other hand, Brust et al. (1985) found that a soil-applied organophosphate insecticide did not suppress soil arthropod predator activity any more in no-till than in conventional tillage treatments. As DSPSC increases the overall sustainability of production systems by taking advantage of natural biological processes for pest management, the evidence reported in this section underscores the importance of keeping pesticide use to a minimum and of using selective molecules so as to minimize any inhibition of biological control processes.

41.6 Discussion

Some disturbance to the environment is a necessary part of agriculture. Just like putting a field under tillage after fallow will alter pest and plant disease problems, so reduced tillage practices, like any change in cultural practices, may increase, decrease, or have no effect on these problems. What will happen depends on the soil type, location, and prevailing environment, as well as on the species of insects or pathogens, as well as plants involved (Bockus and Shroyer, 1998; Kuprinsky et al., 2002).

For instance, the extreme diversity of pedologic situations created by the different geological processes by which the highlands and medium-altitude regions of Madagascar were formed, complicated by the variations in altitude and subtropical climate and the great biodiversity and endemism found in Madagascar, has contributed to biological equilibria that can vary greatly over relatively small distances. These differences in the entomofauna spectrum may account for the apparently contradictory results reported in the introduction to this chapter.

A review of the results of earlier work has shown that the implementation of DSPSC in most situations is associated with reduced pest and disease incidence on crop plants. Evidence has been offered of a wide variety of possible mechanisms accounting for this: direct physical effects of tillage (or the lack of it) and of mulching; changes in pest behavior due to plant diversity; effects of semiochemicals; increased predation, parasitism, or antagonisms; and induced crop resistance through better nutrition.

However, when switching from plowing to no-till, it will probably not be appropriate to continue all other practices in the same way as before. Where farmers have experienced problems with no-till or with direct-drill techniques, this has usually been associated with their failing to adopt other new practices to accompany the changes (Leake, 2001). Some time may be required before new favorable equilibria are reached. Thus, the targeted use of selected chemical inputs, at least for their "starter" effect, can be compatible with the mobilization of soil biological processes underlying DSPSC. This is in line with CIRAD's flexible approach to crop protection (Ratnadass et al., 2003).

The experience of CIRAD and its partners in Madagascar (particularly FOFIFA and TAFA) to alleviating constraints on upland rice production, building on earlier work in Brazil and elsewhere on the African continent, has justified a more holistic approach to soil system management, appropriately adjusting the management of crop, water, and nutrient factors, along with soil management methods that promote more favorable biological conditions for successful crops.

Our organizations have collaborated to provide special opportunities for implementing IPM approaches, evaluating the systems being improved with a focus on the major biotic constraints (white grubs/black beetles, rice blast, and Striga). The main objectives have been: (1) to determine mechanisms involved in the reduction of these pests' adverse impacts in DSPSC; and (2) to minimize externalities of these systems, both in the final product and in the environment, in term of chemicals used for crop protection, so as to enhance production outcomes.

To meet these objectives, we are exploring the potential for using within DSPSC certain biostimulants and concentrated organic fertilizers that can speed up induced resistance, as well as plant-derived insect repellents, along the lines discussed in Chapters 32–35. Although these are assumed to be more environmentally benign than traditional chemical pesticides, we know that their potential unintended side-effects need to be studied and evaluated (Chen et al., 2002; Sonnemann et al., 2002).

This is an area of knowledge generation and practical application where the science is still young, and where there are many opportunities to make improvements. The conclusion from experience so far is that the control of pests and diseases is best pursued in conjunction with knowledge of soil-system management, to take advantage of whatever these associated practices can contribute to making agriculture more reliable and productive.

References

A'Brook, J., The effects of plant spacing on the number of aphids trapped over the groundnut crop, *Ann. Appl. Biol.*, **61**, 289–294 (1968).

Altieri, M.A., The ecological role of biodiversity in agroecosystems, *Agric. Ecosyst. Environ.*, **74**, 19–31 (1999).

Andrianaivo, A.P. et al., *Biologie et Gestion du Striga à Madagascar.* FOFIFA, DPV, GTZ, Antananarivo, Madagascar (1998).

Asman, K., Ekbom, B., and Rämert, B., Effect of intercropping on oviposition and emigration of the leek moth (Lepidoptera: Acroplepiidae) and the diamondback moth (Lepidoptera: Plutellidae), *Environ. Entomol.*, **30**, 288–294 (2001).

Badji, C.A. et al., Impact of deltamethrin on arthropods in maize under conventional and no-tillage cultivation, *Crop Prot.*, **23**, 1031–1039 (2004).

Bengaly, M.P. and Defoer, T., Smallholder perception of the importance of problems caused by *Striga* and its distribution in village hinterlands, *Agric. Dev.*, **13**, 52–57 (1997).

Bockus, W.W. and Shroyer, J.P., The impact of reduced tillage on soilborne plant pathogens, *Annu. Rev. Phytopathol.*, **36**, 485–500 (1998).

Bourguignon, D., *Le ver blanc, dangereux ravageur des rizières pluviales malgaches. Rapport de stage au FOFIFA-CIRAD/Antsirabe*, Ecole Nationale Supérieure Agronomique de Toulouse, France (1997).

Boyer, J., Michellon, R., and Reversat, G., Interactions entre les vers de terre et les nématodes phytoparasites dans diverses cultures de l'Ile de la Réunion, In: *Gestion Agrobiologique des Sols et des Systèmes de Culture, Proceedings of International Workshop, Antsirabe, Madagascar, 23–28 March 1998*, Rasolo, F. and Raunet, M., Eds., CIRAD, Montpellier, France, 323–333 (1999).

Brown, G.G. and Oliveira, L.J., White grubs as agricultural pests and as ecosystem engineers, Abstract for 14th International Colloquium on Soil Zoology and Ecology, 30 August–3 September, Rouen, France (2004).

Brust, G.E., Stinner, B.R., and McCartney, D.A., Tillage and soil insecticide effects on predator-black cutworm (Lepidoptera: Noctuidae) interactions in corn agroecosystems, *J. Econ. Entomol.*, **78**, 1389–1392 (1985).

Byers, R.A., Mangan, R.L., and Templeton, W.C. Jr., Insect and slug pests in forage legume seedlings, *J. Soil Water Conserv.*, **38**, 224–226 (1983).

Chaboussou, F., *Healthy Crops: A New Agricultural Revolution*, Jon Carpenter, Charnley, UK (2004).

Charpentier, H., Semis direct sur couverture végétale dans deux écologies de la Côte d'Ivoire, In: *Gestion Agrobiologique des Sols et des Systèmes de Culture, Proceedings of International Workshop, Antsirabe, Madagascar, 23–28 March 1998*, Rasolo, F. and Raunet, M., Eds., CIRAD, Montpellier, France, 165–177 (1999).

Charpentier, H. et al., *Projet de diffusion de systèmes de gestion agrobiologique des sols et des systèmes cultivés à Madagascar. Rapport de campagne 2000/2001 et synthèse des 3 années du projet*, TAFA, Antanarivo, Madagascar (2001).

Chen, S.K., Subler, S., and Edwards, C.A., Effects of agricultural biostimulants on soil microbial activity and nitrogen dynamics, *Appl. Soil Ecol.*, **19**, 249–259 (2002).

Collier, R., Finch, S., and Davies, G., Pest insect control in organically-produced crops of field vegetables, In: *Proceedings: 53rd International Symposium on Crop Protection*, University of Ghent, Belgium, 259–267 (2001).

Corman, A. et al., Réceptivité des sols aux fusarioses vasculaires: Méthode statistique d'analyse des résultats, *Agronomie*, **6**, 751–757 (1986).

Crovetto Lamarca, C., *Agricultura de Conservacion: El Grano para el Hombre, la Paja para et Suelo*, Mundi Prensa, Madrid (1999).

Davet, P., *Vie Microbienne du Sol et Production Végétale*, National Institute of Agricultural Research (INRA), Paris (1996).

Doupnik, B. and Boosalis, M.G., Ecofallow: A reduced tillage system, and plant diseases, *Plant Dis.*, **64**, 31–55 (1980).

Finch, S. and Collier, R.H., Host-plant selection by insects: A theory based on 'appropriate/ inappropriate landings' by insect pests of cruciferous plants, *Entomol. Exp. Appl.*, **96**, 91–102 (2000).

Finch, S. and Collier, R.H., The effect of increased crop diversity on host-plant selection by insects, In: *2ème Conférence Internationale sur les Moyens Alternatifs de Lutte contre les Organismes Nuisibles aux Végétaux, Lille, 4–7 March 2002*, 567–571 (2002).

Grant, J.F. et al., Invertebrate organisms associated with alfalfa seedlings loss in complete-tillage and no-tillage plantings, *J. Econ. Entomol.*, **75**, 822–826 (1982).

Hoffmann, G. et al., Parasitic plant species of food crops in Africa: Biology and impact study in Mali, *Agric. Dev.*, **13**, 30–51 (1997).

Husson, O. et al. Impacts of cropping practices and direct seeding on permanent vegetal cover (DSPVC) techniques on soil biological activity in Northern Vietnam, In: *II World Congress on Conservation Agriculture, Iguassu Falls, 11–15 August, 2003*, 460–463 (2003).

Khan, Z.R. et al., Intercropping increases parasitism of pests, *Nature*, **388**, 631–632 (1997a).

Khan, Z.R. et al., Utilisation of wild gramineous plants for management of cereal stemborers in Africa, *Insect Sci. Appl.*, **17**, 143–150 (1997b).

Khan, Z.R., Overholt, W.A., and Ng'eny-Mengech, A., Integrated pest management case studies from ICIPE, In: *Integrated Pest Management in the Global Arena*, Maredia, K., Dakouo, D., and Mota-Sanchez, D., Eds., Michigan State University, East Lansing, MI, CAB International, Wallingford, UK, 441–452 (2003).

Knudsen, M.B. et al., Suppressiveness of organically and conventionally managed soils towards brown foot rot of barley, *Appl. Soil Ecol.*, **12**, 61–72 (1999).

Kumar, H. and Mihm, J.A., Fall armyworm (Lepidoptera: Noctuidae), southwestern corn borer (Lepidoptera: Pyralidae) and sugarcane borer (Lepidoptera: Pyralidae) damage and grain yield of four maize hybrids in relation to four tillage systems, *Crop Prot.*, **21**, 121–128 (2002).

Kuprinsky, J.M. et al., Managing plant disease risk in diversified cropping systems, *Agron. J.*, **94**, 198–209 (2002).

Leake, A.R., Integrated pest management for conservation agriculture, In: *World Congress on Conservation Agriculture, Madrid, 1–5 October 2001*, 245–253 (2001).

Litsinger, J.A. and Ruhendi, Rice stubble and straw mulch suppression of preflowering insect pests of cowpea sown after puddle rice, *Environ. Entomol.*, **13**, 509–514 (1984).

Michellon, R., *Modes de gestion écologique des sols et systèmes de culture à base de géranium dans les hauts de l'ouest de la Réunion*, CIRAD-CA, Montpellier, France (1996).

Michellon, R. et al., Influence du traitement des semences et de la date de semis sur la production du riz pluvial en fonction du mode de gestion du sol sur les Hautes Terres, Rapport TAFA/CIRAD/ FOFIFA. Antsirabe, Madagascar (2001).

Michellon, R. et al., Evolution de la faune du sol selon sa gestion: Protection des plantes par traitement des semences, Rapport TAFA/CIRAD/FOFIFA. Antsirabe, Madagascar (1998).

Michellon, R., Séguy, L., and Perret, S., Géranium rosat: Conception de systèmes durables avec couverture herbacée, In: *15th Journées Internationales Huiles Essentielles, APPAM, Digne-les-Bains, 5–7 September 1996*, CIRAD-CA, La Réunion, France (1996a).

Michellon, R., Séguy, L., and Perret, S., Association de cultures maraîchères et du géranium rosat à une couverture de kikuyu (*Pennisetum clandestinum*) maîtrisée avec le Fluazifop-P-butyl, In: *4è colloque: Les substances de croissance, partenaires économiques des productions végétales*, ANPP, Paris, 369–376 (1996b).

Miller, J.R. and Cowles, R.S., Stimulo-deterrent diversion: A concept and its possible application to onion maggot control, *J. Chem. Ecol.*, **16**, 3197–3212 (1990).

Musick, G.J. and Suttle, P.J., Suppression of armyworm damage to no-tillage corn with granular carbofuran, *J. Econ. Entomol.*, **66**, 735–737 (1973).

Naudin, K., *Systèmes de culture sur couverture végétale. Saison 2001–2002. DPGT-Garoua-Cameroun. Rapport d'activité Juin 2001–Février 2002*, CIRAD-Sodecoton, Garoua, Cameroon (2002).

Ou, S.H., *Rice Diseases*, 2nd ed., Commonwealth Mycological Institute, Kew, UK (1985).

Primavesi, A.M., Primavesi, A., and Veiga, C., Influence of nutritional balances of paddy rice resistance to blast, *Agrochimica*, **16**, 459–472 (1972).

Pruett, C.J.H. and Guaman, I., Integrated pest management in no-tillage systems, In: *II World Congress of Conservation Agriculture, Iguassu Falls, 11–15 August 2003*, 96–99 (2003).

Quilici, S., Vercambre, B., and Bonnemort, C., Les insectes ravageurs, In: *Le Géranium Rosat à la Réunion*, CIRAD, Chambre d'agriculture de la Réunion, DAF-SPV, SAFER-Réunion, Chambre d'agriculture, Saint-Denis, La Réunion, France, 79–80 (1992).

Rajaonarison, J.H.J. and Rakotoarisoa, D., Bionomie et contrôle des populations de *Heteronychus arator, H. bituberculatus* et de *H. plebejus* (Coleo. Scarabaeidae: Dynastinae), FOFIFA, Antananarivo, Madagascar (1994).

Ramanantsialonina, H.M., Evolution de la faune et des dégâts aux cultures en fonction du mode de gestion des sols. Mémoire d'ingéniorat en agronomie ESSA, University of Antananarivo. CIRAD/FOFIFA/TAFA, Antananarivo, Madagascar (1999).

Ratnadass, A. et al. IPM experiences of CIRAD-France in developing countries, In: *Integrated Pest Management in the Global Arena*, Maredia, K., Dakouo, D., and Mota-Sanchez, D., Eds., Michigan State University, East Lansing, MI, CAB International, Wallingford, UK, 453–465 (2003).

Schulthess, F. and Setamou, F., *Canavalia ensiformis* et *Mucuna pruriens* plantes-hôtes intermédiaires du ravageur du maïs *Mussidia nigrivenella* Ragonot (Lepidoptera: Pyralidae), *CIEPCA Newslett.*, **4**, 2 (1999).

Séguy, L. et al., La maîtrise de *Cyperus rotundus* par le semis direct en culture cotonnière au Brésil, *Agric. Dev.* **21**, 87–97 (1999a).

Séguy, L., Bouzinac, S., and Maronezzi, A.C., Semis direct et résistance des cultures aux maladies. Potafos: Informaçoes agronômicas, no 88, Piracicaba SP, Brasil, 1–3 (1999b).

Séguy, L., Bouzinac, S., and Pacheco, A., *Les principaux facteurs qui conditionnent la productivité du riz pluvial et sa sensibilité à la pyriculariose sur sols rouges ferrallitiques d'altitude, Goiânia*, IRAT-CIRAD/EMBRAPA, Montpellier, France (1989).

Séguy, L., Notteghem, J.L., and Bouzinac, S., Etude des interactions sol-variétés de riz-pyriculariose, In: *Comptes-rendus du symposium sur la résistance du riz à la pyriculariose*, IRAT-GERDAT, Montpellier, France, 138–151 (1981).

Smith, J.G., Influence of crop background on aphids and other phytophagous insects on Brussels sprouts, *Ann. Appl. Biol.*, **83**, 1–13 (1976).

Sonnemann, I., Finkhauser, K., and Wolters, V., Does induced resistance in plants affect the belowground community?, *Appl. Soil Ecol.*, **21**, 179–185 (2002).

Stinner, B.R. and House, G.J., Arthropods and other invertebrates in conservation-tillage agriculture, *Ann. Rev. Entomol.*, **35**, 299–318 (1990).

Teasdale, J.R. et al., Enhanced pest management with cover crop mulches, *Acta Hortic.*, **638**, 135–140 (2004).

van Bruggen, A.H.C. and Semenov, A.M., In search of biological indicators for soil health and disease suppression, *Appl. Soil Ecol.*, **15**, 13–24 (2000).

Vercambre, B., Equilibre actuel entre la canne à sucre et ses ravageurs à l'Ile de la Réunion (1979–1992), In: *Atelier d'Entomologie Appliquée: Lutte Intégrée Contre les Ravageurs des Cultures*, Girardot, B., Ed., CIRAD, Montpellier, France, 49–58 (1993).

Voland, R.P. and Epstein, A.H., Development of suppressiveness to diseases caused by *Rhizoctonia solani* in soils amended with composted and noncomposted manure, *Plant Dis.*, **78**, 461–466 (1994).

Walker, P.T., *Heteronychus* white grubs on sugar-cane and other crops, *Pest Art. News Summaries*, **14**, 55–68 (1968).

Wamhoff, W. et al., Impact of crop rotation and soil cultivation on the development of pests and diseases of rapeseed, *Zeitschrift für Pflanzenkrankheiten und Pflanzenschutz*, **106**, 57–73 (1999).

Weston, L.A., Utilization of allelopathy for weed management in agroecosystems, *Agron. J.*, **88**, 860–866 (1996).

Wilson-Rummenie, A.C. et al., Reduced tillage increases population density of soil macrofauna in a semiarid environment in central Queensland, *Environ. Entomol.*, **28**, 163–172 (1999).

Workneh, F. et al., Variable associated with corky root and *Phytophtora* root rot of tomatoes in organic and conventional farms, *Phytopathology*, **83**, 581–589 (1993).

42

Revegetating Inert Soils with the Use of Microbes

Gail Papli and Mark Laing

*Department of Microbiology, and Plant Pathology, University of KwaZulu-Natal,
Pietermaritzburg, South Africa*

CONTENTS

"Dead" soils are often found in disturbed sites, such as mine dumps and soil stockpiles, as a result of strip mining or other massive soil system disruption. Even when soil chemistry and physical indicators appear adequate for healthy plant growth, a downward spiral of soil viability ensues once plants are removed, and compaction occurs as a result of soil-moving activities. Reduced pore space lowers both oxygen and water levels, and this combined with the absence of plant root exudates causes soil microbial populations to diminish, previous reduction having contributed to the diminished pore space.

As these interactions occur, the numbers of earthworms and other macrofauna decline and there is even less aeration of the soil. This further reduces the soil's microbial and other populations, which diminishes the production of exopolysaccharides by microbes and impairs the formation of new soil aggregates and pores. Any increase of soil compaction makes it more difficult for plants and microbes to live in these soils. The process is a tight downward spiral of linked activities, with each detrimental outcome compounding soil debility until the soil is biologically dead and extremely difficult to revegetate.

This chapter considers how such soil systems can be restored to productive operation by combining several interventions in an integrated package. This would include the use of containerized seedlings and the treatment of the soil with lime, fertilizers, and organic matter, such as sewage sludge, fly-ash, microbial inoculum, as well as the augmentation of earthworm populations. These interventions enhance soil microbial activity, reduce soil compaction, and render the soil fertile once again for plant growth. To make clearer the effects of these interventions, we review briefly the dynamics of soil systems and their constituent factors. This reprise of relationships considered in Part II presents a short summary of how soil systems function. These relationships are illuminated by considering what happens within soil systems whenever they are absent.

42.1 Challenges and Strategies of Soil Rehabilitation

Soil has been characterized as a "collection of natural bodies on the earth's surface containing living matter and supporting or capable of supporting plants" (Soane and van Ouwwerkerk, 1994). Increasingly we are seeing as a global problem the increase in contaminated and "dead" soils that are unable to support plants and animal life. In recent years, legislation enacted in some countries has made polluters responsible for the rehabilitation of land that they have contaminated. This is focusing more attention on the revegetation of these soils (Glazewski, 2000). Effective revegetation is costly, however, and not necessarily successful unless the knowledge foundations for the interventions are correct and clear.

Soil compaction has been highlighted as a major problem involved in revegetation of strip mining soil dumps by Soane and van Ouwwerkerk (1994). This typically results from the way that the soil has been handled in terms of drainage, tillage, traffic, and use of heavy machinery. Soil compaction reduces the ability of plant roots to penetrate soil layers, resulting in their reduced nutrient uptake. Furthermore, compacted soils commonly have high levels of subsurface soil moisture because water in the soil is unable to drain away freely. This reduces soil oxygen (O_2) levels, further restricting root development and the activities of aerobic micro- and macrobiota (Sopher and Baird, 1982).

Five primary strategies are available for restructuring and remediating compacted and inert soils:

- Physical treatment
- Chemical treatment
- Thermal treatment
- Treatment by stabilization and solidification and
- Biological treatment (Hester and Harrison, 1997).

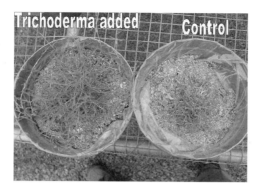

FIGURE 42.1
Response of *Cynodon dactylon* to a probiotic isolate of *Trichoderma harzianum* applied to the roots of grass plugs and planted into "lifeless" mine dump soil (on left) and control with no application.

We propose that the most critical factors in soil restoration are the biological ones. These have been given relatively little attention or emphasis by engineers and soil scientists who have traditionally preferred other kinds of intervention. To examine what is possible with biologically informed strategies of soil recuperation, we focus here on the rehabilitation of degraded mining soils as these are some of the more inert and recalcitrant. Fortunately, there is some promising experience in South Africa of restoring such debilitated soils to productive status, exemplifying many of the principles and practices discussed in preceding chapters. Figure 42.1 shows the difference in plant growth that can be promoted, for example, by soil amendments of Trichoderma, discussed in Chapter 34, to make inert soil more hospitable for plants.

42.2   Disturbed Soils

Mining, especially strip mining of coal, produces large amounts of waste. Waste disposal is often poorly managed, usually creating waste heaps adjacent to mine shafts. After hundreds of years of underground coal mining, the industry has turned to strip mining. This has resulted in the addition of large new areas of land being disturbed by such methods, along with countless unattended spoil heaps at abandoned deep mines. The process of strip mining with use of its heavy earthmoving machinery leads to severe soil problems, creating biologically inert soils. Subsequent attempts at revegetation, often mandated by law, are fraught with difficulties.

42.2.1   Changes in Soil Biota Affect Soil Structure and Chemistry

Mortality rates of $>90\%$ have been observed for earthworms and other macrobiota in disturbed soil as a result of their being buried, crushed, or exposed to predators (Evans et al., 1986). Removed topsoil that is piled up for later return to the area also "dies" as the indigenous populations of aerobic fungi and actinomycetes decrease dramatically when soil is stored for long periods of time. Without plant populations providing nutrients and energy to the soil via root exudates and residues, heterotrophic organisms in the soil system die.

Soil microbial populations, when active, constantly secrete polysaccharides that coat the soil particles around them, keeping these apart and lubricating them from each other.

This increases soil stability and porosity. Without these polysaccharides, the pore spaces between soil particles are severely reduced, resulting in endogenous soil compaction. In such soil, the inorganic soil particles bind together into a hard and impervious matrix (Soane and van Ouwwerkerk, 1994). There can be adequate soil physical and chemical fertility for plant growth, but in the absence of sufficient and appropriate biological components, the soil becomes unable to support plant life, as required for the remediation of these soils.

A further problem for coal mine soils is that they often contain high levels of ferrous iron (FeII) and iron sulfide (FeS_2). When this is oxidized to form ferric iron (FeIII) and iron sulfate ($FeSO_4$), considerable acidity is generated, resulting in a decline in soil pH from 5.0 to 6.5 to 3.0 or even less. Major plant nutrient elements, particularly nitrogen (N) and phosphorus (P), become unavailable at low pH, which exacerbates plant nutrient deficiencies (Hester and Harrison, 1997). Other soil minerals, such as aluminum and manganese, readily dissolve at low soil pH, and are then absorbed into plants at levels that are phytotoxic.

42.2.2 Environmental Impacts of Compacted Soils

Compaction which restricts rooting depth not only reduces the uptake of water, oxygen, and nutrients by plants, but it also decreases soil temperature. This affects microbial activity and decreases the rate of decomposition of soil organic matter (SOM), which reduces the subsequent release of nutrients. Such soil loses its ability to support useful plant and microbial populations (Sopher and Baird, 1982).

The lower levels of O_2 associated with compaction lead to a decrease in the populations of earthworms such as *Aporrectodea longa* and *Lumbricus terrestris* (deep burrowing earthworms). This reduces the number of burrowing holes and diminishes the cycling of organic matter within the soil profile (Hester and Harrison, 1997).

Ammonia can then accumulate in lower soil horizons as the populations of macrobiota and aerobic microbes recycling nitrogenous soil nutrients decline. Soil microbial populations are altered to favor facultative and strict anaerobes at the expense of aerobes (Evans et al., 1986). Since nitrogen-fixing bacteria such as Rhizobium spp. do not flourish in anaerobic, acid soils, these changes reduce the sustainable nitrogen cycle of these soils (Rimmer and Colbourn, 1978).

42.3 Roles of Microbes in Soil Systems

42.3.1 Cycling and Recycling of Nutrients

Microbes are the primary force driving nutrient movement in soil systems. In particular, they drive the carbon (C), nitrogen (N), and sulfur (S) cycles, which are reviewed briefly here to underscore how thoroughly and intimately what are usually analyzed as chemical conversion processes are mediated by or dependent on soil organisms. Any efforts to restore the fertility of soil systems needs to get in place these different sets of biotic actors and associated chemical transformations in order to reestablish plant life on "dead" soils.

42.3.1.1 Carbon Cycle

The carbon cycle is the most important soil cycle from a quantitative standpoint, given that the single largest carbon reservoir on earth is present in the soil, as discussed in the following chapter. Microbes' use of the energy tied up in C–H bonds is the driving force for all other nutrient cycles. Heterotrophs produce carbon dioxide, and chemoautotrophs utilize and fix CO_2 through photosynthesis. The soil carbon pool is about five times the size of the atmospheric pool, which is somewhat larger than the carbon in all living organisms. (Figures on these levels are given in the next chapter.) Humus, a complex mixture of organic materials as discussed in Chapter 6, has particularly high carbon content. Like other carbonaceous materials, humus is inert by itself, needing the activities of microbes to transform its fixed chemical energy into biologically-useful energy.

42.3.1.2 Nitrogen Cycle

Nitrogen is usually the limiting nutrient for plant or microbial growth. Where there is more nitrogen available, both flora and fauna can grow more, until other nutrients become limiting. Many bacteria utilize nitrate (NO_3) as their nitrogen source, in what is known as assimilatory nitrogen reduction. In dissimilatory nitrate reduction, or nitrate respiration, nitrate serves as the final hydrogen acceptor under anaerobic conditions. Nitrate can be further reduced to nitrite by nitrate reductase, a molybdenum-containing enzyme. Many denitrifying bacteria can grow not only with nitrate, but also with nitrite, and some can even grow with nitrous oxide as a hydrogen acceptor. There are many actors and many steps in the nitrogen cycle, sketched in Figure 42.2.

While N_2 fixation is widely recognized as an important soil process, its converse, denitrification, is similarly essential for soil system functioning because this is the primary process that converts fixed nitrogen into molecular nitrogen, which is vital for terrestrial life. Nitrate (NO_3), the end product of nitification under aerobic conditions, is highly soluble and poorly adsorbed on soil particles, so that it is easily leached into lower horizons unless held by soil microbes. This is an important function because NO_3 can be toxic to animals when it accumulates in drinking water, causing cyanosis (Brock and Madigan, 1991). Nitrate and other compounds that can have deleterious effects on flora and fauna are easily lost from "dead" soils, whereas they can be neutralized or immobilized by bacterial activity. At the same time, the utilization of soil nitrogen existing in its various forms depends on microbial transformations.

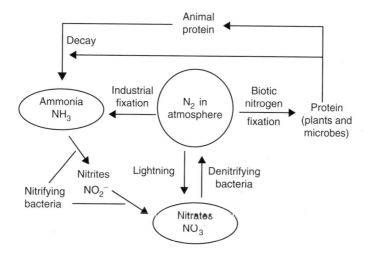

FIGURE 42.2

The N cycle in nature. *Source:* Kimball, J.W., The nitrogen cycle (online: http://users.rcn.com/jkimball.ma. ultranet/BiologyPages/N/NitrogenCycle.html; accessed 03/05/2005) (2005).

42.3.1.3 Sulfate Reduction

Sulfur is another major nutrient necessary for life forms, although potentially harmful. It is very common in contaminated soils, and hydrogen sulfide (H_2S), which is characteristic of damp, anaerobic soils, can be toxic to plants and microbes. This gas is released when soils are disturbed, emitting a "rotten egg" smell. In soils with high concentrations of ferrous iron (Fe^{2+}), sulfides can combine with this iron to form ferrous sulfides, which give the black color characteristic of many anaerobic soils.

Sulfate-reducing bacteria transfer substrate hydrogen to sulfate as the terminal electron acceptor with the reduction of sulfate to sulfide. The process allows electron transport with the participation of cytochrome c, and there is energy gain from electron transport phosphorylation under anaerobic conditions, in a process called dissimilatory sulfate reduction. Most of the hydrogen sulfide produced in nature is due to this reaction.

Sulfate-reducing bacteria are obligate anaerobes, being strictly dependent on anaerobic conditions. They are found in decomposing sediments and black mud where organic materials undergo anaerobic degradation. In soils where sulfur is in reduced form, conditions are inhospitable for most plant growth.

The conversion of sulfur into more tolerable forms depends on soil aeration which can be done mechanically. However, this can also be achieved biologically over time if sufficient aerobic conditions are established, enabling various species of organisms to restore interdependent communities of aerobes that can reverse the dominance of sulfur in reduced forms.

42.3.2 Rhizosphere Balance

The rhizosphere, the zone of soil extending up to 5 mm from the roots, can have microbial populations 10–100 times higher than in the rest of the soil (Brock and Madigan, 1991; Elsas et al., 1997; also Chapter 7). Most of the important soil biological transformations take place in the rhizosphere, especially N_2 fixation and mycorrhizal associations, with numerous chemical transactions occurring with microbial intermediation (Chapters 9 and 12). Roots' excretion of sugar, amino acids, hormones, and vitamins, collectively referred to as root exudates, promotes the growth of bacteria and fungi, which form microcolonies on the roots' surfaces (Pinton et al., 2001). Rhizosphere microorganisms exert strong effects on plant growth and health by nutrient solubilization, N_2 fixation, and the production of plant hormones. The suppression of deleterious microorganisms by protective bacteria also contributes to increases in plant productivity (Smalla et al., 2001). Plants not having such support and protection do not fare well.

Root exudates selectively influence the growth of bacteria and fungi that colonize the rhizosphere by altering the chemistry of soil in the vicinity of plant roots and by serving as selective growth substrates for soil microorganisms (Gershuny and Smillie, 1995). Microorganisms in turn can influence the composition and quantity of root exudation through their effects on root cell leakage, cell metabolism, and plant nutrition (Yang and Crowley, 2000). Variations in the structure and species composition of microbial communities result from differences in root exudates and rhizodeposition, associated with different root locations and also with soil type, plant species, nutritional status, plant age, stress, disease, and other environmental factors (Garland, 1996).

During the growth of new roots, exudates secreted in the zone of elongation behind the root tips (Figure 5.2 in Chapter 5) support in particular the growth of primary root colonizers that utilize easily degradable sugars and organic acids. Around older parts of the root system, carbon is deposited primarily in the form of sloughed cells and consists of more recalcitrant materials, including lignified cellulose and hemicellulose. Fungi and bacteria in these zones are organisms better adapted to crowded, oligotrophic conditions (Yang and Crowley, 2000).

42.3.3 Retardation or Reversal of Soil Compaction

Plant growth is highly affected by the structure of the soil. Compaction reduces the uptake of water by plants: first, by increasing the bulk density of the soil and limiting root penetration; second, by decreasing total pore space; and third, by reducing macropore space and creating fine micropores. The latter increase the permanent wilting coefficient and decrease available water content (Brady and Weil, 2002).

The impact of soil structure on plant growth can be traced to the basic units of soil structure. These are soil aggregates which are generally <10 mm in diameter and composed of solid material; binding agents; and pore space (Sylvia et al., 1999). Aggregate formation is a crucial factor affecting germination, root health, and plant growth (Smalla et al., 2001). The effects of reducing soil compaction are discussed in Chapter 38.

42.3.3.1 *Aggregate Functions*

Aggregates determine the mechanical and physical properties of soil, such as retention and movement of water, aeration, and temperature (Lynch, 1988). When soil particles are bound together to form aggregates, they facilitate the movement of air and water through the soil. Relatively large spaces between aggregates (macropores) allow rapid movement of water and air and permit ready penetration by roots. In contrast, compacted soils lack aggregates and thus structure. Their soil particles function individually with no distinct pore spaces (Herrick and Wander, 1997). Since roots must grow into pores either matching or exceeding their own diameter, many plant species cannot become established in nonaggregated soil (Grayston et al., 1998). The binding forces that attach soil particles in aggregates include chemical reactions with their positive or negative charges and biological components.

Many studies have shown that the formation of stable aggregates depends heavily on the nature and the content of SOM (e.g., Alami et al., 2000). As the stability of aggregates decreases, so too does the level of SOM (Garland, 1996). Plant roots by releasing root exudates indirectly stimulate microbial activity in the rhizosphere, contributing to soil organic material and thereby to soil aggregate stability (Aldén et al., 2001). Microbes affect soil aggregation through their production of exopolysaccharides, which cement small aggregates (Alami et al., 2000). Soil fungi, particularly mycorrhizal fungi, add stability to mid-size aggregates (Elsas et al., 1997). Experimental observations have demonstrated that the amendment of soil with microbial exopolysaccharides results in increased soil aggregation (Alami et al., 2000).

42.3.3.2 *Microbial Roles*

Microbes directly affect the aggregation of root-adhering soil (RAS), which constitutes the immediate physical environment in which roots take up O_2, water, and nutrients for their growth and that of shoots. The importance of O_2 in this zone is often overlooked but cannot be overstated, being crucial to healthy root development and function. Alami et al. (2000) have demonstrated that the exopolysaccharides produced by *Paenibacillus polymyxa* significantly contribute to the aggregation of RAS on wheat roots. A similar effect on wheat was observed after inoculation with *Pantoea agglomerans*, further demonstrating the role of bacteria in the regulation of water and O_2 content of the rhizosphere by improved soil aggregation (Grayston et al., 1998).

Since soil productivity is dependent on aggregate formation, soil microbial populations have a direct influence soil fertility and its productivity. Only if large microbial populations can be restored in compacted soil will aggregates develop and efficient plant growth take place. Soil tillage and crop rotation can help build up such populations by aerating the soil and adding more diversified exudation (Lupwayi et al., 1998). However, unless these physical interventions have the desired biological effects, their benefits are limited.

42.4   Physical Requirements of Soil Microbes

42.4.1   Soil Pore Space

Total pore space is the volume of soil systems occupied by air and water. Soils that have well-structured and stable aggregates have more spaces created and maintained between aggregates. The relative size of the pores is important because macropores are usually filled with air, whereas micropores tend to fill with water (Soane and van Ouwwerkerk, 1994). Most microbes exist on the outside of aggregates and in the small pore spaces between them. Very few reside within aggregates themselves (Sopher and Baird, 1982). The smaller the particles into which the mineral components of soil systems are aggregated, the more relative surface area there is on which microbes can live and function.

The diameter of pore necks (entries) determines how accessible the pores are to microbes. The water content of pores affects this also. Pores with diameters of a very few micrometers easily fill with water and are most suitable for bacteria, while fungi generally require larger pores (Grayston et al., 1998). This is a reason for wanting diversity of soil aggregates. The infertility of compacted soils derives most fundamentally from their having few pore spaces due to the lack of structural aggregates and thus lack of microbial diversity, which in turn hinders plant growth. The downward spiral of compaction accelerates itself. Unfortunately, providing fertilizer cannot compensate for inhospitable physical structure.

42.4.2   Oxygen Content

Oxygen plays an essential role in all living cells. Microbes vary in their need for or tolerance of O_2, being divided into different groups according to their need for O_2, ranging from strict aerobes to strict anaerobes (Chapter 5). It is necessary to provide extensive aeration for the growth of many aerobic microorganisms because O_2 is poorly soluble in water, and the O_2 that is used up by microbes during their growth is not replaced fast enough by the diffusion of O_2 from air. Oxygen is toxic to obligate anaerobes, and this inhibition results from toxic intermediates produced during electron transport (Brock and Madigan, 1991). Their inhibiting effect is noted only in the presence of an electron donor. Accordingly, anaerobes can survive for long periods in oxygenated soil if no substrate is available.

Soil contains 10 to 100 times greater concentrations of CO_2 than of O_2, due to the respiration of roots and other organisms that continuously consume O_2 and produce CO_2. Differences in the pressures of the two gases that are created cause O_2 to diffuse from the atmosphere into the soil, and CO_2 in it to diffuse out (Schlegel, 1993). Because O_2 diffuses slowly in water, poor soil aeration is usually caused more by the presence of water than by the size or amount of pore spaces. Obviously, compaction of soils that eliminates pore space will limit the availability of O_2 in the soil.

42.4.3 Soil Moisture

All organisms require water for their life processes, and water availability is one of the most important factors affecting the growth of microbes in nature (Brock and Madigan, 1991). Water has an electrical dipole, which is responsible for its unique properties. Soil particles also have dipoles that form a strong attraction with water molecules, resulting in the formation of films on the surface of soil particles, called adhesion water. Such water is always present in the soil, but it is not necessarily all available to microbes. Water molecules can also be attracted to one another, forming a film called cohesion water, which is readily available for use by microbes and plant growth in soil micropores (Schlegel, 1993).

Matric potential, the attraction of water to solid surfaces, reduces the free energy of water. Solutes such as salts and sugars in water reduce the free energy of water as well, since these substances are able to absorb water molecules more or less tightly. This bonding is referred to as osmotic potential (Soane and van Ouwwerkerk, 1994). The combined osmotic and matric components of soil water determine the force against which an organism must work to obtain water residing within the soil. Microbial activity is generally optimal at or near -0.01 MPa (megapascals, a measurement of pressure), and it decreases as the soil becomes waterlogged or suffers from drought. Fungi are more tolerant of higher water potentials in drought-prone soils than are bacteria (Sopher and Baird, 1982), which gives them some advantage under these adverse conditions. In compacted soils with little water-holding capacity, water seeps through and out of the soil profile, decreasing microbial and plant growth.

42.4.4 Soil pH

The optimum pH for most plants is in the range of 6.3–6.8. This allows most nutrients to be absorbed by plants and is also the most favorable range for the functioning of most soil bacteria. Fungi can tolerate a wider pH range (Gershuny and Smillie, 1995). At a pH below 5.5, some macro- and micronutrients assume insoluble forms, whereas many other micronutrients and heavy metals become more soluble under acidic soil conditions. The latter include iron (Fe), manganese (Mn), zinc (Zn), copper (Cu), cobalt (Co), aluminum (Al), and lead (Pb) (Baath, 1996). Acidic conditions can induce toxicity with these element and metals, however (Herrick and Wander, 1997). Under very acidic conditions, phosphorus and molybdenum (Mo) may become insoluble, and low levels of calcium (Ca) and magnesium (Mg) will become available to plants (Sopher and Baird, 1982). The activities of nitrogen-fixing bacteria, bacteria that convert ammonium (NH_4) to NO_3, and organic material-degrading bacteria are all diminished in soils of low pH.

Alkaline soils can also have diminished the availability of certain nutrients, especially micronutrients, concurrently with toxic levels of sodium (Na), selenium (Se), and some other minerals that may accumulate. The chemical destruction of SOM which can occur has negative effects on plant growth (Sopher and Baird, 1982). Improper irrigation on alkaline soils, i.e., overapplication of water, leads to a build-up of salts within and on the surface layer, a process known as salinization. Saline soils cause soil crusting, which reduces water infiltration into the soil (Gershuny and Smillie, 1995). This has negative effects on both plants and other soil biota.

42.4.5 Soil Organic Matter

Defined as any material in the soil originating from the growth of plants or animals, living or dead, SOM promotes aggregation of soil particles. Because SOM is less dense than soil minerals, it increases porosity and reduces bulk density. It also releases gases during decomposition (Gershuny and Smillie, 1995). These effects increase soil permeability and generally increase plant-available water. Further, by creating air passages in the soil, SOM leads to increased O_2 diffusion rate (Gelsomino et al., 1999). Chemically, SOM increases the cation exchange capacity of soil and acts as a buffer to pH changes (Wagner and Wolf, 1999).

Soil organic matter, especially humus, is a source of carbon to many soil microbes, which degrade it to obtain nutrients and energy (Chapters 5 and 6). The process by which SOM and humus are broken down in the soil is called mineralization (Sopher and Baird, 1982). The end products of mineralization are available for enhancing plant growth and soil stability. The amount of water still being held in pore spaces after a fully saturated field has been allowed to drain for 24 h is referred to as field capacity, representing how much water that particular soil can retain on the basis of its matric potential (see above). By improving soil structure, organic matter modulates the field capacity of soils that would otherwise be too wet or too dry.

42.5 Amendments to Restore Soil Fertility

42.5.1 Lime

Lime is largely composed of Ca and Mg carbonates which when added to the soil increase its pH. Coal mine dump soils containing high concentrations of sulfates that produce sulfuric acid will have high acidity. By removing hydrogen ions from the soil solution, lime neutralizes Mn and Al in the soil colloids (Sopher and Baird, 1982). Common liming materials are calcitic and dolomitic limestone, burned lime, hydroxide of lime, and suspensions of lime. Several factors affect the choice of lime material, including whether or not the soil also needs magnesium, the reaction speed desired, and the cost of each of the materials based on their relative neutralizing values (Demetz and Insam, 1999).

Adding lime to soil may also affect its resistance to compaction, in addition to having an effect on the production of organic matter (Soane and van Ouwwerkerk, 1994). Studies have shown that lime increases the pH and modifies clay dispersibility in variable-charge soils (Elsas et al., 1997). Liming also reduces the plasticity of clays dominated with montmorillonite, more common in temperate than tropical areas, and increases their sodium (Na) content. The amount of lime to be added depends upon soil pH, acid saturation, and the cation exchange capacity (CEC) of each soil (Soane and van Ouwwerkerk, 1994).

42.5.2 NPK Fertilizers

Fertilizers may be added to the soil to supply those chemical elements required for achieving plant growth that are in short supply. The most appropriate kinds and amounts of fertilizer to be used will vary with the type of soil and nutrient availability (Gershuny and Smillie, 1995). Plants require large amounts of macronutrients to grow normally. In soils that are acidic, some of these nutrients can be unavailable, impeding normal plant growth. Then either this acidity needs to be adjusted through soil system management

(Chapter 23), or fertilizers need to be applied to establish the proper nutrient concentrations for good plant growth (Demetz and Insam, 1999). Because most "dead" soils do not contain appreciable quantities of plant-available nitrogen or phosphorus, managing nitrogen and phosphorus nutrition becomes critical for successful revegetation (Hester and Harrison, 1997).

42.5.3 Sludge and Fly-Ash

Dried human sewage sludge is an organic amendment sometimes added to soils to improve properties such as bulk density (BD), water-holding capacity (WHC), soil structure, and CEC (Daniels and Haering, 1994). All these properties decrease soil compaction of the damaged soils and increase soil fertility. A variety of composted materials can be used similarly. For example, vegetable crops were grown for 3 years consecutively from a single application of spent mushroom substrate compost (Herrick and Wander, 1997). Compost applications tend to have an impact on pH, moving the pH level from acid to near neutral, and on SOM, raising this fraction, in one case from 3.6 to 5% (Haering et al., 2000).

Fly-ash, the particulate waste left in smokestack filters following coal burning, can be a useful waste product to apply to plants and degraded soil as it contains Ca, Mg, and traces of several metals. It has been found to be as effective a mine soil amendment as sewage sludge (Haering et al., 2000).

42.5.4 Earthworms

Soil fertility is increased by earthworms moving through the soil, as well as by their moving the soil itself through their guts, in the process improving the soil's aeration and the transport of nutrients from the subsoil (Sopher and Baird, 1982). Earthworm castings are richer in nutrients and bacteria than the surrounding soil, and earthworms themselves contain high concentrations of microbial cells and nutrients in their gut (Elsas et al., 1997). The addition of earthworms to soil has resulted in an increase of cell concentrations of *Pseudomonas fluorescens* in effluent from intact soil cores by several orders of magnitude (Duah-Yentumi et al., 1998).

The most common earthworm species is *Lumbricus terrestis*. Earthworms are the most prominent of the fauna found in the rhizosphere and play a role in physically breaking down organic materials into smaller pieces and simpler compounds that are then decomposed chemically by bacterial action (Gershuny and Smillie, 1995). Earthworms are the most visible agents assisting in the reduction of soil compaction, thereby increasing its fertility for plant growth. An increase in their populations is both cause and effect of the restoration of vegetation in degraded soils.

42.6 Discussion

As emphasized in the opening chapter, soil systems are hybrids of living and nonliving components, but the living elements are essential for the functioning and success of any soil system. This is very evident when trying to resuscitate bulk soil that has lost its life. A myriad of soil organisms can infuse its volume with chemical acquisitions, accumulations, and exchanges along with physical transport, movement, and structure, all in the course of carrying out their "normal" biological activity. Soil systems are unlike individual organisms in that they can be brought back to life from a "dead" condition.

The most visible evidence of life in the soil is the vegetation that grows on its surface, with roots extended varying distances into the soil below, interacting with the multifarious creatures that are living with and from (on) each other. They multiply and die endlessly so long as their requirements for oxygen (if aerobes) and energy (carbon compounds), as well as other nutrient inputs are satisfied. With such transformation and cycling of energy, the soil becomes a hospitable venue for plant growth.

While plants' requirements for minerals and water are the most evident ones, simply supplying these directly to plants will not restore vegetative systems. Much of the effect from providing chemical fertilizers and water when reviving plant growth in previously inert soils is due to the revival of soil organisms. These in turn create and improve the growth environment for higher flora. Physical interventions are in many ways more important than chemical ones, because reducing compaction is an essential first step for restoring the life to soil systems by getting oxygen and water into soil horizons and getting soil biota to begin the process of building up organic matter in the soil. Happily, once this process gets started, as explainable from the different sections of this chapter, it can become self-sustaining.

References

Alami, Y. et al., Rhizosphere soil aggregation and plant growth promotion of sunflowers by an exopolysaccharide-producing *Rhizobium* sp. strain isolated from sunflower roots, *Appl. Environ. Microbiol.*, **66**, 3393–3398 (2000).

Aldén, L., Demoling, F., and Baath, E., Rapid method of determining factors limiting bacterial growth in soil, *Appl. Environ. Microbiol.*, **67**, 1830–1838 (2001).

Baath, E., Adaptation of soil bacterial communities to prevailing pH in different soils, *FEMS Microbiol. Ecol.*, **19**, 227–237 (1996).

Brady, N.C. and Weil, R.R., *The Nature and Properties of Soils*, 13th ed., Prentice-Hall, Upper Saddle River, NJ (2002).

Brock, T.D. and Madigan, M.T., *Biology of Microorganisms*, 6th ed., Prentice-Hall, Englewood Cliffs, NJ (1991).

Daniels, W.L. and Haering, K.C., Use of sewage sludge for land reclamation in the central Appalachians, In: *Sewage Sludge: Land Utilization and the Environment*, American Society of Agronomy, Madison, WI (1994).

Demetz, M. and Insam, H., Phosphorus availability in a forest soil determined with a respiratory assay compared to chemical methods, *Geoderma*, **89**, 259–271 (1999).

Duah-Yentumi, S., Rønn, R., and Christensen, S., Nutrients limiting microbial growth in a tropical forest soil of Ghana under different management, *Appl. Soil Ecol.*, **8**, 19–24 (1998).

Elsas, J.D. van, Trevors, J.T., and Welligton, E.M.H., *Modern Soil Microbiology*, Marcel Dekker, New York (1997).

Evans, E.J. et al., Comparative studies on the growth of winter wheat on restored opencast and undisturbed soil, *Reclamation Reveg. Res.*, **4**, 223–243 (1986).

Garland, J.L., Patterns of potential C source utilization by rhizosphere communities, *Soil Biol. Biochem.*, **28**, 223–230 (1996).

Gelsomino, A.C. et al., Assessment of bacterial community structure in soil by polymerase chain reaction and denaturing gradient gel electrophoresis, *J. Microbiol. Methods*, **38**, 1–15 (1999).

Gershuny, G. and Smillie, J., *The Soul of Soil: A Guide to Ecological Soil Management*, 3rd ed., AgAccess, Davis, CA (1995).

Glazewski, J., *Environmental Law in South Africa*, Butterworth Publishers, Durban (2000).

Grayston, S.J. et al., Selective influence of plant species on microbial diversity in the rhizosphere, *Soil Biol. Biochem.*, **30**, 369–378 (1998).

Haering, K.C., Daniels, W.L., and Feagley, S.E., Reclaiming mined lands with biosolids, manures, and papermill sludges, In: *Reclamation of Drastically Altered Lands*, Barnhisle, R., Daniels, W.L., and Darmody, R., Eds., Agronomy Society of America, Madison, WI (2000).

Herrick, J.E. and Wander, M.M., Relationships between soil organic carbon and soil quality in cropped and rangeland soils: The importance of distribution, composition and soil biological activity, In: *Soil Processes and the Carbon Cycle*, Lal, R., Ed., CRC Press, Boca Raton, FL, 405–425 (1997).

Hester, R.E. and Harrison, R.M., *Contaminated Land and Its Reclamation*, Issues in Environmental Science and Technology 7, Royal Society of Chemistry UK, 73–81 (1997).

Kimball, J.W., The nitrogen cycle (online: http://users.rcn.com/jkimball.ma.ultranet/BiologyPages/N/NitrogenCycle.html; accessed 03/05/2005) (2005).

Lupwayi, N.Z., Rice, W.A., and Clayton, G.W., Soil microbial diversity and community structure under wheat as influenced by tillage and crop rotation, *Soil Biol. Biochem.*, **30**, 1733–1741 (1998).

Lynch, J.M., Microorganisms in their natural environments: the terrestrial environment, In: *Microorganisms in Action: Concepts and Applications in Microbial Ecology*, Lynch, J.M. and Hobbie, J.E., Eds., Blackwell Scientific Publications, London, 103–131 (1988).

Pinton, R., Varanini, Z., and Nannipieri, P., Eds. *The Rhizosphere: Biochemistry and Organic Substances at the Soil–Plant Interface*, Marcel Dekker, New York (2001).

Rimmer, D.L. and Colbourn, P., Problems in the Management of Soils Forming on Colliery Spoils. Report for the Department of the Environment, London, UK (1978).

Schlegel, H., *General Microbiology*, 7th ed., Prentice-Hall, New York (1993).

Smalla, K. et al., Bulk and rhizosphere soil bacterial communities studied by denaturing gradient gel electrophoresis: Plant-dependent enrichment and seasonal shifts revealed, *Appl. Environ. Microbiol.*, **67**, 4742–4751 (2001).

Soane, B.D. and van Ouwwerkerk, C., *Soil Compaction in Crop Production: Developments in Agricultural Engineering II*, Elsevier Science, Amsterdam, The Netherlands (1994).

Sopher, C.D. and Baird, J.V., *Soils and Soil Management*, 2nd ed. (1982).

Sylvia, D.M. et al., *Principles and Application of Soil Microbiology*, Prentice-Hall, Upper Saddle River, NJ (1999).

Wagner, G.H. and Wolf, D.C., Carbon transformations and soil organic matter formation, In: *Principles and Application of Soil Microbiology*, Sylvia, G.H. et al., Eds., Prentice-Hall, Upper Saddle River, NJ (1999).

Yang, C.H. and Crowley, D.E., Rhizosphere microbial community structure in relation to root location and plant iron nutritional status, *Appl. Environ. Microbiol.*, **66**, 345–351 (2000).

43

Impacts of Climate on Soil Systems and of Soil Systems on Climate

Rattan Lal

The Ohio State University, Columbus, Ohio, USA

CONTENTS

There are many interactions between climate and soil systems. Climate as an active contributor of soil formation affects the rate and intensity of weathering, the accumulation and dynamics of soil organic carbon (SOC) and nitrogen pools, the concentration of soluble salts and soil reactions, the net primary productivity (NPP) of soil systems, the quantity and quality of biomass residue returned, the activity and species diversity of soil biota including microorganisms and macrofauna such as earthworms and termites, the formation of organomineral complexes, and the stability of soil aggregates to natural and anthropogenic perturbations.

In addition to the impact of historical climatic factors on soil genesis, the projected climatic changes caused by an accelerated "greenhouse effect" can alter soil temperature and moisture regimes in the future with strong associated impacts on soil respiration, SOC, and nitrogen pools, soil structure, erosion, salinization, and biomass productivity. Soil processes can, fortunately, be managed to have some reciprocal influence on atmospheric processes because the sink capacity of the soil carbon pool to absorb and sequester atmospheric CO_2 can be enhanced through land-use conversion, restoration of degraded soils and ecosystems, and adoption of recommended management practices (RMPs) for agricultural, pastoral, silvicultural, and restorative land uses.

Soil processes can influence, in addition to CO_2, the fluxes of CH_4 and N_2O in the atmosphere, thus altering radiative forcing (the ability of these gases to retain long-wave radiation emitted by the earth within the atmosphere) and the global warming potential of various trace gases in the atmosphere. Furthermore, the quantity and quality of belowground biomass returned to the soil, especially in terms of the fine root mass, affects the soil's capacity as a carbon sink. This will also affect soil fertility, biomass productivity, and soil systems' ability to moderate environments. However, the net impact of increased CO_2 levels on terrestrial ecosystems is hard to assess because the effects of elevated CO_2 on carbon input to soils — on soil respiration and on plants' use of water and nutrients — often have contrasting responses to changes in microbial processes, NPP, and terrestrial carbon budgets.

The threat of an accelerated greenhouse effect due to anthropogenic emissions of greenhouse gases (GHGs) has enhanced the interest of researchers and policymakers in identifying sinks that can sequester gaseous emissions, especially carbon dioxide (CO_2). The world's soils constitute the third largest pool of carbon on the planet, estimated at 2400 billion metric tons, denoted as 2400 Pg (petagrams). Because this is such a huge number, we will use here the metric terminology and abbreviation, otherwise using more familiar units of measurement.

The oceanic pool of carbon, 38,400 Pg, is many time larger, while the geologic pool beneath the soil is almost twice as large as the carbon found in the soil. Of the 4130 Pg of carbon contained deeper in the earth, 3510 Pg is in coal, 230 Pg in oil, 140 Pg in gas, and 250 Pg in other fossil carbon. The atmospheric pool of carbon, on the other hand, is much smaller, only 760 Pg; and the biotic pool is smaller still, only 560 Pg (Lal, 2004). The volume of carbon in soils, considered to a 1-m depth, is divided between 1550 Pg as SOC and 950 Pg in soil inorganic carbon (SIC). It appears that 60% of the 8 Pg of annual anthropogenic carbon emissions (4.7 Pg), is absorbed by the ocean, land and soil, and unknown terrestrial sinks. This leaves 40% (3.3 Pg), building up in the atmosphere.

There is potential for enhancing the capacity of natural carbon sinks through anthropogenic management that reduces the buildup of carbon in the atmosphere. This chapter discusses interactions between soil system and atmospheric processes, with specific reference to the accelerated greenhouse effect. It reviews land use and soil management options that can enhance SOC sequestration, which will improve soil quality at the same time that it mitigates the accelerated greenhouse effect resulting from increasing CO_2 concentration in the atmosphere.

43.1 Climate and Soil Systems

The impact of climate on soil processes in general, and on soil organic matter (SOM) content in particular, has been recognized for more than two centuries (Jenny, 1980). The principal climatic factors having a strong impact on soil processes include: mean

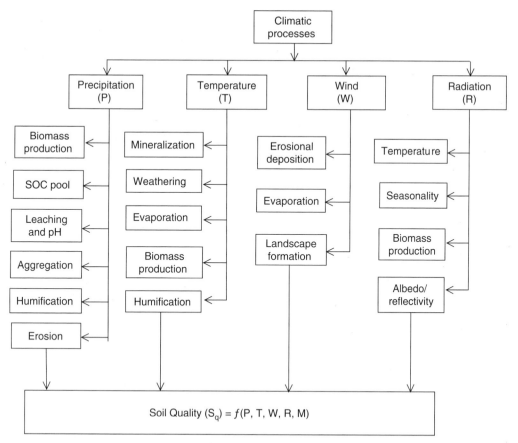

FIGURE 43.1

The effects of climatic factors on soil quality as mediated by anthropogenic perturbation or management. P, precipitation; T, temperature; W, wind; R, radiation; and M, management.

annual precipitation and its seasonal patterns, mean annual temperature and its seasonal variation, wind velocity affecting evaporation, and net solar radiation. Collectively, these four climatic factors affect annual evapotranspiration and the soil moisture budget which moderates the leaching of soluble salts. These factors directly affect the SOC pool, which is determined by NPP, the amount of biomass returned to the soil, and the rate of its decomposition.

The capacity of the soil to produce biomass and to moderate the environment is strongly influenced by climatic processes (Figure 43.1). Total and effective precipitation, minus total evaporation, and mean annual temperature affect both the SOC and SIC pools in the soil. There is also a close link between SOC and nitrogen concentrations, on the one hand, and total population and species diversity of soil biota, on the other. As a general rule, microbial biomass carbon increases with increases in SOC concentration. There is a strong cause-and-effect relationship between macrofauna populations and SOC concentration, earthworms being the predominant species in humid-climate soil systems and termites most numerous in semiarid and arid regions.

Another important impact of climate on soil is its effect on numerous aspects of soil structure. The arrangement of soil particles and voids affects the retention and transmission of fluids in soil, the extent of soil aeration, and the growth and proliferation of plant roots. There is a strong relationship between soil structure and root growth, on the one hand, and soil structure and SOC concentration, as influenced by climatic factors, on the other.

43.2   Effects of Projected Climate Change on Soil Systems

The climate changes currently projected to result from increases in atmospheric concentrations of CO_2 will increase soil temperature and alter patterns of precipitation and evaporation, leading to changes in soil moisture regimes. Such atmospheric changes can have strong impacts on soil processes, soil quality, biomass productivity, and the SOC pool and its dynamics.

43.2.1   Increase in the Atmospheric Concentration of CO_2 and Soil Processes

Given past and present trends, the atmospheric concentration of CO_2 is likely to double by 2100, to 750 parts per million by volume (ppmv). Specific impacts on soil processes of increasing atmospheric concentration of CO_2 are discussed in the following sections.

43.2.1.1   Impacts in the Rhizosphere

Soil faunal population can be strongly affected and in many ways by high atmospheric concentration of CO_2 as seen in Table 43.1. Yeates et al. (1997) observed significant changes in the soil fauna under ryegrass/white clover swards by increasing CO_2 concentration from 350 to 750 ppmv. This was especially evident for populations of nematodes, although the effects were not unidirectional. A marked decrease was seen in the bacteria-feeding *Rhabditis* species, probably due to increases in the populations of omnivores and predacious nematodes. There was an increase in earthworms and Enchytraeid populations, attributable to increases in belowground biotic productivity.

TABLE 43.1

Effects on Soil and Associated Biotic Processes from a Doubling of CO_2 Atmospheric Concentration

Location	Ecosystem	Process	Reference
Laboratory study	Rye grass	Increase in nematodes and earthworms	Yeates et al. (1997)
Estonia	Forest	Increase in forest growth (2–9%)	Kont et al. (2002)
Terrestrial ecosystem	Grassland	Increase in plant utilization of nitrogen	Hu et al. (2001)
Terrestrial ecosystem	Beech-spruce	Stimulation of fine root production	Wiemken et al. (2001)
Norway	Birch-spruce	Increase in mineralization and increase in growth	Solverness (1999)
Amazon	Forest	Increase in carbon storage by 0.2 Pg C/year	Tian et al. (2000)
Canada	Forest	20% increase in NPP	Medlyn et al. (2000)
Norway	Forested catchment	Increase in nitrogen mineralization	Verburg et al. (1999)
U.S.A.	Cropland	Increase in low molecular weight aliphatic quality of SOC	Islam et al. (1999)
Switzerland	Beech-spruce	Increase in stem diameter	Egli et al. (1997)

Closely associated with increases in belowground and root productivity are increases in nutrient mobility (Kont et al., 2002) as well as changes in soil microbial activity. Hu et al. (2001) have observed that increased atmospheric concentration of CO_2 enhances plant nitrogen uptake, microbial biomass carbon, and available carbon for microbes. Consequently, it reduces available soil nitrogen, accentuates nitrogen constraints on microbes, and reduces microbial respiration per unit biomass. Hu and colleagues concluded that increased atmospheric CO_2 concentration can alter the interaction between plants and microbes in favor of plants' greater utilization of nitrogen, thereby slowing microbial decomposition and enhancing the ecosystem carbon pool. However, the net effect of these complex and often counteractive processes on ecosystem carbon is still unclear, specifically with regard to rhizosphere processes. There are both positive and negative feedback loops in the complicated biotic realm underground.

Increased CO_2 concentration can affect mycorrhizal activity and fine root growth. Wiemken et al. (2001) reported that elevated CO_2 concentration stimulated fine root production in the top 10 cm of calcareous and siliceous soils, respectively, by 85 and 43%. Furthermore, the concentration of phospholipid fatty acids (PLFAs), typically reflecting populations of ectomycorrhizal fungi, was significantly higher under conditions of elevated CO_2 in a nutrient-rich calcareous soil. Islam et al. (1999) report that the total SOC concentration did not change in response to elevated CO_2. However, it appears that CO_2 enrichment may favor the accumulation of low-molecular-weight C and more aliphatic varieties, while ozone (O_3) stress can favor forms of carbon with higher molecular weight and more aromatic quality.

43.2.1.2 Mineralization

The rate of mineralization depends not just on the availability of substrate, but also on the availability of nutrients to support microbial populations (Hu et al., 2001). Experiments conducted by Solvernes et al. (1999) have showed that the fertilization effect of CO_2 on Norway spruce (*Picea abies*) and silver birch (*Betula pendula*) was largest when the plants were receiving a good nutrient supply. Yet the chemical composition of the plants was unaffected by the higher CO_2 concentration. An increase in mineralization, along with an accompanying increase in the amount of nitrogen released, was observed only when there was an increase in soil temperature. The increase in total SOC concentration was attributed to a possible increase in the exudation of organic compounds from silver birch roots induced by elevated CO_2 which would have affected soil microbial populations.

43.2.1.3 Net Primary Productivity and Nutrient Availability

Changes in rhizosphere processes, mineralization, and release of nutrients plus the fertilization effect can have a strong impact on the NPP of natural and managed ecosystems, with implications for the global carbon cycle. The general view for temperate forests is that global warming with rising CO_2 concentrations is likely to enhance this ecosystem carbon pool for the next 50–100 years (Nisbet, 2002).

Changes in NPP of the Amazon forest under elevated CO_2 levels have been modeled by Tian et al. (2000). They calculated the ecosystem carbon pool in the Amazon in 1980 to be 127.6 Pg, of which 94.3 Pg was in vegetation and 33.3 Pg in the SOC pool. They estimated that between 1980 and 1995, the ecosystem carbon pool was increased by 3.1 Pg. Of this, 1.9 Pg increase occurred in the vegetation pool and 1.2 Pg in the SOC pool. This means that the undisturbed Amazonian ecosystem accumulated 0.2 Pg C year^{-1} as a result of CO_2 fertilization effect over a 15-year period.

43.2.2   Soil Processes Associated with Increase in Soil Temperature

Atmospheric enrichment of GHGs is projected to increase global temperature over the coming century by 1.4–5.8°C with attendant increase in soil temperature. The projected latter increase is likely to have a strong impact on soil processes and soil quality. It is postulated that global warming may result in: (1) long-term decline in the SOC pool; (2) increase in soil wetness, waterlogging, and flooding during winter in temperate climates; (3) decrease in soil trafficability with adverse effects from tillage on soil aggregation and root survival; (4) increase in the frequency and severity of summer droughts; (5) increase in the mobility, dilution, and in-stream processing of pollutants, with adverse impacts on water quality leading to acidification, eutrophication, and discoloration of water supplies; (6) increase in adverse impacts on fresh water biota; and (7) drastic changes in soil water budgets (Nisbet, 2002).

Increase in soil temperature is an important anticipated impact of the accelerated greenhouse effect. This is likely to affect various soil processes, especially the respiration rate and the emission of CO_2 from soil into the atmosphere. Several experiments have been conducted to assess the impact of projected global warming on soil respiration (Table 43.2). The results of some of these evaluations of the effects of soil warming are briefly described below.

43.2.2.1   *Increased Respiration and CO₂ Emissions*

Increases in soil temperature generally lead to an increase in soil respiration and an attendant increase in CO_2 emissions. However, this simple cause-effect relationship can be drastically altered by other factors. Estimates of future warming are now greater than earlier projections because of expected positive feedback, with greater release of GHGs from soil/terrestrial ecosystems in response to climate warming. Accelerated plant growth would seem to reduce the supply of carbon in the atmosphere. However, this accelerated process can diminish the soil carbon pool while at the same time increasing the stock of carbon in the atmosphere.

Raich et al. (2002) estimated emissions from terrestrial ecosystems for the 15-year period 1980–1994. Mean annual global CO_2 flux over this period was estimated at 80.4 Pg C (range 79.3–81.8 Pg). On a global scale, annual soil CO_2 flux correlated with mean annual temperature, with a slope of 3.3 Pg C year^{-1} °C^{-1}. Raich and colleagues concluded that global warming is likely to stimulate CO_2 emissions from soils. Similar conclusions were arrived at by Kirschbaum (1995), who observed that each 1°C increase in temperature could lead to a loss of 10% of SOC pool at higher latitudes with an annual mean temperature of 5°C, and to a loss of 3% of SOC pool in soils with an annual mean temperature of 30°C. These differences are much greater in absolute amounts because cooler soils contain a larger SOC pool.

Several soil warming experiments have been conducted to test the hypothesis of positive feedback based on the assumption that the observed sensitivity of soil respiration to temperature under present climate conditions would also hold in a warmer climate. Luo et al. (2001) measured soil respiration under tall grass prairie ecosystems in the U.S. Great Plains under artificial warming of about 2°C. Their results indicated that temperature sensitivity of soil respiration decreased because of acclimatization under warming, and that acclimatization would become greater at high temperatures.

Warming also has the potential to stimulate growth and to compensate for the SOC loss from soil. In the eastern Amazon, Cattanio et al. (2002) observed that soil drying caused by warming reduced CO_2 emissions because root growth was lower in dry soil. On the basis of soil warming experiments conducted in Arctic, temperate and tropical soils,

TABLE 43.2

Effects of Soil Warming on Soil Organic Carbon Dynamics

Impact	Location	Land Use	Reference
Decrease in permafrost and soil freezing			
Permafrost degradation and disappearance of lowland birch forest	Central Alaska	Permafrost	Jorgenson et al. (2001)
Decrease in soil freezing	Finland	Soil frost	Venalainen et al. (2001)
Reduce tundra by 50% over 100 years	Iceland	Arctic	Heal (2001)
Increase in CO_2 flux			
CO_2 emission	Great Plains, U.S.A.	Tall grass prairie	Luo et al. (2001)
Increase CO_2 flux	Greenland	Tundra	Mertens et al. (2001)
Increase in emission by 3.3 Pg C with 1°C increase in temperature	World	Global ecosystems	Raich et al. (2002)
Increase in emission of CH_4 and N_2O			
N_2O emission	Germany	Bavarian hills	Kamp et al. (1998)
Emission of CH_4 and N_2O	Germany	Livestock	Seidl (1998)
Change in SOC pool			
Increase in SOC pool	Russia	Diverse region	Stolbovoi and Stocks (2002)
1% reduction of SOC in cool climate with 1°C increase in temperature	Global	Diverse ecosystems	Kirschbaum (1995)
SOC losses more in fine than coarse-textured soils	Argentina	Pampas	Hevia et al. (2003)
Change in mineralization rate			
Nitrogen mineralization	Southern England	Calcareous grassland	Jamieson et al. (1998)
Increase in lignin	Southern Norway	Forested catchment	Verburg et al. (1999)
Increased litter decomposition	Maine, U.S.A.	Red spruce and maple	Rustad and Fernandez (1998)
Increased mineralization (23 kg C m^{-2})	Alaska	Peat soils	Hartshorn et al. (2003)
Response depends on nutrients	Alaska	Boreal forests and Arctic tundra	Hobbie et al. (2002)
Response differs among soils	Norway, Japan, Malaysia	Diverse soils	Bekku et al. (2003)
Increase in decomposition in boreal region	Europe	Diverse ecosystems	Couteaux et al. (2001)
No increase in decomposition	U.K.	Global scale	Glardina and Ryan (2000)

(continued)

TABLE 43.2 (Continued)

Impact	Location	Land Use	Reference
Increase in soil erosion	Malaysia, Indonesia, Hawaii	Forest ecosystem	Lo and Cai (2002)
	Mediterranean basin	Agricultural ecosystem	Alexandrov et al. (2002)
Increase in soil salinity	Cuba	Agricultural ecosystem	Utset and Borroto (2001)
	Cuba	Agricultural ecosystem	Utset et al. (1999)
Adverse effects on plant growth	Argentina	Pampas	Alvarez and Alvarez (2001)
	Eastern Canada	Perennial forages	Belanger et al. (2002)
	Austria	Agricultural ecosystems	Alexandrov et al. (2002)
Biome shifts	U.K.	Forest	Broadmeadow (2000)
	Europe	Agricultural lands	Olesen and Bindi (2002)
Changes in temperature	Germany	Livestock	Seidl (1998)

Impact details:
- Increase in soil erosion
- Increase in soil salinity
- Adverse effects on plant growth: Reduction in biomass; Increase risks of winter injury; Decrease in crop growth duration and yields
- Biome shifts: Distribution on woodland, flora and fauna; Northward expansion of cropping and enhanced productivity
- Changes in temperature: Emission of CH_4 and N_2O

Bekku et al. (2003) have concluded that the response of microbial respiration to climate warming may differ among soils of different latitudes, a not surprising finding.

Hobbie et al. (2002) concluded that increased nutrient mineralization associated with decomposition of peat in northern latitudes will stimulate primary production and ecosystem carbon gain, offsetting or even exceeding the carbon lost through decomposition, a negative feedback to climate warming. A similar observation of negative feedback was made for Russian soils by Stolbovoi and Stacks (2002), who accept that predicted climate warming will enhance the SOC pool in soils of Russia. As reported above, Hu et al. (2001) have observed that carbon accumulation in the terrestrial biosphere could partially offset the effects of anthropogenic emissions because increased CO_2 can alter the interaction between plants and microbes in favor of plant utilization of nitrogen; this would slow microbial decomposition and increase ecosystem carbon accumulation.

43.2.2.2 Nitrogen Mineralization and Soil Warming

In many ecosystems, carbon cycling is closely linked to the cycling of nutrients, particularly nitrogen. Changes in the decomposition rate of SOM due to climate change may affect mineralization of nitrogen, emission of N_2O and NO, and plant growth. In the eastern Amazon, Cattanio et al. (2002) observed that soil drying caused by high temperatures elevated N_2O and NO fluxes. A 21-month field experiment in southern Germany by Kamp et al. (1998) showed that soil warming by 3°C above ambient temperature increased N_2O emissions threefold from heated fallow plots compared with controls during the summer, and increased N_2O emissions from both fallow and wheat plots during the winter.

Nitrogen mineralization experiments in southern England by Jamieson et al. (1998) have suggested that water availability is the main constraint to microbial processes and plant growth. This conclusion accords with results of Cattanio et al. (2002) from the Amazon. Observed treatment effects were attributed to changes in organic carbon and nitrogen input in plant litter resulting from the direct impact of climatic manipulations on plant growth, death, and senescence.

43.2.2.3 Soil Erosion, Salinity, and Water Pollution

It is widely hypothesized that the accelerated greenhouse effect would exacerbate the hazard of soil erosion because of: (a) increase in the frequency and intensity of extreme rainfall events; (b) decreases in SOC concentration with an attendant decline in soil structure and corresponding increase in soil erodability; and (c) decrease in protective ground cover due to a possible increase in litter decomposition. Lo and Cai (2002) based on data from three tropical regions of Malaysia, Indonesia, and Hawaii have predicted that temperature increases would increase organic matter decomposition rates, reduce understory plant cover, and increase susceptibility of tropical forest ecosystems to surface erosion.

Conversion of tropical forests to agricultural ecosystems in rugged topography and steep slopes on highly weathered soils, which already creates a soil erosion hazard, would be accentuated by climate changes. Increasing trends in soil degradation with global warming have been observed in the Mediterranean Basin. Rodolfi and Zanchi (2002) have concluded that predicted temperature increases will lead to stronger and more prolonged drought periods and to decreases in SOC content, decline in soil structure, increases in salinization, and northward extension of arid conditions or desertification. The latter would be exacerbated by reduction in soil water retention and an increase in soil erodability. Rodolfi and colleagues postulate that the interaction of changing land use and agricultural practices with potential climate warming and socioeconomic conditions will determine the trajectory of soil degradation.

Low productivity on marginal lands and their abandonment may exacerbate soil degradation along with massive soil erosion and irreversible conditions for future land use. With reference to some terrestrial ecosystems in Estonia, Kont et al. (2002) observed that soil warming would increase nutrient mobility and enhance nutrient losses through leaching. Similar results were reported from Norway, where Solvernes et al. (1999) found that increased mineralization would leach out nitrogen and aluminum. Therefore, climate warming could adversely affect water quality through increased leaching of pollutants. Utset and Borroto (2001) have concluded that with projected global warming, irrigation in the San Antonio del Sur Valley in southeastern Cuba would increase the hazard of soil salinity within 15 years of the start of irrigation.

43.2.2.4 Agricultural and Forest Productivity

The impact of global warming on biomass productivity is likely to depend on latitude, soil type, and precipitation (Rosenzweig and Hillel, 1998, 2000). At higher latitudes with their shorter growing season, soil warming and the associated increase in mineralization are expected to stimulate NPP and ecosystem carbon gains (Hobbie et al., 2002). Kont et al. (2002) predict a 2–9% increase in harvestable timber in highly productive forest sites in Estonia. A model study on forest ecosystems in Sweden and Australia by Medlyn et al. (2000) showed that a 2°C increase in temperature and CO_2 fertilization would increase NPP by 20% initially, but this later equilibrated at 10–15% on a long-term basis. These responses were similar in both cool and warm climates.

Alexandrov et al. (2002) assessed the vulnerability of major agricultural crops to climate change in northeastern Austria with a projected increase in annual temperatures of 0.4–4.8°C from the 2020s to the 2080s. They observed that warming will decrease the crop growing duration for certain crops, for example, a gradual increase in air temperature for winter wheat would probably reduce grain yield, and although soybean yields might increase with incremental warming and increase in precipitation, they would decrease where climates become drier. Crop yields may also be affected by changes in sowing date and the duration of the growing season.

Olsen and Bindi (2002) predict that warming in Europe will cause a northward expansion of suitable cropping areas leading to increase in productivity and resource-use efficiencies. In eastern Canada, Belanger et al. (2002) predict that an increase of 2–6°C in the minimum temperature during winter months will affect the survival of forage crops. There would be more winter injury to perennial forage crops because of less cold-hardening during autumn and reduced protective snow cover during the winter. This is a somewhat unexpected way in which warmer temperatures could have adverse effects, showing one more way in which climate change can affect the functioning of soil systems above- and belowground.

43.3 Effects of Soil Systems on Projected Climate Change

Soil can be a source or sink for atmospheric CO_2 depending on land use, on residue management and tillage methods, and on cropping systems (Table 43.3 and Figure 43.2). Soil systems are a source of atmospheric CO_2 when natural ecosystems are converted to agricultural ecosystems; when plow-based tillage methods are used for seedbed preparation; when surface and subsurface drains remove excess water; when crop residues are removed or burned; when the nutrients removed in harvested produce are

TABLE 43.3

Anthropogenic Activities That Influence Emission and Sequestration of Greenhouse Gases

Activities That Enhance GHG Emissions	Activities That Sequester GHG Emissions
Soil tillage	No-till farming
Removing crop residue	Residue retention as mulch
Summer fallowing	Winter cover crops
Excessive use of nitrogen and other fertilizers	Integrated nutrient management and BNF
Indiscriminate use of fertilizers	Precision farming
Excessive use of pesticides	Integrated pest management
Deforestation	Afforestation
Soil-degrading land use	Soil-restorative land use
Monoculture	Polyculture and mixed farming
Erosion-promoting farming systems	Conservation-effective measures

not replenished from organic or inorganic sources; and when grazing lands are subjected to uncontrolled and excessive stocking rates.

Soil systems are a source of atmospheric CO_2 whenever the amount of biomass returned above- and belowground is less than the losses of carbon from the ecosystem caused by mineralization, erosion, and leaching. On the other hand, soil systems are a sink for atmospheric CO_2 when degraded soils and marginal lands revert back to natural ecosystems or are converted to forest plantations; when plow-based methods of seedbed preparation are replaced by conservation tillage or no-till farming; when nutrients harvested in farm produce are replenished by practices based on integrated nutrient management; and when conservation-effective measures are adapted to reduce losses by erosion, leaching, and volatilization.

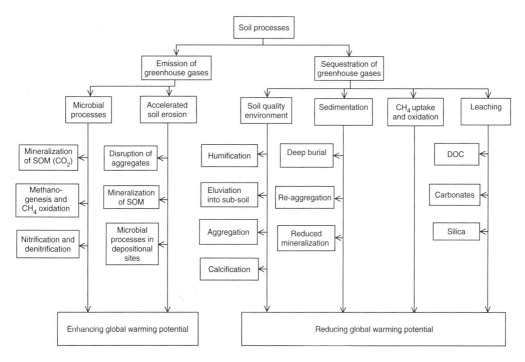

FIGURE 43.2

Processes involved in effects of soils on emission and sequestration of greenhouse gases (DOC, dissolved organic carbon).

43.3.1 Soil Systems as a Source of Greenhouse Gases

Historically, soil, terrestrial, and agricultural ecosystems have been a major source of CO_2, CH_4, N_2O, and NO_x. Agricultural ecosystems, especially rice paddies and livestock production, are principal sources of CH_4. N_2O is emitted are a result of the application of nitrogenous fertilizers, manures, and soil processes that affect nitrification and denitrification.

The impact of anthropogenic activities on CO_2 emission from terrestrial ecosystems into the atmosphere goes back 10,000 years, to the dawn of settled agriculture, and CH_4 emission increases go back 5000 years with the irrigated cultivation of rice and large-scale domestication of livestock (Ruddiman, 2003). The emission of CO_2 from land-use change between 1800 and 1994 has been estimated at 100–180 Pg C, and at 24 ± 12 Pg C just between 1980 and 1999 (Sabine et al., 2004), i.e., rising from <1 Pg C year^{-1} to approximately 1.25 Pg C year^{-1} in recent years. In comparison, the emission of CO_2 from fossil fuel combustion and cement production has been estimated at 244 ± 20 Pg between 1800 and 1994, and 117 ± 5 Pg between 1980 and 1999 (Sabine et al., 2004). Until the 1970s, more CO_2 was emitted from land use conversion and soil cultivation than from fossil fuel combustion, but that balance has now shifted. Between 1990 and 2000, about 20% of the annual anthropogenic emissions are attributed to tropical deforestation, biomass burning, and soil cultivation (IPCC, 2001).

The atmospheric concentration of CH_4 (1745 parts per billion [ppb] in 1998) is increasing at a rate of 7 ppb year^{-1} (IPCC, 2001). Atmospheric CH_4 concentration has increased by about 150% since 1750. The most important terrestrial sources of CH_4 are largely agricultural: wetlands, ruminants, rice agriculture, and biomass burning. For the total atmospheric pool of CH_4 estimated at 4850 million metric tons, annual emissions from all sources are estimated at 598 million metric tons (Prather et al., 2001). In contrast, the total annual sink capacity is about 576 million metric tons. An imbalance or net emission of 22 million metric tons of CH_4 year^{-1} means that atmospheric concentration of CH_4 increases by 0.45% year^{-1}.

The atmospheric concentration of N_2O has increased steadily from 270 ppb in the pre-industrial era to 314 ppb in 1998. This GHG is presently increasing at the rate of 0.8 ppb year^{-1} or 0.25% year^{-1} (IPCC, 2001). Principal agricultural sources of N_2O are soils, biomass burning, and fertilizers. The total atmospheric pool of N_2O is estimated at 1510 million metric tons nitrogen. As the total source is 14.7 million metric tons year^{-1} and total sink is 12.6 million metric tons year^{-1}, this leaves an annual imbalance of 3.8 million metric tons nitrogen (Prather et al., 2001).

The source capacity of soil for all GHGs is accentuated by soil degradation processes, for example, erosion, leaching, acidification, elemental/nutrient imbalance, compaction, and anaerobiosis. The extent and severity of soil degradation are exacerbated by land misuse and soil mismanagement, which are driven by increases in population of human and animals.

43.3.2 Soil Systems as a Sink for Greenhouse Gases

Soil can be a net sink for CO_2 and CH_4 through adoption of land uses and soil restorative measures that enhance soil quality. There are strong interactions between soil and climatic processes with regard to soil quality that affect the SOC pool and its dynamics (Figure 43.3). Pedologic and climatic processes influence soil quality through their effects on humification and SOM dynamics, soil structure and tilth, erosion and deposition, leaching, elemental cycling, and soil moisture regime.

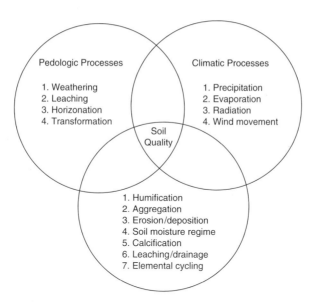

FIGURE 43.3
Interactive effects of soil and climatic processes.

Agricultural activities that sequester emissions and make soils a net sink include: converting from plow tillage to no-till farming in conjunction with crop residue mulch and cover crops; replenishing soil fertility through judicious use of chemical fertilizers and organic manures, for example, compost, green manure, sludge, and biosolids, along with biological nitrogen fixation (BNF); using precision or soil-specific farming practices; restoring agriculturally marginal soils and degraded ecosystems by afforestation; and converting surplus land to perennial culture and conservation-effective measures.

The balance among processes that affect SOC dynamics, sketched in Figure 43.4, determine the sink capacity of soil for atmospheric CO_2. Gains in SOC depend on the quantity and quality of the aboveground biomass and detritus and on the amount of belowground root biomass and exudates returned, plus any deposition of organic

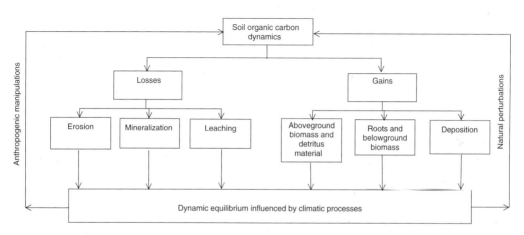

FIGURE 43.4
Processes affecting soil organic carbon dynamics.

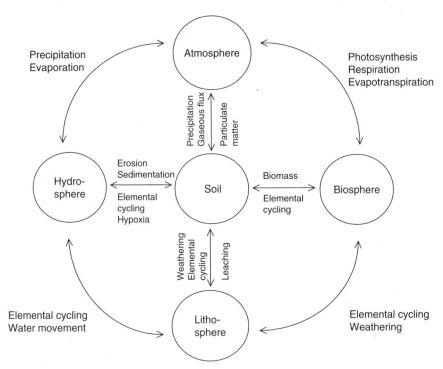

FIGURE 43.5
Interactive processes in soil with its environment.

material, e.g., windblown dust. Sustainable management is achieved by adopting RMPs so that gains exceed losses, leading to net SOC sequestration.

The rates of SOC sequestration achievable through adoption of RMPs are higher for cool and humid climates than for warm and dry ones. In general, rates of SOC sequestration can range from 50–200 kg ha^{-1} year^{-1} for arid regions, and 300–500 kg ha^{-1} year^{-1} for humid regions (Lal, 2004). In some exceptional cases, rates of SOC sequestration in agricultural soils may be as high as 1 t C ha^{-1} year^{-1} (Lal, 2004). Restoration of the SOC pool upon adoption of RMPs follows a sigmoid curve with the maximum rates attained within 5–10 years after land-use conversion.

The total sink capacity of the soil is approximately equal to the historic loss, i.e., 30–50% of the antecedent pool in soils of temperate regions and 50–75% in soils of the tropics. Most agricultural soils have lost 30–50 t C ha^{-1}. It is expected that about two-thirds of this loss can be sequestered through land-use conversion (Lal, 2004).

There is also some potential of sequestration of SIC through the formation of secondary carbonates and the leaching of dissolved carbonates into the groundwater (Figure 43.5). The formation of secondary carbonates is facilitated through biogenic processes accentuated by application of biosolids and other organic amendments. Leaching of carbonates occurs most commonly in soils irrigated with good quality water. The rate of formation of secondary carbonates is low, usually less than 50 kg ha^{-1} year^{-1}.

The sink capacity of the terrestrial biosphere has been estimated as having gone from 61 to 141 Pg C between 1800 and 1994, with another 39 ± 18 Pg C between 1980 and 1999 (Sabine et al., 2004). Pacala and Socolow (2004) have proposed 15 technological options for stabilizing net additions of CO_2 at 6 Pg. Of these 15, the three options related to soil use and management are described in the following subsections.

43.3.2.1 Use of Biofuels

These derive from biomass either for direct combustion or conversion to ethanol or hydrogen, recycling atmospheric CO_2 and offsetting emissions due to fossil fuel combustion. Biofuel production requires establishment of high-yielding plantations ($15 \text{ t ha}^{-1} \text{ year}^{-1}$) of switch grass, willow, poplar, and other site-specific species. Residues of grain crops, e.g., maize, wheat, and soybean, cannot be removed without jeopardizing soil quality and exacerbating soil erosion risks.

Offsetting 1 Pg C year^{-1} in the atmosphere through conversion to biofuels would require production of 34 million barrels of ethanol per day by 2054. This is 50 times larger than the 2004 rate of ethanol production and would require 250 million hectares of land committed to high-biomass plantations (Pacala and Socolow, 2004). Establishing such plantations would enhance SOC sequestration through a decrease in soil erosion risks and an increase in the belowground biomass returned to the soil. Still, huge investments would be required to have much impact on a global atmospheric scale.

43.3.2.2 Afforestation

Conversion of degraded soils and marginal agricultural lands to prospering forest plantations is another option to enhance the terrestrial carbon pool in biomass and soil. Strategies that could offset 1 Pg C year^{-1} in the atmosphere include: reducing tropical deforestation to zero; reforesting 400 million ha in the temperate zone or 250 million ha in the tropics; or establishing new forests on 300 million ha (Pacala and Socolow, 2004). Such conversions would have major resource allocation implications.

43.3.2.3 Management of Agricultural Soils

World soils have a potential of SOC sequestration at the rate of $0.4-1.2 \text{ Pg C year}^{-1}$ (Lal, 2004). This potential can be realized by some combination of: converting conventional plowing to no-till farming (Chapters 22 and 24); eliminating summer fallow; application of biosolids on the soil; incorporating cover crops into rotation cycles (Chapter 30); and restoring degraded soils and ecosystems (Chapter 42).

Each of these three options would also have beneficial effects on soil quality, water quality, biodiversity, NPP, and meso- and macroclimate. Continuous ground cover and return of biomass to the soil will increase soil biotic activity (macro- and microfauna) and the SOC pool, as well as improve soil structure and tilth. These soil processes in turn would have synergistic benefits for NPP and the environment, as well as enhancing agricultural production. This is why soil-system management should be given more attention by scientists and policymakers.

43.4 Soil Systems and Climate

The close links between soil systems and their environment are mapped out in Figure 43.5. Soil is in dynamic equilibrium with the lithosphere through weathering and new soil formation, elemental leaching into the groundwater, and elemental cycling. Soil systems concurrently interact with the hydrosphere. Transport of organic material from soils into the ocean has a strong impact on the coastal ecosystems.

The severe problem of hypoxia in the Gulf of Mexico recognized since 1935 is an example of such an interaction (Ferber, 2001; Snyder, 2001). The area affected varies from year to year, from $<1000 \text{ km}^{-2}$ in 1988, almost $20,000 \text{ km}^{-2}$ in 1999, and 4000 km^{-2} in

2000. The 5-year average of the area affected (1996–2000) was 14,128 km^{-2} (Snyder, 2001). The current goal is to reduce the area to <5000 km^{-2} by 2015. A national plan has been developed to reduce nitrogen coming down the Mississippi by 30% by 2015 (Ferber, 2004). Recently, a new "dead zone" has been discovered off the Oregon coast, attributed to a possible change in sea currents (Service, 2004).

The importance of interaction between soil systems and the biosphere cannot be overstated. All terrestrial life depends on soil systems, and the magnitude of NPP is a function of soil quality. In turn, soil quality depends on the nature of vegetation, the partitioning of the NPP into above- and belowground components, and the diversity and complexity of species mixes. The important concern of achieving and maintaining global food security depends on this close link between soil quality and NPP.

Two issues of importance emerging from the close link between soil and climatic processes include: (1) gaseous exchange between the soil and the atmosphere, which is the main focus of this chapter; and (2) interdependence between soil moisture regime and precipitation.

43.4.1 Soil Systems and Atmosphere

Regarding the link between soil and atmospheric processes, soil processes that have direct effects on atmospheric chemistry and emission/sequestration of GHGs include: humification vs. mineralization, aggregation vs. dispersion of soil, methanogenesis vs. methane oxidation or uptake, denitrification/nitrification vs. BNF, leaching vs. precipitation, erosion vs. new soil formation and deposition, and eluviation vs. illuviation. Land use and soil management practices should be preferred that enhance the SOC pool through increases in humification, soil aggregation, BNF, deposition, and illuviation. Soil restorative processes can be enhanced through sustainable management of soil and water resources based on soil-conserving rather than soil-exploiting or soil-mining practices.

With increasing population, especially in developing countries where it is likely to double between 2000 and 2030, enhancing food production is a supreme concern (Wild, 2003). Decline in soil fertility due to nutrient depletion, serious as it is in Africa and elsewhere in areas of subsistence farming, can be offset by the addition of fertilizers, albeit at often prohibitively high cost. More ecologically compatible and cost-effective alternatives involve the use of biofertilizers and elemental cycling through microbial processes.

Replenishing the depleted SOC pool is a serious challenge especially in tropical climates. Despite the difficulty and slow rate of its enhancement, the SOC pool has to be replenished. There is a little choice in this matter for reasons of both food production and atmospheric stability. The importance of soil microbial processes in enhancing SOC pool for improving soil quality cannot be overstressed. In addition, the SOC pool and its dynamics, NPP and agronomic productivity, elemental/material transport, and numerous other processes also depend on the soil moisture regime. The latter is influenced by a coupling between the soil and atmospheric processes.

43.4.2 Soil Systems and Hydrology

Fresh renewable water is a very scarce resource. Irrigated agriculture, the most productive farming system in arid and some semiarid environments, supplies 40% of the world's food crops (Postel, 1995). In turn, agriculture drastically affects water flow and its quality. About 2.3 billion people live in river basins under water stress with annual per capita water availability below 1700 m^{-3}. Of these, 1.7 billion people reside in river basins with high stress, where water availability already falls below 1000 m^{-3} per capita annually (Johnson et al., 2001). Groundwater resources are also being excessively exploited and heavily polluted. Similar to carbon, freshwater resources are also undervalued and suffer from the "tragedy of the commons."

There is also a close link between soil moisture and the global hydrological cycle mediated through precipitation. Water in soil is eventually returned to the atmosphere through evaporation, where it is the source of water that falls as precipitation. Koster et al. (2004) have reported a strong coupling between soil moisture and precipitation and outlined hotspots in the Sahel and Indian subcontinent with strong local effects on precipitation. Enhancement of soil quality through SOC sequestration with a positive impact on soil moisture regime may also favorably affect precipitation. This is a subject area where research is inadequate, but expanding.

43.5 Discussion

The effects of projected global warming on soil processes are difficult to generalize because they are complex, site-specific, and with uncertain net effects due to the interaction of different factors. Increasing atmospheric concentration of CO_2, with attendant increase in global temperature and change in precipitation, will certainly have a strong impact on numerous soil processes. Rhizosphere processes affected by soil warming include greater activity and species diversity of soil fauna, higher production and turnover rate of fine roots, mineralization of nitrogen, and leaching of elements. These processes will enhance or deplete the ecosystem carbon pool depending on ecoregional characteristics.

Increases in soil temperature caused by projected global warming may also affect soil quality. Soil warming may lead to degradation of permafrost, a decrease in area of soil freezing, an increase in soil respiration and CO_2 emission in some cases, and to decrease in decomposition due to desiccation or reduction in soil moisture in other situations. Positive feedback leading to increased CO_2 emissions can be counterbalanced by an increase in NPP due to enhanced nitrogen mineralization and longer duration of growing seasons. At the same time, the risks of soil erosion and salinization may increase, and increases in leaching of chemicals could adversely affect water quality.

Impacts of global warming on agronomic yield will be invariably site-specific. NPP is likely to increase at higher latitudes due to a longer growing season, while it decreases in the tropics and subtropics primarily due to accentuation of soil moisture stress. A majority of poor countries are in the tropics and subtropics, where yield declines are likely to affect adversely their underdeveloped economies, contributing to growing global inequality.

The world's soils can be either a source or a sink for atmospheric CO_2 depending on land use and management. Improving and sustaining high soil quality through increases in the SOC pool and its residence time is a win–win option. It can increase and maintain food production to meet the food demand of the growing world population, while improving the quality of renewable fresh water resources and mitigating adverse climate change that results from greater atmospheric concentrations of GHGs. The importance of making positive connections between soil and atmospheric processes cannot be overemphasized nor should it be ignored.

Management strategies that can exploit favorable synergism among these processes have not been made sufficiently available to land managers, especially to small landholders in the tropics and subtropics. The literature documents many appropriate technologies that could be taken up, most effectively with participatory approaches. Part III of this book has given detailed evidence on such options. The basic principles of soil quality enhancement and the importance of SOM management apply across international boundaries. The temporal dimensions of technological change cover both short and long time horizons. Adoption of RMPs can have short-term impacts on agronomic yield, but also long-term impacts on NPP, water quality, and net atmospheric emission of CO_2 mitigating future climate change.

Some components of RMP strategies can be adopted immediately, within a short time horizon, while other components can only be adopted after certain biophysical and socioeconomic thresholds have been crossed. The horizontal diffusion of technology is extremely important, especially for resource-poor, small landholders. When a RMP is adopted by some farmers, others can do so by learning from their neighbors. The widespread adoption of improved technology in developing countries will probably depend most on the horizontal spread of innovative technologies.

Projected climate changes are likely in the foreseeable future to cause accelerating declines in soil quality through reduction in the SOC pool, increased risks of soil erosion and salinization, alterations in elemental cycling (N, P, Ca, Mg), and changes in the activity and species diversity of soil fauna. Economic and environmental consequences of these adverse impacts can be minimized through revised management of soil resources, however. The sustainability and effectiveness of land-use choices and management strategies will depend on a sound understanding of the interactions between soil system and atmospheric processes as well as on supportive policies, discussed in the next chapter.

References

Alexandrov, V. et al., Potential impact of climate change on selected agricultural crops in northeastern Alaska, *Global Change Biol.*, **8**, 372–389 (2002).

Alvarez, R. and Alvarez, C.R., Temperature regulation of soil carbon dioxide production in the humid pampa of Argentina: Estimation of carbon fluxes under climate change, *Biol. Fertil. Soils*, **34**, 282–285 (2001).

Bekku, Y.S. et al., Effect of warming on the temperature dependence of soil respiration rate in arctic, temperate and tropical soils, *Appl. Soil Ecol.*, **22**, 205–210 (2003).

Belanger, G. et al., Climate change and winter survival of perennial forage crops in Eastern Canada, *Agron. J.*, **94**, 1120–1130 (2002).

Broadmeadow, M., Ed., *Climate Change: Impacts on U.K. Forests*, Forestry Commission Bulletin No. 125, Edinburgh, UK (2002).

Cattanio, J.H. et al., Unexpected results of a pilot throughfall exclusion experiment on soils emissions of CO_2, CH_4, N_2O and NO emissions in eastern Amazonia, *Biol. Fertil. Soil*, **36**, 102–108 (2002).

Couteaux, M.M. et al., Decomposition of 13-C labelled standard plant material in a latitudinal transect of European coniferous forests: Differential impact of climate on the decomposition of soil organic matter compartments, *Biogeochemistry*, **54**, 147–170 (2001).

Egli, P., Kroner, C., and Kroner, C., Growth responses to elevated carbon dioxide and soil quality in a beech-spruce model ecosystem, *Acta Oecologia*, **18**, 343–349 (1997).

Ferber, D., Keeping the stygian waters at bay, *Science*, **292**, 968–973 (2001).

Ferber, D., Dead zone fix not a dead issue, *Science*, **305**, 1557 (2004).

Glardina, C.P. and Ryan, M.G., Evidence that decomposition rates of organic carbon in mineral soil do not vary with temperature, *Nature*, **404**, 858–861 (2000).

Hartshorn, A.S., Southard, R.J., and Bledsoe, C.S., Structure and function of peatland-forest ecotones in southeastern Alaska, *Soil Sci. Soc. Am. J.*, **67**, 1572–1581 (2003).

Heal, O.W., Potential responses of natural terrestrial ecosystems to arctic climate change, *Buvisindi*, **14**, 3–16 (2001).

Hobbie, S.E., Nadelhoffer, K.J., and Hogberg, P., A synthesis: The role of nutrients as constraints on carbon balances in boreal and arctic regions, *Plant Soil*, **242**, 163–170 (2002).

Hevia, G.G. et al., Organic matter in size fractions of soils of the semi-arid Argentian: Effects of climate, soil texture and management, *Geoderma*, **116**, 265–277 (2003).

Hu, S. et al., Nitrogen limitation of microbial decomposition in a grassland under elevated CO_2, *Nature*, **409**, 188–191 (2001).

IPCC, *Climate Change 2001: The Scientific Basis*, Cambridge University Press, Cambridge, UK (2001).

Islam, K.R., Mulchi, C.L., and Ali, A.A., Tropospheric carbon dioxide or ozone enrichments and moisture effects on soil organic carbon quality, *J. Environ. Qual.*, **28**, 1629–1636 (1999).

Jamieson, N. et al., Soil N dynamics in a natural calcareous grassland under a changing climate, *Biol. Fertil. Soils*, **27**, 267–273 (1998).

Jenny, H., *The Soil Resource: Origin and Behavior*, Springer, NY (1980).

Johnson, N., Revenga, C., and Echeverria, J., Managing water for people and nature, *Science*, **292**, 1071–1072 (2001).

Jorgenson, M.T. et al., Permafrost degradation and ecological changes associated with a warming climate in central Alaska, *Climatic Change*, **48**, 551–579 (2001).

Kamp, T. et al., Nitrous oxide emissions from a fallow and wheat field as affected by increased soil temperatures, *Biol. Fertil. Soils*, **27**, 307–314 (1998).

Kirschbaum, M.U.F., The temperature dependence of soil organic matter decomposition and the effect of global warming on soil organic C storage, *Soil Biol. Biochem.*, **27**, 753–760 (1995).

Kont, A. et al., Biophysical impacts of climate change on some terrestrial ecosystems in Estonia, *GeoJournal*, **57**, 169–181 (2002).

Koster, R.D. et al., Regions of strong coupling between soil moisture and precipitation, *Science*, **305**, 1138–1140 (2004).

Lal, R., Soil carbon sequestration impact on global climate change and food security, *Science*, **304**, 1623–1627 (2004), (www.science.org/cgi/content/full/304/5677/1623/DCI, SOMText).

Lo, K.F.A. and Cai, G.Q., The potential impact of global change in surface erosion from forest lands in Asia, In: *Environmental Changes and Geomorphic Hazards in Forests*, Report 4 of the International Union of Forest Research Organizations, Sidle, R.C., Ed., CAB International, Wallingford, UK, 45–66 (2002).

Luo, Y. et al., Acclimatization of soil respiration to warming in a tall grass prairie, *Nature*, **413**, 622–625 (2001).

Medlyn, B.E. et al., Soil processes dominate the long-term response of forest net primary productivity to increased temperature an atmospheric CO_2 concentration, *Can. J. Forest Res.*, **30**, 873–888 (2000).

Mertens, S. et al., Influence of high temperature on end-of-season tundra CO_2 exchange, *Ecosystems*, **4**, 226–236 (2001).

Nisbet, T.R., Implications of climate change: Soil and water, In: *Climate Change: Impacts on U.K. Forests*, Broadmeadow, M., Ed., Forestry Commission Bulletin No. 125, Edinburgh, UK, 53–67 (2002).

Olsen, J.E. and Bindi, M., Consequences of climate change for European agricultural productivity, land use and policy, *Eur. J. Agron.*, **16**, 239–262 (2002).

Pacala, S. and Socolow, R., Stabilization wedges: Solving the climate problem for the next 50 years with current technologies, *Science*, **305**, 968–972 (2004).

Postel, S., *Where Have All The Rivers Gone?*, World Watch Institute, Washington, DC (1995).

Prather, M. et al., Atmospheric chemistry and greenhouse gases, In: *Climate Change 2001: The Scientific Basis*, Cambridge University, Cambridge, UK, 239–287 (2001).

Raich, J.W., Potter, C.S., and Bhagawati, D., Interannual variability in global soil respiration, 1980–94, *Global Change Biol.*, **8**, 800–812 (2002).

Rodolfi, G. and Zanchi, C., Climate change related to erosion and desertification, In: *Environmental Changes and Geomorphic Hazards in Forests*, Report 4 of the International Union of Forest Research Organizations, Sidle, R.C., Ed., CAB International, Wallingford, UK, 67–86 (2002).

Rosenzweig, C. and Hillel, D., *Climate Change and the Global Harvest: Potential Impacts of the Greenhouse Effect on Agriculture*, Oxford University Press, NY (1998).

Rosenzweig, C. and Hillel, D., Soils and global climate change: Challenges and opportunities, *Soil Sci.*, **165**, 47–56 (2000).

Ruddiman, W., The anthropogenic greenhouse era began thousands of years ago, *Climate Change*, **61**, 261–293 (2003).

Rustad, L.E. and Fernandez, I.J., Soil warming: Consequences for foliar litter decay in a spruce-fir forest in Maine, U.S.A, *Soil Sci. Soc. Am. J.*, **62**, 1072–1080 (1998).

Sabine, C.L. et al., The oceanic sink for atmospheric CO_2, *Science*, **305**, 367–371 (2004).

Seidl, W., Intensive livestock farming: Consequences for the greenhouse effect and soil pollution, *Entwicklung ländlicher Raum*, **32**, 17–19 (1998).

Service, R.F., New dead zone off Oregon coast hints at sea change in currents, *Science*, **305**, 1099 (2004).

Solvernes, K.A., et al., *Effects of Elevated Carbon Dioxide and Increased Temperature on Plants and Soils*, Norwegian Univ. Life Sci., Aas, Norway (1999).

Snyder, C.S., Hypoxia, fertilizer and the Gulf of Mexico, *Science*, **292**, 1485 (2001).

Stolbovoi, V. and Stocks, B.J., Carbon in Russian soils, *Climate Change*, **55**, 131–156 (2002).

Tian, H. et al., Climatic and biotic controls on annual carbon storage in Amazonian ecosystems, *Global Ecol. Biogeogr.*, **9**, 315–335 (2000).

Utset, A. et al., Modeling of transport processes in soils at various scales in time and space, In: *Proc. Intl. Workshop of Eur AgEng's Field of Interest in Soil and Water*, 24–26 Nov. 1999, Leuven, Belgium, Feyen, J., Eds., 563–568 (1999).

Utset, A. and Borroto, M., A modeling-GIS approach for assessing irrigation effects on soil salinisation under global warming conditions, *Agric. Water Manage.*, **50**, 53–63 (2001).

Venalainen, A. et al., The influence of climate warming on soil frost on snow-free surfaces in Finland, *Climatic Change*, **50**, 111–118 (2001).

Verburg, P.S.J., van-Loon, W.K.P., and Lukewille, A., The CLIMEX soil-heating experiment: Soil response after 2 years of treatment, *Biol. Fertil. Soils*, **28**, 271–276 (1999).

Wiemken, V. et al., Effects of elevated carbon dioxide and nitrogen fertilization on mycorrhizal fine roots and the soil microbial community in beech-spruce ecosystems on siliceous and calcareous soil, *Microb. Ecol.*, **42**, 126–135 (2001).

Wild, A., *Soils, Land and Food: Managing the Land during the 21st Century*, Cambridge University Press, Cambridge, UK (2003).

Yeates, G.W., Tate, K.R., and Newton, P.C.D., Response of the fauna of a grassland soil to doubling of atmospheric carbon dioxide concentrations, *Biol. Fertil. Soils*, **25**, 307–315 (1997).

44

Economic and Policy Contexts for the Biological Management of Soil Fertility

Sara J. Scherr

Ecoagriculture Partners/Forest Trends, Washington, USA

CONTENTS

Opportunities for improved biological management of soil fertility have been debated mostly among technical actors — agricultural producers, scientists, field agents, the fertilizer industry, and more recently, land management specialists. While their discussions have focused on whether or how to enhance the role of biological soil components in constructing more sustainable and lower-cost approaches to fertility management for food production, their assessments of technical options and advantages have, either explicitly or implicitly, been responding to changes in economic and policy contexts.

The growing agreement within the scientific community on the desirability of relying more, if not exclusively, on biological factors and dynamics to enhance and maintain soil fertility and soil ecological functions reflects a convergence of policy concerns arising both within and outside the agricultural sector. The promotion of effective biological management of soils (BMS) on a globally-effective scale will require mobilizing political action. This in turn requires explicitly linking the BMS research and development agenda to important policy objectives.

The first major barrier to achieving this linkage is already being overcome, as discussed in the first section of this chapter: economic and policy analysis methods and practices have evolved to where they can more meaningfully incorporate issues of BMS. The second section describes policy challenges that are presently providing momentum moving BMS to center stage in policy debates in four areas: national food security; rural poverty reduction; environmental services; and public health. The concluding section discusses how further development of BMS practices can benefit by making more explicit links with these higher-level policy debates.

44.1 Mainstreaming the Biological Management of Soils in Economic and Policy Analyses

Until the last decade, soil quality issues have been of little interest to economists or policy analysts, largely because no connection was seen with major policy concerns. Soil productivity issues were addressed through fertilizer supply policies, and soil erosion was to be controlled through investments in inert or vegetative barriers or through land-use restrictions. Biological aspects of soil management were almost entirely absent from economic and policy analyses, which required data on soil-productivity or soil-environment connections that did not exist.

However, research on the biological management of soils has in the past decade moved from being done by small groups of scientists motivated by sheer curiosity or by philosophy and values, to involvement of mainstream agronomists and soil scientists who recognize the value of biological management and are using advanced techniques to understand how to do so more effectively (see feature article in *Nature*, April 22, 2004). The communication gap is being bridged further as proponents of biological management strategies, most outside the organic movement, have begun to identify specific soil conditions where supplemental use of inorganic fertilizer may be clearly productive (e.g., Palm et al., 1997). Data are becoming available in forms that are meaningful to economists and policy analysts.

Economists have become more appreciative of biological soil management as research and technologies have created production systems that generate comparable or better levels of yields and income as conventional systems (Chapter 49, Table 49.4; Current et al., 1995). This has occurred especially with shifts from BMS methods, such as composting that are more labor-demanding, to technologies such as the planting of green manures that may save farm labor overall (Bunch, 2003; Chapter 30), others that reduce cash costs, such as tree fallows for soil fertility improvement (Franzel and Scherr, 2002; Chapter 19), and systems that can be much more profitable with higher returns to labor, such as the system of rice intensification (Chapter 28).

While economic policy models commonly used a decade ago might have distinguished among different soil types, they could not address changes in soil quality, and most treated soil as an inert substrate for production. Developments within agricultural and natural resource economics over the last decade have produced policy models and methods that can now address the biological quality of soils (e.g., Wiebe, 2003), although these approaches are still evolving.

44.2 Policy Interest in Biological Management of Soil Fertility

The shift in policy attention toward biological soil fertility management has been driven by diverse factors that are important to policymakers in different sectors:

1. Soil degradation is occurring on a scale that is beginning to threaten national food security and economic development.

2. Persistent poverty is associated with chronically low productivity of low-income households that are farming poor or depleted soils.

3. Threats to human health are increasing from agricultural chemicals used to enhance soil fertility and control crop pests, weeds, and diseases, and the link of soil quality with food quality and human health is becoming better understood.

4. Critical ecosystem functions in agricultural landscapes (biodiversity conservation, watershed conservation, carbon sequestration) are being compromised as a result of detrimental soil management practices.

44.2.1 Addressing National Food Security: Sustainable Land Management

Policymakers concerned with economic development are becoming interested in biological management of soils to protect and improve the economic value of assets critical for national food security. Soil health, soil productivity, and soil erosion as such are not of great intrinsic concern to politicians and administrators, but these considerations impinge on policy calculations if they occur in places and on a scale that will affect aggregate food supplies, food prices, or economic growth.

Until the 1990s, very few policymakers perceived such impacts. But since then, soil specialists have taken advantage of new spatial monitoring, modeling, and survey tools to scale up and improve the quality of their assessments of the subregional and national impacts of soil and land degradation (Bridges et al., 2001). Policy economists have brought a more convincing level of rigor to these analyses, and their work — based on field evidence of large-scale declines in soil productivity in key producing regions — have demonstrated unexpectedly high and adverse impacts of degradation on agricultural productivity and rural economic growth (Scherr, 1999). It could be seen, for example, that maize yields in historically important production areas in Africa had dropped to under 1 ton ha^{-1}, and millions of hectares of farmland in land-hungry Asia were going out of production due to salinization and waterlogging. Soil degradation was documented as significantly affecting productivity in the "bread baskets" and "rice bowls" of Asia, such as the Punjab (Ali and Byerlee, 2001), and China (Huang and Rozelle, 1994).

Policymakers in countries that are highly dependent on agricultural production from degradation-prone lands became suddenly more concerned than before about the emerging evidence of trends of soil-system deterioration. Resulting initiatives are focusing as never before on improving soil health and fertility with biological management. Key concerns have been soil fertility, soil structure and water-holding capacity, as well as basin-wide conservation planning (Penning de Vries et al., 2003). Many developing countries including China, India, and South Africa have put in place large national investment programs to promote sustainable land management and rehabilitate degraded lands (Bridges et al., 2001). Programs for farmers working on soils that are highly susceptible to erosion and structural degradation emphasize investment in soil organic matter and structure, even when inorganic fertilizers are available.

At an international policy level, the Convention to Combat Desertification (CCD) was negotiated in 1992 and came into force in 1996, with a particular emphasis on halting land degradation in the drylands of Africa. Soil and land degradation was still widely seen as a local rather than international concern, however, and little additional funding has been secured to deal with these problems. This may now be changing as:

- In 2003, the governing council of the Global Environment Facility (GEF) added a new Operational Program on Sustainable Land Management that recognized that the international community has an abiding interest in preventing and reversing land degradation (GEF, 2003).

- The World Bank launched a TerrAfrica program in 2004 to support land rehabilitation in Africa, with a strong focus on enhancing agricultural productivity, although as noted in Chapter 25, this was initially more focused on chemical approaches than BMS.

While most of these and other programs are acknowledging and encouraging biological approaches to soil fertility management, strategies for large-scale action in the field are still lacking.

44.2.2   Addressing Rural Poverty: Low-Cost Approaches to Improve Soil Health

Policymakers concerned with rural poverty reduction have become interested in biological soil management in part as a means to improve food security and raise incomes of low-income farm households, particularly those farming so-called marginal lands. Forty years ago, economic development theory predicted and promoted a structural shift in economies — from agriculture-based production and employment through growth of the industrial and service sectors accompanied by urbanization — that reduced the relative importance of rural areas. Investment in higher-potential agricultural areas, defined as having good agricultural soils and climatic conditions and better market infrastructure, was seen as the best strategy for facilitating and financing this transition. It was assumed that populations in lower-potential agricultural areas would seek and find better livelihoods in the cities or in fast-growing, higher-potential rural areas. Official strategies for agricultural development were willing to depend heavily on the use of inorganic fertilizers, especially nitrogen, for wheat, rice, and maize, for which new, input-responsive varieties were specifically bred, as well as on the application of chemical pesticides. Marketing systems were put in place to facilitate use of these external inputs.

Urbanization and rural–rural migration have indeed taken place on a large scale. However, the absolute growth of rural populations along with large-scale migration from over-populated high-potential areas into lower-potential farming areas has continued. Nearly two-thirds of the population in Asia and Africa still live in rural areas, and rural population growth rates, while expected to decline from 2000 to 2030, will still be positive in many subregions, indeed significantly in much of Africa (around 1% or more yr^{-1}). Of total rural population in developing countries, two-thirds now live in the lower-potential areas (Nelson et al., 1997). Many of these lower-potential areas, lacking infrastructure and agricultural investment, have become poverty traps. It is estimated that 75% of the global population who now live on one dollar per day or less are living in rural areas.

Rural poverty in such areas results, in part, from low and declining soil nutrient status and poor soil health, including compaction, low organic matter, and low biological activity. Biologically- and structurally-degraded soils increase risks of crop failure by reducing water-holding capacity and making production more vulnerable to drought. Many soils that have low organic matter often have poor response to fertilizer when applied. The potential for agricultural intensification in many so-called marginal lands thus depends upon first building up the capacity of soils (and associated water management systems) to support intensification sustainably (Scherr, 1999). Biological soil management is now better understood as essential for sustainable production even when fertilizer is available. The highly divisive debates over inorganic fertilizer use that characterized the 1970s and 1980s have softened as the complementarities and even synergies possible with combined use of organic and inorganic sources of nutrients are identified, as reported in many chapters in this volume.

This said, extensive use of synthetic fertilizers (and other agrochemical inputs) is still not an economically viable option for low-value crops or for remote communities where poverty is greatest. In sub-Saharan Africa, road infrastructure serves a much smaller fraction of the rural population than in other parts of the world. Thus, the local price of nitrogen fertilizer is two to four times the world price, due to high transport costs and inefficient marketing systems. Partly as a result, farmers in sub-Saharan Africa apply on average 9 kg of nutrients ha^{-1} of arable land, compared with an average of 107 kg ha^{-1} for all developing countries. Still, even in India — the birthplace of the Green Revolution — half of smallholder farmers do not have affordable economic access (or financial justification) for the purchase of inorganic fertilizers (Swarna et al., 2004). With large numbers of the rural poor living in areas of high climatic risk (where rains fail 2 to 3 years out of every five), the use of scarce cash resources to purchase inorganic fertilizers does not make economic sense at the household level. In many areas, the proportion of farmers purchasing fertilizers and the land area on which they are used has declined in recent years as a result of falling real prices for agricultural commodities, particularly staple foods.

As a result of these trends, the relative economic returns and benefits from protecting and improving biological soil fertility have improved, especially where this can be done using organic inputs available on the plot, on-farm or from nearby forests. The Hunger Task Force of the UN Millennium Project, which has stated as its first goal the reduction of extreme poverty and hunger, places high priority on improving soil health on the holdings of small and marginal farmers, especially in Africa, and in drylands and mountainous regions. It has recommended a major international push for investment with farmers in BMS for soil fertility. While the report also suggested potential merit in fertilizer subsidies, these would be relevant only under limited circumstances (UN Millennium Project, 2005). Farmer movements to invest in soil and land protection and restoration, such as the Landcare movement, the international organic movement (IFOAM), and other ecological agriculture movements, are actively promoting biological soil fertility management (Pretty, 2005).

International initiatives for rural poverty reduction, such as the UNDP's Drylands Development Centre, the World Agroforestry Centre's Trees for Change program, FAO's Farmer Field Schools, the Tropical Soil Biology and Fertility Institute, and NGOs focused on rural poverty reduction, such as OXFAM and CARE, are supporting biological soil management approaches. However, the scale of institutional support to farmers and farming communities for investment, training, and adaptive research on this meets only a small fraction of the present need. The identification of cost-effective, decentralized models for service and information provision will be critical to scaling up for significant poverty-reduction impacts.

44.2.3 Addressing Threats to Human Health: Reduced Agrochemical Use

Policymakers concerned with public health have also begun to take notice of, and to promote, the potential benefits of biological soil management. Their interest is mainly to reduce the negative impacts on human health from pesticides, by promoting alternative crop-production systems that use much lower levels of these chemicals. Public health leaders have been particularly concerned with the impacts of pesticide use on farmworkers occurring through unsafe or unprotected application methods and poor management of equipment, although there is increasing evidence in some places that pesticide residues are accumulating in consumers of foods produced with pesticide (e.g., Schafer et al., 2004, for the U.S.A.) and from pesticides accumulating in fish from waters polluted by agricultural runoff or incorrect washing of equipment or disposal of wastes (Pretty, 2005).

These problems are being addressed primarily through initiatives for safe handling of pesticides, standards for limiting residues in marketed products, and disposal of obsolete stocks. More-developed countries are requiring vegetative or artificial barriers to protect waterways from chemical loads in runoff from fields and farmsteads. Consumers particularly concerned about dangers to themselves or to farm workers have turned to foods certified as organically grown, and governments have put in place certain minimum standards to assure consumer confidence when purchasing organic products (OECD, 2003). These regulatory and consumer trends have spurred greater farmer and agroindustry interest in BMS to help manage crop pests and diseases, including integrated pest management, organic agricultural systems, and best management practices, as described in Clay (2004).

44.2.4 Addressing Ecosystem Threats: Biodiverse Soils for Ecoagriculture

The environmental policy community has become interested in biological management of soils, as it increasingly recognizes the role of biodiverse and biologically healthy soils in the protection of critical ecosystem services, beyond those essential for sustainable agriculture. Analyses of the Pilot Assessment of Global Ecosystems (Wood et al., 2000) and the Millennium Ecosystem Assessment (2005) have concluded that the scale of crop production is ecologically significant. It accounts for at least 30% of land use, on over one-third of global terrestrial area. In some heavily populated areas, such as Europe and South Asia, crop agriculture dominates most landscapes.

Thus a significant share of ecosystem services, including species habitat to maintain biodiversity, watershed protection, and carbon sequestration, must be produced in agricultural landscapes. Innovations are arising from around the world to develop what is now being called ecoagriculture, production systems and landscape mosaics that provide both food and ecosystem services (McNeely and Scherr, 2003). New systems of biological soil management are critical elements of such systems.

44.2.4.1 *Conservation of Biodiversity*

The research and policy communities concerned with biodiversity have begun to expand their original focus on endangered (charismatic) megafauna, on unique and threatened types of ecosystems, and on biodiversity conservation within large undisturbed areas. They are recognizing that a high proportion of global biodiversity consists of microorganisms in biologically-active soils, and that these should be considered in any conservation strategy. These species are important not only in themselves, but also for the roles they play for other species, in terms of habitat, food, waste decomposition, and nutrient recycling. Soil biodiversity is affected by practices for soil preparation and utilization as stressed repeatedly in this book. Organic matter management, plant management, and water, nutrient, and sediment management all have direct implications for biodiversity of soil biota. The biologically-degraded state of many agricultural soils managed using high levels of agrochemical inputs is a major concern, especially in ecosystems where a high proportion of land is under agricultural use.

These new perspectives are reflected in the 1992 Convention on Biological Diversity through its explicit concern with agrobiodiversity. This is defined to include crop and livestock genetic diversity, associated wild species on which agriculture depends (including microorganisms and pollinators), and wild species using agricultural lands as habitat. The GEF's Operational Program 13, established in 2000, focuses on "conservation and sustainable use of biological diversity important to agriculture" (GEF, 2001). Key initiatives have included a research program on belowground biodiversity (CIAT, 2005). Stocking (2003) has documented the benefits for smallholders in East Africa of enhancing and managing soil agrobiodiversity at site, farm, and landscape scales.

44.2.4.2 Watershed Functions

Scientists from diverse fields have concluded that biologically healthy soils are a critical part of the natural infrastructure of well-functioning watersheds maintaining a productive hydrological cycle. Porous, well-drained soils are required to absorb and store rainfall, thereby ensuring percolation to aquifers, as well as to regulate the rate of flow into streams and to filter impurities out of the water supply. The concept of "green water" — as contrasted with "blue water" — has underscored the importance and opportunities of these processes. The latter water is captured, stored, or pumped, and then conveyed to fields through standard irrigation methods. This contrasts with green water that is stored and utilized *in situ*, at much lower cost and reduced losses, as well as with more environmental benefit (Savenije, 1998).

Soil biota play a critical role in the cycling of green water. Earthworms and other organisms in soils are critical to maintaining soil porosity. Soil compaction results at least in part from having inadequate diminished soil organic matter and populations of microflora and -fauna. It can dramatically increase surface runoff during rainfall events, not only reducing the amount of water that can be stored in soils (reducing resilience to drought periods), but also the recharge of groundwater. Soil organic matter helps maintain soil structure and the porosity needed for soils to hold water and enhance infiltration. These are services of huge economic value. Just reducing the losses from flood damage can be worth many millions of dollars.

Biological management can also reduce fertilizer use and fertilizer-laden runoff. Excessive fertilizer nutrient loads from agricultural runoff into waterways, lakes, and coastal areas are now recognized to be a major threat to freshwater and coastal aquatic biodiversity (Revenga et al., 2000). Reduction of fertilizer runoff, through a combination of reduced use and more effective filtering, which can be done through vegetative or other biological means, is now understood to be critical for reducing these impacts.

High-profile issues have motivated policy action. For example, the biological "dead zone" in the Gulf of Mexico, which has been traced to nutrient build-up coming from the heavy use of inorganic fertilizers in the upper Mississippi agricultural region, has induced the governors of several states in the region to commit their states to major programs of nutrient reduction. Nutrient runoff into the Mekong delta is threatening economically important fisheries in Southeast Asia and has prompted a Mekong-wide collaborative policy initiative to manage agricultural runoff.

44.2.4.3 Carbon Sequestration and Storage

Global-level concerns about climate changes due to excessive emission of greenhouse gases have also prompted a rethinking of the handling of agricultural soils. Much research, summarized in the preceding chapter, has made clear that soil management directly affects levels of emission, sequestration, and storage of greenhouse gases. Soils, which currently represent the third largest pool of carbon, have the capacity to store carbon in their organic matter — roots, flora, and fauna, and incorporated residues or other formerly living material. Scientists have estimated that loss of soil organic matter due to soil degradation can account for a significant share of carbon emissions into the atmosphere, with the highest rates coming from tropical soils.

Soils' sink capacity can be enhanced, and fluxes of methane (CH_4) and nitrous oxide (N_2O) can be modified, through management that enhances the biological component of the soil including revegetation (Chapter 42–IPCC, 2003). The UN Framework Convention on Climate Change currently allows developed countries with obligations under the Kyoto Protocols to reduce their carbon emissions to offset some of their emissions through improved soil management within their own borders or abroad. Industrial greenhouse-gas emitters are permitted to compensate for part of their emissions by investing in selected sustainable development activities in the developing countries (which do not yet have obligations under the treaty) through the Clean Development Mechanism (CDM) of the Kyoto Protocols.

The rules governing the CDM currently restrict offset projects to investment in afforestation and reforestation, however, so soil carbon projects are not yet eligible. Options to include this domain are being closely studied, and pilot projects are underway to develop reliable methods of measurement that could demonstrate additionality and permanence from such projects so that they could be included within the scope of the convention.

Pilot projects to increase soil carbon through biological management are being implemented in many parts of the world. Farmer organizations, the soil conservation community, some groups of climate action activities (not all), and NGOs have begun to lobby governments to develop programs for soil investments that address national climate policy goals. While some proposals are more narrowly focused on soil carbon, others seek to support agricultural producers to adopt more sustainable management systems that increase both below- and aboveground carbon pools. Initiatives such as the Community, Climate, Biodiversity Alliance (CCBA) and the BioCarbon Fund are exploring ways to design carbon projects that integrate biodiversity, watershed, and livelihood objectives as well.

44.3 Discussion

Policy imperatives — for national food security, rural poverty reduction, human health, and ecosystem services — have provided the context and stimulus for promotion of biological soil management around the world over the past decade. Yet neither farming communities nor agricultural professionals nor policymakers have fully reoriented their thinking and actions to address these concerns or to work with those farmers and scientists, let alone NGOs, who have pioneered work in this biological domain.

Such engagement is now better grounded in scientific terms and more urgent in terms of practical application. Expanding biological soil management on a large scale, to restore productive resources and enhance agricultural productivity, health, and ecosystem services, calls not only for new research initiatives and new knowledge systems for community innovation, but also more supportive policy frameworks. Proponents of BMS should begin to work more strategically and systematically at the policy level.

While it should be recognized that biological soil management can address all four of the policy issues listed above, in most countries (and even in most international agencies) these issues are treated separately. Support for each must thus be mobilized through different types of evidence and arguments. The case for biological soil management needs to be framed and justified specifically for policymakers in these different sectors.

A valuable way to encourage the incorporation of biological soil management in policy analysis is to seek out collaborative opportunities with policy economists and analysts, and to frame those analyses — including research questions, variables, and indicators, and geographic sampling — so that they address priority policy questions and concerns directly (Scherr, 2000). In communicating with policymakers and in working with them to set priorities, key questions to consider are:

1. Where do poor soil conditions pose a significant impediment for agricultural, social, health, or environmental policy, e.g., for which geographic regions and farming groups?

2. In these situations, is promotion of biological soil management the most appropriate response to this problem? What are competing solutions? On what basis can BMS be justified economically and in terms of implementability?

3. If biological soil management is a preferable solution, what are the most appropriate types of intervention to address those specific policy concerns?

Addressing such sector-specific concerns will not be enough, however. There is an urgent need also to promote cross-sectoral thinking and policy dialogue at both international and national levels. This will document and explore the potential synergies *across* the agriculture, social, health, and environmental sectors attainable from making investments in BMS. In many regions, improved biological management of soils may be a very cost-effective way to meet concurrently the Millennium Development Goals on rural poverty, hunger, water, environment, and health.

Taking advantage of these opportunities will require national and local policy coordination in rural development, sustainable land management, and ecosystem management. At an international level there is also potential to expand financing for such work through carbon emission offset payments for sustainable land management, including BMS. This should be a priority area for negotiation at the UN Framework Convention on Climate Change, as well as for exploiting opportunities to design investments that jointly meet countries' obligations under the conventions for biological diversity and desertification.

References

Ali, M. and Byerlee, D., Productivity growth and resource degradation in Pakistan's Punjab, In: *Response to Land Degradation*, Bridges, E.M., Hannam, I.D., Oldeman, L.R., Penning de Vries, F.W.T., Scherr, S.J., and Sombatpanit, S., Eds., Science Publishers, Enfield, NH, 186–189 (2001).

Bridges, E.M. et al., Eds., *Response to Land Degradation*, Science Publishers, Enfield, NH (2001).

Bunch, R., Note to Hunger Task Force on Green Manure Technology, Submitted to U.N Millennium Project Task Force on Hunger (2003).

CIAT, *Short Description of the Project: Conservation and Sustainable Management of Belowground Biodiversity* (http://www.ciat.cgiar.org/tsbf_institute/bgbd_news.htm) (2005).

Clay, J., *World Agriculture and the Environment: A Commodity-by-Commodity Guide to Impacts and Practices*, Island Press, Washington, DC (2004).

Current, D., Lutz, E., and Scherr, S.J., Eds., *Costs, Benefits and Farmer Adoption of Agroforestry: Project Experience in Central America and the Caribbean*, World Bank Environment Paper 14, World Bank, Washington, DC (1995).

Franzel, S. and Scherr, S.J., Eds., *Trees on the Farm: Assessing the Adoption Potential of Agroforestry Practices in Africa*, CAB International, Wallingford, UK (2002).

GEF, *Program Objectives of Operational Program #13*, Global Environment Facility, Washington, DC (2001).

GEF, *Program Objectives of Operational Program #15*, Global Environment Facility, Washington, DC (2003).

Huang, J. and Rozelle, S., Environmental stress and grain yields in China, *Am. J. Agric. Econ.*, **77**, 246–256 (1994).

IPCC, *Good Practice Guidance for Land Use, Land Use Change and Forestry*, Inter-Governmental Panel on Climate Change, Kanagawa, Japan (2003), http://www.ipcc-nggip.iges.or.jp/public/gpglulucf/gpglulucf_files/0_Task1_Cover/Cover_TOC.pdf.

McNeely, J. and Scherr, S., *Ecoagriculture: Strategies to Feed the World and Save Wild Biodiversity*, Island Press, Washington, DC (2003).

Millennium Ecosystem Assessment, *Millennium Ecosystem Assessment Synthesis Report*, Earth Institute of Columbia University, New York (2005).

Nelson, M. et al., *Report of the Study on CGIAR Research Priorities for Marginal Lands*, Technical Advisory Committee, Consultative Group on International Agricultural Research, and FAO, Rome (1997).

OECD, *Organic Agriculture: Sustainability, Markets and Policies*, CAB International, Wallingford, UK, for Organization for Economic Cooperation and Development (2003).

Palm, C.A., Myers, R.J.K., and Nandwa, S.M., Combined use of organic and inorganic nutrient sources for soil fertility maintenance and replenishment, In: *Replenishing Soil Fertility in Africa*, SSSA Special Publication, Buresh, R.J. and Sanchez, P.A., Eds., Soil Science Society of America and American Society of Agronomy, Madison, WI, 193–217 (1997).

Penning de Vries, F.W.T. et al., *Integrated Land and Water Management for Food and Environmental Security*, CGIAR Comprehensive Assessment Secretariat, Colombo, Sri Lanka (2003).

Pretty, J., Ed., *The Pesticide Detox: Towards a More Sustainable Agriculture*, Earthscan Publications, London (2005).

Revenga, C. et al., *Pilot Analysis of Global Ecosystems: Freshwater Systems*, World Resources Institute, Washington, DC (2000).

Savenije, H.H.G., *The Role of Green Water in Food Production in Sub-Saharan Africa*, Paper prepared for FAO program on Water Conservation and Use in Agriculture (WCA) (http://www.wca-infonet.org) (2003).

Schafer, K. et al., *Chemical Trespass: Pesticides in Our Bodies and Corporate Accountability*, Pesticide Action Network North America, May, http://panna.org/campaigns/docsTrespass/ChemTres Main(screen).pdf (2004).

Scherr, S.J., *Soil Degradation: A Threat to Developing Country Food Security in 2020?*, Food, Agriculture, and the Environment Discussion Paper No. 27. International Food Policy Research Institute, Washington, DC (1999).

Scherr, S.J., Environmental Protection Versus Restoration? A Model for Policy Decisions, Invited paper for the 10th International Soil Conservation Conference on Sustaining the Global Farm, May 1999. Purdue University, West Lafayette, IN (2000).

Stocking, M., Managing biodiversity in agricultural systems, In: *Agricultural Biodiversity in Smallholder Farms of East Africa*, Kaihura, F. and Stocking, M., Eds., United Nations University Press, New York, 20–33 (2003).

Swarna, S.V., *Atlas of the Sustainability of Food Security in India*, M.S. Swaminathan Research Foundation and World Food Programme, Chennai, India (2004).

UN Millennium Project, *Halving Hunger: It Can Be Done*, Summary version of the report of the U.N. Task Force on Hunger. The Earth Institute at Columbia University, New York (2005).

Wiebe, K., Ed., *Land Quality, Agricultural Productivity and Food Security: Biophysical Processes and Economic Choices at Local, Regional, and Global Levels*, Edward Elgar, Cheltenham, U.K. (2003).

Wood, S., Sebastian, K., and Scherr, S.J., *Pilot Analysis of Global Ecosystems: Agroecosystems*, International Food Policy Research Institute, and World Resources Institute, Washington, DC (2000).

45

Village-Level Production and Use of Biocontrol Agents and Biofertilizers

B. Selvamukilan, S. Rengalakshmi, P. Tamizoli and Sudha Nair
M. S. Swaminathan Research Foundation, Chennai, India

CONTENTS

A prime objective of agriculture in the coming decades must be to optimize soil productivity while preserving the capacity of soils to function as healthy systems. This is particularly important for the stressed soils that are so abundant in most developing countries. Increasing productivity in perpetuity without inducing any associated ecological or social ill-effects — achieving what M.S. Swaminathan has called the "evergreen revolution" (Swaminathan, 1996) — is imperative if we are to reduce poverty and hunger without worsening people's current living conditions and future prospects. In this move towards more sustainable agricultural practices, the use of biological inputs will play a key role in the maintenance of soil fertility and in increasing crop production.

Biological inputs such as compost and farmyard manure are well-known and long-standing practices. More novel and innovative are bioinoculants as part of integrated nutrient management (INM) and biocontrol agents and botanicals as part of integrated pest management (IPM). There is growing popular and commercial interest in developing biological materials that are reliable and that can complement if not always replace the chemical inputs already on the market. There is need to multiply and strengthen the units of production and delivery for such materials, and also the connections among them, to be able to ensure a reliable supply of good-quality material at the most appropriate times in rural areas. This need points to the desirability of having such units well-established in rural areas themselves.

This chapter reports on efforts to enhance the availability of such biological inputs for sustainable agriculture practices in rural India, and on some initial solutions devised so far. There is special concern with improving soil quality and increasing crop productivity especially for poor and marginal farmers. Fortunately, we are finding that they themselves can make significant contributions to surmounting these challenges and becoming a major part of the solution.

45.1   Challenges Faced

Inoculation of plants with beneficial bacteria can be traced back many centuries. From experience, Indian farmers knew that when they mixed soil where a previous legume crop had been grown with soil in which they wanted to plant a nonleguminous crop, yields often improved. Towards the end of the 19th century, the practice of mixing "naturally inoculated" soil with seeds to be planted was becoming a recommended method for legume inoculation. At the turn of the century, the first patent (for Nitragin) was registered to promote plant inoculation with Rhizobium sp. (Nobbe and Hiltner, 1896). Eventually, the practice of legume inoculation with Rhizobia became common. For almost 100 years, primarily small companies have produced Rhizobium inoculants around the world, and there has been a progressive increase in the use of Azotobacter, Bacillus, and other species as well.

Two major breakthroughs in plant inoculation technology occurred when Azospirillum was found to enhance nonlegume plant growth by directly affecting plant metabolism (Döbereiner and Day, 1976), and when biochemical agents, mainly from the *Pseudomonas fluorescence* and *P. putida* groups, began to be intensively investigated. Increasingly, in recent years, various other bacterial genera such as Bacillus, Flavobacterium, Acetobacter, and several Azospirillum-related microorganisms have also been evaluated (Tang and Yang, 1997).

The immediate response to the inoculation of soil with associative, nonsymbiotic plant growth-promoting bacteria (PGPB) has varied considerably depending on the microorganism, plant species, soil type, inoculant density, and environmental conditions. This variation has been a discouragement to some, but it has been regarded by others as a challenge to be overcome. Chapters 8, 32–35, and 42 in this volume report on specific efforts to devise practical means of benefiting from the potential contributions that microbes can make to plant health and productivity. Such benefits are detailed in Chapter 9 on mycorrhizal fungi, Chapter 12 on biological nitrogen fixation (BNF), Chapter 13 on phosphorus solubilization, and Chapter 14 on phytohormones.

There is thus growing and evident interest in developing effective bacterial and fungal products for enhancing soil system productivity. Research and subsequent field trials on the plant growth-promotion effects of microbes conducted over the last decade have opened up new horizons for the inoculation industry. While agricultural enterprises in developed countries have been the major promoters of microbial inoculants that are environmentally friendly, one of the countries making the greatest advances is Cuba (Chapters 32 and 33). There, financial constraints have made biologically-based (rather than petroleum-based) inputs particularly attractive to farmers and to the government. Also, in India there has been a large and growing interest in such opportunities. Given that farmers in all developing countries have similar needs and constraints, more attention should be paid to making easy-to-use and inexpensive formulations available. To the extent that production and distribution can be handled at the village level, there will be multiple benefits to rural people, as manufacturers, traders, and users.

Farmers in developing countries now practice mainly some variant of low-input agriculture because fertilizers, pesticides, and machinery are costly and scarce. They have few resources to invest in improved agricultural techniques. Many are engaged in their own agricultural production only part-time as they must also work as laborers to receive a daily wage. Another constraint is that biological inoculations require specialized infrastructure for storing and transporting these products, and such facilities are limited in rural areas. Government extension services and the growers' own formal agricultural education are also limited for these purposes.

There are a variety of reasons for the present low demand for biological inputs: lack of effective promotion and extension work, insufficient publicity, and unavailability of quality products in a timely manner, especially in rural areas. Also, there is need for products that can work under harsh conditions, such as drought or saline soils which many rural producers, especially the poorest, must cope with. Products tend to be developed under and for ideal conditions, which give the largest and most dramatic gains in production. However, in terms of human welfare, smaller but reliable gains achieved by more marginal households will yield more valued benefits.

For inoculation programs to spread based on results and ensuing demand, there is a need to find the best candidate species available, provided in specific and effective formulations. For many applications, irrigation facilities need to be available since the use of inoculants should coincide with sowing into moist soil or the materials are best delivered in conjunction with irrigation. Since the effectiveness of inoculation depends very much upon matching inoculant species with specific plant species and the particular soil conditions such as pH, biological products should be differentiated and matched appropriately. No single bacterial or fungal species will perform best in all situations. This complicates the task of providing effective inoculants because commercial incentives are to keep the number of products small. This runs contrary to the reality of biodiversity among crop species, inoculant species, and soil biotic communities which argues for the production and distribution of a multiplicity of products.

Whenever there is a greater variety of products, of course, there needs to be more infrastructure capacity for storage and transport and more education with farmers. The microbial inoculants themselves can often be produced and marketed fairly inexpensively. However, the system to get them utilized needs to be rather elaborate and multifaceted. With better-organized communication and infrastructure, including some initial support for the uptake and demonstration of these new opportunities, e.g., through agriclinics, farmers can be expected to become regular clients for microbial technologies, especially in the developing world.

45.2 An Integrated Initiative to Support the Production and Use of Biological Agro-Inputs

In keeping with its philosophy and strategy of supporting development that is "pro-nature," the M.S. Swaminathan Research Foundation (MSSRF) based in Chennai, Tamil Nadu state of India, has been promoting an integrated farming systems approach whereby farmers are encouraged to pursue low-external-input agriculture with INM and IPM as core concepts. Wider adoption of traditional practices, which utilize local and natural inputs and that capitalize on indigenous knowledge, is encouraged. As its work has proceeded, the foundation has seen how such practices can be supplemented and improved upon by support from modern scientific endeavors that bring state-of-the-art microbiological knowledge to bear in a synthesis of old and new techniques. During its drives to promote low-cost, sustainable agriculture at village level, the need for improved biological inputs to enhance soil fertility and productivity has become evident. Efforts have been undertaken to introduce such inputs by setting up demonstration plots which followed an integrated farming systems approach.

The institutional platform developed for interaction with villagers was the "biovillage" model that MSSRF has facilitated among rural and tribal communities. This work initially began in a few villages of Pondicherry in the early 1990s, and it has now been extended to other field centers in the states of Tamil Nadu, Kerala, and Orissa. Once the available natural resources, livelihood systems, and human resources in a village have been assessed, appropriate forward and backward linkages are created for the enhancement of natural resources and to establish opportunities for skilled employment.

45.2.1 The Biovillage Model

The biovillage approach seeks to add value to the time and labor of the rural poor, their main assets, through creation of on-farm and nonfarm employment opportunities. It focuses on the capacity of rural communities to blend sustainable natural resource management with secure livelihoods through economically feasible, socially acceptable, ecologically viable, and gender-sensitive initiatives. The approach is to combine traditional knowledge with frontier technologies, developing human resources and building grassroot institutions such as farmers' associations, self-help groups (SHGs), and federations of rural organizations that take up development initiatives within the biovillage framework.

The foundation has integrated the concept of ecoentrepreneurship into the program. This promotes the technical, managerial, and marketing capacities to rural women and men to mobilize and organize themselves as entrepreneurs to produce environment-friendly products and services for sustainable development. "Ecojobs" are defined as employment opportunities in the sectors, which use natural resources efficiently and effectively, without creating environmental instability.

45.2.2 Village-Level Production of Biocontrol Agents and Biofertilizers

Within this framework, the production of biological materials has been taken up by groups of women members who form ecoenterprises to create livelihood opportunities for the socially and economically disadvantaged rural poor and also to build a supply chain at the farmers' field level for good quality bioinoculants at the most appropriate times. The products that can be produced include *Trichoderma viride* and *P. fluorescence*, biofertilizers, such as Azospirillum and Phosphobacter, and arbuscular mycorrhizal fungi (AMF). The MSSRF program has started with production of the first two, to be sure that the organizational, economic, and technical aspects of this provision are all manageable in such rural enterprises.

The first step was to mobilize community interest and understanding of the need for such products and enterprises. The mobilization process started in 1997 in the study area. Since 2000, landless and marginal landholding members, both women and men, have organized themselves into SHGs with 12 to 15 members. These SHGs are recognized by the formal banking sector and are able to get loans to finance their operations. Two groups are presently functioning successfully, with the formation of a third in progress. The first is producing *T. viride* (Figure 45.1), while the second is producing *P. fluorescens*. The third group will go into the production of Azospirillum.

Members of the first and third groups are all landless women agricultural laborers, whereas the second group is composed of landless and marginal landholders. The process of institutional development and capacity building has been carried forward with a participatory approach, with an initial research phase that gives members a better understanding of local conditions and of how they can fine-tune the technology to suit such conditions. The training of these self-formed groups was carried out by interactive learning and learning-by-doing and covered two main aspects: rural entrepreneurship and biological technology.

FIGURE 45.1
Packing of *Trichoderma viride* biofertilizers in a village production center initiated by the M.S. Swaminathan Research Foundation in India.

Initially, all the group members received the same training in skills for rural entrepreneurship. Then, two group members who are semiliterate (the rest are illiterate) received training on production technology for five days at Tamil Nadu Agricultural University. This hands-on, residential training program included the preparation and maintenance of cultures by different methods, handling of the machinery and equipment, and methods for application of bioinoculants (diseases to be targeted, quantities to be used, etc.). The members were also trained in quality control and how to assess quality using simple methods such as a serial dilution technique.

The first group needed about 5 months to get operations going after obtaining its loan from the bank; the second group required only 2 months for its stabilization. During the first 6 months, MSSRF staff interacted with members on a daily basis. This attention was gradually reduced, and now is based on groups' demand, which is mostly concerning marketing issues.

The program is managed from a systems perspective, which means that each group has been exposed to the full range of activities and functions involved in the production and distribution of biological products. Each is coordinating with the government extension service and with private marketing entities, networking at the same time with government technical departments and academic institutions. With the units functioning on their own, the MSSRF role has changed from being an active facilitator to becoming a friendly observer, with group members taking over all management of their activities.

A survey conducted in the area where MSSRF initiated these efforts, to ascertain potential delivery channels for bioinoculants, revealed that several academic institutions and some private companies were already producing bioproducts and supplying them to farmers through the agricultural extension network and a few commercial dealers. However, this system was not able to meet the growing demand among farmers for such products, which has been steadily increasing, given expanding interest in sustainable and also organic agriculture. At present, more than 17 million ha of land in India are under organic farming practices, which consume most of the available ecofriendly biological products. The demand for bioinoculants and botanicals is growing. There is massive adoption of vermicomposting of various postharvest wastes. This practice, discussed in Chapter 31, is being used to create a good carrier material for inoculants.

Before setting up the village-level production units, an assessment was made of potential demand based on a survey of current cropping patterns, the present supply chain for bioinoculants, and estimated unmet demand. An Agricultural Extension Office serving a subdistrict (block) was currently handling only 30 kg of each product in a year. However, considering the total net cultivable area, the demand could exceed 15 to 20 tons in that block. Actual demand will depend on farmers being able to acquire timely, quality products, which can be distributed along with their purchase of seeds.

Units with a production capacity of 1000 kg month^{-1} have been set up using low-cost methods. When drawing up business plans, care is taken to make everything cost-effective, using local materials to set up the units, and sourcing the machinery locally so that maintenance can be easily taken care of with convenient troubleshooting.

Market linkages have been established with the help of MSSRF, e.g., to agricultural input dealers and the agricultural department. Group members have been directly interacting with local farmers and farmers' associations. They are undertaking product promotional activity among farmers by participating in district-level farmers' meetings, state-level agriconferences, and farmers' day programs. They also interact with the local agriculture and horticulture extension departments and the farmers clubs promoted by MSSRF's agriclinics. In addition to these market sources, those NGOs that promote organic farming are emerging as one of the best potential customers for these kinds of products.

The amount of bioproducts being produced by each of these two units already averages 900 to 950 kg month^{-1}, which can cover nearly 10,000 ha yr^{-1}. Each unit creates 100 to 120 days of employment yr^{-1}. Each member receives an additional income of Rs. 1300 to 2000 month^{-1} (US$ 30 to 45).

A major qualitative impact of this initiative has been the change in perceptions of the use of biological products. Local farmers' demand is constantly increasing with farmers coming to the production units to purchase products. This happens mainly through farmer-to-farmer exchange of information as well as by the direct interaction of group members with farmers in association meetings.

To date, nearly 10% of total production is being marketed directly to local farmers. Farmers' associations in the region carried out a survey, from which they estimated that in the previous year, 2000 farm families were already using these kinds of biological inputs. Currently, members are not withdrawing their share of the group's profits, having taken a collective decision that their first priority is to repay the bank loan. This sets a good example for gaining support to expand this kind of local enterprise.

45.3   Discussion

This initiative is in line with a wider, ongoing effort to promote biofertilizers in Indian agriculture through public interventions (Ghosh, 2004). Such policy motivates the private sector to utilize and promote the technology. Increasingly, the long-term implications of this at farm level are being understood, as it is seen that the gains from increased use of biological technologies can benefit other farms and sectors through reduced water pollution now caused by the diffusion of residuals from inorganic fertilization. It has also been recognized that both research and extension and field promotion efforts must be, to the extent possible, specific to local conditions and constraints.

Under these circumstances, decentralized production units at village level will help farmers to get reliable products on time, with suitable effective strains. Supplies of crop-specific and soil-specific strains are receiving wider acceptance from farmers in the local region. Experience reveals that scaling down and dispersing technological improvements in the mass production of biological products to be utilized in small-scale rural ecoenterprises helps create a diversified livelihood system especially benefiting the poor, landless, and marginal landholders.

Critical elements in such a system based on decentralized production units are the mobilization of rural women and men and their creation of groups; establishing need-based and relevant ecoenterprises having forward and backward linkages with different institutions; and capacity-building to manage enterprises as successful business ventures. Creating access to technology through demystification as well as strengthening local organizational capacities brings agronomic, economic, and social benefits to local communities that will help sustain equitable development in rural areas.

Acknowledgments

We would like to thank the brave women members of Elayathendral and Durga Women Self-Help Groups of Kannivadi region, Dindigul District, Tamil Nadu, India, who took on the challenge of becoming first-generation entrepreneurs and who are contributing their energies and leadership to the sustainable agriculture movement; and also the agencies that are supporting this work, the Volkart Stiftung Foundation, the Department of Biotechnology, and the Tata Trust.

References

Döbereiner, J. and Day, J.M., Associative symbioses in tropical grasses: Characterization of micro-organisms and dinitrogen-fixing sites, In: *Proceedings of the First International Symposium on Nitrogen Fixation*, Vol. 2, Newton, W.E. and Nyman, C.J., Eds., Washington State University Press, Pullman, WA, 518–538 (1976).

Ghosh, N., Promoting biofertilizers in Indian agriculture, *Econ. Polit. Weekly*, 5617–5625, December 25 (2004).

Nobbe, F. and Hiltner, L., U.S. patent 570 813. Inoculation of the soil for cultivating leguminous plants (1896).

Swaminathan, M.S., *Sustainable Agriculture: Towards Food Security*, Konark, New Delhi (1996).

Tang, W.H. and Yang, H., Research and application of biocontrol of plant diseases and PGPR in China, In: *Plant Growth-Promoting Rhizobacteria: Present Status and Future Prospects*, Ogoshi, A. et al., Eds., Faculty of Agriculture, Hokkaido University, Sapporo, Japan, 4–9 (1997).

46

Measuring and Assessing Soil Biological Properties

Janice E. Thies

Department of Crop and Soil Sciences, Cornell University, Ithaca, New York, USA

CONTENTS

For decades, measures of soil chemical and physical properties have dominated our views of what constitutes soil fertility. Soil biological fertility has been gaining attention recently, and the soil biota are now better recognized as the key drivers of soil fertility and productivity that they are. New methods employing advances made in microscopy, molecular biology, and biochemistry have revealed a diversity of soil organisms previously unimagined, indicating several orders of magnitude greater richness and abundance. They have confirmed the regular occurrence of biogeochemical transformations, such as dissimilatory nitrate reduction to ammonium, in environments previously thought to be nonconducive for them. The past decade and a half have been particularly rich in discoveries in soil ecology, as seen from preceding chapters. This chapter reviews some of the major advances made in measuring soil biota and assessing their activity that are significantly improving our understanding of soil ecology.

The soil biota are assessed most frequently in terms of their: (1) abundance or density; (2) gross or net rates of specific activities or functions; and (3) diversity or richness,

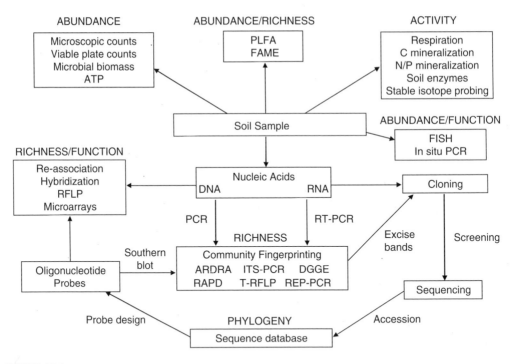

FIGURE 46.1
Overview of analytical methods discussed in this chapter and their relationships to each other. Full descriptions for the abbreviations are given in the text. Updated from Thies and Suzuki (2003).

i.e., the number of different types or groups present. Typically, no single measure is adequate for capturing all of the biological changes that occur in response to soil management or natural soil variation. Hence, a combination of measurements, termed a polyphasic approach, is typically desirable for testing a given hypothesis. The main current approaches to measuring the soil biota and the interrelationships among these approaches are shown in Figure 46.1. Some approaches, as noted, allow the simultaneous measurement of multiple dimensions of the soil community, such as abundance coupled with function, or activity in relation to genetic diversity. This yields richer data sets than were attainable previously.

46.1 Sampling the Soil Biota

One of the reasons why soil biological analyses have lagged behind those of chemical and physical properties is the difficulty in measuring soil biota and their activities and the lack of a unifying framework for interpreting the results. Ironically, responses of the soil biota to changes in abiotic and biotic variables are simultaneously very rapid and very slow, which makes interpreting any data generated difficult. For example, if soil is sampled after a heavy rainfall, soil microbial biomass and respiratory activity would probably be high. However, if the same soil is sampled 2 weeks later after no additional rainfall, the result would probably be quite different. Thus, antecedent factors — those occurring prior to sampling — strongly influence the outcome of most biological measures in the short term and make valid comparisons across sites and regions difficult, if not impossible.

To try to overcome this difficulty, soil is often incubated at a standard moisture and temperature for a period of time following sampling to stabilize populations and their activities. Measurements taken subsequently represent the *potential* of the soil community, under a specified set of conditions, to attain a certain measured population size or to perform a given function. While this improves the ability to compare measured variables between sites and regions, it does not directly reflect ongoing activities in the field, and thus it may easily over- or underestimate a given process in sampled soils. Over a longer term, we can measure the community composition in relation to changes in soil management and find that it takes years to discern significant treatment differences — thus making it difficult to use these measures to assess the potential sustainability of a given management system.

Improved sampling, sample handling, and downstream analyses developed in recent years now allow for greater *in situ* and *ex situ* resolution of soil biological variables and have contributed greatly to our understanding of soil ecology. For any approach in which biological properties of the soil will be measured, however, sampling design, timing, and handling must be considered carefully. The very act of sampling often creates changes in the variables that are measured downstream. Investigations at ICRISAT in India, reported by Boddey et al. (2005), evaluated the effect that sampling methods had on the results with acetylene reduction (AR) techniques used to assess biological nitrogen fixation (BNF) activity in soil samples. "Planted core assays," where a metal core collecting device was inserted around plant roots when they were very young and then carefully removed at the time of analysis 45 days later, were compared with "regular core assays" extracted at the same time by inserting such a device into the soil around plant roots, removing it, and straightaway analyzing it. Researchers found that BNF activity was up to ten times greater in the first set of soil samples, showing how much physical disturbance of the soil could reduce AR results and the derived estimates of BNF. Minimizing sampling effects as a source of variation is thus essential, although it can be extremely difficult. Such influences cannot be eliminated, but they can and should be specified and considered.

What to sample, how to sample, and what to measure will depend on the nature of the question that is being asked in any given study and the resolution that is required to satisfy any given hypothesis test. Wollum (1994), Forster (1995), Fredrickson and Balkwill (1998), and van Elsas et al. (2002) provide guidance for designing robust sampling strategies for different applications. Before sampling, it is also essential to consult a statistician to assure that the relevant statistical assumptions associated with the chosen sampling design will be met. It is also important to remember that all methods, no matter how careful the operator may be, have some degree of bias associated with them. Hence, reasonable efforts should be made to minimize known sources of bias and variation. In all studies, known sources of bias that cannot be eliminated should be acknowledged directly, and the conclusions of a given study should not be projected beyond what is supported by the evidence, once the associated biases are taken into account.

The five main approaches to analyzing the soil biota as individuals, populations, and communities are: microscopy, biochemical assays, physiological assays, molecular analyses, and process-level studies. Recent advances in basic methods and applications of each approach are considered below as they relate to assessing the abundance, activity, and diversity of the soil biota. Standard protocols for a wide range of approaches are given in Weaver et al. (1994), Alef and Nannipieri (1995), Hall (1996), Burlage et al. (1998), Hurst et al. (2002), and Kowalchuk et al. (2004), so the techniques themselves are not treated here in any detail.

46.2   Abundance

Abundance or density refers to the number of individuals (or concentration) of a target group of organisms that are present per unit volume or weight of soil. Traditional methods of measuring abundance have relied on:

1. Culturing organisms on artificial media
2. Counting cells directly in filtered extracts or minimally disturbed samples using various microscopy techniques in conjunction with different dyes, stains, or antigenic markers
3. Extracting specific cell components from cells and measuring their concentrations.

Significant advances in all of these approaches have been made in recent years that have improved estimates of microbial abundance in soil and have in many cases made possible simultaneous measures of activity and diversity.

46.2.1   Culturing

The limitations associated with cell culturing are well-known. Microscopy, even when enhanced with stains to increase accuracy in cell identification, typically yields cell counts one to two orders of magnitude greater than those detected by culturing. However, to understand the physiology, biochemistry, and functional roles of different groups of organisms, it is essential that they can be cultured.

Continuing efforts to "culture the uncultured" (Zengler et al., 2002; Joseph et al., 2003) have allowed numerous groups of organisms to be isolated on artificial media and examined in greater detail. These efforts have involved careful selection of media components frequently guided by knowledge of the functions of target groups of interest, such as the capacity to fix carbon dioxide or atmospheric nitrogen, or degrade specific carbon compounds such as lignin or atrazine.

Very high dilution of soil samples, combined with month- to year-long incubations in liquid or on solid media, as well as simple changes in gelling agents or carbon sources (Joseph et al., 2003) or encapsulation and subsequent incubation (Zengler et al., 2002), have allowed previously intractable organisms to be cultured and studied more rigorously, e.g., members of the phyla Verrucomicrobia, Planctomycetes, and many Actinobacteria (Joseph et al., 2003). Members of these phyla have been shown to be numerically dominant in many soils by use of 16S sequencing approaches (discussed in Section 46.4.2), yet had rarely been isolated previously. Creative approaches to bringing the seemingly vast realm of uncultured soil bacteria into culture is a promising area for extending our understanding of the functional capacity of soil Bacteria and Archaea.

It is well known that our understanding of the abundance (and diversity) of the soil microbial world has been restricted until recently by the limitations of the culturing approach. It is clear that no one medium or set of conditions, or even a host of different media and incubation conditions, is adequate for either enumerating or capturing the extant diversity in soil. New approaches using molecular methods combined with traditional approaches have improved both diversity estimates and, in many cases, abundance estimates as well. Depending on the method to which culturing is compared and how "individuals" are determined, estimates of how much of the soil's microbial diversity is currently known vary between 0.03 and 10%. Clearly, we have much yet to learn, and innovations in both traditional methods and the newer molecular methods are helping us to attain a deeper understanding of soil ecology.

46.2.2 Microscopy

Major advances have been made in recent years in the *in situ* identification and enumeration of as-yet-uncultured bacteria and in determining the composition of complex microbial communities by combining fluorescence microscopy or confocal laser scanning microscopy (CLSM) with the use of methods such as:

- Ribosomal RNA-targeted nucleic acid probes (Amann and Ludwig, 2000)
- A variety of different fluorochromes (Li et al., 2004)
- Marker and reporter genes (Jansson, 2003)
- Microautoradiography (Lee et al., 1999).

CLSM relies on use of a laser scanner in conjunction with an automated stage and image analysis software. Images of a sample are captured by scanning a field of view at varying depths of focus, and the images are then digitally combined to create a composite image of the sample across a large focal plane (Engelhardt and Knebel, 1993; Bloem et al., 1995). This approach allows the relationships between different organisms and between organisms and associated particles to be visualized with high resolution and with vastly less physical disturbance, which facilitates abundance estimates (Bloem et al., 1995).

When CLSM is coupled with molecular techniques such as use of fluorescent probes to specific rRNA sequences (i.e., fluorescent *in-situ* hybridization — FISH) (Amann and Ludwig, 2000; Pernthaler et al., 2001) or marker or reporter genes such as green fluorescent protein (*gfp*) (Errampalli et al., 1999; Lorang et al., 2001; Jansson, 2003), then specific populations can be quantified, and the microsites that they occupy and organisms that they consort with can be revealed. Dazzo and Yanni (Chapter 8, this volume) describe the use of *gfp*-tagged rhizobia to show that these bacteria were entering rice roots at lateral root emergence sites and are multiplying intercellularly within lateral rootlets, qualifying them as true endophytes of rice.

The *gfp* gene for the expression of the GFP protein may also be inserted into recipient cells in association with different promoter sequences, thus making it possible to use transformed cells as biosensors for the detection of the specific elicitor molecules (Bringhurst et al., 2001). Combining CLSM with microarray technology (Dahllof, 2002; Denef et al., 2003) or with microautoradiography (Lee et al., 1999) allows the strengths of multidimensional imaging to be combined with measures of diversity (presence of particular DNA sequences in a sample) or activity (uptake of radiolabeled substrates), respectively. Each approach has particular strengths, weaknesses, and potential biases associated with it that are discussed in each of the articles cited.

46.2.3 Biochemical Approaches

The more common biochemical methods used to assess abundance in soil biology research are:

- Chloroform fumigation followed by sample extraction and analysis of microbial biomass carbon and/or nitrogen, or further incubation and subsequent analysis of carbon dioxide evolved by the microbes recolonizing the soil post-fumigation.

- Analysis of membrane lipids and subsequent classification of the lipids to provide density estimates of specific groups of organisms.
- Detection of specific molecules such as ATP (soil energy charge), ergosterol (a purported marker for fungi), or glomalin (a marker for arbuscular mycorrhizal fungi, AMF).

Use of phospholipid fatty acid (PLFA) or fatty acid methylated ester (FAME) analyses has gained popularity as a means to estimate the abundance of soil organisms, but this technique is limited to separating them into broad categories, such as gram-negative or gram-positive bacteria, fungi, protozoa, or cyanobacteria. The more we discover about the composition of the soil biota, the less clear-cut this interpretive framework has become.

Yet some signature molecules remain very useful for measuring the abundance of key microbial groups, such as:

- 16:1 ω5 as a marker for arbuscular mycorrhizal fungi and cyanobacteria
- cy15:1 which is unique to the clostridia bacteria
- 16:0 and 18:3 ω6 for estimating fungal biomass
- 20:3 ω6 and 20:4 ω6 for protozoa (Paul and Clark, 1996)

The process by which membrane lipids are extracted from samples is sequential, so neutral lipids, glycolipids, and polar lipids can be quantified and characterized individually (White and Ringelberg, 1998; Zelles, 1999). Presence and quantity of one of the neutral lipids, ergosterol, can be used to estimate soil fungal biomass.

More rapid methods for determining total microbial biomass employing microwaving soil samples prior to extracting dissolved organic carbon (Insam and Weil, 1998) or adding chloroform directly to the extraction vials to streamline microbial biomass estimates (Witt et al., 2000) are gaining popularity. Both methods correlate well with the originally published and widely-used chloroform fumigation-extraction method (Vance et al., 1987).

46.2.4   Nucleic Acid Approaches

Enormous advances have been made in our understanding of soil ecology by use of molecular methods used to characterize nucleic acids (DNA and RNA) extracted from soil. Methods for extracting and purifying nucleic acids from varying soil types and environments continue to improve (Kowalchuk et al., 2004), and the availability of nucleic acid extraction kits (MoBio Laboratories, Inc., Carlsbad, CA, and Qbiogene, Inc., Irvine, CA) has streamlined downstream analyses considerably.

The most significant force driving this field forward is the availability of rapidly expanding, public-access nucleic acid sequence databases (GenBank, National Center for Biotechnology Information, Bethesda, MD, and the Ribosomal Database Project II, Cole et al., 2005), which now contain over 120,000 16S rRNA gene sequences. Associated bioinformatics tools enable users to mine the databases for comparative sequence information, develop targeted probes and primers, and construct parsimonious phylogenic trees to examine the relative placement and affiliation of new nucleic acid sequences obtained from the soil. Kowalchuk et al. (2004) provide a definitive collection of protocols for a myriad of molecular approaches useful for characterizing abundance, activity, and diversity of soil microbes.

A molecular approach that has become used widely for estimating abundance of gene sequences in soil samples is real time and/or quantitative polymerase chain reaction (PCR). Real-time PCR is a method that employs fluorogenic probes or dyes to quantify the number of copies of a target DNA sequence in a sample. This approach has been used successfully to quantify target genes that reflect the capacity of soil bacteria to perform given functions. Examples include, respectively, the use of ammonia monooxygenase, nitrite reductase, and methane monooxygenase genes to quantify ammonia oxidizing (Hermansson and Lindgren, 2001), denitrifying (Henry et al., 2004) and methanotrophic (Kolb et al., 2003) bacteria in soil samples. Real-time PCR coupled with primers to specific internal transcribed spacer (ITS) or rRNA gene sequences has also been used to quantify ectomycorrhizal (Landeweert et al., 2003) and endomycorrhizal fungi (Filion et al., 2003) as well as cyst nematodes (Madani et al., 2005) in soil.

46.3 Activity and Function

Activity generally refers to the rate of a given biochemical process in sampled soils. Specific processes are associated with particular ecosystem functions associated with microorganisms, such as BNF, carbon or nitrogen mineralization, ammonia or methane oxidation, or nitrate or nitrite reduction. In typical process-level studies, the microbes themselves are not isolated or identified specifically; instead, their activities are measured. Such studies normally characterize the ability of a soil population to metabolize specific compounds in assays, such as community-level physiological profiles (CLPP, the Biolog™ approach, Zak et al., 1994) or to transform specific molecules from one form to another, such as the conversion of NH_4^+ to NO_3^- during nitrification, or the conversion of NO_3^- to N_2 during the process of denitrification. Stable isotope tracers, such as ^{13}C or ^{15}N, are often used to characterize transformation rates by pool dilution or accumulation over time. The release of specific molecules from a soil sample, such as methane (CH_4), CO_2, hydrogen sulfide (H_2S) or nitrous oxide (N_2O), or the appearance of specific molecules in a soil sample such as NH_4^+, NO_3^- or sulfate ($SO_4^=$), are used to estimate the rate of various biogeochemical processes.

Such studies may also yield estimates of plant nutrient availability and measures of environmental quality. Still, the microbes facilitating these transformations generally remain uncharacterized. Recent advances have been made in CLPP analyses and in the use of stable isotopes in conjunction with nucleic acid analyses or quantifying signature molecules, such as specific PFLAs. This latter approach, called stable isotope probing (section 46.3.2), has provided insight into the microbial populations responsible for catalyzing various biochemical processes and thus has opened many "black boxes" to our view.

46.3.1 Community-Level Physiological Profiles

The ability of bacteria to metabolize specific carbon compounds has long been used as a part of a suite of diagnostic criteria by which they are classified. Biolog, Inc. (Hayward, CA) has developed a miniaturized method to test the capacity of bacterial isolates to metabolize a range of sole carbon substrates, thus enabling their rapid identification. A 96-well microtiter plate containing 95 different carbon sources for testing gram-negative bacteria and another plate for testing gram-positive bacteria have been used for years in diagnostic laboratories around the world. Zak et al (1994) reported the first use of Biolog plates to characterize the metabolic diversity of soil populations, associating the carbon substrate-use patterns obtained with "functional diversity." Since then, Biolog, Inc. has facilitated this application by manufacturing the EcoPlate™ for research use. The EcoPlate™ contains 31 single carbon source substrates, replicated three times on a plate. Limitations of this method include the need to dilute soil samples prior to inoculating the plates, thus potentially diluting out populations of interest. Also, the results are reflective of the activity of only those fast-growing community members that can typically be cultured on laboratory media.

Degens and Harris (1997), and more recently Campbell et al. (2003), have adapted this approach using whole soil to which various sole carbon source substrates are added. The resulting substrate-induced respiratory (SIR) response of the whole community to the added substrate is measured by detecting the release of CO_2 from the soil sample. In the approach of Degens and Harris (1997), soils were contained in separate flasks. Campbell et al. (2003) miniaturized the system by creating a sandwich between a deep-welled, 96-well microtiter plate in which the soil samples and the sole carbon source substrates are placed along with a regular 96-well microtiter plate which contains an agar-based alkali trap to detect CO_2 evolved from the soil. The two plates are sealed together with a silicone rubber gasket with interconnecting holes between the top and bottom wells. The complete assembly is clamped together tightly so that only the CO_2 emanating from each soil sample is detected separately. This microrespiratory system (MicroResp), which has a patent pending, represents a significant improvement to the use of this metabolic fingerprinting approach. Rather than measuring cell growth, as in the Biolog approach, it measures the short-term responsiveness of the community to the addition of a substrate and therefore more closely reflects potential community metabolic activity.

46.3.2   Stable Isotope Probing

Stable isotope probing (SIP) is a method that can be used to identify the microorganisms in a soil sample that use a particular growth substrate (Radajewski et al., 2000; Wellington et al., 2003; Dumont and Murrell, 2005). A selected substrate is first highly enriched with a stable isotope, most frequently ^{13}C, and then incorporated into the soil. Cellular components of interest are then recovered from the soil sample and analyzed for the incorporated stable isotope label. In this way, microbes that are actively using the added substrate can be identified. DNA and rRNA are the biomarkers used most frequently (Radajewski et al., 2003), although PLFAs have also been used successfully (Treonis et al., 2004). The ^{13}C-labeled molecules are purified from unlabeled nucleic acid (or PLFAs) by density-gradient centrifugation. Once separated, labeled nucleic acids can be amplified using PCR and universal primers to Bacteria, Archaea, or Eukarya. Analysis of the PCR products, through cloning and sequencing, for example, allows the microbes that have assimilated the ^{13}C-labeled substrate to be identified.

This approach has been applied successfully to study methanotrophs and methylotrophs (McDonald et al., 2005) and active rhizosphere communities through ^{13}C–CO_2 labeling of host plants (Griffiths et al., 2004). In the latter approach, information about which microbes are assimilating root exudates under a given set of environmental conditions can be obtained. Rangel-Castro et al. (2005) used ^{13}C–CO_2 pulse-labeling, followed by RNA-SIP, to study the effect of liming on the structure of the rhizosphere microbial community metabolizing root exudates in a grassland. Their results indicated that communities in limed soils were more complex and were more active in using recently exuded ^{13}C compounds than were those in unlimed soils.

SIP-based approaches have great potential for linking microbial identity with function (Dumont and Murrell, 2005), but at present a high degree of labeling is necessary to be able to separate labeled from unlabeled marker molecules. This need for high substrate concentrations may bias measures of community responses. Alternatively, use of long incubation times to ensure label incorporation increases the risk of having cross-feeding of ^{13}C from the primary consumers to the rest of the community, which complicates data interpretation.

46.4 Diversity, Richness, and Community Composition

Diversity or richness refers to the number of different kinds of taxonomic, morphological, physiological, or functional groups that are present in a soil sample. Community structure refers typically to the number of different taxonomic groups present and their relationships with each other. Community composition, visualized as DNA fingerprints, has also been used recently to describe how the operational taxonomic units (OTUs) in one sample, generally bands of nucleic acids resolved in an electrophoretic gel, are arranged in comparison to those in another sample. It is well known that for many fingerprinting techniques, a band on an electrophoretic gel is not synonymous with a distinct taxon; instead, since further separation and confirmation of the contents of the band is often not possible, the band becomes the OTU for intersample comparison. This is not a very robust or accurate use of the term "community composition", but its use reflects a rapidly expanding technological base that has enabled researchers to propose taxonomic structuring based on DNA sequencing, or classification based on DNA melting behavior or fragment size. The ecological nomenclature is still catching up.

While culturing, microscopic, and biochemical approaches have been used frequently to characterize population diversity in soil, the most promising advances have been made by use of molecular techniques. These will be the focus of this section. Kowalchuk et al. (2004) provide detailed protocols for many of the techniques used more commonly in soil diversity analyses today. New methods of analyzing nucleic acids to yield diversity data are being published daily, and this chapter cannot discuss them in any detail. Here, a few examples are given so the reader will have an idea of the molecular approaches that have added much to our understanding of the diversity of microbes in soil, at times coupled with their activity and/or function.

Molecular approaches used for estimates of diversity or intersample comparisons of community composition are based on characterizing nucleic acids (DNA and RNA) extracted from soil. While there are many approaches now available, the ones used most frequently in soil microbial ecology are: whole or partial community hybridization experiments (including use of microarrays); different variations on restriction fragment length polymorphism (RFLP) analysis; shotgun cloning and subsequent sequencing of clones from a clone library; and characterizing and/or sequencing of amplified rRNA or other gene targets.

The main strength of the molecular approaches is that they do not rely on the capacity to culture organisms from soil samples. The main drawbacks are the known biases associated with them. For example, when extracting nucleic acids from soil, some cells may be lysed and release their DNA/RNA more readily than others. In cloning, some DNA fragments may be preferentially incorporated into the cloning vector compared with others; and in PCR, some DNA sequences may be preferentially amplified over others (Wintzingerode et al., 1997). These known biases make it difficult, if not impossible, to calculate standard diversity indices reliably. The various molecular techniques can be subdivided into whole community analysis and partial community analysis. The former approach does not involve PCR amplification, while the latter approach does.

In essence, molecular techniques take advantage of the base-pairing rules that are fundamental to replicating new strands of DNA; guanine pairs with cytosine (G:C) and adenine pairs with thymine (A:T), forming three- and two-hydrogen bonds between pairs, respectively. This base-pairing is what allows us to construct specific DNA or RNA oligonucleotides and to hybridize them to their target sequences or to use them in primers in PCR. In this way, the presence and relative abundance of target sequences in appropriately prepared samples can be assessed.

46.4.1 Whole Community Analysis

Whole community analysis methods are those that do not rely on the use of PCR amplification, but instead analyze nucleic acids extracted from soil directly. Torsvik et al. (1990) were the first to describe the use of DNA denaturation/reassociation kinetics to estimate the existing biological diversity in soil samples. Community complexity is inferred from the rate at which complementary sample DNA strands reassociate following heat denaturation. This process can be carried out for a single soil community, or the DNA extracted from two separate soil communities can be mixed and their similarity (or extent of difference) can be estimated by use of this method. For estimating extant diversity in a soil sample, this approach is still considered to be the gold standard. Yet it is rarely used in population studies today because it is extremely time-consuming and requires complex and expensive instruments to perform correctly. Hence, it is not practical for broad-scale or extensive surveys. These are commonly performed by use of other approaches.

Direct cloning and the sequencing of clones from resulting clone libraries is another approach commonly employed to survey the biological diversity and/or richness of a given set of samples. This approach is also labor-intensive and thus is not used as frequently in broad-scale surveys as DNA fingerprinting methods (next section). It has recently become possible to clone very large DNA fragments into *Bac*-vectors (Rondon et al., 2000), thus facilitating the screening of cloned sequences for possible functional genes, such as resistance to antibiotics or specific catabolic capacities (Schloss and Handelsman, 2003).

Microarrays represent a promising new development in the analysis of microbial communities. This technology is based on nucleic acid hybridization and CLSM (46.2.2). The main differences between past protocols and microarrays are that oligonucleotide probes, rather than the DNA or RNA target, are immobilized on a solid surface, and hundreds of probes, bound in a miniaturized matrix, can be tested at the same time (Guschin et al., 1997). A fully developed DNA microchip could include a set of probes that encompasses virtually all natural microbial groupings and thereby can serve to monitor simultaneously the population structure at multiple levels of resolution (Liu and Stahl, 2002) while allowing for high sample throughput.

Ekins and Chu (1999) provide details of how to construct DNA microarrays and the types of arrays that can be developed. Microarrays have great potential for exploring genetic diversity of soil microbial communities (Denef et al., 2003) and for detecting pathogens in soil. However, the sensitivity of detection, which borders on 10^5 copies of the target sequence, is still orders of magnitude away from where it needs to be for use as a robust monitoring or detection tool. As probe design and associated specificity and detection systems improve, microarrays will play a larger and larger role in soil microbial community analysis.

46.4.2 Partial Community Analysis

Polymerase chain reaction-based methods are used to characterize partially complex communities of microorganisms. The most common PCR targets for characterizing soil microbial communities are the ribosomal RNA gene sequences. These are the small subunit (SSU) rRNA genes (16S in prokaryotes or 18S in eukaryotes), the large subunit (LSU) rRNA genes (23S in prokaryotes or 28S in eukaryotes), or the internal transcribed spacer (ITS) regions between the SSU and LSU gene sequences. Other defined targets are genes that code for ecologically significant functions as described in section 46.2.4.

The ribosome is critical to the function of all organisms, and hence, ribosomal genes are very highly conserved. It is this conservation of sequence that has been exploited to examine the evolutionary history and taxonomic relationships of different organisms to each other. The result of such analyses has led to a recasting of the metaphorical tree of life to reflect three broad domains: the Bacteria, the Archaea, and the Eukarya (Pace et al., 1986; Olsen and Woese, 1993).

Knowledge of the variation of the gene sequences between different organisms has been exploited in a variety of ways to characterize microbial communities in soil. The methods used more commonly for soil microbial community analysis are denaturing (or temperature) gradient gel electrophoresis (DGGE, TGGE) analysis (Muyzer and Smalla, 1998), terminal restriction fragment length polymorphism (T-RFLP) analysis (Liu et al., 1997), and cloning and sequencing of amplified SSU rRNA genes (Olsen and Woese, 1993).

When rRNA genes are amplified using universal primers to any of the three domains, generally a single band of a predicted size is obtained on an electrophoretic gel, yielding by itself very little information about the community being analyzed. If cloning and sequencing is then performed, this is not a restriction. However, any other method requires additional analyses to separate amplified fragments in such a way as to yield information about key differences between the soil communities under study. DGGE and T-RFLP are creative means by which to separate SSU rRNA (or other gene) products to yield community-level DNA fingerprints. DNA fingerprints obtained can then be analyzed by use of multivariate statistics to characterize the level of similarity or difference between different soil microbial communities.

46.4.3 Denaturing or Temperature Gradient Gel Electrophoresis (DGGE/TGGE)

DGGE and TGGE are identical in principle. Both techniques impose a parallel gradient of denaturing conditions along an acrylamide gel. Amplified DNA is loaded on electrophoretic gel and, as the DNA migrates, the denaturing conditions of the gel gradually increase. In DGGE, the denaturant is generally urea; in TGGE, it is temperature. As native double-stranded DNA is a compact structure, it migrates faster than partially denatured DNA. The sequence of a fragment determines the point in the gel at which denaturation will start to retard mobility. Sequence affects duplex stability by both percentage G + C content and neighboring nucleotide interactions (e.g., GGA is more stable than GAG).

While the power of DGGE and TGGE to detect diversity among amplified DNA fragments within a single gel is high, the sensitivity of the technique to gel and running conditions makes comparisons between gels very difficult. As with all electrophoretic techniques, the resolving power is limited by the number of bands capable of fitting and being resolved within a single lane on a gel. In practice, no more than 100 to 150 distinct sequence types may be resolved despite the single base-pair sensitivity.

DGGE and TGGE are now frequently being applied in microbial ecology to compare the structures of complex microbial communities and to study their dynamics (Ovreas et al., 1998; Heuer et al., 1999; Rumberger et al., 2004). An important advantage that DGGE analysis has over T-RFLP (section 46.4.4) is that amplicons of interest that are resolved on a DGGE gel can be cut from the gel, reamplified, cloned, and sequenced, thereby obtaining taxonomic and/or phylogenetic information about amplifiable members of the soil community. However, Ovreas et al. (1998) have shown that, compared with reassociation kinetics, DGGE vastly underestimates the existing diversity in soil.

46.4.4   Terminal Restriction Fragment Length Polymorphism (T-RFLP) Analysis

T-RFLP analysis employs a primer pair targeted to the SSU rRNA or other target gene of interest in which one of the primers is tagged with a fluorescent label. Amplified DNA is then digested with a restriction enzyme to yield a family of terminally labeled restriction fragments (TRFs) that can be sized accurately using DNA sequencing technology (Marsh, 1999). The result is again a type of DNA fingerprint that can be used to compare the composition of community DNA extracted from different soil samples. Liu et al. (1997) used this technique to characterize microbial diversity within bioreactor sludge, aquifer sand, and termite guts. Lukow et al. (2000) and Devare et al. (2004) have applied T-RFLP successfully to compare bacterial communities in the rhizosphere of transgenic and non-transgenic crops.

This technique need not be restricted to studying the 16S rRNA gene. T-RFLP can be used as a quick screen with any gene to look at differences between communities in soil samples. Tan et al. (2003) used this technique to examine the effect of nitrogen application and plant genotype on *nif*H gene diversity in the rice rhizosphere. The main drawbacks of the use of T-RFLP are the inability to characterize TRFs further or to obtain sequence information. Further, pseudoterminal fragments can be formed, thus potentially biasing follow-on analyses and comparisons (Egert and Friedrich, 2003).

The strengths of the community-level PCR fingerprinting techniques, particularly the T-RFLP technique, are that they can be used successfully to compare community composition of different samples without culturing, and to compare communities over time and space in response to changes in soil management or environmental factors. A key weakness of these methods is that phylogenetically-related taxa can yield TRF (or electrophoretic types, ET) of the same size. This means that one band does not necessarily represent one taxon, so it is unwise and potentially misleading to use traditional diversity indices such as the Shannon-Weiner index to characterize these types of data. Multivariate statistical approaches, such as principal component analysis (PCA), redundancy analysis, or additive main effect and multiplicative interaction (AMMI) analysis (Gauch, 1992), are the preferred approaches for intersample comparisons.

46.4.5   DNA Cloning and Sequencing

Although it is more time-consuming, cloning and sequencing of amplified rRNA or other target genes is the more robust means to assess community diversity, and this is the means by which the sequence databases so crucial to increasing our understanding of soil ecology continue to grow. Sambrook and Russell (2001) and Kowalchuk et al. (2004) provide detailed methods for cloning and sequencing DNA or reverse-transcribed (RT) RNA. Many community census studies have employed this approach successfully (e.g., Felske et al., 1998; Felske et al., 1999; Poly et al., 2001; Jansa et al., 2002; Carney et al., 2004). Gene sequences obtained in various studies are submitted to and maintained within databases such as the Ribosome Database Project II (http://rdp.cme.msu.edu/html/) or GenBank (http://www.ncbi.nlm.nih.gov/html), which contain over 120,000 prokaryotic sequences. Continued development of databases through DNA sequencing is essential for and is a prerequisite to good primer and probe design.

46.5 Discussion

We have entered an era of great opportunity for making significant discoveries in soil biology and ecology, although many challenges remain. For example, the ability to extract DNA or ribosomal RNA (rRNA) from cells contained within soil samples, and its direct analysis in hybridization experiments or its use in PCR amplification experiments, has allowed us to detect the presence of a vast diversity of microbes previously unimagined.

Yet, beyond deriving their phylogenetic placement and demonstrating that some groups compose a large proportion of the total Bacteria or Archaea in soil, many of their key roles and functions remain an enigma. As methods continue to develop, our ability to probe and to begin to fathom this vast diversity should continue to improve, extending our knowledge and appreciation of the many and varied functions of the soil biota, including their various effects on — and how they are influenced by — soil chemical and physical characteristics.

Recent efforts to combine results of polyphasic studies into a workable assessment framework (Andrews et al., 2004; Lilburne et al., 2004) are helping researchers and land managers alike to determine whether their management strategies are leading to improvements in soil biological condition or if new strategies may be needed. Methods for on-farm monitoring and evaluating soil quality are discussed in the following chapter (Chapter 47).

References

Alef, K. and Nannipieri, P., Eds., *Methods in Applied Soil Microbiology and Biochemistry*, Academic Press, San Diego, CA (1995).

Amann, R. and Ludwig, W., Ribosomal RNA-targeted nucleic acid probes for studies in microbial ecology, *FEMS Microbiol. Rev.*, **24**, 555–565 (2000).

Andrews, S.S., Karlen, D.L., and Cambardella, C.A., The soil management assessment framework: a quantitative soil quality evaluation method, *Soil Sci. Soc. Am. J.*, **68**, 1945–1962 (2004).

Bloem, J., Veninga, M., and Shepherd, J., Fully automatic determination of soil bacterium numbers, cell volumes, and frequencies of dividing cells by confocal laser scanning microscopy and image analysis, *Appl. Environ. Microbiol.*, **61**, 926–936 (1995).

Boddey, R.M. et al., Assessment of bacterial nitrogen fixation in grasses, EMBRAPA-Agroecologia, Saõ Paulo, Brazil, Unpublished paper (2005).

Bringhurst, R.M., Cardon, Z.G., and Gage, D.J., Galactosides in the rhizosphere: Utilization by *Sinorhizobium meliloti* and development of a biosensor, *Proc. Natl Acad. Sci. USA*, **98**, 4540–4545 (2001).

Burlage, R.S. et al., Eds., *Techniques in Microbial Ecology*, Oxford University Press, New York (1998).

Campbell, C.D. et al., A rapid microtiter plate method to measure carbon dioxide evolved from carbon substrate amendments so as to determine the physiological profiles of soil microbial communities by using whole soil, *Appl. Environ. Microbiol.*, **69**, 3593–3599 (2003).

Carney, K.M., Matson, P.A., and Bohannan, B.J.M., Diversity and composition of tropical soil nitrifiers across a plant diversity gradient and among land-use types, *Ecol. Lett.*, 7, 684–694 (2004).

Cole, J.R. et al., The Ribosomal Database Project (RDP-II): Sequences and tools for high-throughput rRNA analysis, *Nucleic Acids Res.*, **33**, D294–D296 (2005), doi: 10.1093/nar/gki038.

Dahllof, I., Molecular community analysis of microbial diversity, *Curr. Opin. Biotechnol.*, **13**, 213–217 (2002).

Degens, B.P. and Harris, J.A., Development of a physiological approach to measuring the catabolic diversity of soil microbial communities, *Soil Biol. Biochem.*, **29**, 1309–1320 (1997).

Denef, V.J. et al., Validation of a more sensitive method for using spotted oligonucleotide DNA microarrays for functional genomics studies on bacterial communities, *Environ. Microbiol.*, **5**, 933–943 (2003).

Devare, M., Jones, C.M., and Thies, J.E., Effects of CRW transgenic corn and tefluthrin on the soil microbial community: Biomass, activity, and diversity, *J. Environ. Qual.*, **33**, 837–843 (2004).

Dumont, M.G. and Murrell, J.C., Stable isotope probing: Linking microbial identity to function, *Nat. Rev. Microbiol.*, **3**, 499–504 (2005).

Egert, M. and Friedrich, M.W., Formation of pseudo-terminal restriction fragments: A PCR-related bias affecting terminal restriction fragment length polymorphism analysis of microbial community structure, *Appl. Environ. Microbiol.*, **69**, 2555–2562 (2003).

Ekins, R. and Chu, F.W., Microarrays: Their origins and applications, *Trends Biotechnol.*, **17**, 217–218 (1999).

Engelhardt, J. and Knebel, W., Leica TCS — the confocal laser scanning microscope of the latest generation: Technique and applications, *Sci. Tech. Inf.*, **10**, 159–168 (1993).

Errampalli, D. et al., Applications of the green fluorescent protein as a molecular marker in environmental microorganisms, *J. Microbiol. Methods*, **35**, 187–199 (1999).

Felske, A. et al., Phylogeny of the main bacterial 16S rRNA sequences in Drentse A grassland soils (The Netherlands), *Appl. Environ. Microbiol.*, **64**, 871–879 (1998).

Felske, A. et al., Searching for predominant soil bacteria: 16S rDNA cloning versus strain cultivation, *FEMS Microbiol. Ecol.*, **30**, 137–145 (1999).

Filion, M., St-Arnaud, M., and Jabaji-Hare, S.H., Direct quantification of fungal DNA from soil substrate using real-time PCR, *J. Microbiol. Methods*, **53**, 67–76 (2003).

Forster, J.C., Soil sampling, handling, storage and analysis, In: *Methods in Applied Soil Microbiology and Biochemistry*, Alef, K. and Nannipieri, P., Eds., Academic Press, San Diego, CA (1995).

Fredrickson, J.K. and Balkwill, D.L., Sampling and enumeration techniques, In: *Techniques in Microbial Ecology*, Burlage, R.S. et al., Eds., Oxford University Press, NY (1998).

Gauch, H.G., *Statistical Analysis of Regional Yield Trials: AMMI Analysis of Factorial Designs*, Elsevier, Amsterdam (1992).

Griffiths, R.I. et al., ($^{13}CO_2$)–C^{13}–pulse labelling of plants in tandem with stable isotope probing: Methodological considerations for examining microbial function in the rhizosphere, *J. Microbiol. Methods*, **58**, 119–129 (2004).

Guschin, D.Y. et al., Oligonucleotide microchips and genosensors for determinative and environmental studies in microbiology, *Appl. Environ. Microbiol.*, **63**, 2397–2402 (1997).

Hall, G.S., Ed., *Methods for the Examination of Organismal Diversity in Soils and Sediments*, CAB International, NY (1996).

Henry, S. et al., Quantification of denitrifying bacteria in soils by *nir*K gene targeted real-time PCR, *J. Microbiol. Methods*, **59**, 327–335 (2004).

Hermansson, A. and Lindgren, P.E., Quantification of ammonia-oxidizing bacteria in arable soil by real-time PCR, *Appl. Environ. Microbiol.*, **67**, 972–976 (2001).

Heuer, H. et al., Polynucleotide probes that target a hypervariable region of 16S rRNA genes to identify bacterial isolates corresponding to bands of community fingerprints, *Appl. Environ. Microbiol.*, **65**, 1045–1049 (1999).

Hurst, C.J. et al., Eds., *Manual of Environmental Microbiology*, 2nd ed., ASM Press, Washington, DC (2002).

Insam, K.R. and Weil, R.R., Microwave irradiation of soil for routine measurement of microbial biomass carbon, *Biol. Fert. Soils*, **27**, 408–416 (1998).

Jansa, J. et al., Intra- and intersporal diversity of ITS rDNA sequences in *Glomus intraradices* assessed by cloning and sequencing, and by SSCP analysis, *Mycol. Res.*, **106**, 670–681 (2002).

Jansson, J.K., Marker and reporter genes: Illuminating tools for environmental microbiologists, *Curr. Opin. Microbiol.*, **6**, 310–316 (2003).

Joseph, S.J. et al., Laboratory cultivation of widespread and previously uncultured soil bacteria, *Appl. Environ. Microbiol.*, **69**, 7210–7215 (2003).

Kolb, S. et al., Quantitative detection of methanotrophs in soil by novel *pmo*A-targeted real-time PCR assays, *Appl. Environ. Microbiol.*, **69**, 2423–2429 (2003).

Kowalchuk, G.A. et al., *Molecular Microbial Ecology Manual*, Vol. 1–2, 2nd ed., Kluwer Academic Publishers, Dordrecht (2004).

Landeweert, R. et al., Quantification of ectomycorrhizal mycelium in soil by real-time PCR compared to conventional quantification techniques, *FEMS Microbiol. Ecol.*, **45**, 283–292 (2003).

Lee, N. et al., Combination of fluorescent *in situ* hybridization and microautoradiography: A new tool for structure–function analyses in microbial ecology, *Appl. Environ. Microbiol.*, **65**, 1289–1297 (1999).

Li, Y., Dick, W.A. and Tuovinen, O.H., Fluorescence microscopy for visualization of soil microorganisms: A review, *Biol. Fert. Soils*, **39**, 301–311 (2004).

Lilburne, L., Sparling, G., and Schipper, L., Soil quality monitoring in New Zealand: Development of an interpretive framework, *Agric. Ecosyst. Environ.*, **104**, 535–544 (2004).

Liu, W.-T. et al., Characterization of microbial diversity by determining terminal restriction fragment length polymorphisms of genes encoding 16S rRNA, *Appl. Environ. Microbiol.*, **63**, 4516–4522 (1997).

Liu, W.-T. and Stahl, D.A., Molecular approaches for the measurement of density, diversity and phylogeny, In: *Manual of Environmental Microbiology*, 2nd ed., Hurst, C.J. et al., Eds., ASM Press, Washington, DC (2002).

Lorang, J.M. et al., Green fluorescent protein is lighting up fungal biology, *Appl. Environ. Microbiol.*, **67**, 1987–1994 (2001).

Lukow, T., Dunfield, P.F., and Liesack, W., Use of the T-RFLP technique to assess spatial and temporal changes in the bacterial community structure within an agricultural soil planted with transgenic and non-transgenic potato plants, *FEMS Microbiol. Ecol.*, **32**, 241–247 (2000).

Madani, M., Subbotin, S.A., and Moens, M., Quantitative detection of the potato cyst nematode, *Globodera pallida*, and the beet cyst nematode, *Heterodera schachtii*, using real-time PCR with SYBR Green I dye, *Mol. Cell. Probes*, **19**, 81–86 (2005).

Marsh, T.L., Terminal restriction fragment length polymorphism (T-RFLP): An emerging method for characterizing diversity among homologous populations of amplications products, *Curr. Opin. Microb.*, **2**, 323–327.

McDonald, I.R., Radajewski, S., and Murrell, J.C., Stable isotope probing of nucleic acids in methanotrophs and methylotrophs: A review, *Org. Geochem.*, **36**, 779–787 (2005).

Muyzer, G. and Smalla, K., Application of denaturing gradient gel electrophoresis (DGGE) and temperature gradient gel electrophoresis (TGGE) in microbial ecology, *Antonie Van Leeuwenhoek Int. J. Gen. Mol. Microbiol.*, **73**, 127–141 (1998).

Olsen, G.J. and Woese, C.R., Ribosomal-RNA: A key to phylogeny, *Fed. Am. Soc. Exp. Biol. J.*, **7**, 113–123 (1993).

Overeas, L. et al., Microbial community changes in a perturbed agricultural soil investigated by molecular and physiological approaches, *Appl. Environ. Microbiol.*, **64**, 2739–2742 (1998).

Pace, N.R., Olsen, G.J., and Woese, C.R., Ribosomal-RNA phylogeny and the primary lines of evolutionary descent, *Cell*, **45**, 325–326 (1986).

Paul, E.A. and Clark, F.E., Eds., *Soil Microbiology and Biochemistry*, Academic Press, San Diego (1996).

Pernthaler, J. et al., Fluorescence *in situ* hybridization (FISH) with rRNA-targeted oligonucleotide probes, *Methods Microbiol.*, **30**, 207–226 (2001).

Poly, F. et al., Comparison of *nif*H gene pools in soils and soil microenvironments with contrasting properties, *Appl. Environ. Microbiol.*, **67**, 2255–2262 (2001).

Radajewski, S. et al., Stable-isotope probing as a tool in microbial ecology, *Nature*, **403**, 646–649 (2000).

Radajewski, S., McDonald, I.R., and Murrell, J.C., Stable-isotope probing of nucleic acids: A window to the function of uncultured microorganisms, *Curr. Opin. Biotechnol.*, **14**, 296–302 (2003).

Rangel-Castro, J.I. et al., Stable isotope probing analysis of the influence of liming on root exudate utilization by soil microorganisms, *Environ. Microbiol.*, **7**, 828–838 (2005).

Rondon, M.R. et al., Cloning the soil metagenome: A strategy for accessing the genetic and functional diversity of uncultured microorganisms, *Appl. Environ. Microbiol.*, **66**, 2541–2547 (2000).

Rumberger, A. et al., Rootstock genotype and orchard replant position rather than soil fumigation or compost amendment determine tree growth and rhizosphere bacterial community composition in an apple replant soil, *Plant Soil*, **264**, 247–260 (2004).

Sambrook, J. and Russell, D.W., *Molecular Cloning: A Laboratory Manual*, Cold Spring Harbor Laboratory Press, Cold Spring Harbor, NY (2001).

Schloss, P.D. and Handelsman, J., Biotechnological prospects from metagenomics, *Curr. Opin. Biotechnol.*, **14**, 303–310 (2003).

Tan, X.Y., Hurek, T., and Reinhold-Hurek, B., Effect of N-fertilization, plant genotype and environmental conditions on *nif*H gene pools in roots of rice, *Environ. Microbiol.*, **5**, 1009–1015 (2003).

Thies, J.E. and Suzuki, K., Amazônian dark earths: Biological measurements, In: *Amazônian Dark Earths*, Lehmann, J., Kern, D., and Falcao, N., Eds., Kluwer Academic Publishers, Dordrecht, Netherlands (2003).

Torsvik, V., Goksoyr, J. and Daae, F.L., High diversity in DNA in soil bacteria, *Appl. Envir. Microb.*, **56**, 782–787 (1990).

Treonis, A.M. et al., Identification of groups of metabolically-active rhizosphere microorganisms by stable isotope probing of PLFAs, *Soil Biol. Biochem.*, **36**, 533–537 (2004).

Vance, E.D., Brookes, P.C., and Jenkinson, D.S., An extraction method for measuring soil microbial biomass C, *Soil Biol. Biochem.*, **19**, 703–707 (1987).

van Elsas, J.D. et al., Methods for sampling soil microbes, In: *Manual of Environmental Microbiology*, 2nd ed., Hurst, C.J. et al., Eds., ASM Press, Washington, DC (2002).

Weaver, R.W. et al., Eds., *Methods of Soil Analysis, Microbiological and Biochemical Properties*, Soil Science Society of America, Pt 2, Madison, WI (1994).

Wellington, E.M.H., Berry, A., and Krsek, M., Resolving functional diversity in relation to microbial community structure in soil: Exploiting genomics and stable isotope probing, *Curr. Opin. Microbiol.*, **6**, 295–301 (2003).

White, D.C. and Ringelberg, D.B., Signature lipid biomarker analysis, In: *Techniques in Microbial Ecology*, Burlage, R.S. et al., Oxford University Press, New York, 255–272 (1998).

Wintzingerode, F.v., Gobel, U.V., and Stackebrandt, E., Determination of microbial diversity in environmental samples: Pitfalls of PCR-based rRNA analysis, *FEMS Microbiol. Rev.*, **21**, 213–229 (1997).

Witt, C. et al., A rapid chloroform-fumigation extraction method for measuring soil microbial biomass carbon and nitrogen in flooded rice soils, *Biol. Fert. Soils*, **30**, 510–519 (2000).

Wollum, A.G., Soil sampling for microbiological analysis, In: *Methods of Soil Analysis, Microbiological and Biochemical Properties*, Pt 2, Weaver, R.W. et al., Eds., Soil Science Society of America, Madison (1994).

Zak, J.C. et al., Functional diversity of microbial communities: A quantitative approach, *Soil Biol. Biochem.*, **26**, 1101–1108 (1994).

Zelles, L., Fatty acid patterns of phospholipids and lipopolysaccharides in the characterisation of microbial communities in soil: A review, *Biol. Fert. Soils*, **99**, 111–129 (1999).

Zengler, K. et al., Cultivating the uncultured, *Proc. Natl Acad. Sci.*, **99**, 15681–15686 (2002).

47

Approaches to Monitoring Soil Systems

David Wolfe

Department of Horticulture, Cornell University, Ithaca, New York, USA

CONTENTS

The terms "soil health" and "soil quality" are becoming increasingly familiar worldwide, reflecting a growing appreciation for soil as a major component of the biosphere and as a fundamental resource that must be monitored and protected for the sustainability of all life on the planet. Doran and Parkin (1994) defined soil quality as: "the capacity of a soil to function, within ecosystem and land use boundaries, to sustain biological productivity, maintain environmental quality, and promote plant and animal health." In general, soil health and soil quality can be considered synonymous. While some scientists are most comfortable with the term "soil quality," at Cornell University we have constituted a Soil Health Program Team as a multidisciplinary effort that involves university researchers, extension educators, and farmer-cooperators. Our terminology reflects an emphasis on new biological, as well as chemical and physical measures of soil quality. The analogy with human health is useful from an educational standpoint. For example:

- Just as a healthy person is resistant to stress and disease and thus uses drugs sparingly, a healthy soil can suppress soil-borne crop diseases, reduce pesticide requirements, and buffer plants from water and nutrient stresses.
- Conversely, much as it takes time and commitment for an unhealthy person, e.g., someone addicted to drugs or bad eating habits, to become healthy again, rehabilitating an unhealthy soil requires patience and commitment by the grower.

- An assessment of human health begins with measurements of vital signs such as heart rate and blood pressure, although these measures by themselves do not permit reliable diagnosis of a problem or guarantee perfect health. Similarly, measurements of pH, organic matter, or a particular microbial activity may be useful "indicators" of soil health, but more comprehensive testing and long-term monitoring are usually required to diagnose and solve serious problems.

There is no universal standard for optimum soil health nor any specific set of soil quality measurements that will suit all purposes. Deciding on exactly what should be measured to evaluate soil health and interpreting the data require *a priori* knowledge of the basic soil type (taxonomic class), intended land use, and management goals. Potential measurements usually fall into two broad categories: (1) soil biological, chemical, and/or physical indicators relevant to specific soil functions, such as nitrogen mineralization rate which would be relevant to assessment of nutrient cycling and soil productivity; and (2) quantification of the soil function or ecosystem service that is of interest, such as disease suppression, productivity, or water quality.

47.1   Identifying Useful Indicators of Soil Health

What makes a good indicator of soil health? A variety of approaches have been suggested for answering this question (e.g., Doran and Parkin, 1996; Andrews et al., 2004; Bending et al., 2004). Criteria being used by the Cornell Soil Health Team to assess possible indicators include: (1) measurability at reasonable cost; (2) sensitivity to changes in management practices; (3) quantifiable effects on crop health, yield, and/or environmental impact; and (4) a useful integration of several soil quality factors, or good correlation with other more costly measures. These criteria are probably generally applicable, but modifications may be necessary to address user priorities and to work within the funding available for monitoring.

47.1.1   Chemical, Physical, and Biological Indicators

Table 47.1 lists a number of candidate chemical, physical, and biological indicators of soil health. For detailed procedures, the reader can consult one of the well-established reference texts available, such as Weaver et al. (1994), Sparks (1996), and Doran and Jones (1996). Many of the chemical, and some of the physical measurements are routinely conducted by university and commercial laboratories on a fee-for-service basis. In contrast, the soil biological measurements are generally not available even for a fee, and interpretation of biological data in relation to crop yield or environmental effects is in its infancy (Section 47.3).

Biological measurements may include actual counts or estimates of the abundance of specific soil organisms or functional groups. Earthworms, saprophytic nematodes, and arthropods are organisms essential to decomposition and nutrient cycling that are sometimes targeted in this way (Blair et al., 1996; Neher, 2001; Parisi et al., 2005). Specific pathogens such as root parasitic nematodes are also often of interest. The arbuscular mycorrhizal fungi — common plant root symbionts that help plants acquire phosphorus and other nutrients — are often of interest, but have been difficult to quantify. Bending et al. (2004) recently described an assay using onions as host plant to determine a soil's potential for mycorrhizal colonization of roots.

TABLE 47.1

Quantitative Chemical, Physical, and Biological Measurements for Soil Health Monitoring[a]

Chemical	Physical	Biological
pH	Soil texture/taxonomy	*Estimations of abundance*:
% OM (loss on ignition)	Bulk density	Microbial biomass
CEC	Soil penetrometer resistance	Earthworm populations
Total C, N	Aggregate stability	Nematode populations (parasitic, saprophytic)
Major nutrients	Water-holding capacity	Arthropod populations
Minor nutrients	Water infiltration rate	Arbuscular mycorrhizal fungi
Electrical conductivity	Depth to hardpan	PLFA profiling
Heavy metals, toxins	Depth to water table	Molecular genetic methods
C, N in POM	Porosity	*Estimations of activity*:
C, N in microbial biomass	Erosion potential	Respiration rate
Glomalin		Potentially mineralizable N Decomposition rate Disease-suppressive assay Soil enzyme activities Pollutant detoxification

[a] This is not an exhaustive list, but is provided as an example of options. References for methods are cited in the text. CEC: cation exchange capacity; OM: organic matter; PLFA: phospholipid fatty acid; POM: particulate OM.

New biochemical and molecular techniques (e.g., Liu et al., 1997) are offering high-tech means for estimating the relative abundance of important functional groups or particular species. Several of these methods have been discussed in Chapter 46. These new methods are revolutionizing our capacity for understanding soil microbial diversity and community structure and hold great promise for monitoring soil health in the future. At this time, however, they are primarily being used for research purposes and are not readily available in many parts of the world.

Measuring soil biological activity rather than abundance is another way to quantify some of the biological components of soil health, as indicated in Table 47.1. Data on activity often relate directly to soil processes that we know are important to crop productivity and environmental effects of farming, e.g., nutrient cycling, disease suppression, and pollutant detoxification. When a measure of microbial activity such as respiration rate is expressed in terms of per-unit microbial biomass, it reflects the metabolic "efficiency" for that activity. Because soils dominated by different functional groups will have different metabolic efficiencies, these measurements can provide indirect information about the overall structure of soil biotic communities.

Some of the methods for measuring biological activity are well established, although modifications for cost-efficiency may be required for their widespread use in soil health monitoring. Measuring soil enzyme activity is viewed as a promising new approach because enzyme activities are often correlated with important functions of soil organisms such as decomposition and C, N, P, and S cycling (Dick et al., 1996; Bending et al., 2004). Most soil enzymes are of microbial origin, and many remain catalytic after cell death in soil solution or after being complexed with clays or organic matter (dehydrogenase is a notable exception). Enzyme activities may thus serve as an "early indicator" of response to soil management practices since they often begin to change much sooner, in 1 to 2 years, than other properties, e.g., soil organic C (Dick et al., 1996). Dehydrogenase, glucosidases, ureases, and phosphatases are examples of enzymes whose activity can now be measured fairly easily.

Some soil chemical and physical measurements listed in Table 47.1 are particularly relevant to biological activities. One example is glomalin, a common glycoprotein found in soils that is important in aggregate stability (Rillig et al., 2002). Glomalin is released in substantial quantities by many types of arbuscular mycorrhizal fungi and can remain in soils for several years. Glomalin levels not only reflect historical abundance and activity of the Glomeromycotan fungi, but also tend to be correlated with aggregate stability because the sticky nature of the compound helps hold soil particles together. There are other substances released by other soil organisms and by roots that help hold aggregates together, and the hyphae of many types of fungi also play a role. This is important to be aware of since aggregate stability has a major influence on water-holding capacity and other soil physical and chemical features. This makes both glomalin and aggregate stability useful as integrators of soil physical, chemical, and biological characteristics.

47.1.2   A Minimum Data Set

In practice, for economic reasons, it is usually necessary to identify a minimum data set (MDS) for monitoring. Several groups (Doran and Parkin, 1996; Bending et al., 2004; Sparling et al., 2004) have attempted to identify such essential lists. Andrews et al. (2004) developed an "expert system" of decision rules to help users select an optimum MDS from a menu of more than 80 options. Users are expected to answer a series of questions in order to categorize their knowledge needs based on inherent soil properties, climate, cropping system, tillage practices, assessment purpose, and management goals (which could range from crop production, to waste management, or maintenance of habitat and biodiversity). This approach recognizes the fact that no one MDS is suitable for all purposes.

Among the chemical measurements listed in Table 47.1, an MDS will often include soil pH, total percent OM, major and minor plant nutrients, and cation exchange capacity (CEC). These have long been considered essential information and are included in standard soil test packages available in many places. Electrical conductivity (EC) has significant effects on soil microbial community structure (Smith and Doran, 1996) and is particularly important in regions where soil salinity is an issue.

Determining the C and N in the small but relatively labile fractions of OM, referred to as particulate OM (POM) or called the light OM fraction, is more time-consuming and expensive than are standard measures of total OM by combustion. However, in a monitoring program it has the benefit of reflecting changes attributable to soil management practices much sooner than will changes in total OM (Magdoff et al., 1996). Also, the C and N in soils that are most readily available to plants and microbes is generally found in POM (and in microbial biomass).

Among the soil physical measurements of interest, bulk density is particularly important. It is required to convert any soil quality data back and forth between a soil dry weight and soil volume basis. In addition, bulk density is correlated with several other soil physical measurements that are often considered informative, such as aggregate stability, porosity, water-holding capacity, and soil compaction (penetrometer resistance).

While chemical and physical measurements have been collected routinely for decades, there is less experience with biological measurements, so determining an MDS for these is comparatively difficult. Total microbial biomass and soil respiration rate are often considered essential (e.g., Doran and Parkin, 1996), but these measurements are particularly variable over time and are difficult to interpret, so they have been left out of the MDS in some monitoring programs (Sparling et al., 2004).

The Cornell Soil Health Program Team is considering a wide range of candidate biological indicators, but has tended to focus on activity measurements, such as N mineralization rate, decomposition rate (using a simple cellulose filter-paper degradation method), and a disease-suppressive assay (evaluated by scoring the roots of young beans grown in soil for root disease symptoms). In addition, we are making counts of both pathogenic (potentially harmful to crops) and saprophytic (beneficial) nematodes, and the ratio of the two (Abawi and Widmer, 2000). We consider aggregate stability to be a biological as well as a physical indicator, and we are measuring glomalin at some sites.

47.1.3 Sampling Protocol

Just as there is no single MDS ideal for all purposes, the sampling protocol will vary depending on goals, variability, acceptable statistical confidence level, and funding constraints. This topic is discussed in various texts, e.g., Carter (1993), so the discussion here can be brief.

In any monitoring program, the basic purpose is to be able to evaluate changes over time. In this book, it has been stressed that biological processes and levels are dynamic and variable. Repeated sampling should of course be carried out at similar times of the year each year, using similar procedures, and with soil at similar conditions, e.g., soil moisture, temperature, and timing relative to tillage, planting, and chemical applications. Ideally, one would like to measure one's MDS of soil-health indicators several times during the year, such as before plowing (for temperate-zone annual cropping), at planting (but before chemical applications), mid-season, at harvest, and in the fall. In tropical settings, the most desirable schedule would be appropriately different.

When restricted to a single sampling time, at-planting is often selected because this provides a "snapshot" at a very critical stage relative to crop establishment and eventual yield. However, pre-plow sampling is best for information on soil physical characteristics, i.e., before disturbance. Biological activity is often at a peak at planting time because of recent plow-down of winter cover-crop residues, but microbial biomass and activity are also highly variable at this time. For this reason, a shift in sampling date by just a few days can give very different results. Accordingly, biological data collected at planting time must be interpreted cautiously. Samples taken later in the growing season may show lower numbers and activity of soil organisms, but results will tend to be less variable and may be preferable for monitoring long-term management effects in some cases.

47.2 In-Field and Subjective Evaluations

47.2.1 In-Field Soil Test Kits

Discussion of soil measurements thus far has focused on a protocol where field soil samples are brought back to the laboratory or greenhouse for measurement under controlled, standardized conditions, with tests often conducted by highly specialized staff. Such data provide the most reliable index of soil "potential" that can be compared across regions or years. However, for locations where laboratory facilities are not available or are too costly for the user, various tools or kits for measuring some basic soil characteristics in the field are available (Liebig et al., 1996; Ditzler and Tugel, 2002).

Even when laboratory facilities are available, on-farm assessments may be preferable because of their educational value, or because research or practitioner objectives require that measurements be taken in the actual field setting rather than in an artificial, standardized, laboratory setting (Sarrantonio et al., 1996). To the extent possible, when conducting in-field measurements, soil moisture and temperature should be similar at the time of data collection, and/or soil moisture and temperature should be measured to make subsequent data adjustments employing available algorithms for standardization.

Some specific soil indicators, such as penetrometer resistance in the root zone, are always measured better directly in the field than in a laboratory. Some techniques for estimating decomposition (Smalla et al., 1998; Knacker et al., 2003) can be easily adapted to field, laboratory, or greenhouse use. Some new in-field methods are being developed, such as a relatively simple permanganate test for active C determination in the field, that could become an alternative to POM-C measurements (Weil et al., 2003).

The Soil Quality Test Kit developed by the U.S. Department of Agriculture (Liebig et al., 1996; Ditzler and Tugel, 2002) enables people to make in-field estimates of two biological indicators (soil respiration and earthworm counts), five physical indicators (bulk density, aggregate stability, water content, infiltration rate, and slaking), and three chemical indicators (pH, EC, and nitrate-N). While the kit can be used directly by farmers, in practice, the primary users have been public and private agriculture consultants and educators. Liebig et al. (1996) have documented that the data collected with the field kit compare well with those from standard laboratory analyses, although respiration rates calculated from the kit were routinely higher than laboratory measurements, presumably due to inclusion of root respiration in the field analysis, whereas roots were absent in lab samples.

47.2.2 Visual Assessments and Soil Health Score Cards

Qualitative measures of soil health involve no special analyses, only the informed scoring or rating of soil characteristics by farmers, agricultural scientists, and/or educators. This is usually carried out by visual assessment, but the smell and feel of soil may also be involved. While this approach is subjective and therefore can reflect user bias, when detailed guidelines and training have been provided, the results can compare well to quantitative measurements (Liebig and Doran, 1999). Training should include information on sampling, standardized verbal descriptions and/or photos that regularize scoring and keep users on track, with sufficient information regarding the interpretation of results.

One of the first soil health score cards was developed by Romig et al. (1995). Since then, many others have been developed by agricultural professionals working with local groups of farmers. The USDA–NRCS produced a "Soil Quality Design Guide" that provides a template and guidelines for modification of a standard soil health card to suit local needs and local terminology for various soil characteristics (Ditzler and Tugel, 2002). Cards developed to date have utilized more than 30 physical indicators and more than ten biological, chemical, and crop indicators of soil health (Ditzler and Tugel, 2002). Wander et al. (2002) discuss the educational and other benefits of a participatory process in developing qualitative soil health monitoring procedures locally.

Soil chemical properties in these score cards are defined by broad, straightforward phrases such as "fertility," with scoring based on the grower's subjective evaluation of chronic nutrient problems or soil test trends, and "salinity," based on observations of stunted growth or saline spots. Soil physical characteristics might be scored for soil "feel," crusting, water infiltration, retention or drainage, and compaction. Soil biological properties might include soil smell (low score for sour, putrid, or chemical odors vs. high score for "earthy," sweet, fresh aroma), soil color and mottling (which reflects balance of aerobic vs. anaerobic bacterial activity, among other things), and earthworm or overall biological activity.

The rating scales used in soil health score cards vary from just a few categories ("poor, fair, or good") to scales of 1 to 10. The descriptions that define categories or rating scales are best based on local terminology and preferences. For example, a score for earthworms might be based on actual counts in a shovel full of soil ($<1 =$ poor; $2-10 =$ fair; $>10 =$ good), or an overall score for biological activity might rely on visual evidence of fungal hyphae, insects, and other organisms including earthworms (poor = no signs; fair = some signs; good = plentiful signs). High quality photographs, such as those used in the field guides developed by Landcare Research in New Zealand (Shepherd, 2000), are an excellent way to train users and standardize scoring.

47.2.3 Measuring Crop or Environmental Responses

For any soil health monitoring effort, in addition to directly evaluating soil characteristics as described above, some measures of crop or ecosystem response should be included in any MDS. These may be measured in a detailed quantitative way, or given a subjective rating. Most of the existing soil health score cards include some of these kinds of measures. In Cornell's Soil Health Program we attempt to quantify several crop responses at all sites. It is not difficult to obtain ratings for (a) early crop establishment, (b) mid-season ratings for crop rooting depth, root health, weed pressure, and aboveground insect and disease pressure, and (c) final yield and quality.

Where environmental goals are important, the monitoring might choose to follow measures such as nitrate leaching and nitrous oxide emissions; toxin levels in surface and groundwater; and indices of biodiversity and stability. Ideally, soil health monitoring should include measures of both environmental and productivity responses. While crop performance and environmental impacts can be influenced by factors other than soil, this information is essential in developing an interpretation of soil biological, chemical, and physical characteristics, as discussed in Section 47.3.

47.3 Integration and Interpretation

Soil health monitoring will be of only limited value to farmers unless they or others can provide meaningful interpretation of the data. What is a "good" level of microbial biomass, or a desirable N mineralization rate? In the case of biological indicators such as these, there is not yet any agreed scientific basis on which to make such determinations. In contrast, for many of the standard chemical tests for plant nutrients, there are well-established yield-response curves available for interpreting results. This is possible because many agricultural scientists have worked together with farmers over several decades on these issues during the 20th century. However, even these conclusions should be reexamined within the context of soil health because now we are no longer interested only in crop response, but also environmental effects (Lilburne et al., 2004).

For nutrients such as N and P, it is now well established that when values are beyond the yield plateau, the curve shifts downward, and there are associated problems such as nitrate leaching into groundwater and phosphate runoff into streams. Such effects should be assessed apart from whatever are the yield results. (In this example, the inclusion of some economic assessment in the monitoring could have the added benefit of demonstrating to farmers that practices causing environmental problems are reducing profit and thus are also economically unwise.)

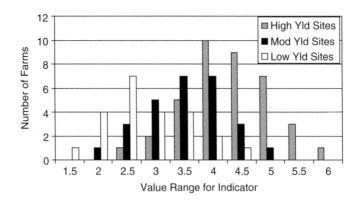

FIGURE 47. 1
Frequency distribution for the range of values for high-yield, moderate-yield, and low-yield, sites for a hypothetical soil health indicator. Compiling data bases such as this will be a first step toward developing better means to interpret soil health measurements.

To help develop better means to interpret soil biological indicators, the Cornell Soil Health Team is building up a large database that shows the range of values observed for specific indicators at low-yield, moderate-yield, and high-yield sites. This is illustrated in the frequency distributions for a hypothetical biological indicator shown in Figure 47.1. With data such as these, we can put the results obtained for a farm site into some context. We may not yet be able to put a simple label such as "acceptable" or "unacceptable" on a test result, but we can tell a farmer where the value obtained from his or her farm fell within the range observed for low-, moderate-, or high-yield sites. A subsequent step could be a workshop process, such as that described by Lilburne et al. (2004). In this case, a team of soil experts established preliminary "acceptable" ranges and response curves using data collected from 500 sites over a 6-year period.

The "acceptable" ranges for most soil indicators will have to be modified based on soil type and other factors. For example, a Ultisol will inherently have a much lower OM content than a Histosol, and this will alter what are "acceptable" levels of microbial biomass for these two different soil classes. We are also exploring various ways to combine specific soil measurements into integrative indicators.

The example in Figure 47.1 suggests a soil indicator where essentially "more is better." Microbial biomass C or percent OM might be in this category. Another kind of indicator is one where there is some clear "optimum" to aim for. Soil pH is a classic example of this category. A third category is one where "less is better," and here pathogenic nematodes or bulk density are examples. In each of these categories, there will be modifying factors that will affect what are the specific "acceptable" levels. For example, a higher level of pathogenic nematodes may be more acceptable in soils with higher microbial biomass, because there are likely to be more potential "natural enemies" of pathogens in a microbe-rich soil.

A particularly sophisticated and comprehensive approach to interpreting soil quality data is used in the Soil Management Assessment Framework recently developed by the USDA–NRCS (Andrews et al., 2004). It involves a three-step process:

1. Identification of an MDS from a menu of over 80 potential indicators, as mentioned above;
2. Interpretation of data by converting all indicator measurements to unitless score values ranging from 0 (undesirable) to 1 (optimum), based on empirically

derived mathematical functions that describe the conversion curves for each indicator; and

3. Integration of indicator scores into a single overall soil health index, calculated as the sum of all indicator score values, divided by the number of indicators in the MDS.

For any particular indicator in step 2, the coefficients used in the conversion function may be modified, or a completely different algorithm may be used, depending on soil class, crop, and other factors. The shapes of the conversion curves used for step 2 will reflect whether the indicator can be characterized as more is better, or less is better, or some optimum can be specified.

The benefits of having a single, integrated index or indicator of overall soil health (step 3 in the framework just described) remain controversial. Some of the qualitative soil health cards (e.g., Shepherd, 2000) ultimately provide a composite single number, usually by weighting the scores of certain indicators more than others before adding up a total.

While some farmers may prefer obtaining a single number to describe the health of their soil, it can be argued that farmers can derive more educational and management benefits from reviewing a set of values that show each soil health indicator separately. Also, when weighting is done, deciding what weighting factors should be applied in calculating an overall soil health index is somewhat arbitrary, not based on any rigorous testing of their relevance for either production or environmental goals.

47.4 Discussion

The Soil Management Assessment Framework developed by USDA and described above reflects the current state-of-the-art with regard to quantitative soil health monitoring. However, as its name implies, it is best viewed as a framework, and none should assume that the many algorithms used for converting measurements of indicators into score values will be applicable to a particular region without first evaluating and modifying them as necessary. Moreover, the parameters can be very different between temperate or tropical agroecosystems, or between humid, subhumid, semiarid, or arid climates.

This has been seen from the overambitious use of the badly-named Universal Soil Loss Equation (USLE) which has had to be considerably revised when used outside the USA. (and even in environments with more slope and rainfall patterns different from where the USLE was originally constructed). Even when well-tested, a sophisticated mathematical approach for soil health assessment will not replace careful interpretation of specific indicator measurements by an expert in consultation with the farmer or land manager. In some situations, simple qualitative soil health cards used by well-trained individuals are likely to provide information that is just as meaningful, as well as cheaper and more timely, than will more sophisticated quantitative approaches.

Acknowledgments

This chapter draws on the work of the Cornell University Soil Health Program Team, supported by grants from the USDA-Northeast Regional SARE, USDA Special Grants-Agricultural Ecosystems Programs, and Cornell University. Team members contributing to the ideas and methods discussed here include George Abawi, Laurie Drinkwater, Janice Thies and Harold van Es.

References

Abawi, G.S. and Widmer, T.L., Impact of soil health management practices on soilborne pathogens, nematodes and root diseases of vegetable crops, *Appl. Soil Ecol.*, **15**, 37–47 (2000).

Andrews, S.S., Karlen, D.L., and Cambardella, C.A., The soil management assessment framework: A quantitative soil quality evaluation method, *Soil Sci. Soc. Am. J.*, **68**, 1945–1962 (2004).

Bending, G.D. et al., Microbial and biochemical soil quality indicators and their potential for differentiating areas under contrasting agricultural management regimes, *Soil Biol. Biochem.*, **36**, 1785–1792 (2004).

Blair, J.M., Bohlen, P.J., and Freckman, D.W., Soil invertebrates as indicators of soil quality, In: *Methods for Assessing Soil Quality*, SSSA Special Publication 49, Doran, J.W. and Jones, A.J., Eds., Soil Science Society of America, Madison, WI, 273–292 (1996).

Carter, M.R., *Soil Sampling and Methods of Analysis*, Lewis Publishers, Ann Arbor, MI (1993).

Dick, R.P., Breakwell, D.P., and Turco, R.F., Soil enzyme activities and biodiversity measurements as integrative microbiological indicators, In: *Methods for Assessing Soil Quality*, SSSA Special Publication 49, Doran, J.W. and Jones, A.J., Eds., Soil Science Society of America, Madison, WI, 247–272 (1996).

Ditzler, C.A. and Tugel, A.J., Soil quality field tool: Experiences of USDA–NRCS soil quality institute, *Agron. J.*, **94**, 33–38 (2002).

Doran, J.W. and Jones, A.J., Eds., *Methods for Assessing Soil Quality*, SSSA Special Publication 49, Soil Science Society of America, Madison, WI (1996).

Doran, J.W. and Parkin, T.B., Defining and assessing soil quality, In: *Defining Soil Quality for a Sustainable Environment*, SSSA Special Publication 35, Doran, J.W. et al., Eds., Soil Science Society of America, Madison, WI, 3–21 (1994).

Doran, J.W. and Parkin, T.B., Quantitative indicators of soil quality: A minimum data set, In: *Methods for Assessing Soil Quality*, SSSA Special Publication 49, Doran, J.W. and Jones, A.J., Eds., Soil Science Society of America, Madison, WI, 25–37 (1996).

Knacker, T. et al., Assessing the effects of plant protection products on organic matter breakdown-litter decomposition test systems, *Soil Biol. Biochem.*, **35**, 1269–1287 (2003).

Liebig, M.A. and Doran, J.W., Evaluation of farmer's perception of soil quality indicators, *Am. J. Altern. Agric.*, **14**, 11–21 (1999).

Liebig, M.A., Doran, J.W., and Gardner, J.C., Evaluation of a field test kit for measuring selected soil quality indicators, *Agron. J.*, **88**, 683–686 (1996).

Lilburne, L., Sparling, G., and Schipper, L., Soil quality monitoring in New Zealand: Development of an interpretive framework, *Agric. Ecosyst. Environ.*, **104**, 535–544 (2004).

Liu, W.-T. et al., Characterization of microbial diversity by determining terminal restriction fragment length polymorphisms of genes encoding 16S rRNA, *Appl. Environ. Microbiol.*, **63**, 4516–4522 (1997).

Magdoff, F.R., Tabatabai, M.A., and Hanlon, E.A., Eds., *Soil Organic Matter: Analysis and Interpretation*, Soil Science Society of America, Madison, WI (1996).

Neher, D., Role of nematodes in soil health and their use as indicators, *J. Nematol.*, **33**, 161–168 (2001).

Parisi, V. et al., Microarthropod communities as a tool to assess soil quality and biodiversity: A new approach in Italy, *Agric. Ecosyst. Environ.*, **105**, 323–333 (2005).

Rillig, M.C., Wright, S.F., and Eviner, V.T., The role of arbuscular mycorrhizal fungi and glomalin in soil aggregation: Comparing effects of five plant species, *Plant Soil*, **238**, 325–333 (2002).

Romig, D.E. et al., How farmers assess soil health and quality, *J. Soil Water Conserv.*, **50**, 229–236 (1995).

Sarrantonio, M. et al., On farm assessment of soil quality and health, In: *Methods for Assessing Soil Quality*, SSSA Special Publication 49, Doran, J.W. and Jones, A.J., Eds., Soil Science Society of America, Madison, WI, 83–105 (1996).

Shepherd, T.G., *Visual Soil Assessment Field Guide*, Horizons.mw Report No. 20/EXT/425, Vol. 1, Horizons.mw and Landcare Research, Palmerston North, New Zealand (2000).

Smalla, K. et al., Analysis of BIOLOG GN substrate utilization patterns by microbial communities, *Appl. Environ. Microbiol.*, **64**, 1220–1225 (1998).

Smith, J.L. and Doran, J.W., Measurement and use of pH and electrical conductivity for soil quality analysis, In: *Methods for Assessing Soil Quality*, SSSA Special Publication 49, Doran, J.W. and Jones, A.J., Eds., Soil Science Society of America, Madison, WI, 169–185 (1996).

Sparks, D.L., Ed., *Methods of Soil Analysis: Chemical Methods*, Pt 3, Soil Science Society of America, Madison, WI, USA (1996).

Sparling, G.P. et al., Soil quality monitoring in New Zealand: Practical lessons from a 6-year trial, *Agric. Ecosyst. Environ.*, **104**, 523–534 (2004).

Wander, M.M. et al., Soil quality: Science and process, *Agron. J.*, **94**, 23–32 (2002).

Weil, R.R. et al., Estimating active carbon for soil quality assessment: A simplified method for laboratory and field use, *Am. J. Altern. Agric.*, **18**, 3–17 (2003).

Weaver, R.W. et al., Eds., *Methods of Soil Analysis: Microbiological and Biochemical Properties*, Pt 2, Soil Science Society of America, Madison, WI (1994).

48

Modeling Possibilities for the Assessment of Soil Systems

Andrew S. Ball[1] and Diego De la Rosa[2]

[1]*School of Biological Sciences, Flinders University of South Australia, Adelaide, Australia*
[2]*Department of Soil-Plant-Atmosphere System Sustainability, Institute for Natural Resources and Agrobiology, Seville, Spain*

CONTENTS

As seen throughout the previous chapters, almost all soils can be used for the growth of crops provided that beneficial soil biological processes are sustained and sufficient nutrients and water are supplied. However, understanding the complex mechanisms of land use change above the plot level so as to make more intelligent resource management decisions requires the integration of scientific data and knowledge from multiple disciplines and diverse landscapes. Planning and management of land resources to maintain sustainable soil systems should be buttressed with tools for assessment and evaluation that permit broad, interactive and informed participation in decision-making processes.

Successful development of such tools requires the integration of spatial and nonspatial information on biological, physical and chemical factors, socio-political and economic analyses, and expert opinion (Sugumaran, 2002). No single technique can address all of these requirements (Fedra, 1995). Fortunately, modern information and analytical technologies are providing unprecedented power and flexibility plus the possibilities to combine them in novel and productive ways. In addition to the methodologies discussed in Chapters 46 and 47, the new technologies include remote sensing, geographic information systems (GIS), multiphase models, user interfaces, database management and knowledge-based systems. This chapter reviews some holistic and analytical techniques that are moving science beyond the assessment of specific soils in certain locations to begin addressing the needs and opportunities for managing different soil systems at various scales.

48.1 Seeking Large-Scale Optimization over Time for Soil-System Productivity

Soil systems are the foundation for any sustainable agricultural development, one of the prime objectives in all countries around the world, whether developed or developing. However, every soil system has its own unique requirements, potentials and limitations. External inputs can be provided in the form of nutrients, energy, labor or capital expenditure. Adverse environmental impacts are external outputs that result from certain practices in managing soil systems; but they can be treated as a kind of input if counted as costs.

Sustainable soil management can maintain and even improve the potential of the soil through the use of beneficial practices such as reduced tillage, organic matter amendments and improved irrigation (Niles et al., 2003). The challenge for sustainable agroecosystems in the future will be to increase crop production on smaller areas of land, and with fewer additions of inorganic nutrients, pesticides and water, relying as much as possible on endogenous processes within soil systems that recycle or recuperate nutrients, enhance biotic populations, protect crops from pathogens or predation, and sustain soil fertility in other ways.

Agroecological land evaluation analysis, such as the assessment of land suitability and land vulnerability, provides an empirical basis for sustainable soil use and management (Dent and Young, 1981). Land evaluation is something different from usual soil quality assessment, however, because soil biological factors are seldom considered when doing land evaluation (Karlen et al., 1997). Moreover, the assessments transcend field units and are undertaken on a landscape scale.

Biological parameters such as soil microbial biomass, respiration, activities of soil enzymes, and detailed organic matter analyses are dynamic parameters that change in response to soil management practices such as non-tillage, application of organic amendments, and crop rotation (Girvan et al., 2003; Girvan et al., 2004). Measurement and use of these parameters at any point in time, let alone over time, as seen in preceding chapters, represents a challenging undertaking. Therefore, land evaluation analysis should be regarded as at most a first step in developing chemical/physical soil quality assessment procedures. To have a good appreciation of soil system potentials, this baseline analysis needs to be followed up by shorter-term monitoring procedures for assessing the biological characteristics of soil (Chapter 47).

Recent developments in the areas of data and knowledge engineering appear to provide opportunities for this technology to be applied to land evaluation. The application phase of land evaluation systems is a process of scaling-up from the representative areas studied during a development phase to implementing the methods across unknown areas.

This application phase can be carried out using computer-assisted procedures that draw on integrated databases, computer models, and spatialization tools that together constitute what are now known as decision-support systems (DSS) (De la Rosa and Van Diepen, 2002). In this chapter, some applications of these technologies to soil assessment systems are reviewed, attempting to ensure that the biological dimensions of soil systems are better integrated into this process.

48.2   Development of a Soil Assessment System

The evolution of a land evaluation system such as the Mediterranean Land Evaluation Information System (MicroLEIS DSS) reflects the various advances made within the computer sector, namely, the initial era of data processing, then the microcomputer era, and now the network era. During the data-processing era, some qualitative and statistical land evaluation models were developed using the DOS environment (De la Rosa et al., 1992). These models moved to Windows software in the late 1990s. These models are now also benefiting from the emergent opportunities that the Internet presents, especially for the rapid dissemination of information and knowledge.

Land evaluation systems are generally developed to assist decision-makers faced with specific agroecological problems. These systems are designed as a knowledge-based approach that incorporates a set of information tools as illustrated in Figure 48.1. Each of these tools is directly linked to another, and applications can be carried out on a wide range of problems related to land productivity and land degradation. They are grouped into the following main modules: (i) basic data warehousing, (ii) land evaluation modeling, and (iii) model application software.

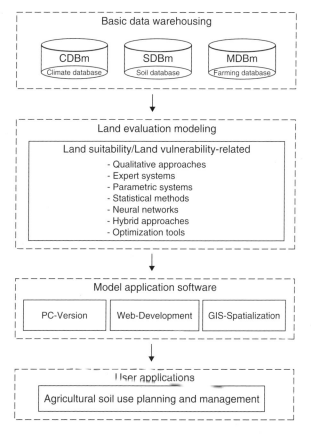

FIGURE 48.1
Conceptual design and component integration of the MicroLEIS DSS land evaluation decision support system.
Source: From De la Rosa et al., *Environ. Model. Software,* **19**, 929–942 (2004). With permission.

48.3 Data Warehousing

The land attributes used in the MicroLEIS DSS system shown in Figure 48.1 correspond to databases on three main factors: soil, climate and farming. There are two kinds of datasets.

1. Controlled, usually from field experiment where the researcher controls the levels of the independent variable.
2. Observed, usually from surveys, where the levels of the independent variables are not controlled, only observed and recorded.

Controlled data provide a more reliable basis for understanding cause-and-effect relationships; however, observed data are an inexpensive method of data acquisition with a wide (numerical) range possible for each causal factor. A limitation on the latter is that nature may not provide the full range of effects that could be imposed in an experimental approach. Soil survey data contain information on the chemical and physical properties of soil, including depth, texture, water-holding capacity, and organic matter content. Because climatic conditions vary from year to year, however, reliable long-term data must be used to predict future events with any degree of confidence. Traditionally, agricultural management practices have been considered as something quite separate from land evaluation. However, increasingly, management factors are being incorporated as input variables. Therefore, factors such as growing season length, rooting depth, tillage operations, and treatment of residues are now often considered as land characteristics.

For each of these main factors, a database has been constructed with inter-connectivity among the three databases, respectively, for climate, soil and farming system. Development of an inter-connected database management system to facilitate the integrated consideration of land attributes has been crucial in the development of decision-support systems.

48.4 Land Evaluation Modeling

Generally in soil assessment models, land evaluation analysis focuses on agricultural land use, planning, and management for soil protection purposes. However a few land evaluation studies have focused on land productivity through the modeling of crop systems (Jones et al., 2003). Land evaluation models are usually based on nonspatial single areas, where the land suitability or vulnerability is determined without any influence from the surrounding areas considered, and without explicit reference to its actual geographic location. Such a modeling phase uses basic information from representative areas. The traditional FAO scheme of land evaluation analysis (FAO, 1976) is still generally followed, along with use of the terminology suggested by Rossiter (1996), e.g., land suitability, land characteristics, land quality, land utilization type, land-use requirements and severity level.

The modeling phase involves the following main stages.

1. Definition of relevant land-use requirements or limitations, land-use response or degradation level.
2. Matching of land attributes with land-use requirements, identifying cause-and-effect relationships.
3. Selection of land attributes, land characteristics, and associated land qualities.
4. Validation of the developed algorithms in other representative areas.

TABLE 48.1

Set of Input Land Characteristics Considered in the MicroLEIS DSS Models

Factor Type	Input Land Characteristics
Land productivity	
Site/Soil	Latitude; altitude; physiographic position; parent material; slope gradient; useful depth; stoniness; texture; clay content; structure; color; reaction; organic matter content; carbonate content; salinity; sodium saturation; cation exchange capacity; free iron; bulk density; drainage; water retention; hydraulic conductivity
Climate	Monthly precipitation; monthly maximum temperature; monthly minimum temperature
Crop/Management	Growing season length; maximum rooting depth; specific leaf area; cropping intensity coefficient; coefficient of efficiency
Land degradation	
Site/Soil	Latitude; altitude; physiographic position; parent material; slope gradient; slope form; slope aspect; land cover; useful depth; stoniness; texture; clay content; structure; organic matter content; carbonate content; salinity; sodium saturation; cation exchange capacity; bulk density; drainage; water retention; hydraulic conductivity
Climate	Monthly precipitation; monthly maximum precipitation; monthly maximum temperature; monthly minimum temperature
Crop/Management	Land use type; growing season length; leaf situation; leaf duration, plant height; maximum rooting depth; sowing date; tillage practice; tillage depth; row spacing; artificial drainage; conservation technique; residues treatment; crop rotation; operation sequence; implement type; material input type; material input rate; wheel load; tire inflation pressure

Source: From De la Rosa et al., *Environ. Model. Software*, **19**, 929–942 (2004). With permission.

The selection of land attributes (climate, crop/management factors, and site/soil) as input variables for the predictive models is an essential part of the land evaluation analysis. Table 48.1 lists the main land characteristics used in evaluation modeling. This broad set of indicators is grouped according to the developed land suitability and land vulnerability approaches. It is expected that soil biological indicators discussed in prior chapters could be integrated into the first set of factors considered regarding site and soil.

To match land characteristics with land-use requirements, a number of parameters that assess the quality of the land are used (Table 48.2). In the land vulnerability models, environmental impacts such as increased N and P concentrations in water resources, loss of wildlife habitat, presence of pesticides, sedimentation of waterways, and their effects on crop productivity are also considered.

TABLE 48.2

Main Land Qualities Considered in the MicroLEIS DSS Models

Land Evaluation Approach	Land Quality Parameters
Land suitability	Plant water-use efficiency; water- and air-filled pore space; nutrient availability; plant root penetration; water infiltration; crop growth
Land vulnerability	Runoff and leaching potential; erosion resistance; soil structure; cover protection; pesticide absorption and mobility; subsoil compaction; soil workability

Source: From De la Rosa et al., *Environ. Model. Software*, **19**, 929–942 (2004). With permission.

48.5   Modeling Approaches

Models are considered as simplified representations of the real world that can be expressed in a wide variety of statistical or deterministic mathematical models. The two principal modeling approaches are: (i) empirical (also called statistical) modeling, and (ii) dynamic simulation. In land evaluation, empirical-based modeling has moved on from simple qualitative approaches to other strategies that are more sophisticated and based on artificial intelligence techniques. The basic idea of statistical modeling for land evaluation is that observed relations once quantified and analyzed can be made applicable for predicting future situations. This will not work, however, unless there are sufficient data on which to base the statistical inferences. So this methodology is not appropriate for new land uses or for areas from which sufficient samples have not been taken. For land evaluations of established land uses with sufficient historical or experimental data, such analyses can be useful and are often the preferred method (Van Lanen, 1991).

48.5.1   Empirical or Statistical Modeling

Advantages of using statistical methods for land evaluation include the following.

1. They employ actual observations, either random or imposed.
2. The more observations are made, the more reliable their predictions should become, in a predictable and quantifiable manner.
3. Because predictions are made on a continuous scale, fine distinctions can be made when there are enough data.
4. Input levels as well as natural resources can be considered as predictor variables.
5. There is a rich set of analytical methods available.
6. Each prediction comes with an estimate of its reliability.

However, some disadvantages also need to be considered.

1. The form of the statistical relationship is not obvious *a priori*, and several different forms may give similar results in terms of their goodness-of-fit.
2. The assumptions that are made about the distribution of the predictor and response variables may be impossible to verify.
3. Extrapolation to values of the independent variable outside the domain of calibration is not justified, and it may not be obvious which independent variables should be included in the relation.
4. Because it is not obvious how to define the sample space, errors may invalidate the inferences.
5. The precision of the statistical relationship may not be great enough for meaningful predictions, especially with observational data.
6. Multicollinearity and unreliable coefficients can limit the value and validity of modeling results.
7. These models are often "black boxes," not providing any insight into the mechanisms and interactions involved. This is less of a disadvantage with simulation models that can be manipulated for this purpose.

The linking of land characteristics with land-use requirements or limitations may be as simple as making statements about land suitability for particular uses, or lands may be grouped subjectively into a small number of classes or grades of suitability. In many qualitative approaches, quantification is achieved by the application of the rule that the most-limiting land quality determines the degree of land suitability or vulnerability.

This, however, assumes some knowledge of optimum land conditions and of the consequences of deviations from this optimum (Verheye, 1988), an assumption that is often not tenable. Relatively simple systems of land evaluation depend largely on experience and intuitive judgment. They are really empirical models, and no quantitative expressions of either inputs or outputs are normally given. The USDA Land Capability System (1961) and its diverse adaptations is an example of this kind of analytical model that has been widely used around the world.

48.5.2   Expert Systems, Simulation Modeling, and Land Evaluations

Expert systems are computer programs that simulate the problem-solving skills of human experts in a given field. They provide solutions to a problem, expressing inferential knowledge through the use of decision trees, such as discussed in Chapter 18. In land evaluation, decision trees give a clear expression of the comparison between land use requirements and land characteristics. Expert decision trees are based on scientific background information and on discussions with human experts, thereby reflecting available expert knowledge. Where suitable data on practical experience are available, statistical decision-tree analysis can be used to generate land evaluation models with good prediction rates.

Between qualitative and quantitative methods there lie semi-quantitative land evaluations, derived from the numerically inferred effects of various land characteristics. Parametric methods can be considered a transitional phase between qualitative methods, based entirely on expert judgment and mathematical models. They account for interactions between the most significant factors by the multiplication or addition of single-factor indexes (Riquier, 1974). Multiplicative systems assign separate ratings to each of several land characteristics, and then take the product of all factor ratings as the final rating index.

These systems have the advantage that any important factor can control the rating. The most widely known method to include specific, multiplicative criteria for rating land productivity inductively was developed in 1933 by Storie (1978). Certain land degradation analytical systems such as the Universal Soil Loss Equation (USLE) and its adaptation (Wischmeier and Smith, 1965) have a very similar form to that of the Storie Index, operated by multiplying the factor values.

In additive systems, various land characteristics are assigned numerical values according to their inferred impact on land use. These numbers are either summed, or subtracted from a maximum rating of 100, to derive a final rating index. Additive systems have the advantage of being able to incorporate information from more land characteristics than can multiplicative systems. The agro-climatic zoning project developed by FAO (1978) represents a milestone in the development of land evaluation, introducing a new approach to land-use systems analysis (Driessen and Konijn, 1992).

The combination of dynamic simulation models and empirically-based land evaluation techniques is currently producing good scientific and practical results, improving the accuracy and applicability of the models. For example, simulation modeling focusing on soil/plant-growth/contamination systems is well advanced at the local scale, e.g., process measurement sites, experimental stations, and small catchments (Jones et al., 2003). However, extrapolation to a regional scale is still a major priority.

Land evaluation decision-support systems for land users and policy-makers are expected to select optimal use and management decisions. For such purposes, optimization tools based on land evaluation models are very important in formulating decision alternatives. The possibilities for utilizing land evaluation models in decision-making through the development of model application software are significant. This phase of development will allow the practical use of information and knowledge gained during the prior phase of building evaluation models (Antoine, 1994).

When land evaluation models are expressed in a form that can be understood by a calculating device, their algorithms become computer programs. To put the models to use in practical applications, i.e., to automate the application of land evaluation models, a library of PC-based software must be developed. A graphical interface is also helpful. This user interface is considered a very important component because, to the user, it is the system.

The model computer programs can also be implemented on the Internet through a WWW server, so that users can apply the models directly if they have access to a Web browser. It is not necessary to download and install the PC software. Open-access WWW applications offer several advantages, including their use by a wider audience, enabling their suitability to be checked to improve the systems. Upgrades are immediately made available on the WWW server. The Website is the center of activity in developing operative decision-support systems.

48.5.3   Neural Networks and Artificial Intelligence

Neural networks are artificial intelligence-based technologies that have grown rapidly over the past few years and that show an ability to deal with nonlinear multivariate systems. An artificial neural network is a computational mechanism that is able to acquire, represent and compute a weighting (or mapping) from one multivariate space of information to another, given a set of data representing that mapping. Very important, it can identify patterns in input training data that may be missed by conventional statistical analysis. In contrast to regression models, neural networks do not require knowledge of the functional relationships between the input and the output variables. Also, these techniques are nonlinear and thus may handle complex data patterns that make simulation modeling unattainable.

48.6   Geographic Information Systems

Regionalization analysis includes the use of spatial techniques to expand land evaluation results from points to geographic areas, using soil survey and other related maps. The use of geographical information system technology has led to the rapid generation of thematic maps and area estimates. It enables many of the analytical and visualization operations of land evaluation to be carried out in a spatial format, by combining different sets of information in various ways that produce overlays and interpretable maps. Furthermore, digital satellite images can be incorporated directly into many GIS packages.

This technology is already a prerequisite for managing the datasets required for spatial land evaluation application. A simple map subsystem, e.g., ArcView, is all that is required to show basic data and model results on a map, or to extract information from maps to be used in a land evaluation model. At a regional scale, the assessments are made from a very broad and generalized perspective. However, this level of assessment is where policy decisions are usually made (Davidson et al., 1994).

Remote sensing and GIS technology are thus being effectively utilized in sustainable agricultural development and management decision-making, including cropping system analysis, agroecological zonation, quantitative assessment of soil carbon dynamics and land productivity, soil erosion inventories, and integrated agricultural drought assessment and management.

48.7 Discussion

Land evaluation systems such as the MicroLEIS DSS system focus on soil protection by improving agricultural soil use and its management. Soil-use planning has derived mainly from the application of the results from land suitability-related models and from soil-use management from the land vulnerability-related results. The MicroLEIS DSS models assess the impacts of soil loss on crop productivity, allow identification of areas with soil erosion and contamination problems, and enable the selection of management practices to minimize soil erosion.

The application of such systems can be seem by observing the results from users of MicroLeiss DSS from 1990 until 2003. Out of the total of 1,963 users, the highest number of users were from Europe (1,248) and Central and South America (583), together about 93% of users. According to the type of activity, teaching and private sector use are the main user activities. Farmers have shown relatively little interest in this system so far (De la Rosa et al., 2004). Various reasons have been suggested to explain this passing over of information technology by farmers (Ascough et al., 1999). However, there are likely to be significant changes in the ways that information technology is used in agriculture. It can be anticipated that farmers will begin to want to apply information technology to support many different operational aspects of farming in the future, e.g., real-time decision-support systems (Thysen, 2000).

With a modular framework such as that used in the MicroLEIS decision-support system, the components can be easily used as required for a particular application. Owing to the wide range of data types required for most of the models, the use of the databases is normally the initial step of any application project. Overall, the knowledge-based, decision-support system approach appears to be a very useful method for responding to the need to bring agriculture and land resources sciences together for decision-makers.

At the beginning of the 21st century, the Web is becoming the center of activity in developing decision-support systems. The future generation of technologically-advanced users will expect more functionality in decision-support technology. Senior managers and executives will get more directly involved in problem solving, decision-making and planning. This means that decision-support tools are likely to play a more central role in this rapidly changing environment.

References

Antoine, J., *Linking Geographical Information Systems (GIS) and FAO's Agroecological Zone (AEZ) Models for Land Resource Appraisal 1978*, World Soil Resources Report 75. FAO, Rome (1994).

Ascough, J.C. et al., Computer use in agriculture: An analysis of great plains producers, *Comput. Electron. Agric.*, **23**, 189–204 (1999).

Davidson, D., Theocharopoulos, S.P., and Bloksma, R.J., A land evaluation project in Greece using GIS and based on Boolean and fuzzy set methodologies, *Int. J. Geogr. Inf. Syst.*, **8**, 369–384 (1994).

De la Rosa, D. et al., A land evaluation decision support system (MicroLEIS DSS) for agricultural soil protection, *Environ. Model. Software*, **19**, 929–942 (2004), http://www.microleis.com.

De la Rosa, D. et al., MicroLEIS: A microcomputer-based Mediterranean land evaluation information system, *Soil Use Manage.*, **8**, 89–96 (1992).

De la Rosa, D. and Van Diepen, C., Qualitative and quantitative land evaluation, In: 1.5: Land Use and Land Cover, *Encyclopedia of Life Support Systems* (EOLSS–UNESCO), Verheye, W., Ed., Eolss Publisher, Oxford (2002), http://www.eolss.net.

Dent, D. and Young, A., *Soil Survey and Land Evaluation*, George Allen and Unwin, London (1981).

Driessen, P.M. and Konijn, N.T., *Land-Use Systems Analysis*, Wageningen Agricultural University, Wageningen (1992).

FAO, *A Framework for Land Evaluation*, Soils Bull. 32. FAO, Rome (1976).

FAO, *A Framework for Land Evaluation: Report on the Agro-Ecological Zones Project*, World Soil Resources Report 48. FAO, Rome (1978).

Fedra, K., Decision support for natural resources management: Models, GIS and expert systems, *AI Appl.*, **9**, 3 (1995).

Girvan, M.S. et al., Soil type is the primary determinant of the composition of the total and active bacterial communities in arable soils, *Appl. Environ. Microbiol.*, **69**, 1800–1809 (2003).

Girvan, M.S. et al., Monitoring of seasonal trends in the soil microbial community of an agricultural field, *Appl. Environ. Microbiol.*, **70**, 2692–2701 (2004).

Jones, J.W. et al., The DSSAT cropping system model, *Eur. J. Agron.*, **18**, 235–265 (2003).

Karlen, D.L. et al., Soil quality: A concept, definition and framework for evaluation, *Soil Sci. Soc. Am. J.*, **61**, 4–10 (1997).

Niles, J.O. et al., Potential carbon mitigation and income in developing countries from changes in use and management of agricultural and forest lands, In: *Capturing Carbon and Conserving Biodiversity: The Market Approach*, Swingland, I.R., Ed., Royal Society Press, London, 70–89 (2003).

Riquier, J., A summary of parametric methods of soil and land evaluation, In: *Approaches to Land Classification*, Soils Bulletin 22. FAO, Rome (1974).

Rossiter, D.G., A theoretical framework for land evaluation (with discussion), *Geoderma*, **72**, 165–202 (1996).

Storie, R.E., *Storie Index Soil Rating (Revised)*, University of California Division of Agricultural Science Special Publication 3203. University of California, Davis, CA (1978).

Sugumaran, R., Development of an integrated range management decision-support system, *Comput. Electron. Agric.*, **37**, 199–205 (2002).

Thysen, I., Agriculture in the information society, *J. Agric. Eng. Res.*, **76**, 297–303 (2000).

USDA, *Land Capability Classification*, USDA Agriculture Handbook 210. U.S. Government Printing Office, Washington, DC (1961).

Van Lanen, H.A.J., Qualitative and quantitative physical land evaluation: An operational approach, Ph.D. thesis, Wageningen Agricultural University, Wageningen, The Netherlands (1991).

Verheye, W., The status of soil mapping and land evaluation for land use planning in the European community, In: *Agriculture: Socio-Economic Factors in Land Evaluation*, Boussard, J.M., Ed., Office for Official Publications of the EU, Luxembourg (1988).

Wischmeier, W.H. and Smith, D.D., *Predicting Rainfall Erosion Based from Cropland East of the Rocky Mountains*, USDA Agriculture Handbook 282. U.S. Government Printing Office, Washington, DC (1965).

49

Opportunities for Overcoming Productivity Constraints with Biologically-Based Approaches

Norman Uphoff
Cornell University, Ithaca, New York, USA

CONTENTS

Ten years ago, a member of our editorial team, speaking to the 15th World Congress of Soil Science, proposed that it was time to move soil science toward a "second paradigm" in order to meet agricultural production needs in the tropics, and indeed in the world more generally (Sanchez, 1994; 1997). Sanchez summarized the prevailing paradigm which had grown out of 150 years of research and practice as: "Overcome soil constraints through the application of fertilizers and amendments to meet plant requirements." This conception, which currently guides most soil science and "modern" management, focuses primarily on production goals, and gave little attention to ecological functions, he noted. Research and applied efforts have been directed primarily to managing inputs that are *exogenous* to the processes of plant growth and production rather than to *endogenous* processes and potentials that exist within soil systems.

The utilization of external inputs has been very successful, but that does not necessarily mean that this approach must or will remain forever the guiding paradigm. Sanchez suggested an alternative paradigm for soil science and management because of his particular concern that the benefits of modern agricultural development were by-passing millions of rural households around the world. These depend on poorly endowed lands that, given their present agronomic productivity, are considered "marginal." These areas have not been served well by the Green Revolution of the preceding four decades because the technologies it offered were not appropriate or not accessible. In the last 10 years, what has become increasingly evident is that these technologies, in fact, are no longer serving the better-endowed agricultural areas as well as previously. This is seen in Section 49.1. This conclusion adds to the reasons why agricultural scientists, policy-makers, and practitioners should take biologically-based development strategies more seriously.

This chapter begins by reviewing the agricultural sector globally. It then considers the alternative paradigm that Sanchez proposed in 1994, which has acquired more definition, urgency, and scientific justification in the meantime. The chapter then summarizes the evidence reported in this volume on the productivity potentials of technologies and methods associated with this second paradigm. The prevailing assumption that to raise agricultural productivity, producers need to rely mostly on exogenous inputs, is contradicted by an abundance of data for a large number of crops in a great variety of locations around the world. This does not mean that external inputs must or should be abandoned; they have established an important place in agricultural systems and strategies and will continue to have a role in future production efforts. However, the preoccupation with inputs that has shaped agricultural thinking and practice for the past half century of subsidized agriculture is now challenged by efforts, grounded in scientific research and farmer innovation, to redirect production strategies for the agricultural sector based on a deeper and practical appreciation of soil systems.

49.1 Agricultural Production Trends and Current Stagnation

The first paradigm's height of achievement, and possibly its culmination, was the Green Revolution during the latter third of the 20th century. Between 1966 and 1996, cereal production worldwide was raised by 90% — from 991 to 1882 million metric tons. Grain availability per capita during this period of rapid population growth went up by 21% — from 290 to 326 kg person^{-1}. Since 1996, however, cereal production has been declining, absolutely as well as relatively. In fact, world grain production over the last 7 years has averaged only 1870 million metric tons yr^{-1}, with grain production per capita declining more than this, averaging 310 kg person during these 7 years; in 2003 per capita grain production was 9% below its peak in 1996–1997 (all data are from FAO and USDA sources.) Statistics for 2004 indicate some resumption in growth of grain production, but the factors contributing to the preceding stagnation have not changed significantly. While cereals are not the only agricultural products that should be considered, they are indicative of major trends and have the most impact on people's well-being, as well as on public and policy-makers' perceptions of agricultural sector performance.

Production depends on many factors, of course. The recent reduction in grain output has been in part a response to a decline in market prices, which can be seen as an indication of the success of Green Revolution approaches. Between 1966 and 1996, world food prices overall, not just grain prices, fell by 46% (in constant dollars), an epochal accomplishment. More grain could surely have been produced with conventional modern methods if price incentives had been more favorable. However, such a strategy for raising food production, through higher prices, would do nothing to reduce the remaining extent of hunger and poverty in the world, where at least 850 million people are still undernourished despite the surpluses produced by the Green Revolution (Sanchez and Swaminathan, 2005). Inducing more production by higher prices would make food even less accessible to poor households, which rely most on basic grains for their subsistence. Rather than use the stimulus of higher prices to promote food output, a better way to improve the situations of both poorer and richer households is to raise the productivity of the land, labor, capital, and water employed in agriculture. Achieving gains in factor productivity is the preeminent challenge for both the first and second paradigms.

There is no agreement on how much of the decline in productivity gains now being seen with Green Revolution technologies is attributable to attrition in soil system fertility or, further, how much of this loss is due to erosion of topsoil, global warming, changes in agronomic practice, or to changes in soil biology and biodiversity. Some expert assessments have sought a solution for declining yields through overcoming genetic limitations (e.g., for rice, see Peng et al., 1998). This implies that plant breeding can overcome all of these constraints. Given what is known about the functioning of soil systems, however, as presented in Part II, it is unlikely that the current stagnation in agricultural production gains can be reversed and made sustainable simply by improving varieties and by continuing to add more exogenous nutrients to the soil. Greater attention will have to be paid to soil fertility, and particularly to enhancing the life in the soil.

The current area of degraded soil worldwide is estimated to be approximately 2 billion hectares, according to Cassman (1998), who adds that this does not include "less obvious forms of degradation [that] may become an increasingly important constraint to food production capacity in the next century." Acknowledging this, however, the solutions proposed to reverse the current yield stagnation are really "first paradigm," focusing on how to meet plants' nutrient needs better by applying inorganic fertilizers in optimizing, more efficient ways. The two suggestions that Cassman considers to deal with yield stagnation are: site-specific nutrient management (Dobermann et al., 2002), and precision agriculture. Neither of these approaches for changing the present unsatisfactory situation assigns any role to altering plant, soil, water, and nutrient management practices so that the abundance and diversity of soil biota will be enhanced.

As seen in Table 49.1, the expansion of inorganic fertilizer and pesticide use has been slowing since the peak of the Green Revolution. There has been little or no increase in fertilizer use since 1996, and exports of pesticides, a reflection of rates of use, have decreased since that year. (Unfortunately, data on pesticide production and consumption are not complete or consistent enough to permit comparisons over time; so we have to consider as data the international trade in pesticides as in indicator of use, even though this does not include domestic production.) The drop-off seen in growth of pesticide exports over time has been great enough that these figures are probably reflective of worldwide trends.

The slow-down in use of fertilizer and agrochemical inputs, such as that observed in agricultural production, reflects many factors. The collapse of the Soviet Union's agricultural sector, which previously relied heavily on chemical inputs, caused a big drop in world demand for fertilizer. Little connection has been drawn between the decline of Soviet agriculture and its heavy use of agrochemicals — except in a region like Kazakhstan

TABLE 49.1

Changes in Rates of Increase by Decade, 1960–2001, in Percent

Decade	World Grain Production, by Volume (%)	World Grain Production Capita^{-1} (%)	Fertilizer Use, by Volume (%)	Pesticide Exports, by Value (%)
1961–1971	48.3	20.6	135.5	93.9
1971–1981	25.3	4.8	57.5	163.7
1981–1991	14.8	−3.0	17.4	34.7
1991–2001	11.1	−3.1	2.2	15.5

Source: Analysis of data from FAO and USDA, adapted from Buck et al., *Ecoagriculture: A Review and Assessment of its Scientific Foundations,* Cornell International Institute for Food, Agriculture and Development, Ithaca, NY, 2004. With permission.

where excessive input applications have visibly contributed to the shrinking of the Aral Sea and to the loss of large areas of formerly productive land.

Farmers in more developed countries have been scaling back their use of agrochemicals, either because of environmental quality regulations or because such use is becoming less profitable with higher prices and lower returns. A more positive reason for the decline is the advent of conservation agriculture (Chapters 22 and 24) which has been spreading in the U.S., South America, South Asia, and elsewhere. While many versions of conservation agriculture rely heavily on herbicides, at the same time that they reduce the use of fertilizer and energy for tillage, crop rotations are now being developed that reduce and even eliminate herbicide applications (Petersen et al., 1999; Calegari, 2002). There are no comprehensive statistics available, but there appears to more interest among farmers than a decade ago in the reduction, if not cessation of use of agrochemical inputs as they see their soil systems "rebound" with more environmentally-friendly management strategies (Pretty, 2002). The costs of petrochemical-based inputs are unlikely to decrease in the future and are most likely to increase.

49.1.1 Problems with External Input-Dependent Agricultural Development

What is emerging is a picture of diminishing returns to agrochemical inputs. The global pattern for fertilizer use and productivity over 50 years is summarized in Table 49.2. In most developed countries and emerging economies and for many individual farmers, fertilizer is still productive and profitable at the margin. But on a global scale, there has been some disengagement from agrochemical-led production in recent years.

For one of the world's main crops, rice, diminishing returns to further fertilizer inputs, in particular nitrogen (N), have become apparent at an aggregate level. It has been projected that, given the current (and declining) efficiency of N fertilizer use, to achieve the increase of 60% in the production of rice estimated to be needed by 2030, average total rates of N fertilizer use would need to be approximately *tripled* (Cassman and Harwood, 1995). The infeasibility of this prompted the authors to argue that alternatives need to be

TABLE 49.2

World Grain Production and Fertilizer Use (in million metric tons, Mt) and Grain:Fertilizer Response, 1950–2001

	Grain Production (Mt)	% Increase Over Previous Decade	Fertilizer Use (Mt)	% Increase Over Previous Decade	Incremental Response Ratio[b]
1950	631	—	14	—	—
1961	805	+174 (28%)	31	+17 (121%)	10.2:1
1969–1971[a]	1116	+311 (39%)	68	+37 (113%)	8.4:1
1979–1981[a]	1442	+326 (29%)	116	+48 (70%)	6.8:1
1981–1991[a]	1732	+290 (20%)	140	+24 (21%)	12.1:1
1999–2001	1885	+153 (9%)	138	−2 (−1.4%)	Not calculable

Note: During the 1980s, efforts began to be made to increase the use-efficiency of fertilizer as its cost was increasing and as regulations increased; these efforts continued during the 1990s.

[a] Averages.

[b] This ratio is the number of tons of additional grain production per ton of additional fertilizer applied.

Source: Worldwatch Institute, *State of the World 1994*, Norton, New York, 1994, and data from UNFAO, International Fertilizer Industry Association, and USDA.

found to the current heavy reliance on inorganic N fertilizer applications. In China, the marginal productivity of N fertilizer has fallen sharply as farmers have increased their applications, partly to compensate for its declining productivity. In 1958–1963, rice producers could obtain 15–20 kg more rice for each 1 kg of N fertilizer that they added to their fields; this ratio has now come down to about 5 additional kg of rice per 1 kg of fertilizer added (Peng et al., 2004). Further increases bring little additional production, with however serious effects on soil systems, water quality, and crop response, e.g., vulnerability to disease, insects and lodging.

The suggestion of tripling N fertilizer applications was a purely agronomic projection, taking account of declining yield response. Of course, farmers could not afford to triple their inputs of N fertilizer unless food shortages in the future drive up rice prices considerably. Moreover, such large increases could hardly ever be acceptable environmentally. Researchers are raising more urgent questions whether soil and water systems can continue absorbing higher levels of inorganic, i.e., reactive, N without serious ecological damage. The former chief executive of the Natural Environmental Research Council in the UK, John Lawton, recently described the rising use of N fertilizer as "the third major threat to our planet, after biodiversity loss and climate change" (*Nature*, 24 February 2005), referring just to the impacts of reactive N on water quality and aquatic ecosystems.

Levels of nitrate in groundwater are already reaching toxic or near-toxic levels in many locations. Further applications, especially at high rates, could prove very harmful. In China, for example, N fertilizer applications for rice production are currently 180–240 kg ha^{-1}, which is 2–4 times the agronomic optimum of 60–120 kg ha^{-1} (Peng et al., 2004). Chapter 39 reported current *average* application rates of 450 kg ha^{-1} with the maize–wheat cropping system in northern China. Groundwater levels of nitrate have already reached >300 mg NO_3^- l^{-1} in parts of China where fertilizer-N applications have reached 500–1900 kg ha^{-1} (Hatfield, 2004), as farmers try to compensate for diminishing returns to fertilizer by applying greater quantities. The current U.S. Environmental Protection Agency threshold, beyond which there are health concerns over nitrate in groundwater, is 50 mg NO_3^- l^{-1}. While there is no clear consensus on how great the health risks are from high concentrations of nitrate in water supply and what are acceptable levels (see Wilson and Ball, 1999), levels over 300 mg NO_3^- l^{-1} can hardly be desirable, and such levels continue to rise.

Some other forms of inorganic fertilization present fewer problems and often are well justified for agricultural production, e.g., phosphorus where it is in short supply. Nitrogen, the nutrient most emphasized in the agronomic literature and in agricultural programs, is the most disruptive ecologically, although excess P can also have adverse effects on aquatic ecosystems. The conclusion here is not to dismiss all use of fertilizer and agrochemicals but to call attention to the economic and environmental problems associated with expecting that doing "more of the same" for some or many decades to come will meet agricultural sector objectives.

49.1.2 Paradigms as Guides to Practice

Paradigms are ways of thinking that organize ideas and evidence (Kuhn, 1970). They privilege some factors and explanations over others, while they screen some factors and explanations out of consideration. Although paradigms are themselves immaterial, being ideas, they have very concrete impacts on people's lives by framing and influencing decisions. How resources and trends are understood depends on prior concepts and assumptions about cause-and-effect. Since decision-makers, whether farmers, officials or donors, invariably have multiple objectives, actual strategies of production are usually composites or compromises, not reflections of pure paradigms. This makes paradigms important as guides for thinking and acting, rather than as blueprints for action. The impact of paradigms can be seen from current agricultural policies and practices. These are shaped by the present dominant paradigm that focuses on exogenous inputs to soil systems and pays much less attention to endogenous processes within them.

The editors and contributors to this volume have seen evidence accumulating from many disciplines and parts of the world which calls attention to the productive potential of endogenous soil-system processes that should be better understood to be used more widely, even if not exclusively. The trends reviewed above reinforce this concern. Fortuitously, evidence of new opportunities comes at a time when rising economic costs, plus negative environmental externalities, make continued heavy reliance on petrochemical-based inputs less profitable and less sustainable. These create incentives for following up on the new thinking that is arising from research and experimentation across many fields. Until it has been more thoroughly investigated, evaluated and tried, we cannot know how far it will supersede external-input dependence.

49.2 The Second Paradigm and Postmodern Agriculture

Paradigms strive for simple, coherent explanations and predictions that make complex realities more comprehensible and actionable. On the other hand, to obtain best results, practical strategies need to utilize the best knowledge available from any and all sources. So while knowledge is invariably shaped by and gets incorporated into paradigms, strategies themselves aim at some kind of optimization, combining resources and ideas that are best suited to particular situations.

The second paradigm that Sanchez articulated 10 years ago is scale-neutral, applying to good farmlands as well as poorer ones. However, it particularly addressed the problems of farmers who are managing marginal lands with a combination of biophysical and socioeconomic constraints: nutrient depletion, aluminum toxicity, low nutrient reserves, low water-holding capacity, high phosphorus fixation, steep slopes, little access to markets, and disadvantageous policies. For such farmers, the prescriptions of the Green Revolution, with its reliance on external inputs, have not been working. The economic costs and logistical problems involved in procuring fertilizers and agrochemicals are prohibitive, and few of these farmers have access to irrigation, which makes them dependent on rainfall with its uncertainties and insufficiencies. External inputs give little benefit unless the water requirements of the plants and the soil, i.e., of the organisms living within it, are met.

The emergent "second paradigm" articulated by Sanchez to deal with the constraints of small and marginalized farmers is: "Rely more on biological processes by adapting germplasm to adverse soil conditions, enhancing soil biological activity, and optimizing nutrient cycling to minimize external inputs and maximize the efficiency of their use." The first point sought to redress the plant breeding strategy of selecting for varieties that perform well under the most favorable soil conditions — with the expectation that less-than-ideal soils will be modified by soil amendments to meet the needs of the new varieties. Over the last decade, breeding efforts have increasingly been directed to addressing certain soil constraints, such as selecting cultivars for their aluminum or drought tolerance. The second and third points that Sanchez articulated are addressed by the various chapters in this book.

It is becoming evident that this second paradigm is relevant not only to marginal farms and farmers, but also to many producers who are better endowed with factors of production. While they can be successful with external input-dependent production practices, the rising costs of inputs in real terms, negative impacts on environmental quality, and diminishing returns are prompting some rethinking of modern agricultural practices. More attention is being paid by "modern" farmers to soil biology where previously, mostly soil chemistry or physical qualities were considered. All three sets of factors are important and need to be managed jointly.

The rapid spread of conservation agriculture (no-till or reduced till) in the USA, Europe, Latin America, and South Asia over the past 20 years (Table 24.1, Chapter 24) reflects this realization, as does the parallel spread and adoption of integrated pest management (IPM) practices that enable farmers, including large commercial ones, to reduce their use of chemical pesticides for crop protection. This is carried out primarily by changing plant, soil, water, and nutrient management practices. As experience has been gained with IPM and as more research has been carried out on interactions among crops, pests, pathogens, soil, and water, including the ways that some populations of bacteria, fungi, or arthropods keep others in check, a broader strategy has evolved. Increasingly, IPM is viewed not just as a means to control certain pests with less use of chemicals, but as an entry point for comprehensive ecosystem management employing multiple-function interventions that capitalize on synergies within natural processes and enhance production as part of the effort to curb losses to pests (Lewis et al., 1997; Chapter 41).

This experience suggests that the second paradigm is not a "second-best" approach limited to the poorly endowed, to be followed if one cannot afford the inputs endorsed by the first paradigm. Rather, the second paradigm offers productive opportunities in its own right. Inasmuch as the first paradigm has been associated with industrial agriculture, building on the knowledge advances made in chemistry and engineering beginning approximately 150 years ago, and improved over the last 50 years by advances in genetics and biotechnology, the second paradigm represents a kind of "postmodern agriculture."

This new orientation does not reject science, as "postmodernism" does in the arts and humanities. Rather, it is grounded on recent advances in knowledge made in the biological realm, particularly in microbiology but also in ecology. These two fields, at opposite ends of the spectrum from nano to mega, are both important for a better understanding and practice of agriculture, including their combination into the emergent field of microbial ecology (Alexander, 1971; Metting, 1994; Atlas and Barth, 1998; Wardle, 2002). As the frontiers of knowledge in plant molecular biology are advancing, some of the mechanisms for these synergistic effects are being clarified, as seen in Chapters 8, 14, and 15.

In common with the "postmodernist" perspective in the humanities and social sciences, postmodern agriculture takes account of the fact that our patterns and premises of thinking affect our perceptions and choices. It also accepts the importance of values, such as poverty reduction and environmental conservation, rather than promulgate a value-neutrality that is unrealistic since neutrality tends to reinforce the status quo and thus in practice it favors some interests over others.

Rather than view plants from an engineering perspective as "living machines," to be designed to certain specifications and then grown by the investment of planned inputs, as implied by the first paradigm, the second paradigm regards plants as organisms with their own capabilities and strategies for growth and survival. Management efforts should comprehend and capitalize upon these. Plants are best treated not as isolated species and specimens to be grown separately from the rest of nature, but rather as existing inter-dependently with other organisms. Plants serve the needs of belowground biota and are served by them in return as part of complex energy and nutrient webs (Chapters 5, 6, and 7).

With such a vision of plants and their associations with other biota, the activities that Sanchez (1994) enumerated — enhancing soil biological activity, optimizing nutrient cycling, and maximizing the efficiency of all nutrient use — acquire different meanings and greater promise than with more simplified, mechanistic conventional ways of thinking

TABLE 49.3

Nutrient Balances in Tropical Rainforest and Smallholder Agroecosystems (kg ha^{-1} yr^{-1})

	Amazon Rainforest[a]			Kenya Farms[b]		
	N	P	K	N	P	K
Inputs						
Atmospheric deposition	6.1	0.2	10.6	6	1	4
Nitrogen fixation	16.2	0	0	8	0	0
Organic manure	0	0	0	14	5	25
Mineral fertilizers	0	0	0	17	12	2
Total	22.3	0.2	10.6	55	18	31
Outputs						
Crop harvest removal	0	0	0	55	10	43
Crop residue removal	0	0	0	6	1	13
Runoff and erosion	0	0	0	37	10	36
Leaching	14.1	0	4.6	41	0	9
Denitrification	2.9	ND	ND	28	0	0
Total	17.0	0	4.6	167	21	101
Balance	+5.3	0	6.0	−112	−3	−70

Source: Sanchez, *Tropical Soil Fertility Research: Towards the Second Paradigm*, Mexican Soil Science Society, Chapingo, Mexico, 1994, 65–68. With permission. ND, Negligible or not determined.
[a] Oxisol in Venezuela (Jordan, 1989).
[b] Kisii District (Smaling, 1993).

about crops and soil science. In presenting the second paradigm, Sanchez compared the nutrient balances in an undisturbed tropical rain forest with those in an intensively cultivated small-farm agroecosystem in East Africa (Table 49.3). With intensive modern agriculture in the USA or Europe, there would be still more fertilizer inputs than in the latter, but also considerably higher removal of nutrients from harvest — and also, with conventional soil tillage, from runoff and erosion.

"Postmodern" soil management strategies try to refashion agricultural practices so that they can mimic the biodynamics of forest systems, as suggested in Chapters 2 and 18–24. In such strategies, one has intensive nutrient cycling, accessing of nutrients from lower soil horizons (more available in boreal than humid tropical forests), and the creation of favorable conditions for soil bacteria and fungi to contribute to plants' nutrition and health in such systems. A movement away from monocropping toward more agro-ecological management capitalizes on synergies among plant and animal species that are matters for empirical investigation and validation.

One of the characteristics of the Green Revolution was to seek technologies that could be applied on a very broad scale, with minimal adaptation to local soil, climatic, and other conditions. Cultivars were selected that could succeed widely despite variations in local context. The second paradigm, on the other hand, is very attuned to location-specific variability, responding to particular conditions of soil, climate, economy, and even policy. Industrial agriculture regarded these factors as constraints, to be overridden, whereas postmodern approaches seek to understand and adapt to place-specificity, expecting that this can achieve optimal production and greater sustainability beyond the reach of standardized technologies that invariably must settle for some suboptimization over time even if short-run profitability is maximized.

A postmodern perspective in agriculture does not propose the abandonment of all external inputs. As seen in many preceding chapters and as discussed in the concluding chapter, such inputs can be very beneficial within the second paradigm, promoting species and cycles that are endogenous to soil systems in a kind of "pump-priming" activity. There can be synergy between organic and inorganic inputs, as demonstrated by Palm et al. (1997), as seen in agroforestry evaluations (Chapters 19, 20, and 21) or as seen with the zaï holes in West Africa (Chapter 26) and P solubilization strategies (Chapter 37). Supplying inorganic nutrients to the soil can help, for example, to increase the production of biomass which in turn provides more mulching material to protect soil surfaces and more organic matter with which to enrich the soil and nurture soil biota.

This newer view, however, proposes that the impacts of exogenous inputs on endogenous processes be carefully considered and evaluated, so that positive effects get enhanced and negative ones are minimized. It is well established that applying inorganic N and P inhibits processes of biological N fixation and bacterial and mycorrhizal P solubilization. Within the second paradigm, there is a burden of proof expected for the use of exogenous inputs that is absent in the first. The first regards agrochemical inputs as intrinsically beneficial, having seen positive returns from their initial use in smaller amounts than are now being utilized. However, there are diminishing returns from such inputs. This is where strategic considerations come in: how to combine elements from several paradigms to achieve some optimizing result.

The second paradigm addresses another consideration that has received little attention in the first: sustaining productivity over time, not just maximizing this in the short run. Short-term production gains can commonly be achieved with external inputs, provided they are available and affordable. However, their use, especially heavy use, can compromise a soil system's capacity to maintain its productivity because of adverse impacts on soil ecology. This is a concern that has gained salience in recent years as negative externalities, e.g., groundwater pollution and water-body eutrophication from use of N fertilizer, have become more evident with the intensification of first-paradigm practices.

From the perspective of the second paradigm, the internal capacity of soil systems deriving from the functioning of diverse populations of flora and fauna is valued for its longer-term productive benefits. To be sure, when cropping systems have been heavily reliant on agrochemical inputs, there may at first be little or no benefit from changing to more biologically-dependent methods — or even some loss during a transition period while the abundance and diversity of soil organisms are restored. Actions that build up this basic form of biological capital for agricultural production are assigned more value in the second paradigm than in modern agriculture. The first paradigm assumes that all or at least most constraints can be dealt with by applying the right kinds and amounts of external inputs. Some of the current efforts guided by the second paradigm are seeking to make transition quicker or less costly. Some of the innovations reported in this volume can indeed match modern agriculture results without any transition.

As noted in the previous section, the use of external inputs has indeed been productive, and it has had many social as well as economic benefits, often underestimated by critics of the Green Revolution. Reducing the average price of food (by 46% in real terms over a 30-year period) has been one of the most powerful anti-poverty strategies imaginable. But the question is now – as the returns to external input use are becoming less and as adverse impacts on human and environmental health are growing concerns – where does agricultural research and practice go from here?

The second paradigm expands upon the first-paradigm thinking that in recent decades has shaped decision-making about agriculture from the cabinet level to the farm level.

It would not discount or discard the use of external inputs, but instead it changes presumptions and introduces broader concerns such as equity, resilience, and sustainability. Here, we consider how these approaches can contribute to increased agricultural productivity. Sustainability issues are addressed in the concluding chapter.

49.3   Evidence of Productivity

One of the main arguments in favor of the first paradigm has been that exogenous inputs are needed to achieve the higher outputs that are required for meeting world food needs. In fact, evidence is accumulating that substantial increases in production are attainable with sound biologically-driven production practices, and often with lower costs of production so that farmers' profitability is enhanced. Biological interventions can also be more accessible to small and poor farmers by reducing their requirements for purchased inputs.

Proponents of the first paradigm have justified its methods by pointing to increases in the marginal value of production that are greater than the increased cost of additional inputs. By expending more, an even higher return is possible. This follows the standard investment logic that some investment is required to get more income: one employs more inputs to get even more output. Less attention is focused on how using resources *differently* can also raise output.

Biological processes are more complex than simple engineering logic allows. Where living organisms are involved, relatively small inputs with the right growing conditions can produce large outputs (recall the adage about little acorns and mighty oaks). On the other hand, even large inputs can yield little or no output when growing conditions are unfavorable. This variability means that under the right circumstances, *less can produce more*. This possibility is excluded by an engineering paradigm of production, as well as by the economic maxim that tradeoffs are unavoidable — that there are "no free lunches." The realm of biology has potential to diverge from the logic and dynamics of these two domains.

While soil systems are subject to entropic forces like everything else when operating within closed systems, soil systems are *open* to the extent that solar energy is continually being injected into them by pathways discussed in Chapter 6. The complicated networks and webs for processing this energy can, however, be impaired by external interventions that inhibit or distort biological processes. As noted in Chapter 5, these relationships are inherently rather inefficient, with approximately 90% of the energy that exists at each trophic level being lost when taken up at the next higher level. Yet even so, given the vast energy received from the sun on a continuous basis, high levels of productivity can be sustained despite the vast energy dissipation, with equilibria established through mechanisms that have resilience as well as productivity.

49.3.1   Comparisons in Agronomic Terms

In this book, readers have seen how these dynamics of symbiotic interdependence and biodiversity produce positive-sum outcomes. Even ostensibly negative processes such as predation, and especially decomposition, paradoxically contribute to plant growth and to sustainable systems. In Table 49.4, we see the gains that have been demonstrated for more than a dozen crops, with many different kinds of biological interventions around the world. Some are the results of agronomic trials, where the objection that these were not large-scale is offset by the fact that there was careful measurement with appropriate

TABLE 49.4

Summary of Representative Agronomic Results from Biologically Based Agricultural Interventions

Crop	Intervention	Yield Attained ($t\ ha^{-1}$) and/or Yield Increase (in %)	Comparison Practices	References
Rice				
Egypt	Inoculation with endophytic *Rhizobia leguminosarum*	+3.5 to 30.3%	Farmer yields on adjoining fields	Chapter 8; Yanni et al. (2001)
Brazil (rainfed)	Direct-seeded, mulch-based cropping system	7 to 8 $t\ ha^{-1}$ in 2000; +250 to 300%	2 $t\ ha^{-1}$ in 1986	Chapter 22; Séguy et al. (1998)
Vietnam	Water and plant management on acid sulphate soils	3.5 to 4.0 $t\ ha^{-1}$ now spread to 120,000 ha	<1.0 $t\ ha^{-1}$	Chapter 23; Husson (1998)
Madagascar	Soil smouldering: On ferralitic soils; On volcanic soils	+92 to 177%; 3.6 $t\ ha^{-1}$[a]; 3.9 $t\ ha^{-1}$[a]; 2.5 to 2.8 $t\ ha^{-1}$[b]	Standard practices: 1.3 $t\ ha^{-1}$[a]; 2.9 $t\ ha^{-1}$[a]; 1.3 $t\ ha^{-1}$[b]	Chapter 23; Michellon et al. (2004)
Madagascar	System of rice intensification (SRI)	+127 to 246%; 5-year average: 8.55 $t\ ha^{-1}$ with SRI methods	Modern methods: 3.77 $t\ ha^{-1}$; Farmer practices: 2.47 $t\ ha^{-1}$	Chapter 28; Hirsch (2000)
India; China; Nepal	SRI methods	8.73 $t\ ha^{-1}$[c]; 11.44 $t\ ha^{-1}$[d]; 7.85 $t\ ha^{-1}$[e]; +28 to 133%	6.31 $t\ ha^{-1}$[c]; 8.13 $t\ ha^{-1}$[d]; 3.37 $t\ ha^{-1}$[e]	Chapter 28; sources in footnotes[c, d, e]
Colombia	*A. chroococcum, A. brasiliense,* and *Penicilium bilaii* with a 40% reduction in N fertilizer	8.14 $t\ ha^{-1}$; +2.3% with less cost	Standard practices: 7.96 $t\ ha^{-1}$	Chapter 33; Martínez Viera et al. (2001)
Wheat				
Egypt	Inoculation with endophytic *Rhizobia leguminosarum*	+16.2 to 29.2%	Farmer yields on adjoining fields	Chapter 8; Yanni et al. (2001)

(continued)

TABLE 49.4 (Continued)

Crop	Intervention	Yield Attained (t ha^{-1}) and/or Yield Increase (in %)	Comparison Practices	References
Pakistan	Inoculation with auxin-producing *Azotobacter*	+21.3%	Control plots	Chapter 14; Khalid et al. (1999, 2003)
Pakistan	Zero-till wheat compared normal tilled wheat	*41% average increase* for 6 locations, with different dates of planting	Paired plots: 6–8 passes = 2.6 t ha^{-1} vs. zero-till = 3.7 t ha^{-1}	Chapter 24; Hobbs and Gupta (2003)
Pakistan	Surveys of zero-till wheat adopters	1991–1992, +13% 1995–1996, +16% 2000–2001, +18% 2001–2002, +14%	Farmer practice with plowing, compared to zero-tillage	Chapter 24; Khan and Hashmi (2004)
India	Zero-till wheat over 3 years, 3 sites in permanent plots	Zero-till went from 5.2 to 5.4 t/ha with cost-saving	Normal-tilled wheat went from 4.9 to 5.1 t/ha	Chapter 18; Singh et al. (2002)
Maize				
Zimbabwe	Integrated soil fertility management	+663 to 1188 kg ha^{-1} more than controls	Standard practices	Chapter 18; Nhamo (2001)
East, West and Southern Africa countries	Application of organic resources of varying quality	Produced nutrient effect equivalent to 55% of N fertilizer (4 to 139% range)	Comparison with results using N fertilizer	Chapter 18; Vanlauwe et al. (2002)
Côte d'Ivoire, Nigeria, Benin and Togo	Applications of organic resources with or without fertilizer application	1.6–3.7 t ha^{-1} with fertilizer + organic resources *+ 85–100%*	0.8–2.0 t ha^{-1} with fertilizer only	Chapter 18; Vanlauwe et al. (2001)
Zambia	Fertilizer trees: Sesbania Tephrosia Gliridicia	*1st yr* / *2nd yr* 3.9 t ha^{-1}[b] / 1.9 t ha^{-1}[b] 4.3 t ha^{-1}[b] / 2.6 t ha^{-1}[b] 4.1 t ha^{-1}[b] / 2.9 t ha^{-1}[b]	*1st yr* / *2nd yr* Sesb. 1.7 t ha^{-1}[b] / 1.4 t ha^{-1}[b] Sesb. + fert. 5.9 t ha^{-1}[a] / 3.4 t ha^{-1}[a]	Chapter 19
Zambia	Cajanus Sesbania Tephrosia	*1st yr* / *2nd yr* 2.8 t ha^{-1} / 1.8 t ha^{-1} 5.4 t ha^{-1} / 3.3 t ha^{-1} 3.2 t ha^{-1}[b] / 2.3 t ha^{-1}	*1st yr* / *2nd yr* Sesb. 1.1 t ha^{-1}[b] / 1.2 t ha^{-1}[b] Sesb. + fert. 4.0 t ha^{-1}[a] / 3.6 t ha^{-1}[a]	Chapter 19; Kwesiga et al. (1999)

Brazil	Groundnut or Mucuna cropping in rotation, instead of fallow between cropping seasons	+ *14.5%* after groundnut, + *35.8%* after Mucuna + value of crop	Comparison with yield after fallow only	Chapter 27; Okito et al. (2004)
Honduras	Association with Mucuna as cover crop/green manure	2.5 t ha^b 3.2 t ha^a with 70 kg ha^{-1} urea + *194 to 276%*	National average: 850 kg ha^{-1}	Chapter 30; Flores and Estrada (1992)
Kenya	On-farm trials with legume fallow (*Crotalaria grahamiana*)	+*40%* first year +*30%* average for several years, with high variation	Maize monocrop	Chapter 37; Bünemann et al. (2004)
USA	Thermally-composted swine manure	+*10%* grain yield +*15%* aboveground biomass	Yield with fresh manure	Chapter 31; Loecke et al. (2004)
Sorghum				
Burkina Faso	Zaï holes Zaï + compost Zaï + compost + NPK	200 kg ha^{-1} 654 kg ha^{-1} 1704 kg ha^{-1}	Control standard practice: 150 kg ha^{-1}	Chapter 26
Soybean				
Brazil	Direct-seeded, mulch-based cropping system	4.5 t ha^{-1} in 2000 +*125 to 165%*	1.7 to 2.0 t ha^{-1} (in 1986)	Chapter 22; Séguy et al. (1998)
Sugar Cane				
Brazil	Inoculation with endophytic *G. diazotrophicus*	+*28%* average, up to 50%	Control plots	Chapter 12; Baldini et al. (2000)
Cassava				
Madagascar	Associating Brachiaria with cassava on acid ferralitic soils	25 to 30 t ha^{-1b} +*150 to 200%* + cattle forage	8 to 12 t ha^{-1b}	Chapter 23
Potato				
Madagascar	One year after soil smoldering: On volcanic soils On ferralitic soils	+*50 to 108%* 30 t ha^{-1a} 25 t ha^{-1a}	20 t ha^{-1a} 12 t ha^{-1a}	Chapter 23; Michellon et al. (2004)
South Africa	Trichoderma applications (Eco-T®)	21–22.5 kg yield from treated plots (4 replics.) +22 to 31%	17.2 kg yield from untreated plots (4 replications)	Chapter 34; Mark Laing, unpubl. results

(continued)

TABLE 49.4 (Continued)

Crop	Intervention	Yield Attained (t ha^{-1}) and/or Yield Increase (in %)	Comparison Practices	References
Tomato USA	Hairy vetch mulch with 100 kg N ha^{-1} (50% reduction)	+*12 to 57%* +*10 to 48%*	Control: plastic mulch with 200 kg N ha^{-1}	Chapter 15; Abdul-Baki et al. (2002)
Citrus Cuba Orange Grapefruit	Application of *A. chroococcum* with 50% reduction in recommended N fertilization	+*9.8 to 32.4%* 48.0 t ha^{-1} 73.0 t ha^{-1}	With recommended 100% N: 36.25 t ha^{-1} 66.5 t ha^{-1}	Chapter 32; Martínez Viera et al. (2001)
South Africa Navel oranges	Deep profile modification down to 80 cm, 1 m from trees, both sides	+44% in yield[b] and +52% in large fruit[b]	32-year-old orchard with soil compaction	Chapter 38; Abercrombie and Hoffman (1996)
Multiple Crops Turkey — Tomato, maize, pepper, cotton, eggplant, soy, sunflower	Application of *A. chroococcum* with 30% less N fertilizer	36% *average* increase in yield across 7 crops with cost savings	Controls with 100% of recommended N fertilization	Chapter 32; Dibut and Martínez Viera, 2003
Bolivia — Cotton, maize, wheat, soybean, bean, sunflower	Application of Ecomic® (mycorrhizal fungi) on high-input farms	+*39.6% average* for 10 trials (ave. farm size 160 ha)	Controls cultivated with standard practices	Chapter 33; INCA (1999)
Colombia — Rice, cotton, maize, bean	Application of Ecomic® (mycorrhizal fungi) on low-input farms	+*59.7% average* for 7 trials (ave. farm size 1.5 ha)	Controls cultivated with standard practices	Chapter 33; INCA (1999)
Cuba — Rice, bean, maize, soy, groundnut	Application of Ecomic® (mycorrhizal fungi) on low-input farms	+*36.5% average* for 6 trials (ave. farm size 1.6 ha)	Controls cultivated with standard practices	Chapter 33; INCA (1999)

India — Sorghum, pigeon pea, cotton, cowpea, and maize	Mixed cropping system employing no chemical inputs, as described in Chapter 35	5-year total yield: T1: 18.14 t, T2: 16.36 t with only organic inputs +2.3–13.4% > control	5-yr yield: T3: 16.00 t for same crops, same size plots, and with chemical fertilizer + biocides	Chapter 35; Rupela et al. (2005)
Brazil — Cowpea, rice, moon bean, soy bean, alfalfa, common beans, carrots, peas, maize, pasture	Bio-char applications, as explained in Chapter 36	*+60% average increase* in biomass for 24 trials[f] — range from *+16 to 224%* according to the crop	Control plots	Chapter 36
Madagascar — Typical upland crop rotation followed by a Crotalaria fallow	Natural fallow cut-and-mulched (no burning of fallow), with application of 80 kg ha^{-1} P as guano-phosphate form	Difference over 4 seasons 1st: Rice −38% 2nd: Beans +224% 3rd: Ginger +95% 4th: Crotalaria +266%	Control plots	Chapter 29; Styger (2004)
Beef Production Colombia	On-station trials introducing of exotic tropical grasses adapted to acid soils (mostly Brachiaria spp) with low P inputs	*10–15-fold increases* in animal productivity	Natural savanna pastures, without any fertilization	Chapter 37; Oberson et al. (1999)

[a] With fertilizer.

[b] Without fertilizer.

[c] Data from 2003-2005 seasons, 1,525 on-farm comparison trials, average 0.4 ha each, monitored and measured by the Andhra Pradesh extension service (Dr. A. Satyanarayana, Director of Extension, ANGRAU, pers. comm.).

[d] Report of 2004 summer season results from 8 on-farm comparison trials, average 0.2 ha each with hybrid rice varieties, monitored and measured by the Sichuan Academy of Agricultural Sciences (Dr. Zheng Jiaguo, director, Crop and Soil Research Institute, pers. comm.).

[e] Report of 2004 summer season results from 22 on-farm comparison trials monitored and measured by the District Agricultural Development Office, Morang (Rajendra Uprety, District Agricultural Extension Officer, pers. comm.).

[f] Crop yields are in the same order of magnitude, but have been determined for a subset of trials only, since some experiments were pasture trials or bioassays that do not produce grains or tubers.

controls; other data are from larger-scale, in-field comparisons, where measurements were less complete or exact, but where there is compensating realism in the results.

These results are presented not as a definitive assessment of biological approaches, since many of the innovations are still in the early stages of development or refinement. Rather, they are representative of the kinds of production increases that are being achieved by following the logic of the second paradigm. Because biological processes are involved, results can vary considerably from place to place, or from season to season, or even from initial results to subsequent ones as soil systems become adjusted to more organic and fewer inorganic inputs. Precise numbers are not as important or revealing as are patterns and orders of magnitude, across countries and over time. No summary quantification is possible or justified, given that the practices themselves are so varied, but most effects can be seen to be in the range of 40 to 100%, and some are much greater.

Some change only one practice such as microbial inoculation (Chapters 8, 32, 33, and 34), while others change whole systems of production, e.g., direct-seeded, mulch-based cropping systems (Chapter 22) or the System of Rice Intensification (SRI) (Chapter 28). Table 49.4 provides a variety of data that are appropriate for meta-analysis rather than strictly comparable data with which one could test a hypothesis. No single hypothesis-test can either prove or disprove the propositions offered here about biologically-informed agricultural innovations since we are dealing here with an emerging paradigm for understanding soil systems and for intervening in them. Paradigms are as valid and central to scientific inquiry as is hypothesis testing.

Some of the case studies in this book have not been particularly concerned with increased agricultural output that can be associated with biologically-based innovations. Rather, they have focused mostly on achieving changes in parameters of soil fertility, e.g., Chapters 18 and 35–40. Others are more concerned with reducing farmers' costs of production by substituting biological inputs for chemical or mechanical ones, e.g., conservation farming in South Asia (Chapter 24), zero-till with legumes in Brazil (Chapter 27), and managed fallows in Madagascar and other countries (Chapter 29).

In some other cases, what are reported are long-term improvements rather than the kind of direct increases in productivity shown in Table 49.4. This is evident in the former Machakos district in Kenya (Chapter 25), where soil systems there were virtually written off by colonial authorities in the mid-1930s as "exhausted." Yet over the next six decades, through various productive changes in the management of plants, animals, soil, water, and nutrients, the value of agricultural output ha^{-1} was increased tenfold (at constant prices) even while the density of population km^{-2} was increasing sixfold, making the person:land ratio ever more unfavorable (Tiffen et al., 1994). To be sure, this transformation involved some external inputs, but not just the use of chemical fertilizer so much as labor and capital investments financed by income earned in Nairobi and then from rising incomes from agriculture. This is a good example of where exogenous and endogenous inputs of many kinds were combined and adjusted to dramatically advance farmers' production-possibility frontiers.

To put into context the increases reported in Table 49.4, we should consider the gains in production currently being achieved within the first paradigm. The site-specific nutrient management methodology being developed by scientists associated with the International Rice Research Institute (IRRI) has been generating yield increases of 7–11% compared with current farmer fertilizer practice in the same year, or preceding year. This assessment comes from evaluations carried out at 179 sites in eight irrigated rice areas of six Asian countries. The increase in profitability was only 12% (Dobermann et al., 2002).

In contrast, evaluations of SRI based on random samples of farmers in Cambodia, Sri Lanka, and India who were using SRI methods (incompletely), compared with farmers who were not, have showed profitability increasing by 68–100 + %. This was greater than their increase in yields because costs of production were reduced by approximately 20% invariably (Anthofer, 2004; Namara et al., 2003; Sinha and Talati, 2005).

The development of hybrid rice has certainly been a boon for many farmers. However, its resulting yield advantage is 7–10%, according to Cassman (1998), while others consider it as much as 15%. Since farmers have higher costs from purchasing hybrid seed, their profitability gains will be less than their yield increase. In Andhra Pradesh, India, and Sichuan, China, average SRI yields in on-farm yield evaluations using whatever rice varieties farmers were already planting are 10.0–10.5 t ha^{-1}. These yields are 35–40% higher than on adjacent control plots using modern inputs and getting 7.0–7.5 t ha^{-1} (data and sources reported in Table 49.4).

That the current rates of increase using first-paradigm approaches are not matching those being attained with more biologically-based innovations does not diminish the achievements previously made. However, there is growing evidence that second-paradigm strategies offer attractive opportunities, for many farmers if not all. These innovations are not getting the acceptance or credit that the data justify, however, because of biases in perception established by first-paradigm thinking. Second-paradigm alternatives are at a cognitive disadvantage, not because there is a lack of evidence but rather because of preconceptions that increased inputs are necessary to produce more outputs — rather than changing management practices to mobilize agroecosystem potentials.

Several evaluations of agroecological innovations have demonstrated very favorable agronomic and economic returns, augmented by additional benefits that are not monetizable including reduced costs to the environment (e.g., Pretty and Hine, 2001; Uphoff, 2002). These innovations have received less attention than warranted by the available evidence, at least in part because their results are so markedly at variance from standard thinking and expectations. Recognizing that there is "a paradigm problem," not just a matter of data, this volume has presented the scientific foundations in Part II that can explain the Part III results that are accumulating in favor of the second paradigm.

There is no empirical basis for the prejudgment that external inputs are necessary for high yield and that biologically-based, agroecological methods will invariably reduce agricultural productivity. Both paradigms will have more and less successful applications, according to differences in soil capabilities, crops grown, climatic and other factors, and management skill. The best of one will always surpass poor examples of the other. Valid comparisons need to compare production practices and systems on the same terms.

Readers should carefully consider the comparisons between biologically-based and chemically-driven cropping systems made at ICRISAT over a 6-year period (Chapter 35). These demonstrated productivity being at least comparable so that the economic and environmental benefits accruing from the more novel system represent a bonus. In most cases summarized in Table 49.4, the advantages of second-paradigm approaches are substantial and obvious.

49.3.2 Difficulties of Evaluation in a Systems Context

When assessing the impacts of changes in the management of soil systems, however, it must be acknowledged that assigning precise numbers to before- and after-intervention results of new practices can be very difficult, especially in on-farm contexts. Michael Mortimore, who contributed Chapter 25, tried to calculate such effects from published data for the Kano Close Settled Zone of Northern Nigeria, which he and colleagues have evaluated over time. The multi-variate nature of soil systems became very evident as he tried to sort out the effects of new practices, given the essential covariance among factors. The effects of varying just a single factor in a complex system will be relevant only for identical conditions. Such conditions are artificial because farmers strive for system optimization, continuously altering numerous parameters at the same time rather than just one.

Data on organic and inorganic inputs and on the resulting plant biomass produced around Kano were available for 2 years, 1993 and 1994 (Harris, 1998; Mortimore et al., 1999). The data were averages from 12 fields managed by three farmers, with high inter-field variability. There was an average inter-year increase of $2.6–2.8\,\mathrm{t\,ha^{-1}}$ of plant biomass, including harvested crops, associated with an average increase of $0.84\,\mathrm{t\,ha^{-1}}$ of organic inputs (dry manure and compost). This was quite substantial, but other factors were simultaneously operative.

Between 1993 and 1994 farmers increased their use of inorganic fertilizer by $18\,\mathrm{kg\,ha^{-1}}$, starting from low absolute levels, so this contributed something to the increased output. There was also an increase of $\sim 85\,\mathrm{mm}$ in total seasonal rainfall. This affected farmers' behavior as well as their water availability. Given the increased rainfall, not only were farmers more willing to invest in inorganic nutrient inputs, but they also planted more leguminous crops in 1994, and this increased soil N by $14\,\mathrm{kg\,ha^{-1}}$. Thus, one cannot say how much of the recorded yield increase came from farmers' greater input of organic matter, because the increased output was linked with other associated changes in inputs.

This example indicates the difficulty of separating biological from chemical or physical factors when trying to explain variance in yield under actual farm conditions. Mortimore's conclusion was that his analytical exercise, though inconclusive, was instructive because it underscored that evaluating soil biological approaches involves more than shifting emphasis from one group of factors to another (personal communication). To assess real farming systems, analysts need to embrace the complexity of the systems as they are actually operating. This need not invalidate all comparisons. When differences are large, as generally seen in Table 49.4, it is not so necessary to know what they are precisely, as to understand patterns of association and to look for consistency in cause-effect relations even if the mechanisms are not all identified.

49.4   Discussion

No final conclusions can be, or need to be, drawn at this time about the second paradigm because it is still emerging. There are by now sufficient reasons to undertake more research on and experimentation with alternative approaches departing from the first, input-dependent paradigm. These reasons include constraints being encountered with the standard current approaches to agricultural development, particularly economic and environmental costs, and attractive opportunities already demonstrated both in well-conducted research evaluations and on farmers' fields. However, we need to consider not just the potentials for productivity gains, but also, given our concerns and responsibilities as scientists and as human beings, questions of sustainability, which have fortunately gained salience in development circles in recent years. In a concluding chapter, the editors have reflected on some issues that pertain to the sustainability of soil systems, the central concern of this book.

References

Abdul-Baki, A.A. et al., Marketable yields of fresh-market tomatoes grown in plastic and hairy vetch mulch, *HortScience*, **37**, 878–881 (2002).

Abercrombie, R.A. and Hoffman, J.E., The effect of alleviating soil compaction on yield and fruit size in an established navel orange orchard, *Proc. Int. Soc. Citriculture*, **2**, 979–983 (1996).

Alexander, M., *Microbial Ecology*, Wiley, New York (1971).

Anthofer, J. et al., An Evaluation of the System of Rice Intensification (SRI) in Cambodia, Report to the German Agency for Development Cooperation (GTZ), Phnom Penh, Cambodia (2004).

Atlas, R.M. and Barth, R., *Microbial Ecology: Fundamentals and Applications*, 4th ed., Addison-Wesley, New York (1998).

Baldini, J.I. et al., Biological nitrogen fixation (BNF) in non-leguminous plants: The role of endophytic diazotrophs, In: *Nitrogen Fixation: From Molecules to Crop Productivity, Proceedings of the 12th International Congress on Nitrogen Fixation*, Pedrosa, F.O. et al., Eds., Kluwer, Dordrecht, The Netherlands, 397–400 (2000).

Buck, L.E. et al., *Ecoagriculture: A Review and Assessment of its Scientific Foundations, SANREM CRSP Report*. Cornell International Institute for Food, Agriculture and Development, Ithaca, NY (2004).

Bünemann, E. et al., Maize productivity and nutrient dynamics in maize-fallow rotations in western Kenya, *Plant Soil*, **264**, 195–208 (2004).

Calegari, A., The spread and benefits of no-till agriculture in Paraná state, Brazil, In: *Agroecological Innovations: Increasing Food Production with Participatory Development*, Uphoff, N., Ed., Earthscan, London, 187–202 (2002).

Cassman, K.G., Ecological intensification of cereal production systems: Yield potential, soil quality, and precision agriculture, *Proc. Natl Acad. Sci. U.S.A.*, **96**, 5952–5959 (1998).

Cassman, K.G. and Harwood, R.R., The nature of agricultural systems: Food security and environmental balance, *Food Policy*, **20**, 439–454 (1995).

Dibut, B. and Martínez Viera, R., Biofertilizantes y bioestimuladores: Métodos de inoculación, *Manual de Agricultura Orgánica Sostenible*, 17–22. UN Food and Agriculture Organization, Havana (2003).

Dobermann, A. et al., Site-specific nutrient management for intensive rice cropping systems in Asia, *Field Crop Res.*, **74**, 37–66 (2002).

Flores, M. and Estrada, N., Estudio de caso: La utilizacion del frijol abono (Mucuna spp) como alternativa viable para el sostenimiento productivo de los sistemas agricolas del Litoral Atlantico, Paper presented to Center for Development Studies, Free University of Amsterdam (1992).

Harris, F.M.A., Farm-level assessment of the nutrient balance in northern Nigeria, *Agric. Ecosyst. Environ.*, **71**, 201–214 (1998).

Hatfield, J.L., Nitrogen over-use, under-use and efficiency. Paper presented to 4th International Crop Science Congress, Brisbane, Australia, September (2004).

Hirsch, R., La riziculture malgache revisitée: Diagnostic et perspectives (1993–1999), Agence Française de Développement, Antananarivo (2000).

Hobbs, P.R. and Gupta, R.K., Resource-conserving technologies for wheat in rice–wheat systems, In: *Improving the Productivity and Sustainability of Rice–Wheat Systems: Issues and Impact*, ASA Special Publication 65, Ladha, J.K., Hill, J. et al., Eds., Agronomy Society of America, Madison, WI, 149–171 (2003).

Husson, O., Spatio-temporal variability of acid sulphate soils in the Plain of Reeds, Vietnam: Impact of soil properties, water management and crop husbandry on the growth and yield of rice in relation to microtopography, Ph.D. Thesis, Wageningen Agricultural University, The Netherlands (1998).

INCA, *Efecto de las aplicaciones del biofertilizante Ecomic (HMA) en cultivos de interés económico, durante el periodo 1990–1998*, INCA Research Report. National Institute of Agricultural Sciences, Havana (1999).

Jordan, C.F., Ed., *An Amazonian Rain Forest*, Man and Biosphere Series, Vol. 2. UNESCO, Paris (1989).

Khalid, A., Arshad, M., and Zahir, Z.A., Growth and yield response of wheat to inoculation with auxin producing plant growth promoting rhizobacteria, *Pak. J. Bot.*, **35**, 483–498 (2003).

Khalid, M. et al., Azotobacter and L-tryptophan application for improving wheat yield, *Pak. J. Bio. Sci.*, **2**, 739–742 (1999).

Khan, M.A. and Hashmi, N.I., Impact of no-tillage farming on wheat production and resource conservation in the rice–wheat zone of Punjab, Pakistan, In: *Sustainable Agriculture and the Rice–Wheat System*, Lal, R., Hobbs, P.R., Uphoff, N., and Hansen, D.O., Eds., Marcel Dekker, New York, 219–228 (2004).

Kuhn, T.S., *The Structure of Scientific Revolutions*, 2nd ed., University of Chicago Press, Chicago (1970).

Kwesiga, F.R. et al., *Sesban sesbania* improved fallows in eastern Zambia: Their inception, development and farmer enthusiasm, *Agroforest. Syst.*, **47**, 49–66 (1999).

Lewis, W.J. et al., A total system approach to sustainable pest management, *Proc. Natl Acad. Sci.*, **94**, 12243–12248 (1997).

Loecke, T.D. et al., Corn growth responses to composted and fresh solid swine manures, *Crop Sci.*, **44**, 177–184 (2004).

Martínez Viera, R. et al., Trascendencia internacional de los biofertilizantes cubanos, In: *Memorias del V. Congreso Latinoamericano de las Ciencias del Suelo*. Sociedad Latinoamericana de la Ciencia del Suelo, Varadero, Cuba, 265–274 (2001).

Metting, E.B., *Soil Microbial Ecology*, Marcel Dekker, New York (1994).

Michellon, R. et al., Projet d'appui a la diffusion des techniques agroecologiques a Madagascar. Rapport de campagen 2002–2003, TAFA/GSDM/CIRAD, Antananarivo, Madagascar (2004).

Mortimore, M., Harris, F.M.A., and Turner, B., Implications of land use change for the production of plant biomass in densely populated Sahelo-Sudanian shrub-grasslands in north-east Nigeria, *Global Ecol. Biogeogr.*, **8**, 243–256 (1999).

Namara, R., Weligamage, P., and Barker, R., *Prospects for Adoption of System of Rice Intensification in Sri Lanka: A Socio-Economic Assessment*, Research Report 75. International Water Management Institute, Colombo, Sri Lanka (2003).

Nhamo, N., An evaluation of the efficacy of organic and inorganic fertilizer combinations in supplying nitrogen to crops, M. Phil. Thesis, University of Zimbabwe, Zimbabwe (2001).

Oberson, A. et al., Phosphorus status and cycling in native savanna and improved pastures on an acid low-P Colombian Oxisol, *Nutr. Cycl. Agroecosys.*, **55**, 77–88 (1999).

Okito, A. et al., N_2 fixation by groundnut and velvet bean and residual benefit to a subsequent maize crop, *Pesquisa Agropecuária Brasileira*, **39**, 1183–1190 (2004).

Palm, C.A., Myers, R.J.K., and Nandwa, S.M., Combined use of organic and inorganic nutrient sources for soil fertility maintenance and replenishment, In: *Replenishing Soil Fertility in Africa*, SSSA Special Publication No. 51, Buresh, R.J., Sanchez, P.A., and Calhoun, F., Eds., Soil Science Society of America, Madison, WI, 193–217 (1997).

Peng, S. et al., Yield potential trends of tropical rice since the release of IR8 and the challenge of increasing rice yield potential, *Crop Sci.*, **39**, 1552–1559 (1998).

Peng, S. et al., Improving fertilizer-nitrogen use efficiency of irrigated rice: Progress of IRRI's RTOP Project in China, Paper for International Conference on Sustainable Rice Production, China National Rice Research Institute, Hangzhou, October (2004).

Petersen, P., Tardin, J.M., and Marochi, F., Participatory development of no-tillage systems without herbicides for family farming: The experience of the center-south region of Paraná, *Environ. Dev. Sustainability*, **1**, 235–252 (1999).

Pretty, J., *Agri-Culture: Reconnecting People, Land and Nature*, Earthscan, London (2002).

Pretty, J. and Hine, R., *Reducing Food Poverty with Sustainable Agriculture: A Summary of New Evidence*, University of Essex, Colchester, UK (2001).

Rupela, O.P. et al., Lessons from non-chemical input treatments based on scientific and traditional knowledge in a long-term experiment, In: *Agricultural Heritage of Asia, Proceedings of an International Conference, Hyderabad, 6–8 December 2004*, Nene, Y.L. et al., Eds., Asian Agri-History Foundation, Secunderabad, India (2005).

Sanchez, P.A., Tropical soil fertility research: Towards the second paradigm, In: *Transactions of the 15th World Congress of Soil Science, Acapulco, Mexico*. Mexican Soil Science Society, Chapingo, Mexico, 65–88 (1994).

Sanchez, P.A., Changing tropical soil fertility paradigms: From Brazil to Africa and back, In: *Plant–Soil Interactions at Low pH*, Moniz, A.C. et al., Eds., Brazilian Soil Science Society, Lavras, Brazil, 19–28 (1997).

Sanchez, P.A. and Swaminathan, M.S., Cutting world hunger in half, *Science*, **307**, 357–359 (2005).

Séguy, L., Bouzinac, S., and Maronezzi, A.C., Les plus récents progrès technologiques réalisés sur la culture du riz pluvial de haute productivité et à qualité de grain supérieure, en systèmes de semis direct: Ecologies des forêts et cerrados du Centre Nord de l'état du Mato Grosso, CIRAD internal document. Montpelier, France: CIRAD (1998).

Singh, S. et al., Long-term effect of zero-tillage sowing technique on weed flora and productivity of wheat in rice–wheat cropping zones of the Indo-Gangetic Plains, In: *Proceedings of International Workshop on Herbicide-Resistance Management and Zero-Tillage in Rice–Wheat Cropping System, 4–6 March 2002*, Malik, R.K. et al., Eds., Haryana Agricultural University, Hisar, 155–158 (2002).

Sinha, S.K. and Talati, J., Impact of System of Rice Intensification (SRI) on rice yields: Results of a new sample study in Purulia district, India, research paper for 4th IWMI-Tata Annual Partners' Meet, Anand, February 24–26, International Water Management Institute India Program, Ahmedabad (2005).

Smaling, E., An Agroecological Framework for Integrated Nutrient Management with Special Reference to Kenya, PhD thesis, Wageningen Agricultural University, Wageningen, The Netherlands (1993).

Styger, E., Fire-Less Alternatives to Slash-and-Burn Agriculture (tavy) in the Rainforest Region of Madagascar, PhD thesis, Department of Crop and Soil Sciences, Cornell University, Ithaca, NY (2004).

Tiffen, M., Mortimore, M., and Gichuki, F., *More People, Less Erosion: Environmental Recovery in Kenya*, Wiley, Chichester, UK (1994).

Uphoff, N., Ed., *Agroecological Innovations: Increasing Food Production with Participatory Development*, Earthscan Publications, London (2002).

Vanlauwe, B. et al., Maize yield as affected by organic inputs and urea in the West-African moist savanna, *Agron. J.*, **93**, 1191–1199 (2001).

Vanlauwe, B. et al., Organic resource management in sub-Saharan Africa: Validation of a residue quality-driven decision support system, *Agronomie*, **22**, 839–846 (2002).

Wardle, D.A., *Communities and Ecosystems: Linking the Aboveground and Belowground Components*, Princeton University Press, Princeton, NJ (2002).

Wilson, W.S. and Ball, A.S., Eds., *The Environmental, Agricultural and Medical Aspects of Nitrogen Chemistry*, The Royal Society of Chemistry, Cambridge, UK (1999).

Worldwatch Institute, *State of the World 1994*, Norton, NY (1994).

Yanni, Y.G. et al., The beneficial plant growth promoting association of *Rhizobium leguminosarum* bv. *trifolii* with rice roots, *Aust. J. Plant Physiol.*, **28**, 845–870 (2001).

50

Issues for More Sustainable Soil System Management

Norman Uphoff, Andrew S. Ball, Erick C.M. Fernandes, Hans Herren, Olivier Husson, Cheryl Palm, Jules Pretty, Nteranya Sanginga and Janice E. Thies

CONTENTS

Assessing sustainability is more difficult than evaluating productivity because it depends on future evidence, which by definition cannot be known in the present. Certainly sustainability is an aspiration for both the first and second paradigms for soil system management. As seen in the previous chapter, there are reasons for questioning the sustainability of Green Revolution technologies, with their heavy dependence on external inputs. Nobody can know the future prices for petroleum, which will influence the cost of energy for mechanized production and of inorganic fertilizers and many agrochemicals, but recent data give no grounds for an optimistic view of agricultural input prices. Biotechnology advances could possibly overcome the stagnation of cereal yields in most major producing countries at some time in the future, but this is uncertain.

There are no long-term or aggregate data to support any claims of sustainability for second-paradigm approaches since these are relatively new. No better claims can be made for first-paradigm agriculture. What time-series data are available on specific innovations treated in Part III indicate that higher yield levels with biologically-based management are sustainable so long as farmers can maintain their inputs of biomass to soil systems that support and enhance levels of soil organic matter. Reduced use of inorganic fertilizers and of agrochemicals together with increased application of organic matter to the soil should have positive effects on soil biota and on associated agricultural production so long as essential nutrient levels can be sustained in soil systems. This is an empirical issue to be addressed and we consider it in section 50.2.

50.1 Processes Contributing to Sustainability

As stated above and discussed below, second-paradigm approaches are not necessarily "organic" in that they do not reject the use of inorganic inputs. Rather, there is a positive emphasis on mobilizing and managing biological processes so as to minimize the need for inorganic inputs. Table 49.4 in the preceding chapter showed that yields as high or higher than those from first-paradigm practices can be attained through second-paradigm approaches, contradicting the often-assumed superiority of "modern agricultural practices." How sustainable alternative agricultural systems will be, and how their long-term productivity can be enhanced, are both important questions, not currently answerable.

As relatively little investment has been made in researching the alternatives to date, these questions remain to be addressed explicitly and thoroughly. Evaluations made will be more illuminating if they are undertaken not just in terms of certain hypotheses to be tested, but are linked to broader questions of how to understand soil systems and their sustainable productivity. On the basis of both research and experience, optimizing patterns and rationale for resource use should be developed and also changed over time as knowledge and feedback from practice accumulate.

It is difficult to sum up in single numbers the changes in soil systems' fertility and capabilities as these include contingent qualities such as resilience when confronted with biotic or abiotic stresses. Of particular importance for sustainable agriculture is the enhancement of soil water-holding capacity and drainage. This is very dependent on the kinds of soil biological activity that lead to better particle aggregation, creating soil that can be both better aerated and infused with water at the same time. The ability of soil systems to absorb rain runoff — to capture what Savenije (1998) has characterized as "green water," i.e., water stored and used *in situ* — will become more and more essential in this century as variability in the timing and amount of precipitation is likely to become more extreme, which has dreadful effects on most agriculture. Acquiring and distributing "blue water" from surface flows or groundwater reserves with all of its costs and inefficiencies in conveyance will become ever more costly. By contrast, improving soil characteristics through biological activity and management will store water, the most essential resource for agriculture, in soil horizons and root zones where it is most needed, and at lower cost.

The practices presented and evaluated in this book are recent enough that no conclusions can be firmly drawn about their sustainability. But the biological processes and effects that are being intensified or enhanced are ones that have been occurring for ages. The results of sustaining mutually productive associations between flora and fauna — specific and diverse microbial populations in the rhizosphere and in plant roots themselves — have been the production of growth-promoting hormones (Chapters 14 and 15), beneficial meso- and macrofauna activity in the soil (Chapters 10 and 11), biological nitrogen fixation and phosphorus solubilization (Chapters 12 and 13), the build-up of carbon in the soil (Chapter 6), mycorrhizal "infection" of roots (Chapter 9), induced systemic resistance of plants to damage by pathogens (various chapters), diversified root systems in the soil that can access a larger proportion of its volume (Chapter 39), bringing up nutrients from lower horizons of the soil to distribute on the surface and in upper layers (Chapter 22), and reducing leaching, to list some of the most prominent.

Any one of these processes, if occurring to an extreme, can have deleterious effects, much like the overuse of inorganic fertilizer. The complexity of natural systems includes mechanisms and feedback loops for curbing excesses. Agricultural practices that fit into these flows and these mechanisms, not truncating this complexity, have better prospects for sustainability than ones that seek to set their own parameters independent of what existing systems would support. This is the challenge presented by the second paradigm.

50.2   Nutrients in a Soil System Context

Agricultural systems lose carbon and nutrients through the off-take of crops, but there are multiple mechanisms for restoring elements in deficit. Plants have been "exploiting" soil resources for millions of years without depleting them because of efficient cycling and little nutrient loss. Depletion of available nutrient supplies in soil systems has been more a consequence of their management and off-take of nutrients through harvest than of natural processes.

Nutrient constraints need to be understood and remedied in terms of the supply of "available" nutrients. Most soil systems have large stocks of nutrients in the soil that are currently "unavailable," being bound up in recalcitrant chemical complexes or physically inaccessible. The issue for agricultural practice becomes whether rates of utilization of these "unavailable nutrients" can meet production needs and expectations and are sustainable.

- Carbon is continuously restored to the soil through processes of photosynthesis and root exudation and through litterfall. The share of photosynthate exuded into the rhizosphere is difficult to measure and certainly varies, but plants commonly put about 10–20% of the carbon they acquire from the atmosphere into the soil (Pinton et al., 2001). The amount of litterfall and crop residues returned to the system varies with system composition and management.

- Nitrogen also from the atmosphere is fixed by organisms in, on and around plant roots and even on their leaves, in what is referred to as the phyllosphere, so N is restored to the soil through multiple pathways. Nitrogen is certainly abundant; the question is whether sufficient amounts in available forms can be maintained in the soil to meet crop needs, offsetting losses through leaching and denitrification as well as crop removal. As seen in Chapter 12, N fixation is not limited to leguminous species. Also the contributions that protozoa and nematodes make to N available in plant root zones (Chapter 10) have seldom been given the attention this process deserves.

- Phosphorus, often identified as a key constraint to crop production, is actually abundant in most soils, with much less than 10% of the total supply "available" at any one time. As discussed in Chapter 13, there is much potential for P solubilization and mobilization through biological processes. Turner and Haygarth (2001) in discussing their evidence that microorganisms significantly increase the P available in soil when it is alternately wetted and dried suggest that the same mechanisms probably apply for other nutrients, but these have not been studied. It remains to be determined to what extent these processes can provide P and other nutrients at the levels and rates needed to meet crop demand and if it can be done sustainably. Chapter 37 provides considerable evidence of this potential.

There is a legitimate concern about "where the nutrients will come from" if exogenous inputs are reduced. The ICRISAT study reported in Chapter 35 addressed that question and showed that organic inputs could match inorganic input-dependent practices in terms of yield, with a concomitant build-up of soil resources both chemically and biologically. A concern whether farmers can access sufficient organic resources for such production is valid but may be solvable. Relatively little scientific research and experimentation have gone into producing biomass rather than just yield. Plant breeding efforts over recent decades, aiming to maximize the Harvest Index, have sought to reduce the biomass, which can feed soil microorganisms as well as livestock. Chapter 22 showed biomass production being raised from $6-8$ t ha^{-1} to 25 t ha^{-1} in Brazil by introducing a calculated variety and sequence of plants into the system.

Significant research and experimentation have been done on N-fixing trees and on green manures and cover crops grown *in situ* (Chapters 19 and 30). If research and extension efforts comparable with those that went into the Green Revolution were focused on the production of biomass within agricultural systems as well as on otherwise nonarable land, this nutrient and biomass constraint could, it seems likely, be alleviated creating more scope for biomass-based soil fertility management. Work would need to be done on implements such as cutting tools, shredders, and equipment for transport that could raise labor productivity when handling biomass for agricultural purposes. A combination of enhanced productivity and reduced costs based on innovations that alleviate this constraint could make second-paradigm practices more profitable and would present producers with a different incentive structure in the future.

There will in most soil systems be some nutrient constraints, following von Liebig's "law of the minimum," based on the concept that there will always be some nutritionally-limiting factor operative in the soil (van der Ploeg et al., 1999). This is why the second paradigm is better characterized as "biological" than as "organic," since it does not reject the use of inorganic nutrient inputs.

Justus von Liebig, the first major contributor to our knowledge of soil fertility, considerably expanded his thinking by the end of his scientific career. Rather than focus on particular chemical elements in a reductionist manner, he advocated a more holistic view of soil systems and paid more attention to their living components. In 1865, reflecting on his life's work, von Liebig wrote:

> In the years 1840 to 1842, I proposed that the natural sources which deliver to plants the nitrogen they need are not sufficient for the [production] objectives of agriculture. A series of observations as well as continuous reconsideration have indicated to me, however, that this view is not correct...
>
> For millennia, millions of people have believed, and millions believe it still, that the sun revolves around the earth because this is what they perceive. In the same way, many thousands of farmers have believed, and thousands still believe, that the practice of agriculture revolves around nitrogen, even though this belief has never been scientifically validated, and never will be scientifically supported because all progress and indeed all improvements in agriculture revolve around the soil (republished in Liebig, 1995: 12–13; translation by Uphoff).

Von Liebig's conclusion from a lifetime of research devoted to understanding soil fertility was to emphasize the biological factors and processes within soil.

This perspective does not make chemical elements less important but rather puts them into a living context. The interactions of the components and processes of biological systems that have evolved over millennia provide a framework for comprehending and managing soil systems. This does not suggest that "nature" cannot be improved upon. But the admonition of Leonardo da Vinci cited by Dazzo and Youssef at the end of Chapter 8: "Look first to Nature for the best design before invention," not only has a certain logic; a growing body of scientific evidence is explaining the merits attainable from complex relationships and biodiversity within soil systems when they are enlisted on behalf of agricultural production.

How sustainable any particular set of practices will be remains an empirical question that deserves close and continuous study. Sometimes physical interventions, such as profile modification (Chapter 38), will enhance the soil's biological processes and capabilities. On the other hand, research has shown that one biologically-oriented practice, zero-tillage, by itself is not always the best practice (Govaerts et al., 2004). Reduced tillage needs to be coupled with the use of mulch to create conditions for plant, microbial and macrofauna growth that are optimal. While these can contribute to denitrification, they contribute also N fixation. In complex systems, one seeks net positive results, as many contradictory and offsetting processes are likely to be involved. There is no reasonable basis for being opposed to "chemical interventions" in soil systems as all of the processes discussed here involve chemicals, in various forms. While some adverse effects of certain interventions can be identified, there are at the same time various chemical interventions that can be supportive, and in many situations essential, for well-functioning agricultural soil systems.

We began this book by affirming that all soil systems have these three interactive facets — chemical, physical and biological. These are not subsystems or components, but basically coequal *dimensions* that are conjoined in time and space. Our focus and emphasis on this third facet has not tried to make it supreme but rather to restore balance to soil system analyses and prescriptions, compensating for past neglect. Everyone must recognize that there are some dynamic forces driving soil systems that come from outside, particularly climate and human interventions, while appreciating explicitly the animation of soil systems that is endogenous.

50.3 Some Issues for Biologically-Driven Soil System Management

50.3.1 Optimizing the Use of Organic and Inorganic Inputs

One of the most important issues for the next decade or two as agricultural systems move toward more biologically-framed management practices will be how to optimize the use of inorganic soil amendments so that there is a positive-sum effect on agricultural productivity. This issue was addressed particularly in Chapter 18, with data from evaluations by the Tropical Soil Biology and Fertility Program. The principle is that of "pump-priming," where utilizing a small amount of resources can elicit a much larger flow of desired resources. Where available soil nutrients are deficient, the practice of adding inorganic fertilizers has been conceived initially as zero-sum, compensating for a deficiency.

When inorganic nutrients are introduced into soil systems, unless sufficient organic matter is supplied to feed the soil biota and maintain levels of soil organic matter, there is often over time a depression of soil biotic communities and their processes that support many aspects of soil fertility. In such cases, plants in the soil become increasingly dependent on inorganic inputs, because organic inputs are diminished. A substantial amount of the nitrogen taken up by plants, by some estimates 20–40%, is cycled through nematodes and other fauna that occupy middle ranks of the soil food web (Badalucco and Kuikman, 2001; Bonkowski, 2004; Chapter 10). Much of this is forgone when inorganic amendments are made. On the other hand, there can be positive-sum dynamics when organic inputs are combined with inorganic amendments that maintain the nutrients, soil organic matter and its biological processes that underpin soil fertility.

Short-term benefits from inorganic soil amendments are common. There can also be long-term benefits such as the residual effect of phosphorus and lime applications. But the long-term productivity ensuing from such amendments should be assessed empirically rather than simply assumed. Often when soil systems have been primarily managed with inorganic nutrients (first-paradigm approaches) are switched to more biologically based systems (second-paradigm approaches), there can be a reduction in yields until the soil biological system has been redeveloped, including stocks of soil organic matter and the diversity as well as abundance of soil organisms. How to reduce the length and magnitude of such "transitions" in agricultural systems is one of the most important and practical research questions for plant and soil scientists in the years ahead. The answers will vary, probably widely, among soil systems and for different crops. The results reported in Chapter 35 from ICRISAT are encouraging in this regard. But this remains a thoroughly empirical question. Some rough generalizations can be formulated, but actual practices need to be evaluated with both data and with sensitivity to the variability and surprises inherent in the biological realm.

50.3.2 Applicability to Commercial Agriculture

The impetus for most of the work that is reported in Part III was to identify agricultural production practices and systems that could benefit particularly the kind of impoverished, food-insecure rural households for whom Sanchez's "second paradigm" was explicitly formulated. The needs and opportunities of these fellow citizens of the world have motivated most of the work of the editorial group throughout our lifetimes, as it has the research and practice of most of the contributors. However, what has been learned about biological approaches to enhancing soil system fertility and sustainability is similarly relevant, with appropriate modifications, to large-scale farmers practicing industrialized agriculture.

High-input farming systems are of limited benefit for a majority of the world's current farmers who have low incomes and are often isolated, relating to "the market" intermittently and seldom on very favorable terms because they lack information, bargaining power, and the essential infrastructure and institutions necessary for effective market participation. Modes of production that reduce their dependence on capital inputs and thus lower costs of production give them more opportunity to engage in market exchanges on terms that benefit them.

The vision for second-paradigm agriculture is not perpetual subsistence cultivation. Instead, it points toward various kinds of intensification that are more remunerative as well as environmentally benign, toward what Conway (1999) has dubbed "the doubly green revolution." Recently, UN secretary-general Kofi Annan has called for "a uniquely African 'green revolution' for the 21st century" (Annan, 2004). We expect that this will depend heavily on the kinds of innovations discussed in this book. When looking for good examples of productive new approaches for Part III, it was gratifying to see that more than half of the chapters written were based entirely or in large part on work going on in Africa. This is an encouraging statistic. Moreover, for small-scale farmers the increases in profitability accompanying more biologically-based production methods can be even greater than the changes in output, so the socio-economic benefits from these methods can be more than the agronomic and environmental ones.

Larger commercial farmers are, at the same time, experiencing cost-price squeezes that are eroding the profitability even of large-scale operations. The globalization of commodity markets is making even big producers subject to the vagaries of the market. These larger producers are supported by over \$1 billion day^{-1} of governmental subsidies, meaning that taxpayers in the richer countries are encouraging and paying for these inefficiencies. They also have serious negative externalities for small-scale producers in poorer countries.

As global climatic influences become more variable and extreme, the vagaries of weather further complicate those of market forces. The more capital that farmers have invested in their operation, the more vulnerable they become to shifts that benefit consumers at the expense of producers. Cultivating robust soil systems that can better withstand the effects of drought and flooding, through better water-holding capacities and better aggregation, will become more and more relevant to commercial agriculturists, whose large capital investments and debt hold them hostage to climatic stresses.

More immediately, environmental and health considerations are likely to begin shifting current calculations as regulations constrain the timing, amounts and kind of inorganic fertilization that can be used, and the application of pesticides. Consumers' concerns about their exposure to agrochemicals and residues have made organic agriculture the fastest growing part of the agricultural sector. Worldwide demand for organic products is rising approximately 20% per annum, as reported in *Nature* (April 22, 2004). While scientific evidence on the health benefits of organic food products is still mixed and thus contested, the main uncertainty is over whether benefits are as great as proponents claim, with no support for the converse conclusion that food grown with synthetic inputs is better for human health than that produced "organically." As long as there is rising consumer demand and it is profitable to move toward sustainable agriculture practices — as some large commercial producers such as Dole and Unilever have begun to do — the appeal of more biologically-based agriculture will continue to grow.

50.3.3 Some Constraints to Be Addressed

50.3.3.1 Labor Intensity

One limitation on many biologically-based practices has been their relative labor-intensity, although some like direct seeding through permanent vegetative cover (Chapter 22) and green manures and cover crops (Chapter 30) are labor-saving from the start. Mechanized, energy-intensive agriculture was developed to enhance farm profitability by reducing labor requirements. The cost of labor is rising around the world; but so are the prices for fuel and agrochemical inputs. The low prices for petroleum that supported agricultural as well as industrial expansion in the latter part of the 20th century are probably now "history."

The economic logic of technical change charted by Hayami and Ruttan (1985) will sooner or later begin reflecting the greater relative scarcity of productive land and the rising costs of fossil fuel-based inputs, and their transportation, even as labor costs continue to rise. Especially in developing countries, relative availability of labor will continue to influence factor markets.

By expanding units of production in the past, profits could be enhanced by economies of size, not just of scale. (Economies of size derive from economic advantages that are due to greater bargaining power in the market; economies of scale reflect gains from more efficient resource use.) Indeed, the productivity of land usually declines in larger-scale operations as soil systems were less carefully managed, as labor and capital are applied across larger areas.

In this century, as population continues to grow, even if the rate of growth is slowing, previous strategies that use land profligately will become less viable as the relative availability of good land per capita diminishes. Actually, if the current system of agricultural subsidies in rich countries were eliminated, this would rapidly transform the capital- and chemical-intensive nature of their agricultural production, which has been favored by policies rather than by market-determined price structures. In the future, it will probably become economic to apply more labor to land, provided this is done productively, rather than continuing to rely on land-extensive strategies. Various methods and systems that have been documented in Part III can help with such a transition.

Some biological innovations are not necessarily more labor-intensive, e.g., soil inoculations (Chapters 8, 32 and 33) and direct seeding with no-till (Chapters 22 and 24). The latter saves fuel as well as the labor and other costs of plowing and weeding, which is why it is spreading in Latin America, North America, Europe and South Asia. Others methods like the System of Rice Intensification (SRI) (Chapter 28) and composting (Chapter 31) may require more labor, but to the extent that they give higher per-hour returns to labor, they are economically attractive, and with time and experience as well as mechanical innovations, their labor time is reduced.

Most of the technologies heretofore identified as labor-intensive did not offer high returns to labor. Labor has been used abundantly, even excessively where it was cheap, to make land, capital or water more productive. This necessarily diminished labor productivity. Because second-paradigm methods are mobilizing resources from the soil or atmosphere through essentially free (unpaid) biological activity, it is possible to have higher labor productivity at the same time that labor inputs are increased, i.e., with greater labor intensity. If this is a more profitable use of labor, it can become attractive to farm households and to investors even if more labor is required. The limitation will then be whether sufficient labor supply is available to take advantage of the opportunities (Moser and Barrett, 2003).

Biologically-oriented methods are not necessarily limited to a small scale. SRI, for example, is being practiced on a larger scale, not limited to smallholdings, now that its methods are being better understood. Good organization of cultivation practices is required, but an increased requirement for labor could benefit both the farmer and hired laborers. Actually, the SRI methodology is proving to be labor neutral or even labor-saving once farmers gain familiarity with its techniques (Anthofer, 2004; Li et al., 2005; Sinha and Talati, 2005). Whether alternative technologies will in the long run require more labor is still an open question. Not all of them will be equally dependent on more labor inputs. In any case, what is more important is whether and how much they can raise labor productivity.

Many of the innovations reported in this book as they are scaled up and as farmers, scientists and extensionists gain experience with them, will have labor-saving modifications that diminish this constraint to wider adoption. As seen in Chapters 22 and 24, a main constraint for the spread of conservation agriculture (no-till) has been the availability of suitable implements. As the designs for tools, equipment and implements become better suited to farmers' conditions and as the production and supply of these are ramped up, with concomitant reductions in price, the acceptability and spread of biological innovations should be hastened.

50.3.3.2 *Biomass*

A constraint that can be critical for many of these biologically-driven innovations is the availability of biomass for keeping soil energy and nutrient stocks sufficient to support higher levels of biological activity. The use of fast-growing leguminous trees and cover crops, reviewed in Chapters 19–21, 27 and 30, presents varied opportunities for increasing biomass and also enhancing N supplies in the soil. Finding ways and means to grow more abundant biomass on presently uncultivated land, with techniques that are environmentally-benign and labor-efficient, is one of the most important areas for research in support of various biologically-driven approaches. Complementing this are the production and use of bioproducts such as compost and inoculants that enhance the productivity of organic and other inputs (Chapters 31–35).

As noted above, little thought and little investment have been devoted to reducing biomass production as a constraint. As long as the returns to making organic inputs are moderate to low, there is little incentive for researchers or farmers to tackle this problem. But the kind of productivity and profitability gains that are documented in this book and in Table 49.4 should make this an attractive area for experimentation, including the design and production of tools and transport equipment that can enhance labor productivity. The work of CIRAD and its partners in Brazil and Madagascar (Chapters 22 and 23) has shown, for instance, that there are some plants that can grow very well in dry or cold seasons and have aggressive rooting systems that improve soil structure. These can produce large amounts of biomass when there are no crops being grown. This means that there is little or no opportunity cost in terms of agricultural output and, instead, a substantial augmentation of production when these plants are utilized to increase soil organic matter and improve physical characteristics.

Further, as noted above, inorganic nutrients can often be productively used to increase biomass output. This is part of the direct-seeding strategy discussed in Chapter 22. No opposition or mutual exclusion between organic and inorganic inputs should be erected that leads to a suboptimization that is not in the interest of farmers or of sustainable soil systems. Soil and climatic constraints have been the major physical limitations on agricultural production in the past. Inorganic inputs that help to increase organic outputs can diminish both constraints. By creating better soil conditions and root systems, they can even offset some of the constraints of rainfall and temperature by holding water and buffering heat or cold.

The production of compost on a commercial basis, and especially of vermicompost, is expanding in India and other countries. The production of biofertilizers and agents for biological control of pests and diseases is being taken down to the village level, as reported in Chapter 45, so that employment is created at the same time that farmers using these bioproducts get higher output and profits. Unlike the production of biomass, manufacturing these products does not require any access to land. The same is true for production of bacterial or fungal inoculants, which have the effect of "producing more land" by raising the productivity of existing cultivated area. These are innovations well suited for the 21st century. Biologically-based agriculture will, it appears, be increasingly integrated into commercial production activities at both large and small scale as part of future agriculture.

50.3.3.3 *Training*

Because these new approaches are knowledge-intensive, and require some changing of mindset as well as having factual information, a principal constraint for their spread is a lack of understanding not just of techniques but also of rationale and principles. The practices being proposed often go against what has been taught in schools and universities for 150 years. Yet they are supported by a huge amount of research and now-spreading practice.

Training is probably too narrow a concept for what is needed, though the substance of biologically-based agricultural thinking should be incorporated into training programs around the world. These new practices represent a shift in paradigm, from input-dependent, exogenously-focused production systems to ones that are soil-system-based and endogenously-focused. They adopt an ecological perspective that appreciates the interactions among organisms, seeking to maximize positive synergies and to control or eliminate negative effects. The expanding field of biotechnology can become compatible with this perspective if it becomes less preoccupied with manipulating the genotypes of individual species and appreciates more the interaction among species. Genes are of course important, but a genocentric view of biology is being superseded by concerns with G × E interdependence, studying genetic interactions with environment. Thus, the relearning is not just for farmers but also for scientists and extension workers.

50.4   New Directions for Agriculture in the 21st Century

Brazil is a country held up as a paragon of modern agriculture, with a dynamic and productive agricultural sector, expanding through many large-scale operations. Yet as seen in this book, it is also a country where some of the most interesting large-scale applications of the new, biologically-based thinking about agricultural improvement can be found (Boddey et al., 2003; Chapters 21, 22, 24 and 27). Brazil is a country where agriculture is not subsidized as in North America and Europe. Indeed, it faces significant disadvantages of transportation costs given its global location and the location of many of its farming areas. Still, Brazil is becoming more and more competitive in the world market.

The area under the newer systems of production described in Part II has now reached at least 22 million ha in Brazil, growing by 1–2 million ha yr^{-1}. Use of conservation agriculture techniques has spread even faster in the USA, more than doubling between 1997 and 2003 and now covering an area 50% more than in Brazil, according to Derpsch and Benites (2004). Some of the "negative externalities" of Brazil's modern agriculture, employing a high degree of mechanization and soil tillage along with heavy inputs of mineral fertilizers and agrochemicals, are becoming too great to be ignored, with adverse impacts on soil quality and microclimates, as mentioned in Chapter 2. The rising economic costs of conventional production methods are pushing Brazilian farmers to reevaluate their technical options.

Research on different approaches to soil system management is cumulating across many countries, building upon decades of basic research conducted while most investigations were still being carried out within the context of the first paradigm. It is surprising to see how many of the seminal scientific studies published on phytohormones, mycorrhizal associations, and nitrogen cycling through protozoa and nematodes, to take just three examples, were done in the 1950s, 1960s and 1970s, with little attention paid to them. However, this work has persisted and matured over the past 50 years, strengthened now by new analytical techniques, many at the molecular level (Chapter 46), giving clearer outlines and more specificity to the actors and processes in the soil food web that have been amorphous and inexact. Their consequences for plant/crop performance and for agroecosystem functioning are becoming better known.

50.4.1   Rationale for New Directions

The factors making the second paradigm more salient and attractive are numerous, going beyond the accumulation of scientific knowledge that offers explanations for the beneficial effects observed and measured. These include the following.

- Changing factor proportions — as discussed in section 1.6, section 49.1 and section 50.3.3.1 of this chapter. Land per capita ratios will require raising land productivity through more intensive, i.e., less extensive, production strategies as labor supply relative to land will continue to increase. Whether labor in the agricultural sector becomes more productive will depend on the technological and institutional configurations that this century evolves. The scientific basis for enhanced land and labor productivity through intensive management of plants, soil, water and nutrients is available and growing. Especially more productive utilization of freshwater resources will become imperative in many countries and regions, making the enhancement of root growth and soil biotic communities more essential.

- Calculations of real cost are changing as environmental "externalities" get figured into societal if not always individual assessments (Pretty et al., 2000; Pretty, 2005; Tegtmeier and Duffy, 2005). Groundwater and soil contamination from N fertilizers and agrochemicals is increasingly subject to regulation while the economic costs of their use become relatively greater. Economic and environmental considerations are favoring movement toward management strategies relying more upon intrinsic biological and ecological processes.

- The disgrace of poverty and hunger that still afflict too many people on our planet suggests that new approaches are called for in the agricultural sector, where, ironically, most of the world's hunger is still concentrated. External input-dependent technologies continue to by-pass the poor. As discussed in Chapter 45, biological technologies can be adapted to the conditions of resource-limited households and can be made to benefit producers and consumers.

The argument that synthetic external inputs are necessary to "feed the planet" (Avery, 1995) is contradicted by the evidence presented in this volume, summarized in Table 49.4. Biologically-based agriculture that combines the use of organic and inorganic inputs can match or surpass first-paradigm agriculture, as seen in Chapter 35 and many others. It should be possible to produce more surpluses to meet the food needs of the urban poor with more intensive methods given their demonstrated potentials.

Also, the lower costs of production with these methods can make them more profitable for the poor and also reduce risks of loss when less capital investment is required. This was seen in an evaluation by IWMI-India of SRI adoption by impoverished farmers in West Bengal. Even with only partial use of the recommended methods, yields for 110 farmers using both methods were 32% higher on their SRI plots, while their net profits ha^{-1} were 67% higher because costs of production were lower (Sinha and Talati, 2005). Concerns with poverty reduction and food security add to the rationale for taking the second paradigm seriously.

50.4.2 Paradigm Change

As stated at the outset of the chapter, we do not expect one paradigm to replace the other. That is not how the history of interaction between ideas and practice has occurred. However, as new paradigms gain credence and influence based on accumulating evidence in their favor, eventually new combinations of practice emerge that suit both the ideas and the objective conditions as well as the needs of those people who are engaged in the application of available knowledge. The second paradigm is still in its early stages of development, although it is now much more fleshed out and robust than when it was proposed 10 years ago.

There is need for continuing and expanded research to validate and vary the principles and conclusions upon which it is based. But this is not a case where science will first create new opportunities, and then technological applications will be derived from the emergent knowledge. Much of the knowledge base for the second paradigm has been emerging from practice, with scientists then investigating the new ideas and opportunities with their standard methods of analysis. A two-track rather than a sequential approach is indicated for this domain, as there is already enough evidence and scientific justification for application and refinement of second-paradigm thinking, supported by government extension services, NGOs and farmer organizations. Concurrently, researchers have a huge and promising research agenda before them framed by the second paradigm. Pursuing the questions it raises should give higher returns to research investments at the margin than continuing with more thoroughly investigated questions deriving from the first paradigm.

50.4.3 Knowledge-Driven Change

This book was written for researchers who have a direct interest in practice and for practitioners who appreciate the fruits of research. We have tried to meet high academic and scientific standards, but the motivating concern was to produce and share knowledge of practical value. The contributions have been intended to speak to the interests and needs of persons who are seeking to get most, and most sustainable, benefits from the resources available to the agricultural sector and for the producers within it, including particularly the most resource-limited ones.

More productive and sustainable strategies will derive from a thorough knowledge and appreciation of the biological nature of soil systems. These are shaped by physical characteristics, and their transactions are made in the coin of chemistry. But the fundamental determinants of fertility remain the soil biota that have coevolved symbiotically with plant root systems and with plant shoots and animals above ground for 400 million years.

This reality can be overlooked but it cannot be repealed. It can be compensated for, if undercut, by use of external inputs. Such a strategy can be and often has been successful, and it will continue to be beneficial in many places for many farmers, but its productivity is abating (Section 49.1). Fortunately, the opportunities that postmodern agriculture is opening up in the 21st century are not limited to either richer or poorer farmers. The ubiquity and synergy of genetic potentials in plants and soil organisms is widely and freely available to all those who understand and respect them.

References

Annan, K., Africa's Green Revolution: A Call to Action. Opening remarks at high-level U.N. meeting on Innovative Approaches to Meeting the Hunger Millennium Goal in Africa, Addis Ababa, July 5 (Press release: SG/SM/9405 AFR/988) (2004).

Anthofer, J., The potential of the System of Rice Intensification (SRI) for poverty reduction in Cambodia. Report for GTZ/Cambodia. Deutscher Tropentag, Berlin, November (http://www.tropentag.de/2004/abstracts/full/399.pdf) (2004).

Avery, D.T., *Saving the Planet with Pesticides and Plastic: The Environmental Triumph of High-Yield Farming*, Hudson Institute, Indianapolis, IN (1995).

Badalucco, L. and Kuikman, P., Mineralization and immobilization in the rhizosphere, In: *The Rhizosphere: Biochemistry and Organic Substances at the Soil–Plant Interface*, Pinton, R., Varanani, Z., and Nannipieri, P., Eds., Marcel Dekker, New York, 159–196 (2001).

Boddey, R.M. et al., Brazilian agriculture: The transition to sustainability, *J. Crop Prod.*, **9**, 593–621 (2003).

Bonkowski, M., Protozoa and plant growth: The microbial loop in soil revisited, *New Phytol.*, **162**, 616–631 (2004).

Conway, G.L., *The Doubly Green Revolution: Food for All in the 21st Century*, Cornell University Press, Ithaca, New York (1999).

Derpsch, R. and Benites, J.F., Situation of conservation agriculture in the world, In: *Proceedings of the Second World Congress on Conservation Agriculture: Producing in Harmony with Nature, Iguassu Falls, Paraná, Brazil, August 11–15, 2003*. Food and Agriculture Organization, Rome (2004), published on CD.

Govaerts, B., Sayre, K.D., and Deckers, J., Stable high yields with zero tillage and permanent bed planting?, *Field Crops Res.* **94**, 33–42 (2004).

Hayami, Y. and Ruttan, V.W., *Agricultural Innovation*, Johns Hopkins University Press, Baltimore, MD (1985).

Li, X.Y., Xu, X.L., and Li, H., A socio-economic assessment of the System of Rice Intensification (SRI): A case study from Xinsheng Village, Jianyang County, Sichuan Province, College of Humanities and Rural Development, China Agricultural University, Beijing (2005).

Liebig, Jv., Der Stickstoff im Haushalt der Natur und in der Landwirtschaft (Nitrogen in the Realm of Nature and in Farming), *Naturegesetz im Landbau: Es ist ja dies die Spitze meines Lebens* (Natural Laws in Agriculture: Essays at the Culmination of my Life). Stiftung Ökologie Landbau, Bad Dürkheim (1995).

Moser, C.M. and Barrett, C.B., The disappointing adoption dynamics of a yield-increasing, low external-input technology: The case of SRI in Madagascar, *Agric. Syst.*, **76**, 1085–1100 (2003).

Pinton, R., Varanini, Z., and Nannipieri, P., Eds., *The Rhizosphere: Biochemistry and Chemical Substances at the Soil–Plant Interface*, Marcel Dekker, New York (2001).

Pretty, J., Ed., *The Pesticide Detox*, Earthscan, London (2005).

Pretty, J. et al., An assessment of the total external costs of UK agriculture, *Agric. Syst.*, **65**, 113–136 (2000).

Savenije, H.H.G., The role of green water in food production in sub-Saharan Africa. Paper prepared for FAO program on Water Conservation and Use in Agriculture (WCA) (http://www.wca-infonet.org) (1998).

Sinha, S.K. and Talati, J., Impact of System of Rice Intensification (SRI) on rice yields: Results of a new sample study in Purulia District, India, Research Paper for 4th IWMI-Tata Annual Partners' Meeting, Anand, February 24–26. International Water Management Institute India Program, Ahmedabad (2005).

Tegtmeier, E.M. and Duffy, M.D., External costs of agricultural production in the United States, *Int. J. Agric. Sustain.*, **2**, 155–175 (2005).

Turner, B. and Haygarth, P., Phosphorus solubilization in rewetted soils, *Nature*, **411**, 258 (2001).

van der Ploeg, R.R., Böhm, W., and Kirkham, M.B., On the origin of the theory of mineral nutrition of plants and the Law of the Minimum, *Soil Sci. Soc. Am. J.*, **63**, 1055–1062 (1999).

Biography

Norman Uphoff, managing editor for this collaborative book project, is professor of government and international agriculture at Cornell University, and retiring director of the Cornell International Institute for Food, Agriculture and Development (CIIFAD). Uphoff joined the Cornell faculty in 1970 after completing a Ph.D. in political science at the University of California, Berkeley in that year.

During most of the next 20 years, he served as chair of the interdisciplinary Rural Development Committee in Cornell's Center for International Studies while teaching and doing research on social science issues in agricultural and rural development. When CIIFAD was established in 1990 with private gift funding to enable Cornell faculty and students to become more engaged with problems of sustainable agricultural and rural development, Uphoff was appointed as its first director.

Uphoff's interest in agroecological approaches and sustainable soil systems grew out of his involvement with the System of Rice Intensification (SRI) developed 20 years ago in Madagascar. This methodology greatly increases rice yields on soils that tests evaluate as being "very poor," at the same time that it reduces farmers' inputs of seed and water and makes inorganic fertilizer and agrochemicals unnecessary (Chapter 28).

Over the past 10 years, Uphoff has delved into the literature of multiple disciplines to get a better understanding of biological dimensions of soil and crop management so as to understand how changing the management of plants, soil, water, and nutrients can achieve better and more sustainable production. He has found many other cases where such changes have produced remarkable improvements.

Through CIIFAD, it has been possible to develop an international network of professional associates who from their own work and vantage points have developed similar perspectives. Such a network has made this book possible, as many colleagues in 28 countries have contributed scientific expertise and field experience far beyond Uphoff's own.

Editors and Contributors

Lead contributions to chapters are indicated by boldfaced numbers

Editorial Team

Andrew S. Ball (1, **6**, **48**, 50): Foundation Chair in environmental biotechnology, School of Biological Sciences, Flinders University of South Australia, Australia; previously reader in microbiology in the Dept. of Biological Sciences, University of Essex, UK, and director of the University of Essex's Center for Environment and Society (0088ball@flinders.edu.au)

Erick C.M. Fernandes (1, **21**, 29, 50): Land management advisor, The World Bank, Washington, DC, USA; previously associate professor in Dept. of Crop and Soil Sciences, Cornell University, USA, and formerly coordinator for the Alternatives to Slash-and-Burn research program at ICRAF for the Consultative Group for International Agricultural Research (efernandes@worldbank.org)

Hans Herren (1, 50): President of the Millennium Institute, Arlington, VA; previously director-general of the International Centre of Insect Physiology and Ecology (ICIPE), Nairobi, Kenya; co-chair of International Assessment of Agricultural Science and Technology for Development (IAASTD); recipient of World Food Prize in 1995 for work on cassava mealybug biological control (hherren@threshold21.com)

Olivier Husson (1, 22, **23**, 50): Agronomist from CIRAD research unit on direct seeding, based in Madagascar as deputy director of GSDM (Direct Seeding Group of Madagascar), coordinating research, training and extension under various agroecological conditions; has worked for 17 years with farmers in Africa and Southeast Asia developing and testing different agroecological techniques based on better understanding of soil systems' functioning (olivier.husson@cirad.fr)

Mark Laing (16, 34, **42**): Director of African Centre for Crop Improvement and chair of Dept. of Plant Pathology, University of KwaZulu-Natal, South Africa; chairperson of PlantBio Trust (South African National Plant Biotechnology Innovation Centre) and director of Plant Health Products (Pty) Ltd; research areas cover classical and innovative biological control, microbial augmentation, and plant breeding of African food crops (laing@ukzn.ac.za)

Cheryl Palm (1, 50): Senior research scientist in the Earth Institute at Columbia University, and associate director of its Center for Globalization and Sustainable Development, USA; previously senior research scientist with the Tropical Soil Biology and Fertility Program, Nairobi, Kenya (cpalm@iri.columbia.edu)

Jules Pretty (1, 50): Professor of environment and society, and head of the Dept. of Biological Sciences, University of Essex, UK; deputy chair of UK Government's Advisory Committee on Releases to the Environment; author of *The Pesticide Detox* (2005), *Agri-Culture* (2002), *The Living Land* (1998), and *Regenerating Agriculture* (1995); Fellow of the Royal Society of Arts and Fellow of the Institute of Biology (jpretty@essex.ac.uk)

Pedro Sanchez: Director of the Tropical Agriculture Program in the Earth Institute at Columbia University, and professor emeritus of soil science and forestry, North Carolina State University, USA; previously director-general, World Agroforestry Centre (ICRAF), Nairobi, Kenya: co-chair of the U.N. Millennium Project Task Force on Hunger; recipient of World Food Prize in 2002 for work on agroforestry and tropical soils (sanchez@iri.columbia.edu)

Nteranya Sanginga (1, 18, 50): Soil microbiologist and director-general, Tropical Soil Biology and Fertility Institute (TBSF) of the International Center for Tropical Agriculture (CIAT), based in Nairobi, Kenya; research in integrated soil fertility management, with emphasis on biological nitrogen fixation in tropical cropping systems; current focus is on the integration of natural resource management with livelihood creation and sustainability (n.sanginga@cgiar.org)

Janice E. Thies (1, **5**, 31, **46**, 50): Associate professor in Dept. of Crop and Soil Sciences, Cornell University, USA; faculty leader for Soil Health Working Group supported by the Cornell International Institute for Food, Agriculture and Development (CIIFAD) (jet25@cornell.edu)

Norman Uphoff (managing editor; **1, 28, 49,50**): Professor of government and international agriculture, Cornell University, USA, and retiring director of Cornell International Institute for Food, Agriculture and Development (CIIFAD); editor and contributor for *Agroecological Innovations: Increasing Agricultural Production with Participatory Development* (2002); has worked with the System of Rice Intensification since 1994 (ntu1@cornell.edu)

Contributors

Aref Abdul-Baki (15): Sustainable Agricultural Systems Laboratory, Henry A. Wallace Beltsville Agricultural Research Center, U. S. Dept. of Agriculture, USA; develops alternative systems in the production of vegetables and management of orchards by using cover crops and no-tillage with focus on reducing chemical inputs, improving soil fertility, and reducing soil erosion and compaction (Abdul-Ba@ba.ars.usda.gov)

Bruno J.R. Alves (12, 27): Soil fertility and plant nutrition specialist at EMBRAPA-Agrobiologia, the National Research Centre of Agrobiology Research of Brazil's national agricultural research agency; specializing in the use of isotopes and related techniques for the study of C and N cycles in agroecosystems (bruno@cnpab.embrapa.br)

Muhammad Arshad (14): Professor of soil microbiology and biochemistry, Institute of Soil and Environmental Sciences, University of Agriculture, Faisalabad, Pakistan;

previously director of the Centre for Agriculture, Biochemistry and Biotechnology at University of Agriculture, Faisalabad (bio@fsd.comsats.net.pk)

Joeli Barison (28): Rural development manager for BAMEX project, Chemonics International, Tananarive, Madagascar; thesis research in Madagascar on the system of rice intensification for first degree in agriculture from the University of Antananarivo Faculty of Agricultural Sciences (ESSA), and for MSc degree in crop and soil sciences, Cornell University (JOE@chemonics.mg)

Hameeda Bee (35): Research scholar in the Dept. of Microbiology, Osmania University, Hyderabad, carrying out research at the International Crops Research Institute for the Semi-Arid Tropics (ICRISAT), India; studying plant-growth-promoting bacteria and recycling of crop residues for sustainable agriculture (dr_hami2002@yahoo.com)

Suzette R. Bezuidenhout (16): Researcher on weed science in the Dept. of Agriculture and Environmental Affairs, KwaZulu-Natal, South Africa, with specialization on allelopathic effects (suzette.bezuidenhout@dae.kzntl.gov.za)

Charles L. Bielders (26): Lecturer on soil, land and water conservation in the Dept. of Environmental Sciences and Land Use Planning, Université de Louvain, Belgium; works on evaluation of the extent and severity of land degradation processes and on development of environmentally-friendly soil and water conservation technologies (bielders@geru.ucl.ac.be)

Robert M. Boddey (12, 27): Research scientist at EMBRAPA's Agrobiologia Center in Brazil; has worked for 25 years in Brazil on the quantification of biological N_2 fixation inputs to rice and sugarcane as well as legumes, working with stable isotope and other techniques; now studying nutrient fluxes and carbon dynamics in various tropical agroecosystems (bob@cnpab.embrapa.br)

Stéphane Boulakia (23): Agronomist since 1992 with the French International Center for Research on Agricultural Development (CIRAD) working on programs in the highlands of Madagascar, Gabon, Vietnam, and currently Cambodia on direct-seeding technologies and their adaptation and adoption in environments varying physically and socio-economically (stephane.boulakia@cirad.fr)

Serge Bouzinac (22): Agronomist working since 1984 with CIRAD program in Central Brazil on mechanized, no-till systems of production (serge.bouzinac@cirad.fr)

Lijbert Brussaard (11, 26): Professor of soil biology and biological soil quality in Dept. of Soil Quality, Wageningen University, Netherlands; co-editor-in-chief of *Applied Soil Ecology*, 1993-1997; books or special journal issues on soil structure-soil biota relationships (1993), soil ecology of conventional and integrated arable farming systems (1994), soil ecology in sustainable agricultural systems (1997), earthworm management in tropical agroecosystems (1999), and soil biodiversity in Amazonian and other Brazilian ecosystems (in press) (Lijbert.Brussaard@wur.nl)

Else K. Bünemann (37): School of Earth and Environmental Sciences, University of Adelaide, Australia; previously with Institute of Plant Sciences, Swiss Federal Institute of Technology, Switzerland (else.bunemann@adelaide.edu.au)

Roland Bunch (30): Coordinator of sustainable agriculture and rural livelihoods programs for World Neighbors, USA; previously coordinator for COSECHA (Association of Consultants for Sustainable, Ecological and People-Centered Agriculture), Honduras; author of *Two Ears of Corn: A Guide to People-Centered Agricultural Improvement* (1982), translated into 10 languages; member of U.N. Millennium Project Task Force on Hunger (rolandbunchw@yahoo.com)

Frank B. Dazzo (8): Professor of microbiology in Dept. of Microbiology and Molecular Genetics, Michigan State University, USA; research on soil microbial ecology and beneficial plant-rhizobium associations, and on microscopy and image analysis of microorganisms seeking to strengthen microcoscopy-based methods for understanding soil microbial ecology (dazzo@msu.edu)

Diego De la Rosa (48): Head of the Soil Evaluation Group (www.microleis.com) in the Department of Soil-Plant-Atmosphere System Sustainability, Instituto de Recursos Naturales y Agrobiologica de Sevilla (IRNAS), Spain; research focuses on integrated studies of soil-water-plant relationships, agricultural use of residues, and development of models and expert systems for the evaluation of soil use capacity and risks of degradation (diego@irnase.csic.es).

Hanadi El-Dessougi (4): Agricultural chemist and soil fertility scientist in program on improved land management to combat desertification, International Center for Agricultural Research in the Dry Areas (ICARDA), Syria; research on community-based nutrient and integrated soil fertility management in crop-livestock systems in Central and West Asia and North Africa (h.dessougi@cgiar.org)

Bernardo Dibut Alvarez (32): Senior researcher in the Institute for Basic Research in Tropical Agriculture (INIFAT), Cuba; long experience in practical use and management of biofertilizers and biostimulators and development of industrial technology (bdibut@inifat.co.cu)

Dougbedji Fatondji (26): Senior scientific officer with International Crops Research Institute for the Semi-Arid Tropics (ICRISAT), Niger; research on integrated systems for water harvesting, use of perennials to improve soil fertility, combat wind erosion and provide firewood, fruit trees to increase rural incomes, and annuals in rotation to sustain soil fertility and enhance livelihoods; long-term interest in legume physiology and agronomy (d.fatondji@cgiar.org)

Felix Fernandez (33): Senior researcher and head of arbuscular mycorrhizal group in National Institute of Agricultural Sciences (INCA), Cuba; long experience in practical use and management of biofertilizers and development of inoculant technology; in 1999, received patent for mycorrhizal product EcoMic® widely used now in both low-input and high-input agricultural systems in Cuba with use spreading to other countries (felixfm@inca.edu.cu)

Dennis K. Friesen (37): Senior scientist with the International Center for Soil Fertility and Agricultural Development (IFDC) and International Center for Improvement of Wheat and Maize (CIMMYT) in Ethiopia; previously with International Center for Tropical Agriculture (CIAT), Colombia; research on nutrient cycling, dynamics and requirements of crop-pasture-fallow production systems in acid soil tropical lowland agroecologies (d.friesen@cgiar.org)

Emmanuel Frossard (13, 37): Professor and head of the group on plant nutrition, Institute of Plant Sciences, Swiss Federal Institute of Technology, Switzerland; research on the biotic and abiotic processes controlling the release of mineral elements from the soil and their transport to the root surface and in the plant; developing new concepts and tools for quantifying fluxes of elements between compartments of the soil/plant system (emmanuel.frossard@ipw.agrl.ethz.ch)

Zhang Fusuo (39): Dean and professor in College of Resources and Environmental Sciences, China Agricultural University, Beijing, China; research fields include plant nutrition, plant physiology, and soil ecology; special interests in plant-soil interactions, rhizosphere nutrition, and integrated nutrient resource management of various agro-ecosystems in China; chair of 15th International Plant Nutrition Colloquium (zhangfs@cau.edu.cn)

Silas Garcia (21): Weed ecologist at the Western Amazon Agroforestry Center (CPAA) of EMBRAPA near Manaus, Brazil; has worked extensively on secondary succession in cultivated areas and abandoned pastures in the western Amazon; currently leader of EMBRAPA-CPAA's Agroforestry Program for Rehabilitating Degraded Pasture Lands in the Amazon.

C.L.L. Gowda (35): Global theme leader for crop improvement and management for the International Crops Research Institute for the Semi-Arid Tropics (ICRISAT), India; 30 years of research as plant breeder on improving chickpea varieties; coordinator of the Cereals and Legumes Asia Network (CLAN) since 1987 (c.gowda@cgiar.org)

A. Stuart Grandy (3): Visiting scientist in the Dept. of Crop and Soil Sciences and Kellogg Biological Station at Michigan State University, USA; research interests include the ecology and biology of managed ecosystems, particularly microbial and physical controls over carbon and nitrogen cycling and their response to agricultural management (Grandya1@kbs.msu.edu)

Julie M. Grossman (5): Postdoctoral fellow in Dept. of Crop and Soil Sciences, Cornell University, USA; previous research at The Ohio State University; focus on microbial ecology of tropical soils, nitrogen fixation in coffee agroforestry systems in Mexico, and farmer knowledge of below-ground processes; advisor for Cornell New World Agriculture and Ecology Group (jmg225@cornell.edu)

Raj Gupta (24): Facilitator of the Rice-Wheat Consortium for South Asia, which is based in India with membership of IRRI, CIMMYT and ICRISAT and the national agricultural research systems (NARS) of Bangladesh, India, Nepal and Pakistan as well as other institutions such as Cornell University (r.gupta@cgiar.org)

Mitiku Habte (9): Professor of soil microbiology in Dept. of Tropical Plant and Soil Science, University of Hawaii, USA; research focus on microbial ecology, plant nutrition, and the ecology of plant-soil microorganism interaction with emphasis on arbuscular mycorrhizal symbiosis (mitiku@hawaii.edu)

Ahmed Hassanali (40): Principal scientist and head of the Behavioral and Chemical Ecology Dept., International Centre of Insect Physiology and Ecology (ICIPE), Kenya;

research on chemical ecology of locusts, tsetse fly, ticks, mosquitoes, and Striga (a.hassanali@icipe.org)

Peter Hobbs (**24**): Adjunct professor in the Dept. of Crop and Soil Sciences, Cornell University, USA; previously with International Center for the Improvement of Wheat and Maize (CIMMYT), South Asia, and International Rice Research Institute (IRRI), Bangladesh; co-facilitator and founder of the Rice-Wheat Consortium, an eco-regional program of the CGIAR, and promoter of zero-tillage crop establishment in Indo-Gangetic Plains of South Asia (ph14@cornell.edu)

Allison L.H. Jack (**31**): Recently completed PhD in Dept. of Plant Pathology, Cornell University, USA; became interested in composting through work as waste reduction educator in AmeriCorps VISTA program; has been volunteer Master Composter through Cornell Cooperative Extension and is active in Cornell chapter of the New World Agriculture and Ecology Group (alh54@cornell.edu)

Deon Joubert (38): Area manager for the Eastern Cape, Unifrutti SA Pty Ltd, South Africa; after completing degrees in Plant Pathology at University of Pretoria, joined the South African Citrus Exchange as an extension officer specializing in production; later joined Tambankulu Estates in Swaziland as citrus production manager, and then in 2000 joined Unifrutti, large international fruit producer and exporter, responsible for managing 550+ ha farm (deon@dunbrodyestates.co.za)

Azeem Khalid (14): Lecturer doing teaching and research in the Institute of Soil and Environmental Sciences, University of Agriculture, Faisalabad, Pakistan; research focused on plant-microbial interactions and on recycling of organic wastes (azeemuaf@yahoo.com)

Zeyaur Khan (**40**): Principal scientist and leader of Habitat Management Program, International Centre of Insect Physiology and Ecology (ICIPE), Kenya, focusing on new strategies for suppressing insect pests and parasitic weeds; previous work with the International Rice Research Institute (IRRI), University of Wisconsin, and Kansas State University (zkhan@mbita.mimcom.net)

Ulrike Krauss (20): Tropical plant pathologist and acting director of CAB International's program for the Caribbean and Latin American region, based at the Centro Agronómico Tropical de Investigación y Enseñanza (CATIE), Costa Rica; experience in ecology of plant-microbe-soil interactions, control of soil-borne plant pathogens, IPM and biocontrol in semi-arid and humid tropics, and participatory research and extension approaches (cabi-catie@cabi.org)

Elias Kuntashula (19): Lecturer in the Dept. of Agricultural Economics, University of Zambia; worked with the Zambia program of the World Agroforestry Centre (ICRAF), 2000-2005, on sustainable soil improvement in vegetable production systems using leguminous leafy biomass transfer (Kelias@mailcity.com)

Nico Labuschagne (38): Associate professor in the Dept. of Microbiology and Plant Pathology, University of Pretoria, South Africa; research on etiology and control of soil-borne diseases, especially Fusarium, Phytophthora and Pythium; phytonemato-logy; plant disease control using chemical and biological interventions; evaluation of plant disease resistance; and integrated control of diseases in hydroponic systems (nlabusch@postino.up.ac.za)

Rattan Lal (42): Professor of soil science and director of the Carbon Management and Sequestration Centre, The Ohio State University, USA; previously research soil scientist for 18 years at the International Institute of Tropical Agriculture (IITA) in Nigeria; research interests include soil erosion and its control, soil carbon dynamics in relation to land use and management, watershed management, soil degradation, conservation tillage and mulch farming, sustainable agricultural systems, tropical agriculture, and world food security (Lal.1@osu.edu)

Johannes Lehmann (36): Head of soil fertility and soil biogeochemistry program in Dept. of Crop and Soil Sciences, Cornell University, USA; research ranges from carbon dynamics in microaggregates to global nutrient and carbon budgets; particular attention to carbon, nitrogen, phosphorus and sulphur cycles in soil and how long-term human interventions influence soil functions and the reversibility of soil degradation (cl273@cornell.edu)

Li Long (39): Professor in the College of Resources and Environmental Sciences, China Agricultural University, Beijing, China; heading research group on biodiversity and resource utilization; research on interspecific interactions between different species and mechanisms behind interspecific facilitation/competition regarding nutrients; exploring root communication underground between plants in natural and agricultural ecosystems (lilong@cau.edu.cn)

Paramu L. Mafongoya (19): Senior soil scientist responsible for coordination of soil fertility research in the Southern Africa regional program of the World Agroforestry Centre (ICRAF), and country representative for ICRAF's agroforestry project in Zambia; ongoing research on mechanisms by which trees contribute to nutrient cycling and soil fertility improvement in agroforestry systems (p.mafongoya@cgiar.org)

Abdoulaye Mando (11, 26): Coordinator of program for sustainable integrated soil management practices for smallholder farms in sub-Saharan Africa, International Center for Soil Fertility and Agricultural Development (IFDC), Togo; previous research on agro-silvo-pastoral land use and management with Wageningen University, Netherlands (amando@ifdc.org)

Rafael Martinez Viera (32): Senior researcher and head of the Biotechnology Division in the Institute for Basic Research in Tropical Agriculture (INIFAT), Cuba; previously director of the Cuban Soil Science Institute (vieramartinezrafael@yahoo.com)

Christopher Martius (26): Soil biologist and agroecologist in the Center for Development Research (ZEF), University of Bonn, Germany; currently coordinator of German-Uzbek project on economic and ecological restructuring of land and water use in drylands of the Aral Sea basin; studies of soil fauna and soil fertility interactions in central Asia, sub-Saharan Africa, and Amazonia (c.martius@uni-bonn.de)

Autar K. Mattoo (15): Sustainable Agricultural Systems Laboratory, Henry A. Wallace Beltsville Agricultural Research Center, U. S. Dept. of Agriculture, USA; research on the fundamental principles in hormonal signaling in plant inter-organ communication, stress biology, and chloroplast-chromoplast metabolism; devising biotechnological and biochemical approaches for food security, with focus on fruits and vegetables (mattooa@ba.ars.usda.gov)

Craig Meisner (24): Agronomist with International Center for Soil Fertility and Agricultural Development (IFDC) and adjunct associate professor of crop and soil sciences, Cornell University, USA; previously with the International Center for Maize and Wheat Improvement (CIMMYT) working on holistic problem-oriented agricultural research, mostly in Bangladesh (cmeisner@ifdc.org)

Roger Michellon (23, 41): CIRAD agronomist currently based in Madagascar on assignment with the NGO TAFA (Land and Development); involved in the creation, evaluation and dissemination of cropping systems based on direct seeding through mulch cover; previously engaged in the improvement of flower cultivation, diversification of vegetable crops, and direct-seeding cropping systems in the Indian Ocean region (michellon@cirad.mg)

Michael Mortimore (25): Drylands Research, UK; previously with Cambridge University and Overseas Development Institute, UK; research on the management of African dryland ecosystems by small farmers and on the sustainability of natural resources and livelihoods in the course of demographic, economic, policy and environmental change (mikemortimore@compuserve.com)

Sudha Nair (45): Microbiologist currently serving as program director for the Ecotechnology Division of the M. S. Swaminathan Research Foundation, India; current work on the diversity of beneficial organisms in the rhizosphere and on taking biotechnologies to rural areas and applying them for income generation among the resource-poor; resource person for UNESCO in the Asia-Pacific region on developing science-based programs for women (sudhanair@mssrf.res.in)

Brendon Neumann (34): Postdoctoral student at the University of Kwazulu-Natal, South Africa; research on interactions between Trichoderma and both biotic and abiotic (environmental) influences in the rhizosphere; currently developing new biological control products for a variety of pathogen and pest problems (942411714@ukzn.ac.za)

Günter Neumann (7): Senior scientist and lecuturer at the Institute of Plant Nutrition, Hohenheim University, Germany; research on rhizosphere processes, regulatory aspects of root exudation, role of root exudates in plant-microbial interactions, nutrient mobilization, and interactions with toxins in the rihizosphere (gd.neumann@t-online.de)

Astrid Oberson (13, 37): Senior scientist and lecturer in the Institute of Plant Sciences, Swiss Federal Institute of Technology, Switzerland; research to improve understanding of nitrogen and phosphorus cycles in agroecosystems in the temperate and tropical regions, with emphasis on microbial-mediated processes in soil and their response to agricultural production systems (astrid.oberson@ipw.agrl.ethz.ch)

Elisée Ouédraogo (11): Sustainable agriculture program officer and chief of Agroecology Dept. in Albert Schweitzer Center for Ecology, Burkina Faso; previous research on soil quality improvement and management in semi-arid West Africa, with focus on soil organic matter, soil fauna-mediated processes and nutrient dynamics under auspices of Wageningen University, Netherlands, and Burkina Faso's national research institute (INERA) (oelisee@hotmail.com)

Gail Papli (42): Soil microbiologist and senior laboratory technician in Dept. of Microbiology, University of KwaZulu-Natal, South Africa; studying the revegetation of mine dumps using probiotics, in particular Trichoderma; more generally working on the rehabilitation of disturbed soils and assessments of environmental impact (Papli@ukzn.ac.za)

Tony Pattison (10): Senior nematologist in Queensland Dept. of Primary Industries, Australia; research on management and ecology of soil-dwelling nematodes to improve the sustainability of tropical agricultural systems; has developed successful integrated nematode management system currently being implemented in tropical banana production to reduce losses due to plant parasitic nematodes (Tony.Pattison@dpi.qld. gov.au)

Alice N. Pell (17): Professor in Dept. of Animal Science, Cornell University USA, and now director of Cornell International Institute for Food, Agriculture and Development (CIIFAD); principal investigator for research project on coupled human and natural systems in smallholder systems in the Kenyan highlands funded by NSF Biocomplexity Program (ap19@cornell.edu)

Rogerio Perin (21): Livestock and pasture management specialist at EMBRAPA's Western Amazon Agroforestry Center (CPAA) near Manaus, Brazil; has developed a range of pasture management techniques for intensifying traditional extensive pasture systems in the Amazon and has collaborated with CIAT-Colombia in testing a range of shrubby and herbaceous forage legumes with potential to improve productivity and sustainability of Amazonian pastures.

John Pickett (40): Professor and head of the Biological Chemistry Division, Rothamsted Research, UK; research focus on how volatile natural products (semiochemicals) affect the behavior and development of animals, pests and other organisms; for his contributions to the field of chemical ecology, Pickett was elected as a Fellow in the Royal Society in 1996 and was honored by the British government in 2004 when appointed CBE (john.pickett@bbsrc.ac.uk)

Ana Primavesi (2): Professor emeritus of soil management and plant nutrition, University of Santa Maria, Rio del Sul, Brazil; previously director of that university's soil chemistry and biological and microbiological laboratories; currently scientific counselor for the Mokiti Okada Foundation in Brazil; recipient of 26 awards from Ibero-American countries for her contributions to tropical soil and plant science (sindritai@hotmail.com)

Joshua J. Ramisch (18): Social science officer, Tropical Soil Biology and Fertility Institute (CIAT), Kenya; research on connections between natural and social sciences within domain of human ecology; currently visiting scholar at Land Tenure Center, University of Wisconsin, with research focusing on local ecological knowledge and dynamics of community-based learning for improved soil fertility management (j.ramisch@cgiar.org)

Richard Randriamanantsoa (41): Entomologist with the Sustainable Farming and Rice Cropping Systems cooperative research unit established by CIRAD, the University of Antananarivo, and the Madagascar government's agency for rural development research (FOFIFA); national coordinator of ASARECA participatory research network (PRIAM); previously in charge of a Swiss-funded IPM rice research program in the Lac Alaotra region (r.randiam@blueline.mg)

Robert Randriamiharisoa (28): Formerly head of the Dept. of Agriculture and director of research in the Faculty of Agricultural Sciences (ESSA), University of Antananarivo, Madagascar; former director of the National Centre for Applied Pharmaceutical Research, Antananarivo; supervised a number of research studies on the System of Rice Intensification before his untimely death in August 2004.

Idupulapati M. Rao (37): Plant nutritionist and physiologist in Tropical Soil Biology and Fertility Institute, International Center for Tropical Agriculture (CIAT), Colombia; research on crop and forage adaptation to abiotic stress factors, and nutrient cycling in crop-pasture-fallow systems in the tropics (i.rao@cgiar.org)

Alain Ratnadass (41): CIRAD entomologist currently heading cooperative research unit on Sustainable Farming and Rice Cropping Systems in Madagascar; previously coordinator of ICRISAT-CIRAD joint regional sorghum program in Mali, and of a European Union-funded IPM project in Mali and Burkina Faso; president of the IPM working group of the West and Central African Sorghum Research Network (WCASRN) (ratnadass@cirad.mg)

Veronica M. Reis (12): Research scientist at EMBRAPA's Agrobiologia Center near Rio de Janiero, Brazil; has worked for 10 years on the microbial aspects of biological N_2 fixation, especially in sugar cane, rice, grasses and maize (veronica@cnpab.embrapa.br)

S. Rengalakshmi (45): Agronomist working with the M. S. Swaminathan Research Foundation, India; research on application of participatory approach to the conservation of agro-biodiversity and natural resource management; adaptations to be made against climate variability; and sustainable agriculture, rural development, and multiple livelihoods.

Ramon Rivera (33): Research director of the National Institute of Agricultural Sciences (INCA), Cuba; previously chief of INCA's Biofertilizers and Plant Nutrition Dept., and before that leader of its Coffee Nutrition and Fertilization Research Program; leader of Cuban Network on Mycorrhizal Management and president of Cuban Soil Science Society (rrivera@inca.edu.cu)

G. Philip Robertson (3): Professor of soil ecology and director of the Long-Term Ecological Research Program, Kellogg Biological Station and Dept. of Crop and Soil Sciences, Michigan State University, USA; research on biogeochemistry and ecology of field crop ecosystems (Robertson@kbs.msu.edu)

Volker Römheld (7): Professor in the Institute for Plant Nutrition at the University of Hohenheim, Germany, focusing on the rhizosphere and nutrient transactions; project leader for study of irrigation and fertigation strategies for water-saving and optimized nutrient supply for litchis and mangoes; delegated organizer for the 7th international symposium of the International Society for Root Research in 2006 (roemheld@ uni-hohenheim.de)

Marco Rondon (36): Leader of the Climate Change Program of the Tropical Soil Biology and Fertility (TSBF) Institute of the International Center for Tropical Agriculture (CIAT), Colombia; working on land-use impacts on trace gas emissions of carbon dioxide, methane and dinitrous oxide, and carbon trading aspects of land-use and land-cover change (m.rondon@cgiar.org)

O.P. Rupela (35): Principal scientist in microbiology at the International Crops Research Institute for the Semi-Arid Tropics (ICRISAT), India; contributions mainly in areas of identification and evaluation of agriculturally-beneficial microorganisms from natural environments; maintenance and supply of microbial germplasm; crop residue management for crop production; biological approaches for crop production and protection (o.rupela@cgiar.org)

Sara J. Scherr (**44**): Director for ecosystem services, Forest Trends, and president of Ecoagriculture Partners, both Washington, DC; member of the U.N. Millennium Project Task Force on Hunger; previously with International Food Policy Research Institute (IFPRI) and World Agroforestry Center (ICRAF); leading international effort to establish convergence of sustainable agriculture and biodiversity conservation (ecoagriculture) (SScherr@ecoagriculturepartners.org)

Götz Schroth (**20**): Advisor for agroforestry and land use practices, integrating agricultural production and biodiversity conservation, Conservation International, Washington, DC, USA; previously with Center for International Forestry Research (CIFOR); research on tropical land use focusing on ecological, institutional and socioeconomic aspects of complex land use systems and biodiversity conservation in human-dominated landscapes (g.schroth@conservation.org)

Lucien Séguy (**22**, 23, 41): Agronomist based in Brazil coordinating worldwide CIRAD network on non-till practices in tropical regions of Asia, Africa and Brazil; working with conservation-agriculture farming systems for a range of farmers, from smallholders to mechanized producers, since 1984 (lucien.seguy@cirad.fr)

B. Selvamukilan (45): Biochemist working with the M. S. Swaminathan Research Foundation, India, for the last five years, promoting multiple livelihoods among resource-poor families.

Gudeta Sileshi (19): Agroforestry pest management scientist with World Agroforestry Centre (ICRAF) program in Zambia; formerly lecturer at Alamaya University, Ethiopia; research interests on integrated pest management and soil biodiversity (gsileshi@ zamtel.zm)

Paul C. Smithson (37): Assistant professor of environmental chemistry, Berea College, USA; previously senior soil scientist with the World Agroforestry Centre (ICRAF), Nairobi, Kenya; analytical laboratory specialist addressing soil fertility issues in agroforestry systems (Paul_Smithson@berea.edu)

Erika Styger (29): Agronomist consultant with the World Bank, Washington, DC; previously did research on managed fallows in Madagascar for PhD in crop and soil sciences, Cornell University, USA; before that, worked with the World Agroforestry Center (ICRAF); field research on farmer-based solutions for improved agricultural production and agroforestry systems in semi-arid, humid and highland farming systems in Africa (estyger@worldbank.org)

M.S. Swaminathan (Foreword): Chairman of the M. S. Swaminathan Research Foundation in Chennai, India, chairman of the Government of India's National Commission on Farmers, co-chair of the U.N. Millennium Project's Task Force on Hunger, and President of the Pugwash Conferences on Science and World Affairs; formerly director-general of the International Rice Research Institute, and first recipient of World Food Prize for his leadership of IRRI and in India's Green Revolution

P. Tamizoli (45): Anthropologist with M. S. Swaminathan Research Foundation, India; research expertise in the areas of cultural ecology, participatory research and rural development, and gender and tribal development; currently concerned with operationalizing biovillage models in different ecosystems.

Richard J. Thomas (4): Director of program on improved land management to combat desertification, International Center for Agricultural Research in the Dry Areas (ICARDA), Syria; research interests include integrated soil, water and nutrient management in crop-livestock systems in Latin America, Africa, West and Central Asia (r.thomas@cgiar.org)

Ashraf Tubeileh (4): Agronomist and specialist on plant-soil-water relations, International Center for Agricultural Research in the Dry Areas (ICARDA), Syria; current research focuses on soil and water management in fruit tree systems (a.tubeileh@cgiar.org)

Benjamin L. Turner (13, 37): Staff scientist with Smithsonian Tropical Research Institute, Panama; previously with Dept. of Soil and Water Science, University of Florida, USA, the Institute of Grassland and Environmental Research in UK, and the USDA-ARS Northwest Irrigation and Soils Research Laboratory; studies on nutrient biogeochemistry in soil and its ecological significance for both above- and belowground communities (turnerbl@si.edu)

Segundo Urquiaga (12, 27): Research scientist at the National Research Center of Agrobiology Research (EMBRAPA-Agrobiologia) in Brazil; has worked for 21 years on the quantification of biological N_2 fixation to both graminaceous and leguminous crops, and in nutrient cycling in different cropping systems, working with stable isotopes and other techniques; now studying N and C dynamics in tropical agroecosystems (urquiaga@cnpab.embrapa.br)

Bernard Vanlauwe (18): Soil fertility management specialist, Tropical Soil Biology and Fertility Institute (TBSF) of the International Center for Tropical Agriculture (CIAT), Kenya; research on nutrient cycling, interactions between organic and mineral nutrient inputs, and soil organic matter dynamics, with integration of legumes into farming systems; current research focuses on long-term effects of organic resource quality, and adapting soil fertility management strategies to heterogeneous farming systems (b.vanlauwe@cgiar.org)

Elisa Wandelli (21): Forest ecologist at EMBRAPA's Western Amazon Agroforestry Center (CPAA) near Manaus, Brazil; previously professor of ecology at the Federal University of Amazonas in Manaus; worked on the impact of deforestation and edge effects on sub-canopy palm species; co-leader of program on agroforestry systems for rehabilitating degraded lands sponsored by the Large-Scale Biosphere Atmosphere program and NASA's Terrestrial Ecology program.

S.P. Wani (35): Regional coordinator for global theme on agroecosystems management at the International Crops Research Institute for the Semi-Arid Tropics (ICRISAT), India; with training in soil microbiology, has focused on community watershed development and management to improve and sustain ecosystem productivity and enhance rural livelihoods (s.wani@cgiar.org)

David Wolfe (**47**): Professor in Dept. of Horticulture, Cornell University, USA, with state-wide extension responsibilities for soil and water management in vegetable cropping systems; co-chair of Cornell Soil Health Program; author of award-winning book on soil ecology for general audiences, *Tales from the Underground: A Natural History of Subterranean Life* (2001) (dww5@cornell.edu)

Liu Xuejun (39): Associate professor in the College of Resources and Environmental Sciences, China Agricultural University, China; research focuses on biogeochemical cycles and the management of nutrients, especially nitrogen, ranging from the rhizosphere to global scales, with particular attention to N deposition and its impact on ecosystems in China (liu310@cau.edu.cn)

Youssef G. Yanni (8): Chief researcher in soil microbiology, Sakha Agricultural Research Station, Egypt; research on development and utilization of biofertilizers for soil reclamation and field production of cereal and legume crops; several national awards for scientific contributions in soil science and applied microbiology (yanni244@yahoo.com)

Gregor W. Yeates (**10**): Soil zoologist, Landcare Research, New Zealand; research on the ecology, biology and taxonomy of plant and soil nematodes, with research showing positive correlation between nematode abundance and primary production; work includes interactions between soil microfauna and microflora, interactions between land use and soil animals, and biodiversity (YeatesG@landcareresearch.co.nz)

Zahir Ahmad Zahir (14): Soil microbiologist in the Institute of Soil and Environmental Sciences, University of Agriculture, Faisalabad, Pakistan; pioneer in the utilization of microbially-produced plant growth regulators for the improvement of crop performance; major focus on development of biofertilizers for sustainable agricultural production (bio@fsd.comsats.net.pk)

Robert Zougmoré (26): Agronomist and soil scientist in the National Institute for Environment and Agricultural Research, Burkina Faso; research on soil and water conservation at plot and watershed levels, integrated water and nutrient management in arid and semiarid sub-Saharan Africa; use of water-harvesting techniques and soil fertility management practices (including legume cover crops) to reduce runoff and soil erosion and improve crop productivity (robert.zougmore@messrs.gov.bf)

Acronyms and Abbreviations

ABA	Abscisic acid (phytohormone)
ADEOS	Advanced Earth Observing Satellite, joint project of NASA and National Space Development Agency of Japan
AM or AMF	Arbuscular mycorrhiza or arbuscular mycorrhizal fungi
AR	Acetylene reduction, for analysis of BNF
ATP	Adenosine triphosphate
BMS	Biological management of soils
BNF	Biological nitrogen fixation
C3	Most common pathway for photosynthesis, in 95% of plants
C4	Alternative pathway for photosynthesis, more efficient in strong sunlight, e.g., found in maize and sugarcane
CA	Conservation agriculture
CAM	Crassulacean acid metabolism, another pathway for photosynthesis, found mostly in succulents
CBD	Convention on Biological Diversity
CEC	Cation exchange capacity
CIAT	International Center for Tropical Agriculture, Colombia
CIFOR	Centre for International Forestry Research, Indonesia
CIIFAD	Cornell International Institute for Food, Agriculture and Development, USA
CIMMYT	International Maize and Wheat Improvement Center, Mexico
CIRAD	International Center for Research on Agricultural Development, France
CLPP	Community-level physiological profiles
CLSM	Confocal laser scanning microscopy
C:N or C/N	Carbon-nitrogen ratio
C:P or C/P	Carbon-phosphorus ratio
CT	Conventional tillage
DGGE	Denaturing gradient gel electrophoresis
DSPSC	Direct seeding on permanent soil cover
DSS	Decision support system
EMBRAPA	Brazilian Agency for Agricultural Research
FAME	Fatty acid methylated ester
FAO	U.N. Food and Agriculture Organization
FOFIFA	Madagascar government agricultural research organization
GA	Gibberellin (phytohormone)
GC	Gas chromatography
GDSM	Direct Seeding Group, Madagascar, unit of CIRAD
GEF	Global Environmental Facility
GFP	green fluorescent protein; *gfp* denotes the gene for this protein
GHG	Greenhouse gas(es)
GIS	Geographic information system
GMCC	Green manure/cover crop

GS	Glutamate synthetase, an enzyme for protein synthesis
HMW	High molecular weight
IAA	Indole acetic acid, an auxin (phytohormone)
ICARDA	International Center for Agricultural Research in the Dry Areas, Syria
ICIPE	International Centre for Insect Physiology and Ecology, Kenya
ICRAF	International Centre for Research in Agroforestry, now World Agroforestry Centre, Kenya
ICRISAT	Institute for Crop Research in the Semi-Arid Tropics, India
IFDC	International Center for Soil Fertility and Agricultural Development
IFPRI	International Food Research Policy Institute, USA
IGP	Indo-Gangetic plains, South Asia
IITA	International Institute for Tropical Agriculture, Nigeria
INCA	National Institute for Agricultural Science, Cuba
INIFAT	Institute for Basic Research in Tropical Agriculture, Cuba
INM	Integrated nutrient management
INRM	Integrated natural resource management
IPM	Integrated pest management
IRRI	International Rice Research Institute, Philippines
ISFM	Integrated soil fertility management
ISR	Induced systemic resistance
ITP	Internal transcribed spaces
IWMI	International Water Management Institute, Sri Lanka
KCSZ	Kano Close-Settled Zone, Nigeria
LER	Land equivalent ratio
LEISA	Low external-input sustainable agriculture
LF	Light fraction of SOM
LMW	Low molecular weight
MPN	Most probable number
MS	Mass spectrometry
MSSRF	M. S. Swaminathan Research Foundation, India
MO	Microorganism(s)
MWD	Mean weight diameter, a measure of soil porosity
NARS	National agricultural research system(s)
NASA	National Aeronautics and Space Agency, USA
NGO	Non-governmental organization
NPK	Nitrogen-phosphorus-potassium fertilizer
NPP	Net primary productivity
OM	Organic matter
OTU	Operational taxonomic unit
PCA	Principal components analysis
PCR	Polymerase chain reaction
PGP	Plant growth promoter
PGPR	Plant growth-promoting rhizobium
pH	Measure of acidity/basicity of the soil solution; $< 7 =$ acid, $> 7 =$ basic
PGPB	Plant growth-promoting bacteria
PLFA	Phospholipid fatty acids
PM	Plastic mulch
POM	Particulate organic matter
PSS	Penetration soil strength
RAS	Root-adhering soil
RMP	Recommended management practices

RPR	Root-pulling resistance
SBD	Soil bulk density
SE	Standard error
SED	Standard error of difference
SHG	Self-help group, India
SIC	Soil inorganic carbon
SIP	Stable isotope probing
SM	Straw mulch
SML	Soil moisture level
SOC	Soil organic carbon
SOM	Soil organic matter
SPM	Soil profile modification
SPR	Soil penetration resistance
SRI	System of Rice Intensification
SSA	Sub-Saharan Africa
TF	Traditional flooding
TGGE	Temperature gradient gel electrophoresis
T-RFLP	Terminal restriction fragment length polymorphism
TRP	Tryptophan, a precursor for auxin (phytohormone)
TSBF	Tropical Soil Biology and Fertility program, now unit of CIAT, Kenya
UNCCD	United Nations Convention to Combat Desertification
UNEP	United Nations Environment Program, Kenya
USDA	US Department of Agriculture
USLE	Universal soil loss equation
UV	Ultra-violet
V	Base saturation level
VAM	Vesicular arbuscular mycorrhizas (fungus)
WHC	Water-holding capacity of soil
Z	Zeatin, an important cytokinin (phytohormone)
ZT	Zero-tillage, form of conservation agriculture

Chemical Notations

Ag	Silver
Al	Aluminum
B	Boron
Ca	Calcium
Cd	Cadmium (heavy metal)
CO_2	Carbon dioxide (gas)
Co	Cobalt
Cr	Chromium (heavy metal)
Cu	Copper
Fe	Iron
FeS	Iron sulfide
$FeSO_4$	Iron sulfate
H_2S	Hydrogen sulfide (gas)
Hg	Mercury (heavy metal)
K	Potassium
K_2O	Potassium oxide, known as potash
Mg	Magnesium
Mn	Manganese

Mo	Molybdenum
N_2	Dinitrogen (gas)
N_2O	Nitrous oxide (gas)
NH_3^+	Ammonia (gas)
NH_4^+	Ammonium
NO	Nitric oxide
NaCl	Sodium chloride, salt
Ni	Nickel
O	Oxygen
O_2	Dioxygen (gas)
O_3	Ozone (gas)
P	Phosphorus
P_2O_5	Phosphorus pentoxide (= 44% elemental P), referred to as phosphate
Pb	Lead (heavy metal)
S	Sulfur
SO_4^{2-}	Sulfate
Se	Selenium (heavy metal)
Si	Silicon
Zn	Zinc

Units of Measurement

cm	centimeter = .39 inch
ha	hectare = 2.54 acres
kg	kilogram = 2.2 pounds
km	kilometer = 1.61 miles
kPa	kilopascal, unit of pressure
m	meter = 39.37 inches
Mg	million grams = 1 metric tonne
mm	millimeter = 1 thousandth of a meter
MPa	megapascal, unit of pressure
μm	micron = 1 millionth of a meter
Pg	petagram = 1 billion metric tonnes
t	metric tonne = 1,000 kg

Index

This book focused on soil biology was not designed as a comprehensive soil science text. However, its chapters cover most subjects addressed in soil chemistry and physics as well as many other subjects related to soil and its management. The page numbers below refer to where in the book the respective subjects are discussed or referred to, either in some depth or in passing. This index will enable readers to track subjects throughout the book. Where terms are defined, page numbers are given in *italics*, while chapters or sections that are devoted to the subject are indicated in **boldface**.